D1116200

INTRODUCTION TO

ELECTRICITY AND ELECTRONICS

CONVENTIONAL CURRENT VERSION

INTRODUCTION TO
ELECTRICITY AND ELECTRONICS

SECOND EDITION

CONVENTIONAL CURRENT VERSION

ALLEN MOTTERSHEAD
Department of Electronics Technology
Cypress College, California

JOHN WILEY & SONS
New York   Chichester   Brisbane   Toronto   Singapore

Cover and text were designed by Raphael Hernandez. Design was supervised by Ann Marie Renzi. Production was supervised by Cindy Funkhouser. Editing was supervised by Tamara Lee.

Photo Credits
Cover: R. Laird/Leo deWys.
Openers for Part 1 and Chapters: Pat Lanza Field/Bruce Coleman.
Openers for Part 2 and Chapters: Douglas Mesney/Leo deWys.
Openers for Part 3 and Chapters: John Launois/Black Star.

Copyright © 1986, John Wiley & Sons, Inc.

All rights reserved. Published simultaneously in Canada.

Reproduction or translation of any part of
this work beyond that permitted by Sections
107 and 108 of the 1976 United States Copyright
Act without the permission of the copyright
owner is unlawful. Requests for permission
or further information should be addressed to
the Permissions Department, John Wiley & Sons.

Library of Congress Cataloging in Publication Data:

Mottershead, Allen.
    Introduction to electricity and electronics : conventional current version.
    Includes index.
      1. Electric circuits.   2. Electronic circuits.
I. Title.
TK454.M67   1986   621.319'2   85-26306
ISBN 0-471-80654-4

Printed in the United States of America

10 9 8 7 6 5 4 3 2 1

PREFACE

In the five years since the first edition of this text was written, I have had the opportunity to receive comments from a great many instructors who have taught from the text. Those comments provided a working plan for improving the clarity of the presentation and making the book more useful as a learning tool for students.

There are three areas in which the improvements to this second edition are concentrated: making the book easier to read, adding important new material, and adding to the scope of the various applications.

First, the text has been almost completely rewritten to provide a more realistic reading level. Shorter sentences combined with a more conversational style will help students to better understand the material, including the more difficult concepts.

Second, more than half of the chapters have had new material added, expanded, or clarified:

Chapter 3—Work and power topics have been streamlined. More information on modern resistors has been included.

Chapter 5—More examples and problems on series circuits.

Chapter 6—More examples and problems on parallel circuits.

Chapter 7—Many more examples and problems on series-parallel circuits, including ladder schematics.

Chapter 9—Various interconnections of cells and the effects on internal resistance and capacities have been made more clear.

Chapter 10—Nodal analysis and additional examples and problems on Thévenin's and Norton's techniques have been added.

Chapter 11—An optional section on nonlinear magnetic circuits that require the use of B-H curves for analysis has been added.

Chapter 15—More information on true rms values and nominal line voltage and frequencies in other countries are presented.

Chapter 18—Consideration of transformers with multiple secondaries.

Chapter 19—Material on exponential equations for solving dc *RL* circuits has been moved from the Appendix to an optional section.

Chapter 21—Additional information on electrolytic capacitors.

Chapter 22—Material on exponential equations for solving dc *RC* circuits has been moved from the Appendix to an optional section.

Chapter 23 and 24—Added material on voltage division in a series *RC* circuit.

Chapter 27—Added information on phase-shifting circuits (magnitude and phase) and the use of complex numbers in Norton's Theorem.

Continuing with a major strength of the first edition, many new applications have been provided in the following chapters:

Chapter 2—Application of electrostatics to Xerography.
Chapter 5—An illustration of a series control circuit in a swimming pool heater.
Chapter 7—Series-parallel ladder schematic of relay controls in a lathe.
Chapter 10—Thévenin's and Norton's theorems applied to a transistor equivalent circuit.
Chapter 23 and 24—Resistor-capacitor circuits used in audio amplifier tone controls.
Chapter 24—Series *RLC* circuit theory applied to operation of relays and solenoids at lower than normal voltage.
Chapter 25—Use of capacitors as voltage dropping devices to operate elapsed time indicators at higher voltages with low power consumption.
Chapter 28—Use of a half-wave rectifier to extend the life of an incandescent lamp.

I hope that these changes will make the book even more understandable, informative, and enjoyable for students of electricity and electronics.

Allen Mottershead

PREFACE

TO THE FIRST EDITION

Everyone knows that the world is becoming more complex. From the cars we drive to the calculators, computers, and entertainment equipment familiar to all, we take for granted the advances that electronics has brought. And yet, the fundamental principles that made many of these devices possible first had to be understood before such devices could be developed. This book provides the fundamentals of electricity, magnetism, and electronics to help you understand the world around you. Many everyday applications are described to show these fundamentals at work. Also, completion of the material in this book will allow you to study in detail the advanced theory and applications of electrical and electronic devices, circuits, and systems.

The book is written for the first-year course in an electricity/electronics program at a community college, vocational institute, or technical college. It covers the traditional material of dc and ac circuits, including magnetism, that is required for both electronic and electrical majors. It assumes no previous knowledge of electricity or electronics. Only a basic understanding of algebra is required, and right-angle trigonometry is introduced only when it is necessary to understand alternating current.

There are three parts to the book. In Part One, after introducing calculators, electrical meters, and the basic electrical quantities, series and parallel circuits are covered followed by series-parallel circuits and their applications in loaded voltage dividers. A comprehensive chapter on voltage sources contains up-to-date information on the latest secondary batteries as well as fuel and solar cells. The important effect of internal resistance on maximum power transfer is also considered. A chapter on network analysis covers all the important methods such as branch currents, loop, superposition, Thévenin's, Norton's, and delta-wye transformations. Magnetism is then covered so that its applications in dc ammeters, voltmeters, and other dc measuring instruments can be shown.

Part Two begins with the generation of alternating current (ac) and direct current (dc), and continues with a detailed coverage of alternating voltage and current. This includes a chapter on ac measuring instruments, concentrating on the oscilloscope and its use. The oscilloscope can then be referred to during the following chapters on inductance and capacitance where their effects in both dc and ac circuits are considered. A whole chapter on transformers is included. After treating various combinations of resistance, inductance, and capacitance in alternating circuits, power and resonance in these circuits are also covered. Part Two ends with an optional chapter on complex numbers.

Part Three introduces the broad field of electronics. How a PN junction diode can cause rectification is explained, and its use to convert ac to dc in power supplies is followed by a discussion of filters. Finally, the bipolar junction transistor is introduced, and its use is demonstrated in amplifiers and oscillators.

The book has a number of features. First, theory and principles are made easier to understand by presenting applications whenever possible. For example, following a theoretical discussion of magnetism, we see how loudspeakers and dc motors operate, how solenoids and relays are used in a typical home heating/cooling system, and how the Hall effect is used in a clamp-on ammeter to measure both ac *and* dc.

In the chapter on electromagnetic induction we see the basic principles applied to a microphone, a magnetic tape recorder, practical alternators and dc generators, and a voltage regulator for an automotive battery-alternator system.

In the chapter on transformers, applications to power distribution and residential wiring systems are made, including the safety features of a ground connection and a ground fault circuit interrupter. Other applications include the conventional and electronic automotive ignition system, the role of an inductor in a fluorescent lamp circuit, the use of capacitors in photoflash units, differentiating and delay circuits, and power factor correction applications. The chapter on resonance shows how the principles of resonance are applied to a superheterodyne AM receiver, and the chapter on diodes applies the principles of rectification to the three-phase output of an automobile's alternator.

Second, every opportunity is taken to show how measurements are properly made by the latest in instrumentation. Three chapters are devoted to dc and ac measuring instruments. The conventional meters to measure current, voltage, resistance, and power, both analog and digital, are examined, in addition to the following: Gaussmeter, potentiometer for voltmeter and ammeter calibration, Wheatstone bridge, frequency and period meter, instrument transformers, impedance bridge, capacitance and inductance meters, power factor meter, vector impedance and voltage meters, and curve tracers. The use of the oscilloscope is stressed in the measurement of time constants, inductance, capacitance, phase angle, quality factor, and so on.

Third, every chapter has a summary, self-examination, review questions, and problems. Answers to the self-examination questions and to the odd-numbered problems are at the back of the book. There is also a comprehensive glossary containing all the book's important terms, as well as the symbols and units of each electrical quantity. The appendixes contain 14 sections that cover tables of information on resistors, wire gauges, determinants, trigonometry ratios, derivations, and so on.

Fourth, the text is arranged so that material needed for a concurrent laboratory course is given early in the text. A laboratory manual is available that was written to accompany the text. The laboratory manual has one laboratory experiment with many self-contained sections, for each chapter in this book. It is designed to be used with most standard laboratory equipment and supplies.

Finally, this book is available in both the *conventional-current version* and the *electron-flow version*. Although the SI system recommends the eventual adoption of conventional current, many will prefer to remain with the more familiar electron flow. Either way, it is the hope of the author that this book will make electricity and electronics more understandable and enjoyable to both reader and instructor.

Allen Mottershead

ACKNOWLEDGMENTS

I am grateful for the comments and suggestions that were provided by many users of the first edition as I prepared this second edition. I especially thank the following reviewers for their detailed recommendations:

A. L. Abbott
Chaffey Community College

David J. Bradley
Prince Georges Community College

Dr. William Perkins
ITT Technical Institute

Mel Jansen
Madion Area Technical College

Ronald Cohen
Grumman Data Systems Institute

Linda J. Wagner
Wisconsin School of Electronics

R. Max Steadman
Weber State College

James Vandernald
DeVry Institute of Technology,
Los Angeles

Maurice E. Ryan
Santa Barbara City College

Phillip D. Anderson
Muskegon Community College

Irving L. Brooks
Suburban Tech

Robert T. Campbell
North Shore Community College

Larry A. Oliver
Columbus Technical Institute

Stanley W. Lawrence
Utah Technical College,
Salt Lake City

Special thanks are also owed to James Smith, Wendy Ford, Jack Klasey, Karen Beam, and Joy Sherrill of White River Press for their valuable contributions to the revision.

I acknowledge the help, patience, and advice of the John Wiley staff who brought this book into production for the second time. It has been a pleasure to work with them once again.

A.M.

TO THE INSTRUCTOR

This book may be used in a number of different ways.

1   For a conventional two-semester dc/ac course:

Semester 1: Part I, DC Circuits: Chapters 1 to 13.
Semester 2: Part II, AC Circuits: Chapters 14 to 27.

2   An alternate two-semester dc/ac course may use the following sequence:

Semester 1: DC Circuits: Chapters 1-10, 21, 22, 11, 17, 19, 12, and 13.
Semester 2: AC Circuits: Chapters 14, 15, 18, 16, 20, and 23–27.

3   For a one-semester combination dc/ac course (no analysis) that may include none-lectrical/electronics majors, the following chapters and sections are recommended:

WEEK	DC CIRCUITS
1	Chapter 1—read lightly
	Chapter 2
2	Chapter 3
3	Section 4–1 and Chapter 5
4	Chapter 6
	Section 7-1
5	Chapter 8—read lightly
	Chapter 9
6	Chapter 11—Omit Section 11-15
7	Chapter 12—Omit Section 12-4
	Chapter 13—Omit Sections 13-4 and 13-6

WEEK	AC CIRCUITS
8	Chapter 14—All plus Appendix F
9	Chapter 15
	Sections 16-1 and 16-6
10	Chapter 17—Omit Sections 17-2, 17-5.3, and 17-5.4
	Chapter 18—Omit Sections 18-1.4 and 18-6
11	Chapter 19—Omit Sections 19-4, 19-6.1, 19-6.2 and 19-10
	Chapter 20—Omit Sections 20-4, and 20-6 through 20-9
12	Chapter 21—Omit Sections 21-4 and 21-7.1
13	Chapter 22—Omit Sections 22-6, 22-8, and 22-9

Chapter 23—Omit Sections 23-3.3 through 23-6
14   Chapter 24—Omit Sections 24-4.1, 24-6.1, 24-9, 24-10, and 24-11

WEEK   **ELECTRONICS**
15   Chapter 28—Omit Sections 28-12 and 28-13
16   Chapter 29

4 For a one-semester dc/ac analysis course that concentrates on areas such as network theorems, magnetic circuits, exponential equations for L and C circuits, and complex numbers and is designed to follow the course in No. 3 above, the following chapters, sections, and appendixes are recommended. (It is assumed that coverage of sections omitted in the first course will be accompanied by a general review of pertinent material in the appropriate chapter.)

WEEK
1   General review plus Chapter 4
2   Chapter 7
3   Chapter 8
4,5   Chapter 10 and Appendix D
6   Review plus Section 11-15 plus Appendix K
7   Review of Chapter 14 plus Appendixes E, G, and H
8   Chapter 16
9   Review of Chapter 17 plus Sections 17-2, 17-5.3, and 17-5.4
         Review of Chapter 18 plus Sections 18-1.4 and 18-6
10   Sections 19–4, 19–6.1, 19-6.2, and 19-10
         Sections 20-4, 20-6 through 20-9 plus Appendix I
11   Sections 21-4 and 21-7.1 plus Appendix J
         Sections 22-6, 22-8, and 22-9
12   Sections 23-3.3 through 23-6 plus Appendix L
         Review Chapter 24 Plus Sections 24-4.1, 24-6.1, 24-9, 24-10, and 24-11
13   Chapter 25
14   Chapter 26
15,16   Chapter 27

Individual needs vary from one instructor and course to another so that many variations of the above are possible. For example, Chapter 27 on complex numbers may appear more logically (for some) immediately after Chapter 24 on ac circuits. Others may wish for this material to appear even earlier, before reactance. Or they may feel it is too advanced and omit it altogether. For this reason, Chapter 27 stands on its own as optional material at the end of Part Two.

It is assumed that a laboratory session constitutes part of each semester course, using a minimum of three hours lecture and three hours laboratory per week. Topics have been arranged with this in mind so that, for example, Chapter 16 on the oscilloscope precedes capacitance and inductance. This means that the oscilloscope can be used in the laboratory experiments on these two topics, allowing the observation and measurement of time constants.

Students can check their own progress by doing the review questions, self-examinations, and problems. In general, problems become more difficult toward the end of each chapter, and even more challenging ones are found among the even numbers. A Solutions Manual for all problems is available to instructors.

A.M.

CONTENTS

INTRODUCTION TO

ELECTRICITY AND ELECTRONICS

CONVENTIONAL CURRENT VERSION

PART 1

DC CIRCUITS

CHAPTER 1

INTRODUCTION TO CALCULATORS, CIRCUITS, AND METERS

To study electricity and electronics in any detail, you will have to perform calculations. They are needed in solving problems to become familiar with a subject or to predict results using equations. The electronic calculator has made ordinary arithmetic computations a much less tedious task. One purpose of this chapter is to help you choose and learn to use this essential tool for everyday calculations. Another is to introduce you to a simple electric circuit and the symbols used to represent that circuit in a schematic diagram. Still another purpose of this chapter is to help you learn to correctly read the scales of multirange meters used to test circuits and components.

1-1   THE ELECTRONIC CALCULATOR

Except when very large or very small numbers are involved, even the lowest-cost calculators will allow you to enter and process the four basic arithmetic operations ($+$, $-$, $\times$, $\div$). These calculators provide only an eight-digit readout. Consider the problem of multiplying the number 50,000,000 by 2. The result, of course, is 100,000,000, but such calculators cannot display all the digits. On some calculators, the 1 disappears off the left end of the display and only zeros are shown. On one calculator model, the result is shown as 10,000,000, but is indicated as incorrect

FIGURE 1-1

A typical scientific calculator. (Courtesy of Texas Instruments.)

by a flashing display. Still other calculators may show that an answer is too long for the display by placing decimal points after every digit.

A "scientific" calculator or "electronic slide rule" calculator, of the type shown in Fig. 1-1, is a worthwhile investment. Besides the capability of displaying 10 or more digits, these calculators have many useful functions. Most include the trigonometric functions of *sin, cos, tan,* and their inverse, as well as $\log_{10}$ and ln. If you plan to continue in electronics technology, the calculator should be capable of converting from rectangular to polar notation, and vice-versa. The calculator you choose may have some functions you do not immediately use, but you will learn to understand and use them as you advance. Depending on your budget, you may choose a scientific calculator that is programmable, or even one with printout capability.

1-2   SCIENTIFIC NOTATION

The scientific calculator handles very large and very small numbers by using *scientific notation,* which is a form of "shorthand notation." It involves writing a given number as a number between 1 and 10 (but not including 10), multiplied by 10 raised to a suitable exponent. The exponent may be positive or negative. For example:

$$34{,}275 = 3.4275 \times 10^4$$
$$1100 = 1.1 \times 10^3$$
$$56{,}000{,}000 = 5.6 \times 10^7$$
$$0.00678 = 6.78 \times 10^{-3}$$
$$0.000{,}0016 = 1.6 \times 10^{-6}$$

Assume that you have a scientific calculator with a 10-digit display and want to do this calculation:

$$5{,}000{,}000{,}000 \times 2.2$$

If you do the problem by hand, using scientific calculation, you obtain

$$
\begin{aligned}
5{,}000{,}000{,}000 \times 2.2 &= 5 \times 10^9 \times 2.2 \\
&= 11 \times 10^9 \\
&= 1.1 \times 10^{10} \\
&= 11{,}000{,}000{,}000
\end{aligned}
$$

Obviously, the 11-digit result is too long to be displayed by your calculator. To overcome this problem, the calculator will display the result in scientific notation as 1.1 10. This is the same as the number written in the form 1.1×10^{10}. On some calculators, an E or EE may be displayed between 1.1 and 10.

You can *enter* numbers in scientific notation by pressing the *scientific exponent* key (often labeled EE). For example:

Multiply: $(7.6 \times 10^3) \times (6.5 \times 10^4)$

Enter	Press	Display	Purpose
	C	0	Clears the calculator
7.6	EE 3	7.6 03	Enters first number in scientific notation
	×	7.6 03	Instruction to multiply
6.5	EE 4	6.5 04	Enters second number in scientific notation
	=	4.94 08	Answer

Answer: **4.94 × 10⁸**

> **NOTE** If you wish to enter only a power of 10 (such as 10⁴), the calculator may not respond unless you enter a 1 before pressing the EE key. Your entry would be 1 EE 4, or 1 × 10⁴. See Problem 1-10.

Some calculators can convert from scientific notation to decimal notation, as follows:

Enter	Press	Display	Purpose
4.94 08	2nd E̶E̶	494,000,000	Converts scientific notation to decimal form

If the decimal form has more digits than the calculator can display, the display may continue to show the entered numbers, or may flash as described earlier.

Some calculators also can convert from decimal notation to scientific notation. For example:

Enter	Press	Display	Purpose
494,000,000		494,000,000	
	EE	494,000,000 00	Converts decimal form to scientific notation
	=	4.94 08	

Answer: **4.94 × 10⁸**

Finally, consider $\dfrac{3.04 \times 10^{-2}}{21.6 \times 10^4}$

Enter	Press	Display	Purpose
3.04	EE +/− 2	3.04-02	To multiply 3.04 by 10^{-2}, use the +/- key to enter a negative exponent
	÷	3.04-02	Indicates division as the next operation
21.6	EE 4	21.6 04	Enters the denominator in scientific notation
	=	1.407407407-07	
	2nd E̶E̶	.0000001407	

Answer: **1.407407407 × 10⁻⁷ (1.41 × 10⁻⁷)**

By using the appropriate key, operations involving addition, subtraction, squaring or square root can be performed in a similar way on numbers in scientific notation.

The method of entering data may vary depending on the make of the calculator. Some calculators use algebraic notation; others a method known as "reverse Polish notation." The instruction booklet for your calculator will tell you how data should be entered and usually will provide examples.

1-3 ACCURACY AND ROUNDING OFF NUMBERS

The answer to the last example, 1.407407407, raises the important question, "How many digits are significant in the final answer?" This depends on how many *significant figures* are in the original numbers used in the problem.

Significant figures are those that provide information on how accurately a number is known. If you weigh an object and state its weight to be 125 pounds, you are indicating *three* significant figures of accuracy. That is, the weight is closer to 125 pounds than 124 or 126 pounds. If the measuring instrument had sufficient *precision,* you might state that the weight is 125.0 pounds. This is now a *four* significant figure accuracy, stating the weight is closer to 125.0 than 124.9 or 125.1 pounds. In the first case, the weight is known to within ±1 pound, but in the second case, the weight is known to within ±0.1 pound.

Although higher accuracy generally requires more precise measurements, accuracy is not the same as precision. It is quite possible for a precision digital voltmeter to indicate a value such as 1.578 volts, implying the value is known to the nearest 0.001 volt. But if the instrument is defective, or has not been recently calibrated against a *standard* (a known true value of voltage), the reading itself may not be *correct* or *true*. That is, you may have made a highly precise reading that is inaccurate. In short, precise implies "sharply defined to within narrow limits"; accurate implies "correctness" or "truth."

It should be noted that a measured value of 0.00782 is known to *three* significant figures. The zeros are not significant in determining accuracy; they are used only to show the size of the number. And finally, a measured value of 1000 is considered to have only one significant digit, but the number 1000.0 has five significant figures.

The importance of significant figures is seen when expressing a final answer for a computation that involves *measured* data.

In the computation ($3.04 \times 10^{-2} \div 21.6 \times 10^4$) both numbers contain *three* significant figures, so the answer can have no more than three significant figures. The calculator answer of $1.407407407 \times 10^{-7}$ must be rounded off to 1.41×10^{-7}.

If the original numbers had been quoted as 3.039×10^{-2} and 21.58×10^4, these would be accurate to four significant figures and the calculator result of $1.408248378 \times 10^{-7}$ would be written as 1.408×10^{-7}.

In general, a calculation that involves multiplication, roots, and division of *measured* data provides a result that is only as accurate as the *least* accurate information used, and must be rounded off accordingly.

To illustrate this, imagine that you want to calculate the average miles per gallon (mpg) of fuel used by your car. The odometer reading shows a distance of 221.7 miles for a gas pump reading of 11.5 gallons. Using your calculator, you get this result:

$$\frac{221.7 \text{ miles}}{11.5 \text{ gallons}} = 19.27826087 \text{ mpg}$$

Because the gas pump reading was the *least* accurate information (containing only three significant figures), the rounded-off answer is 19.3 mpg.

It should be noted that addition of measured data must be treated differently as far as rounding off is concerned. If two objects were weighed on two different scales, with one measured to be 12 pounds, and the other to be 0.0005 pounds, the computation involving their sum (12 + 0.0005) pounds must be rounded off to 12 pounds. That is, although 0.0005 is known to only one significant figure, this does not determine the accuracy of the addition. The weight of 12 pounds is known only to the nearest pound, so the addition of 0.0005 pounds is not meaningful.

The above discussion has emphasized computations involving *measured data* where some limitation on accuracy is always in effect. In this case, final answers should always be rounded off although intermediate calculations should not.

However, in many problems, where data are *given*, it is assumed in this text to be exact, so that computations such as 9/4 = 2.25 will not be rounded off to 2.

Neither will the computation be shown as 9.00/4.00 = 2.25, since the numbers 9 and 4 are assumed to have whatever accuracy is needed for the final answer. Where results are not exact, rounding off will be done to reflect the given data's decimal place accuracy.

1-4 THE SI SYSTEM

Although the United States has not officially adopted the metric system, it is used in most other countries of the world. For this reason, many American industries that market products in other nations have chosen to use metric units in their drawings and specifications.

There are three systems of units, differing basically in the units used for length and mass. (All three systems measure time in seconds.) Table 1-1 compares the units used in the three systems. Note that the name of each system stems from the first letter of each unit of measurement for length, mass, and time.

In 1965, the Institute of Electrical and Electronic Engineers (IEEE) adopted the International System of Units, better known as the "SI" system from the initial letters of its name in French (Système International d'Unites). The SI (or metric) system is based on the fundamental units of the meter, kilogram, second, and ampere and is sometimes referred to as the "MKSA" system. Other metric units of measurement are derived from these fundamental units, and are thus called "derived units."

TABLE 1-1
Physical Units

System	Length	Mass	Time
MKS	Meter (m)	Kilogram (kg)	Second (s)
CGS	Centimeter (cm)	Gram (g)	Second (s)
FPS	Foot (ft)	Pound-mass (lb_m)	Second (s)

1-5 CONVERSION FACTORS

Sometimes, it is necessary to make *conversions* from the units of one system of measurement to the units of another system. To help you make such conversions, a list of some of the common *conversion factors* is provided in Table 1-2.

In Table 1-2, the unit abbreviations for the conversion factors are also given in *ratio* form. This is obtained by dividing both sides of the equation by the *same* unit. Since the ratio represents a factor of 1, multiplying or dividing a quantity by a conversion ratio does not change the quantity.

EXAMPLE 1.1

Convert 3.67 feet to meters.

Solution

Since 1 foot = 0.3048 m, then $\dfrac{0.3048 \text{ m}}{1 \text{ ft}} = 1$ (see Table 1-2).

TABLE 1-2

Common Conversion Factors, in Equation and Ratio Form

1 inch = 2.54 centimeters; $\dfrac{2.54}{1 \text{ in.}}$

1 foot = 0.3048 meter; $\dfrac{0.3048 \text{ m}}{1 \text{ ft}}$

1 mile = 1.609 kilometers; $\dfrac{1.609 \text{ km}}{1 \text{ mi}}$

1 gallon = 3.785 liters; $\dfrac{3.785 \text{ l}}{1 \text{ gal}}$

1 ounce = 28.35 grams; $\dfrac{28.35 \text{ g}}{1 \text{ oz}}$

1 pound = 0.4536 kilogram; $\dfrac{0.4536 \text{ kg}}{1 \text{ lb}_m}$
(mass)

1 pound = 4.45 newton; $\dfrac{4.45 \text{ N}}{1 \text{ lb}_f}$
(force)

1 horsepower = 746 watts; $\dfrac{746 \text{ W}}{1 \text{ hp}}$

1 horsepower = 550 ft lb/s; $\dfrac{550 \text{ ft lb/s}}{1 \text{ hp}}$

$$3.67 \text{ ft} \times 1 = 3.67 \cancel{\text{ft}} \times \frac{0.3048 \text{ m}}{1 \cancel{\text{ft}}}$$

$$= 3.67 \times 0.3048 \text{ m}$$

$$= 1.118616 \text{ m (by calculator)}$$

$$= \mathbf{1.12 \text{ m}} \text{ (rounded to three significant figures)}$$

Note that the units of feet in this example *cancel*, leaving the desired units of meters.

Some calculators have conversion factors stored in memory, and can make the conversions when you supply a code (these codes may be printed on the back of the calculator). One popular scientific calculator uses the code 02 for feet to meters conversion:

Enter	Press	Display
	C	0
3.67		3.67
	2nd	3.67
	02 (Code)	1.118616

Answer: **3.67 ft = 1.12 m**

1-6 UNIT PREFIXES

Standard *prefixes* are attached to metric *unit* names to identify multiples and submultiples of that unit. The prefixes are used with both fundamental (length, mass, time) units and derived units. You will see them frequently in the units of electrical measurement used later in this book. Table 1-3 lists the common prefixes and the symbols used

TABLE 1-3

SI Unit Prefixes

Value	Scientific Notation	Prefix	Symbol
1,000,000,000,000	10^{12}	tera	T
1,000,000,000	10^{9}	giga	G
1,000,000	10^{6}	mega	M
1,000	10^{3}	kilo	k
0.001	10^{-3}	milli	m
0.000,001	10^{-6}	micro	μ
0.000,000,001	10^{-9}	nano	n
0.000,000,000,001	10^{-12}	pico	p
	10^{-15}	femto	f
	10^{-18}	atto	a

when abbreviating the unit (the milligram, or 1/1000th gram, is abbreviated mg).

With this system, any quantity that is much larger or much smaller than the basic unit is referred to in a compact form that includes the name of the basic unit. A kilometer, for example, can be written as 1×10^3 m, or in the simple abbreviated form, 1 km. It is important to distinguish between the lowercase *m,* which stands for milli, and the capital *M,* which stands for mega. There's a considerable difference between 1 MV (1,000,000 volts) and 1 mV (0.001 volt)!

EXAMPLE 1-2

Convert 0.0005 seconds to:
a. milliseconds (ms)
b. microseconds (μs)

Solution

a. 0.0005 s = 0.5×10^{-3} s = **0.5 ms**
b. 0.0005 s = 500×10^{-6} s = **500 μs**

EXAMPLE 1-3

Convert 0.035 nanoseconds to:
a. picoseconds (ps)
b. microseconds (μs)

Solution

a. 0.035 ns = 0.035×10^{-9} s = 35×10^{-12} s
     = **35 ps**
b. 0.035 ns = 0.035×10^{-9} s = $0.000,035 \times 10^{-6}$ s
     = 0.000,035 μs = **3.5 $\times$ 10$^{-5}\mu$s**

1-7 A SIMPLE ELECTRIC CIRCUIT

An electric circuit consists of various components connected together in a way that will provide a complete path for the flow of electric current. These components include *passive* devices that do not amplify or provide a source of energy. The three basic passive components in electrical circuits are the resistor *(R),* the inductor *(L),* and the capacitor *(C).* Also included in many circuits are such electrical *devices* as buzzers, lamps, relays, transformers, and

motors. All passive devices can be reduced to circuit component combinations of *R, L,* and *C.*

Sources of electrical energy (and devices that amplify) are known as *active* devices. These include generators, chemical batteries, and solar cells.

A distinction must be made between *electric* and *electronic* circuits. An *electric* "power" circuit usually involves the transfer of relatively large amounts of energy to produce heat, light, motion, and so on. An *electronic* circuit involves the transfer of much smaller amounts of energy. In addition to the three basic passive devices, electronic circuits involve additional active devices such as semiconductors (transistors and integrated circuits) and electron tubes.

Most electric circuits have four main parts:

1. A *source* of electrical energy such as a chemical battery, generator, or solar cell.
2. A *load* or output device such as a lamp, motor, or loudspeaker.
3. *Conductors* to transport the electrical energy from the source to the load. The conductors are usually copper or aluminum wires.
4. A *control* device such as a switch, thermostat, or relay. This device controls the flow of energy to the load.

The energy source may be either direct current (dc) or alternating current (ac). *Direct* current flows in only one direction and is generally constant in value; *alternating* current continuously changes in value and periodically reverses direction of flow (polarity). Figure 1-2 represents the ac and dc sources in graphic form, along with the symbols used to represent them in circuit diagrams.

The source applies an *electromotive force* (emf) or *potential difference* to the circuit. The emf is measured in *volts* (V) and is related to the *work* that a source can do to move an electrical *charge* through a circuit. This flow of an electrical charge is called *current,* and is measured in *amperes* (A). The *frequency* of an alternating current is the number of cycles (changes in polarity) per second, and is measured in *hertz* (Hz). In the United States and Canada, the typical domestic supply voltage is 120/240 volts ac at a frequency of 60 Hz. (Potential difference and current will be discussed in greater detail in Chapter 2.)

A flashlight is a good example of a simple *series* dc circuit. To represent such a circuit, a *pictorial diagram* may be used. In such a diagram, the physical appearance of each component is shown. Engineers, electricians, and technicians, however, prefer the method of representing a circuit called a *schematic diagram.* This type of diagram,

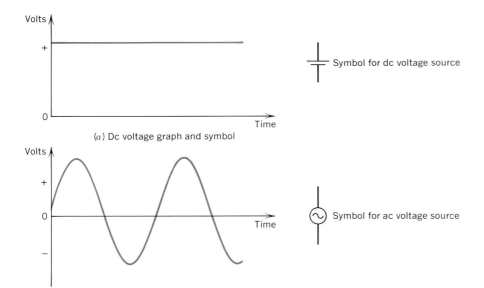

(a) Dc voltage graph and symbol

Symbol for dc voltage source

(b) Ac voltage graph and symbol

Symbol for ac voltage source

FIGURE 1-2
Graphs showing how voltage is related to time for dc and ac voltages.

often referred to simply as a "schematic," uses interconnected symbols that *represent* electrical components. A schematic diagram is much easier to draw than a pictorial diagram, especially for a complicated circuit. The flashlight circuit is shown in both pictorial and schematic form in Fig. 1-3. Note that the components are not labeled in the schematic diagram. Symbols for various types of components are standardized for easy recognition.

Also note the length of the lines in the battery symbols. Each cell is represented by a long line for the *positive* terminal and a short line for the *negative* terminal. The polarity labels (+ and −) on the diagram are optional.

A *series* circuit, like the one shown, provides only one path for current flow. This is the definition of a series circuit. (A *parallel* circuit provides two or more paths.) In this circuit, the *source* of electrical energy consists of two

$1\frac{1}{2}$-V dry cells (batteries) connected in series to supply a total emf of 3 V. The *load* is a 3-V lamp, and the *control device* is a knife-blade switch connected between the source and the load. The conductors in this example are provided by the metal case of the flashlight. The same circuit could be constructed using copper wire for the conductors.

When the switch is open ("off"), as shown in Fig. 1-3, no current flows, and thus the lamp does not light. When the switch is closed ("on"), a complete current path exists, and electrical energy flows from the cells to light the lamp. High resistance to current flow in the lamp's filament causes it to heat up to a white hot incandescence. In the lamp, electrical energy is converted to both light and heat energy.

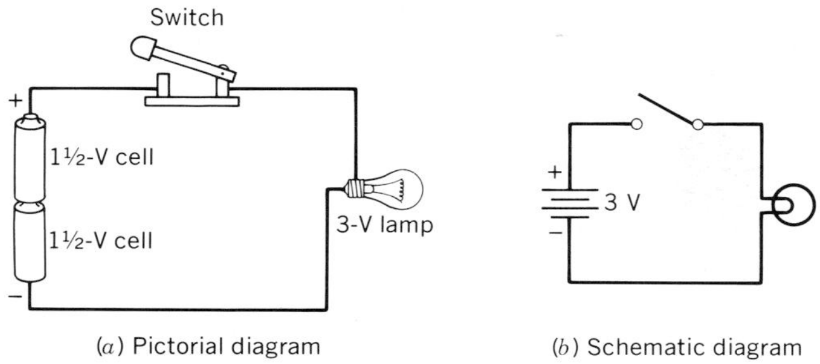

1-8 ELECTRICAL SYMBOLS FOR RESISTANCE, INDUCTANCE AND CAPACITANCE, AND ELECTRICAL DEVICES

The symbols used in electrical and electronic schematic diagrams for the three basic passive components (resistors, inductors, and capacitors) are shown in Table 1-4. Photographs of some representative types of these components are shown in Fig. 1-4. For a more complete list of electrical and electronic components and devices, see Appendix A.

Switch

+

$1\frac{1}{2}$-V cell

$1\frac{1}{2}$-V cell

3-V lamp

−

(a) Pictorial diagram

+

3 V

−

(b) Schematic diagram

FIGURE 1-3
Simple electric circuit.

TABLE 1-4

Schematic Symbols for Resistors, Inductors, and Capacitors

Type	Symbol		Property	Unit
Fixed resistor	⎯\/\/\/⎯	R	Limits or opposes	Ohm
Variable resistor	⎯\/\/\/⎯	R	current	
Fixed inductor or coil (air core)	⎯ℓℓℓ⎯	L	Opposes *change* of current	Henry
Iron core inductor	⎯ℓℓℓ⎯	L		
Variable inductor	⎯ℓℓℓ⎯	L		
Fixed capacitor	⎯⎬⎬⎯	C	Opposes *change* of voltage	Farad
Polarized capacitor	⎯⎬⎬+	C		
Variable capacitor	⎯⎬⎬⎯	C		

FIGURE 1-4

Typical resistors, inductors, and capacitors. (Components are mounted on bases for breadboarding in a circuit.)

(*a*) Resistors

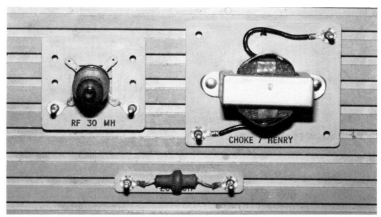

(*b*) Inductors (coils)

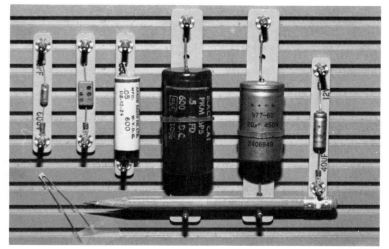

(*c*) Capacitors

Electric current naturally encounters some resistance (opposition to flow) as it flows through any circuit. In some cases, it is desirable to *limit* the current by intentionally increasing the amount of resistance. This is done by inserting devices known as *resistors* that have known resistance values. Resistance, symbolized by R, is measured in *ohms*. The Greek letter omega (Ω) is used as the abbreviation for ohms. Resistors are available in a wide range of standard resistance values. Some typical values are $R = 220 \ \Omega$, $4.7 \ k\Omega$, and $1 \ M\Omega$. The value of a resistor is often identified by a series of colored stripes around its body (See Section 3-11.1).

An inductor is a coil of wire that may be wound on an iron core, and has the property of opposing any *change* of current through it. This property is called inductance, and is symbolized by L. The basic unit of measurement for inductance is the henry (H). Typical inductance values are $L = 100 \ \mu H$, $30 \ mH$, and $7 \ H$.

Capacitors consist of conductors separated by an insulating material or *dielectric*. Capacitance, symbolized by C, is the property of being able to store an electrical charge and oppose any *change* in voltage. The basic measuring unit for capacitance is the *farad* (F), but more practical values are the microfarad (μF) and picofarad (pF). Typical values are $C = 270 \ pF$, $0.1 \ \mu F$, and $100 \ \mu F$.

1-9 PICTORIALS AND SCHEMATICS

A pictorial wiring diagram for a power supply and the corresponding schematic diagram are shown in Fig. 1-5.

There are a number of points to observe when comparing the diagrams.

1. Terminal strips used as solder tie-points are shown in the pictorial diagram, but *do not* appear on the schematic.
2. The connection dots shown on the schematic *do not* necessarily coincide with the tie-points in the pictorial.
3. Straight horizontal and vertical connecting lines are used in the schematic. They bear little resemblance in location and length to the wires in the actual circuit.
4. The schematic is usually read left-to-right, following the current flow.
5. For reference purposes, each component on the schematic is identified by a letter and number such as R_1, R_2, S_1, C_1, and so on.

6. Only the *values* and *ratings* of the components are shown on the schematic; no other labeling is used.

The pictorial diagram is used for component recognition and wiring purposes. The schematic diagram helps to show how a circuit works and is used as a reference for tracing signals through the circuit during troubleshooting. Frequently, the schematic is the only diagram available. When a schematic is not available, it is often worthwhile to trace the circuit and draw your own schematic as an aid in troubleshooting. (You will be given a chance to practice drawing schematics at the end of the chapter.)

In the absence of a pictorial diagram, it is also possible to use a schematic as a guide to wiring a circuit. In the case of laboratory "breadboarding," this is most easily done by laying out the components on the circuit board in the same relative positions shown in the schematic. As each wiring connection is made, you should check off or cross out that line on the schematic to keep track of your work.

1-10 SAFETY

One of the reasons for introducing the basic electric circuit and the units describing emf, current, and resistance in this chapter is to help make you aware of some important precautions in working with electricity.

If you touch a high-voltage ac or dc source, your body's resistance may be low enough to allow a lethal amount of current to flow through it. In effect, you become part of the electrical circuit. The amount of current needed to disrupt the regular pumping of your heart is in the order of a few milliamperes. It is, of course, difficult to get exact values of this current since figures vary widely. You can get some idea of the effects of electrical currents passing through the body by studying Fig. 1-6.

If you are working in a wet location or are in contact with a grounded metal water pipe, the normal 120-V domestic supply voltage can easily kill you. Always follow these precautions:

1. **If at all possible, avoid working on live circuits.**
2. **When working on a circuit carrying 120 V or higher, use only one hand to make voltage measurements. Keep the other hand in your pocket or behind your back. This will prevent creation of a complete path for current to flow**

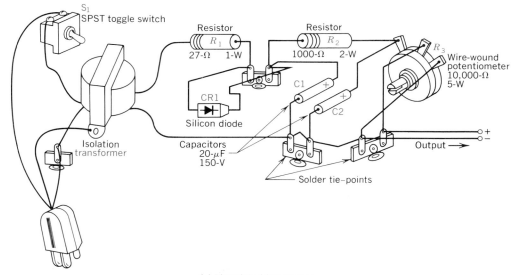

(a) Pictorial wiring diagram

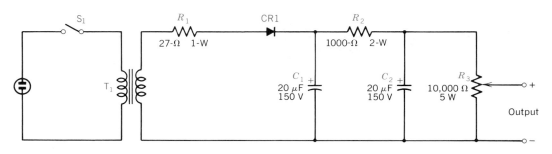

(b) Schematic diagram

FIGURE 1-5
Pictorial and schematic diagrams of a power supply. (Courtesy of McGraw-Hill.)

through your heart and lungs. (It does not, however, avoid the possibility of burns on the hand being used.)

3. Before working with electricity, always remove rings, wrist watch, and any other metal jewelry that could come in contact with a conductor. A belt buckle may also provide contact, so it should be moved to the side or back.

4. If the circuit includes a high-voltage electrolytic capacitor, remove the charge from the device by shorting it with a screwdriver. A capacitor can hold a dangerous charge for hours after the power has been turned off.

5. Be careful to avoid contact with components that become quite hot during circuit operation, such as resistors, vacuum tubes, or power transistors. Let them cool sufficiently before attempting to remove them.

6. Do not work with power cords that are burned or frayed. They present a serious shock hazard.

7. Know the location of the nearest fire extinguisher designed for use on *electrical* fires, and how to use it properly.

8. When working on high voltage, try not to work alone. Make sure the other person knows the circuit breaker location.

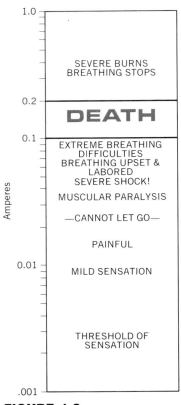

FIGURE 1-6
The effects on the body of electric currents.
(Diagram courtesy of Graymark International, Inc.,
Irvine, Calif.)

9. Concentrate on what you are doing, and treat all
    voltages with the proper respect.

Fortunately, most electronic circuits (those involving
transistors and integrated circuits) seldom involve voltages
in excess of 15 V. But when these voltages are obtained
from a 120-V (or higher) supply, precautions are always
in order.

1-11   READING SCALES OF MULTIRANGE METERS

1-11.1   Types of Meters

Although direct-reading digital meters are now common,
you are likely to use deflection (or analogue) meters at
some time. Your first electrical experiments will probably

FIGURE 1-7
Typical VOM. (Courtesy of Triplett Corporation.)

involve readings of voltage, amperes, and ohms. These
may be made on separate instruments (voltmeter, ammeter, or ohmmeter), but can also be made on a single instrument called a *multimeter*.

A typical multimeter has a *function* switch used to select the mode of operation, and a *range* switch to provide
a readable deflection for a wide range of measured quantities. The *VOM* (volt-ohm-milliammeter) shown in Fig. 1-
7 combines the function and range switches into a single
selector. This portable meter can measure ac or dc voltages, resistance, and dc currents.

Another common multirange instrument is the electronic voltmeter (EVM) or vacuum-tube voltmeter
(VTVM). Solid-state versions are available, but these voltmeters are usually powered from a 120-V outlet. The
EVMs can measure ac and dc volts and resistance. Some
solid-state instruments can also measure ac and dc current.
(See Fig. 1-8.)

A multirange milliammeter that can measure current
from 1 mA to 1 A is shown in Fig. 1-9. This type of meter
generally has a lower resistance than is found in the milliammeter portion of a multimeter.

FIGURE 1-8

Typical electronic voltmeter or VTVM. (Courtesy of Heath Company.)

A digital multimeter, capable of measuring ac and dc voltage as well as ac and dc current and ohms, is shown in Fig. 1-10. This is a $4\frac{1}{2}$-digit meter. This means that it has a five-digit display, with the first digit only able to indicate a 1 when necessary.

Detailed information on the operation and use of meters

FIGURE 1-9

Typical multirange dc milliammeter. (Courtesy of Hickok Teaching Systems, Inc., Woburn, Mass.)

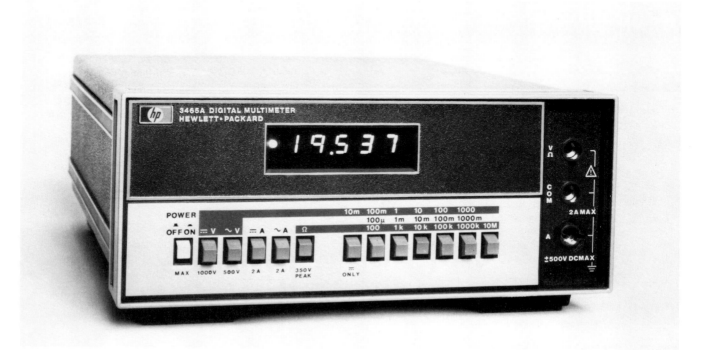

FIGURE 1-10

A $4\frac{1}{2}$-digit five-function multimeter. (Courtesy of Hewlett-Packard.)

will be presented in later chapters. However, the methods used to connect a voltmeter, ammeter, and ohmmeter are shown in Fig. 1-11. Note that polarity must be observed when connecting a voltmeter or ammeter in a dc circuit. **The ohmmeter can be used to measure resistance *only* if the voltage is first removed from the circuit.**

1-11.2   Reading the Scales

If you use a multirange ammeter or voltmeter, it is important to know the function of the range switch. That switch selects the amount of current or voltage that will cause *full-scale deflection* (FSD) of the meter. When you measure an unknown quantity, begin with the highest range and reduce to lower ranges step-by-step until you obtain a deflection somewhere between mid-scale and full-scale. This method will protect the meter from damage and help ensure accurate readings.

Meters do not generally have as many scales on the face as there are ranges. Instead, scales are used to read several different ranges. This is usually done by multiplying or dividing the scale numbers by 10 or 100. To read a multirange meter, check the setting of the range switch, then determine which scale shows a full-scale deflection value that corresponds to the range setting. Read the number off the scale at the point where the pointer has come to rest.

Two examples follow to demonstrate this method.

EXAMPLE 1-4

Determine the readings of the multirange milliammeter in Fig. 1-12 when the range switch is on:
a.   50 mA
b.   100 mA

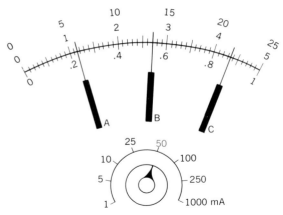

FIGURE 1-12

Multirange dc milliammeter scale with range switch on 50 mA.

Solution

a.   With the range switch on 50 mA, the FSD must be 50 mA. Therefore, use the 0–5 scale with each number multiplied by 10.
At A: reading is between 10 and 20 = **11.5 mA**
At B: reading is between 20 and 30 = **27 mA**
At C: reading is between 40 and 50 = **43.5 mA**
b.   With the range switch on 100 mA, the FSD must be 100 mA. Therefore, use the 0–1.0 scale with the decimal point moved to the right two places.
At A: reading is between 20 and 40 = **23 mA**
At B: reading is between 40 and 60 = **54 mA**
At C: reading is between 80 and 100 = **87 mA**

NOTE   The only direct reading scales are those corresponding to ranges of 1, 5, and 25 mA.

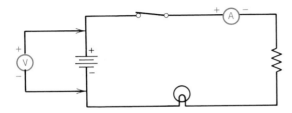

(*a*) Connection of voltmeter in parallel with the source. The ammeter is in series with the load.

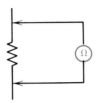

(*b*) The ohmmeter is connected across the component with no applied voltage

FIGURE 1-11

Proper methods of connecting voltmeter, ammeter, and ohmmeter to measure voltage, current, and resistance.

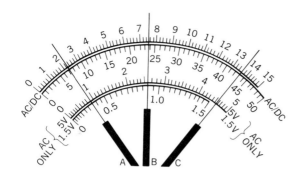

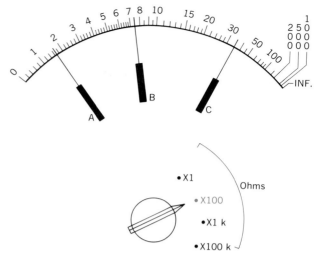

FIGURE 1-13

Multirange ac/dc voltmeter scales with range and function switches set to 5-V dc.

EXAMPLE 1-5

Determine the readings of the multirange voltmeter in Fig. 1-13 for the following settings:

a. Function: dc; Range: 5 V
b. Function: ac; Range: 150 V

Solution

a. With the range switch on 5 V dc, the FSD must be 5 V on the ac/dc scale. Therefore, we must use the 0–50 scale with each number divided by 10.
   At A: reading is between 0.5 and 1.0 = **0.72 V**
   At B: reading is between 2.0 and 2.5 = **2.37 V**
   At C: reading is between 4.0 and 4.5 = **4.30 V**
b. With the range switch on 150 V ac, the FSD must be 150 V on the ac/dc scale. Therefore, we must use the 0–15 scale with each number multiplied by 10.
   At A: reading is between 20 and 30 = **23 V**
   At B: reading is between 70 and 80 = **75 V**
   At C: reading is between 130 and 140 = **136 V**

NOTE The direct reading 1.5-V and 5-V scales may be used for ac voltages only.

FIGURE 1-14

Multirange ohmmeter set on ×100 range.

On an ohmmeter, the scale always has *zero* at one end and *infinity* at the other. Figure 1-14 shows the type of scale normally found on EVMs, with zero on the left and infinity on the right. On VOMs the scale is usually reversed, with infinity on the left and zero on the right. In both cases, however, the method of reading is the same. The value read from the scale is multiplied by the range switch setting ($\times 1$, $\times 100$, and so on).

EXAMPLE 1-6

Determine the readings of the multirange ohmmeter in Fig. 1-14 on the following ranges:

a. ×100
b. ×100 k

Solution

a. At A: reading = 1.7 × 100 = **170 Ω**
   At B: reading = 7.5 × 100 = **750 Ω**
   At C: reading = 35 × 100 = **3.5 kΩ**
b. At A: reading = 1.7 × 100 k = **170 kΩ**
   At B: reading = 7.5 × 100 k = **750 kΩ**
   At C: reading = 35 × 100 k = **3.5 MΩ**

In a digital multimeter, infinity or overrange is usually indicated by the display 1. _____. That is, the display is blank except for the first digit and a decimal point. Some digital meters may show a flashing or blinking series of zeros for infinite resistance or an overrange reading.

Unlike the analogue meters, the measurement of resistance does not require the multiplication of the display by the range selected. For example, in Fig. 1-10, with the function switch on ohms, and the range switch on 10 kΩ, the displayed number of 19.537 is interpreted as 19.537 kΩ. If the resistance being measured exceeds 19.999 kΩ, the display will revert to 1. _____, and the next range must be selected. Similarly, on the 100-kΩ range, a display of 58.75 would be read as 58.75 kΩ, with a maximum reading possible of 199.99 kΩ.

Other features of digital meters are covered in detail in Section 13-8.

Multirange analogue VOMs and EVMs may have all or some combination of the three sets of scales (shown in Figs. 1-12, 1-13, and 1-14) included on the face of the meter. Great care must be taken to determine which scale and set of numbers must be used to correspond with the settings of the function and range switches.

SUMMARY

1. The size of numbers that a low-cost calculator can process is limited by the calculator's display capability.

2. Larger numbers can be processed by a scientific calculator that makes use of scientific notation. This allows the writing of a given number, regardless of size, as a number between 1 and 10 (but not including 10) and multiplied by 10 raised to a positive or negative exponent.

3. For measured data, the least accurate information fed into the calculator when multiplying, dividing, or taking roots, determines the accuracy of the result, which should be rounded off accordingly.

4. Conversion of English (customary) units to metric units of the MKS system can be done by using the conversion ratios given in Table 1-2.

5. Prefixes attached to the metric unit name are used to indicate multiples and submultiples. Table 1-3 lists the prefixes.

6. An electric circuit consists of interconnected components that form a complete path for the flow of electric current.

7. The four main parts of most electric circuits are the voltage source, load, conductors, and a control device.

8. The three basic passive electrical circuit components are the resistor, the inductor, and the capacitor. Symbols for these components are shown in Table 1-4.

9. A pictorial diagram shows the appearance of the components; a schematic diagram uses symbols to represent components.

10. A small current (only a few milliamperes) passing through the body can be fatal. Treat all voltages with respect and always follow safety precautions.

11. To read an analogue multirange meter correctly, you must zero the meter, identify the function, the range, and the appropriate scale.

12. On a voltmeter or ammeter, the range selector is used to choose the amount of voltage or current that will result in full-scale deflection (FSD) of the meter pointer.

13. The resistance reading from the scale of an analogue ohmmeter must be multiplied by the range setting to find the proper value. This is not necessary in a digital ohmmeter where a high enough range must be selected to avoid an overrange indication.

SELF-EXAMINATION

Answer true or false
(Answers at back of book)

1-1. The scientific notation for 1215 is 1.215×10^3. _____

1-2. The scientific notation for 0.0794 is 7.94×10^{-1}. _____

1-3. The standard decimal form for 8.98×10^2 is 898. _____

1-4. The standard decimal form for 3.64×10^{-3} is 0.00364. _____

1-5. The number in question 1 is accurate to four significant figures. _____

1-6. The number in question 2 is accurate to four significant figures. _____

1-7. If 0.494 is rounded-off to two significant figures, the result is 0.49. _____

1-8. If 0.494 is rounded-off to one significant figure, the result is 0.5. _____

1-9. A conversion ratio may be used to convert a quantity given in the FPS system to the MKS system. _____

1-10. One millisecond is one millionth of a second. _____

1-11. A time interval of 27 µs equals 0.027 ms. _____

1-12. A time interval of 300 ps equals 3 ns. _____

1-13. A schematic diagram uses pictures of the components, but a pictorial diagram uses symbols. _____

1-14. Symbols are graphic drawings that represent electrical components. _____

1-15. When a switch is opened in a circuit, the current stops flowing. _____

1-16. A resistor opposes current whereas an inductor opposes a change of current. _____

1-17. The unit of resistance is the henry, and for inductance it is the ohm. _____

1-18. A capacitor opposes change of voltage; its unit of measurement is the farad. _____

1-19. Only a few milliamperes of current through the body is sufficient to be painful. _____

1-20. A multimeter may be of the VOM type that can measure voltage, current, and resistance. _____

1-21. An EVM is portable, but a VOM needs a 120-V ac source. _____

1-22. A range switch in a multirange voltmeter or ammeter selects the quantity required to cause full-scale deflection. _____

1-23. When using an ohmmeter, it is necessary to multiply the ohms scale reading by the range setting. _____

1-24. If, in Fig. 1-12, the range switch is on 1 mA, the current indicated by the pointer at position C is 8.7 mA. _____

1-25. If, in Fig. 1-13, the voltmeter is set to 5 V ac, the voltage indicated by the pointer at position C is 4.7 V. _____

1-26. If, in Fig. 1-14, the ohmmeter range is set to × 1k, the resistance indicated by the pointer at position C is 350 kΩ. _____

1-27. An ammeter is always connected in series with the component whose current it is going to measure. _____

1-28. A voltmeter is always connected in parallel with the component in order to measure the voltage across the component. _____

REVIEW QUESTIONS

1. What is the largest number that a 10-digit readout calculator can display if it does not use scientific notation?
2. What are the largest and smallest (nonzero) numbers that a 10-digit readout calculator can display if it does use scientific notation?
3. Why is scientific notation useful?
4. What does it mean when a given calculator display blinks repeatedly or, on another calculator, decimal points appear after each number?
5. What is one way of converting a decimal number to scientific notation on some calculators?
6. What is the rule for rounding off a calculated number?
7. What are the three systems for measuring physical units?
8. What conversion factor would you use to convert inches to centimeters?
9. Why does multiplying a quantity by a conversion ratio not change that quantity?
10. What do the symbols T, M, m, μ, p stand for when used as unit prefixes?
11. Draw pictorial and schematic diagrams of two electric circuits and indicate the four main parts in each.
12. What is the main difference between dc and ac voltage?
13. Which passive electrical component has the property of opposing any change of current?
14. What is the main use of a pictorial diagram?
15. Give at least three differences between a schematic and a pictorial.
16. Describe how you would "breadboard" a circuit from a schematic diagram and how you would check it.
17. List the important safety precautions in working with electricity.
18. In using a multirange meter, what range should be used when making a measurement of an unknown voltage or current?
19. What precaution *must* be observed when using an ohmmeter to measure resistance in a circuit?
20. Explain the relation between the ranges on a range switch and the various scales on the meter face.
21. What is the main difference between a VOM and an EVM?
22. What other difference is there between a VOM and an EVM as far as the resistance scales are concerned?
23. Which of the two meters in Question 21 generally measures current?
24. What does it mean to say that polarity must be observed when dc voltmeters and ammeters are used to make voltage and current measurements?
25. In what way (series or parallel) are voltmeters and ammeters connected in a circuit?

PROBLEMS

(Answers to odd-numbered problems at back of book)

1-1. Convert to scientific notation:
    a. 571,000
    b. 0.0000645
    c. 2306

1-2. Convert to scientific notation:
  a. 49,960
  b. 0.007825
  c. 1,100,000

1-3. Round off the numbers in Problem 1-1 to two significant figures.

1-4. Round off the numbers in Problem 1-2 to three significant figures.

1-5. Round off the numbers in Problem 1-1 to one significant figure.

1-6. Round off the numbers in Problem 1-2 to two significant figures.

1-7. Express the following in standard decimal form:
  a. 4.79×10^5
  b. 1.135×10^{-3}
  c. 9.079×10^2

1-8. Express the following in standard decimal form:
  a. 3.57×10^4
  b. 2.07×10^{-4}
  c. 8.064×10^2

1-9. Carry out the following computations using a calculator, expressing the answers in scientific notation and rounded off to show the appropriate significant figures:
  a. $(1.357 \times 10^4) \times (7.25 \times 10^{-3})$

  d. $\dfrac{6.3 \times 8.40}{0.45 + 3.39}$

  b. $\dfrac{1.357 \times 10^4}{7.25 \times 10^{-3}}$

  e. $\sqrt{7.4} + \sqrt{23.1}$

  c. $\dfrac{\sqrt{1.357 \times 10^4}}{(7.25 \times 10^{-3})^2}$

  f. $\dfrac{5.42\sqrt{8.77}}{13.2 - 19.9}$

1-10. Repeat Problem 1-9 for the following:
  a. $(8.079 \times 10^2) \times (3.574 \times 10^{-7})$

  b. $\dfrac{3.574 \times 10^{-7}}{8.079 \times 10^2}$

  c. $\dfrac{\sqrt{3.574 \times 10^{-7}}}{(8.079 \times 10^2)^3}$

  d. $44.3 \times 10^2 \times 10^3$

  e. $\dfrac{927 \times 10^{-4} \times 10^1}{10^6 \times 10^{-1}}$

1-11. Using the conversion factors in Table 1-2 (or conversion code on your calculator if available), make the following conversions:
  a. 13.5 in. to cm
  b. 2.89 ft to m
  c. 125 mi to km
  d. 8.75 gal to l

1-12. Repeat Problem 1-11 for the following:
  a. 32 oz to g
  b. 165 lb_m to kg
  c. 150 lb_f to N
  d. 75 hp to kW

1-13. a. Convert 50 mA to A.
  b. Convert 50 mA to μA.

1-14. a. Convert 0.15 A to mA.

  b. Convert 0.15 A to μA.

1-15. a. Convert 2500 V to kV.

  b. Convert 0.35 V to mV.

1-16. a. Convert 465 kV to V.

  b. Convert 50 mV to V.

1-17. a. Convert 1200 Ω to kΩ.

  b. Convert 0.75 Ω to mΩ.

  c. Convert 1500 kΩ to MΩ.

1-18. a. Convert 5.1 kΩ to Ω.

  b. Convert 270 kΩ to MΩ.

  c. Convert 2.2 MΩ to kΩ.

1-19. a. Convert 50 ns to ps.

  b. Convert 50 ns to μs.

1-20. a. Convert 50,000 μF to F.

  b. Convert 30 mH to H.

  c. Convert 1500 pF to μF.

1-21. a. Convert 213 ps to ns.

  b. Convert 0.001 F to μF.

  c. Convert 0.5 mH to μH.

1-22. Given the pictorial diagram of the one-transistor radio in Fig. 1-15, draw the corresponding schematic. Refer to Appendix A for symbols.

1-23. Determine the readings of the multirange milliammeter in Fig. 1-12 when the range switch is on:

  a. 1 mA

  b. 25 mA

  c. 1000 mA

1-24. Repeat Problem 1-23 for the following ranges:

  a. 5 mA

  b. 10 mA

  c. 250 mA

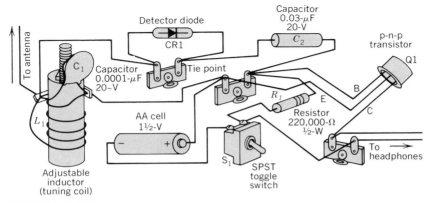

FIGURE 1-15

Pictorial of a transistor radio for Problem 1-22. (Courtesy of McGraw-Hill.)

1-25. Determine the readings of the multirange voltmeter in Fig. 1-13 for the following range and function settings:
   a.   1.5 V ac
   b.   1.5 V dc
   c.   50 V dc
   d.   150 V ac

1-26. Repeat Problem 1-25 for the following range and function settings:
   a.   5 V ac
   b.   15 V dc
   c.   500 V ac
   d.   1500 V dc

1-27. Determine the readings of the multirange ohmmeter in Fig. 1-14 on the following ranges:
   a.   ×1
   b.   ×1 k

1-28. What range (and function) settings would you use to make the following measurements on a multirange multimeter as in Figs. 1-12, 1-13, and 1-14?
   a.   6-V battery
   b.   120-V ac outlet
   c.   A current of 28 mA dc
   d.   3.15 V ac
   e.   10-Ω resistor
   f.   8.2-kΩ resistor
   g.   4.7-MΩ resistor

CHAPTER 2

BASIC ELECTRICAL QUANTITIES

The word *electricity* is derived from the Greek word *elektron,* which means amber. When amber (a natural resin) is rubbed with fur, the amber becomes capable of attracting lightweight particles such as tiny pieces of paper. This *electrification by friction* is caused by the transfer of electrons from the fur to the amber. When this occurs, the two materials are said to be *oppositely* charged—a *potential difference* exists between the two materials. This potential difference has an effect similar to that between the two electrodes of a battery. Often referred to mistakenly as a "pressure," this potential difference or electromotive force (emf) causes electrons to flow through a metallic conductor connected across the electrodes. This flow of electrons is an electric *current,* which can perform such useful functions as producing heat or light. In the case of a battery, a chemical action maintains the potential difference and thus, the flow of current.

In this chapter, you will learn about the basic units of electricity, including resistance—the opposition to the flow of current.

2-1 STATIC ELECTRICITY

Your first experience with static electricity probably took place on a dry winter day when you walked across a carpet in your rubber-soled shoes, then touched a metal door-knob. The spark and mild shock you experienced was a result of your body becoming "electrified." As you walked across the room, friction between your rubber-soled shoes and the carpeting caused your body to accumulate excess electrons. This is called *static* electricity because the charges are at rest. When you touched the door-knob, a brief flow of current took place and you felt a shock.

In the same way, if amber, ebonite, or a hard rubber rod is rubbed with wool or fur, the amber, ebonite, or rubber accumulates excess electrons at the expense of the wool or fur. Benjamin Franklin, the eighteenth-century statesman and scientist, is credited with first referring to the rubber rod as *negatively charged* and the fur as *positively charged*. (Franklin, who did not know about electrons, speculated that the fur *gained* some charge and thus became *positive,* while the rubber *lost* some charge and became *negative*.)

The opposite effect takes place when a glass rod is rubbed with silk. The glass rod becomes positively charged because some electrons have been transferred to the silk, making the cloth negative. These effects are shown in Fig. 2-1, with positive and negative signs placed around the two rods.

Both rods were originally *neutral* or uncharged. After being rubbed with fur, the rubber rod is said to have a *surplus* of electrons, while the glass rod, after being rubbed with silk, is said to have a *deficiency* of electrons.

The opposite charges of the two rods can be demonstrated by briefly *touching* them to two lightweight pith balls suspended from fine threads. Each ball will become charged to the same *polarity* as the rod it is touched by. The effect is shown in Fig. 2-2c. The two balls are at-

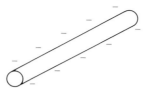 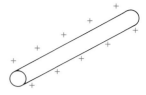

(a) Rubber rod rubbed with fur   (b) Glass rod rubbed with silk

FIGURE 2-1
Two different rods oppositely charged by friction.

tracted to each other because they are oppositely charged. If the same rod (rubber *or* glass) is used to touch both balls, they will have *like* charges and be repelled from each other, as shown in Figs. 2-2a and 2-2b. This demonstrates an important rule of electrical *forces:*

Like charges *repel; unlike* charges *attract*

2-2 COULOMB'S LAW

In 1785, the French physicist Charles Coulomb determined how much force of attraction or repulsion is exerted between two charged bodies separated by a given distance. The force may be calculated by using Coulomb's law (see Fig. 2-3):

$$F = k \frac{Q_1 Q_2}{r^2} \quad \text{newtons} \quad (2\text{-}1)$$

where: F is the force in newtons (N)*

---

*One newton is the force required to give a mass of 1 kg an acceleration of 1 m/s².

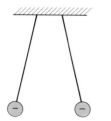   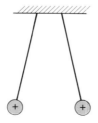   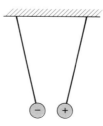

(a) Repulsion produced by rubber rod touching both balls   (b) Repulsion produced by glass rod touching both balls   (c) Attraction produced by oppositely charged pith balls

FIGURE 2-2
Pith ball demonstration of like charges repelling, unlike charges attracting.

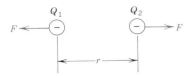

FIGURE 2-3
Quantities and forces involved in Coulomb's law.

Q_1, Q_2 are the charges of the two bodies in coulombs (C)

r is the separation of the two bodies in meters (m)

k is a constant, equal to 9×10^9 Nm²/C²

Note that all quantities in the equation must be expressed in *basic* units, regardless of the units in which they are given.

Note also that Eq. 2-1 involves the product of two charges. If both are positive or both are negative, F will be positive. A *positive* force means *repulsion*. If only *one* charge is negative, then F will be negative. A *negative* force must be interpreted as a force of *attraction*.

Coulomb's law is being considered at this time to determine what units are used to measure *charge*. You have already seen how charge depends upon the accumulation (or deficiency) of a number of electrons. Andrew Millikan, in experiments carried out during the years 1909–1913, determined that the combined charge of 6.24×10^{18} electrons should be called 1 coulomb (1 C).

Now it should be clear why Q is used to represent charge. **A coulomb is a *quantity* of electrical charge equal to the combined charge of 6.24×10^{18} electrons. Therefore, 6.24×10^{18} electrons = 1 coulomb, 1 C (of charge).**

> **NOTE** A coulomb does not have any mass, since it is not a particle but rather a *number*. It represents a *quantity* of charge. Note also that Q is the *letter symbol* for charge, but the coulomb, the unit in which charge is *measured*, is abbreviated C.

Evidently the charge of one electron (what Millikan actually determined) is given by:

charge of 1 electron

$$= \frac{1}{6.24 \times 10^{18}} \text{ C} = 1.60 \times 10^{-19} \text{ C}$$

A coulomb is a rather *large* quantity of charge. Typical electrostatic values obtained in the laboratory seldom exceed a few millionths of a coulomb, so the microcoulomb (μC) is often used. (See Example 2-1.)

EXAMPLE 2-1

Two pith balls, separated by 3 cm, are each charged positively with 0.25 μC.
a. Calculate the force of repulsion in newtons.
b. Convert the force to pounds.

Solution

a. $F = k\dfrac{Q_1 Q_2}{r^2}$ (2-1)

$= 9 \times 10^9 \dfrac{\text{Nm}^2}{\text{C}^2}$

$\times \dfrac{(+0.25 \times 10^{-6} \text{ C}) \times (+0.25 \times 10^{-6} \text{ C})}{(3 \times 10^{-2} \text{ m})^2}$

$= 9 \times 10^9 \dfrac{\text{Nm}^2}{\text{C}^2} \times \dfrac{0.0625 \times 10^{-12} \text{ C}^2}{9 \times 10^{-4} \text{ m}^2}$

$= \textbf{0.625 N}$

b. $F = 0.625 \text{ N} \times \dfrac{1 \text{ lb}}{4.45 \text{ N}}$ (see the conversion factors in Table 1-2.)

$= \textbf{+0.14 lb}$ (repulsion)

EXAMPLE 2-2

Calculate the force of attraction between an electron and the positive nucleus of an atom having a charge 29 times that of an electron if the distance between them is 1×10^{-10} m.

Solution

$F = k\dfrac{Q_1 Q_2}{r^2}$

$= 9 \times 10^9 \dfrac{\text{Nm}^2}{\text{C}^2}$

$\times \dfrac{(-1.6 \times 10^{-19} \text{ C}) \times (+29 \times 1.6 \times 10^{-19} \text{ C})}{(1 \times 10^{-10} \text{ m})^2}$

$= \textbf{--6.68} \times \textbf{10}^{-7} \textbf{ N}$ (attraction)

2-3   APPLICATIONS OF STATIC ELECTRICITY (ELECTROSTATICS)

A number of modern applications in business and industry make use of the attractive force of charged particles for oppositely charged surfaces. Three are mentioned here, but they are only a few among the many beneficial uses of electrostatic forces.

Anyone who has done ordinary spray painting knows that a lot of paint remains suspended in the air as a vapor cloud and is wasted. Typically, ordinary spray painting is only 65% efficient, and requires a good ventilation system to clear the air of the suspended paint particles.

An *electrostatic* spray gun (Fig. 2-4) is up to 98% efficient, so as little as 2% of the paint is lost to the air and painting can be done without a face mask. The electrostatic painting system uses a powder, or resin material, that is mixed with the paint and carried to the gun by compressed air. A high-voltage lead (several thousand volts) at the tip of the spray gun imparts a positive charge to the particles of powder as they are sprayed. The part to be painted is grounded and therefore at a lower potential, and attracts the positively charged particles. (In effect, the part to be painted is connected to the negative side of the high voltage and the spray gun to the positive side.)

The electrostatic painting technique provides uniform paint coverage as thin as 1 mil (0.001 in.) on cold parts and up to 20 mil in thickness on preheated parts. The paint application is usually followed by heat curing in an oven. Electrostatic painting is economically competitive with ordinary liquid spray painting.

Another application of electrostatics is in air filtration. An *electrostatic* air cleaner can capture microscopic particles that would normally pass right through an ordinary air filter. The air passes through a series of plates that impart a positive charge to the dust and dirt particles it carries. The air then passes through a negatively charged filter, which attracts and traps the positively charged particles. A typical home air cleaning system may involve potential differences of about 2 kV.

A third application is *xerography,* or electrostatic printing, widely used for office copiers.

A plate coated with the light-sensitive element selenium is first given a positive electrostatic charge, then light reflected from the material to be copied is projected onto the plate through a lens system. When the light strikes the selenium plate, it sensitizes the selenium (makes it more conductive) and produces a positively charged hidden image of the material to be copied. Next, the plate is dusted with *toner,* a negatively charged black powder that is attracted to the positively charged image. A sheet of paper with a negative charge is placed against the plate, and the powder image is attracted to it. The image is "fixed" on the paper by heat, which melts the powder and produces a perfect copy of the original material.

2-4 POTENTIAL DIFFERENCE

What would happen to the two oppositely charged rods shown in Fig. 2-1 if they were joined together by a conductive metallic wire? (See Fig. 2-5.)

In a very short time, the two rods are no longer oppositely charged. Since like charges repel and opposite charges attract, electrons travel from the location of electron surplus (the negatively charged rod) through the metallic conductor to the location of electron deficiency (the positively charged rod). **This flow of electric charges is called *electric current*. And the ability (or *potential*) to cause this flow between two *differently* charged bodies is called *potential difference*.** The negatively charged side is said to have a *negative* potential; the positively charged side a *positive* potential. These terms are *relative*—if there are two negatively charged bodies, the *less* negatively charged one would be considered to have a *positive* potential with respect to the *more* negatively charged body. (See Fig. 2-6.)

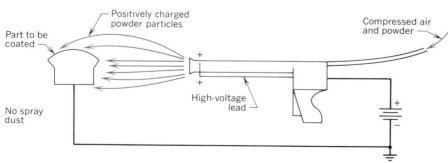

FIGURE 2-4
Electrostatic spray gun.

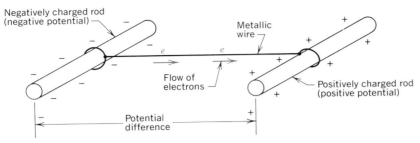

FIGURE 2-5

Two differently charged bodies establish a potential difference that causes electrons to flow between the two. *Note:* the charges actually exist on the *surface* of the rods.

If the two negatively charged bodies in Fig. 2-6 are connected with a copper wire, electrons will flow from left to right until equilibrium is established. In other words, electrons will flow until there is no longer a potential difference.

Before the wire is attached to the rods, there is a *static electricity* situation: the electric charges are at rest. When the rods are connected with the wire, there is a brief current (charge flow) called a *discharge* current. The current will flow only while the electrons are rearranging themselves.

This discharge current is what you feel when you touch the doorknob after walking across a carpet on a dry day. (The stinging "shock" sensation is actually caused by the concentrated flow of electrons leaving your skin at a point.) If the room is dark, you can sometimes see a blue spark as the air breaks down and allows the flow of

charges or current between your skin and the doorknob. If you held a key tightly in your hand as you walked across the floor, you would see the spark jumping from the key to the doorknob, but would not feel the stinging sensation.

2-4.1 The Volt

As mentioned in Section 1-7, the *potential difference* of a source is related to the *work (W)* that the source can do to move the *charge (Q)* around a complete circuit. **The basic measuring unit of potential difference is the volt (V). The potential difference, *V*, is defined as the *ratio* of the work done to the charge transferred.** That is,

$$V = \frac{W}{Q} \quad \text{volts} \qquad (2\text{-}2)$$

where: *V* is the potential difference in volts (V)

  W is the amount of work or energy expended in joules (J)*

  Q is the charge in coulombs (C)

From Eq. 2-2, we may say:

1 volt = 1 joule/coulomb (1 V = 1 J/C)

A potential difference of 1 V exists when 1 J of work is done moving 1 C of charge between two points.

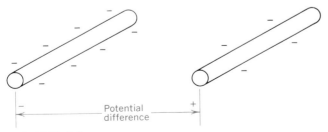

FIGURE 2-6

Two negatively charged bodies establish a potential difference. The rod on the left has more negative charges than the one on the right.

*One joule is the work done by a force of 1 N acting through a distance of 1 m.

EXAMPLE 2-3

Determine the potential difference between two points if 24 joules of work must be done to move 0.2 C of charge between the two points.

Solution

$$V = \frac{W}{Q} \qquad (2\text{-}2)$$
$$= \frac{24 \text{ J}}{0.2 \text{ C}}$$
$$= 120 \text{ J/C} = \mathbf{120\ V}$$

NOTE *V* is the letter *symbol* for potential difference, while V is an abbreviation for the basic *unit,* the volt.

EXAMPLE 2-4

Calculate the amount of work done by a 12-V battery that causes 50 C of charge to be transported from source to load.

Solution

$$V = \frac{W}{Q} \qquad (2\text{-}2)$$
$$W = V \times Q$$
$$= 12 \text{ V} \times 50 \text{ C}$$
$$= 600 \text{ J/C} \times \text{C} = \mathbf{600\ J}$$

2-4.2 Electromotive Force

A term used to refer to the potential difference of a source is *electromotive force* (emf). This term makes it possible to distinguish between the potential difference of a *source* and other potential differences that exist in a circuit (called *voltage drops*). Voltage drops occur due to a flow of current through a component; an emf does not require a flow of current to exist. Both emf and voltage drops, however, are measured in volts. In fact, when we refer to the *voltage* between two points, we may be referring to either the emf or some other potential difference in the circuit. This ''voltage'' may be in millivolts (mV), microvolts (μV), or kilovolts (kV).

In this chapter, the concept of potential difference (PD) was introduced in terms of electrostatic charges. It should

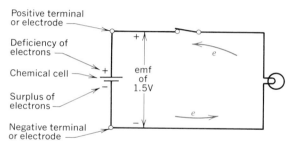

FIGURE 2-7

Electric circuit showing how electrons flow from the negative terminal to the positive terminal.

be clearly understood, however, that this PD is equivalent to the potential difference produced by a battery. In a battery (such as a carbon-zinc flashlight cell that produces 1.5 V), chemical activity causes one electrode to have a surplus of electrons, while the other has a deficiency. When electrons are made to flow through an external circuit to light a lamp, for example, the chemical energy in the cell is continuously converted to electrical energy. Thus the cell tries to maintain the potential difference and a continuous flow of current. (See Fig. 2-7.) This continuous flow is in contrast to the very brief flow of electrons that occurs when *electrostatic charges* equalize, as described in Section 2-3.

In Fig. 2-5, of Section 2-3, the material used to connect the two charged bodies so that a charge could flow between them was a *metallic wire.* **Metallic material, which readily permits the transfer of charges, is called a *conductor.*** If the material used to connect the charged bodies in Fig. 2-5 had been nonmetallic (such as a piece of string or plastic fishing line), there would have been no current or charge flow. **Materials which do not readily permit the transfer of charges are called nonconductors, or *insulators.*** To understand the different electrical properties of materials—why some are conductors and others are insulators—it is necessary to consider their atomic structure.

2-5 ATOMIC STRUCTURE

In 1913, the physicist Niels Bohr proposed that an atom consists of a central nucleus surrounded by orbiting electrons (somewhat like our solar system, with planets orbit-

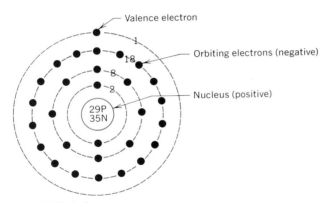

FIGURE 2-8
Bohr concept of an atom of copper.

TABLE 2-2
Mass and Charge for Atomic Particles

Particle	Mass, kg	Charge, C
Electron	9.11×10^{-31}	-1.6×10^{-19}
Proton	1.67×10^{-27}	$+1.6 \times 10^{-19}$
Neutron	1.67×10^{-27}	0

2-5.1 Valence Electrons

In all atoms of all elements, the outermost electron shell is called the *valence* shell. This shell determines the chemical and electrical behavior of the material. In a copper atom, at room temperature, the single valence electron in the outer shell is very loosely bound to the nucleus, for two reasons:

1. The outer shell is not completely filled (it could hold up to 32 electrons). If the outer shell were filled, a very *stable* electron configuration would result. Such a configuration tightly binds the electrons together and does not permit any one electron to be removed easily.

2. Because it is *further away* from the nucleus than the rest of the electrons, the single electron in the outer shell is only weakly bound to the nucleus.
   Coulomb's law (Section 2-2) states that the force of attraction between opposite charges varies inversely as the square of the distance between them.

These valence electrons, readily given up by metallic atoms such as those of copper, account for the ability of metallic materials to readily conduct current. In a material that is an *insulator,* such as plastic, mica, or glass, the valence shells of the atoms are either completely filled or almost full. Electrons in the valence shells are not readily given up, so these materials do not conduct electricity.

There are also materials called *semiconductors,* which will allow the conduction of current but do so less readily than metallic materials. The semiconductors, such as silicon, have four electrons in their valence shells. Semiconductors like silicon share the four electrons in their valence shell with four other atoms, in a manner called *covalent bonding.* Each atom then effectively has eight electrons, on a *shared* basis, in its valence shell. This is a stable situation, since it means that two *subshells* within the third shell are completely filled. The atoms of semiconductors do not *readily* give up electrons. However, at room tem-

ing around the sun). Although there are more modern theories, Bohr's concept of the atom is well suited to explaining the electrical properties of materials.

An atom of copper (Fig. 2-8) has a nucleus consisting of 29 positively charged protons and 35 neutrons. The neutrons have no charge, so the charge of the whole nucleus is positive. The amount of the charge of a proton is the same as that of an electron (1.6×10^{-19} C). Therefore, there must be 29 negatively charged electrons orbiting around the nucleus to make a *neutral* atom of copper. This is provided by the electrons filling up "shells" or orbits from the center outward, until there is a total of 29. Electrons are allowed to exist at only certain energy levels, and each shell is usually filled before another shell can be started. As shown in Table 2-1, the charge of a whole atom of copper is neutral.

Table 2-2 shows technical data for the mass and charge of the particles making up an atom.

TABLE 2-1
Atomic Structure of Copper

Shell	Number of Electrons	Charge
1st	2 (Filled)	
2nd	8 (Filled)	
3rd	18 (Filled)	
4th	1 (Could hold 32)	
Total:	29 electrons	Negative
Nucleus:	29 protons	Positive
	35 neutrons	Neutral

perature, *some* of the electrons do gain enough energy to escape from the atoms, accounting for some *conductivity*. For this reason the materials are called "semiconductors."

2-6 FREE ELECTRONS

The valence electrons that acquire sufficient energy to leave their parent atoms are called *free electrons.* These electrons are free to wander from atom to atom, moving in a random manner through the material, colliding with atoms but not attached to any of them. A copper atom that loses its valence electron becomes a *positively charged ion,* since it now has 29 positively charged protons in the nucleus, but only 28 negatively charged electrons.

Copper has approximately 8.5×10^{28} free electrons per cubic meter, and thus is an excellent conductor. A piece of copper wire, with no emf connected across it, has this tremendous number of free electrons randomly moving within it. However, the *net* combined average motion of *all* the electrons to the left or right is *zero*. When a battery is connected across the copper wire, a force acts upon these free electrons and begins moving them in a *definite* direction. (See Fig. 2-9.) **It is the motion of free electrons, in a** *definite overall direction,* **that constitutes an electric current.** An electric current is a *flow* of electrical charges.

2-7 CONDUCTORS AND INSULATORS

Most metals are good conductors of electricity. The more free electrons a material has, the better it will allow electrical charges to flow through it (conduct electricity). Thus

the metal silver, which has approximately 5% more free electrons (for the same volume) than annealed copper, is a slightly better conductor than copper. Aluminum, however, has only 61% of the free electrons that copper has available. It is not as good a conductor.

For the reasons explained in Section 2-5, insulators have very few free electrons. A perfect insulator, of course, would have *no* free electrons. In practice, a good insulator like the plastic polystyrene, may have as many as 6×10^{16} free electrons per cubic meter. This is a large number, but still is relatively small compared to copper (8.5×10^{28} free electrons per cubic meter). Most of the free electrons in polystyrene come from impurities in the material or moisture it has absorbed. It is these free electrons that account for the very small leakage currents through, or on the surface of, most insulators.

2-8 CURRENT

Current is the *rate* of flow of electrical charges and is measured in terms of how *much* charge flows *per second*. If you think of cars on a busy expressway, the *rate* of traffic flow would be measured in the number of cars passing a given point in a given period of time.

The basic unit used to measure current is the *ampere* (A), which may be defined by this equation:

$$I = \frac{Q}{t} \qquad \text{amperes} \qquad (2\text{-}3)$$

where: I is the current in amperes (A)
Q is the charge in coulombs (C)
t is the time in seconds for the charge to flow
Thus 1 ampere = 1 coulomb/second (1 A = 1 C/s)

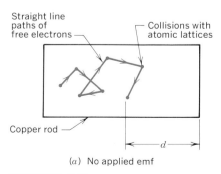

(*a*) No applied emf

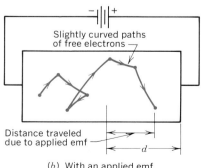

(*b*) With an applied emf

FIGURE 2-9
Schematic representation of a few paths of an electron in copper.

EXAMPLE 2-5

In an electroplating circuit, it is determined that 60 C of charge have been transferred at a steady rate in 4 min.
a. How much current is flowing in the circuit?
b. At this rate, how many coulombs will be transferred in 10 min?
c. How many electrons does this represent?

Solution

a. $I = \dfrac{Q}{t}$ (2-3)

$$= \frac{60 \text{ C}}{4 \text{ min} \times \dfrac{60 \text{ s}}{\text{min}}} = \frac{60 \text{ C}}{240 \text{ s}} = 0.25 \text{ C/s} = \mathbf{0.25 \text{ A}}$$

b. $Q = It = 0.25 \text{ C/s} \times 10 \text{ min} \times \dfrac{60 \text{ s}}{\text{min}} = \mathbf{150 \text{ C}}$

c. Since the charge of 6.24×10^{18} electrons makes up a charge of 1 C, the number of electrons represented by 150 C of charge is given by:

$$150 \text{ C} \times 6.24 \times 10^{18} \frac{\text{electrons}}{\text{C}}$$
$$= \mathbf{9.36 \times 10^{20} \text{ electrons}}$$

Note, from Example 2-5b, that 1 coulomb = 1 ampere-second (1 C = 1 A · s). Current is easier to measure than charge, so one coulomb (charge of 6.24×10^{18} electrons) is defined as the charge that flows past a given point in a circuit in a period of one second, due to a steady current of one ampere.

In many circuits, however, the ampere is too large a unit to be used conveniently. Recall from Section 1-6;
$$1 \text{ mA} = 1 \times 10^{-3} \text{ A} = 0.001 \text{ A}$$
$$1 \text{ } \mu\text{A} = 1 \times 10^{-6} \text{ A} = 0.000{,}001 \text{ A}$$

EXAMPLE 2-6

Convert: a. 0.075 A to mA
b. 0.35 mA to μA
c. 2300 mA to A
d. 0.1 A to μA
e. 40 μA to mA

Solution

a. $0.075 \text{ A} = 75 \times 10^{-3} \text{ A} = \mathbf{75 \text{ mA}}$
b. $0.35 \text{ mA} = 0.35 \times 10^{-3} \text{ A}$
$= 350 \times 10^{-3} \times 10^{-3} \text{ A}$
$= 350 \times 10^{-6} \text{ A} = \mathbf{350 \text{ } \mu\text{A}}$
c. $2300 \text{ mA} = 2300 \times 10^{-3} \text{ A}$
$= 2.3 \times 10^{3} \times 10^{-3} \text{ A}$
$= \mathbf{2.3 \text{ A}}$
d. $0.1 \text{ A} = 0.1 \times 10^{0} \text{ A}$
$= 0.1 \times 10^{6} \times 10^{-6} \text{ A}$
$= 0.1 \times 10^{6} \text{ } \mu\text{A} = \mathbf{1 \times 10^{5} \text{ } \mu\text{A}}$
e. $40 \text{ } \mu\text{A} = 40 \times 10^{-6} \text{ A}$
$= 40 \times 10^{-3} \times 10^{-3} \text{ A}$
$= 40 \times 10^{-3} \text{ mA}$
$= \mathbf{0.04 \text{ mA}}$

Alternatively:

a. Since $1 \text{ A} = 10^{3} \text{ mA}$, $\dfrac{10^{3} \text{ mA}}{1 \text{ A}} = 1$

Then $0.075 \text{ A} = 0.075 \text{ A} \times \dfrac{10^{3} \text{ mA}}{1 \text{ A}}$
$= 0.075 \times 10^{3} \text{ mA}$
$= \mathbf{75 \text{ mA}}$

b. Since $1 \text{ mA} = 10^{3} \text{ } \mu\text{A}$, $\dfrac{10^{3} \text{ } \mu\text{A}}{1 \text{ mA}} = 1$

Then $0.35 \text{ mA} = 0.35 \text{ mA} \times \dfrac{10^{3} \text{ } \mu\text{A}}{1 \text{ mA}}$
$= 0.35 \times 10^{3} \text{ } \mu\text{A}$
$= \mathbf{350 \text{ } \mu\text{A}}$

c. Since $10^{3} \text{ mA} = 1 \text{ A}$, $\dfrac{1 \text{ A}}{10^{3} \text{ mA}} = 1$

Then $2300 \text{ mA} = 2300 \text{ mA} \times \dfrac{1 \text{ A}}{10^{3} \text{ mA}}$
$= \dfrac{2300 \text{ A}}{10^{3}}$
$= \mathbf{2.3 \text{ A}}$

d. Since $1 \text{ A} = 10^{6} \text{ } \mu\text{A}$, $\dfrac{10^{6} \text{ } \mu\text{A}}{1 \text{ A}} = 1$

Then $0.1 \text{ A} = 0.1 \text{ A} \times \dfrac{10^{6} \text{ } \mu\text{A}}{1 \text{ A}}$
$= 0.1 \times 10^{6} \text{ } \mu\text{A}$
$= \mathbf{1 \times 10^{5} \text{ } \mu\text{A}}$

e. Since $10^{3} \text{ } \mu\text{A} = 1 \text{ mA}$, $\dfrac{1 \text{ mA}}{10^{3} \text{ } \mu\text{A}} = 1$

Then $40 \text{ } \mu\text{A} = 40 \text{ } \mu\text{A} \times \dfrac{1 \text{ mA}}{10^{3} \text{ } \mu\text{A}}$
$= \dfrac{40 \text{ mA}}{10^{3}}$
$= 40 \times 10^{-3} \text{ mA} = \mathbf{0.04 \text{ mA}}$

2-9 CONVENTIONAL CURRENT AND ELECTRON FLOW

Current is defined as the rate of charge flow. In a metallic conductor, this flow consists entirely of electrons—a flow of negative charges. When electrons flow from the negative terminal of the source and around the external circuit toward the positive terminal, this direction is called *electron flow*.

In semiconductors, there is another type of current carrier that is a *positive* charge (called a *hole*). Holes flow in a direction *opposite* that of electrons. "Hole current" in this case is a flow of positive charges. This subject will be covered in detail in Chapter 28.

Early experimenters in electricity were not aware of the actual mechanism of current flow. They were aware of its effect long before they understood its nature. They reasoned that, like water flowing from a higher level to a lower level, *current* flowed from a point of high potential (+) toward a point of lower potential (−). Accordingly, the experimenters agreed on a *convention* for the direction of current flow that everyone would use. And the letter *I* (from the French word "intensité") was used to designate current, shown flowing from positive to negative. This is called *conventional current direction*. (See Fig. 2-10.) It was not until years later that the flow of electrons (in metallic conductors, at least) was determined to be from negative to positive. This *electron flow* is usually designated by the letter *e*, as shown in Fig. 2-10.

It should be noted that when an electric current flows through a conductive liquid, it is the result of the movement of both positive *and* negative charges. The same holds true when current flows through conductive gases or plasmas. Electron flow (the flow of negative charges alone) is found only in solid metallic conductors. Conse-

quently, there is as much justification for the conventional current direction as for the direction of electron flow.

Engineers, physicists, and other scientists have continued to use *conventional* current flow in developing laws, rules, and terminology. This includes symbols for modern semiconductors, *all* of which have arrows pointing in the direction of *I* (conventional current flow) through the devices, as shown in Fig. 2-11.

Finally, notation in the SI metric system requires the use of conventional current *I* as an international standard for all electronic devices. You will find the conventional current direction an advantage when considering the term *voltage drop* and its mathematical meaning in Chapter 5.

2-10 DRIFT VELOCITY OF CURRENT

As you read in Section 2-6, free electrons move about randomly in a metallic conductor with no electromotive force (emf) applied. The free electrons move about, colliding with atoms at speeds approaching the speed of light, but have no overall definite direction.

When an emf is applied, a steady-state *drift speed* is *superimposed* upon the random motion of the free electrons. As a given free electron collides with one positive ion after another, it makes a slow but definite progress toward a point of more positive potential. The relation between the electron drift speed and the current flowing in a metallic conductor is given by this equation:

$$I = Aenv \quad \text{amperes} \quad (2\text{-}4)$$

where: *I* is the current in amperes (A)

   A is the conductor's cross-sectional area in square meters (m^2)

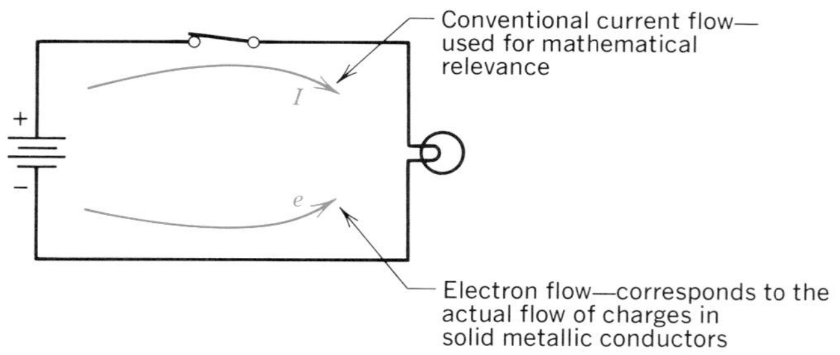

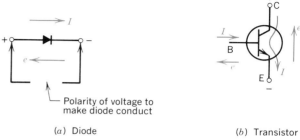

(a) Diso

(b) Transistor

FIGURE 2-10

Opposite directions of electron flow and conventional current flow.

FIGURE 2-11

Graphic symbols for diode and transistor, showing the directions of e, electron flow, and I, conventional current.

e is the charge of an electron $= 1.6 \times 10^{-19}$ C

n is the number of free electrons per cubic meter of conductor

v is the velocity of electron drift in meters per second (m/s)

EXAMPLE 2-7

A number 14 gauge copper wire, of diameter 0.064 in., is carrying a current of 15 A. Given the number of free electrons per cubic meter is 8.5×10^{28} for copper, calculate:

a. The drift velocity of an individual electron in the conductor.

b. The distance in inches an individual electron moves in the conductor in 1 min.

Solution

a. Diameter $= 0.064$ in.

$$= 0.064 \text{ in.} \times 2.54 \frac{\text{cm}}{\text{in.}} \times \frac{1 \text{ m}}{100 \text{ cm}}$$

$$= 1.63 \times 10^{-3} \text{ m}$$

$$\text{Area } A = \frac{\pi d^2}{4} = \frac{\pi}{4} \times (1.63 \times 10^{-3} \text{ m})^2$$

$$= 2.09 \times 10^{-6} \text{ m}^2$$

$$I = Aenv \qquad (2\text{-}4)$$

Therefore, $v = \dfrac{I}{Aen}$

$$= 15 \text{ A}/(2.09 \times 10^{-6} \text{ m}^2) \times (1.6 \times 10^{-19} \text{ C}) \times (8.5 \times 10^{28} \text{ m}^{-3})$$

$$= \mathbf{5.28 \times 10^{-4} \text{ m/s}}$$

b. $d = vt$

$$= \left(5.28 \times 10^{-4} \frac{\text{m}}{\text{s}}\right) \times 1 \text{ min} \times \frac{60 \text{ s}}{\text{min}}$$

$$= 3.17 \times 10^{-2} \text{ m}$$

$$= 3.17 \times 10^{-2} \text{ m} \times 10^2 \frac{\text{cm}}{\text{m}} \times \frac{1 \text{ in.}}{2.54 \text{ cm}}$$

$$= \mathbf{1.25 \text{ in.}}$$

The low charge velocity, as shown in Example 2-7b, is 1.25 in./min, which accounts for the use of the term *drift*. The drift velocity is so slow because of the tremendous number of free electrons available in the conductor. A current of 15 A corresponds to 15 C/s, or about 1×10^{20} electrons per second. Since there are so many electrons, they do not have to move very quickly to transport 15 C of charge each second. If a smaller conductor is used, or if the conductor is a metal with fewer free electrons (such

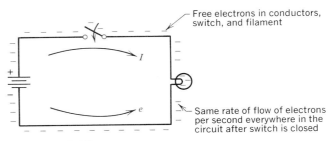

FIGURE 2-12

A circuit has free electrons distributed throughout the circuit even before the switch is closed.

as aluminum), the drift velocity of the charges will increase.

If the charges move so slowly, why does a lamp seem to come on immediately when you close a switch to complete the circuit?

As you can see in Fig. 2-12, free electrons are distributed throughout the circuit (the lamp filament and the wires) even *before* the switch is closed. When you close the switch, you complete the circuit, allowing the potential of the source ($V = W/Q$) to do work on all those free electrons. A force is exerted on all the electrons and starts them moving. Figure 2-13 shows a trough of marbles that illustrates the action of the emf on the free electrons. As a force is exerted on the left end of the row of marbles, a marble at the right end is immediately displaced because all the marbles are set in motion.

The flow of electrons is the *same* everywhere in the circuit shown in Fig. 2-12, so you don't have to wait for an electron to travel all the way from the source to the lamp before the lamp lights up (this could take hours). **Current is not something that squirts out of a battery and gets all used up by the time it reaches the other terminal.**

The effect of the current starting to move throughout the circuit is near the speed of light. This is called the impulse or *propagation velocity* and is almost 3×10^8 m/s.

2-11 RESISTANCE AND CONDUCTANCE

When current flows in an electric circuit, the free electrons are involved in constant collisions. They collide with each other, and with the bound (captive) electrons of the atoms in the conductor. **These collisions impede the progress of the free electrons, and the effect is called the *resistance* (R) of the circuit.** This opposition to the flow of current

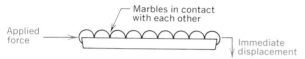

FIGURE 2-13
Marble analogy for immediate effect of current.

is measured in the basic unit of **ohms,** symbolized by the Greek letter omega (Ω).

A circuit is said to have a resistance of 1 ohm whenever a potential difference of 1 volt is applied to the circuit and a current of 1 ampere flows:

$$1 \ \Omega = 1 \ V/1 \ A$$

To describe larger values of resistance, prefixes are used with the unit symbol:

$$1,000 \ \Omega = 1 \ k\Omega \ (1 \ \text{kilohm})$$
$$1,000,000 \ \Omega = 1 \ M\Omega \ (1 \ \text{megohm})$$

Thus 470,000 Ω of resistance can be written as 470 kΩ; in the same way, 22,000,000 Ω is written as 22 MΩ. Small values of resistance may be expressed in milliohms (mΩ), but generally the decimal form is sufficient. For example, 0.030 Ω can be expressed as 30 mΩ.

In the next chapter, factors affecting the resistance of conductors will be covered in detail. Generally, however, resistance increases with conductor length or a smaller cross-sectional area. *Both* these factors increase the possibility of electron collisions, and thus, opposition to the flow of current.

In some ways, resistance is similar to the friction found in a mechanical system. Friction may be undesirable in bearings, for example, because of wear and losses that could slow down a rotating shaft. In the same way, resistance in a conductor can be considered detrimental because it wastes electrical energy in the form of unwanted heat. Good conductors should have a very *low* resistance to efficiently transport energy from source to load.

On the other hand, the friction in a car's braking system is beneficial—it allows the driver to stop the car. In a similar manner, resistance can be beneficial in some circuits— the heat produced by resistance is used to toast bread, iron clothing, and warm up a cold room. Heating elements in such appliances are purposely made to have a *high* resistance by suitable selection of the material and cross-sectional area of the material used.

An *insulator,* of course, must have a very high (ideally, infinite) resistance to oppose the flow of current.

A term that will be useful as you continue your study of electricity is *conductance (G).* Conductance is the *reciprocal* of resistance and is an indication of how well a conductor *permits* the flow of current.

$$G = \frac{1}{R} \quad \text{siemens} \quad (2\text{-}5)$$

where: G is the conductance in siemens (S)*
R is the resistance in ohms (Ω)

EXAMPLE 2-8

a. A circuit has a resistance of 20 Ω. What is its conductance?
b. What resistance does a conductance of 1×10^{-6} S represent?

Solution

a. $G = \dfrac{1}{R}$ (2-5)

$\quad = \dfrac{1}{20 \ \Omega} = 0.05 \ \text{S} = \textbf{50 mS}$

b. $R = \dfrac{1}{G}$

$\quad = \dfrac{1}{1 \times 10^{-6} \ \text{S}} = 1 \times 10^{6} \ \Omega = \textbf{1 M}\boldsymbol{\Omega}$

---

*Siemens is the SI unit for conductance. It replaces mho and its symbol ($\mho$) for the unit of conductance.

SUMMARY

1. Static electricity results from the transfer of electrons from one body to another.
2. A body that is negatively charged has a surplus of electrons; a positively charged body has a deficiency of electrons.
3. Like charges repel; unlike charges attract.
4. Coulomb's law states that the force exerted between two charges is proportional to the product of the two charges and is inversely proportional to the square of the distance separating the two charges.

5. A coulomb is the combined electrical charge of 6.24×10^{18} electrons.
6. Static electricity has many useful applications. Three such applications are the electrostatic spray gun, the electrostatic air cleaner, and xerography (electrostatic printing).
7. A potential difference exists between two differently charged bodies; it represents the ability to cause a flow of charge between the two bodies.
8. Potential difference is measured in volts. One volt equals one joule per coulomb.
9. Bohr's concept of the atom involves a positively charged nucleus surrounded by negatively charged orbiting electrons. In a *neutral* atom, the positively and negatively charged particles are equal in number.
10. A free electron is the outermost or *valence* electron, capable of leaving its parent atom and moving freely through the material.
11. Conductors are materials with many free electrons. They readily permit the flow of electric charge.
12. Insulators have few or no free electrons, and do not readily conduct electricity.
13. Current is the *rate* of flow of charge, measured in coulombs per second (amperes). One ampere equals one coulomb per second.
14. In a solid metallic conductor, electrons (identified by the letter e) flow from negative to positive. Conventional current direction (identified by the letter I) is considered to flow from positive to negative.
15. Electron drift is the overall movement through a conductor made under the influence of an applied emf. Drift velocity is very low, in the order of a few centimeters per minute.
16. At the instant a circuit is completed, the effect of current moves at almost the speed of light because all the electrons start to move at once. This is called the propagation velocity.
17. Resistance is the opposition to current, caused by electron collisions. Resistance is measured in ohms (one ohm is equal to one volt per ampere).
18. Conductors exhibit a very low resistance value; insulators, a high resistance value.
19. Conductance is the reciprocal of resistance and is measured in siemens.

SELF-EXAMINATION

Answer true or false
(Answers at back of book)

2-1. Friction between two materials causes the transfer of electrons from one material to the other. This is called electrification by friction. _____

2-2. A body is said to be negatively charged if it loses electrons and positively charged if it gains electrons. _____

2-3. Like charges repel; unlike charges attract. _____

2-4. Coulomb's law determines how much force of attraction or repulsion occurs between two charged bodies. _____

2-5. The amount of force is directly proportional to the square of the distance and inversely proportional to the product of the charges. _____

2-6. Electrical charge is measured in coulombs. _____

2-7. Applications of static electricity rely upon particles becoming electrically charged and being attracted to oppositely charged surfaces. _____

2-8. A potential difference exists between two differently charged electrodes.
_____

2-9. It is necessary to have a positively charged body and a negatively charged body to produce a potential difference. _____

2-10. Potential difference is a measure of the ability to cause charge to flow between two differently charged bodies. _____

2-11. Potential difference is measured in volts. _____

2-12. An electromotive force is a source, measured in volts. _____

2-13. Both emf and voltage drops are measured in volts. _____

2-14. When a cell discharges, chemical energy is converted to electrical energy as the cell tries to maintain the potential difference. _____

2-15. A material that readily permits the flow of charges is called an insulator. _____

2-16. The nucleus of every atom is positive. _____

2-17. There are as many electrons in orbit around the nucleus as there are neutrons in the nucleus. _____

2-18. The outer electron shell is the valence shell of an atom. _____

2-19. In a copper atom the valence electron is loosely bound to the nucleus. At room temperature this electron gains sufficient energy to become a free electron and wanders from atom to atom. _____

2-20. Free electrons have a random thermal motion when no emf is applied to a conductor. _____

2-21. Under the influence of an applied emf the free electrons drift toward the negative terminal of the emf. _____

2-22. A conductor has a large number of free electrons. _____

2-23. Current is the rate of flow of charge, measured in coulombs per second or amperes. _____

2-24. A current of 0.05 A is equal to 5 mA. _____

2-25. A current of 1000 μA is equal to 1 mA. _____

2-26. Conventional current is shown as flowing from positive to negative. _____

2-27. In a solid metallic conductor, current is the flow of electrons from negative to positive. _____

2-28. The arrows in the symbols for solid-state devices indicate the direction of electron flow. _____

2-29. The drift velocity of current in a metallic conductor is very slow because there are a tremendous number of free electrons available. _____

2-30. Propagation velocity refers to the near-instantaneous effect of current when a circuit is first completed. _____

2-31. Resistance in a wire is the opposition to current caused by electron collisions. _____

2-32. A resistance of 1200 Ω equals 12 kΩ. _____

2-33. A resistance of 1200 kΩ equals 1.2 MΩ. _____

2-34. A resistance of 50 Ω has a conductance of 20 mS. _____

REVIEW QUESTIONS

1. a. When two initially neutral bodies, such as rubber and fur, are rubbed together, what can you say about the sign and magnitude of the charge on each?
   b. What prevents *all* the electrons from moving from one body to the other?

2. What is the law of electric forces? To what law is this similar?

3. What factors determine the size of forces between charged bodies? Would this same force exist in a vacuum?

4. How many protons are necessary to produce a positive charge of 1 coulomb?

5. If opposite charges attract, why don't the electrons in an atom move toward, and come in contact with, the nucleus?

6. Describe how static electricity may be used in the manufacture of sandpaper.
7. What do you understand by the term *potential difference?*
8. Explain how it is possible for two electrodes, both positively charged, to have a potential difference between them.
9. Explain the difference between the potential differences set up by two electrostatically charged rods and the potential difference from a dry cell.
10. Name five conducting materials and five insulators.
11. What is a free electron?
12. How many free electrons would a perfect insulator have? What characteristic would a perfect conductor have?
13. What is covalent bonding? Name a material that has this type of bonding. How does this material's conductance compare with that of copper or an insulator?
14. How is it possible for free electrons to be moving in different directions in a copper wire when there is no emf applied? How does an applied emf change this situation?
15. Is current necessarily the flow of electrons or negative charge?
16. Why is the term *drift* appropriate in describing current in a metallic conductor?
17. What happens to the drift velocity in a series circuit where the cross-sectional area of the conductor decreases? Why is the drift velocity low in a metallic conductor? Would you expect it to be higher or lower (for a given current) in a semiconductor of the same cross-sectional area? Why?
18. What is one ampere-second?
19. Distinguish between electron flow and conventional current. Which direction is shown on solid-state symbols? Why?
20. If the drift velocity of current is in the order of only a few centimeters per minute, explain how a light comes on immediately after closing a switch to complete the circuit.
21. Describe in your own words what resistance to current is in an electric circuit. What factors affect the amount of resistance and why?
22. Why is it desirable for a conductor to have a low resistance? How would you describe the property of an insulator in terms of conductance?

PROBLEMS

(Answers to odd-numbered problems at back of book)

2-1. In an experiment to verify Coulomb's law two metal spheres, separated by a distance of 10 cm, are each charged to 2.5×10^{-8} C. Calculate:
   a. The force of repulsion in newtons.
   b. The force in pounds.
2-2. What must the separation between an electron and a proton be if the force of attraction between them is to be 1×10^{-6} N?
2-3. How many electrons would have to be removed from each of the spheres in Problem 2-1 if the spheres were initially neutral?
2-4. How many electrons would have to be transferred from one sphere to another, if they were both initially neutral, so that they would experience an attractive force of 1×10^{-6} N when they are 1 cm apart?
2-5. If it takes 30 J of chemical energy to move 20 C of electric charge between the positive and negative terminals of a battery, what is the emf between the terminals?

2-6. What is the emf of a source that moves 600 mC of charge from source to load with the expenditure of 3.6 J of energy?

2-7. If it takes 1 min for the charge in Problem 2-5 to be transported, what is the current?

2-8. What is the current in Problem 2-6 if the time required is 10 s?

2-9. In an electroplating system a steady current of 20 A flows for 10 min. What quantity of charge flows during this period?

2-10. How much charge is transported by a steady current of 50 mA flowing for 15 s?

2-11. How much energy does a 12-V battery supply to a load when the load draws a current of 4.5 A for 5 min?

2-12. A current of 350 mA causes a PD of 20 V across a resistor. How much energy is dissipated in the form of heat in 1 min?

2-13. How long does it take for a source to deliver 50 C of charge to a load if the current is 100 mA?

2-14. How long does it take for 1×10^6 electrons to pass by a given point in a circuit when the current is 10 mA?

2-15. How many electrons are transferred by a current of 10 μA flowing for 10 s?

2-16. What is the current in a circuit if one electron passes by in 1 ns?

2-17. Convert:
a. 0.0025 A to mA
b. 0.075 mA to μA
c. 5000 mA to A

2-18. Convert:
a. 15,000 μA to mA
b. 6500 μA to A
c. 1300 pA to nA

2-19. A copper wire carries a current of 10 A. Determine the electron drift velocity if the wire has a diameter of
a. 1 mm
b. 2 mm
c. 4 mm

2-20. Repeat Problem 2-19 using an aluminum wire with $n = 5.2 \times 10^{28}$ free electrons per cubic meter of aluminum.

2-21. How far does an individual electron move in 1 min for each of the diameters in Problem 2-19?

2-22. How far does an individual electron move in 1 min for each of the diameters in Problem 2-20?

2-23. Convert:
a. 1500 Ω to kΩ
b. 4700 kΩ to MΩ
c. 0.045 Ω to mΩ

2-24. Convert:
a. 100 Ω to kΩ
b. 1.2 kΩ to Ω
c. 2.2 MΩ to kΩ

2-25. Find the conductance of each of the resistances in Problem 2-23.

2-26. Find the conductance of each of the resistances in Problem 2-24.

2-27. Find the resistance represented by the following conductances:
a. 40 mS
b. 0.4545 μS
c. 1 S

CHAPTER 3

OHM'S LAW, POWER, AND RESISTORS

The three basic quantities in an electrical circuit are voltage, current, and resistance. In this chapter, you will examine *Ohm's law,* which relates these three variables to each other. By using Ohm's law, you can determine any of the three variables if the other two are known.

Power is the *rate* of doing work, or consuming energy. *Power* and *energy* are not the same—it is important to distinguish between them. The amount of electrical energy consumed depends on both the power *and* the length of time the power is being used.

Resistors are devices that introduce a known amount of opposition to current flow *(resistance)* into a circuit. These devices may be of the *fixed* type or of the *variable* type. On the body of many resistors, color bands are printed to identify the resistance value. These bands follow a standard color code. Resistors are available in many standard values, depending on the accuracy of their nominal values. Various power ratings, based on physical size of the device, indicate how much heat can be safely dissipated.

3-1 CONNECTION OF CIRCUIT METERS

Figure 3-1 shows a circuit with a variable emf that includes a resistor whose value can be varied. Meters connected to this circuit allow you to measure the voltage (a *voltmeter*) and the current flow (an *ammeter*). If the meters used to make such measurements are of the multirange type, their ranges should generally be adjusted to give readings as close to full scale deflection (FSD) as possible. This will help minimize reading errors, and improve accuracy.

The voltmeter is polarity-sensitive since it measures a dc voltage. The positive lead of the meter must be connected to the positive terminal of the source, and the negative lead to the negative terminal. If the leads are reversed, the meter will read backwards and could be damaged. To help in making proper connections, the positive lead of the meter is usually colored red and the negative lead black. Note that the voltmeter is only temporarily connected to the circuit—it can be disconnected and the circuit will continue to function. For this reason, the connections are shown in Fig. 3-1 with arrows, rather than dots (which would signify a more permanent connection).

The ammeter, however, forms a part of the circuit. The switch should be opened before the meter is inserted or removed. Like the voltmeter in this example, the ammeter is polarity-sensitive. (Note the polarity signs on the meter in Fig. 3-1.) Connected as shown, the meter will obtain a positive reading. For both meters, the conventional current must enter the positive terminal (or, in electron flow terms, the electrons must enter the negative terminal.)

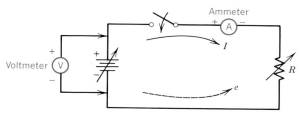

FIGURE 3-1

Circuit showing the use of voltmeter and ammeter for the measurement of applied voltage and current.

tance *(R)*. As shown, a doubling of the voltage, from 10 V to 20 V causes the current to double from 1 A to 2 A. If the voltage increases to 30 V, the current increases to 3 A, and so on.

Ohm originally expressed his findings in a form known as Ohm's law (Eq. 3-1):

$$R = \frac{V}{I} \quad \text{ohms} \quad (3\text{-}1)$$

where: *R* is the resistance in ohms (Ω)
        V is the potential difference in volts (V)
        I is the current in amperes (A)

Many texts use the letter symbol *E* to represent the emf voltage and *V* for potential differences in the rest of the circuit. The SI system, however, uses the symbol *V* for *all* voltages. The letter *E* is used to designate electrical field intensity.

Equation 3-1 allows you to calculate the resistance of a circuit if you know the voltage and the current. The following examples illustrate this relationship.

3-2 OHM'S LAW

In 1827, the relationship between applied voltage *(V)*, current *(I)*, and resistance *(R)* was observed by a German physicist, Georg Simon Ohm. He found that for a fixed value of resistance, a given current will flow when a certain voltage is applied. If the voltage is doubled, the current will double; if voltage is tripled, current will triple. **That is, current is *directly proportional* to voltage, if resistance is kept *constant*.**

Figure 3-2 shows this relationship graphically. The current *(I)* is plotted against the voltage *(V)* for a fixed resis-

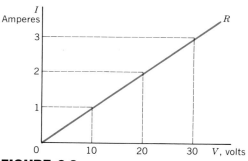

FIGURE 3-2

Ohm's law in graphical form.

EXAMPLE 3-1

Calculate the value (amount of resistance) of the resistor represented by Fig. 3-2 for voltages of:

a.   10 V
b.   20 V
c.   30 V

Solution

a.   From the graph, when $V = 10$ V, $I = 1$ A

then
$$R = \frac{V}{I} \tag{3-1}$$

$$= \frac{10 \text{ V}}{1 \text{ A}} = \mathbf{10 \ \Omega}$$

b.   When $V = 20$ V, $I = 2$ A

$$R = \frac{V}{I} \tag{3-1}$$

$$= \frac{20 \text{ V}}{2 \text{ A}} = \mathbf{10 \ \Omega}$$

c.   When $V = 30$ V, $I = 3$ A

$$R = \frac{V}{I} \tag{3-1}$$

$$= \frac{30 \text{ V}}{3 \text{ A}} = \mathbf{10 \ \Omega}$$

Figure 3-2 is a straight line graph, relating current *(I)* to voltage *(V)*. Example 3-1 shows that it represents a circuit with a constant resistance of 10 Ω. Put another way, the resistance of the circuit is constant if the *ratio* of voltage to current in the circuit is constant.

EXAMPLE 3-2

A heating element of pure resistance draws a current of 8 A when an emf of 120 V is applied to it. Calculate:

a.   the resistance of the element
b.   how much resistance a *different* heating element should have if the *same* current is to be drawn when a potential difference of 240 V is connected across the element

Solution

a.   From the question, I (current) $= 8$ A and V (voltage) $= 120$ V.

$$R = \frac{V}{I} \tag{3-1}$$

$$= \frac{120 \text{ V}}{8 \text{ A}} = \mathbf{15 \ \Omega}$$

b.
$$R = \frac{V}{I} \tag{3-1}$$

$$= \frac{240 \text{ V}}{8 \text{ A}} = \mathbf{30 \ \Omega}$$

Note that, in Example 3-2, you must *double* the resistance if the *same* current is to be drawn with *double* the voltage.

EXAMPLE 3-3

What resistance will allow a current of 5 mA to flow when a potential difference of 12 V exists across the resistor?

Solution

From the question, current $I = 5 \text{ mA} = 5 \times 10^{-3}$ A and voltage $V = 12$ V.

$$R = \frac{V}{I} \tag{3-1}$$

$$= \frac{12 \text{ V}}{5 \times 10^{-3} \text{ A}}$$

$$= 2.4 \times 10^3 \ \Omega = \mathbf{2.4 \ k\Omega}$$

EXAMPLE 3-4

How much resistance is necessary to limit the current flowing through a resistor to 4 μA when the resistor is connected across an emf of 40 mV?

Solution

From the question, $I = 4 \ \mu\text{A} = 4 \times 10^{-6}$ A when $V = 40 \text{ mV} = 40 \times 10^{-3}$ V.

$$R = \frac{V}{I} \tag{3-1}$$

$$= \frac{40 \times 10^{-3} \text{ V}}{4 \times 10^{-6} \text{ A}}$$

$$= 10 \times 10^3 \ \Omega = \mathbf{10 \ k\Omega}$$

Note that, in the last two examples, you had to convert the given units (mA, μA, mV) to basic units (A and V) to obtain the resistance in ohms.

3-3 GRAPHICAL PRESENTATION OF OHM'S LAW

Ohm's law represents a *linear* relationship between current and voltage for any *constant* resistance. This simply means that a graph of a change in current plotted against a change in voltage will result in a straight line for a *fixed* resistor, as shown in Fig. 3-2 and Example 3-1. If graphs of *I* versus *V* are plotted for a *family* of fixed resistors, they have the appearance of Fig. 3-3. Note that a high resistance has a more horizontal slope and a low resistance has a more vertical slope, but all the graphs shown are *straight* lines, indicating linear or *constant* resistance.

If the data for plotting the graphs in Fig. 3-3 are obtained experimentally (for example, from a circuit such as the one in Fig. 3-1), the following points should be noted:

1. The *independent* variable (the one that is directly varied)—voltage—is the *abscissa* and is plotted horizontally.
2. The *dependent* variable (the one that changes indirectly)—current—is the *ordinate* and is plotted vertically.
3. The graphs are drawn as straight lines through the best *average* positions of the plotted data and through the origin. To allow for experimental error, as in reading of the meters, plotted data on such a graph are seldom connected directly (point-to-point).

When a resistance is not constant, as in the filament of a lamp, a graph of *I* versus *V* will result in a *curved* line because the resistance is *nonlinear*. (See Fig. 3-4.) This means that the current is not directly proportional to the

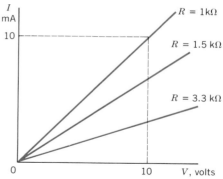

FIGURE 3-3

Graphs of *I* versus *V* for three different resistors showing a low slope for high resistance (low conductance) and high slope for low resistance (high conductance).

voltage: it does not double, for example, when voltage doubles. In this case, the circuit does not obey Ohm's law. It *is* possible, however, to apply Ohm's law to any *pair* of data for *I* and *V* and calculate the resistance *(R)* at those *points*. (See Example 3-5.)

EXAMPLE 3-5

Using the graph shown in Fig. 3-4, calculate the resistance of the lamp filament at points 1 and 2 on the curve. Explain why the overall graph does not obey Ohm's law.

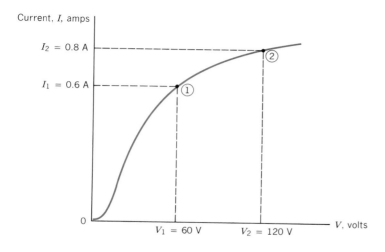

FIGURE 3-4

Graph of *I* versus *V* for a nonlinear incandescent lamp filament.

Solution

At point 1, $R_1 = \dfrac{V_1}{I_1}$ (3-1)

$= \dfrac{60 \text{ V}}{0.6 \text{ A}} =$ **100 Ω**

At point 2, $R_2 = \dfrac{V_2}{I_2}$ (3-1)

$= \dfrac{120 \text{ V}}{0.8 \text{ A}} =$ **150 Ω**

The curve does not obey Ohm's law since a doubling of voltage from 60 V to 120 V does not result in a corresponding doubling of *current*. That is, the resistance is not constant—it increases at higher currents due to a heating effect. (See Section 4-4.)

3-4 OTHER FORMS OF OHM'S LAW

The original statement of Ohm's law (Eq. 3-1) can be algebraically transposed and written in these two forms:

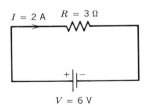
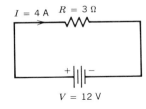

(a) A doubling of voltage causes a doubling of current if resistance is constant

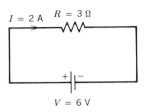
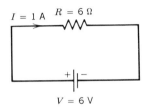

(b) A doubling of resistance causes the current to be cut in half if voltage is constant

FIGURE 3-5

Illustrating Ohm's law: $I = \dfrac{V}{R}$.

$I = \dfrac{V}{R}$ amperes   (3-1a)

$V = IR$ volts   (3-1b)

Equation 3-1a implies that the current in a circuit is *directly* proportional to the voltage and *inversely* proportional to the resistance.

This simply means that a higher voltage across a resistance will cause a higher current to flow. If the circuit obeys Ohm's law (if the resistance is *constant*), the increase in current will be in the same proportion as the increase in voltage, as shown in Fig. 3-5a. But if the resistance is *increased* for a given voltage, the current will *decrease* by the same proportion. (See Fig. 3-5b.)

Equation 3-1a allows you to calculate the current flowing in a circuit if you know the voltage and the resistance.

In the same way, Eq. 3-1b can be used to calculate the "voltage drop" that occurs across a resistor due to the current flowing through that resistor. If there is an increase in current or resistance (or both), an increased voltage drop will occur. The use of Eqs. 3-1a and 3-1b is shown in the following examples.

EXAMPLE 3-6

A 150-Ω resistor has an emf of 120 V applied across it.
a. Calculate how much current flows through the resistor.
b. If the resistance is reduced to 50 Ω, calculate the new emf required to maintain the same current as in part (a).
c. To what value must the resistance be increased if a current of 50 mA must flow when the emf is 60 V?

Solution

a. $I = \dfrac{V}{R}$ (3-1a)

$= \dfrac{120 \text{ V}}{150 \text{ Ω}} =$ **0.8 A**

b. $V = IR$ (3-1b)

$= 0.8 \text{ A} \times 50 \text{ Ω} =$ **40 V**

c. $R = \dfrac{V}{I}$ (3-1)

$= \dfrac{60 \text{ V}}{50 \text{ mA}} = \dfrac{60 \text{ V}}{50 \times 10^{-3} \text{ A}}$

$= 1.2 \times 10^3 \text{ Ω} =$ **1.2 kΩ**

EXAMPLE 3-7

A 2.2-MΩ resistor has a current of 6 μA flowing through it.

a. Calculate the potential difference across the resistor.
b. The resistor is changed until its conductance is 1×10^{-6} S. What is the new current if the potential difference is now 12 V?

Solution

a. $V = IR$ (3-1b)

$= 6 \text{ μA} \times 2.2 \text{ MΩ}$

$= 6 \times 10^{-6} \text{ A} \times 2.2 \times 10^{6} \text{ Ω} = \textbf{13.2 V}$

b. $I = \dfrac{V}{R}$ (3-1a)

$= V \times \dfrac{1}{R} = V \times G$

$= 12 \text{ V} \times 1 \times 10^{-6} \text{ S} = 12 \times 10^{-6} \text{ A} = \textbf{12 μA}$

3-5 LOAD RESISTANCE

The amount of resistance in the circuit shown in Fig. 3-1 could be measured by using an *ohmmeter*. As mentioned in Section 1-11, you would first have to open the switch or remove the resistor from the circuit, zero the ohmmeter, then connect the meter leads across the resistor. (In this case, it would not matter which way the leads are connected across the resistor, since it is not polarity sensitive.) **An ohmmeter must *never* be used in a "live" circuit (one with voltage applied to it). The ohmmeter has its own battery, and any external voltage may damage the meter and/or the ohmmeter circuitry.**

It is important to realize that the resistor in the circuit (Fig. 3-1) can represent an *electrical load,* such as an incandescent lamp or a heating element. It is also important to note that an *increase* in load represents a *decrease* in resistance. This is really an increase in load *current,* commonly referred to merely as "load." Throughout this text, references to an increase in *load* mean that the *current* is increased. In short, *load* implies *current.*

3-6 WORK AND ENERGY

Work is done whenever a force is exerted over a distance in the direction of the force. For example, you do mechanical work against gravity whenever you walk up a flight of stairs. When you slide a heavy box along the floor, you do work in overcoming friction. *Work* requires an expenditure of *energy.* In fact, work and energy can be thought of as *equivalent;* they are measured in the same units.

In an electrical circuit, work involves moving electrical charge *(Q)* through a potential difference *(V).* From the definition of the volt in Section 2-4 *(V = W/Q),* work in an electrical circuit is calculated using:

$$W = V Q \quad \text{joules} \qquad (3\text{-}2)$$

where: *W* is work done or energy expended in joules (J)
  V is the potential difference in volts (V)
  Q is the charge in coulombs (C)

Since *Q = It* (Eq. 2-3), the following equation can be used to determine the amount of work done in a length of time *(t).*

$$W = VIt \quad \text{joules} \qquad (3\text{-}3)$$

where: *W* is the work done or energy expended in joules (J)
  V is the potential difference in volts (V)
  I is the current in amperes (A)
  t is time in seconds (s)

EXAMPLE 3-8

An emf of 120 V causes a current of 3 A to flow in a circuit. Calculate how much energy is supplied by the source in 12 s.

Solution

$W = VIt$ (3-3)

$= 120 \text{ V} \times 3 \text{ A} \times 12 \text{ s} = \textbf{4230 J}$

In the SI (MKS) system, the joule is the measuring unit used for both electrical and mechanical energy. There is, in fact, little difference between electrical and mechanical energy. The energy exists in different *forms* and calculations are made with different equations, but the measuring units are the same. An alternative unit exists, however, for measuring large quantities of electrical energy. It will be introduced in Section 3-10.

3-7 POWER

Power is the *rate* at which work is done, or the rate at which energy is converted or consumed. Thus, power not only involves *how much* work is done, but *how long* it takes for the work to be accomplished.

$$\text{power} = \frac{\text{work done}}{\text{time taken to do the work}} = \frac{\text{energy}}{\text{time}}$$

$$P = \frac{W}{t} \qquad (3\text{-}4)$$

where, in the MKS system:

   P is power in joules per second (J/s) or watts (W)
   W is work or energy in joules (J)
   t is time in seconds (s)

Note that one joule per second equals 1 watt. (1 J/s = 1 W)

In the FPS (foot-pound-second) system, work is measured in foot pounds, or ft lb. Thus, power in the FPS system uses units of ft lb/s. James Watt, a Scottish engineer who sought a way to measure the power produced by his improved steam engine, decided to call 550 ft lb/s one *horsepower* (hp). In this measurement, he expressed the equivalence between his steam engine and the average power that a horse could develop over some interval of time.

$$1 \text{ hp} = 550 \text{ ft lb/s}$$

EXAMPLE 3-9

If the work done by a 160-lb man running up a 20-ft flight of stairs in 8 seconds is 3200 ft lb or 4340 J, calculate the power he developed in:
a.   watts
b.   hp

Solution

a. $P = \dfrac{W}{t}$ $\qquad (3\text{-}4)$

   $= \dfrac{4340 \text{ J}}{8 \text{ s}} = 542.5 \text{ J/s} \approx$ **543 W**

b. $P = \dfrac{W}{t}$ $\qquad (3\text{-}4)$

   $= \dfrac{3200 \text{ ft lb}}{8 \text{ s}} = 400 \text{ ft lb/s}$

   $= 400 \text{ ft lb/s} \times \dfrac{1 \text{ hp}}{550 \text{ ft lb/s}} =$ **0.73 hp**

Electrical power is the rate at which work is done, or the rate at which energy is converted or consumed, in an electrical circuit. Thus

$$P = \frac{W}{t} = \frac{V \times Q}{t} = \frac{V \times It}{t}$$

Therefore, $P = VI$ watts $\qquad (3\text{-}5)$
where: P is power in watts (W)
   V is potential difference in volts (V)
   I is current in amperes (A)

EXAMPLE 3-10

A lamp operating at 120 V draws a current of 0.5 A. Calculate:
a.   the amount of power used by the lamp
b.   the rate at which energy is converted into heat and light
c.   the amount of energy used by the lamp in one minute

Solution

a.   Power used,
   $P = VI$ $\qquad (3\text{-}5)$
   $= 120 \text{ V} \times 0.5 \text{ A} =$ **60 W**
b.   The rate at which energy is converted is power. Therefore,
   $P = VI$ $\qquad (3\text{-}5)$
   $= 120 \text{ V} \times 0.5 \text{ A} =$ **60 W or 60 J/s**
c.   Energy used,
   $W = VIt$ $\qquad (3\text{-}3)$
   $= 120 \text{ V} \times 0.5 \text{ A} \times 60 \text{ s} =$ **3600 J**

Thus, every second it is lit, a 60-W lamp uses 60 J of energy. Example 3-10 clearly shows how a lamp consumes energy and that power is the *rate* of energy consumption. The lamp consumes 60 W of power (60 J/s of energy) all the time it is on, whether for 1/10 second or one week.

In some applications, the watt is a very small unit of power. In those applications, the *kilowatt* is a more convenient unit.

$$1 \text{ kilowatt} = 1 \text{ kW} = 1000 \text{ W}$$

EXAMPLE 3-11

a. What maximum power can be delivered from a 120-V outlet if it is fed from a 20-A circuit breaker?

b. What voltage must be applied to a 3.3-kW heating element if the current must not exceed 15 A?

c. What current is drawn by a 40-W lamp operating at 120 V?

Solution

a. $P = VI$ (3-5)
$$= 120 \text{ V} \times 20 \text{ A} = 2400 \text{ W} = \textbf{2.4 kW}$$

b. $V = \dfrac{P}{I}$
$$= \frac{3.3 \text{ kW}}{15 \text{ A}} = \frac{3300 \text{ W}}{15 \text{ A}} = \textbf{220 V}$$

c. $I = \dfrac{P}{V}$
$$= \frac{40 \text{ W}}{120 \text{ V}} = 0.33 \text{ A} = \textbf{330 mA}$$

3-8 EFFICIENCY

To convert energy from one form to another, a machine is generally used. One example is an electric motor, which takes electrical energy and converts it to mechanical energy in the form of a rotating shaft. In the conversion process, some energy is inevitably lost (in the example of the motor, part of the loss is in the form of heat due to winding resistance). The ratio of the *useful* output work to the total input work is called the *efficiency*. This is usually represented by the Greek letter eta (η).

Efficiency may also be expressed in terms of useful output *power* and total input *power*. Thus

$$\eta = \frac{P_{\text{out}}}{P_{\text{in}}} \qquad (3\text{-}6)$$

where: η is efficiency
P_{out} is useful output power in watts (W or hp)
P_{in} is total input power in watts (W or hp)

Because it is impossible to get more power out of a machine than is put into the machine, η must be between 0 and 1.

Efficiency is generally expressed as a percentage by multiplying Eq. 3-6 by 100%, as shown in the following example.

EXAMPLE 3-12

A dc power supply converts 50 W of ac input power to 30 W of dc output power. Calculate the conversion efficiency of this power supply.

Solution

$$\eta = \frac{P_{\text{out}}}{P_{\text{in}}} \times 100 \% \qquad (3\text{-}6)$$
$$= \frac{30 \text{ W}}{50 \text{ W}} \times 100\% = \textbf{60\%}$$

The output power of a motor is often expressed in horsepower. For problems involving power given in hp, the following equivalence is useful:

$$\textbf{1 hp} = \textbf{746 W}$$

Consider a 1-hp electric motor. If this motor were 100% efficient, 746 W of electrical input power would develop 1 hp of mechanical power at the shaft. But in a practical motor, more than 746 W of input power is needed to develop 1 hp of mechanical power, due to losses. The following example will illustrate this point.

EXAMPLE 3-13

How much current must be supplied to a 75%-efficient, 1½-hp motor operating at 120 V?

Solution

$$\eta = \frac{P_{\text{out}}}{P_{\text{in}}} \qquad (3\text{-}6)$$

Therefore, $P_{\text{in}} = \dfrac{P_{\text{out}}}{\eta}$
$$= \frac{1.5 \text{ hp}}{0.75} = 2 \text{ hp}$$
$$= 2 \text{ hp} \times \frac{746 \text{ W}}{1 \text{ hp}} = 1492 \text{ W} \approx \textbf{1500 W}$$

$$I = \frac{P}{V} \qquad (3\text{-}5)$$
$$= \frac{1492 \text{ W}}{120 \text{ V}} = \textbf{12.4 A}$$

Notice that the 1½-hp motor draws almost 1½ kW from the supply. This equivalence of 1 kW per hp is a useful rule of thumb for motors of this size, assuming an efficiency of approximately 75%.

Example 3-9 showed the calculation of mechanical power in both horsepower and watts. In the United States, the common practice is to use hp when referring to mechanical power and watts when referring to electrical power. But in Europe, it is not unusual to give the power rating of a mechanical device, such as a gasoline-engine vehicle, in kW. If such a vehicle were rated at 40 kW, the conversion 746 W = 1 hp can be used to find the horsepower:

$$40 \text{ kW} \times \frac{1000 \text{ W}}{1 \text{ kW}} \times \frac{1 \text{ hp}}{746 \text{ W}} = \textbf{53.6 hp}$$

3-9 OTHER EQUATIONS FOR ELECTRICAL POWER

When calculating the power dissipated (converted to heat) in a resistor, the known quantities may involve the resistance (R). In such a case, the following two equations are useful. They result when $V = IR$ and $I = V/R$ are separately substituted in Eq. 3-5.

$$P = I^2R \quad \text{watts} \tag{3-7}$$
$$P = V^2/R \quad \text{watts} \tag{3-8}$$

where: P is power dissipated in watts (W)
R is resistance in ohms (Ω)
I is current in amperes (A)
V is potential difference in volts (V)

EXAMPLE 3-14

a. The element in an electric toaster has a resistance of 10 Ω and carries a current of 12 A. How much power is dissipated in the element?
b. How much power is dissipated in a 10-kΩ resistor connected to a 12-V supply?

Solution

a. $P = I^2R$ $\hspace{3cm}$ (3-7)
$\hspace{1.2cm} = (12 \text{ A})^2 \times 10 \text{ }\Omega$
$\hspace{1.2cm} = 1440 \text{ W} = \textbf{1.44 kW}$

b. $P = \dfrac{V^2}{R}$ $\hspace{3cm}$ (3-8)

$\hspace{1cm} = \dfrac{(12 \text{ V})^2}{10 \times 10^3 \text{ }\Omega}$

$\hspace{1cm} = 14.4 \times 10^{-3} \text{ W} = \textbf{14.4 mW}$

EXAMPLE 3-15

Calculate:
a. The maximum current that a 100-Ω, 2-W resistor can handle without overheating.
b. The resistance of a resistor that dissipates 50 mW of power when an emf of 40 V is connected across the resistor.

Solution

a. $\hspace{2cm} P = I^2R$ $\hspace{2cm}$ (3-7)

Therefore, $I^2 = \dfrac{P}{R}$

$\hspace{1.5cm} I = \sqrt{\dfrac{P}{R}}$

$\hspace{1.5cm} = \sqrt{\dfrac{2 \text{ W}}{100 \text{ }\Omega}} = \sqrt{0.02} \text{ A}$

$\hspace{1.5cm} = 0.14 \text{ A} = \textbf{140 mA}$

b. $\hspace{2cm} P = \dfrac{V^2}{R}$ $\hspace{2cm}$ (3-8)

Therefore, $R = \dfrac{V^2}{P}$

$\hspace{1cm} = \dfrac{(40 \text{ V})^2}{50 \times 10^{-3} \text{ W}} = \dfrac{1600}{50} \times 10^3 \text{ }\Omega$

$\hspace{1cm} = 32 \times 10^3 \text{ }\Omega = \textbf{32 k}\Omega$

Note again that all substitutions in Eq. 3-7 and Eq. 3-8 are made in *basic* units.

3-10 COST OF ELECTRICAL ENERGY

In Section 3-6, you saw how energy could be calculated in joules by the use of Eq. 3-3.

$$W = VIt \quad \text{joules} \tag{3-3}$$

And, in Example 3-10, you saw how a 60-W lamp, used for just one minute, consumed 3600 J of energy. It is obvious that the joule is too small a unit to conveniently measure large quantities of energy. The preferred unit of electrical energy is the *kilowatt-hour*. This is obtained by recognizing that *VI* in Eq. 3-3 is equivalent to power *(P)*. Thus

$$W = Pt \qquad \text{kilowatt-hours} \qquad (3\text{-}9)$$

where: *W* is work or energy, in kilowatt-hours (kWh)
         P is power in kilowatts (kW)
         t is time in hours (h)

The kilowatt-hour is the basic unit the utility company uses to calculate your monthly electric bill. The watthour *meter* indicates the amount of energy used (in kilowatt-hours) during the month or other billing period. The number of kilowatt-hours for the period is then multiplied by cost per kilowatt-hour established by the utility company. Rates vary widely, depending on the method used to generate the electricity and other factors. Typical rates in the mid-1980s range from 1.6 cents per kilowatt-hour in the northwestern United States to 16 cents per kilowatt-hour in the northeastern part of the country.

EXAMPLE 3-16

A 250-W lamp is used 4 h a day for 30 days. Calculate:
a. The amount of energy used by the lamp.
b. The cost of the energy at 5¢/kWh.
c. The cost of the energy at 10¢/kWh.

Solution

a. $W = Pt$ 　　　　　　　　　　　　　　　(3-9)

$$= \left(250\ W \times \frac{1\ kW}{1000\ W}\right) \times \left(4\ \frac{h}{day} \times 30\ days\right)$$

$$= 0.25\ kW \times 120h$$

$$= \mathbf{30\ kWh}$$

b. $Cost = kWh \times ¢/kWh$

$$= 30\ kWh \times \frac{5¢}{kWh} = \mathbf{\$1.50}$$

c. $Cost = 30\ kWh \times \frac{10¢}{kWh}$

$$= \mathbf{\$3.00}$$

EXAMPLE 3-17

How much will it cost to run the 1½-hp motor in Example 3-13 for 1 h a day, for 30 days, at 5¢/kWh?

Solution

$$W = Pt \qquad\qquad\qquad\qquad (3\text{-}9)$$

$$= \left(1492\ W \times \frac{1\ kW}{1000\ W}\right) \times \left(1\ \frac{h}{day} \times 30\ days\right)$$

$$= 44.76\ kWh$$

$$\text{Therefore, Cost} = 44.76\ kWh \times \frac{5¢}{kWh} = \mathbf{\$2.24}$$

Note that a low-powered device can consume a large amount of energy if it is operated for a long period of time.

3-11   PRACTICAL RESISTORS

So far, resistance has been considered as an inherent property of a circuit, with values over which you have little control. But there are some applications in which resistance is intentionally inserted in a circuit by adding devices called *resistors*. These devices may be used to:

1. *Limit current* to a safe value.
2. *Drop voltage* to a required value.
3. *Divide voltage* to different values from a single source.
4. *Discharge energy* from a capacitor.

Resistors are usually described in terms of:

1. Their electrical resistance (in ohms).
2. Their ability to dissipate heat (in watts).
3. Their construction (carbon composition, wire-wound, film).
4. Their functional makeup (fixed or variable).
5. The percentage tolerance of their nominal resistive value.

3-11.1   Fixed Resistors

The lowest-cost fixed-value (nonvariable) resistors are of the carbon composition type. These resistors are commonly used in electronic circuits where high accuracy is not critical. Carbon composition resistors, Fig. 3-6*a*, consist of powdered carbon and insulating materials molded into the form of a cylinder with axial leads at each end. Insulation and mechanical protection are provided by a bakelite case.

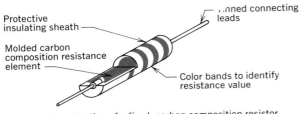

(a) Construction of a fixed, carbon composition resistor

(b) Symbol for resistor

FIGURE 3-6
Commercial carbon composition resistor with symbol.

Carbon-film or metal-film resistors provide more accurate resistance values. They are made by depositing a carbon or nickel-chromium alloy spiral film on a ceramic base (substrate). Resistors of this type are available in standard values up to 22 MΩ. (Others are available with values from 100 MΩ to 1 million MΩ.)

TABLE 3-1

A Portion of Appendix B Showing Standard Resistance Values for Commercial and Military Type Resistors with Various Tolerances

±1%	±2%	±5%	±10%	±20%
182		18	18	
187	187			
191				
196	196			
200		20		
205	205			
210				
215	215			
221		22	22	22
226	226			
232				
237	237			
243		24		
249	249			
255				
261	261			
267				
274	274	27	27	

3-11.2 Standard Values and Tolerances

The *nominal* values available depend upon the *tolerance* of the resistor. The tolerance is a percentage figure that indicates how close the actual resistance value should come to the nominal (stated) value of the resistor. Standard resistance values for commercial resistors are listed in Appendix B; a portion of that appendix is reproduced in Table 3-1.

Resistors with 5% or 10% tolerances are available from 0.22 Ω; 1% and 2% tolerance resistors are available from 2.26 Ω (See Table 3-1). Multiples and submultiples of the numbers shown in Table 3-1 and in Appendix B also can be obtained. For example, resistors with 5% or 10% tolerances are available with values of 0.22, 2.2, 22, 220, 2200 Ω, and so on. You should note that the resistance values for 5%, 10%, and 20% tolerance resistors have only two significant figures; 1% and 2% tolerance resistors have three significant figures, such as 2.26, 22.6, 226 Ω, and so on.

3-11.3 Standard Color Code

Colored bands (stripes) around the resistor body show the resistance and tolerance. The standard color code for the bands is shown in Table 3-2.

The resistance value and tolerance for 5%, 10%, and 20% tolerance resistors is shown by the use of *four* colored bands, as shown in Fig. 3-7a. To read the coding, hold the resistor so that the bands are at the left end and read the sequence from left to right. The first two colors represent the first two significant digits of the resistance value. (Note that black is not used as the first color band). This two-digit number is then multiplied by the factor represented by the third colored band. (See Table 3-2.) The fourth band indicates the tolerance.

In some resistors produced to military specifications, a fifth band is used. The color of this band indicates the *failure rate,* given as the average number of failures per 1000 resistors in a standard 1000-h test. A resistor is considered a failure when its ohmic value falls outside of its

TABLE 3-2

Standard Color Code for Resistors

Color	First and Second Stripes Digit	Third Stripe Multiplier	Fourth Stripe Tolerance
Black	0	$10^0 = 1$	$\pm 1\%$*
Brown	1	$10^1 = 10$	$\pm 2\%$*
Red	2	$10^2 = 100$	—
Orange	3	$10^3 = 1000$	—
Yellow	4	$10^4 = 10,000$	—
Green	5	$10^5 = 100,000$	—
Blue	6	$10^6 = 1,000,000$	—
Violet	7	10^7	—
Gray	8	10^8	—
White	9	10^9	—
Gold	—	$10^{-1} = 0.1$	$\pm 5\%$
Silver	—	$10^{-2} = 0.01$	$\pm 10\%$
No color	—	—	$\pm 20\%$

*Tolerance is indicated by the fifth stripe in 1% and 2% tolerance resistors, since they have three significant digits and a multiplier.

tolerance range. For example, if the fifth (or sixth) band is violet (7), this means an average failure rate of 7 resistors per 1000. Alternatively, the *reliability* may be expressed as 993 per 1000 or 0.993.

EXAMPLE 3-17

What are the resistance values of resistors coded as follows:

a. Red, violet, brown, silver.
b. Brown, black, yellow, none.
c. Red, red, gold, gold.
d. Blue, gray, green, silver.

Solution

a.	red	violet	brown	silver		
	2	7	10^1	$\pm 10\%$ =	**270 Ω ± 10%**	
b.	brown	black	yellow	none		
	1	0	10^4	$\pm 20\%$ =	**100 kΩ ± 20%**	
c.	red	red	gold	gold		
	2	2	0.1	$\pm 5\%$ =	**2.2 Ω ± 5%**	
d.	blue	gray	green	silver		
	6	8	10^5	$\pm 10\%$ =	**6.8 MΩ ± 10%**	

To better understand the *tolerance* concept, consider the 270 Ω ± 10% resistor in the preceding example. The actual value of this resistor may be anywhere between 270 Ω ± 27 Ω—that is, between 297 Ω and 243 Ω. If this spread is too wide for a critical application, a resistor with a smaller tolerance (1%, 2%, 5%) may be specified. The smaller tolerance resistors are more expensive, however.

Figure 3-7b shows the use of five colored bands for 1% and 2% tolerance resistors. The colors of the first three bands represent the first three significant digits of the resistance value; the fourth band is the multiplier, and the fifth shows the tolerance.

EXAMPLE 3-18

What are the resistance values of resistors with the following codings:

a. Brown, green, gray, black, black.
b. Red, brown, green, brown, brown.
c. Violet, orange, red, yellow, black.

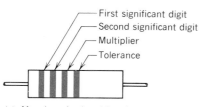

(a) Meaning of colored bands on resistor body for 5% and 10% tolerance (note orientation)

- First significant digit
- Second significant digit
- Multiplier
- Tolerance

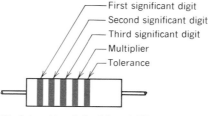

(b) Colored bands for 1% and 2% tolerance resistors

- First significant digit
- Second significant digit
- Third significant digit
- Multiplier
- Tolerance

FIGURE 3-7

Resistor color coding. *Note:* A wide band for the first digit indicates a wire-wound resistor.

d. Blue, yellow, white, silver, brown.

e. Gray, red, black, orange, black.

Solution

a.	brown	green	gray	black	black	
	1	5	8	10^0	$\pm 1\%=$	**158 Ω ± 1%**
b.	red	brown	green	brown	brown	
	2	1	5	10^1	$\pm 2\%=$	**2.15 kΩ ± 2%**
c.	violet	orange	red	yellow	black	
	7	3	2	10^4	$\pm 1\%=$	**7.32 MΩ ± 1%**
d.	blue	yellow	white	silver	brown	
	6	4	9	10^{-2}	$\pm 2\%=$	**6.49 Ω ± 2%**
e.	gray	red	black	orange	black	
	8	2	0	10^3	$\pm 1\%=$	**820 kΩ ± 1%**

3-11.4 Power Ratings

Standard *power ratings* of the resistors discussed previously range from ⅛ W to 2 W. The power rating of a resistor is determined by its physical size (length and diameter). A 2-W resistor, for example, is larger and has more surface area than a 1-W resistor, allowing it to dissipate more heat. There is no color coding to identify a resistor's power rating; experience will eventually allow you to recognize power ratings by the size of the resistor. Figure 3-8 shows resistors with various power ratings in their actual size.

Resistors with power ratings greater than 2 W are generally of the wire-wound type and are covered with a porcelain coating. Resistors of this type, available with power ratings in the hundreds of watts, can operate at temperatures as high as 300°C without damage. Carbon composition resistors, with power ratings of 2 W or less, should

operate at temperatures of 85°C or lower, since excessive temperatures produce changes in resistance value. High temperatures may even cause such resistors to open (lose continuity, breaking the circuit).

Low-powered resistors are also available in packages that resemble integrated circuits, as shown in Fig. 3-9. These may be of the single in-line package (SIP) or dual in-line package (DIP) type. These packages are fabricated from thin-film and thick-film metals and are often laser-trimmed to very close tolerances. They may contain eight fully isolated resistors of equal value (typically from 100 Ω to 100 kΩ), or some resistor network combination. They are especially useful where matched resistor pairs within ±0.005% are required.

Also available are fixed resistors with a tolerance to ± 0.001%. They consist of a resistive element of metal foil mounted on a ceramic substrate. The metal foil is etched to the required resistance and hermetically sealed to maintain that tolerance. These resistors are available with values from 5 Ω to 3.3 MΩ, and can dissipate from 0.2 W to 2.5 W at maximum working voltages from 250 V to 600 V. They are used in laboratories for high-precision calibration purposes and in aerospace applications.

For miniaturization projects, thick film chip resistors only 3.2 mm long, 1 mm wide, and 0.6 mm thick are available. They have resistance values from 10 Ω to 3.3 MΩ, at ⅛ W dissipation and ±1% tolerance.

3-11.5 Variable Resistors

Resistors that allow their resistive value to be changed are called *variable resistors*. They usually have three terminals, compared to the two terminals (leads) found on fixed-value resistors. **Strictly speaking, if all three terminals**

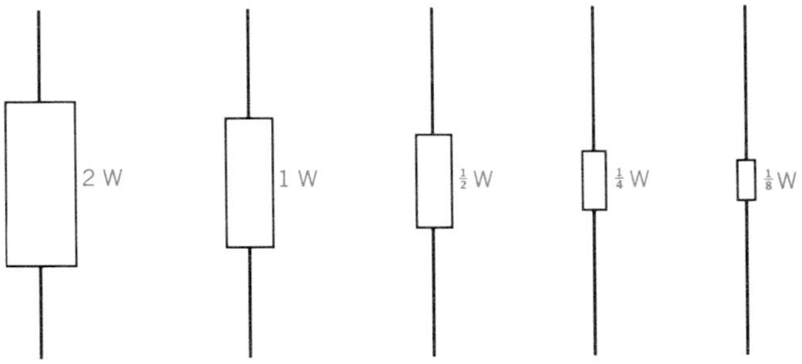

FIGURE 3-8
Full-size drawings of resistors of various power ratings.

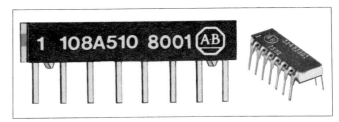

FIGURE 3-9
Resistors in single in-line package (SIP) and dual in-line package (DIP). (Courtesy of Allen Bradley Co.)

are actively used, the device is a *potentiometer*, used for varying *voltage*. (This will be explored in Chapter 5.) If there are only two active terminals, the device is a variable resistor or *rheostat*, used for varying *current*. Any three-terminal potentiometer can be wired as a rheostat. However, three-terminal variable resistors designed as rheostats have a wiper terminal (see Fig. 3-10a) with a large current rating. When designed as a potentiometer, the wiper terminal current-rating is typically extremely small. Potentiometers are damaged by excessive wiper currents when they are used as rheostats.

Figure 3-10*a* shows a typical wire-wound variable resistor. Such resistors allow the varying of resistance values from 1 Ω to 10 kΩ, and are suitable for power ratings from 2 to 300 W. They are typically used as rheostats. The ends of the resistance wire are connected to terminals 1 and 3; terminal 2 is connected to a wiper that slides and makes contact with the wire to vary the resistance. The wiper moves when the shaft is rotated.

Low-power variable resistors, often called "controls," are of the carbon composition type. They are available in power ratings from ¼ W to 2 W, in standard values of 100, 250, and 500 Ω, 1, 2.5, 5, 10, 25, 50, 100, and 500 kΩ, and 1 MΩ. They are used primarily as potentiometers, as in the volume controls of audio devices, or as occasionally adjusted variable resistance devices called "trimmers." Wire-wound types, with slightly higher power ratings, are also available.

Precision trimming potentiometers are usually rectangular in shape and have an adjusting screw (rather than a shaft turned by finger pressure). They are capable of very fine resistance adjustments, since the screw can be moved as many as 26 complete turns, allowing variation from zero to maximum resistance (in the region of 25 kΩ).

Two methods of wiring a potentiometer as a rheostat are shown in Figs. 3-10*c* and 3-10*d*. For example, if the rheostat is 100 Ω, the total resistance from Terminal 1 to

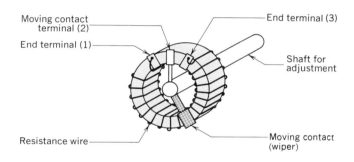

(a) Typical wirewound variable resistor
(potentiometer)

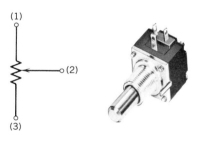

(b) Symbol for a potentiometer
and photograph

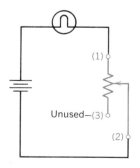

(c) Potentiometer used as a rheostat
(no connection to terminal 3)

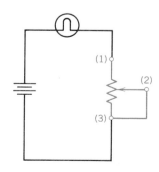

(d) Potentiometer used as a
rheostat (only two active
terminals)

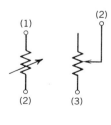

(e) Alternate symbols
for variable resistor
or rheostat

FIGURE 3-10

Variable resistors. [Photograph in (b) courtesy of Allen Bradley Co.]

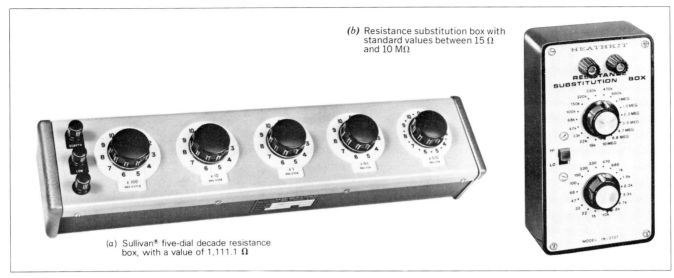

(b) Resistance substitution box with standard values between 15 Ω and 10 MΩ

(a) Sullivan® five-dial decade resistance box, with a value of 1,111.1 Ω

FIGURE 3-11
Adjustable resistors. [Photograph in (a) courtesy of James G. Biddle Co.; photograph in (b) courtesy of Heath Co.]

3 is 100 Ω and does not change. (See Fig. 3-10c.) By rotating the shaft, Terminal 2 is moved toward Terminal 1, decreasing the resistance between those terminals toward zero. The lamp will have maximum current and brightness. Rotating the shaft the opposite way will move Terminal 2 away from Terminal 1, increasing the resistance (to a maximum of 100 Ω) and reducing the current. The light will dim.

The advantage of the connection shown in Fig. 3-10d is that it provides an alternate current path from Terminal 1 to 3. That is, if a "worn spot" (high resistance contact) should occur between the sliding contact and the resistance element, the current flow would not be completely interrupted as it would in the circuit of Fig. 3-10c.

A type of variable resistor that is very useful in the laboratory is the *decade resistance box*. (See Fig. 3-11a.)

Some seven-dial decade resistance boxes can produce resistance values in a range from 1 Ω to 10 MΩ. The values can be obtained in 1-Ω steps, often with an accuracy to within 0.1%, by using the dials. Each dial controls up to 10 series-connected resistors, which can be connected in series with the other dialed resistors.

When using a decade box, be careful to avoid exceeding the current capability of the highest-valued resistance. Maximum current values are usually shown on each dial.

Another method used to obtain different resistance values is the *resistance substitution box*. Figure 3-11b shows a substitution box that can provide standard resistance values from 15 Ω to 10 MΩ, with a tolerance of ±10%. This device is useful when you are experimenting to determine which standard resistor value to use in a circuit.

SUMMARY

1. A voltmeter connected *across* the source indicates the voltage applied to a circuit, and an ammeter *in series* with the load indicates the load current.
2. Ohm's law states the relationship between voltage, current, and resistance in a circuit; for a fixed resistance, the ratio of potential difference to current is a constant called the resistance ($R = V/I$).
3. The linear relationship between voltage and current for any constant resistance is

a statement of Ohm's law. It can be shown graphically as a straight line graph of I plotted against V.

4. Other forms of Ohm's law allow the voltage or current to be determined, given the other two variables $(I = V/R$ and $V = IR)$.

5. A resistor symbol in a circuit can represent an electrical load, such as an incandescent lamp or a heating element. A reduction in resistance is an increase in load.

6. Electrical work and energy are measured in joules, using either $W = VQ$ or $W = VIt$ for calculations.

7. Power is the *rate* at which work is done or energy is converted $(P = W/t)$.

8. The unit of mechanical power in the FPS system is the foot pound per second or the horsepower (550 ft lb/s $= 1$ hp). In the MKS system, the unit is the joule per second or watt (1 J/s $= 1$ W).

9. Electrical power may be calculated by using $P = VI$, $P = I^2R$ or $P = V^2/R$. The basic unit of measurement is the watt (1 kW $= 1000$ W, 1 hp $= 746$ W).

10. Efficiency can be defined as follows: $\eta = P_{out}/P_{in}$. It is usually expressed as a percentage.

11. The practical unit for measuring large quantities of electrical energy is the kilowatt-hour, using the equation $W = Pt$.

12. The cost of electrical energy is the product of the kilowatt-hours used and the cost per kilowatt-hour.

13. Practical resistors are devices that have a known amount of resistance (in certain standard values). They can be used to intentionally introduce resistance to a circuit.

14. Resistors may be fixed in value and identified by a standard color banding code. (See Table 3-2.)

15. Variable resistors generally have three terminals. They are known as potentiometers when used to vary voltage in a circuit, and as rheostats when used to vary current in a circuit.

16. A decade resistance box is a special type of variable resistance that allows almost any desired value to be "dialed" to within 0.1% accuracy.

SELF-EXAMINATION

Answer true or false, or fill in the correct answer
(Answers at back of book)

3-1. A voltmeter is connected in parallel, an ammeter in series. _____

3-2. A voltmeter may be removed from an operating circuit without affecting the circuit, an ammeter cannot. _____

3-3. In a circuit with constant resistance, an increase in voltage results in a proportional increase in current. _____

3-4. If applied voltage is held constant, an increase in resistance results in an increase in current. _____

3-5. If the current through a resistor drops, the voltage drop across the resistor becomes smaller. _____

3-6. A resistor of 2.2 MΩ represents a load that is 1000 times larger than a 2.2-kΩ load. _____

3-7. Work and energy are essentially the same thing. _____

3-8. Power and energy are essentially the same thing. _____

3-9. Mechanical and electrical work are measured in the same units in the MKS system. _____

3-10. Efficiency is the ratio of input work or power to useful output work or power. _____

3-11. Both mechanical and electrical power involve the rate of doing work or converting energy. _____

3-12. Mechanical and electrical power are both measured in watts in the MKS system. _____

3-13. A 100% efficient 2-hp motor would require 1492 W of electrical input power. _____

3-14. Where current flows through a resistor the power dissipated in the resistor depends upon the square of the current. _____

3-15. If the voltage applied to a constant resistance doubles, the power dissipated in the resistance also doubles. _____

3-16. Electrical energy may be measured in joules or kilowatt-hours. _____

3-17. If 500 kWh of energy are used, the total cost at 5¢/kWh is $250. _____

3-18. The resistance of a red, yellow, and green resistor is 2.4 MΩ. _____

3-19. The color stripes for a 100-Ω resistor are brown, black, black. _____

3-20. A rheostat is a two-terminal device that varies resistance and current. _____

3-21. A 120-V appliance that draws 12 A has a resistance of _____.

3-22. A 220-V oven with a resistance of 11 Ω draws a current of _____.

3-23. A 5-mA current through a 10-kΩ resistor causes a voltage drop of _____.

3-24. A 120-V, 360-W television receiver draws a current of _____.

3-25. A 6.6-kW oven that requires a current of 30 A must operate at a voltage of _____.

3-26. A 10-Ω resistor with 2 A of current through it must be able to dissipate a power of _____.

3-27. If a 220-V source must deliver 2.2 kW to a heating element, the resistance of the element must be _____.

3-28. A 200-W lamp used for 6 h consumes an energy of _____.

3-29. If energy costs 20¢/kWh, the cost of running a 100-W lamp for 10h/day for 10 days is _____.

3-30. A 470-kΩ, ±5% resistor, has color stripes of _____.

REVIEW QUESTIONS

1. State Ohm's law in your own words.

2. Draw a graph of I versus V for a 2-kΩ resistor when the voltage is varied in 2-V steps from 0 to 10 V. Find the slope of this graph (rise/run) using the units on the axes. How is this number representing the slope related to 2 kΩ?

3. On the graph drawn in Question 2 draw a second graph for a 4-kΩ resistor. Does it have a larger or smaller slope than that for the 2-kΩ? Explain why the current through a 3-kΩ resistor at 10 V is *not* given by a point on the above graph located *midway* between the 2- and 4-kΩ lines at 10 V. *Hint:* Draw a graph of I versus R for a constant voltage of 10 V with R varying from 0 to 5 kΩ.

4. Explain why a decrease in load resistance should be considered an increase in load.

5. Explain the difference between work and power.
6. Explain how efficiency can be given in terms of work *or* power ratios.
7. Which is a larger unit of power, the horsepower or the kilowatt?
8. Explain how the *power* used by a 100-W lamp is the same no matter whether the lamp is lit for a second, a minute, or an hour; but the *energy* used is different.
9. Under what conditions would a 100-W lamp use more energy than a 1500-W toaster?
10. Which is the larger unit of energy, the joule or the watt-hour?
11. What is meant by the nominal value of a resistor?
12. Explain why brown, black, black does not represent a 100-Ω, 5% resistor. How many bands would be necessary to represent a 1-million-ohm resistor if a colored band were used for each digit? What is the largest theoretical value of resistance the standard color code can represent? What is the smallest?
13. Explain how a three-terminal potentiometer can be wired as a rheostat with two active terminals. What is the advantage of connecting the center terminal to one of the end terminals?
14. What maximum value of resistance can be dialed by a five-dial decade resistance box, whose smallest increments are one tenth of an ohm? Which resistance will determine the current capability of such a value?

PROBLEMS

(Answers to odd-numbered problems at back of book)

3-1. What is the resistance of a circuit that draws 1.5 A from a 220-V source of emf?
3-2. What resistance will limit the current in a circuit to 3.75 A when connected to a 120-V source?
3-3. A voltmeter across a resistor indicates 15 V when an ammeter in series with the resistor indicates 2.5 mA. What is the resistance of the resistor?
3-4. What resistance will allow 25 μA of current when connected across 40 mV?
3-5. A voltmeter connected across a precision 1-kΩ resistor indicates 2.2 V. What is the current through the resistor?
3-6. How much current will flow through a 2.4-MΩ resistor when it is connected across a 1.5-V cell?
3-7. What voltage drop occurs across a 680-Ω resistor that has a current of 450 mA through it?
3-8. What voltage must be applied to a 470-kΩ resistor to obtain a current of 150 μA?
3-9. If the surface resistance of an insulator is 1000 MΩ, what is the current that flows on the surface when an emf of 500 kV is connected?
3-10. If a fault on a 220-kV line consists of a 10-Ω path to ground, how much current initially flows?
3-11. A typical lightning flash consists of 25,000 amperes of current due to a potential difference of 30 MV. What is the average resistance of the ionized air during a lightning flash?
3-12. The resistance of a dc relay coil is 10 kΩ and requires a current of 12 mA for its operation. What potential difference is necessary across the coil to make the relay operate?

3-13. An electric heating element operating at 120 V draws 10 A to boil a kettle of water in 5 min. Calculate how much electrical energy is supplied during this time:
   a.   in joules
   b.   in kWh
3-14. An electric clothes dryer operating at 220 V draws 15 A to dry a load of clothes in half an hour. Calculate how much electrical energy is supplied:
   a.   in joules
   b.   in kWh
3-15. If the water in the kettle in Problem 3-13 absorbs 300,000 J of energy in the form of heat, determine the efficiency of the kettle.
3-16. If the efficiency of the dryer is 60%, how much energy in the form of heat is absorbed by the clothes in Problem 3-14?
3-17. Calculate the cost of operating the kettle in Problem 3-13 if energy costs 10¢/kWh.
3-18. Calculate the cost of operating the clothes dryer in Problem 3-14 if energy costs 10¢/kWh.
3-19.   a.   How many joules of energy are equivalent to 1 kWh of energy?
   b.   What is 1 J equal to in kWh?
3-20. The annual energy use of the world is 0.3 million billion megajoules (MJ).
   a.   Express this quantity in kWh.
   b.   If the total content of proved fossil fuel reserves (oil, coal, and gas) is 25 million billion MJ, how long will this last at the present rate of consumption?
3-21. A steam iron operates at 120 V and draws 12 A. Calculate its power rating in
   a.   kW
   b.   hp
3-22. What current is drawn from a 220-V source by a 6-kW heating element?
3-23. At what voltage must a 55-W soldering iron operate if it uses 0.47 A?
3-24. A 220-V motor drawing 12 A develops 3 hp. What is its efficiency?
3-25. What current must be supplied to an 85%-efficient 25-hp motor operating at 550 V?
3-26. Which motor costs more to run, a 60%-efficient 1½-hp motor or an 80%-efficient 2-hp motor?
3-27. What is the maximum current that a 10-kΩ, ½-W resistor can handle without overheating?
3-28. What is the maximum current that a 2.2-MΩ, ¼-W resistor can handle without overheating?
3-29. What maximum voltage can safely be connected across a 680-Ω, 1-W resistor?
3-30. What maximum voltage can safely be connected across a 10-Ω, 2-W resistor?
3-31. What must be the nearest standard power rating of a 470-Ω resistor that has a current of 50 mA flowing through it?
3-32. What is the minimum power rating of a 5.6-kΩ resistor that has a voltage drop of 50 V across it?
3-33. What are the resistance values and tolerances of resistors coded as follows:
   a.   Orange,   c.   Red, red, green
        orange, orange
        silver
   b.   Green, brown,   d.   Blue, gray,
        brown, gold   gold, gold

e. Brown, black,
black, silver

f. Brown, red, yellow,
black, black

g. Orange, black, brown,
brown, brown

h. Green, red, orange,
orange, black

3-34. What are the resistance values and tolerances of resistors coded as follows:

a. Gray, black, blue,
silver, black

b. White, green,
orange, blue, brown

c. Red, yellow, white,
gold, black

d. Yellow, red, red,
silver, brown

e. Green, blue,
silver, gold

f. Yellow, violet,
red, gold

g. Violet, green,
brown, gold

h. Orange, white,
green, silver

3-35. What are the color bands for the following resistors:

a. 11 kΩ, $\pm 5\%$
b. 820 Ω, $\pm 10\%$
c. 15 MΩ, $\pm 20\%$
d. 3 Ω, $\pm 5\%$
e. 270 kΩ, $\pm 5\%$
f. 196 Ω, $\pm 2\%$
g. 5.11 Ω, $\pm 1\%$
h. 8.25 MΩ, $\pm 2\%$

3-36. What are the color bands for the following resistors:

a. 10.7 Ω, $\pm 1\%$
b. 205 kΩ, $\pm 2\%$
c. 3.92 MΩ, $\pm 1\%$
d. 1.13 kΩ, $\pm 1\%$
e. 2.61 Ω, $\pm 2\%$
f. 16 Ω, $\pm 5\%$
g. 4.7 MΩ, $\pm 10\%$
h. 56 Ω, $\pm 10\%$

3-37. Select the nearest standard value of 10% resistor for the following:

a. 16 kΩ
b. 290 kΩ
c. 5.9 MΩ

3-38. Repeat Problem 3-37 using 5% resistors.

3-39. Select the nearest standard value of 2% resistor for the following:

a. 238 kΩ
b. 372 Ω
c. 6.7 Ω

3-40. Repeat Problem 3-39 using 1% resistors.

CHAPTER
4

RESISTANCE

Good conductors are those materials whose atomic electronic structures readily provide free electrons. This ability to provide free electrons is more easily identified by knowing a property called the material's *resistivity.* The smaller the resistivity value, the better the material is as a conductor. When resistivity is combined with the factors of length and cross-sectional area, the result is an equation that allows you to determine the resistance of any conductor at a specific temperature of 20°C. At any other temperature, a correction is applied. This correction requires the defining of the material's *temperature coefficient* of resistance.

In this chapter, you will also consider *linear* resistors—those that obey the proportional relationship of V to I in Ohm's law—and *nonlinear* resistors, such as thermistors. Nonlinear resistors will be examined in some of their current-surge limiting applications.

4-1 FACTORS AFFECTING RESISTANCE

As you learned in Chapter 2, resistance to current is caused by the collisions of free electrons within the conductive material. Anything that increases the number of such collisions obviously will contribute to an increase in resistance. Figure 4-1 and Eq. 4-1 show the factors that determine the resistance of a conductor at a given temperature.

$$R = \frac{\rho l}{A} \quad \text{ohms} \qquad (4-1)$$

where: R is the resistance of the material in ohms (Ω)
l is the length of the material in meters (m)
A is the cross-sectional area of the material in square meters (m^2)
ρ is the resistivity (specific resistance) of the material in ohm meters ($\Omega \cdot m$) at a specified temperature

Equation 4-1 implies:

1. Resistance is *directly* proportional to length (more collisions occur in a longer wire).
2. Resistance is *inversely* proportional to cross-sectional area (a smaller cross-section increases the current density and the number of collisions that will occur).
3. Resistance depends on the type of material, as indicated by the specific resistance or *resistivity*, designated by the Greek letter rho (ρ).

Table 4-1 lists the resistivities at room temperature for some common conductors, semiconductors, and insulators. These numbers actually give the resistance of a *specific volume* of material (''specific resistance'' is another name for resistivity). For example, ρ for copper $= 1.72 \times 10^{-8}$ ohm meter ($\Omega \cdot m$) at 20°C. This means that a piece of copper 1 m long with a cross-sectional area of 1 m^2 has a resistance of $1.72 \times 10^{-8}\ \Omega$. Note in Example 4-1 how the units for resistivity (in $\Omega \cdot m$) give the correct unit for resistance in Eq. 4-1.

TABLE 4-1
Resistivities at Room Temperature, 20°C

Substance		$\rho\ (\Omega \cdot m)$
Conductors		
Metals	Silver	1.47×10^{-8}
	Copper	1.72×10^{-8}
	Gold	2.45×10^{-8}
	Aluminum	2.63×10^{-8}
	Tungsten	5.51×10^{-8}
	Nickel	$7.8\ \times 10^{-8}$
Alloys	Manganin	$44\ \ \times 10^{-8}$
	Constantan	$49\ \ \times 10^{-8}$
	Nichrome	$100\ \ \times 10^{-8}$
Semiconductors		
Pure	Carbon	$3.5\ \times 10^{-5}$
	Germanium	0.60
	Silicon	2300
Insulators		
	Wood	$10^{8} - 10^{11}$
	Paper	10^{10}
	Nylon	8×10^{12}
	Lucite	10^{13}
	Glass	$10^{10} - 10^{14}$
	Rubber	$10^{13} - 10^{14}$
	Amber	5×10^{14}
	Mica	$10^{11} - 10^{15}$
	Sulfur	10^{15}
	Teflon	10^{15}
	Porcelain	10^{16}
	Polystyrene	10^{16}
	Quartz (fused)	75×10^{16}

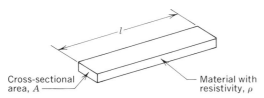

FIGURE 4-1

Factors that determine the resistance of a material at a given temperature.

EXAMPLE 4-1

The type of copper wire used for house wiring has a diameter of approximately 2 mm.
a. What resistance will a 100-m-long reel of copper wire have at room temperature?
b. What is the resistance if the wire has the same dimensions but is made of aluminum?

Solution

a. Area $A = \dfrac{\pi d^2}{4}$

$$= \frac{\pi}{4} \times (2 \times 10^{-3}\ m)^2 = 3.14 \times 10^{-6}\ m^2$$

$$R = \frac{\rho l}{A} \qquad (4\text{-}1)$$

$$= \frac{1.72 \times 10^{-8}\ \Omega \cdot m \times 100\ m}{3.14 \times 10^{-6}\ m^2} = \mathbf{0.55\ \Omega}$$

b. For aluminum, $\rho = 2.63 \times 10^{-8}\ \Omega \cdot m$. Therefore,

$$R = R_{Cu} \frac{\rho_{Al}}{\rho_{Cu}}$$

$$= 0.55\ \Omega \times \frac{2.63 \times 10^{-8}\ \Omega \cdot m}{1.72 \times 10^{-8}\ \Omega \cdot m} = \mathbf{0.84\ \Omega}$$

NOTE In the solution of Example 4-1b, it is not necessary to resubstitute for l and A in solving the problem. Since the only factor that changed is ρ, and the new value of ρ for aluminum is higher than copper, the resistance of copper is multiplied by an *increased* ratio of resistivity.

EXAMPLE 4-2

A 100-W incandescent lamp uses a tungsten filament that is approximately 3 cm long. If its cold (20°C) resistance is 10 Ω, what is its diameter? ($\rho_{tungsten} = 5.5 \times 10^{-8}\ \Omega \cdot m$.)

Solution

$$R = \frac{\rho l}{A} \qquad (4\text{-}1)$$

Therefore, $A = \dfrac{\rho l}{R}$

$$= \frac{5.5 \times 10^{-8}\ \Omega \cdot m \times 3 \times 10^{-2}\ m}{10\ \Omega}$$

$$= 1.65 \times 10^{-10}\ m^2$$

$$A = \frac{\pi d^2}{4}$$

Therefore, $d = \sqrt{\dfrac{4A}{\pi}}$

$$= \sqrt{\frac{4 \times 1.65 \times 10^{-10}\ m^2}{\pi}}$$

$$= 1.45 \times 10^{-5}\ m = \mathbf{0.0145\ mm}$$

NOTE In both Examples 4-1 and 4-2, regardless of the units in which the information is given, it is first necessary to convert these units to basic units before substitution in the appropriate equation.

4-2 WIRE GAUGE TABLE

To avoid the need for calculating resistance (using Eq. 4-1) each time for a given length and diameter of copper wire, tables have been worked out for *standard* wire sizes. An example is the table in Appendix C. It is based on a system called the American Wire Gauge (AWG) or Brown and Sharpe (B&S) gauge. Although metric equivalents are also given in the table, the AWG system is still widely used in the United States. This system is actually based on measuring the cross-sectional area of conductors in units called *circular mils*.

$$\text{CM area} = (\text{diameter in mils})^2 \qquad (4\text{-}2)$$

where: CM is a *measure* of the area in circular mils
   1 mil = 0.001 in.

   Thus, a 0.030-in. diameter wire has a diameter of 30 mils and a circular mil area of 30^2 or 900 CM.

Appendix C shows the largest size of wire to be gauge No. 0000. Wires larger than this are indicated by their cross-sectional areas given in thousands of circular mils. Thus a wire that has a cross-sectional area of 250,000 CM is referred to as having a gauge of 250 MCM, where M represents 1000.

   It can be shown that a decrease in gauge number by 3 (from 14 to 11, for example) approximately *doubles* the area. Each decrease by a single gauge number increases the area by a factor of approximately 1.26.

The larger the diameter of the wire, the smaller the gauge numbers.

   Residential wiring varies from No. 12 or No. 14 for lighting circuits (20 A to 15 A) to No. 8 for electric stoves and other heavy loads. An electronic circuit may use wire as fine as No. 22 gauge, capable of carrying about 1 A.

EXAMPLE 4-3

a. Use Appendix C to calculate the circular mil area of a No. 14 gauge wire
b. Calculate the resistance of 150 ft of No. 14 gauge wire using Appendix C
c. Calculate the approximate resistance of 150 ft of No. 8 gauge wire, *without* using Appendix C

Solution

a. From Appendix C, the diameter of No. 14 gauge wire is 64.1 mils (0.0641 in.)

$$\text{CM area} = (\text{diam. in mils})^2 \qquad (4\text{-}2)$$

$$= 64.1^2 = \mathbf{4109\ CM}$$

b. From C, resistance of No. 14 gauge wire is 2.52 Ω/1000 ft.

$$\text{Resistance} = \Omega/1000 \text{ ft} \times \text{length in feet}$$
$$= \frac{2.52 \ \Omega}{1000 \text{ ft}} \times 150 \text{ ft}$$
$$= \frac{2.52 \times 150 \ \Omega}{1000} = \mathbf{0.378 \ \Omega}$$

c. No. 8 gauge represents a decrease of six gauge numbers from No. 14. Since three gauge numbers cause an increase in cross-sectional area by a factor of 2, a change of six gauge numbers will result in four times the area. This will cause the resistance to *decrease* to one-fourth the value of No. 14 gauge:

$$\text{Resistance of No. 8 gauge} = \frac{\text{Resistance of No. 14}}{4}$$
$$= \frac{0.378 \ \Omega}{4} = \mathbf{0.095 \ \Omega}$$

TABLE 4-2
Average Breakdown Voltages for Typical Insulators

Material	Breakdown Voltage (kV/mm)
Air	4
Porcelain	8
Transformer oil	16
Plastic	16
Paper	20
Rubber	28
Glass	28
Teflon	60
Mica	100

4-3 INSULATORS AND SEMICONDUCTORS

As shown in Table 4-1, insulators have very high resistivity values. Insulators such as rubber, plastic, or paper are used to coat or cover conductors to prevent unwanted contact (a *short*) with other wires or a metal chassis. However, if a high enough voltage is connected across it, an insulator can *break down* and conduct current. Under these conditions, an extremely powerful force acts upon the valence electrons to break their bonds and rupture the insulator. The resulting current that flows is usually accompanied by a large amount of heat. In the case of solid insulators, this leads to their destruction.

Air is a fairly poor insulator compared to mica, for example. An air gap 1 mm wide will break down with only 4000 V across it, while a mica insulator 1 mm thick can withstand up to 100,000 V before breaking down. Average breakdown voltages for some typical insulating materials are listed in Table 4-2. Typical household wiring is insulated for 600 V, but some electronic and power circuits must be capable of withstanding many thousands of volts without breaking down.

The relationship between an insulator's thickness and the voltage it can withstand is given by Eq. 4-3.

Voltage = Breakdown Voltage × Thickness (4-3)

A consistent set of units is used, as shown in Example 4-4.

EXAMPLE 4-4

a. If the thickness of an average sheet of paper is 0.01 cm, what potential difference will "puncture" the sheet of paper?
b. What is the minimum thickness necessary for the plastic coating around a piece of copper wire if the wire must be able to carry 600 V without the insulation breaking down?
c. What material should be used to insulate two metal plates having a potential difference of 250 kV, if the plates are 0.5 cm apart? Use a safety factor of 2 (twice the minimum breakdown voltage required).

Solution

a. Voltage = Breakdown Voltage × Thickness (4-3)
From Table 4-2, breakdown voltage for paper is 20 kV/mm.

$$\text{Voltage} = 20 \ \frac{\text{kV}}{\text{mm}} \times 0.01 \text{ cm} \times \frac{10 \text{ mm}}{\text{cm}}$$
$$= 2 \text{ kV} = \mathbf{2000 \ V}$$

b. $$\text{Thickness} = \frac{\text{Voltage}}{\text{Breakdown Voltage}}$$
From Table 4-2, breakdown voltage for plastic is 16 kV/mm.

$$\text{Thickness} = \frac{600 \text{ V}}{16 \text{ kV/mm}} = \frac{600 \text{ V}}{16 \times 10^3 \text{ V/mm}}$$
$$= 37.5 \times 10^{-3} \text{ mm}$$
$$= \mathbf{0.0375 \ mm}$$

c. For a safety factor of 2, the material should be able to withstand 2 × 250 kV = 500 kV.

Breakdown Voltage $= \dfrac{\text{Voltage}}{\text{Thickness}}$

$$= \dfrac{500 \times 10^3 \text{ V}}{0.5 \text{ cm} \times 10 \text{ mm/cm}}$$

$$= 100 \times 10^3 \text{ V/mm}$$

$$= 100 \text{ kV/mm}$$

From Table 4-2, the required material is **mica.**

The resistivity of silicon is appreciably higher (in its pure state) than all good conductors, but very much less than the resistivity of insulators. Thus silicon is known as a *semiconductor*. When used in solid-state devices, silicon is modified or *doped* by adding other materials that increase its conductivity at least a thousand times. It is the way in which doping changes the conductivity of silicon that makes this material so important in the manufacture of such semiconductor devices as diodes and transistors.

4-4 TEMPERATURE COEFFICIENT OF RESISTANCE

For most metals, an increase in temperature will increase the resistance of the conductor in a fairly linear manner over a range of a few hundred degrees Celsius (°C). This increased resistance is due to the increase in molecular activity, and thus, the number of collisions of free electrons, as a result of heat imparted to the metal. The higher temperature raises the electron energy of the electrons in the valence and conduction shells of the atoms. The more pronounced vibrations caused by this *thermal agitation* increases the chances of collisions with the free electrons that are trying to move through the material.

Figure 4-2 shows a typical variation of resistance with temperature for a metal conductor. As shown, the graph becomes *nonlinear* at very high and very low temperatures, approaching zero resistance at −273°C (absolute zero). Some special alloys exhibit a property called *superconductivity,* in which their resistance abruptly drops to zero at temperatures within a range of 0.1 and 20°C of absolute zero. Once a current is established in a superconducting circuit, it will continue almost indefinitely without the need for an emf (electromotive force).

Equation 4-4 allows you to calculate the resistance R_2 at a different temperature T_2, if you know the resistance R_1 at a known temperature T_1 (usually 20°C or room temperature).

$$R_2 = R_1 \left[1 + \alpha_1 (T_2 - T_1) \right] \quad \text{ohms} \quad (4\text{-}4)$$

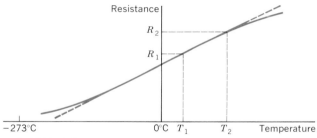

FIGURE 4-2

Variation of resistance with temperature for a metal conductor.

where: R_2 is the final resistance (at temperature T_2) in ohms (Ω)

R_1 is the initial resistance (at temperature T_1) in ohms (Ω)

T_2 is the final temperature in °C

T_1 is the initial temperature, 20°C

α_1 is the temperature coefficient of resistance, per °C, quoted for the material at room temperature, $T_1 = 20°C$

Temperature coefficients for some common materials are listed in Table 4-3. Note that most are positive temperature coefficients (PTC), meaning that they exhibit an increase in resistance as a result of a rise in temperature. Carbon, seimconductors, and most insulators have a negative temperature coefficient (NTC), and exhibit a *drop* in resistance as a result of a rise in temperature. This takes place because, in NTC materials, the current carriers come from electrons *freed* from the valence shells by heat. In insulators and other NTC materials, much more energy is needed to free an electron from the valence shell (making it available for conduction) than is needed in a metal conductor. Therefore, any increase in temperature results in a greater *availability* of current carriers.

EXAMPLE 4-5

A No. 2-gauge copper wire is used in a 10-km-length overhead transmission line. What will be its resistance at:

a.   20°C?

b.   100°F?

c.   −40°F?

TABLE 4-3

Temperature Coefficients of Resistance for Common Materials at 20°C

Material	α_1, per °C
Aluminum	0.0039
Brass	0.0020
Carbon	−0.0005
Constantan (Cu 60, Ni 40)	+0.000002
Copper	0.00393
Gold	0.0034
Iron	0.0050
Lead	0.0043
Manganin (Cu 84, Mn 12, Ni 4)	0.000000
Mercury	0.00088
Nichrome	0.0004
Nichrome II	0.00016
Nickel	0.006
Platinum	0.003
Silver	0.0038
Tungsten	0.0045

Solution

a. From Appendix C, resistance for No. 2 copper = 0.511 Ω/km, at 20°C. Therefore, resistance for 10 km = 10 km × 0.511 Ω/km = **5.11 Ω** at 20°C.

b. $100°F = (100 - 32)\frac{5}{9}°C = 37.8°C$

$$R_2 = R_1[1 + \alpha_1(T_2 - T_1)] \qquad (4\text{-}4)$$

$$= 5.11\ \Omega\left[1 + \frac{0.00393}{°C}(37.8 - 20)°C\right]$$

$$= 5.11\ \Omega\ [1 + 0.07] = \mathbf{5.47\ \Omega}$$

c. $-40°F = (-40 - 32)\frac{5}{9}°C = -40°C.$

$$R_2 = R_1[1 + \alpha_1(T_2 - T_1)] \qquad (4\text{-}4)$$

$$= 5.11\ \Omega\left[1 + \frac{0.00393}{°C}(-40 - 20)°C\right]$$

$$= 5.11\ \Omega[1 - 0.236] = \mathbf{3.90\ \Omega}$$

EXAMPLE 4-6

The resistance of a 100-W tungsten filament lamp increases from 10 Ω at room temperature to 144 Ω when lit. Assuming that this increase in resistance with temperature is linear, calculate the temperature of the hot filament.

Solution

$$R_2 = R_1[1 + \alpha_1(T_2 - T_1)] \qquad (4\text{-}4)$$

Solve for T_2:

$$\alpha_1(T_2 - T_1) = \frac{R_2}{R_1} - 1$$

Hence, $T_2 = \dfrac{1}{\alpha_1}\left(\dfrac{R_2}{R_1} - 1\right) + T_1$

$$= \frac{1°C}{0.0045}\left(\frac{144\ \Omega}{10\ \Omega} - 1\right) + 20°C$$

$$= 2978°C + 20°C = 2998°C \approx \mathbf{3000°C}$$

This calculation compares favorably with typical filament temperatures of 2500°C. Of course, the calculation assumed that the increase in resistance would be completely linear over a wide temperature range, which is not the case.

4-5 LINEAR RESISTORS

In Section 3-3, you learned that a resistor with a constant V/I ratio is known as a *linear* resistor. When current flows through a resistor, electrical energy is converted to heat, which tends to cause a slight increase in resistance in most common conductors. However, this change in resistance is so small over the usual temperature range that such materials as copper and aluminum are considered to be linear resistors.

The *carbon composition* resistors discussed in Section 3-11 are also considered to be linear resistors. As shown in Fig. 4-3, these resistors tend to show a slight increase in resistance for any *appreciable* change in temperature from room temperature.*

Metal film resistors, often used for higher precision resistance values (2% or less), exhibit very high temperature stability. This is given, for example, as 50 PPM/°C (50 parts per million per °C, or a change of only 0.005% per °C). In very accurate measuring instruments and similar critical applications, a material called *constantan* is often

---

*This characteristic is necessary to prevent "thermal runaway." If the resistance were to decrease as temperature increases, this would draw more current. This, in turn, would increase the temperature, further decreasing resistance, and leading to the eventual destruction of the resistor. A *positive* temperature coefficient provides *thermal stability*. A *negative* temperature coefficient leads to thermal *instability*.

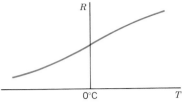

(a) An ordinary metal

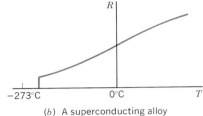

(b) A superconducting alloy

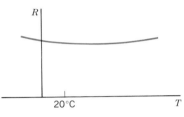

(c) A carbon composition resistor

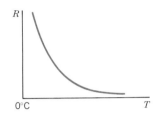

(d) A semiconductor (thermistor)

FIGURE 4-3
Variation of resistance with temperature for various materials.

used. As the name implies, this material has a temperature coefficient of almost zero. It is used to maintain a practically constant resistance when the temperature changes.

4-6 NONLINEAR RESISTORS

The tungsten filament of an incandescent lamp experiences a large change in resistance over its operating range of temperature, as was shown by Example 4-6. A graph of I versus V is not a straight line, so tungsten is a nonlinear resistance in this application. The low initial (cold) resistance of the lamp causes an inrush of current at the instant when it is switched on, as shown in Example 4-7.

Solution

a. Initial inrush, $\quad I = \dfrac{V}{R} \qquad\qquad$ (3-1a)

$$= \frac{120 \text{ V}}{18 \ \Omega} = \textbf{6.7 A}$$

b. Final current, $\quad I = \dfrac{P}{V} \qquad\qquad$ (3-5)

$$= \frac{60 \text{ W}}{120 \text{ V}} = \textbf{0.5 A}$$

c. Hot resistance, $R = \dfrac{V}{I} \qquad\qquad$ (3-1)

$$= \frac{120 \text{ V}}{0.5 \text{ A}} = \textbf{240 } \Omega$$

See Fig. 4-4.

EXAMPLE 4-7

A 120-V, 60-W incandescent lamp has a cold (20°C) resistance of 18 Ω. Calculate:
a. The initial inrush of current when connected to a 120-V source.
b. The steady operating current of the lamp.
c. The hot resistance of the lamp.

If the lamp takes 1 ms to reach its operating temperature, sketch a graph of the current versus time.

Because the very small mass of the filament quickly becomes hot enough to reach its *hot resistance* of 240 Ω, the lamp's current surge is very short-lived. However, this inrush of current causes "thermal shock," in which the sudden increase of temperature is accompanied by a rapid expansion of the metal, causing the filament to bend and flex. This flexing of the element each time the lamp is switched on contributes to the eventual rupture of the filament. Lamps often "burn out" (suffer a filament rupture) at the instant they are switched on.

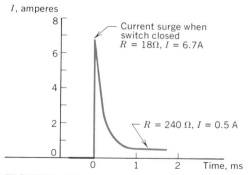

FIGURE 4-4

Graph of the inrush current of an incandescent lamp for Example 4-7.

4-6.1 Thermistors

One method of eliminating current inrush, and thus extending the life of an incandescent lamp, is to place a *thermistor* in series with the lamp. (See Fig. 4-5.)

A thermistor is a semiconductor made from metallic oxides, and has a very large negative temperature coefficient (NTC). It is a nonlinear "thermal resistor," with charac-

teristics shown in Fig. 4-3*d*. These devices are manufactured in many different shapes. (See Fig. 4-6.)

At room temperature, the thermistor may have a resistance of approximately 150 Ω, so that the total resistance of the circuit in Fig. 4-5*a* would be 168 Ω, allowing an initial current of 0.7 A:

$$\frac{120 \text{ V}}{168 \text{ }\Omega} = 0.7 \text{ A}$$

As this current flows through both the thermistor and the lamp filament, the resistance of the thermistor *decreases* while the resistance of the filament *increases*. Eventually, the thermistor's resistance will drop to approximately 1 Ω, a negligible amount compared to the resistance of the filament. (See Fig. 4-5*b*.) The electrical characteristics of the thermistor allow the filament to heat up much more gradually, over a period of several seconds. This avoids "thermal shock" and greatly extends the life of the lamp. Thermistors are now available commercially in disc form, which allows them to be dropped into a lamp socket before inserting the lamp. They last the lifetime of the socket, and are suitable for single-wattage lamps of up to 300 W.

Thermistors are used to prevent current inrush in radio

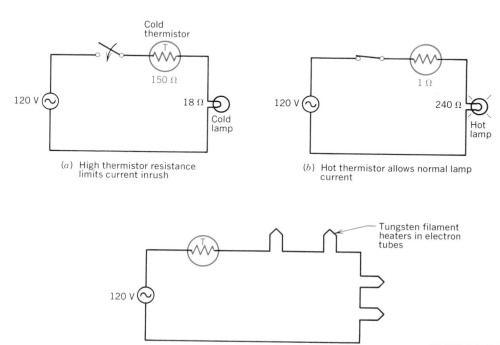

(a) High thermistor resistance limits current inrush

(b) Hot thermistor allows normal lamp current

(c) Thermistor controls the initial current surge in series-connected, electron-tube heaters

FIGURE 4-5

Using thermistors to limit current inrush.

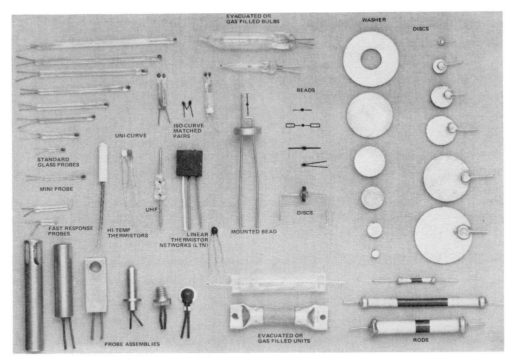

FIGURE 4-6
Various types of thermistors. (Courtesy of Fenwal Electronics.)

and TV broadcast equipment that makes use of vacuum tubes. In these applications, the tungsten heater elements of the tubes are connected in series and operated from a 120-V ac line. (See Fig. 4-5c.) The thermistor in the circuit reduces the current inrush, allowing the heaters to warm up slowly.

It should be noted that the cast-iron or other alloy heating elements used in electric stoves take many seconds to reach their final operating temperature. If tungsten were used, the current inrush would tend to blow fuses or trip circuit breakers. For this reason, the element is usually manufactured from an alloy (such as Nichrome II) that has a relatively low temperature coefficient. This virtually eliminates the current inrush problem, since there is not a very large change between the cold and hot resistance.

4-6.2 Varistors

When lightning "strikes" a power line, a momentary voltage surge or transient is created. On a residential 120-V line, this surge may reach 2–3 kV and occasionally as high as 6 kV. High-voltage transients are also produced when

power equipment containing coils—such as motors, generators, or solenoids—is turned off. Transients as high as 600 V have been recorded upon engine shutdown in an automobile. These transient high voltages, lasting only a few microseconds, can shorten the useful life of semiconductor devices such as transistors or integrated circuits, or even cause them to fail.

A *varistor,* also known as a VDR or voltage-dependent resistor, can be used to suppress high voltage surges. A varistor is a nonlinear resistor whose resistance decreases as voltage increases.

Metal-oxide varistors are produced by pressing and sintering (heating) zinc oxide-based powders into ceramic discs. This produces the semiconductor-like characteristic shown in the *VI* curve in Fig. 4-7a. When the voltage across the varistor exceeds a given value, there is an increase in the breaking of valence bonds (the value may be any standard voltage from 4 V to more than 4 kV, depending on wafer thickness). The breaking of the valence bonds results in a sharp decrease in resistance, allowing transient pulses of current to flow through the varistor. The current may range from 100 A for small varistors to 70,000 A for the bulk type of varistor. The voltage across the varistor is

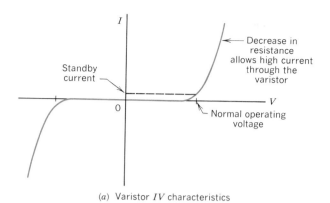

(a) Varistor *IV* characteristics

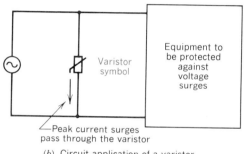

(b) Circuit application of a varistor

FIGURE 4-7
How a varistor reduces voltage surges.

FIGURE 4-8
Typical varistors. (Reprinted with permission of General Electric Company.)

thus "clamped" at some relatively low value. Since the *VI* curve in Fig. 4-7a is symmetrical, it can be used for both ac and dc protection.

When a varistor is connected in parallel with a piece of equipment, as shown in Fig. 4-7b, momentary high voltage surges that could damage the equipment are suppressed. The extra current passes through the varistor, instead of the protected equipment.

A typical varistor used to protect kitchen appliances operating at 120 V would be 7 mm in diameter. It would have a "standby" current of less than 1 mA, but would be capable of bypassing a single current pulse of 1200 A (provided the pulse is less than 20 μs in duration).

The range of metal-oxide varistors available, suitable for continuous operation from 50 V to 2800 V, is shown in Fig. 4-8. Before metal-oxide was introduced as a material for varistors, silicon carbide was used for high-voltage-surge arrestors. The silicon carbide type of varistor, known by the trade name *Thyrite,* was developed to provide lightning protection on power transmission lines.

SUMMARY

1. For a given temperature, the resistance of a material is directly proportional to length, inversely proportional to the cross-sectional area, and dependent upon the type of material. $R = \rho l/A$
2. The resistive property of a material is indicated by its *resistivity* or specific resistance (ρ), ohm meters, as given in Table 4-1.
3. A low resistivity value indicates a more conductive material.
4. The circular mil area of a conductor is found by squaring the diameter in mils. 1 mil = 0.001 in.
5. The wire gauge table is a list of standard wire sizes based on circular mil area. A

smaller gauge number indicates a conductor with a larger cross-sectional area and higher current-carrying capacity.

6. The resistivities of semiconductors are between those of conductors and insulators, but closer to conductors.

7. Insulators are rated in terms of their breakdown voltages, in volts per meter, as given in Table 4-2.

8. The resistance of most metals increases with a rise in temperature as a result of a larger number of electron collisions.

9. The temperature-related change in a material's resistance depends on that material's temperature coefficient, which has units of per °C, as listed in Table 4-3.

10. Carbon, semiconductors, and insulators have a negative temperature coefficient (NTC) because any increase in temperature results in an increase in the availability of current carriers (free electrons).

11. If the resistance and temperature coefficient of resistance are given at one temperature, the resistance at a different temperature may be calculated, using $R_2 = R_1 [1 + \alpha_1(T_2 - T_1)]$

12. Linear resistors, those with a constant V/I ratio, are materials whose resistance changes very little over the range of temperatures involved in a particular application.

13. The nonlinear nature of the resistance of a tungsten lamp filament over a wide temperature range accounts for the large inrush current when the lamp is first switched on.

14. A thermistor is a nonlinear resistor with a very large negative temperature coefficient. When connected in series with a lamp filament, the characteristics of the thermistor allow it to suppress the inrush current.

15. The varistor is a nonlinear resistor whose resistance decreases as voltage increases. It can ''clamp'' voltage at a relatively low value to prevent high-voltage transients that would damage transistors or other semiconductor devices.

SELF-EXAMINATION

Answer true or false or a, b, c, d
(Answers at back of book)

4-1. If the diameter of a given conductor is doubled with the length and temperature remaining the same, the new resistance will be
   a. doubled
   b. halved
   c. quadrupled
   d. reduced to one-fourth

4-2. If the radius of a given conductor is tripled and the length is also tripled, the new resistance at the same temperature will be
   a. the same
   b. reduced to one-third
   c. tripled
   d. nine times as large

4-3. The resistivity of a given material is the resistance of a specific volume of the material at a given temperature. _____

4-4. Aluminum is a better conductor than copper because it has a higher resistivity than copper. _____

4-5. 100 ft of No. 18 gauge copper wire has a resistance of approximately
  a. 6.5 Ω
  b. 0.65 Ω
  c. 1.6 Ω
  d. 4 Ω
4-6. A 0.020-in.-diameter wire has a circular mil area of
  a. 400
  b. 40
  c. 200
  d. 0.004
4-7. A 2-mm-thick layer of paper will break down when the following voltage is applied:
  a. 20 kV
  b. 10 kV
  c. 40 kV
  d. anything above 2 kV
4-8. A material that has an increase in electron collisions at higher temperatures has a positive temperature coefficient. _____
4-9. A negative temperature coefficient material is one in which an increase in heat reduces the number of current carriers. _____
4-10. A reel of copper wire has a resistance of 1 kΩ at 20°C. If its temperature coefficient is 0.004 per °C, the resistance at 120°C will be
  a. 1004 Ω
  b. 1.4 kΩ
  c. 1040 Ω
  d. 1000.4 Ω
4-11. A graph of I versus V for a material has a constant slope. This indicates a linear resistance. _____
4-12. A thermistor is an example of a nonlinear resistor, since its resistance decreases significantly with an increase in temperature. _____

REVIEW QUESTIONS

1. What are all the factors that affect the resistance of a conductor?
2. What would be the units for resistivity if the length was measured in feet and the area in CM (circular mils)?
3. Justify, in your own words, why an increase in length increases resistance, but an increase in area decreases resistance.
4. A No. 14 gauge copper wire is recommended to carry a maximum of 15 A in house wiring applications. Why? Does this mean that it is impossible for more than 15 A to pass through a No. 14 wire? Under what conditions, if any, could you allow a No. 14 wire to carry more than 15 A?
5. Describe what happens when an insulator breaks down.
6. Explain why the resistance of common metals increases with temperature, but the resistance of semiconductors and insulators decreases.
7. Given an accurate resistance measuring device and the means to vary temperature, describe how you would determine the temperature coefficient of resistance for a copper coil. Why would α at 30°C be different from the value obtained at 20°C? Why is α usually specified at 20°C?

8. Sketch, on the same set of axes, graphs of R versus T for carbon, constantan, and copper. Which one of carbon and copper is more constant in resistance as the temperature changes? At a given temperature which one has the largest resistance for a given length and cross-sectional area?

9. What causes the current inrush problems in a tungsten filament lamp? Under what conditions could tungsten be considered a linear resistance?

10. Sketch a graph of I versus V for a 100-W incandescent lamp. How does it differ from a graph for a linear resistor? What effect will a series-connected thermistor have upon the V–I graph for the tungsten filament?

PROBLEMS

(Answers to odd-numbered problems at back of book)

4-1. The line cord to a table lamp uses a 2-m (4-m total) length of copper wire that has a diameter of 1.27 mm. Calculate:
  a. The resistance of the lamp cord at 20°C.
  b. The total voltage drop along the wire when a 120-V, 300-W lamp is in use.
  c. The closest standard wire gauge size of the copper wire.

4-2. Repeat Problem 4-1 assuming that aluminum wire is used.

4-3. A relay coil is to be wound using copper magnet wire that has a radius of 0.4 mm. What length of wire is necessary to produce a 100-Ω coil?

4-4. Repeat Problem 4-3 assuming that aluminum wire is used.

4-5. A two-wire copper feeder line is to supply a 1500-W load 500 ft away from a 120-V source. What must be the minimum size of wire if a 6% maximum voltage drop can be tolerated at the load? (Specify the minimum cross-sectional area and nearest wire gauge number.)

4-6. A stove heating element is 1.5 m long and has a cold resistance of 26 Ω. If the element contains Nichrome II wire, determine the diameter of the wire.

4-7. What length of copper wire, 1 mm in diameter, will have the same resistance as 10 cm of gold wire that has a diameter of 1.2 mm?

4-8. A 10-m length of wire having a diameter of 1.8 mm has a resistance of 1.93 Ω. What material is used for this wire?

4-9. a. What air spacing is necessary to provide a minimum breakdown voltage of 2500 V?
  b. What spacing would be necessary if Teflon is used instead of air?

4-10. The typical air gap found in an automotive spark plug is 0.030 in. What minimum voltage is necessary to cause a spark over this distance?

4-11. The copper winding of a motor has a cold resistance of 0.2 Ω at 20°C. After the motor has been in operation, the winding temperature rises 50°C. What is the new resistance?

4-12. A No. 4 gauge copper wire is used in a 400-km length overhead transmission line.
  a. What is the resistance of the line at 20°C?
  b. What is the voltage drop along this line when a current of 100 A is carried by the line?
  c. Express this voltage as a percentage of the applied voltage of 230 kV.
  d. Repeat parts (a), (b), and (c) for a temperature of 120°F.
  e. Repeat parts (a), (b), and (c) for a temperature of −30°F.

4-13. If the resistance of a nickel wire is 50 Ω at 300°C, what is its resistance at room temperature, 20°C?

4-14. If the tungsten filament of a 120-V, 250-W lamp is 2800°C, calculate its resistance at room temperature, 20°C.

4-15. A Nichrome II heating element is designed to produce 2100 W at 240 V. If the cold resistance at 20°C is 26 Ω, calculate the temperature of the hot element.

4-16. The inrush current to a 120-V lamp is 15 A. The current drops to 1.25 A after the tungsten filament has reached white heat. Calculate the operating temperature of the filament.

4-17. Early incandescent lamps used carbon filaments and operated at a lower temperature with a more reddish light.

a.   If a 120-V, 60-W carbon lamp operated at 1500°C, what was the resistance of the lamp at room temperature, 20°C?

b.   What advantage would such a lamp have, compared to a tungsten filament?

4-18. A coil of wire of unknown length and area has a resistance at 20°C of 25 Ω. The resistance increases to 37 Ω when the coil is lowered into boiling water. What is the material of the wire coil?

4-19. At 20°C a 1-kΩ metal film resistor has a temperature coefficient of 60 PPM/°C. What is its resistance at 80°C?

4-20. a.   What is the cold resistance of a thermistor required to limit the inrush current of a 120-V, 150-W lamp to a value equal to that of the lamp's final steady current? The cold resistance of the lamp is 8 Ω, and the hot resistance of the thermistor is 1 Ω.

b.   What is the power rating of the thermistor?

CHAPTER 5

SERIES CIRCUITS

When several components are connected in a circuit that provides only one path for current, the components are said to be in *series*. If any one element fails, it interrupts current through all the elements of the circuit. In this sense, the elements depend on each other for continued flow of current through the circuit.

The relationship between the applied electromotive force (emf) and the *voltage drops* that occur in such a circuit is governed by *Kirchhoff's voltage law.* This law also provides a way to obtain the total circuit resistance and/or current.

Open circuits (which have infinite resistance) and *short circuits* (which have zero or very low resistance) can be located in a series circuit by making voltage measurements and applying Kirchhoff's voltage law.

A process called *voltage division* can be used to provide different voltages in a series circuit. A *potentiometer* provides *continuous* voltage division in a series circuit by means of a sliding contact. The potentiometer is a three-contact device used to vary voltage.

Either the positive or the negative side of a dc supply may be connected to a metal chassis to provide a common return path in a circuit. This common path is referred to as *ground*. Voltages throughout the circuit may have a positive or negative potential *with respect to* this ground.

5-1 VOLTAGE DROPS IN A SERIES CIRCUIT

In Section 4-6, you saw a circuit (Figs. 4-5a and 4-5b) in which both a thermistor and a lamp had the *same* current flowing through them. Such an arrangement is called a *series* circuit.

The distinguishing feature of a series circuit is that the current is the *same* throughout all the elements in that circuit.

Figure 5-1a shows three resistors in series, connected to a voltage supply. As the current flows through R_1, a "voltage drop" (called V_{R_1}) occurs across R_1, as shown in Fig. 5-1b. This voltage drop causes point C to be at a lower potential or voltage than point B. This is shown by the negative sign ($-$) at C with respect to the positive sign ($+$) at B. When current moves from a point that is positive to a point that is, by comparison, negative, there is a *reduction* in the potential (or voltage) to do work. A *voltage drop* has taken place. Because point C is closer to the neg-

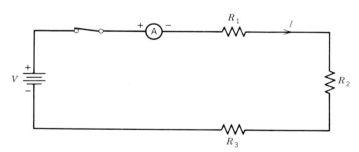

(a) Series circuit showing a single path for current

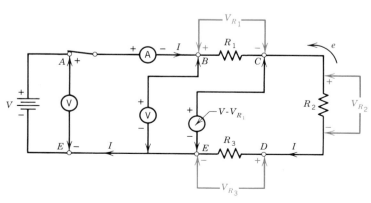

(b) Series circuit showing voltage drops across resistors

FIGURE 5-1
Series circuits.

ative side of the applied voltage source, it is at a lower voltage than point B. This is shown by placing a negative sign at point C.

To determine the polarity signs of voltage drops with conventional current, use the following rule:

Current flows *toward the positive side* of a voltage drop and *away from the negative side* of a voltage drop.

Using this rule, a *positive* sign ($+$) is placed at the end of a resistor where the current enters, and a *negative* sign ($-$) at the end of the resistor where the current leaves.

To further understand the meaning of the term *voltage drop*, consider Fig. 5-1*b*. If a voltmeter is connected from B to E, the reading would indicate the full voltage supply V (ignoring any negligible voltage drop across the ammeter, which has extremely low resistance).

If the *positive lead* of the meter is moved from B to C, the voltmeter would indicate a *lower* voltage, equal to $V - V_{R_1}$. This measurement shows that a drop in voltage occurred across R_1, *reducing* the voltage (emf) available to the rest of the circuit. Likewise, the voltage from D to E is equal to $V - V_{R_1} - V_{R_2}$ (because of the additional voltage drop across R_2), and this voltage is equal to the voltage drop across R_3.

Voltage drops exist across the resistors only when *current flows* in the circuit. This is the reason why a voltage drop across a resistor is sometimes called an *IR* (voltage) drop. (See Eq. 3-1*b*.) If the switch in Fig. 5-1*b* is open, a voltmeter connected from A to E would still indicate that the source voltage exists. The resistor voltage drops would be zero, however, since there is no current flowing. **A current does not have to flow for an emf to exist.** (Consider a ''live'' 120-V wall outlet with no load connected.)

The term *IR* voltage drop also helps to distinguish between a voltage drop across a resistor and the voltage drop of a battery that is being recharged. When a battery is recharged, current flows *into* the positive terminal rather than *out* of that terminal. Therefore, the battery's voltage is also a voltage drop, but *not* an *IR* voltage drop. Figure 5-2 illustrates a situation in which V_{R_1} *is an IR* voltage drop and V_2 is a voltage drop (a ''bucking'' emf) that must be overcome by supply voltage V_1 to reverse the current through V_2.

5-2 KIRCHHOFF'S VOLTAGE LAW

As you saw in Section 5-1, the reading of a voltmeter decreases as you progressively subtract the voltage drops across each resistor in the circuit. That is, in Fig. 5-1*b*:

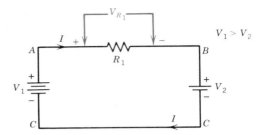

FIGURE 5-2
Battery V_2 is being recharged by V_1 so that V_2 is a voltage drop and V_{R_1} is an *IR* voltage drop.

voltage from B to $E = V$
voltage from C to $E = V - V_{R_1}$
voltage from D to $E = V - V_{R_1} - V_{R_2}$
voltage from E to $E = V - V_{R_1} - V_{R_2} - V_{R_3}$

In the last step, the positive and negative leads of the voltmeter are both connected to the same point, E. Since the voltmeter is no longer connected across any source of emf or across any resistance that could cause a voltage drop, it must now read *zero*. Thus

$$V - V_{R_1} - V_{R_2} - V_{R_3} = 0$$

This is one form of Kirchhoff's voltage law:

Around any *complete* circuit, the *algebraic* sum of the voltages equals zero.

''Complete circuit'' means that you must end up tracing the current at the point where you started. ''Algebraic'' means that some voltages are positive and others are negative. Thus V (the voltage source) is a *positive* voltage, while V_{R_1}, V_{R_2}, and V_{R_3} (the voltage drops) are *negative*.

The term *algebraic* can be avoided by rearranging the equation:

$$V = V_{R_1} + V_{R_2} + V_{R_3} \quad \text{volts} \qquad (5\text{-}1)$$

and restating Kirchhoff's law as follows:

Around any complete circuit, the sum of the voltage *rises* is equal to the sum of the voltage *drops*.

If you use the rule in Section 5-1 to determine the polarity signs for the *IR* voltage drops, you can use the following to determine whether a voltage is a voltage *rise* or a voltage *drop*.

1. A voltage *rise* occurs if, as you move around a circuit in the direction of the *conventional* current, you proceed from a negative sign $(-)$ at one end of a component to a positive sign $(+)$ at the other end.

2. A voltage *drop* occurs if, as you move around a circuit in the direction of conventional current, you proceed from a positive sign $(+)$ at one end of a component to a negative sign $(-)$ at the other end.

Note that the polarity signs referred to may be those across a resistor *or* a voltage source. For example, if you apply Kirchhoff's voltage law to the circuit in Fig. 5-2, a voltage rise occurs between C and A, and is equal to V_1. As you move from A to B in the direction of the current, you proceed from a positive sign to a negative sign across R_1. Thus V_{R_1} is a voltage drop. As you continue from B to C, you again move from a positive sign to a negative sign across V_2. Thus V_2 is also a voltage drop. You have now returned to the starting point at C, and thus have traveled a *complete* loop. Therefore, by Kirchhoff's voltage law:

$$\text{sum of voltage rises} = \text{sum of voltage drops}$$
$$V = V_{R_1} + V_2$$

Now, apply Kirchhoff's voltage law to the circuit shown in Fig. 5-3a.

After you determine the polarity signs across the resistors, it is clear that

$$V = V_{R_1} + V_{R_2} + V_{R_3} \qquad (5\text{-}1)$$

But if you apply Ohm's law to each section of the circuit:

$$V_{R_1} = IR_1, \qquad V_{R_2} = IR_2, \qquad V_{R_3} = IR_3 \quad (3\text{-}1b)$$

Thus

$$\begin{aligned} V &= V_{R_1} + V_{R_2} + V_{R_3} \\ &= IR_1 + IR_2 + IR_3 \\ &= I(R_1 + R_2 + R_3) \\ &= IR_T \end{aligned}$$

where

$$R_T = R_1 + R_2 + R_3 \ \text{ohms} \ (\Omega) \qquad (5\text{-}2)$$

is called the *total* circuit resistance. Thus, as far as V is concerned, the *same* current (I) would flow if only one resistor, R_T, is connected across V.

Equation 5-2 proves an important relationship:

In a series circuit, the *total* resistance of the circuit is the *sum* of the individual resistances.

EXAMPLE 5-1

A source of 18 V is connected across three series resistors having values of $R_1 = 3.3 \ \text{k}\Omega$, $R_2 = 1 \ \text{k}\Omega$, and $R_3 = 4.7 \ \text{k}\Omega$, as in Fig. 5-4.
Calculate:

a. The total circuit resistance.
b. The current in the circuit.
c. The voltage drops across the resistors to verify Kirchhoff's voltage law.
d. The reading of a voltmeter connected between points C and D in Fig. 5-4. Show this by two different methods.

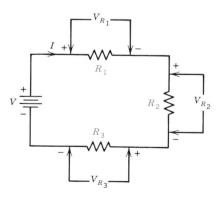

(a) Series circuit

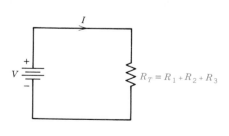

(b) Equivalent circuit with same current

FIGURE 5-3
The use of Kirchhoff's voltage law to obtain an equivalent circuit.

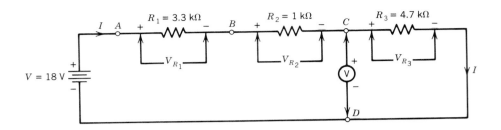

FIGURE 5-4
Circuit for Example 5-1.

Solution

a. Total circuit resistance,
$$R_T = R_1 + R_2 + R_3 \qquad (5\text{-}2)$$
$$= 3.3 \text{ k}\Omega + 1 \text{ k}\Omega + 4.7 \text{ k}\Omega$$
$$= \mathbf{9\ k\Omega}$$

b. Circuit current,
$$I = \frac{V}{R_T} \qquad (3\text{-}1a)$$
$$= \frac{18 \text{ V}}{9 \text{ k}\Omega} = \frac{18 \text{ V}}{9 \times 10^3 \ \Omega}$$
$$= 2 \times 10^{-3} \text{ A} = \mathbf{2\ mA}$$

c. Voltage drops,
$$V_{R_1} = IR_1 \qquad (3\text{-}1b)$$
$$= 2 \text{ mA} \times 3.3 \text{ k}\Omega$$
$$= 2 \times 10^{-3} \text{ A} \times 3.3 \times 10^3 \ \Omega = \mathbf{6.6\ V}$$
$$V_{R_2} = IR_2$$
$$= 2 \text{ mA} \times 1 \text{ k}\Omega$$
$$= 2 \times 10^{-3} \text{ A} \times 1 \times 10^3 \ \Omega = \mathbf{2\ V}$$
$$V_{R_3} = IR_3$$
$$= 2 \text{ mA} \times 4.7 \text{ k}\Omega$$
$$= 2 \times 10^{-3} \text{ A} \times 4.7 \times 10^3 \ \Omega = \mathbf{9.4\ V}$$

But
$$V_{R_1} + V_{R_2} + V_{R_3} = V \qquad (5\text{-}1)$$
$$6.6 \text{ V} + 2 \text{ V} + 9.4 \text{ V} = 18 \text{ V}$$
Note that the above values verify Kirchhoff's voltage law.

d. The voltmeter indicates:
$$V - V_{R_1} - V_{R_2} = 18 \text{ V} - 6.6 \text{ V} - 2 \text{ V}$$
$$= \mathbf{9.4\ V}$$
or directly, the voltmeter indicates $V_{R_3} = \mathbf{9.4\ V}$

In Example 5-1, note how the product of current in mA and resistance in kΩ results in a voltage drop in volts. This occurs because the equal positive and negative exponents cancel. In general, (1 mA) × (1 kΩ) = 1 V.

In the circuit shown in Fig. 5-4, the wires connecting the three resistors are assumed to have no (zero) resistance, and therefore, no voltage drops across them. For example, observe that the wire joining R_2 and R_3 has a negative sign on one end and a positive sign on the other. This does not mean that there is any voltage difference between the two ends. The negative sign indicates that this wire is at a lower *potential* than the wire on the left side of R_2, which has a positive sign. In the same way, the positive sign indicates that this wire (in fact, any point along this wire) is at a higher potential than the right side of R_3, which has a negative sign. **Polarity markings are only relative to one another when they occur on opposite ends of the same component.** Whether something in a circuit is positive or negative depends upon the point of reference. In Fig. 5-4, for example, point B is positive with respect to point C, but negative with respect to point A.

5-3 POWER IN A SERIES CIRCUIT

The *total* power dissipated in a series circuit is the *sum of the individual powers dissipated by each individual resistor in the circuit.*

Total power $P_T = P_1 + P_2 + P_3$ watts (5-3)

where P_1 (etc.) may be calculated by using
$$P_1 = I^2 R_1 \qquad (3\text{-}7)$$
$$\text{or } P_1 = V_{R_1} I \qquad (3\text{-}5)$$
$$\text{or } P_1 = \frac{V_{R_1}^2}{R_1} \qquad (3\text{-}8)$$

$$\text{Thus } P_T = I^2 R_T = VI = \frac{V^2}{R_T} \text{ watts} \qquad (5\text{-}4)$$

where: P_T is the total power in watts (W)
        I is the current in amperes (A)
        R_T is the total resistance in ohms (Ω)
        V is the applied voltage in volts (V)

EXAMPLE 5-2

For the circuit and data in Example 5-1 calculate:
a. The power dissipated individually in each resistor.
b. The total power delivered by the source, using three different methods.

Solution

a. $\quad P_1 = V_{R_1}I$ (3-5)

$\quad\quad = 6.6\ \text{V} \times 2\ \text{mA}$

$\quad\quad = 6.6\ \text{V} \times 2 \times 10^{-3}\ \text{A}$

$\quad\quad = 13.2 \times 10^{-3}\ \text{W} = \textbf{13.2 mW}$

$\text{or } P_1 = I^2R_1$ (3-7)

$\quad\quad = (2 \times 10^{-3}\ \text{A})^2 \times 3.3 \times 10^3\ \Omega$

$\quad\quad = 13.2 \times 10^{-3}\ \text{W} = \textbf{13.2 mW}$

$\text{or } P_1 = \dfrac{V_{R_1}^2}{R_1}$ (3-8)

$\quad\quad = \dfrac{(6.6\ \text{V})^2}{3.3 \times 10^3\ \Omega} = \textbf{13.2 mW}$

$\quad P_2 = V_{R_2}I$ (3-5)

$\quad\quad = 2\ \text{V} \times 2\ \text{mA} = \textbf{4 mW}$

$\quad P_3 = V_{R_3}I$

$\quad\quad = 9.4\ \text{V} \times 2\ \text{mA} = \textbf{18.8 mW}$

b. $\quad P_T = P_1 + P_2 + P_3$ (5-3)

$\quad\quad = 13.2\ \text{mW} + 4\ \text{mW} + 18.8\ \text{mW}$

$\quad\quad = \textbf{36 mW}$

$\text{or } P_T = VI$ (5-4)

$\quad\quad = 18\ \text{V} \times 2\ \text{mA} = \textbf{36 mW}$

$\text{or } P_T = I^2R_T$ (5-4)

$\quad\quad = (2 \times 10^{-3}\ \text{A})^2 \times 9 \times 10^3\ \Omega$

$\quad\quad = 36 \times 10^{-3}\ \text{W} = \textbf{36 mW}$

Note in Example 5-2 that the product of voltage in volts and current in mA yields power directly in milliwatts (mW).

EXAMPLE 5-3

You wish to use a 9-V transistor battery to operate a portable radio that is designed to use two AA (1.5-V) cells and requires a current of 75 mA. Determine:
a. How much resistance must be connected in series with the 9-V battery and the radio to allow proper operation.

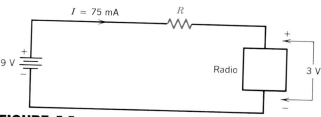

FIGURE 5-5
Circuit for Example 5-3.

b. The power rating of the resistor.
c. The power used by the radio.

Solution

Refer to Fig. 5-5.
a. By Kirchhoff's voltage law, the voltage drop caused by the series resistor must be:

$$9\ \text{V} - 3\ \text{V} = 6\ \text{V}$$

The same current of 75 mA passes through the resistor as through the radio. Therefore

$R = \dfrac{V}{I}$ (3-1)

$\quad = \dfrac{6\ \text{V}}{75 \times 10^{-3}\ \text{A}} = \textbf{80 }\boldsymbol{\Omega}$

b. Power dissipated in R is given by

$P = VI$ (3-5)

$\quad = 6\ \text{V} \times 75\ \text{mA}$

$\quad = \textbf{450 mW}$

A standard 82-Ω ± 5%, 1-W rating resistor would provide a safety factor of 2 to avoid overheating. That is, the power dissipation in the resistor could double without the resistor's rating being exceeded.
c. Power used by the radio is given by:

$P = VI$ (3-5)

$\quad = 3\ \text{V} \times 75\ \text{mA}$

$\quad = \textbf{225 mW}$

In the above example, twice as much power is dissipated in the resistor as is delivered to the radio. This waste of power is a disadvantage of operating a lower-voltage device from a higher-voltage source, especially when a battery is used to supply power.

5-4 OPEN CIRCUITS

An *open* circuit results whenever a circuit is broken or is incomplete. In a series circuit, this means that there is no path for current, and so, no current flows. Another way to put it is that there is no *continuity* in the circuit.

Assume that the circuit shown in Example 5-1 was connected in the laboratory, but no current flow was indicated on the ammeter. (See Fig. 5-6.) How could you determine the location of the break in the circuit?

The break can be located by using a voltmeter set to a range that can accommodate the supply voltage. The meter is connected, in turn, across each of the connecting wires in the circuit. If one of the wires is open, as in Fig. 5-6a, the full supply voltage will be indicated on the voltmeter. Remember that in the absence of current, there is no voltage drop across any of the resistors. Therefore, the voltmeter must be reading the full supply voltage across the open portion of the circuit. That is,

$$\text{Voltmeter reading} = 18 \text{ V} - V_{R_1} - V_{R_2} - V_{R_3}$$
$$= 18 \text{ V} - 0 \text{ V} - 0 \text{ V} - 0 \text{ V}$$
$$= 18 \text{ V}$$

If the open circuit condition is due to a defective resistor (resistors usually open when they burn out) as shown in Fig. 5-6b, the voltmeter would indicate 18 V when connected across the defective resistor (R_2).

The location of the open in the circuit can also be found with an ohmmeter. With the voltage removed, the ohmmeter will show *no* continuity (infinite resistance) when connected across the open wire or resistor.

EXAMPLE 5-4

Figure 5-7 shows a typical series control circuit for a gas-fired swimming pool heater with a standing (constantly burning) pilot light. The *thermopile* in the circuit is made up of a number of series-connected *thermocouples*. A thermocouple consists of two different metals (such as iron and constantan) and generates an emf of approximately 30 mV when placed in a flame. A typical thermopile generates an emf of 750 mV. If the ON-OFF switch is closed, the thermostat is closed ("calling" for heat), the high-temperature limit switch is closed, and the water flow switch is closed (indicating that water is flowing), then the main gas solenoid valve will be energized. Gas will flow to the burners and be ignited by the pilot flame to heat the water.

Of course, if the pilot light is not burning, the thermocouple will not generate an emf and no current will flow. As a result, the gas valve will not open.

If the thermostat malfunctions and water temperature becomes too high, the high-temperature limit switch will open, breaking the circuit. This will de-energize the solenoid, and gas flow will be cut off.

If the water flow is interrupted for any reason, the low-pressure water flow switch will open. Again, the circuit will be broken, de-energizing the solenoid and cutting off gas flow.

Determine the voltage readings shown on a voltmeter when connected between points *A-B*, *B-C*, *C-D*, *D-E*, *E-F*, and *F-A* for the following conditions:

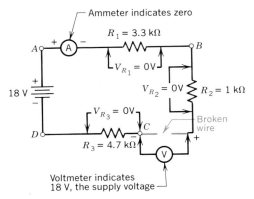

(a) Open circuit due to break between R_2 and R_3

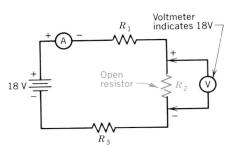

(b) Open circuit due to open resistor R_2

FIGURE 5-6

Open circuits.

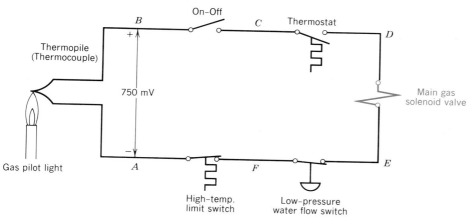

FIGURE 5-7

Typical low-voltage series control circuit for a pool heater.

a. Pilot not lit.
b. Pilot lit, but thermocouple defective (open, not generating an emf).
c. Thermopile generates 750 mV, ON-OFF switch closed, thermostat closed but insufficient water flowing.
d. Same conditions as in part (c), but with high-temperature limit switch stuck open.
e. Thermopile generates 750 mV, ON-OFF switch defective (open), thermostat, high-temperature limit switch and low-pressure switch all closed.
f. Same conditions as in part (e), but with ON-OFF switch repaired and closed.

Solution

a. If the pilot is not lit, no emf will be generated and all voltages will be zero.
b. If the thermocouple is defective, all voltages will be zero (no emf generated).
c. A-B = 750 mV (negative)
   B-C = 0 V
   C-D = 0 V
   D-E = 0 V
   E-F = 750 mV
   F-A = 0 V

NOTE V_{AB} is negative because B is *positive* with respect to A.

d. A-B = 750 mV (negative)
   B-C = 0 V
   C-D = 0 V
   D-E = 0 V

E-F = 0 V
F-A = 0 V

NOTE   E-A would indicate 750 mV.

e. A-B = 750 mV (negative)
   B-C = 750 mV
   C-D = D-E = E-F = F-A = 0 V
f. A-B = 750 mV (negative)
   B-C = C-D = 0 V (closed switch has zero resistance, so no voltage drop occurs even when current flows).
   D-E = 750 mV
   E-F = F-A = 0 V

5-5 SHORT CIRCUITS

A short circuit is a path of zero or very low resistance compared to the normal circuit resistance. All short circuits cause an increase in current. If one of the elements in a series circuit is short-circuited, the result is an *increase* in current that may or may not be damaging. (In a simple *parallel* circuit, the result is usually a blown fuse or tripped circuit breaker.)

The circuit shown in Fig. 5-8 is the same one you saw in Example 5-1, and should have a normal current of 2 mA. Assume that the ammeter in Fig. 5-8 reads higher than 2 mA, and that you suspect one of the components is "shorted out" (short-circuited). How could you tell which component is shorted by making *voltage* measurements?

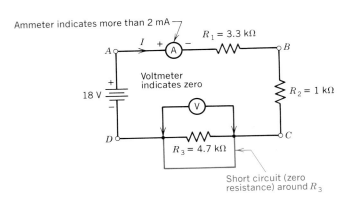

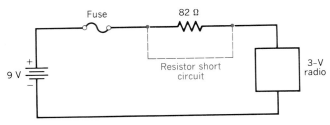

FIGURE 5-8
Short circuit across R_3 results in a current increase.

FIGURE 5-9
Fuse provides protection if resistor burns out.

If a voltmeter is connected across each of the resistors in turn, it will indicate 0 V for the defective resistor. Although current is flowing through the short circuit, no voltage drop occurs if there is zero resistance. That is,

$$I = \frac{V}{R_T} \tag{3-1a}$$

$$= \frac{18 \text{ V}}{4.3 \text{ k}\Omega} = 4.2 \text{ mA}$$

and

$$V_3 = IR$$
$$= 4.2 \text{ mA} \times 0 \ \Omega = 0 \text{ V}$$

Even if the defective resistor, R_3, amounted to a few ohms (say 10 Ω), this would still be effectively a short circuit. This is because R_T would be 4.31 kΩ and the voltmeter would try to indicate

$$V_3 = IR_3 \tag{3-1b}$$
$$= 4.2 \text{ mA} \times 10 \ \Omega = 42 \text{ mV (essentially zero)}$$

EXAMPLE 5-5

Assume that you must connect a fuse in series with the 82-Ω resistor and 3-V radio operating from a 9-V battery shown in Fig. 5-9. The fuse size must be chosen to provide protection for the radio in the event that the resistor burns out and short-circuits. Calculate:
a. The normal current flowing in the circuit in Fig. 5-9.
b. The current that will flow if the 82-Ω resistor shorts out.
c. The size of fuse needed to protect the radio from damage.

Solution

a. Under normal operation, the 82-Ω resistor should drop 6 V across itself. Therefore the normal circuit current is given by:

$$I = \frac{V}{R} \tag{3-1a}$$

$$= \frac{6 \text{ V}}{82 \ \Omega}$$

$$= 0.0732 \text{ A} \approx \textbf{73 mA}$$

b. The effective resistance of the radio under normal operation is given by:

$$R = \frac{V}{I} \tag{3-1}$$

$$= \frac{3 \text{ V}}{73.2 \times 10^{-3} \text{ A}} = \textbf{41 } \mathbf{\Omega}$$

Assuming this stays constant when the voltage across the radio increases to 9 V, the current that will flow is given by:

$$I = \frac{V}{R}$$

$$= \frac{9 \text{ V}}{41 \ \Omega}$$

$$= 0.2195 \text{ A} \approx \textbf{220 mA}$$

Note that this current is three times the normal current, since the applied voltage is three times higher.

c. Since the current will rise suddenly from 73 mA to 220 mA if the resistor shorts, select the nearest standard fuse above the normal operating current: **0.1 A.**

5-6 VOLTAGE DIVISION

If you refer back to Example 5-1 and examine the way in which voltage drops are related to resistor values, you will find that the *highest* resistance has the *largest* voltage drop across it, and the *lowest* resistance has the *smallest* voltage drop. Because the circuit current is common, the voltage drops are in direct proportion to the resistance values.

Consider the circuit in Fig. 5-10. The voltage across R_2 could be obtained as follows:

$$I = \frac{V}{R_T} = \frac{V}{R_1 + R_2 + R_3} \qquad (3\text{-}1a)$$

$$V_{R_2} = IR_2 = \frac{V}{R_1 + R_2 + R_3} \times R_2 \qquad (3\text{-}1b)$$

Therefore,

$$V_{R_2} = V \times \frac{R_2}{R_1 + R_2 + R_3} = V\frac{R_2}{R_T} \quad \text{volts} \quad (5\text{-}5)$$

Equation 5-5 is called the *voltage division rule*. It states that, in a series circuit, the voltage across the resistors "divides up" in a way that is proportional to the ratio of the resistances.

Similarly,

$$V_{R_1} = V \times \frac{R_1}{R_1 + R_2 + R_3} = V\frac{R_1}{R_T} \quad \text{volts} \quad (5\text{-}6)$$

and

$$V_{R_3} = V \times \frac{R_3}{R_1 + R_2 + R_3} = V\frac{R_3}{R_T} \quad \text{volts} \quad (5\text{-}7)$$

The voltage division rule can be applied to *any* number of series-connected resistors.

Since the denominators are equal in size for Equations 5-5 through 5-7, it is clear that *the largest voltage drop occurs across the largest resistance.* In the case of *equal*

value resistors, each resistor has an *equal* voltage across it.

EXAMPLE 5-6

Resistors of 22 kΩ, 100 kΩ, and 56 kΩ are connected in series across a 12-V supply. What is the voltage drop across the 56-kΩ resistor?

Solution

$$V_{56\ k\Omega} = V \times \frac{R_3}{R_1 + R_2 + R_3} = V\frac{R_3}{R_T} \qquad (5\text{-}7)$$

$$= 12\ V \times \frac{56\ k\Omega}{22\ k\Omega + 100\ k\Omega + 56\ k\Omega}$$

$$= 12\ V \times \frac{56\ k\Omega}{178\ k\Omega}$$

$$= 12\ V \times 0.315 = \textbf{3.78 V}$$

EXAMPLE 5-7

A 200-W, 120-V lamp and a 100-W, 120-V lamp are connected in series with a 240-V source (see Fig. 5-11). Assuming that the resistances of the lamps are the same as when operating at their normal voltages, calculate:

a. The voltage across each lamp.
b. The power dissipated by each lamp.
c. The visual effect.

Solution

a. The operating resistances of the lamps are given by

$$P = \frac{V^2}{R} \qquad (3\text{-}8)$$

$$R_{200} = \frac{V^2}{P}$$

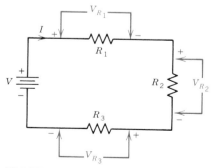

FIGURE 5-10
Voltage division across three series-connected resistors.

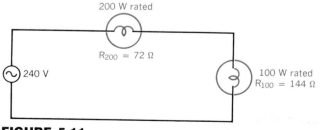

FIGURE 5-11
Circuit diagram for Example 5-7.

$$= \frac{(120\ V)^2}{200\ W} = 72\ \Omega$$

$$R_{100} = \frac{(120\ V)^2}{100\ W} = 144\ \Omega$$

By the voltage divider rule, and referring to Fig. 5-11, the voltage across the 200-W lamp is given by

$$V_{200} = V \times \frac{R_{200}}{R_T} \qquad (5\text{-}5)$$

$$= 240\ V \times \frac{72\ \Omega}{72\ \Omega + 144\ \Omega}$$

$$= 240\ V \times \frac{72}{216}$$

$$= 240\ V \times \frac{1}{3} = \mathbf{80\ V}$$

Similarly,

$$V_{100} = V \times \frac{R_{100}}{R_T} \qquad (5\text{-}5)$$

$$= 240\ V \times \frac{144\ \Omega}{144\ \Omega + 72\ \Omega}$$

$$= 240\ V \times \frac{2}{3} = \mathbf{160\ V}$$

b. Power dissipated in each lamp is given by

$$P = \frac{V^2}{R} \qquad (3\text{-}8)$$

$$P_{200} = \frac{(V_{200})^2}{R_{200}}$$

$$= \frac{(80\ V)^2}{72\ \Omega} = 88.9\ W \approx \mathbf{89\ W}$$

$$P_{100} = \frac{(160\ V)^2}{144\ \Omega} = 177.8\ W \approx \mathbf{178\ W}$$

c. The 200-W lamp will be approximately half as bright as normal; the 100-W would be much brighter (almost twice as bright) than normal and would probably burn out in a few minutes.

NOTE From Example 5-7, you can see that only lamps of equal wattage, each of 120 V, should be series-connected across a 240-V line. In that case, the lamps would have equal resistance and the voltage would divide evenly, giving each lamp its normal 120 V.

5-7 THE POTENTIOMETER

The potentiometer is a three-terminal device with a sliding contact that can be moved from one end to the other. It is shown in schematic form in Fig. 5-12a, with the upper

portion of the resistance given by R_1 and the rest of its resistance by R_2.

This may represent the condition in which the potentiometer shaft is turned to somewhere near the center of its travel. This condition may be thought of as two separate resistors, as shown in Fig. 5-12b. The voltmeter indicates the output of the potentiometer between the moving contact (B) and the lower terminal (C). The input voltage is connected between the terminals of the potentiometer (A, C).

Consider the voltage division rule:

$$V_{out} = V_{R_2} = V_{in} \times \frac{R_2}{R_1 + R_2} \qquad (5\text{-}5)$$

If the shaft of the potentiometer is turned so that B is in contact with A (as in Fig. 5-12c), $R_1 = 0$ and $R_2 =$ full resistance of the potentiometer. Then

$$V_{out} = V_{in} \times \frac{R_2}{R_1 + R_2} = V_{in} \times \frac{R_2}{0 + R_2} = V_{in}$$

and the *full supply* voltage will be indicated on the voltmeter. This is because the voltmeter is now effectively connected directly across the input supply (V_{in}).

Conversely, if B is moved to make contact with C (as in Fig. 5-12d), $R_2 = 0$. Then

$$V_{out} = V_{in} \times \frac{R_2}{R_1 + R_2} = V_{in} \times \frac{0}{R_1 + 0} = 0\ V$$

V_{out} is zero because the voltmeter is now connected across two points with zero resistance between them. Even though current is flowing through the wire connecting the bottom of R_1 to C, there is no voltage drop because the resistance is zero ($V = IR = I \times 0\ \Omega = 0\ V$). This means that a potentiometer allows a variable voltage to be developed between one of its end terminals and the sliding contact.

It should be noted that, as the output voltage is varied from minimum to maximum, there is no change in the current (I) drawn from the supply voltage (V_{in}). Assuming that the voltmeter has a very high resistance compared to the resistance of the potentiometer, **there is *no change* in resistance between A and C as the potentiometer contact (B) is moved.**

In practice, potentiometers are used in applications where a *load* is connected between B and C. If this load draws any appreciable current compared with I, then the current from V_{in} will vary as the potentiometer is varied (by moving the sliding contact). An analysis of this type of circuit will be presented in Chapter 7. Potentiometers

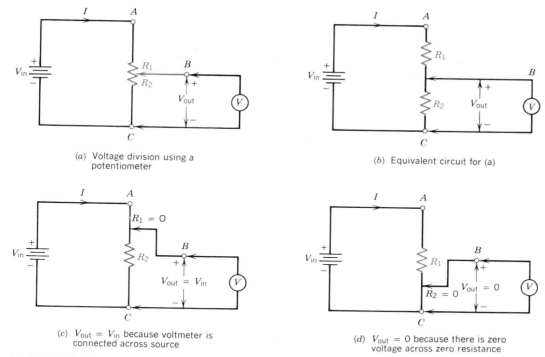

(a) Voltage division using a potentiometer

(b) Equivalent circuit for (a)

(c) $V_{out} = V_{in}$ because voltmeter is connected across source

(d) $V_{out} = 0$ because there is zero voltage across zero resistance

FIGURE 5-12

How a potentiometer provides a variable voltage output.

are generally used, however, with very light loads that draw a small current.

One example is the volume control in an audio amplifier, as shown in Fig. 5-13. The input signal, which may come from a magnetic tape head, is first made larger by a pre-amplifier. Then, it is applied to the potentiometer. By rotating the potentiometer shaft, the amount of audio signal voltage applied to the power amplifier may be increased or decreased (more or less volume). The input resistance to the power amplifier (the load on the potentiometer) is usually high, so very little current is drawn from the potentiometer.

The change in resistance produced by turning the potentiometer shaft may be related to shaft rotation or position in a *linear* or *nonlinear* manner.

If the resistance of the potentiometer (between its center terminal and one end) varies in a linear relationship with its position, the potentiometer is said to have a linear *taper*. Thus, a potentiometer with a linear taper and a resistance of 10 kΩ from one end to the other will have a resistance of 5 kΩ between the center terminal and one end when the shaft is at the center of its travel.

Potentiometers with various nonlinear tapers are available, as well. An example is the volume control shown in

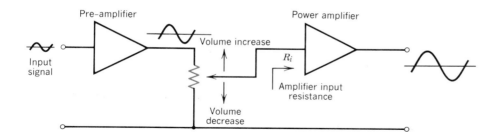

FIGURE 5-13

A potentiometer used as a volume control in an audio amplifier.

Fig. 5-13, which has an *audio* taper. This taper provides smaller changes in resistance at low volume settings than at the high end, permitting a finer degree of adjustment at low volume.

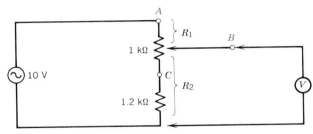

FIGURE 5-14

Circuit diagram for Example 5-9.

EXAMPLE 5-8

A 5-kΩ potentiometer with a linear taper has 6 V connected across its end terminals. Calculate:
a. The range of output voltage that can be developed.
b. The two possible output voltages when the potentiometer is in its three-quarter position.
c. The current through the potentiometer in this position.
d. The potentiometer's power rating.

Solution

a. Minimum = **0 V**
   Maximum = **6 V**
b. $V_{out} = V_{in} \times \frac{3}{4}$
   $= 6 \text{ V} \times \frac{3}{4} = $ **4.5 V**
   or $\quad V_{out} = V_{in} \times \frac{1}{4}$
   $= 6 \text{ V} \times \frac{1}{4} = $ **1.5 V**
c. $I = \dfrac{V}{R_T}$ (3-1a)

   $= \dfrac{6 \text{ V}}{5 \text{ k}\Omega} = $ **1.2 mA**
d. $P = I^2 R$ (3-7)
   $= (1.2 \text{ mA})^2 \times 5 \text{ k}\Omega$
   $= (1.2 \times 10^{-3} \text{ A})^2 \times 5 \times 10^3 \text{ }\Omega$
   $= $ **7.2 mW** (10 mW minimum)

EXAMPLE 5-9

Given the circuit in Fig. 5-14, calculate:
a. The maximum and minimum voltage that the voltmeter will indicate as the potentiometer is varied to its extremes.
b. Assuming a linear taper, the voltage indicated when the potentiometer is in its center position.

Solution

a. A maximum voltage will result when the potentiometer wiper is moved to bring B into contact with A. This is equal to **10 V.**

   For a minimum voltage indication, B is moved into contact with C. The output voltage is now the voltage drop across the 1.2-kΩ fixed resistor, given by

$$V_{out} = V_{in} \times \frac{R_2}{R_1 + R_2} \quad (5\text{-}5)$$

$$= 10 \text{ V} \times \frac{1.2 \text{ k}\Omega}{1 \text{ k}\Omega + 1.2 \text{ k}\Omega}$$

$$= 10 \text{ V} \times \frac{1.2}{2.2} = 5.45 \text{ V} \approx \mathbf{5.5 \text{ V}}$$

b. When the potentiometer is in its center position, the voltmeter is connected across a total resistance R_2 of 500 Ω + 1.2 kΩ = 1.7 kΩ.

$$V_{out} = V_{in} \times \frac{R_2}{R_1 + R_2} \quad (5\text{-}5)$$

$$= 10 \text{ V} \times \frac{1.7 \text{ k}\Omega}{0.5 \text{ k}\Omega + 1.7 \text{ k}\Omega}$$

$$= 10 \text{ V} \times \frac{1.7}{2.2} = 7.73 \text{ V} \approx \mathbf{7.7 \text{ V}}$$

The above circuit might be used as a volume control in a battery-operated radio. It would not be possible, however, to turn the volume down to zero. The low-level sound would be a reminder to turn off the radio when not in use, to prevent running down the batteries.

5-8 POSITIVE AND NEGATIVE GROUNDS

In an automotive electrical system, one side of the battery is connected to the metal chassis of the auto and called the *ground* side. This connection allows the metal chassis to be used as the *return path* for any circuit, avoiding the need for an extra wire. On most U.S.-made vehicles, a *negative* ground is used, but some European vehicles use a *positive* ground, instead. (See Fig. 5-15.) The positive ground system is believed to reduce corrosion problems.

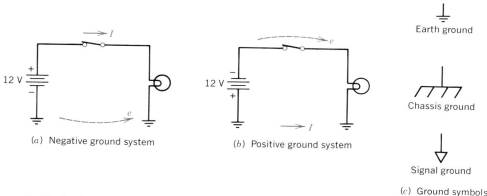

FIGURE 5-15
Possible grounding systems.

In a negative ground system (Fig. 5-15*a*), all wiring is at a positive potential with respect to the chassis. In a positive ground system (Fig. 5-15*b*), all potentials are negative. Current flows in opposite directions in the two systems. But in *both* systems, the metal chassis serves as the common reference point for stating the value of voltage at any point in the system.

In a strict sense, the use of the word *ground* for the metal chassis is not correct; *chassis ground* is better (see symbol in Fig. 5-15*c*). This is because the word *ground*, by itself, usually implies a connection to *earth ground*, such as a cold water pipe. One side of a domestic 120-V ac outlet—the neutral side—is connected to earth ground. (In a car, the chassis is insulated from the ground by its rubber tires.)

Most electronic equipment is built on a metal chassis, which may serve as a convenient tie point for components and a common return to one side of a power supply. Sometimes, this chassis may also be connected to the 60-Hz supply (earth) ground, and may require the use of polarized plugs to ensure the proper connection of neutral and grounded chassis.

Another type of grounding system is the *signal ground*, shown by the symbol in Fig. 5-15*c*. This may simply be a conductive strip running around a printed circuit board that serves as a common connection for the returning signal. It is not necessarily connected to the chassis *or* to the earth ground. In some equipment a movable shorting strap is provided to allow operation of the signal ground at a different voltage from the chassis or earth ground.

Note that, if one side of a battery or power supply is to be grounded, there is no requirement that the *negative* side

be used. **There is nothing about the positive side of a battery that prevents it from being connected to ground.**

5-9 POSITIVE AND NEGATIVE POTENTIALS

In some cases, a circuit may require both positive *and* negative potentials. This makes necessary a different placing for the common or ground connection. (See Fig. 5-16.)

In Fig. 5-16*a*:

$$V_A = V_{R_1} = 12 \text{ V} \times \frac{R_1}{R_1 + R_2} \tag{5-6}$$

$$= 12 \text{ V} \times \frac{100 \ \Omega}{100 \ \Omega + 220 \ \Omega} = \textbf{3.75 V}$$

$$V_B = -V_{R_2} = -12 \text{ V} \times \frac{R_2}{R_1 + R_2} \tag{5-5}$$

$$= -12 \text{ V} \times \frac{220 \ \Omega}{100 \ \Omega + 220 \ \Omega} = \textbf{-8.25 V}$$

Thus, *A* is positive and 3.75 V higher in potential than *C*, while *B* is −8.25 V with respect to the ground point *(C)*. Since it is understood that all voltages are stated with respect to the common or ground at *C*, you could simply say *A* = +3.75 V, *B* = −8.25 V.

A very common method used to obtain dual operating voltages for integrated circuits, such as operational amplifiers, is two series-connected batteries or power supplies with a common ground. (See Fig. 5-16*b*.) The advantage

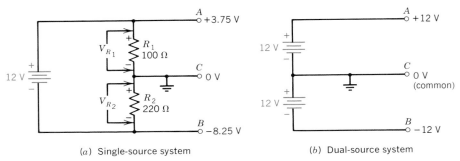

(a) Single-source system   (b) Dual-source system

FIGURE 5-16
Two methods of obtaining positive and negative potentials.

of this system is a more constant output voltage when a load is connected, compared to using a voltage-divider resistance network like the one shown in Fig. 5-16a. It should be noted that both these systems require the use of batteries or a "floating" power supply. For example, if the two 12-V supplies in Fig. 5-16b had their negative sides grounded, it would not be possible to connect them in series.

A commercial power supply that provides both positive and negative potentials with respect to the common (COM) terminal is shown in Fig. 5-17. In this device, the common is completely isolated from the 120-V ac earth ground (the power supply converts the alternating current to direct current), so no ground symbol is shown. A *tracking control*

FIGURE 5-17
Commercial power supply that provides both positive and negative potentials with respect to the common (COM) terminal. (Courtesy of Hewlett-Packard Company.)

permits simultaneous adjustment of both the positive and negative outputs.

5-10 VOLTAGE SOURCES CONNECTED IN SERIES

The series interconnection of cells (batteries) is covered in detail in Chapter 9. However, two general statements can be made concerning the total or net voltage that results from two or more voltage sources being connected in series.

When the cells are connected as shown in Fig. 5-18a, they are said to be *series-aiding*, with total voltage given by

$$V_T = V_1 + V_2 \quad \text{volts} \tag{5-8}$$

The circuit current, which is the same through both batteries and the resistor, is

$$I = \frac{V_T}{R} = \frac{V_1 + V_2}{R}$$

When the cells are connected as in Fig. 5-18b or Fig. 5-18c, the voltage sources are referred to as being *series-opposing*. The net voltage of the combination is

$$V_{\text{NET}} = V_1 - V_2 \quad \text{volts} \tag{5-9}$$

where V_1 is assumed to have the larger emf. Note that the resultant polarity of V_{NET} is the same as the larger emf. Similarly, any number of series-connected emfs may be reduced to a single equivalent voltage. It should be noted that the internal resistance associated with each cell (as discussed in Section 9-1) has an additive effect, whether the cells are aiding or opposing.

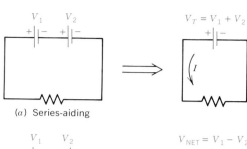

(a) Series-aiding

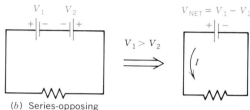

(b) Series-opposing

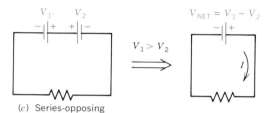

(c) Series-opposing

FIGURE 5-18
Series-connected voltage sources.

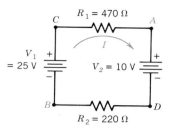

(a) Original circuit

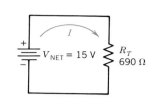

(b) Equivalent circuit

FIGURE 5-19
Circuits for Example 5-10.

EXAMPLE 5-10

For the circuit given in Fig. 5-19a, calculate:
a. The current.
b. The voltage at A with respect to B.

Solution

a. $V_{NET} = V_1 - V_2$ (5-9)
$\quad = 25\ V - 10\ V = 15\ V$
$R_T = R_1 + R_2$ (5-2)
$\quad = 470\ \Omega + 220\ \Omega = 690\ \Omega$

$I = \dfrac{V_{NET}}{R_T}$ (3-1a)

$\quad = \dfrac{15\ V}{690\ \Omega} = 0.0217\ A = \textbf{21.7 mA}$

b. $V_{AB} = V_2 + V_{R_2}$
$\quad = 10\ V + 21.7 \times 10^{-3}\ A \times 220\ \Omega$
$\quad = 10\ V + 4.77\ V = \textbf{14.8 V}$

or $\ V_{AB} = V_1 - V_{R_1}$
$\quad = 25\ V - 21.7 \times 10^{-3}\ A \times 470\ \Omega$
$\quad = 25\ V - 10.2\ V = \textbf{14.8 V}$

SUMMARY

1. In a series circuit, there is only one path for current, so the current is the *same* throughout all the elements of the circuit.
2. The failure of any one element in a series circuit will interrupt the current for all elements in the circuit.
3. An *IR* voltage drop occurs as current passes through a resistor.
4. Kirchhoff's voltage law states that the applied emf equals the sum of the voltage drops around any complete circuit.
5. The total resistance of a series circuit is the sum of the individual resistances.
6. The total power dissipated in a series circuit is the sum of the individual powers dissipated in each resistor.

7. In a series circuit, an open circuit creates infinite resistance due to a break in the circuit. The full supply voltage appears across such an open circuit.

8. A short circuit exhibits very low or zero resistance, resulting in increased current and zero voltage drop across the short circuit.

9. In a series circuit involving more than one resistance, the voltage divides. The highest voltage appears across the largest resistance, which also dissipates the highest power.

10. The potentiometer is a three-terminal device with a sliding contact. It is used to vary voltage from zero to the full supply voltage.

11. A ground is a common return path for a number of components connected to the same supply. Either the positive or the negative side may be grounded, and may or may not be connected to earth ground.

12. A power supply may consist of two emfs connected in series, with their common interconnection grounded and used as a reference to provide both positive and negative potentials to a circuit.

13. Voltage sources may be connected as series-aiding or series-opposing, with the total voltage equal to the sum or the difference, respectively.

SELF-EXAMINATION

Answer true or false or a, b, c, or d
(Answers at back of book)

5-1. No matter where an ammeter is connected in a series circuit the reading is always the same. _____

5-2. Although the current is not used up in a series circuit, the voltage is. _____

5-3. The sum of all the voltage drops in a series circuit may slightly exceed the applied voltage. _____

5-4. As electrons pass through a resistor, we still say that there is a voltage drop across the resistor even though electrons enter the negative side and leave the positive side. _____

5-5. An emf can exist only if there is a complete circuit for current to flow. _____

5-6. If there are two emfs in a series circuit, the net applied voltage may be either the sum or the difference of the two emfs. _____

5-7. Three resistors of 1 kΩ, 2 kΩ, and 7 kΩ are connected in series with a 30-V supply. The total resistance and current are
    a. 10 kΩ, 3 A   c. 10 kΩ, 3 mA
    b. 10 kΩ, 300 mA   d. 5 kΩ, 6 mA

5-8. The voltage drops across the three resistors in Question 5-7 are
    a. 1 V, 2 V, 7 V   c. 3 V, 6 V, 21 V
    b. 2 V, 4 V, 14 V   d. 3 mV, 6 mV, 21 mV

5-9. The powers dissipated in each of the resistors in Question 5-7 are
    a. 9 mW, 36 mW, 441 mW   c. 9 W, 36 W, 441 W
    b. 9 mW, 18 mW, 63 mW   d. 9 μW, 18 μW, 63 μW

5-10. The total power supplied to the whole circuit in Question 5-7 is
    a. 90 mW   c. 9 W
    b. 90 μW   d. 900 mW

5-11. If an open circuit occurs between the 2-kΩ and 7-kΩ resistors in Question 5-7, a voltmeter connected across the 7-kΩ resistor will indicate:

    a. 30 V   c. 4 V

    b. 21 V   d. 0 V

5-12. If an open circuit occurs between the 1-kΩ and 2-kΩ resistors in Question 5-7, a voltmeter connected between these two resistors indicates:

    a. 30 V   c. 4 V

    b. 1 V   d. 0 V

5-13. If a short circuit occurs across the 7-kΩ resistor in Question 5-7, the current will be

    a. 3 mA   c. 3 A

    b. 10 mA   d. 10 A

5-14. A voltmeter connected across the shorted 7-kΩ resistor in Question 5-7 indicates:

    a. 30 V   c. 0 V

    b. 21 V   d. 9 V

5-15. Three resistors are connected in series across a 27-V supply. The second resistor has twice the resistance of the first; the third resistor has three times the resistance of the second. The voltage across the third resistor is

    a. 3 V   c. 9 V

    b. 6 V   d. 18 V

5-16. A 10-kΩ potentiometer is connected across an 18-V supply. When the shaft is rotated to the $\frac{2}{3}$ position the output voltage and current through the potentiometer are

    a. 6 V and 1.8 mA   c. 12 V and 1.2 mA

    b. 12 V and 1.8 mA   d. either (a) or (b)

5-17. Current flows in the opposite direction in a negative ground system compared with a positive ground system. _____

5-18. If two 15-V supplies are series-connected, it is possible to connect loads that operate at $+15$ V, -15 V, and 30 V. _____

5-19. If two 12-V batteries are series-connected, the total available voltage may be either 24 V or 0 V. _____

5-20. In order to connect two batteries series-opposing, the two negative terminals must be connected to one another. _____

REVIEW QUESTIONS

1. In an experimental verification of the Kirchhoff voltage law, the sum of the measured voltage drops was slightly less than the applied emf. How could you explain this? Why may the sum have been slightly larger? What equipment would you need for an exact verification?

2. Why do the polarity signs across a resistor seem more suggestive of a voltage *drop* when tracing conventional current than when tracing electron flow?

3. What is an *IR* drop? Why is this term useful?

4. Under what conditions would two emfs in a series circuit provide zero current?

5. Give an equation for the total power in a series circuit in terms of the individual resistances and the voltage drops across each.

6. A series circuit is suspected to have an open circuit. How could you verify this to be the case and how would you locate the break?

7. One of the components in a series circuit is shorted-out. How would you locate the fault using the following equipment?
   a. A voltmeter.
   b. An ohmmeter.

8. Assume a series circuit with a number of resistors connected to a power supply, one side of which is grounded. What difficulty would you have in locating an open resistor if you used a voltmeter that had its negative lead also grounded (as many VTVMs have)? Show by means of a diagram the readings you would obtain for three resistors, the middle one of which is open.

9. Assume a series circuit connected to a 9-V battery. No current is flowing and an open resistor is suspected. It is suggested that an ohmmeter be used to locate the open resistor. Is this a good idea? Will an infinite reading be indicated when connected across the open resistor? If an ohmmeter is to be used, what must be done first?

10. In a series circuit, justify why the largest resistance will have the highest voltage drop.

11. Use the voltage division rule (Eq. 5-6) to justify why a shorted resistor will have zero voltage across it.

12. When an ohmmeter is connected across the outer two terminals of a potentiometer, why is there no change in resistance reading as the shaft is rotated?

13. How is it possible for a potentiometer to reduce the voltage to zero without causing a short circuit and drawing a large current from the supply?

14. If a potentiometer with a *linear* taper varies the resistance linearly as the shaft is rotated, what is a *logarithmic* taper?

15. Why do you think it is unimportant to a headlight in a car whether the electrical system has a positive or negative ground system but very important to a transistor radio?

16. What does it mean to have a "floating" power supply? Draw a diagram to show why it is impossible to obtain positive and negative supply potentials from two power supplies, both of which have their negative sides grounded to the same point.

17. Explain how it would be possible, in a series circuit, for the number of electrons-per-second flowing at two points to be the same but the electron drift velocity at the two points to be different.

18. Under what conditions can two 120-V lamps be safely connected in series across a 240-V supply?

19. What do you understand by the terms *series-aiding* and *series-opposing?*

20. If two batteries are connected in a series circuit so that both batteries are discharging, in what way have the batteries been connected?

PROBLEMS

(Answers to odd-numbered problems at back of book)

5-1. Given $R_1 = 150 \ \Omega$, $R_2 = 300 \ \Omega$, and $R_3 = 50 \ \Omega$, find the total resistance when connected in series.

5-2. Given $R_1 = 5.1 \ k\Omega$, $R_2 = 6.8 \ k\Omega$, and $R_3 = 4.7 \ k\Omega$, find the total resistance when connected in series.

5-3. The three resistors in Problem 5-1 are connected to a 90-V source of emf. Calculate:

    a. The circuit current.

    b. The voltage drop across each resistor.

5-4. The three resistors in Problem 5-2 are connected to a 24-V source of emf. Calculate:

    a. The circuit current.

    b. The voltage drop across each resistor.

5-5. Calculate:

    a. The power dissipated in each resistor in Problem 5-3.

    b. The total power supplied by the source.

5-6. Calculate:

    a. The power dissipated in each resistor in Problem 5-4.

    b. The total power supplied by the source.

5-7. Refer to Fig. 5-4.

    a. What will a voltmeter indicate when connected from *B* to *D?*

    b. What is the voltage at *C* with respect to *A?*

5-8. Refer to Fig. 5-6*a*. Determine the reading of a voltmeter for the following connections:

    a. Between *A* and *B*.

    b. Between *B* and *C*.

    c. Between *C* and *D*.

    d. Between *B* and *D*.

5-9. Three resistors, when connected in series to a 50-V supply, draw a current of 5 mA. Two resistors are equal in value to each other; the third has a resistance of 2 kΩ. What is the value of each of the other two resistors?

5-10. Three resistors are connected in series to a 20-V source of emf. A current of 2 mA passes through the first resistor; a 10-V drop occurs across the second resistor; and the third resistor has a resistance of 2.2 kΩ. Determine the resistance of the first two resistors.

5-11. Given the circuit in Fig. 5-20, calculate the resistance of R_4, R_3, and R_1.

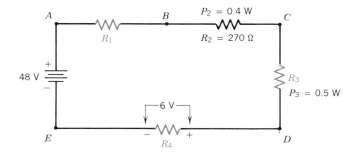

FIGURE 5-20

Circuit diagram for Problems 5-11 and 5-12.

5-12. For the circuit given in Fig. 5-20, determine:

    a. The voltage at *C* with respect to *E*.

    b. The voltage at *C* with respect to *A*.

    c. The voltage at *D* with respect to *A*.

5-13. Four resistors are connected in series to a 25-V supply that provides 100 mW of power. The first two resistors have equal resistance and a combined voltage drop across the pair of 10 V. If the fourth resistor has half the resistance of the third resistor, calculate the resistance of all four resistors.

5-14. Refer to Fig. 5-8. Determine the reading of a voltmeter for the following connections:

    a.   Between A and B.

    b.   Between A and C.

    c.   Between C and D.

    d.   Calculate the reading of an ammeter connected between C and D.

5-15. The following three resistors are connected in series: 330 Ω, 2 W; 100 Ω, 1 W; 33 Ω, ½ W. What is the maximum voltage that can safely be connected across this series combination?

5-16. A series Christmas tree light set consists of twenty-five 7½-W lamps operating at 120 V. What is the total hot resistance of the circuit?

5-17. A 14.8 -Vsource of dc is to be used to recharge a battery as in Fig. 5-2. If the battery is 11.2 V, calculate:

    a.   The resistance R_1 must have to limit the initial charging current to 6 A.

    b.   The current that will flow with R_1 in the circuit if the polarity of the battery is mistakenly reversed.

5-18. What resistance must be connected in series with a 6-V car radio if it is to operate from a 13.2-V battery? The radio uses 3 A at 6 V. Determine the power rating of the resistor.

5-19. A 100-W, 120-V lamp, and a 60-W, 120-V lamp are connected in series to a 240-V supply. Assuming that the resistances of the lamps are constant, use the voltage division rule to determine the voltage across each lamp, the power dissipated by each lamp, and the visual effect.

5-20. A 2.2-MΩ resistor is in a series circuit whose total resistance is 8.5 MΩ. How much voltage appears across the 2.2-MΩ resistor when 40 V is applied to the circuit? Calculate the power dissipated in this resistor.

5-21. A 5-kΩ linear taper potentiometer is connected in series with a 3.3-kΩ resistor across a 50-mV source as shown in Fig. 5-21. Calculate the voltmeter reading for the following locations of contact B:

    a.   Moved fully upward to A.

    b.   Moved fully downward to C.

    c.   Moved halfway between A and C.

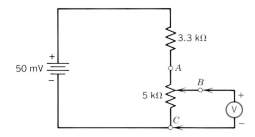

FIGURE 5-21
Circuit diagram for Problems 5-21, 5-23, and 5-25.

5-22. Repeat Problem 5-21 using the values and circuit shown in Fig. 5-22.

5-23. Calculate the current flowing in the circuit of Fig. 5-21 as the potentiometer control is varied.

5-24. Repeat Problem 5-23, using Fig. 5-22.

5-25. If the applied voltage in Fig. 5-21 is 50 V, calculate the required power rating of the potentiometer.

FIGURE 5-22

Circuit diagram for Problems 5-22, 5-24, and 5-26.

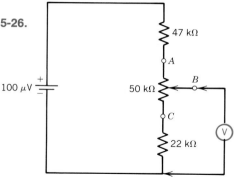

5-26. If the applied voltage in Fig. 5-22 is 100 V, calculate the required power rating of the potentiometer.

5-27. Given the circuit in Fig. 5-23, determine the voltage across the lamp and the power dissipated in the lamp, if the lamp's resistance is 6 Ω.

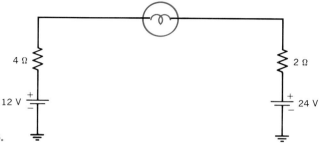

FIGURE 5-23

Circuit diagram for Problems 5-27, and 5-28.

5-28. Repeat Problem 5-27 with the polarity of the 24-V source reversed.

5-29. Refer to Fig. 5-16*a*. It is required to produce +9 V at A and −6 V at B with respect to ground *(C)*. Determine:
   a.   The necessary applied emf.
   b.   The values of R_1 and R_2 if the no-load current drawn from the applied emf is to be 100 mA.
   c.   The potential at B with respect to A.
   d.   The potential at C with respect to A.

5-30. Refer to Fig. 5-19, with $V_1 = 20$ V, $V_2 = 45$ V, $R_1 = 15$ kΩ, and $R_2 = 10$ kΩ. Calculate:
   a.   The current.
   b.   The voltage at C with respect to D.
   c.   The voltage at B with respect to A.

5-31. Repeat Problem 5-30 using $V_1 = 50$ V and $V_2 = 30$ V.

5-32. Given the circuit in Fig. 5-24, calculate:
   a.   The current flowing in the circuit.
   b.   The power dissipated in the 22-kΩ resistor.
   c.   The potential at A with respect to B.

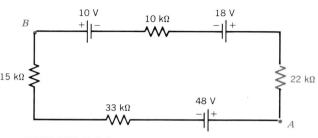

FIGURE 5-24

Circuit diagram for Problems 5-32 and 5-33.

5-33. Repeat Problem 5-32 with the polarity of the 48-V source reversed.

CHAPTER 6

PARALLEL CIRCUITS

A *parallel* circuit provides two or more paths for current but has a common applied voltage. This is the method used in wiring homes, schools, and other buildings, since it allows each load to operate independently of the other loads. *Kirchhoff's current law* states, essentially, that the total current supplied by a source equals the sum of the parallel *branch* currents. This law is used to derive a general expression for the total resistance of any number of parallel-connected resistors. The total resistance of such a circuit is always less than the smallest branch resistance. Finally, two special equations will be presented: one that applies to any *two* resistors in parallel, the other for any number of *equal* resistors in parallel.

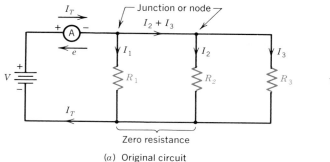 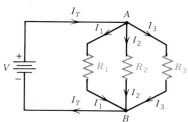

(a) Original circuit

(b) Original circuit redrawn

FIGURE 6-1
Alternate forms of a parallel-connected circuit.

6-1 CHARACTERISTICS OF A PARALLEL CIRCUIT

In a series circuit, the failure of any one element (or any of the connecting wires) interrupts current for *all* elements in that circuit, as you learned in Section 5-1. This is a useful feature when a fuse or circuit breaker opens under overload conditions to prevent damage to the other elements of the circuit. But, as you know if you have ever tried to locate the defective lamp in a series-connected string of Christmas tree lights, the series circuit is *not* a very suitable method for supplying power to a number of different loads. A more practical method is to connect the loads in parallel with each other, then connect the combination to the voltage supply.

The distinguishing feature of a parallel circuit is that it provides more than one path for the flow of current. (A *series* circuit provides only *one* path.) Also, in a true parallel circuit, all the components have the *same* (common) voltage across them.

The current drawn by each component, load, or *branch* will be determined by the resistance of that branch. This is illustrated by the circuit containing three parallel-connected resistors, shown in Fig. 6-1.

In a parallel circuit, the total current (I_T) from the source divides at each *junction* (also called a *node*). The largest current passes through the branch with the smallest resistance. The currents through the branches recombine (as at junction B) to provide the same total current (I_T) flowing back into the source.

6-2 KIRCHHOFF'S CURRENT LAW

In its simplest form, Kirchhoff's current law states:
The current flowing into a junction (or node) must equal the current flowing out of the junction.

Strictly applied, Kirchhoff's current law is the law of conservation of charge. It means that the flow of charge at any junction must be accounted for. Kirchhoff's current law can be applied to junction A in the circuit shown in Fig. 6-1*b*.

Mathematically, this law states:

$$I_T = I_1 + I_2 + I_3 \quad \text{amperes} \qquad (6\text{-}1)$$

EXAMPLE 6-1

A toaster drawing 10 A is connected in parallel with two lamps, one drawing ½ A, the other drawing 1½ A. As shown in Fig. 6-2*a*, a 120-V source is connected across this combination. Calculate:
a. The total current supplied by the source.
b. The voltage across the lamp that draws ½ A.
c. The combined resistance (equivalent resistance) of all three loads.
d. The resistance of a fourth load, connected in parallel with the first three, that will increase the total current to 15 A.
e. The combined resistance of all four loads.

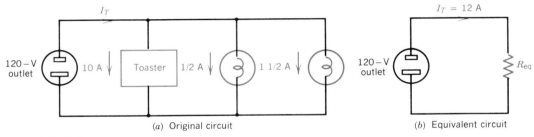

FIGURE 6-2

Circuits for Example 6-1.

Solution

a. $I_T = I_1 + I_2 + I_3$　　　　　　　　　　　(6-1)

     $= 10 + 0.5 + 1.5$ A

     $= \textbf{12 A}$

b. The voltage across the lamp drawing ½ A is the same as the source voltage across all the loads, **120 V.**

c. By Ohm's law (see Fig. 6-2b):

$$R_{eq} = \frac{V}{I_T}$$　　　　　　　　　　(3-1)

$$= \frac{120 \text{ V}}{12 \text{ A}} = \textbf{10 } \Omega$$

d. $I_T = I_1 + I_2 + I_3 + I_4$　　　　　　　(6-1)

Therefore: $I_4 = I_T - (I_1 + I_2 + I_3)$

     $= 15$ A $- 12$ A

     $= 3$ A

$$R = \frac{V}{I}$$　　　　　　　　　　(3-1)

$$= \frac{120 \text{ V}}{3 \text{ A}} = \textbf{40 } \Omega$$

e. New combined equivalent resistance of all four loads:

$$R_{eq} = \frac{V}{I}$$　　　　　　　　　　(3-1)

$$= \frac{120 \text{ V}}{15 \text{ A}} = \textbf{8 } \Omega$$

Example 6-1 shows an interesting result—*adding* a resistor to this parallel circuit has *reduced* the total circuit resistance (the combined resistance) connected to the source. Note that this is the *opposite* of what happens when you add a resistor to a series circuit.

In the next section, you will learn how to find the total equivalent resistance in a parallel circuit by using the individual resistance values.

6-3 EQUIVALENT RESISTANCE FOR A PARALLEL CIRCUIT

If you apply Ohm's law to the resistance in each branch of Fig. 6-3, you obtain:

$$I_1 = \frac{V}{R_1}, \quad I_2 = \frac{V}{R_2}, \quad I_3 = \frac{V}{R_3}$$

Thus, Eq. 6-1 may be rewritten as:

$$I_T = I_1 + I_2 + I_3$$　　　　　　　(6-1)

$$I_T = \frac{V}{R_1} + \frac{V}{R_2} + \frac{V}{R_3}$$　　　　(6-1a)

or

$$I_T = V \left(\frac{1}{R_1} + \frac{1}{R_2} + \frac{1}{R_3} \right)$$　　　(6-2)

Now let R_{eq} stand for the equivalent combined resistance of all three resistors in parallel. (See Fig. 6-3b). That is, R_{eq} can take the place of R_1, R_2, and R_3. The battery would still supply the same current (I_T) as it did to the three resistors. Expressed in equation form:

$$I_T = \frac{V}{R_{eq}} = V \left(\frac{1}{R_{eq}} \right)$$　　　　(6-3)

For the current (I_T) to be the same in Eqs. 6-2 and 6-3, it is required that:

$$V \left(\frac{1}{R_{eq}} \right) = V \left(\frac{1}{R_1} + \frac{1}{R_2} + \frac{1}{R_3} \right)$$

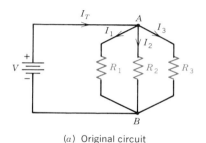

(a) Original circuit

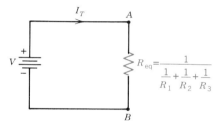

(b) Equivalent circuit

FIGURE 6-3
Parallel resistor circuit.

Therefore,

$$\frac{1}{R_{eq}} = \frac{1}{R_1} + \frac{1}{R_2} + \frac{1}{R_3}$$ (6-4)

Equation 6-4 also may be written as:

$$R_{eq} = \frac{1}{\dfrac{1}{R_1} + \dfrac{1}{R_2} + \dfrac{1}{R_3}} \quad \text{ohms}$$ (6-5)

EXAMPLE 6-2

Three resistors—2 Ω, 5 Ω, and 10 Ω—are connected in parallel across a 25-V source. Calculate:
a. The total (equivalent) resistance of the three resistors.
b. The total current drawn from the source.

Solution

a. $\dfrac{1}{R_{eq}} = \dfrac{1}{R_1} + \dfrac{1}{R_2} + \dfrac{1}{R_3}$ (6-4)

$$= \frac{1}{2} + \frac{1}{5} + \frac{1}{10}$$

$$= \frac{5}{10} + \frac{2}{10} + \frac{1}{10}$$

$$= \frac{8}{10}$$

Therefore,

$$R_{eq} = \frac{10}{8} = \textbf{1.25 Ω}$$

Alternatively, if the given quantities do not have a simple common denominator, use their decimal equivalents for the reciprocals.

$$R_{eq} = \frac{1}{\dfrac{1}{R_1} + \dfrac{1}{R_2} + \dfrac{1}{R_3}}$$ (6-5)

$$= \frac{1}{\dfrac{1}{2} + \dfrac{1}{5} + \dfrac{1}{10}}$$

$$= \frac{1}{0.5 + 0.2 + 0.1}\ \Omega$$

$$= \frac{1}{0.8}\ \Omega$$

$$= \textbf{1.25 Ω}$$

b. $I = \dfrac{V}{R_{eq}}$ (3-1a)

$$= \frac{25\ V}{1.25\ \Omega} = \textbf{20 A}$$

The use of a calculator greatly simplifies the work involved to find equivalent resistance using Eq. 6-5. The calculator automatically stores each reciprocal and adds it to the next, as shown below for Example 6-2.

Enter	Press	Display	Purpose
	C	0.	Clears the calculator.
2	1/x	0.5	Finds reciprocal of first term.
	+	0.5	Instruction to add next reciprocal.
5	1/x	0.2	Finds reciprocal of second term.
	+	0.7	Instruction to add next reciprocal.
10	1/x	0.1	Finds reciprocal of third term.
	=	0.8	Gives total of reciprocals.
	1/x	1.25	Gives reciprocal of total— final answer.

Answer: **1.25 Ω**

A very important conclusion that you can draw from Example 6-2 is that the equivalent resistance (R_{eq}) of a parallel circuit is *always smaller* than the smallest *branch* resistance. In this case, $R_{eq} = 1.25\ \Omega$, while the smallest branch resistance was $2\ \Omega$. Adding *any* resistor in parallel with the 2-Ω resistor provides another parallel path for current, allowing more *total* current to flow from the source. This means that a *drop* in equivalent resistance has taken place. ($R = V/I$)

Since conductance is the reciprocal of resistance, $G = 1/R$ (Section 2-11), you can also write Eq. 6-4 as:

$$\frac{1}{R_{eq}} = \frac{1}{R_1} + \frac{1}{R_2} + \frac{1}{R_3} \qquad (6\text{-}4)$$

$$G_T = G_1 + G_2 + G_3 \quad \text{siemens} \qquad (6\text{-}6)$$

The total conductance of a parallel circuit is the *sum* of its *branch* conductances. If G_T is found first, then $R_{eq} = 1/G_T$.

EXAMPLE 6-3

Two resistors, one with a resistance of 400 Ω and the other with a resistance of 2 kΩ, are connected in parallel with a load that has a conductance of 1 mS, and a 10-V source. (See Fig. 6-4a.) Calculate:
a. The total conductance of the circuit.
b. The combined (equivalent) resistance of the circuit.
c. The total current (I_T) drawn from the source.

Solution

a. $G_T = G_1 + G_2 + G_3$ $\qquad\qquad\qquad$ (6-6)

$$= \frac{1}{400\ \Omega} + \frac{1}{2 \times 10^3\ \Omega} + 1 \times 10^{-3}\ \text{S}$$
$$= 0.0025\ \text{S} + 0.0005\ \text{S} + 0.001\ \text{S}$$
$$= 2.5\ \text{mS} + 0.5\ \text{mS} + 1\ \text{mS}$$
$$= \mathbf{4\ mS}$$

b. $R_{eq} = \dfrac{1}{G_T}$ $\qquad\qquad\qquad\qquad\qquad$ (2-5)

$$= \frac{1}{4 \times 10^{-3}\ \text{S}}$$
$$= 0.25 \times 10^3\ \Omega = \mathbf{250\ \Omega}$$

c. $I_T = \dfrac{V}{R_{eq}}$ $\qquad\qquad\qquad\qquad\qquad$ (3-1a)

$$= \frac{10\ \text{V}}{250\ \Omega} = 0.04\ \text{A} = \mathbf{40\ mA}$$

or

$$I_T = VG_T$$
$$= 10\ \text{V} \times 4 \times 10^{-3}\ \text{S}$$
$$= 40 \times 10^{-3}\ \text{A} = \mathbf{40\ mA}$$

Note, from Example 6-3, how connecting *more* resistors in parallel *increases* the total conductance. And, since

$$I_T = \frac{V}{R} = V \times \frac{1}{R} = V \times G_T,$$

this causes *more* total current to flow from the source as more resistors are added in parallel.

6-4 POWER IN A PARALLEL CIRCUIT

As is the case in a series circuit, **the total power dissipated in a parallel circuit is the sum of the individual powers dissipated by each load.**

$$\text{Total power} \quad P_T = P_1 + P_2 + P_3 \quad \text{watts} \qquad (5\text{-}3)$$

where P_1, and so on, may be calculated using

$$P_1 = I_1^2 R_1 \qquad\qquad (3\text{-}7)$$
$$\text{or} \quad P_1 = VI_1 \qquad\qquad (3\text{-}5)$$
$$\text{or} \quad P_1 = \frac{V^2}{R_1} \qquad\qquad (3\text{-}8)$$

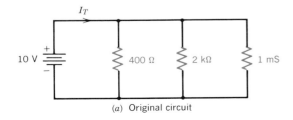

(a) Original circuit

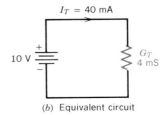

(b) Equivalent circuit

FIGURE 6-4

Circuits for Example 6-3.

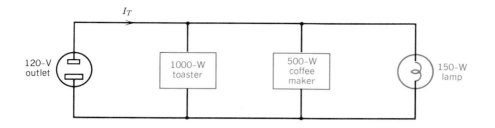

FIGURE 6-5

Circuit for Example 6-4.

Thus $P_T = I_T^2 R_{eq}$ (6-7)

or $P_T = VI_T$ (6-8)

or $P_T = \dfrac{V^2}{R_{eq}}$ (6-9)

$$= \frac{120 \text{ V}}{13.75 \text{ A}} = \textbf{8.73 } \boldsymbol{\Omega}$$

or $P_T = \dfrac{V^2}{R_{eq}}$ (6-9)

$$R_{eq} = \frac{V^2}{P_T}$$

$$= \frac{(120 \text{ V})^2}{1650 \text{ W}} = \textbf{8.73 } \boldsymbol{\Omega}$$

or $P_T = I_T^2 R_{eq}$ (6-7)

$$R_{eq} = \frac{P_T}{I_T^2}$$

$$= \frac{1650 \text{ W}}{(13.75 \text{ A})^2} = \textbf{8.73 } \boldsymbol{\Omega}$$

EXAMPLE 6-4

A 120-V kitchen outlet (wall receptacle) supplies a 1-kW toaster, a 500-W coffee maker, and a 150-W lamp, as shown by the circuit diagram in Fig. 6-5. Calculate:
a. The total power demanded from the outlet.
b. The total current drawn from the source.
c. The effective equivalent resistance of the three parallel-connected loads.

EXAMPLE 6-5

A source of 24 V is connected across three resistors in parallel having values of $R_1 = 2.2$ kΩ, $R_2 = 1$ kΩ, and $R_3 = 4.7$ kΩ, as shown in Fig. 6-6. Calculate:
a. The current drawn by each resistor.
b. The total current from the source.
c. The combined equivalent circuit resistance using b.
d. The equivalent resistance using Eq. 6-4.
e. The total current from the source using the resistance from d.
f. The reading of an ammeter between R_1 and R_2.
g. The power dissipated in each resistor.

Solution

a. $P_T = P_1 + P_2 + P_3$ (5-3)

         $= 1000 + 500 + 150$ W

         $= \textbf{1650 W}$

b. $P_T = VI_T$ (6-8)

         $I_T = \dfrac{P_T}{V}$

         $= \dfrac{1650 \text{ W}}{120 \text{ V}} = \textbf{13.75 A}$

c. $R_T = \dfrac{V}{I_T}$ (3-1)

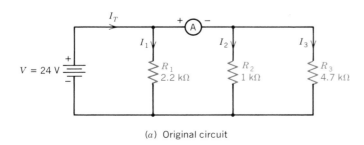

(a) Original circuit

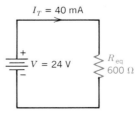

(b) Equivalent circuit

FIGURE 6-6

Circuits for Example 6-5.

h. The total power delivered by the source.
i. The voltage across R_3 if an open circuit develops in R_2.

Solution

a. $I_1 = \dfrac{V}{R_1}$ (3-1a)

       $= \dfrac{24 \text{ V}}{2.2 \text{ k}\Omega} = \textbf{10.9 mA}$

     $I_2 = \dfrac{V}{R_2}$

       $= \dfrac{24 \text{ V}}{1 \text{ k}\Omega} = \textbf{24.0 mA}$

     $I_3 = \dfrac{V}{R_3}$

       $= \dfrac{24 \text{ V}}{4.7 \text{ k}\Omega} = \textbf{5.1 mA}$

b. $I_T = I_1 + I_2 + I_3$ (6-1)

       $= 10.9 + 24.0 + 5.1 \text{ mA} = \textbf{40 mA}$

c. $R_{eq} = \dfrac{V}{I_T}$ (3-1)

       $= \dfrac{24 \text{ V}}{40 \text{ mA}} = \dfrac{24 \text{ V}}{0.04 \text{ A}} = \textbf{600 }\boldsymbol{\Omega}$

d. $\dfrac{1}{R_{eq}} = \dfrac{1}{R_1} + \dfrac{1}{R_2} + \dfrac{1}{R_3}$ (6-4)

       $= \dfrac{1}{2.2 \text{ k}\Omega} + \dfrac{1}{1 \text{ k}\Omega} + \dfrac{1}{4.7 \text{ k}\Omega}$

       $= \dfrac{1}{2.2 \times 10^3 \, \Omega} + \dfrac{1}{1 \times 10^3 \, \Omega} + \dfrac{1}{4.7 \times 10^3 \, \Omega}$

       $= 0.455 \times 10^{-3} + 1 \times 10^{-3} + 0.213 \times 10^{-3} \text{ S}$

       $= 1.668 \times 10^{-3} \text{ S}$

Therefore, R_{eq} $= \dfrac{1}{1.668 \times 10^{-3} \text{ S}} = \textbf{600 }\boldsymbol{\Omega}$

Note that this is essentially the same as using Eq. 6-6 with a total circuit conductance of 1.668 mS. See Fig. 6-6b.

e. $I_T = \dfrac{V}{R_{eq}}$ (3-1a)

       $= \dfrac{24 \text{ V}}{600 \, \Omega} = 0.04 \text{ A} = \textbf{40 mA}$

f. The ammeter will indicate:

       $I_T - I_1 = 40 \text{ mA} - 10.9 \text{ mA} = \textbf{29.1 mA}$

       or $I_2 + I_3 = 24 \text{ mA} + 5.1 \text{ mA} = \textbf{29.1 mA}$

g. $P_1 = \dfrac{V^2}{R_1}$ (3-8)

       $= \dfrac{(24 \text{ V})^2}{2.2 \times 10^3 \, \Omega} = \textbf{262 mW}$

$$P_2 = \dfrac{V^2}{R_2}$$

       $= \dfrac{(24 \text{ V})^2}{1 \times 10^3 \, \Omega} = \textbf{576 mW}$

$$P_3 = \dfrac{V^2}{R_3}$$

       $= \dfrac{(24 \text{ V})^2}{4.7 \times 10^3 \, \Omega} = \textbf{123 mW}$

NOTE This is preferable to using $P = I^2 R$, since I is a result of a previous calculation and has been rounded off.

h. $P_T = P_1 + P_2 + P_3$ (5-3)

       $= 262 + 576 + 123 \text{ mW}$

       $= \textbf{961 mW}$

       or $P_T = I_T^2 R_{eq}$ (6-7)

          $= (0.040 \text{ A})^2 \times 600 \, \Omega = \textbf{960 mW}$

       or $P_T = V I_T$ (6-8)

          $= 24 \text{ V} \times 0.040 \text{ A} = \textbf{960 mW}$

i. If R_2 is removed (due to an open circuit), the voltage across R_3 will remain at 24 V even though the total source current will drop from 40 to 16 mA. This is because all the resistors are connected in parallel with the 24-V source.*

As you have seen once again (in Example 6-5), the **equivalent resistance (R_{eq}) of a parallel circuit is smaller than the smallest branch resistance.** In this case, $R_{eq} = 600 \, \Omega$, while the smallest branch resistance was $1000 \, \Omega$ (1 kΩ).

In fact, if a 600-Ω resistor is added to the circuit, in parallel with the other three resistors, the equivalent resistance of the circuit would drop to just one-half of the original amount, or 300 Ω. This takes place because another parallel path for current has been added, permitting *more* current to flow and resulting in a *drop* in equivalent resistance. Another way to look at this situation is to view it as an increase in *conductance* — G_T has increased, and $R_{eq} = 1/G_T$ has decreased.

---

*In practice, the 24-V source would have some internal resistance that would cause the terminal voltage available for the circuit to depend somewhat upon the total load current. A drop in this current would probably cause some increase in voltage across the parallel circuit. In our circuit, in Fig. 6-6, we have assumed this internal resistance to be zero. We shall consider this topic in detail in Chapter 9.

6-5 OTHER EQUIVALENT RESISTANCE EQUATIONS

Two further equations are useful for parallel resistance networks (circuits with resistors wired in parallel).

For the special case in which the circuit contains *only* two parallel-wired resistors:

$$R_{eq} = \frac{R_1 R_2}{R_1 + R_2} \tag{6-10}$$

$$= \frac{\text{product (of the 2 } R\text{s in parallel)}}{\text{sum (of the 2 } R\text{s in parallel)}} \quad \text{ohms}$$

And for *N equal-valued* resistors, each of resistance R, connected in parallel:

$$R_{eq} = \frac{R}{N} = \frac{\text{resistance of one resistor}}{\text{number of resistors}} \tag{6-11}$$

A notation used to show that resistors are wired in parallel consists of two vertical parallel lines drawn between the resistors involved. For example, if three resistors (R_1, R_2, and R_3) are wired in parallel, their equivalent resistance may be represented by $R_{eq} = R_1 \| R_2 \| R_3$.

For the special case of two parallel-connected resistors:

$$R_{eq} = R_1 \| R_2 = \frac{R_1 \times R_2}{R_1 + R_2}$$

This notation is especially useful when you work with series-parallel circuits, which you will learn about in Chapter 7.

EXAMPLE 6-6

Calculate the equivalent resistance of a 2.2-kΩ resistor and a 4.7-kΩ resistor in parallel with each other, as in Fig. 6-7a.

Solution

$$R_{eq} = R_1 \| R_2 = \frac{R_1 \times R_2}{R_1 + R_2} \tag{6-10}$$

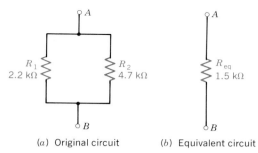

(a) Original circuit   (b) Equivalent circuit

FIGURE 6-7
Circuits for Example 6-6.

$$= \frac{2.2 \text{ k}\Omega \times 4.7 \text{ k}\Omega}{2.2 \text{ k}\Omega + 4.7 \text{ k}\Omega} = \frac{10.34 \text{ (k}\Omega)^2}{6.9 \text{ k}\Omega} = \textbf{1.5 k}\boldsymbol{\Omega}$$

Note that if this resistance is now combined with a 1-kΩ resistor in parallel, using the same equation, we obtain:

$$R_{eq} = \frac{R_1 R_2}{R_1 + R_2} \tag{6-10}$$

$$= \frac{1 \text{ k}\Omega \times 1.5 \text{ k}\Omega}{1 \text{ k}\Omega + 1.5 \text{ k}\Omega} = 0.6 \text{ k}\Omega = \textbf{600 }\boldsymbol{\Omega}$$

This is the same result as in Example 6-5. That is, we can repeatedly use Eq. 6-10 for any number of parallel resistors, provided that we consider them two at a time.

EXAMPLE 6-7

Three 600-Ω resistors are parallel-connected across a 60-V source of emf, as in Fig. 6-8a. Calculate:
a. The total circuit resistance.
b. The total current drawn from the supply.
c. The current in each parallel branch.

Solution

a. $$R_{eq} = \frac{R}{N} \tag{6-11}$$

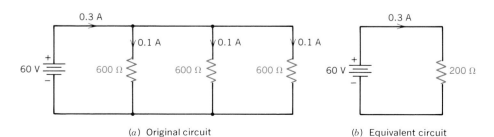

(a) Original circuit   (b) Equivalent circuit

FIGURE 6-8
Circuits for Example 6-7.

$$= \frac{600 \ \Omega}{3} = \mathbf{200 \ \Omega}$$

b. $\quad I_T = \dfrac{V}{R_{eq}} \qquad\qquad\qquad\qquad\qquad$ (3-1a)

$$= \frac{60 \ V}{200 \ \Omega} = \mathbf{0.3 \ A}$$

c. The total current of 0.3 A must divide equally among the three 600-Ω branches, so **0.1 A** must flow in each branch. See Fig. 6-8*a*.

6-6 CURRENT DIVISION

If you examine Example 6-5 again, you will find that the current through the 2.2-kΩ resistor is approximately *double* the current flowing through the 4.7-kΩ resistor with which it is connected in parallel. This should not be surprising, because of the differing resistance values, but it does suggest a formula by which the current divides in a circuit with two parallel resistances. Consider a circuit in which there are two resistors in parallel and a total current (I_T) that flows into those two resistors. (See Fig. 6-9).

Although the source voltage is not shown, you could obtain it by

$$V = I_T R_{eq} = I_T \times \frac{R_1 R_2}{R_1 + R_2}$$

Then,

$$I_1 = \frac{V}{R_1} = \frac{I_T}{R_1} \times \frac{R_1 R_2}{R_1 + R_2}$$

Therefore,

$$I_1 = I_T \times \frac{R_2}{R_1 + R_2} \quad \text{amperes} \qquad (6\text{-}12)$$

and

$$I_2 = I_T \times \frac{R_1}{R_1 + R_2} \quad \text{amperes} \qquad (6\text{-}13)$$

The current division rules given in Eqs. 6-12 and 6-13 apply only to two resistors connected in parallel. Note how *opposite* subscripts are used for the resistance numerator and for the branch current being determined.

EXAMPLE 6-8

A 1-kΩ resistor (R_1) and a 3.3-kΩ resistor (R_2) are in parallel with each other and have a combined current of 16 mA through them. What is the current through each?

Solution

$$I_1 = I_T \times \frac{R_2}{R_1 + R_2} \qquad\qquad (6\text{-}12)$$

$$= 16 \ mA \times \frac{3.3 \ k\Omega}{1 \ k\Omega + 3.3 \ k\Omega}$$

$$= 16 \ mA \times \frac{3.3 \ k\Omega}{4.3 \ k\Omega}$$

$$= 16 \ mA \times 0.767 = \mathbf{12.3 \ mA}$$

$$I_2 = I_T \times \frac{R_1}{R_1 + R_2} \qquad\qquad (6\text{-}13)$$

$$= 16 \ mA \times \frac{1 \ k\Omega}{4.3 \ k\Omega}$$

$$= 16 \ mA \times 0.233 = \mathbf{3.7 \ mA}$$

PROOF $I_1 + I_2 = I_T$
12.3 + 3.7 = **16 mA**

Note the way in which the current divides, so that approximately *one-quarter* of the current passes through the 3.3-kΩ resistor. (See Fig. 6-10.)

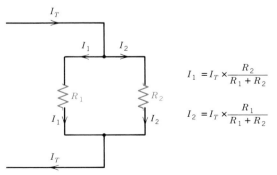

FIGURE 6-9
Current division in two parallel-connected resistors.

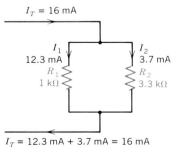

$I_T = 12.3 \ mA + 3.7 \ mA = 16 \ mA$

FIGURE 6-10
Circuit for Example 6-8.

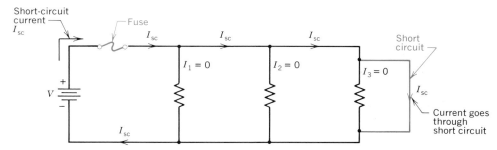

FIGURE 6-11

Short circuit in a parallel network protected by a
fuse.

6-7 SHORT CIRCUITS

As noted in Section 6-6, the *opposite* subscripts were used
in Eqs. 6-12 and 6-13 for the resistance numerator and for
the branch current being determined. The correctness of
this can be shown by assuming that $R_1 = 0$ and determining I_1:

$$I_1 = I_T \times \frac{R_2}{R_1 + R_2} \qquad (6\text{-}12)$$

$$= I_T \times \frac{R_2}{0 + R_2} = I_T \times 1 = I_T$$

That is, if the resistance of R_1 is zero, all of the current
(I_T) will flow through it, and *no* current will flow through
R_2:

$$I_2 = I_T \times \frac{R_1}{R_1 + R_2} \qquad (6\text{-}13)$$

$$= I_T \times \frac{0}{0 + R_2} = \frac{0}{R_2} = 0$$

This condition, of course, represents a *short circuit*. In
a parallel circuit, a short circuit in one branch will always
result in *no* current flowing in the other branches. The current
will follow the path of least resistance. It is important
to note that the current flowing in a short circuit could be
high enough to damage the circuit and source components.
It would be limited only by the current capability of the
source and its internal resistance. Figure 6-11 shows how
a fuse or circuit breaker is connected in series with the
source to prevent a large current flow if a short circuit
occurs.

SUMMARY

1. Components connected in parallel have the same voltage across them.
2. The current drawn by each component in a parallel circuit is determined by the
   resistance of the branch containing that component.
3. Each branch in a parallel circuit is independent of the other branches, unless a
   short circuit occurs.
4. Kirchhoff's current law states that the current into a junction equals the current
   out of the junction.
$$I_T = I_1 + I_2 + I_3$$
5. The equivalent resistance of a number of parallel-connected resistors may be obtained
   from $1/R_{eq} = 1/R_1 + 1/R_2 + 1/R_3$.
6. The equivalent resistance of a circuit containing a number of parallel resistors is
   always less than the smallest individual resistance.

7. For a circuit containing only two parallel-connected resistors, the equivalent resistance is given by $R_{eq} = R_1R_2/(R_1 + R_2)$.
8. For N resistors of equal value connected in parallel, each of resistance R, the equivalent resistance is given by $R_{eq} = R/N$.
9. The total conductance of a parallel circuit is given by $G_T = G_1 + G_2 + G_3$, and the equivalent resistance is given by $R_{eq} = 1/G_T$.
10. The total power dissipated in a parallel circuit is given by $P_T = P_1 + P_2 + P_3$.
11. For two parallel-connected resistors with a total current I_T, the current-division rule gives the current through each resistor by $I_1 = I_T \dfrac{R_2}{R_1 + R_2}$ and $I_2 = I_T \dfrac{R_1}{R_1 + R_2}$.
12. A short circuit in any one branch of a parallel circuit causes a large current to flow from the source through that branch, reducing the current in the other branches to zero.

SELF-EXAMINATION

Answer true or false or a, b, c, d
(Answers at back of book)

6-1. A junction requires the connection of three or more wires. _____
6-2. Kirchhoff's current law applies only to a parallel circuit. _____
6-3. In Fig. 6-3a, the branch currents will be equal to each other only if all the branch resistances are equal. _____
6-4. In a parallel circuit the combined equivalent resistance is always larger than the smallest resistance. _____
6-5. If, in Fig. 6-6, R_2 is twice as large as R_1 and only half as large as R_3, the smallest current flows in:
   a. R_1
   b. R_2
   c. R_3
6-6. For the same conditions as in Question 6-5, the largest current flows in:
   a. R_1
   b. R_2
   c. R_3
6-7. For the same conditions as in Question 6-5, the ammeter in Fig. 6-6 will have a reading compared with the current through R_1 that is
   a. larger
   b. smaller
   c. the same
6-8. Two 20-kΩ resistors and a 10-kΩ resistor are connected in parallel across a 50-V source. Their combined equivalent resistance is
   a. 30 kΩ
   b. 50 kΩ
   c. 10 kΩ
   d. 5 kΩ
6-9. The branch currents in Question 6-8 are
   a. all 2½ mA
   b. all 1 mA

    c. 2½ mA, 2½ mA, 5 mA

    d. 1.25 mA, 1.25 mA, 5 mA

6-10. The total source current in Question 6-8 is

    a. 7½ mA

    b. 3 mA

    c. 10 mA

    d. 1 mA

6-11. The total power delivered by the source in Question 6-8 is

    a. 375 mW

    b. 150 mW

    c. 0.5 W

    d. 50 mW

6-12. One million 1-MΩ resistors, if connected in parallel, would have a combined resistance of:

    a. 1 MΩ   c. 1 kΩ

    b. 1 mΩ   d. 1 Ω

6-13. A 4-Ω and an 8-Ω resistor in parallel have a combined resistance of

    a. 6 Ω

    b. 2⅔ Ω

    c. 2⅓ Ω

    d. 12 Ω

6-14. If the two resistors in Question 6-13 have a total current of 24 mA flowing through them, the current in the 8-Ω resistor is

    a. 8 mA

    b. 16 mA

    c. 12 mA

    d. 24 mA

6-15. If a short circuit develops in the 4-Ω resistor of Question 6-14, the current in the 8-Ω resistor will be

    a. 24 mA

    b. 0 mA

    c. 16 mA

    d. 8 mA

REVIEW QUESTIONS

1. What advantages does a parallel circuit have compared with a series circuit?
2. In what ways is a parallel circuit ''opposite'' to a series circuit? (Name at least three.)
3. In what ways is a parallel circuit similar to a series circuit? (Name at least two.)
4. Justify why the equivalent resistance of a parallel circuit must be less than the smallest branch resistance.
5. Why does ''adding'' a resistor in a parallel circuit decrease the total combined resistance but increase the total in a series circuit?
6. Justify why it is possible to add conductances to obtain the total conductance in a parallel circuit, but not in a series circuit.
7. What electrical equations are the same for a parallel circuit as for a series circuit? Why is this so?

8. Derive the equation for two resistors in parallel, Eq. 6-10, from the reciprocal equation, Eq. 6-4.
9. Use the current division equations (6-12 and 6-13) to determine what current flows in each resistor when an open circuit occurs in R_1 $(R_1 = \infty)$.
10. Why can a short circuit cause damaging currents in a parallel circuit whereas it is less likely to happen in a series circuit?

PROBLEMS

(Answers to odd-numbered problems at back of book)

6-1. What is the equivalent resistance of a parallel circuit consisting of $R_1 = 100\ \Omega$, $R_2 = 220\ \Omega$, and $R_3 = 330\ \Omega$?
6-2. What is the equivalent resistance of a parallel circuit consisting of $R_1 = 100$ kΩ, $R_2 = 220$ kΩ, and $R_3 = 330$ kΩ?
6-3. Repeat Problem 6-1 using $R_1 = 10$ kΩ, $R_2 = 6.8$ kΩ, and $R_3 = 1$ MΩ.
6-4. What is the total conductance of the parallel circuit in Problem 6-2?
6-5. A resistor of conductance 750 μS is connected in parallel with a component having a conductance of 50 mS. What is the equivalent resistance of the combination?
6-6. What is the total current when the parallel circuit of Problem 6-2 is connected across a 40-V source of emf?
6-7. What is the total power dissipated in the three resistors of Problem 6-1 when the parallel circuit is connected across a 120-V supply?
6-8. Three 120-V lamps, rated at 60 W, 100 W, and 200 W, are connected in parallel to a 120-V supply. Find the equivalent hot resistance of this combination in the following two ways:
   a. Find the individual resistances and combine them.
   b. Find the total power and solve for R.
6-9. Given the circuit in Fig. 6-12, calculate:
   a. The missing values.
   b. The total power delivered by the source.

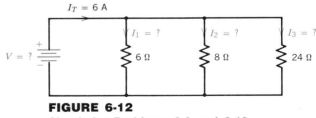

FIGURE 6-12
Circuit for Problems 6-9 and 6-10.

6-10. What single resistance should be added to the circuit in Fig. 6-12 to:
   a. Double the current?
   b. Triple the current?
6-11. A 120-V source that can deliver a maximum of 20 A is connected to a 1500-W electric frying pan and a 600-W coffee maker. Calculate:
   a. The smallest resistance that may be connected in parallel with the two given loads without exceeding the maximum supply current.
   b. The total resistance of the parallel circuit before and after connecting the resistance.
6-12. Given the circuit in Fig. 6-13, find the missing values.
6-13. What resistance must be connected across a 2.2-kΩ resistor to produce a total resistance of 1.5 kΩ?
6-14. What resistance must be placed in parallel with a 33-kΩ resistor to produce a total conductance of 50 μS?

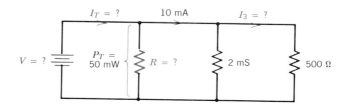

FIGURE 6-13

Circuit for Problem 6-12.

6-15. Three resistors in parallel draw a total current of 7 mA. The first resistor dissipates 50 mW; the second resistor has a voltage drop of 20 V; and the third resistor has a resistance of 10 kΩ. Determine the resistance of the first two resistors.

6-16. Three resistors in parallel draw a total current of 10 mA and dissipate a total power of 500 mW. If the first resistor draws 5 mA and the second and third resistors are equal in value, determine the resistance of all three resistors.

6-17. A 47-kΩ and a 51-kΩ resistor are connected in parallel and draw a total of 12 mA.

    a. How much current passes through each resistor?

    b. What must be the minimum power rating of each?

6-18. What is the largest resistance that can be connected in parallel with a 10-kΩ, 1-W resistor if the combination must handle a total of 25 mA? What is the power rating of this resistor?

6-19. Given the circuit in Fig. 6-14, find the missing values.

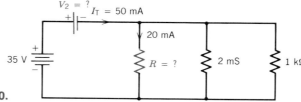

FIGURE 6-14

Circuit for Problems 6-19 and 6-20.

6-20. If V_2 in Fig. 6-14 is shorted, what is the new current through the 1-kΩ resistor?

6-21. Given the circuit in Fig. 6-15, find the missing values.

6-22. How much resistance must be connected in parallel with a 22-kΩ and a 10-kΩ resistor to produce a resistance of 5 kΩ?

6-23. If the three resistors in Problem 6-1 are each rated at 2 W, what is the highest voltage that can safely be applied to the parallel combination?

6-24. How many parallel-connected 25-W Christmas tree lamps can be supplied from a 120-V source fused at 15 A?

6-25. The total current drawn by the three resistors in Problem 6-2 is 3 mA. What is the current through each resistor?

6-26. Six equal resistors are to be connected in parallel with a 100-kΩ resistor to drop the overall resistance to 40 kΩ. What must be the value of each of the six resistors?

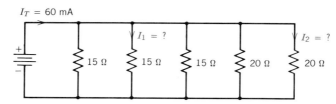

FIGURE 6-15

Circuit for Problem 6-21.

CHAPTER 7

SERIES-PARALLEL AND VOLTAGE-DIVIDER CIRCUITS

A practical circuit is seldom a truly series or truly parallel circuit. For example, when a fuse is inserted in series with a power supply and its parallel-connected loads, the result is a *series-parallel* circuit. Similarly, the internal resistance of a voltage source or the resistance of feeder lines results in series-parallel circuits. All of the rules and laws developed separately for series and parallel circuits, however, can be applied to the series-parallel combination.

An important use of series and parallel principles is in voltage-divider circuits. These circuits permit a load to operate at a voltage different from the immediately available supply voltage. Sometimes, this is accomplished simply by the use of a *series dropping resistor*. A *bleeder resistor* may be added to improve *voltage regulation* with varying loads. In some transistor and integrated circuit power supplies, both positive and negative potentials with respect to ground often must be provided. In this chapter, you will learn how to select the resistors for such a *loaded voltage-divider network*.

7-1 SERIES-PARALLEL RESISTOR CIRCUITS

When a parallel bank of resistors is connected in series with another resistor or parallel group of components, a *series-parallel* circuit is formed. (See Fig. 7-1a.) To solve for the currents and voltage drops in such a circuit, you must combine the methods you learned in Chapters 5 and 6. In general, the series circuit rules are applied to the series portions of the circuit, and the parallel circuit rules to the parallel portions. Some general observations can be made about the series-parallel circuit shown in Fig. 7-1.

1. Resistors R_1 and R_4 are in series with each other because they both have the same current. (The current that leaves the battery and passes through R_1 must return to the battery through R_4.)
2. Resistors R_2 and R_3 are in parallel with each other because there is the same voltage across them. (A voltmeter connected from B to C indicates the same voltage as from B' to C'.)
3. The parallel *combination* of R_2 and R_3 is in *series* with R_1 and R_4. (The equivalent resistance of R_2 and R_3 carries the same current as R_1 and R_4. See Fig. 7-1b.)
4. The current through either R_2 or R_3 is *less* than the current through R_1. (The current through R_1 must

split up or *divide* into two smaller currents at junction B.)
5. The sum of the two currents through R_2 and R_3 must equal I_T. That is, the total current must equal the sum of the branch currents (Kirchhoff's current law).
6. Only *half* as much current flows through R_2 as flows through R_3 because the resistance of R_2 is *twice* that of R_3 (current division equation).
7. The sum of the voltage drops around either loop (*ABCD* or *AB'C'D*) equals the applied voltage, as described by Kirchhoff's voltage law.

NOTE The voltage drop across a parallel circuit is counted only *once* when determining the total of the voltage drops around a complete circuit.

The current and voltages in Fig. 7-1a will be found in Example 7-2.

EXAMPLE 7-1

Using the circuit shown in Fig. 7-2, calculate:
a. The total resistance of the circuit.
b. The current drawn from the source.
c. The current through the 60-Ω resistor.
d. The voltage across the 24-Ω resistor.

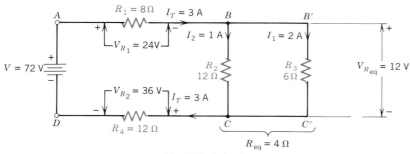

(a) Original circuit

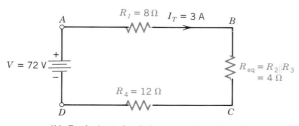

(b) Equivalent circuit for parallel portion = R_{eq}

(c) Final equivalent circuit

FIGURE 7-1
Simplification of a series-parallel circuit.

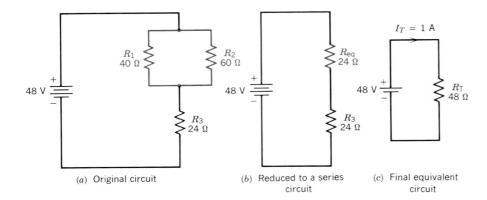

(a) Original circuit   (b) Reduced to a series circuit   (c) Final equivalent circuit

FIGURE 7-2

Circuits for Example 7-1.

Solution

a. The equivalent resistance of the two parallel-connected resistors is

$$R_{eq} = R_1 \| R_2 = \frac{R_1 \times R_2}{R_1 + R_2} \qquad (6\text{-}10)$$

$$= \frac{40\ \Omega \times 60\ \Omega}{40\ \Omega + 60\ \Omega}$$

$$= \frac{2400}{100}\ \Omega = 24\ \Omega$$

See Fig. 7-2b.
The total resistance of the series circuit is

$$R_T = R_{eq} + R_3$$
$$= 24\ \Omega + 24\ \Omega = \mathbf{48\ \Omega}$$

See Fig. 7-2c.

b. $I_T = \dfrac{V}{R_T} \qquad (3\text{--}1a)$

$$= \frac{48\ V}{48\ \Omega} = \mathbf{1\ A}$$

c. $I_2 = I_T \times \dfrac{R_1}{R_1 + R_2} \qquad (6\text{-}13)$

$$= 1\ A \times \frac{40\ \Omega}{40\ \Omega + 60\ \Omega}$$

$$= 1\ A \times \frac{40\ \Omega}{100\ \Omega} = 1\ A \times 0.4 = \mathbf{0.4\ A}$$

d. $V_{R_3} = I_T\,R_3 \qquad (3\text{--}1b)$
$$= 1\ A \times 24\ \Omega = \mathbf{24\ V}$$

EXAMPLE 7-2

For the series-parallel resistor circuit shown in Fig. 7-1, calculate:

a. The total resistance "seen" by the 72-V source.
b. The current delivered by the battery.

c. The voltage drops across the resistors.
d. The currents through the two parallel-connected resistors.

Solution

a. The total resistance of the whole circuit is given by

$$R_T = R_1 + R_2 \| R_3 + R_4$$
$$= 8\ \Omega + \frac{12\ \Omega \times 6\ \Omega}{12\ \Omega + 6\ \Omega} + 12\ \Omega$$
$$= 8\ \Omega + 4\ \Omega + 12\ \Omega = \mathbf{24\ \Omega}$$

See Fig. 7-1c.

b. The current delivered by the battery is

$$I_T = \frac{V}{R_T} \qquad (3\text{-}1a)$$
$$= \frac{72\ V}{24\ \Omega} = \mathbf{3\ A}$$

c. The voltage drop across any resistance is given by

$$V = IR \qquad (3\text{-}1b)$$

$$V_{R_1} = I_T R_1 = 3\ A \times 8\ \Omega = \mathbf{24\ V}$$
$$V_{R_{eq}} = I_T R_{eq} = 3\ A \times 4\ \Omega = \mathbf{12\ V}$$
$$V_{R_4} = I_T R_4 = 3\ A \times 12\ \Omega = \mathbf{36\ V}$$

Note that $V_{R_1} + V_{R_{eq}} + V_{R_4} = \mathbf{72\ V}$.

d. The branch currents are

$$I_2 = \frac{V_{R_{eq}}}{R_2} = \frac{12\ V}{12\ \Omega} = \mathbf{1\ A}$$

$$I_3 = \frac{V_{R_{eq}}}{R_3} = \frac{12\ V}{6\ \Omega} = \mathbf{2\ A}$$

and $\qquad\qquad I_2 + I_3 = \mathbf{3\ A}$

or $I_2 = I_T \times \dfrac{R_3}{R_2 + R_3}$ (6-12)

$= 3 \text{ A} \times \dfrac{6 \ \Omega}{12 \ \Omega + 6 \ \Omega}$

$= 3 \text{ A} \times \dfrac{6}{18} = 3 \text{ A} \times \dfrac{1}{3} = \textbf{1 A}$

$I_3 = I_T \times \dfrac{R_2}{R_2 + R_3}$

$= 3 \text{ A} \times \dfrac{12 \ \Omega}{12 \ \Omega + 6 \ \Omega}$

$= 3 \text{ A} \times \dfrac{12 \ \Omega}{18 \ \Omega} = 3 \text{ A} \times \dfrac{2}{3} = \textbf{2 A}$

EXAMPLE 7-3

Using the circuit shown in Fig. 7-3, calculate:
a. The total resistance connected across the 40-V source.
b. The current drawn from the source.
c. The total power delivered by the source.
d. The power dissipated by R_4.
e. The voltage drop across R_1.

Solution

a. Resistors R_2 and R_3 are in series:
$R_{eq \ 1} = R_2 + R_3$
$= 2 \text{ k}\Omega + 4 \text{ k}\Omega = 6 \text{ k}\Omega$

See Fig. 7-3b.
$R_{eq \ 1}$ is in parallel with R_4:

$R_{eq \ 2} = R_{eq1} \| R_4$

$= \dfrac{6 \text{ k}\Omega \times 3 \text{ k}\Omega}{6 \text{ k}\Omega + 3 \text{ k}\Omega}$

$= \dfrac{18}{9} \text{ k}\Omega = 2 \text{ k}\Omega$

See Fig. 7-3c.

$R_{eq \ 3} = R_{eq \ 2} + R_1$
$= 2 \text{ k}\Omega + 1 \text{ k}\Omega = 3 \text{ k}\Omega$

See Fig. 7-3d.

$R_T = R_{eq \ 3} \| R_5$

$= \dfrac{3 \text{ k}\Omega \times 5 \text{ k}\Omega}{3 \text{ k}\Omega + 5 \text{ k}\Omega}$

$= \dfrac{15}{8} \text{ k}\Omega = \textbf{1.875 k}\Omega$

See Fig. 7-3e.

b. $I_T = \dfrac{V}{R_T}$ (3-1a)

$= \dfrac{40 \text{ V}}{1.875 \text{ k}\Omega} = \textbf{21.3 mA}$

c. $P_T = V_T I_T$ (5-4)

$= 40 \text{ V} \times 21.3 \text{ mA}$

$= \textbf{852 mW}$

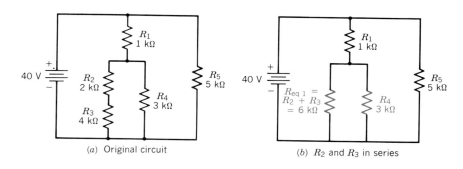

(a) Original circuit

(b) R_2 and R_3 in series

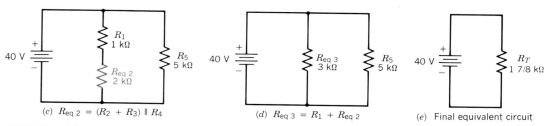

(c) $R_{eq \ 2} = (R_2 + R_3) \| R_4$

(d) $R_{eq \ 3} = R_1 + R_{eq \ 2}$

(e) Final equivalent circuit

FIGURE 7-3
Circuits for Example 7-3.

d. To find the power dissipated in R_4, you must find the current through R_4. First, find the current through $R_{eq\,3}$:

$$I = \frac{V}{R_{eq\,3}} \qquad (3\text{-}1a)$$

$$= \frac{40\text{ V}}{3\text{ k}\Omega} = 13.3\text{ mA}$$

Then

$$I_{R_4} = I \times \frac{R_{eq\,1}}{R_{eq\,1} + R_4} \qquad (6\text{-}12)$$

$$= 13.3\text{ mA} \times \frac{6\text{ k}\Omega}{6\text{ k}\Omega + 3\text{ k}\Omega}$$

$$= 13.3\text{ mA} \times \frac{6}{9} = 8.9\text{ mA}$$

$$P_{R_4} = I_{R_4}^2 \times R_4 \qquad (3\text{-}7)$$

$$= (8.9 \times 10^{-3}\text{ A})^2 \times 3 \times 10^3\ \Omega$$

$$= 237 \times 10^{-3}\text{ W} = \mathbf{237\ mW}$$

e. The current through R_1 is the same as through $R_{eq\,3}$, which was found to be 13.3 mA.

$$V_{R_1} = IR_1 \qquad (3\text{-}1b)$$

$$= 13.3\text{ mA} \times 1\text{ k}\Omega$$

$$= 13.3 \times 10^{-3}\text{ A} \times 1 \times 10^3\ \Omega = \mathbf{13.3\ V}$$

Example 7-4 shows a circuit with two parallel groups that are in series with a resistor and two opposing voltage sources.

EXAMPLE 7-4

Given the series-parallel resistor network in Fig. 7-4a, calculate:

a. The total circuit resistance.
b. The total supply current.
c. The current through R_1 and R_2.
d. The voltage drop across R_2.
e. The voltage drop across R_3.
f. The voltage drop across R_5.
g. The current through R_4, R_5, and R_6.
h. The voltage at B with respect to A.
i. The power delivered by V_2.
j. The total power dissipated by all the resistors.
k. The power recharging voltage source V_1.

Solution

a. For R_1 and R_2 in parallel in Fig. 7-4a,

$$R_{eq\,1} = R_1 \| R_2 = \frac{R_1 R_2}{R_1 + R_2} \qquad (6\text{-}10)$$

$$= \frac{1\text{ k}\Omega \times 2.2\text{ k}\Omega}{1\text{ k}\Omega + 2.2\text{ k}\Omega} = 688\ \Omega$$

For R_4, R_5 and R_6 in parallel in Fig. 7-4a,

$$\frac{1}{R_{eq\,2}} = \frac{1}{R_4} + \frac{1}{R_5} + \frac{1}{R_6} \qquad (6\text{-}4)$$

$$= \frac{1}{4.7\text{ k}\Omega} + \frac{1}{5.1\text{ k}\Omega} + \frac{1}{6.8\text{ k}\Omega}$$

$$= 0.213 \times 10^{-3}\text{ S} + 0.196 \times 10^{-3}\text{ S}$$

$$+ 0.147 \times 10^{-3}\text{ S}$$

$$= 0.556 \times 10^{-3}\text{ S}$$

Therefore, $R_{eq\,2} = \dfrac{1}{0.556 \times 10^{-3}\text{ S}} = 1799\ \Omega$

Using Figs. 7-4b and 7-4c, we obtain

$$R_T = R_{eq\,1} + R_3 + R_{eq\,2} \qquad (5\text{-}2)$$

$$= 688 + 3,300 + 1,799 = \mathbf{5787\ \Omega}$$

b. From Fig. 7-4c,

$$I_T = \frac{V}{R_T} \qquad (3\text{-}1a)$$

$$= \frac{V_2 - V_1}{R_T}$$

$$= \frac{30 - 10\text{ V}}{5.787\text{ k}\Omega} = \frac{20\text{ V}}{5.787\text{ k}\Omega} = \mathbf{3.46\ mA}$$

This value of I_T is the same as that shown in Figs. 7-4a and 7-4b.

c.
$$I_1 = I_T \times \frac{R_2}{R_1 + R_2} \qquad (6\text{-}12)$$

$$= 3.46\text{ mA} \times \frac{2.2\text{ k}\Omega}{1\text{ k}\Omega + 2.2\text{ k}\Omega} = \mathbf{2.38\ mA}$$

$$I_2 = I_T \times \frac{R_1}{R_1 + R_2} \qquad (6\text{-}13)$$

$$= 3.46\text{ mA} \times \frac{1\text{ k}\Omega}{1\text{ k}\Omega + 2.2\text{ k}\Omega} = \mathbf{1.08\ mA}$$

CHECK $I_T = I_1 + I_2 = 2.38 + 1.08 = \mathbf{3.46\ mA}$

d. From Fig. 7-4a,

$$V_{R_2} = I_2 R_2 \qquad (3\text{-}1b)$$

$$= 1.08\text{ mA} \times 2.2\text{ k}\Omega = \mathbf{2.38\ V}$$

CHECK
$$V_{R_1} = I_1 R_1$$

$$= 2.38\text{ mA} \times 1\text{ k}\Omega = \mathbf{2.38\ V}$$

e. From Fig. 7-4a,

$$V_{R_3} = I_T R_3 \qquad (3\text{-}1b)$$

$$= 3.46\text{ mA} \times 3.3\text{ k}\Omega = \mathbf{11.42\ V}$$

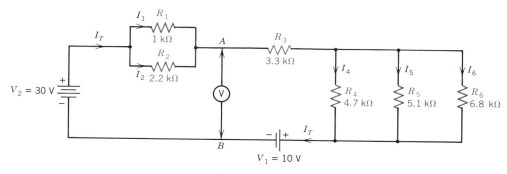

(a) Original circuit

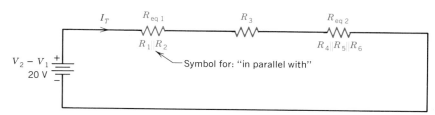

(b) Simplified equivalent series circuit

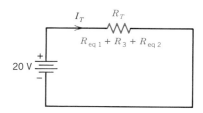

(c) Total equivalent resistance R_T

FIGURE 7-4

Circuits for Example 7-4.

f. From Fig. 7-4b,
$$V_{R_5} = I_T R_{eq\ 2} \qquad \text{(3-1b)}$$
$$= 3.46 \text{ mA} \times 1799\ \Omega = \textbf{6.2 V}$$

g. $I_4 = \dfrac{V_{R_4}}{R_4}$ (3-1a)

$$= \dfrac{6.2 \text{ V}}{4.7 \text{ k}\Omega} = \textbf{1.32 mA}$$

$$I_5 = \dfrac{V_{R_5}}{R_5} \qquad \text{(3-1a)}$$

$$= \dfrac{6.2 \text{ V}}{5.1 \text{ k}\Omega} = \textbf{1.22 mA}$$

$$I_6 = \dfrac{V_{R_6}}{R_6} \qquad \text{(3-1a)}$$

$$= \dfrac{6.2 \text{ V}}{6.8 \text{ k}\Omega} = \textbf{0.92 mA}$$

CHECK $I_T = I_4 + I_5 + I_6 = 1.32 + 1.22 + 0.92 = \textbf{3.46 mA}$

h. $V_{AB} = V_2 - V_{R_2}$

$$= +30 - 2.38 = \textbf{+27.62 V}$$

Thus point B is $\textbf{-27.62 V}$ with respect to point A.

i. Power delivered by V_2 is

$$P_2 = V_2 \times I_T \qquad \text{(5-4)}$$
$$= 30 \text{ V} \times 3.46 \text{ mA} = \textbf{104 mW}$$

j. Power dissipated by all the circuit

resistors $= I_T^2 R_T$ (5-4)

$$= (3.46 \times 10^{-3} \text{ A})^2 \times 5787\ \Omega$$
$$= \textbf{69.3 mW}$$

k. Power delivered to V_1 in recharging it

$$P_1 = V_1 \times I_T \qquad (5\text{-}4)$$
$$= 10\text{ V} \times 3.46 \times 10^{-3}\text{ A}$$
$$= \mathbf{34.6\text{ mW}}$$

CHECK The power dissipated in all the resistors

$$= P_T - P_1 = (104 - 34.6)\text{ mW}$$
$$\approx \mathbf{69.4\text{ mW}}$$

Note that there are various methods for checking the validity and accuracy of calculations, as shown in Example 7-4.

7-1.1   Series-Parallel Ladder Circuits

A method of drawing a series-parallel circuit widely used in industry is the "ladder" schematic diagram. This type of schematic shows the circuit component symbols on hor-izontal lines between two vertical lines that are connected to the source. (See Fig. 7-5.) The reduction of such a circuit to a single resistance is done in the same way shown earlier. Example 7-5 illustrates the method.

EXAMPLE 7-5

For the ladder circuit shown in Fig. 7-5a, find the total resistance and the total current supplied by the source.

Solution

Refer to Fig. 7-5b:

$$R_{\text{eq }1} = R_2 \| R_3 = \frac{3\text{ k}\Omega}{2} = 1.5\text{ k}\Omega$$

$$R_{\text{eq }2} = R_5 \| R_6 \| R_7 = \frac{6\text{ k}\Omega}{3} = 2\text{ k}\Omega$$

Refer to Fig. 7-5c:
$$R_{\text{eq }3} = R_1 + R_{\text{eq }1} + R_4$$

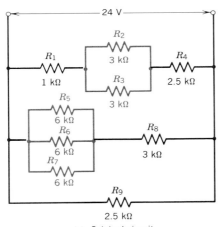

(a) Original circuit

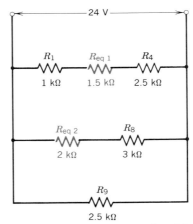

(b) $R_{\text{eq }1} = R_2 \| R_3$; $R_{\text{eq }2} = R_5 \| R_6 \| R_7$

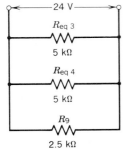

(c) Reduction to parallel circuit

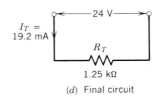

(d) Final circuit

FIGURE 7-5

Example of a series-parallel ladder circuit.

$$= 1 \text{ k}\Omega + 1.5 \text{ k}\Omega + 2.5 \text{ k}\Omega = 5 \text{ k}\Omega$$
$$R_{eq\ 4} = R_{eq\ 2} + R_8$$
$$= 2 \text{ k}\Omega + 3 \text{ k}\Omega = 5 \text{ k}\Omega$$

Refer to Fig. 7-5*d*:

$$R_T = R_{eq\ 3} \| R_{eq\ 4} \| R_9$$
$$= 5 \text{ k}\Omega \| 5 \text{ k}\Omega \| 2.5 \text{ k}\Omega$$
$$= \frac{5 \text{ k}\Omega}{2} \| 2.5 \text{ k}\Omega$$
$$= 2.5 \text{ k}\Omega \| 2.5 \text{ k}\Omega$$
$$= \frac{2.5 \text{ k}\Omega}{2} = \mathbf{1.25 \text{ k}\Omega}$$

$$I_T = \frac{V}{R_T}$$
$$= \frac{24 \text{ V}}{1.25 \text{ k}\Omega} = \mathbf{19.2 \text{ mA}}$$

Ladder diagrams are used extensively for drawing *control* circuits. In these circuits, a low voltage (commonly 24 V) is used to energize relays. A *relay* is a magnetically operated switch that controls the flow of current from a higher voltage (usually 120 V or 240 V) source to a motor or other type of load. Relays are discussed in Chapter 11. Figure 7-6 shows a simple example of a ladder diagram involving both high-voltage loads and low-voltage control.

A transformer "steps down" the 120-V supply to 24 V for the control circuit. When switch S is closed and the ON button is pressed, a path is completed for current to flow through the relay coil *A* (the OFF button is in a "normally closed" condition). The flow of current through the relay coil turns it into an electromagnet ("energizes" it), causing it to close two sets of switch contacts. These contacts are shown by the symbol ⊣ ⊢.

One set of these switch contacts is connected in parallel with the ON button, providing a path for current to continue to flow through coil *A* after the spring-loaded ON button is released. This is called a *holding* or *maintained contact circuit*.

The second set of contacts completes a circuit to a high-voltage load (in this case, the lathe motor). The motor will continue to run until the OFF switch is momentarily depressed, opening the circuit and de-energizing coil *A* (thus opening the switch contacts controlled by that coil).

The *overload* switch (marked O/L on the diagram) is designed to open the circuit if the motor overloads and begins to overheat. The heat generated by excessive motor current will open the overload switch contacts, de-energizing coil *A* and stopping the motor.

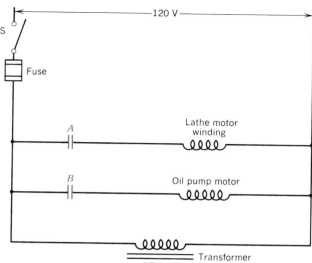
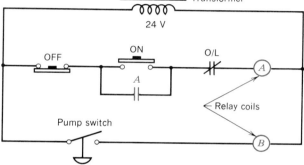

FIGURE 7-6
Example of a series-parallel ladder schematic.

In this circuit, relay coil *B* controls the lathe's oil pump motor. The pump switch shown on the diagram closes whenever the level of the cutting oil in a reservoir becomes too low. The closing of the switch allows current to flow to relay coil *B*, energizing it and completing the circuit to the oil pump motor. The motor will operate until the oil level in the reservoir rises enough to open the switch.

In this circuit, the only loads are the motors and relay coils. The current paths provided by the opening and closing of switches make it an example of a series-parallel circuit.

7-1.2 Practical Series-Parallel Circuit

In practice, most circuits are of the series-parallel type. Every voltage source, for example, has some *internal* resistance, as shown in Fig. 7-7.

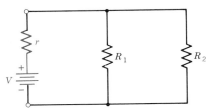

FIGURE 7-7
A series-parallel circuit due to the internal resistance of the voltage source.

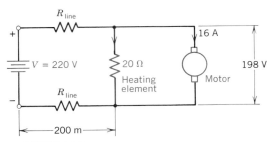

FIGURE 7-8
Circuit for Example 7-6.

This means that, although a number of loads may be connected in parallel across the source, the internal resistance is in *series* with those parallel-connected loads. (The internal resistance is *inside* the source, and cannot be separated from the emf of the source.)

A *truly* parallel circuit is possible only when the internal resistance of the source is negligible. In the same way, the circuit can be truly parallel only if the leads from the source to load have negligible resistance. If the source and the load are some distance apart, the resistance of the feeder line must be taken into account, as shown in Example 7-6.

EXAMPLE 7-6

An electric motor and an electric heater are to be operated in parallel with each other at a distance of 200 m from a 220-V source of emf. The current drawn by the motor is 16 A and the resistance of the heater is 20 Ω. If the total voltage drop in the line must not exceed 10% of the source voltage, calculate:

a. The minimum wire gauge to be used for the copper feeder line.
b. The power dissipated in the feeder line.
c. The power delivered to the load.
d. The power supplied by the source.

Assume that the source has negligible internal resistance. See Fig. 7-8.

Solution

a. Total permissible voltage drop

$$= 10\% \text{ of } 220 \text{ V} = 22 \text{ V}$$

Load voltage $= 220 \text{ V} - 22 \text{ V} = 198 \text{ V}$

Heater current $= \dfrac{V}{R}$ (3-1a)

$$= \frac{198 \text{ V}}{20 \text{ Ω}} = 9.9 \text{ A}$$

Total line current $= 9.9 \text{ A} + 16 \text{ A} = 25.9 \text{ A}$
Maximum resistance of total

feeder line $= \dfrac{V}{I_T}$ (3-1)

$$= \frac{22 \text{ V}}{25.9 \text{ A}} \approx 0.85 \text{ Ω}$$

Maximum resistance of each line,

$$R_{line} = \frac{0.85 \text{ Ω}}{2} = 0.425 \text{ Ω}$$

$$R_{line} = \frac{\rho l}{A} \qquad (4\text{-}1)$$

Therefore, $A = \dfrac{\rho l}{R_{line}}$

$$= \frac{1.72 \times 10^{-8} \text{ Ω} \cdot \text{m} \times 200 \text{ m}}{0.425 \text{ Ω}}$$

$$= 8.09 \times 10^{-6} \text{ m}^2$$

$$A = \frac{\pi d^2}{4}$$

so that $d = \sqrt{\dfrac{4A}{\pi}}$

$$= \sqrt{\frac{4 \times 8.09 \times 10^{-6} \text{ m}^2}{\pi}}$$

$$= 3.21 \times 10^{-3} \text{ m}$$

Thus the *minimum* diameter of the wire is **3.21 mm.** Referring to Appendix C, we find that the minimum AWG size of wire is **gauge number 8** (3.26 mm).

NOTE If we had selected the wire according to its current-carrying capacity, a number 10 gauge wire (30 A without overheating) would have been too small and would have caused too high a voltage drop.

Also, after determining that the maximum resistance should be 0.425 Ω for 200 m, we know that the resistance is 0.425 Ω × 5 or 2.125 Ω/km, maximum. From the wire gauge table, Appendix C, we see that number 8 gauge has a resistance of 2.06 Ω/km, and may be used.

b. Since the diameter of the number 8 gauge wire is practically equal to the calculated value, we can use the voltages and currents calculated in part (a). Power dissipated in the line

$$= I_T^2 \times 2R_{line} \qquad (3\text{-}7)$$
$$= (25.9\ A)^2 \times 2 \times 0.425\ \Omega$$
$$= \textbf{570 W}$$

c. Power delivered to the load

$$= V \times I_T \qquad (3\text{-}5)$$
$$= 198\ V \times 25.9\ A$$
$$= \textbf{5128 W}$$

d. Power supplied by source $= 570\ W + 5128\ W$
$$= \textbf{5698 W}$$

or power $= V \times I_T \qquad (3\text{-}5)$
$$= 220\ V \times 25.9\ A = \textbf{5698 W}$$

Note how 10% of the power supplied by the source is "lost" in the line. This is because we have permitted a 10% voltage drop to occur due to the resistance of the line.

7-1.3 Thermistor-Resistor Series-Parallel Circuit

An interesting series-parallel circuit is a string of lights in which each series-wired lamp has a thermistor *in parallel* with it. (See Fig. 7-9.)

When a thermistor is cold, its resistance is high and very little current flows through it. This means that the string of lights is an essentially *series* circuit, if all the lamps are operating properly. But if a lamp burns out, the current must pass through the thermistor wired in parallel with it. The increased current flow will raise the temperature of the thermistor, causing its resistance to decrease (ideally, to a value close to that of a hot lamp filament). This allows the other lamps in the circuit to continue operating at normal voltage and current. With this type of circuit, it is easy to find and replace the burned-out lamp. In a true *series* circuit, of course, the failure of one lamp causes all the others to stop operating. This makes locating the failed lamp difficult.

7-2 VOLTAGE DIVIDERS

A common problem in electronic circuit design is to provide power to a load at a voltage different from the source voltage. In some cases, a number of loads—each with different voltage and current requirements—must operate from a common supply. The following examples illustrate the use of principles you have learned in the preceding chapters to design three typical voltage-divider networks.

7-2.1 Series-Dropping Resistor

EXAMPLE 7-7

A 9-V transistor radio is to be operated from a 12-V automobile battery. If the average load current of the radio is 100 mA, calculate the value and the power rating of

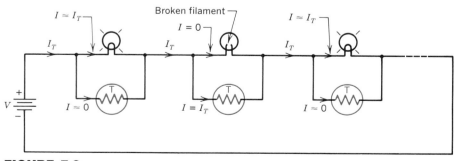

FIGURE 7-9
Series-parallel circuit with a thermistor across each lamp.

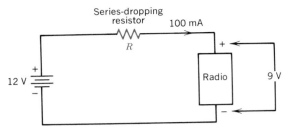

FIGURE 7-10
Circuit for Example 7-7.

the required series-dropping resistor (using the nearest commercial values).

Solution

Draw the circuit shown in Fig. 7-10, entering all the known values. The voltage dropped across R must be $V_R = 12\text{ V} - 9\text{ V} = 3\text{ V}$. Using Ohm's law, you obtain

$$R = \frac{V_R}{I} \tag{3-1}$$

$$= \frac{3\text{ V}}{0.1\text{ A}} = 30\ \Omega \text{ (use a \textbf{33-}}\Omega \text{ resistor)}$$

$$P_R = V_R I$$

$$= 3\text{ V} \times 0.1\text{ A} = 0.3\text{ W} \tag{3-5}$$

(use a power rating of ½ **W**)

The nearest standard value resistor that would be suitable would be 33 Ω, ±10% with a power rating of ½ **W**.

The disadvantage of the circuit in Example 7-7 is that it allows load voltage to vary with load current. For example, if the radio's volume is increased, the load current will increase—possibly to as much as 200 mA. As a result, the voltage drop across the series-dropping resistor would increase to 6 V (with a 200-mA current), leaving only 6 V for the load.

This reduced voltage could cause distortion in the audio section of the radio. If the current is reduced, on the other hand, the load voltage will increase toward a maximum of 12 V at no load.

7-2.2 Bleeder Resistor

It is possible to overcome this problem to some degree by adding a *bleeder resistor* to the circuit. This resistor is placed in parallel with the load, and should draw about

10% to 25% of the total current from the source. (See Example 7-8.)

EXAMPLE 7-8

Design a simple voltage divider with a series-dropping resistor and a bleeder resistor to provide 9 V at a load current of 100 mA from a 12-V supply. Use a bleeder current of (a) 30 mA, (b) 100 mA, determining the no-load voltage in each case. Use commercial values for all resistors. See Fig. 7-11.

Solution

a. $I_S = I_B + I_L$ (6-1)

$= 30\text{ mA} + 100\text{ mA} = 130\text{ mA}$

$V_{R_S} = 12\text{ V} - 9\text{ V} = 3\text{ V}$

Therefore, $R_S = \dfrac{V_{R_S}}{I_S}$ (3-1)

$= \dfrac{3\text{ V}}{130\text{ mA}} = \textbf{23 }\Omega$

$P_{R_S} = V_{R_S} I_S$ (3-5)

$= 3\text{ V} \times 0.13\text{ A} = \textbf{0.39 W}$

Use a 22-Ω, 5%, 1-W resistor, for R_s.

$V_{R_B} = 9\text{ V}, I_B = 30\text{ mA}$

Therefore, $R_B = \dfrac{V_{R_B}}{I_B}$ (3-1)

$= \dfrac{9\text{ V}}{30\text{ mA}} = \textbf{300 }\Omega$

$P_B = V_{R_B} I_B$ (3-5)

$= 9\text{ V} \times 0.03\text{ A} = \textbf{0.27 W}$

Use a 300-Ω, 5%, ½-W resistor for R_B. With the load disconnected

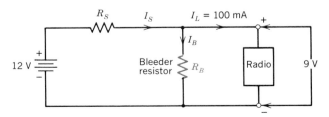

FIGURE 7-11
Circuit for Example 7-8.

$$V_{R_B} = V \times \frac{R_B}{R_B + R_S} \qquad (5\text{-}5)$$

$$= 12 \text{ V} \times \frac{300 \ \Omega}{300 \ \Omega + 22 \ \Omega} = \mathbf{11.2 \ V}$$

b. Using $I_B = 100$ mA, the required values are $R_S = $ **15 Ω, 5%, 0.6 W** (1 W) and $R_B = $ **90 Ω, 0.9 W** (91 Ω, 5%, 2 W).

$$\text{unloaded } V_{R_B} = V \times \frac{R_B}{R_B + R_S} \qquad (5\text{-}5)$$

$$= 12 \text{ V} \times \frac{91 \ \Omega}{91 \ \Omega + 15 \ \Omega} = \mathbf{10.3 \ V}$$

As shown in Example 7-8, a larger bleeder current provides an unloaded output voltage of 10.3 V, which is closer to the loaded value of 9 V. This provides better *voltage regulation*, but draws more power from the source and wastes it in the form of heat in the divider resistors.

A better form of regulation is obtained by connecting a *Zener diode* in place of R_B. Zener diodes, which are often used in situations where wide variations in load current might occur, are covered briefly in Section 28-7.1. Where the load current is relatively constant, however, the bleeder resistor circuit shown in Fig. 7-11 is adequate.

7-2.3 Positive and Negative Potentials

It is possible to design a voltage divider that provides both positive and negative output potentials. Such a divider is shown in Example 7-9.

EXAMPLE 7-9

Determine the resistance and power values of R_1, R_2, and R_3 in Fig. 7-12 to provide $+25$ V at 10 mA for load A, $+15$ V at 50 mA for load B, and -15 V at 20 mA for load C from a 40-V source of emf with a total current of 100 mA.

Solution

To find the resistances of R_1, R_2, and R_3, we need the voltage drop across each and the current through each.

At junction W, $\qquad I_T = I_A + I_{R_1} \qquad (6\text{-}1)$

Therefore, $I_{R_1} = I_T - I_A$
$\qquad\qquad = 100 \text{ mA} - 10 \text{ mA} = 90 \text{ mA}$

At junction X, $\qquad I_{R_1} = I_B + I_{R_2} \qquad (6\text{-}1)$

Therefore, $I_{R_2} = I_{R_1} - I_B$
$\qquad\qquad = 90 \text{ mA} - 50 \text{ mA} = 40 \text{ mA}$

At junction Y, $\qquad I_T = I_C + I_{R_3} \qquad (6\text{-}1)$

Therefore, $I_{R_3} = I_T - I_C$
$\qquad\qquad = 100 \text{ mA} - 20 \text{ mA} = 80 \text{ mA}$

NOTE At junction Z, 40 mA must be flowing "in" from the ground connection. This is because 60 mA total is "returning" from loads A and B, but 20 mA is flowing "out" to load C. The ground is positive, as far as the top side of load C is concerned, with respect to Y.

Clearly, from Fig. 7-12,

$$V_{R_3} = 15 \text{ V and } V_{R_2} = 15 \text{ V}$$

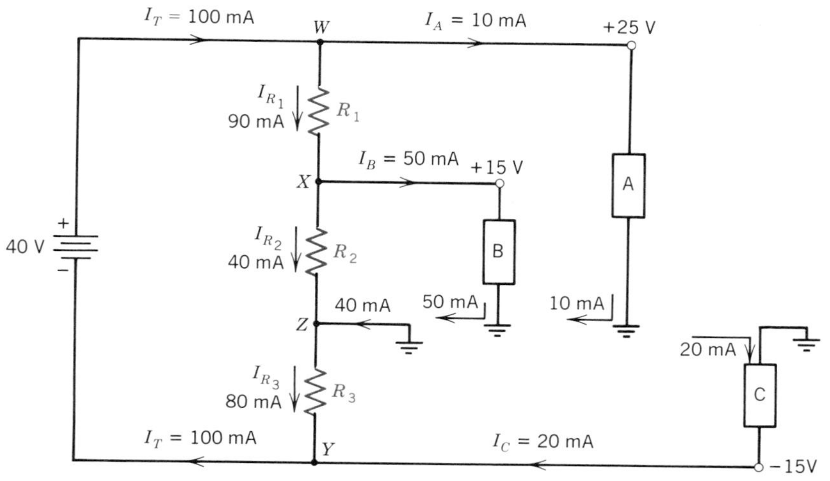

FIGURE 7-12
Voltage-divider circuit for Example 7-9.

V_{R_1} = load A voltage − load B voltage

= 25 V − 15 V = 10 V

$$R_1 = \frac{V_{R_1}}{I_{R_1}} \tag{3-1}$$

$$= \frac{10\ \text{V}}{90\ \text{mA}} = \mathbf{111\ \Omega}$$

$$P_1 = V_{R_1}I_{R_1} \tag{3-5}$$

= 10 V × 0.09 A = **0.9 W**

(110 Ω, 5% 2 W)

$$R_2 = \frac{V_{R_2}}{I_{R_2}} \tag{3-1}$$

$$= \frac{15\ \text{V}}{40\ \text{mA}} = \mathbf{375\ \Omega}$$

$$P_2 = V_{R_2}I_{R_2} \tag{3-5}$$

= 15 V × 0.04 A = **0.6 W**

(390 Ω, 5%, 1 W)

$$R_3 = \frac{V_{R_3}}{I_{R_3}} \tag{3-1}$$

$$= \frac{15\ \text{V}}{80\ \text{mA}} = \mathbf{187.5\ \Omega}$$

$$P_3 = V_{R_3}I_{R_3} \tag{3-5}$$

= 15 V × 0.08 A = **1.2 W**

(180 Ω, 5%, 2 W)

NOTE When selecting resistors, it is good practice to select power ratings approximately double their calculated values. Thus, in Example 7-9, R_1 should be selected as a 2-W resistor, allowing it to safely handle the expected 0.9-W dissipation. Choosing a 2-W rating, rather than 1 W, will increase the reliability of the resistor, since it will not be working near its 1-W limit.

SUMMARY

1. A series-parallel circuit includes both series- and parallel-connected components, with series and parallel rules applying to their respective portions of the circuit.
2. Most practical circuits are actually series-parallel, due to internal resistance in the source or resistance in the connecting leads.
3. A string of series-connected lights, with thermistors connected in parallel with each lamp, will continue to operate if one of the lamps fails. The thermistor provides a low-resistance path for current to flow.
4. Voltage-divider circuits allow a load to operate at a voltage different from the immediately available supply voltage.
5. A single resistor can be connected in series between the source and the load to drop voltage to a suitable value, depending upon the load current.
6. If load current varies widely, a series-dropping resistor voltage divider is not the best choice for good voltage regulation. It allows large changes in load voltage as load current varies.
7. A bleeder resistor can be connected in parallel with the load to improve the voltage regulation of a circuit that employs a series-dropping resistor voltage divider.
8. The current drawn by a bleeder resistor should be from 10% to 25% of the total current from the source.
9. The design of a loaded voltage divider network to provide both positive and negative potentials involves the use of Kirchhoff's current law to determine the current through each of the required resistors.

SELF-EXAMINATION

Answer T or F or, in the case of multiple choice, a, b, c, or d
(Answers at back of book)

7-1. In Fig. 7-1a, resistors R_1 and R_4 are in series with each other. _____

7-2. In Fig. 7-1a, resistors R_1, R_2, and R_4 are in series with each other. _____

7-3. In any series-parallel circuit, the largest current always flows in the smallest resistance. _____

7-4. The total resistance of a series-parallel circuit is always less than the smallest parallel resistor. _____

7-5. In Fig. 7-4a, the following resistors are in series:
a. R_1 and R_3 c. Both of the above.
b. R_2 and R_3 d. None of the above.

7-6. Refer to Fig. 7-4a.
a. Resistors R_1, R_3, and R_4 are in series.
c. Resistors R_4, R_5, and R_6 are in parallel.
b. Resistors R_2, R_3, and R_4 are in series.
d. All the above.

7-7. Refer to Fig. 7-4. Assume that R_1 and R_2 are each twice as large as R_3 and assume that R_4, R_5, and R_6 are each three times as large as R_3. The total combined circuit resistance is equal to
a. $6 R_3$ c. R_3
b. $3 R_3$ d. $R_3/3$

7-8. If, in Fig. 7-9, the thermistors are ideally matched with the lamps, there will be no change in the total current drawn from the source regardless of how many lamps burn out. _____

7-9. All complex circuits can be thought of as some combination of series-parallel circuits. _____

7-10. If a load requires one-half of the available supply voltage, a series-dropping resistor must have the same resistance as the load. _____

7-11. A disadvantage of a voltage divider that uses a single-series dropping resistor is the wide change in load voltage with the change in load current. _____

7-12. In a single-series dropping resistor voltage divider, an increase in load causes an increase in load voltage. _____

7-13. A bleeder resistor voltage divider improves voltage regulation because it causes some current to always flow through the series-dropping resistor, even when the load is removed. _____

7-14. The larger the bleeder current, the better is the voltage regulation. _____

7-15. To obtain positive and negative potentials from a single supply in a voltage-divider network requires the selection of 0 V or ground at a point different from the negative or positive side of the single supply. _____

7-16. It is not possible to obtain two different positive potentials and two different negative potentials from a single-supply voltage-divider network. _____

REVIEW QUESTIONS

1. Explain why the circuit in Fig. 7-7 is really an example of a series-parallel circuit.
2. Can you devise any rule concerning the total resistance of a series-parallel circuit?
3. In Fig. 7-4 show all the points at which Kirchhoff's current law can be applied.
4. Apply Kirchhoff's voltage law to the circuit in Fig. 7-4.
5. Draw a circuit that contains two sources of emf and three resistors in which the resistors do not form a series, a parallel, or a series-parallel combination.
6. When is the size of feeder line determined by other factors than the necessary current-carrying capacity?

7. a. Identify which components are parallel-connected and series-connected in Fig. 7-12.
   b. Write the equation for the total resistance between W and Y in Fig. 7-12, in terms of the voltage-divider resistors and load resistances.

8. a. What is a disadvantage of the single-series dropping resistor voltage divider?
   b. How is this disadvantage overcome?

9. What can be done to improve the voltage regulation of a voltage divider?

10. What is the power rating of a single-series dropping resistor compared with the load if the load operates at half the supply voltage?

11. Explain why a larger bleeder current improves the voltage regulation of a voltage divider when the load varies.

12. Under what conditions might you use a power rating for the voltage-divider resistors double the nearest standard sizes to the calculated values?

13. Would it be possible to obtain $+25$ V and -15 V supplies from a voltage-divider circuit with a single emf of 45 V? Explain.

PROBLEMS

7-1. Refer to Fig. 7-1. Given $R_1 = 1$ kΩ, $R_2 = 2.2$ kΩ, $R_3 = 3.3$ kΩ, $R_4 = 4.7$ kΩ, and $V = 40$ V, calculate:
   a. The total circuit resistance.   c. The current through R_3.
   b. The total source current.   d. The voltage drop across R_4.

7-2. For the values given in Problem 7-1 refer to Fig. 7-1 and calculate:
   a. The voltage drop across R_2.
   b. The total power dissipated in all the resistors.
   c. The voltage at C with respect to A.

7-3. Refer to Fig. 7-4. Let $R_1 = 5.6$ kΩ, $R_2 = 3.3$ kΩ, $R_3 = 1$ kΩ, $R_4 = 10$ kΩ, $R_5 = 6.8$ kΩ, $R_6 = 12$ kΩ, and $V_2 = 9$V, $V_1 = 22.5$ V.
   Calculate:
   a. The total supply current.
   b. The current through R_5.
   c. The power dissipated in R_3.

7-4. For the values given in Problem 7-3 refer to Fig. 7-4 and calculate:
   a. The current through R_1.
   b. The power dissipated in R_6.
   c. The voltage at A with respect to B.

7-5. An electric motor (effective resistance 12 Ω) and a combined lighting load of resistance 18 Ω are to be operated in parallel with each other at a distance of 300 m from a 230-V source of emf. If the total voltage drop in the line must not exceed 8% of the source voltage, calculate:
   a. The minimum wire gauge to be used for the copper feeder line.
   b. The power dissipated in the feeder line.
   c. The power supplied by the source.
   Assume that the source has negligible internal resistance.

7-6. Calculate the actual voltage at which the load in Problem 7-5 operates if the selected wire gauge is used.

7-7. Refer to Fig. 7-13. What is the resistance between terminals A and B under the following conditions?

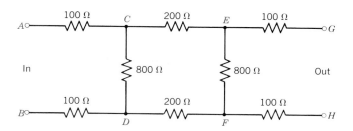

FIGURE 7-13
Circuit for Problems 7-7, 7-8, and 7-9.

a. The output terminals have no load (open-circuit).
b. The output terminals are short-circuited.
c. A load of 600-Ω is connected at the output.

7-8. Refer to Fig. 7-13. Assume that a 600-Ω load is connected to the output and a 6-V source at the input. Calculate:
a. The voltage between C and D.
b. The voltage between E and F.
c. The current in the 600-Ω load.

7-9. Refer to Fig. 7-13. Assume that a 600-Ω load is connected to the output and a 10-V source at the input. Calculate:
a. The current and power in the load.
b. Which resistor has the largest power dissipated in it.

7-10. Refer to Fig. 7-14.
a. What voltage exists across the open switch?
b. Find the current through the switch when it is closed.

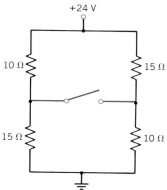

FIGURE 7-14
Circuit for Problem 7-10.

7-11. Using the circuit shown in Fig. 7-15, find the total resistance between A and B.

7-12. For the circuit shown in Fig. 7-15, determine the power dissipated in the 5-Ω resistor when a 120-V source is connected between A and B.

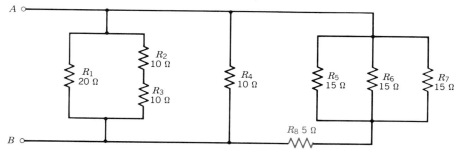

FIGURE 7-15
Circuit for Problems 7-11 and 7-12.

7-13. Refer to Fig. 7-16. Calculate, with the switch *open:*
  a. The current flowing through R_6.
  b. The power dissipated in R_8.
  c. The power supplied by V_1.

7-14. Repeat Problem 7-13 with the switch *closed.*

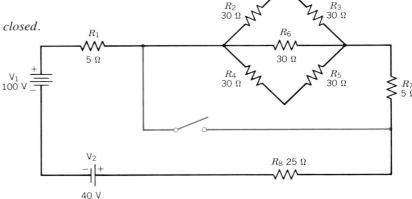

FIGURE 7-16
Circuit for Problems 7-13 and 7-14.

7-15. Refer to Fig. 7-17 and calculate:
  a. The total resistance "seen" by the 100-V source.
  b. The power supplied by the 100-V source.
  c. The potential difference across R_3.
  d. The current through R_4.

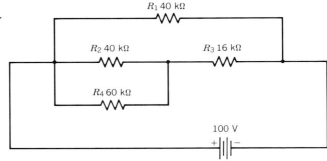

FIGURE 7-17
Circuit for Problem 7-15.

7-16. Using the circuit shown in Fig. 7-18, calculate the total resistance between *A* and *B:*
  a. With the switch open.
  b. With the switch closed.

7-17. Repeat Problem 7-16, but change the value of all the resistors in Fig. 7-18 to 10 Ω.

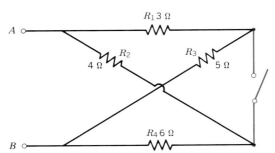

FIGURE 7-18
Circuit for Problems 7-16 and 7-17.

7-18. Find the total resistance of the circuit in Fig. 7-19 and the total current supplied by the source.

7-19. Refer to Fig. 7-19 and calculate:
   a. The current through R_{13}
   b. The voltage across R_6.
   c. The current through R_2.

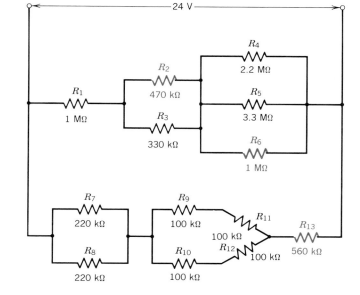

FIGURE 7-19
Circuit for Problems 7-18 and 7-19.

7-20. For the circuit shown in Fig. 7-20, calculate:
   a. The total resistance "seen" by the source.
   b. The total current supplied by the source.
   c. The voltage across R_3.
   (*Hint:* Redraw the circuit in simpler form.)

FIGURE 7-20
Circuit for Problem 7-20.

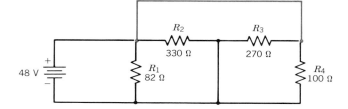

7-21. It is required to operate a 6-V transistor radio from a 12-V automobile battery. If the average load current of the transistor radio is 200 mA calculate:
   a. The series dropping resistor and its power rating. (Use commercial values.)
   b. The voltage at the radio when the automobile battery is being recharged with a terminal voltage of 14.1 V. (Assume that the load current increases to 250 mA.)
   c. The voltage at the terminals of the radio when it is turned off and the battery voltage is 13.2 V.

7-22. A series-dropping resistor had been inserted in the line from a 12-V battery to supply a 9-V tape deck operating at 200 mA. It is required now to replace the tape deck with a 6-V radio operating at 100 mA.
   a. What resistance must be connected in parallel with the operating radio to provide 6 V? (Use the nearest commercial value.)
   b. What is the power rating of this resistance?
   c. How much power is the series-dropping resistor now dissipating? How does this compare with its initial dissipation?
   d. What voltage appears at the terminals of the radio when the radio is turned off?

7-23. Design a simple voltage divider with a series-dropping resistor and a bleeder resistor to provide 1.5 kV at a load current of 5 mA from a 2-kV supply. Use a bleeder current of 1 mA and determine the no-load voltage. (Use the nearest commercial value.)

7-24. Repeat Problem 7-23 using a bleeder current of 2 mA.

7-25. a. What nearest commercial bleeder resistor is required to complete a voltage divider to provide 250 V at 40 mA from a 400-V source of emf if the series dropping resistor is 2.2 kΩ?
   b. What are the power ratings necessary for these two resistors?
   c. What is the no-load voltage?

7-26. Repeat Problem 7-25 using a 3.3-kΩ series-dropping resistor.

7-27. A 12-V power supply is fused at 750 mA.
   a. What should be the minimum (commercial value) bleeder resistor to provide a 500-mA load with 9 V?
   b. What must be the value of the series-dropping resistor?
   c. How much power is dissipated in the series and bleeder resistors?
   d. If 5-W power ratings are chosen for the resistors to what maximum voltage can the source increase before either of the resistors overheats? (Assume that the fuse is replaced by a higher value.)
   e. What will be the load voltage for the maximum input voltage? (Assume that the load resistance is constant.)

7-28. Repeat Problem 7-27 using an initial power supply of 13 V fused at 1 A.

7-29. Design a voltage divider to provide 90 V at 15 mA and 250 V at 20 mA from a 350-V dc power supply with a bleeder current of 5 mA in parallel with the 90-V load. (Use the nearest commercial values and power ratings.)

7-30. Repeat Problem 7-29 but use a total current from the 350 V supply of 50 mA instead of the 5-mA bleeder current.

7-31. Design a voltage divider to provide ±20 V at 20 mA each and ±12 V at 50 mA each from a 40-V supply with a total current of 100 mA. (Use the nearest commercial values and power ratings.)

7-32. Repeat Problem 7-31 using a 45-V supply.

7-33. Given the circuit in Fig. 7-21, find:
   V, V_1, V_{AD}, V_{BF}, V_{CD}, and V_{FD}.

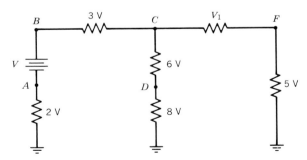

FIGURE 7-21
Circuit for Problem 7-33.

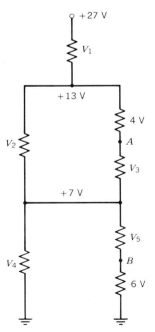

FIGURE 7-22
Circuit for Problem 7-34.

7-34. Given the circuit in Fig. 7-22, find:
   V_1, V_2, V_3, V_A, V_4, V_5, V_B, and V_{AB}.

7-35. Given the circuit shown in Fig. 7-23, find: I_{R_1}, I_{R_2}, and V_N.

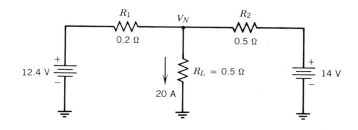

FIGURE 7-23
Circuit for Problem 7-35.

CHAPTER 8

VOLTAGE SOURCES

Because of the large amount of portable equipment built with solid-state circuitry, batteries have many important everyday applications, ranging from toys to calculators to radios to cardiac pacemakers. All these devices, and many more, rely on a power source that converts chemical energy into electrical energy. This process involves the separation of charges between two terminals. There are two basic types of cells (or batteries), classified according to the type of chemical activity: primary cells and secondary cells. **In a *primary* cell, the chemical materials are used up as electric energy is produced. It is discarded when its active material is depleted and it no longer produces electrical energy. In a *secondary* cell, a *reversible* chemical reaction is used to separate charges so that the cell can be returned to its original chemical state (recharged) many times.**

The most common primary cell is the *carbon–zinc* cell or common "flashlight battery." Carbon–zinc cells are low in cost and are best suited to intermittent use. When more continuous and heavier loads are involved, *alkaline* batteries will give longer and more satisfactory service than carbon–zinc cells. For small, solid-state devices such as hearing aids or watches, higher energy-density "button-type" primary cells are used. These cells are usually referred to as *mercury–oxide* or *silver–oxide* batteries.

A *secondary* cell is usually used for situations requiring larger amounts of energy at high current. The most common example is the automotive *lead–acid* battery, which can supply several hundred amperes at 2 V per cell. Strictly speaking, a *battery* consists of several interconnected *cells,* each of which has an emf from 1.2 V to 2.2 V (depending on type). However, the term *battery* is used interchangeably with *cell* for the common AA, C, and D sizes.

8-1 SIMPLE COPPER–ZINC CELL

Although it is not available in commercial form, the copper–zinc cell can be used to clearly illustrate the chemical reactions that produce electricity. In the form called a *voltaic cell,* it demonstrates how charges are separated by the motion of ions.

As shown in Fig. 8-1, a voltaic cell consists of a strip of copper (Cu) and a strip of zinc (Zn) immersed in an *electrolyte* of dilute sulfuric acid (H_2SO_4). When the cell functions, there is a great deal of activity around the zinc strip (called an *electrode*) as it is "eaten away" by the

acid. The copper electrode is not "eaten away," but has many small bubbles of a gas (hydrogen) clinging to its surface.

If a voltmeter is connected between the two electrodes, it will show a reading of approximately 1 V, with the copper strip positive and the zinc strip negative. If a low-resistance load is connected between the electrodes, a current will flow until the zinc electrode is completely eaten away. The chemical process is described in detail below.

Each molecule of sulfuric acid (the electrolyte) *dissociates* or separates into two positive hydrogen ions ($2H^+$) and one negative sulfate ion (SO_4^{2-}).

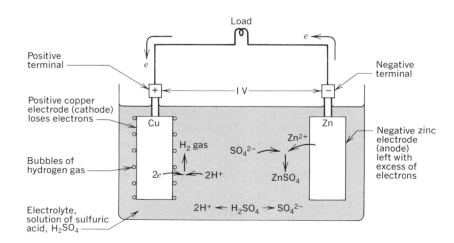

FIGURE 8-1
Simple copper–zinc cell.

$$H_2SO_4 \rightarrow 2H^+ + SO_4^{2-} \qquad (8\text{-}1)$$

Equation 8-1 simply means that a molecule of sulfuric acid is initially neutral, with its components held in an ionic bond. When the two hydrogen ions break away, they leave their valence electrons with the sulfate molecule. This gives the sulfate ion a negative charge of two electrons.

Since atomic zinc (Zn^0) is more chemically active than atomic copper (Cu^0), it dissolves more readily in the acid. Some of the zinc atoms (Zn^0) form Zn^{2+} ions and go into solution, reacting with the sulfate ions to form zinc sulfate.

$$Zn^{2+} + SO_4^{2-} \rightarrow ZnSO_4 \qquad (8\text{-}2)$$

The zinc is said to have *replaced* the hydrogen when it combines with the sulfate ion to form zinc sulfate. The zinc ion has the equivalent charge of the two positively charged hydrogen ions it replaces. But the *doubly charged* positive ion of zinc is an atom that is *deficient* in two valence electrons. Thus an ion of zinc departing from the electrode leaves behind an *excess* of two electrons. The zinc electrode, initially neutral, becomes *negatively* charged due to the excess of electrons.

The hydrogen (H^+) ions, meanwhile, each lack one valence electron. They pick up an electron from the readily available supply of free electrons in the copper electrode, forming neutral atomic hydrogen (H^0).

$$2H^+ + 2e \rightarrow H_2 \uparrow \qquad (8\text{-}3)$$

In this reaction, each *pair* of hydrogen atoms becomes *one* molecule of hydrogen gas. The gas forms a visible coating of bubbles on the copper electrode and also rises to the surface of the electrolyte. The copper strip, initially neutral, becomes *positively charged* due to the electrons it has given up to the hydrogen.

The overall reaction is represented by the following chemical equation. The equation does not show, however, the manner in which the charge is separated.

$$Zn + H_2SO_4 \rightarrow ZnSO_4 + H_2 \uparrow \qquad (8\text{-}4)$$

A potential difference of approximately 1 V is established by the chemical action. The potential of the copper electrode is *raised* by 0.5 V and the potential of the zinc electrode is *lowered* by 0.5 V.

If a load is connected across the terminals, electrons flow from the negative zinc electrode through the external circuit to the positive copper electrode. (See Fig. 8-1.) At the copper electrode, the electrons are available to combine with hydrogen ions and form more hydrogen gas. Additional positive zinc ions enter the solution to replace the lost positive hydrogen ions. This process continues until the zinc electrode is completely eaten away or the acid solution becomes zinc sulfate. Thus, there is a motion, in solution, of positive ions toward the positive (copper) electrode and of negative ions toward the negative (zinc) electrode.

Although the chemical action tries to maintain a difference in potential of approximately 1 V, the voltaic cell is *not*, in practice, a very successful 1-V source. This is due to accumulation of hydrogen gas around the copper electrode, a process called *polarization*. Since hydrogen is a nonconductor, polarization results in an insulating effect around the electrode, increasing the cell's *internal resistance*. It also reduces the *active surface area* available for chemical action, and the cell's terminal voltage soon decreases. In practical primary cells, a *depolarizer* is often used to decrease this *insulating* effect.

When referring to a cell, the terms *anode* and *cathode* are frequently used. The *anode* is the terminal that *loses* electrons to the external circuit during normal operation (discharge). The *cathode* is the terminal that *gains* electrons. In the cell shown in Fig. 8-1, the negative zinc electrode is the anode and the positive copper electrode is the cathode. Note that while the *electron flow* is from anode to cathode in the external circuit, *conventional current* is from the cathode to the anode.

8-2 CARBON–ZINC CELL

The most common type of primary cell is the carbon–zinc cell, also known as the *Leclanché* cell. Since it uses a paste electrolyte rather than a liquid, it is known as a *dry cell*. Since the electrolyte will not spill out, a dry cell can be used in any position. The main construction features of a carbon–zinc dry cell are shown in Fig. 8-2.

In this cell, the negative electrode is the zinc can, while the positive electrode is a carbon rod in the center. The electrolyte paste is made up of zinc chloride and ammonium chloride (NH_4Cl, also called "sal ammoniac"). To prevent *polarization*—the formation of insulating hydrogen gas around the carbon electrode—manganese dioxide (MnO_2) is mixed into the electrolyte paste to act as a *depolarizer*. Magnesium dioxide is rich in oxygen, which combines with the hydrogen to form water.

$$MnO_2 + 2H_2 \rightarrow Mn + 2H_2O \qquad (8\text{-}5)$$

To prevent direct contact between the zinc can and the

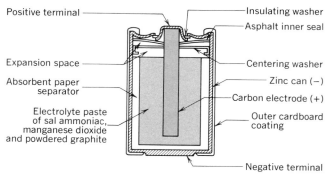

Positive terminal
Expansion space
Absorbent paper separator
Electrolyte paste of sal ammoniac, manganese dioxide and powdered graphite
Insulating washer
Asphalt inner seal
Centering washer
Zinc can (−)
Carbon electrode (+)
Outer cardboard coating
Negative terminal

FIGURE 8-2

Construction of a typical carbon–zinc dry cell.

electrolyte, an absorbent moist paper liner is used. This liner does not impede the passage of ions, but helps to reduce chemical reactions between the zinc and impurities in the electrolyte. These reactions form small voltaic cells (a process called *local action*) that would use up the zinc and shorten the cell's life. To reduce local action due to impurities in the zinc itself, a coating of mercury is placed on the inside surface of the zinc can during manufacture. This process, called *amalgamation,* effectively places the zinc and its other metallic impurities at the same electrical potential.

The motion of ions (similar to that described for the copper–zinc cell) produces an emf of approximately 1.5 V. Carbon–zinc cells range in size from the small AAA penlight cell to the six-inch-tall No. 6 cell, but all produce an emf of approximately 1.5 V. The physical size determines the amount of *current* the cell can deliver and the amount of energy stored. The smaller cells can deliver continuously a maximum current of only a few milliamperes; the No. 6 cell can deliver $\frac{1}{4}$ A (250 mA). A "D" flashlight cell can provide a load current of 50 mA for approximately 15 hours before the terminal voltage drops to about 1.3 V. To provide higher voltages, dry cells can be interconnected to form *batteries* of convenient multiples of 1.5 V (6 V, 9 V, etc.).

8-3 THE ALKALINE–MANGANESE CELL

Dry cells of the alkaline–manganese type are capable of providing heavy current for long periods. They require no

"rest periods" to recover after continuous use, as do carbon–zinc cells.

As shown in Fig. 8-3, this cell consists of an outer steel jacket (which does not take part in the chemical reaction) and an inner steel case which serves as the "cathode collector." These electroplated steel cases greatly decrease the possibility of leakage, compared to carbon-zinc cells. Note, in Fig. 8-3, that the cell includes a venting system to prevent case rupture in the event of a sustained short circuit (which would cause excessive gas buildup).

The cathode of an alkaline–manganese cell is a compressed mixture of manganese dioxide and graphite; the anode is compressed powdered zinc. The two chemical components are separated by a sleeve of porous synthetic fiber impregnated with the electrolyte, potassium hydroxide (KOH). The inner steel case, or cathode collector, is in contact with the cathode material and the positive terminal at the top of the cell. A metal anode collector in the center of the cell is in contact with the anode material and the negative terminal at the bottom of the cell.

Like the carbon–zinc cell, the alkaline–manganese cell produces a nominal 1.5 V. It can produce that emf, however, for about four times *longer* than the same size carbon–zinc cell. For example, a "D" alkaline cell can provide a load current of 50 mA for approximately 56 hours before terminal voltage drops to 1.3 V. A carbon–zinc cell, under the same conditions, would last about 15 hours. The *capacity* of the alkaline–manganese cell can be calculated as:

$$50 \text{ mA} \times 56 \text{ h} = 2800 \text{ mAh (milliampere hours)}$$
$$= 2.8 \text{ Ah (ampere hours)}$$

The *capacity* of a cell (expressed in ampere hours) is a measure of the total charge (and energy) that the cell can deliver under normal operating conditions. In the example above, the cell would be said to have a capacity of 2.8 Ah (if a terminal voltage of 1.3 V is the lowest acceptable value for the given application).

The alkaline cell also exhibits much longer *shelf life* than the carbon–zinc cell. Shelf life is the length of time that the cell retains its energy while stored. A carbon–zinc cell may retain 80% of its capacity for only 6 to 12 months, but an alkaline cell may be stored for 30 to 36 months. The loss of energy for either type of cell is partially due to the electrolyte drying out, so shelf life can be extended by storing the cells at *cooler* temperatures, as shown by the curves in Fig. 8-4.

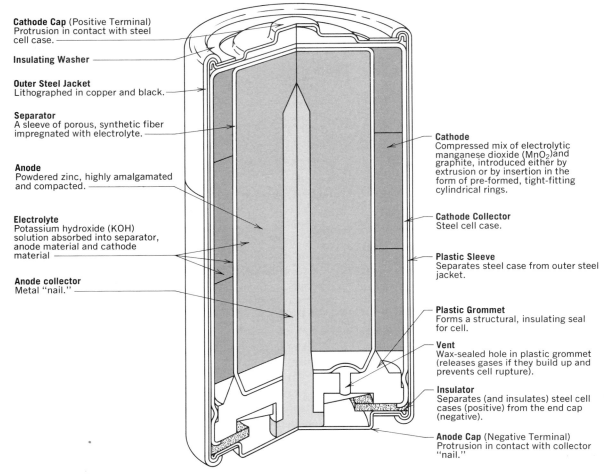

Cathode Cap (Positive Terminal)
Protrusion in contact with steel cell case.

Insulating Washer

Outer Steel Jacket
Lithographed in copper and black.

Separator
A sleeve of porous, synthetic fiber impregnated with electrolyte.

Anode
Powdered zinc, highly amalgamated and compacted.

Electrolyte
Potassium hydroxide (KOH) solution absorbed into separator, anode material and cathode material

Anode collector
Metal "nail."

Cathode
Compressed mix of electrolytic manganese dioxide (MnO_2)and graphite, introduced either by extrusion or by insertion in the form of pre-formed, tight-fitting cylindrical rings.

Cathode Collector
Steel cell case.

Plastic Sleeve
Separates steel case from outer steel jacket.

Plastic Grommet
Forms a structural, insulating seal for cell.

Vent
Wax-sealed hole in plastic grommet (releases gases if they build up and prevents cell rupture).

Insulator
Separates (and insulates) steel cell cases (positive) from the end cap (negative).

Anode Cap (Negative Terminal)
Protrusion in contact with collector "nail."

FIGURE 8-3

Cross section of an alkaline manganese cell. (Courtesy of Duracell International Inc.)

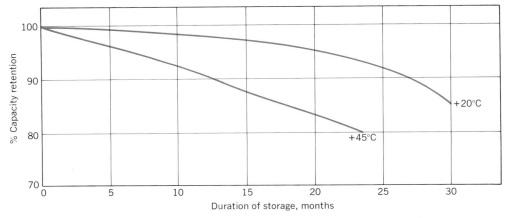

FIGURE 8-4

Typical capacity retention curves for an alkaline cell.

EXAMPLE 8-1

A fresh alkaline D-cell has a capacity of 3 Ah in a given application. Determine the available capacity of the cell after being stored for 20 months at a temperature of
a. 45°C
b. 20°C

Solution

a. From Fig. 8-4, capacity retention after 20 months at 45°C is 84%.

$$\text{capacity} = 3 \text{ Ah} \times 0.84 = \textbf{2.52 Ah}$$

b. After 20 months at 20°C, capacity retention is 96%.

$$\text{capacity} = 3 \text{ Ah} \times 0.96 = \textbf{2.88 Ah}$$

Example 8-1 shows the benefit of storing dry cells at reduced temperatures; Example 8-2 demonstrates how this prolongs the active life of the cell.

EXAMPLE 8-2

Determine the length of time the cell in Example 8-1 could deliver a current of 50 mA after being stored for 20 months at a temperature of
a. 45°C
b. 20°C

Solution

a. At 45°C, capacity = 2.52 Ah. Since

$$\text{time} = \frac{\text{capacity}}{\text{current}} \text{ or } t = \frac{Q}{I}$$

then $t = \dfrac{Q}{I} = \dfrac{2.52 \text{ Ah}}{50 \times 10^{-3} \text{ A}} = \textbf{50.4 h}$

b. At 20°C, capacity = 2.88 Ah.

$$t = \frac{Q}{I} = \frac{2.88 \text{ Ah}}{50 \times 10^{-3} \text{ A}} = \textbf{57.6 h}$$

8-4 OTHER PRIMARY CELLS

The development of *mercury* cells has made possible many electronic applications. These cells are essentially alkaline cells with mercuric oxide (HgO) replacing manganese dioxide as the positive electrode. Mercury cells have 50% more energy for a given volume than the alkaline–manganese cell and three times as much energy as the carbon–zinc cell.

They are available as either 1.35-V or 1.4-V cells. The 1.35-V cells are used primarily for instrumentation applications, the 1.4-V cells for general use in such devices as electronic watches. There are two different structures of the 1.4-V cell available. One form (the type used in watches) provides a low continuous current for a year or more. The second form is suitable for short-duration, high-surge current applications like photographic flash units. General-use mercury cells are usually of the flat disc or "button" type. Mercury cells can be series-connected to produce batteries with higher output voltages, such as the 12.6 V needed for some smoke alarm circuits.

The *silver oxide* cell is similar to the mercury cell, but uses silver oxide (Ag_2O) as the positive electrode. Silver cells develop an emf of 1.5 V and have four times more energy than a carbon–zinc cell. Like the mercury cell, they are usually "button" cells, and are widely used for watches.

An adaptation of the carbon–zinc cell is the *heavy-duty zinc–chloride* cell. It is similar in construction to the carbon–zinc cell, but uses only zinc chloride—rather than a combination of zinc chloride and ammonium chloride—as the electrolyte. This provides vigorous chemical action and keeps polarization to a minimum at high currents. Heavy-duty zinc–chloride cells usually last twice as long as carbon–zinc cells. They are available in all standard sizes.

The primary cell with highest energy density of all is the *lithium* cell. A lithium cell, with a nominal emf of 3.0 V, has the energy of 30 carbon–zinc cells. These cells retain 95% of their original capacity after a 5-year period, and operate well in low-temperature conditions. Lithium cells are used for continuous or standby power in low-drain integrated circuit memory or microprocessor applications.

The ultrathin "paper" battery is the most recent development of dry cell technology. It operates in much the same way as any other dry cell, but uses a stainless steel plate (sheet) instead of a carbon rod as the positive electrode. The electrolyte consists mainly of zinc perchlorate, which will not attack the steel plate as would ammonium chloride or zinc chloride. The steel plate electrode makes possible a flat, thin cell shape. A typical cell, which develops 1.5 V and has a 27 mAh capacity, is only 2.75 × 0.78 × 0.03 in. and weighs only 0.06 ounce. One major application is in camera film magazines, where a 6-V "pa-

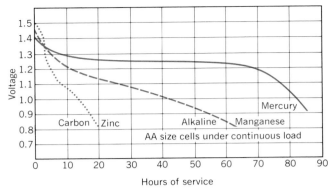

FIGURE 8-5
Comparison of discharge characteristics for primary cells.

per'' battery provides power for the camera's motorized film advance.

8-5   COMPARISON OF PRIMARY CELLS

The basic operating differences in the various types of primary cells is in their discharge characteristics, as shown in Fig. 8-5. Carbon–zinc and alkaline–manganese cells have a *sloping* discharge curve; mercury, silver, and lithium cells have a *flat* discharge curve. And, because of their different energy densities, an alkaline cell takes three times as long to drop to 0.8 V as a carbon–zinc cell, and a mercury cell takes four times as long.

EXAMPLE 8-3

Determine the hours of service for a continuous load to reduce the terminal voltage of AA size cells by 20% for the following types:
a. carbon–zinc
b. alkaline
c. mercury

Assume that the discharge curves conform to Fig. 8-5.

Solution

a. Nominal voltage of carbon–zinc cell = 1.5 V.

20% reduction = $1.5 \text{ V} \times \dfrac{20}{100} = 0.3 \text{ V}$

final terminal voltage = 1.2 V
From Fig. 8-5, the time required to reach 1.2 V
= **5 h.**
b. Final terminal voltage of alkaline = 1.2 V.
From Fig. 8-5, the time required to reach 1.2 V
= **10 h.**
c. Nominal voltage of mercury cell = 1.4 V.

20% reduction = $1.4 \text{ V} \times \dfrac{20}{100} = 0.28 \text{ V}$

final terminal voltage = 1.12 V.
From Fig. 8-5, the time required to reach 1.12 V
= **75 h.**

Table 8-1 lists other electrical characteristics of the common primary cell types and a summary of construction materials. Note how alkaline cells may be used at much lower operating temperatures than carbon–zinc cells. Shelf life (and capacity) of all types of dry cells is improved by storage at low temperatures, but the output of a cell during *operation* is greatly improved at *higher* temperatures. This output improvement is due to *increased* chemical activity. (See Fig. 8-6.)

8-6   LEAD–ACID SECONDARY STORAGE CELLS

The most common *secondary* storage cell is the lead–acid wet cell. Used most commonly for starting automobiles and other internal-combustion-engine vehicles, the lead–acid cell was introduced by Gaston Planté in 1860. The earliest automobiles were powered by electricity, rather than internal combustion engines, and used Planté cells. (Today's electric vehicles use batteries with higher energy density and lower weight. See Section 8-8.3.) Lead–acid batteries are still widely used in the telephone industry, where they serve as standby power sources. A lead–acid cell can be *recharged* several hundred times, and will last a number of years with proper use and maintenance.

A fully charged lead–acid cell has a lead peroxide (PbO_2) *positive* electrode, reddish-brown in color, and a *negative* electrode of gray spongy lead (Pb). With an approximately 27% electrolyte of sulfuric acid (specific gravity of 1.3), the cell produces about 2.2 V. Three or six of these cells are connected in series to obtain the common 6-V or 12-V battery used in autos or other applications.

TABLE 8-1
Comparison of Common Primary Cells

	Carbon–Zinc	Alkaline–Manganese	Mercury	Silver Oxide
Negative, anode	Zinc	Zinc	Zinc	Zinc
Positive, cathode	Carbon	Manganese dioxide	Mercuric oxide	Silver oxide
Electrolyte	Ammonium chloride	Potassium hydroxide	Potassium hydroxide	Potassium hydroxide
Nominal voltage, volts	1.5	1.5	1.35 or 1.4	1.5
Maximum rated current amperes	2–30	0.05–20	0.003–3	0.1
Energy output				
Watt-hr/lb	22	35	46	50
Watt-hr/in.^{3}	2.0	3.5	6.0	8.0
Temperature range				
Storage, °F	−40 to 120	−40 to 120	−40 to 140	−40 to 140
Operating, °F	20 to 130	−5 to 160	−5 to 160	−5 to 160
Shelf life in months at 68°F to 80% initial capacity	6–12	30–36	30–36	30–36
Shape of discharge curve	Sloping	Sloping	Flat	Flat

8-6.1 Chemical Action

The diagrams in Fig. 8-7 show the processes that take place during charging and discharging of a lead–acid cell. During *discharge,* the lead in both electrodes (plates) reacts with the sulfuric acid electrolyte. Hydrogen is displaced and lead sulfate is formed. The lead sulfate, a whitish material, is somewhat insoluble, and partially coats both the plates (Eq. 8-6). Since both plates are changing chemically toward the same material (lead sulfate, $PbSO_4$), the potential difference begins to decrease. Also, the oxygen in the lead peroxide combines with the hydro-gen ions of the electrolyte, forming water (Eq. 8-6). The sulfuric acid solution becomes weaker (more dilute, with its specific gravity approaching 1.0) as the cell delivers energy to a load. The combined effect of decreased potential difference and more dilute electrolyte causes the voltage developed by the cell to drop off as the cell loses its charge. Also, the internal resistance of the cell rises due to the lead sulfate coating on the plates.

Fortunately, this chemical process is reversible. If a battery charger is connected as shown in Fig. 8-7b (positive to positive, negative to negative), the direction of current and ionic flow is reversed. Electrical energy causes the lead sulfate to recombine with the hydrogen ions in the electrolyte. The effect is to remove excess water from the electrolyte and return the plates to their original states (lead peroxide and sponge lead). Since lead sulfate will harden into an insoluble salt over a period of time, it is wise to fully recharge a battery if it will not be used for some time.

The chemical action described above can be represented in this reversible equation:

$$Pb + PbO_2 + 2H_2SO_4 \underset{\text{charge}}{\overset{\text{discharge}}{\rightleftharpoons}}$$

$$2\,PbSO_4 + 2H_2O + \text{electrical energy} \qquad (8\text{-}6)$$

$$\text{lead} + \text{lead peroxide} + \text{sulfuric acid} \underset{\text{charge}}{\overset{\text{discharge}}{\rightleftharpoons}}$$

$$\text{lead sulfate} + \text{water} + \text{electrical energy}$$

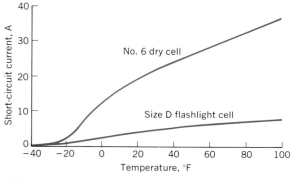

FIGURE 8-6
Typical effects of temperature on cell activity.

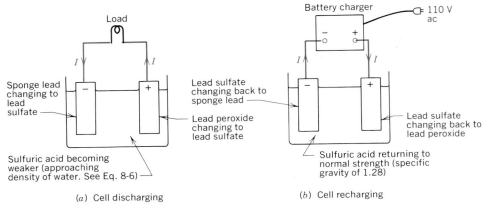

FIGURE 8-7
Chemical actions in a lead–acid cell under conditions of discharge and recharge.

An automotive charging system requires approximately 14.1 V to recharge a 12-V battery at currents up to 30 A. Recharging at excessively high currents can cause "boiling" of the electrolyte. This lowers the liquid level and can buckle and crumble the plates, shortening battery life.

8-6.2 Construction

The diagram in Fig. 8-7 has been simplified by showing each electrode as a single plate. However, in practical batteries, a number of positive and negative plates are interleaved and separated by porous rubber sheets. (See Fig. 8-8a.) This layering provides greater surface area and current availability. All of the positive plates are electrically connected, as are all the negative plates. These connections yield a parallel (higher-current) arrangement for a single cell developing approximately 2 V. Three such cells are series-connected, positive to negative, to form a 6-V battery, as shown in Fig. 8-8b. Since hydrogen gas is produced during recharging, vents are provided to allow hydrogen and water vapor to escape from the battery. The vents are part of *filler caps,* which can be removed to add distilled water. Water is added to make up for evaporation from the electrolyte during operation.

There are two basic ways of manufacturing the battery. Both start with plates made of lead antimony and cast in the form of grids with holes or spaces in the surface. In the *dry charge* method of manufacturing, lead peroxide is pressed into the holes of the positive plate grids and sponge lead into the negative plate grids. When the electrolyte (dilute sulfuric acid) is added, the battery is ready for use, without the need for charging.

In the *wet charge* method, lead oxide is pressed into both positive and negative plates. After electrolyte is added, the battery must be charged. In the charging process, the current *forms* the plate material into lead peroxide and sponge lead.

In recent years, low-maintenance and maintenance-free batteries have been developed. The low-maintenance type seldom need water added; maintenance-free batteries are sealed and cannot be opened to add water.

This advance in technology was made possible by the discovery that the amount of "gassing" (hydrogen production) that takes place during charging is related to the antimony content of the plates. Antimony is alloyed with lead to make casting of the grids easier, and ordinary lead–acid batteries have up to 4% antimony content. Low-maintenance batteries are produced by reducing the antimony content to 2%. This reduces the need for adding water, since very little is "boiled off" during charging. The plates of maintenance-free batteries contain no antimony. Instead, a lead–calcium or lead–calcium–tin combination is used. Since this eliminates "gassing," there is no need for venting, and the battery can be sealed completely. Once the battery is sealed, no electrolyte can evaporate from the cells. A very small vent is provided to relieve pressure due to altitude changes.

8-6.3 Specific Gravity

The state of charge of a lead–acid cell can be checked by determining the *specific gravity* of the electrolyte. Specific gravity is a ratio of the weight of a given volume of the electrolyte to the same volume of water. At a temperature

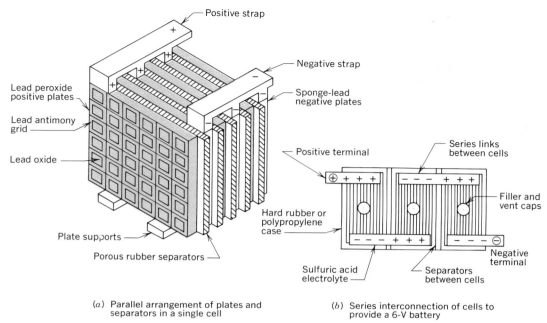

(a) Parallel arrangement of plates and separators in a single cell

(b) Series interconnection of cells to provide a 6-V battery

FIGURE 8-8

Construction of a lead–acid battery.

of 68°F, a fully charged cell should have a specific gravity (abbreviated sp. gr.) of 1.28; a fully *discharged* cell, 1.13. Both figures are related to the specific gravity of pure water, which is 1.0.

Specific gravity can be measured with a *hydrometer*. This is a glass tube containing a weighted, calibrated float. The numbers on the scale of the float range from 1280 on the bottom to 1120 at the top. (The scale multiplies the specific gravity reading by 1000.) When some electrolyte is drawn into the tube by means of a squeeze bulb on its top, the position of the float shows the specific gravity of the liquid. The lower the specific gravity of the electrolyte (the closer it is to water), the lower the float will sink, and the smaller will be the number visible on the scale.

A fully discharged cell will freeze if exposed to temperatures of 20°F or lower. But a fully charged cell can operate at temperatures between −76 and +140°F, because the higher concentration of acid depresses the freezing point of the liquid. (The freezing point of pure water is 32°F or 0°C.)

8-6.4 Ampere-Hour Capacity

A battery's current rating is usually given in units of ampere-hour (Ah) capacity, based on an 8-h discharge period.

During that period, the cell's output voltage must not drop below 1.7 V. For example, a 60-Ah battery (typical of those used for smaller vehicles) should deliver 7.5 A for 8 hours without any cell dropping below 1.7 V. Similarly, the battery should be able to deliver 5 A for 12 hours. It is unlikely, however, that the battery could actually deliver 60 A for 1 hour—cells are less efficient at higher discharge currents. Batteries with capacities of up to several hundred Ah are available.

Like primary cells, the lead–acid cell's capacity decreases significantly at lower temperatures. It will lose approximately 0.7% of capacity for each decrease of 1°F. At 0°F (−18°C), the lead–acid cell's capacity is only 60% of the value at 60°F (15.6°C). For automotive starting applications, a "cold cranking power" is usually specified. This is stated in the number of amperes available for 30 seconds at 0°F, with figures ranging from 300 to over 400 A.

8-7 THE NICKEL–CADMIUM CELL

For portable devices such as calculators, walkie-talkies, or electric tools, the most widely used rechargeable dry cell is the *nickel–cadmium* or nicad cell. A single nicad cell develops an open-circuit voltage of approximately 1.2 V. These cells are available in AA, C, and D sizes, and also in 6.1-V, 9.7-V, and 12.2-V output voltages. A nickel–cadmium cell has a lower energy content than a corresponding carbon–zinc cell, but a greater energy content than a lead-acid cell. Its major advantage is *rechargeability*—although the initial cost is higher than any primary cell, its overall cost is lower because it can be recharged up to 2000 times.

8-7.1 Construction

The base material for the nickel–cadmium cell electrodes is flexible nickel-plated steel sheets. Nickel powder is sintered onto the sheets at high temperature to provide a porous surface layer. On the negative electrode, this layer contains *metallic cadmium;* on the positive electrode, nickel hydroxide is present. A separator consisting of absorbent material saturated with an electrolyte—(basic) potassium hydroxide—is sandwiched between the plates. The two plates and separator are then rolled into a tight cylinder and sealed inside a nickel-plated steel case. A resealing safety vent relieves any excess internal pressure on the cell caused by overcharging or reverse charging.

8-7.2 Charging

Table 8-2 shows the nominal capacity for some standard-size nicad cells. Others are available with capacities of up to 10 Ah.

The rate at which sealed nickel–cadmium cells can be safely charged is the 8-hour or $C/8$ rate, where C is defined as the capacity in ampere-hours. For example, if C is 4 Ah, a maximum charging current of 0.5 A should be used. At this current, it will take 12 hours (because of internal losses) to restore a fully discharged cell to full charge. It is permissible, at this rate, to continue charging for *an indefinite period* without damaging the cell. Depending on size, each cell has its maximum continuous charge current. Some of these are given in Table 8-2.

The oxygen that is produced at the positive electrode when a nicad cell is overcharged reacts quickly with the cadmium in the charged negative electrode. The oxygen is reused continually according to the following chemical reaction:

$$O_2 + 2H_2O + 2Cd \rightarrow$$
$$Cd(OH)_2 + \text{electrical energy} \qquad (8-7)$$
oxygen + water + cadmium →
 cadmium hydroxide + electrical energy

TABLE 8-2

Capacities and Maximum Charge Rates for Nickel–Cadmium Cells

Size	Nominal Capacity, C Ah	Maximum Continuous Charge Rate, mA
AA	0.5	65
C	2.0	250
D	4.0	500

The overall chemical reaction in the nickel–cadmium cell is:

$$Cd + 2NiOOH + 2H_2O \underset{\text{charge}}{\overset{\text{discharge}}{\rightleftharpoons}} 2Ni(OH)_2$$
$$+ Cd(OH)_2 + \text{electrical energy} \qquad (8-8)$$
cadmium + nickel hydroxide + water $\underset{\text{charge}}{\overset{\text{discharge}}{\rightleftharpoons}}$
nickel hydroxide + cadmium hydroxide + electrical energy

Thus, the cadmium hydroxide produced by the overcharging is available to be reconverted to cadmium and water to repeat the cycle.

8-7.3 Quick-Charging

Sometimes, batteries must be charged more rapidly than normal. When this is the case, you must be careful not to exceed the recommended times given in Table 8-3. The times in the table assume that the cell has been completely discharged and that it is at a reasonable temperature (20° to 45°C, 68° to 113°F).

Some automatic battery chargers sense both temperature and voltage of the cell; when either reaches the set limit, an electronic switch reduces the charging rate to a trickle charge of $C/8$.

EXAMPLE 8-4

It is required to charge a nickel–cadmium C-cell (that has been fully discharged) in approximately 30 min.
a. What is the maximum current that can be used and the maximum time the cell can be left on charge?

TABLE 8-3

Normal and Maximum Charge Times
for Various Charge Rates for Nickel–Cadmium Cells

Charge Rate, A	Normal Charge Time	Maximum Charge Time
$C/8^a$	12 h	Indefinite
$C/4$	5 h	6 h
$C/2$	$2\frac{1}{4}$ h	$2\frac{1}{2}$ h
$C/1$	1 h	$1\frac{1}{4}$ h
2C	27 min	30 min
4C	12 min	12 min
8C	5 min	5 min

$^a C$ is the capacity in ampere hours (Ah).

b. If the cell is to be left on trickle charge, to what must the current be reduced?

Solution

a. From Table 8-2, a C-cell has a capacity of 2 Ah.

From Table 8-3, a charge rate (in amperes) of $2\,C$ is required to fully charge in 27 min.

Maximum charge current $= 2\,C = 2 \times 2\,A = $ **4A**
The maximum charge time at this rate is **30 min.**

b. Trickle charge rate $= C/8 = \frac{2}{8} = $ **0.25 A**

8-7.4 Disadvantages

Compared to other secondary cells, nickel–cadmium cells have three disadvantages.

1. *"Memory effect."* If the cell is operated at certain *low* discharge levels for short, repetitive periods, it becomes conditioned to this level and "forgets" its original design capacity. Thus it delivers only this low output, even when the demand for power is increased.
2. *Shelf life.* A fully charged cell, if it is allowed to stand unused at 20°C, will drop to 80% of its capacity in only 21 days. If the storage temperature is 30°C, the same result will occur in only 10 days.
3. *Cost.* Nickel–cadmium cells cost about twice as much as a lead–acid battery for the same amount of energy.

8-8 OTHER SECONDARY CELLS

8-8.1 Gelled-Electrolyte Lead–Acid Cells

These cells provide all the advantages of the wet-cell lead–acid battery, but use a gelled electrolyte to avoid the problems presented by the liquid electrolyte. They are completely sealed and can be mounted in any position, since the electrolyte will not spill. A relief valve opens to release internal pressure if it rises too high during charging, and automatically closes as pressure drops to an acceptable level. For maximum cell life, a gelled-electrolyte cell should not be left on continuous charge.

These batteries are available from 2 to 12 V, with capacities ranging from 0.9 to 20 Ah, based on a 20-h discharge rate. The maximum current ranges from 40 to 200 A. Gelled-electrolyte lead–acid batteries are used in portable tools, portable TV sets, and a variety of industrial applications.

8-8.2 Silver–Zinc and Silver–Cadmium Cells

These cells have an energy output, per pound, that is two to three times greater than nickel–cadmium cells. They are used as power sources for portable television cameras, videotape recorders, missile guidance systems, and similar applications. Battery packs are available with capacities ranging from 0.1 to 750 Ah.

The silver–cadmium cell produces 1.1 V; the silver–zinc, 1.5 V. Both are more expensive than nickel–cadmium cells and their recharge life cycles are much shorter. However, they do not suffer from the "memory effect" disadvantage of the nickel–cadmium cell.

8-8.3 Cells of the Future

Interest in producing practical, efficient electric vehicles has promoted development of a battery with higher energy density and lower weight than the present lead-acid battery. One recent development is the zinc–nickel oxide battery, which weighs less than half as much as a lead–acid battery and produces equal energy. A vehicle powered by these batteries can travel at up to 50 miles per hour, with a range of 100 miles before recharging. Battery life is 30,000 miles. The zinc–nickel oxide battery can be fully

FIGURE 8-9
The General Motors breakthrough on electric vehicle battery technology is illustrated here in a size comparison between the conventional lead–acid batteries in the foreground and the zinc–nickel oxide batteries in the rear. The battery packs have equal energy, but the zinc–nickel is only half as large and at 900 lb weighs less than half as much as the 2000-lb lead–acid pack. (Courtesy of General Motors Corporation.)

recharged in 10 to 12 hours from a 120-V ac source, or 6 to 7 hours from a 220-V ac source. (See Fig. 8-9.)

The generation of batteries now under development includes a lithium–iron sulfide cell (Fig. 8-10), with even greater energy density than the zinc–nickel oxide cell, and a sodium–sulfur cell that has four times the energy density of a lead–acid cell. A problem to be overcome with the sodium–sulfur cell is that both sodium and sulfur are very corrosive, and must be kept at temperatures of 300 to 400°C to maintain the chemical reaction.

8-9 COMPARISON OF SECONDARY CELLS

The differences in terminal voltage, discharge characteristic, and relative energy content for the four types of secondary cells discussed in Section 8-8 are shown in Fig. 8-11. Other electrical characteristics of the four types of cells are shown in Table 8-4.

FIGURE 8-10
Comparison of three battery technologies. The three battery cells shown here have the same energy-storage capacity. At the left is a conventional lead–acid battery. The next two are experimental batteries under development at General Motors Research Laboratories, Warren, Mich. In the center is the zinc–nickel oxide battery, about ⅓ the size and weight of the lead–acid; at the right is the lithium–iron sulfide, about ⅙ the size and weight. (Courtesy of General Motors Corporation.)

8-10 OTHER SOURCES OF EMF

The different types of sources available to produce electrical energy are shown in Table 8-5. All the cells described so far in this chapter are *chemical* sources, which *store* electrical energy in the form of chemical energy. All the other sources listed in Table 8-5 are *converters*—they can-

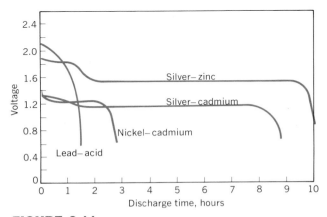

FIGURE 8-11
Discharge characteristics for lead–acid, nickel–cadmium, silver–cadmium, and silver–zinc cells of equal weight and current drain.

TABLE 8-4

Comparison of Common Secondary Cells

	Lead–Acid, Wet and Dry	Nickel–Cadmium	Silver–Cadmium	Silver–Zinc
Negative, anode	Sponge lead	Nickelic hydroxide		Zinc
Positive, cathode	Lead peroxide	Cadmium		Silver oxide
Electrolyte	Sulfuric acid	Potassium hydroxide	Potassium hydroxide	Potassium hydroxide
Nominal voltage, V	2.2	1.2	1.1	1.5
Maximum rated current	High (100–40,000 A)	Medium (0.1–100 A)	Medium	Medium
Energy output				
Watt-hours/lb	10 to 14	12 to 16	22 to 34	40 to 50
Watt-hours/in.^{3}	0.9 to 1.1	1.2 to 1.5	1.5 to 2.7	2.5 to 3.2
Cycle life	200 to 500	500 to 2,000	150 to 300	80 to 100
Temperature range				
Storage, °F	−76 to 140	−40 to 140	−85 to 165	−85 to 165
Operating, °F	−76 to 140	−20 to 140	−10 to 165	−10 to 165
Shelf life at 68°F to 80% initial capacity, months	8 (with lead-calcium grids)	$\frac{1}{2}$ to 1	3	3
Internal resistance, Ω	Low (0.001 to 0.03)	Low (0.003 to 0.5)	Very low	Very low
Shape of discharge curve	Sloping	Flat	Flat	Flat
Relative cost; 1 = lowest	1	2	4	3

not store electrical power, but must produce and deliver it at the instant it is needed.

8-10.1 Fuel Cells

A *fuel cell* is a chemical source that provides, but does not store, electrical energy. It generates an emf only as long as the oxygen and hydrogen fuels are fed to it. (See Fig. 8-12.)

TABLE 8-5

Different Sources of emf

Type of Energy Source	Typical Device
Chemical	Voltaic batteries, fuel cell
Photovoltaic (light)	Solar and photocells
Mechanical	Alternators and generators
Thermoelectric (heat)	Thermocouple
Piezoelectric (pressure)	Crystal

A fuel cell generates electricity by *reversing* the process known as *electrolysis*. In electrolysis, a dc voltage is applied to two platinum electrodes immersed in water. The current that flows between the electrodes decomposes the water into its components: hydrogen and oxygen. The hydrogen gas accumulates around the negative electrode

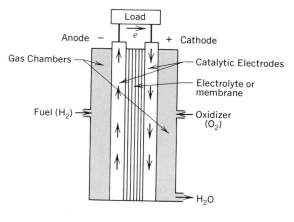

Arrangement of an oxygen–hydrogen fuel cell.

(cathode) and the oxygen gas around the positive electrode (anode). In chemical terms:

$$2H_2O + electricity \rightleftharpoons 2H_2 + O_2 \tag{8-9}$$

Equation 8-9 is reversible. This means that the combination of hydrogen and oxygen, in some suitable reaction, should produce water *plus* electricity. That is,

$$2H_2 + O_2 \rightleftharpoons 2H_2O + electrical\ energy \tag{8-10}$$

In the fuel cell (Fig. 8-12), a fuel containing hydrogen (such as ammonia, methyl alcohol, or natural gas) is introduced at one side. When it comes in contact with a nickel boride catalyst around the anode, the hydrogen ionizes and forms H^+ ions. The anode picks up the electrons and becomes negatively charged. The H^+ ions move through the membrane or potassium hydroxide electrolyte and react with the oxygen gas introduced at the other side of the cell. A silver catalyst at the cathode promotes this reaction, at the expense of electrons, and the cathode becomes positively charged. The potential difference between the two electrodes produces an emf. Neither the electrolyte nor the electrodes are affected by the operation. A by-product of the reaction is water (Eq. 8-10), which is useful in certain applications. In manned spacecraft, for example, the fuel cells used for power supplies also produce water for drinking and cooling purposes.

FIGURE 8-13
Fuel cells under construction that operate from natural gas and develop 40 kW of electric power. (Courtesy of Gas Research Institute, Chicago, Illinois. Photography by Bruce Quist.)

A single fuel cell can generate from 0.5 to 1 V. By suitable series-parallel interconnection, cells can make up batteries with a typical output of 2 kW for spacecraft applications. The high efficiency (over 40%), low noise, low pollution, and low waste characteristics of fuel cells have led industry to develop a 40-kW power plant for potential commercial use. (See Fig. 8-13.) A number of these plants are now in operation, supplying up to 40 kW of electric power, and an equivalent thermal output in the form of hot water and steam, providing an overall efficiency of 80%. This is an example of a cogeneration plant, where the electricity may be used directly by the owner or may be sold back to the electric generating utility. In these fuel cell plants, natural gas provides the hydrogen and air is the oxygen source. Phosphoric acid is used as the electrolyte and presently needs periodic replenishment as the fuel cell operates. The direct current output is converted to three-phase 120/208-V, 60-Hz alternating current by solid-state inverters, suitable for connection to the local power grid. Fuel cell power plants with higher outputs are now under consideration, since costs are competitive with conventional and nuclear generating facilities.

8-10.2 Solar Cells

The electrical power in unmanned satellites is usually developed by *silicon solar cells,* which convert light energy from the sun *directly* into electrical energy.

Basically, the solar cell is a *PN* junction, semiconductor device similar to the diodes used for rectification. (Chapter 28 will present a detailed discussion of *P*-type and *N*-type silicon.) When light energy strikes the solar cell, it passes through a thin transparent layer of *P*-type silicon to generate electrons and holes in the *N*-type silicon.* These carriers cross over the junction depletion layer, as shown in Fig. 8-14a, to produce an emf of approximately 0.5 V on open circuit.

Depending on its surface area, each solar cell produces only a few milliamperes. Commercially available solar cells, 3 in. in diameter, produce approximately 120 mA at 0.45 V. Using series-parallel interconnections, cells can be combined into panels producing 5 W at 12 V. Such panels are used with telephone repeater amplifiers in remote desert locations. During daylight hours, the panels recharge batteries that provide power through the night. A typical solar module is shown in Fig. 8-14b.

A *hole* is the vacancy left in a covalent bond when an electron breaks free. Holes contribute to current in pure silicon. (See Chapter 28.)

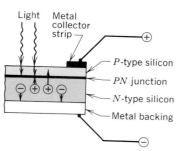

(a) Basic construction of a silicon photovoltaic cell

(b) Typical solar module. Thirty-six 90-mm-diameter silicon cells on this 44 × 17 in. module produce 31W of power to charge a 12-V battery at 13.8V and 2.25 A.

FIGURE 8-14

Features of a solar cell. [Photograph in (b) courtesy of the Solar Power Corporation.]

FIGURE 8-15

Solar installation at Mt. Laguna, California. (Courtesy of the Solar Power Corporation.)

The most efficient cells produced thus far can convert 28% of the light falling on them into electrical energy. (Optical losses in focusing the energy reduce this to an effective value of 25%.) In spite of this relatively low figure (thermal power plants are about 40% efficient) and the relatively high cost of the electricity produced, solar cells are being manufactured and installed around the world.

One such installation provides for the electrical needs of an Arizona Indian village with 96 residents. The village is too far away from existing power lines to make it economically feasible to bring in electricity by transmission lines. The installation consists of an array of three 64-foot rows of panels containing a total of 8064 three-inch cells. The array produces 120-V dc. Excess electrical energy is stored in a bank of lead–acid batteries for use at night and on cloudy days. The batteries can deliver full power to the village for up to 10 days at a time. The solar photovoltaic array delivers 3.5 kW of peak power at noon on a clear day, and is used to power 47 fluorescent lights, a 2-hp water pump, and 15 small refrigerators. Figure 8-15 shows another large solar array.

8-10.3 Alternators

Most of the electrical power used in homes and industry is generated by using *mechanical* energy to move a conductor through a magnetic field. This process, called *electromagnetic induction*, is the basic principle of all dc generators and ac alternators. Electromagnetic induction will be covered in detail in Chapter 14.

8-10.4 Thermocouples

There are a number of *thermoelectric* effects that can use temperature differentials as a means of generating an emf. One such device is the *thermocouple*. It consists of two dissimilar metals (such as iron and copper) twisted together at one end and subjected, at that junction, to high temperature. A small potential difference of a few millivolts is produced between the conductors at the *cold* end. Thermocouples are most widely used in temperature measurement (from $-450°F$ to more than $3000°F$), and in heating systems. In a heating system, the voltage generated by the thermocouple indicates that the pilot flame is burning, and thus available to light the main gas burner. (See Section 11-3.4.)

8-10.5 Piezoelectrics

An emf can be generated by pressure through what is known as the *piezoelectric* effect. Certain quartz crystals, when subjected to a physical force or pressure, will generate a voltage across the crystal face. They are used primarily as *transducers* (converters of mechanical to electrical energy) in phonograph pickups, microphones, and devices to measure sound and pressure.

SUMMARY

1. A voltaic cell requires two electrodes, immersed in an electrolyte, to produce an emf.
2. A copper–zinc cell produces approximately 1 V due to ions from the zinc electrode ''replacing'' hydrogen in the electrolyte and causing a separation of charge through motion of the ions.
3. Polarization is the accumulation of insulating hydrogen gas around a cell's electrode. It causes an increase in internal resistance and a reduction in generated voltage.
4. The carbon–zinc cell is a dry cell that produces 1.5 V using an ammonium chloride electrolyte. Manganese dioxide is a depolarizer material used to remove accumulated hydrogen gas from the electrode.
5. An alkaline–manganese cell uses a potassium hydroxide electrolyte, zinc anode, and manganese cathode to produce 1.5 V.
6. Shelf life is the length of time a stored cell will retain 80% of its energy. It can be extended by storage at cool temperatures. An alkaline cell has three to six times the shelf life of a zinc–carbon cell.

7. Mercury– and silver–oxide cells have a very high energy density and a flat discharge curve. They are most suitable for low-current applications.
8. A secondary cell, unlike a primary cell, can be returned to its original condition by recharging. Recharging makes use of a reversible chemical reaction.
9. A fully charged lead–acid cell has a lead peroxide positive plate, a spongy lead negative plate and an electrolyte of sulfuric acid. It develops approximately 2.2 V.
10. The lead sulfate produced during discharge of a lead–acid cell can be converted back to lead peroxide and sponge lead by the external connection of a charger (positive to positive, negative to negative).
11. A hydrometer can be used to determine the state of charge of a lead–acid cell by measuring the specific gravity of the electrolyte. A reading of 1.13 indicates a discharged cell; 1.28 indicates a fully charged cell.
12. The ampere-hour (Ah) capacity of a battery is the product of current and time. It determines how much current the cell can deliver, based on an 8-hour discharge period.
13. The nickel–cadmium cell is the most popular rechargeable dry cell for use in portable equipment. It develops 1.2 V and can be recharged up to 2000 times. The nickel–cadmium cell can be left on trickle charge for an indefinite period, but loses its charge rapidly when on standby.
14. Gelled-electrolyte lead–acid cells are portable versions of the lead–acid automotive battery. They have a much smaller capacity and should not be left on continuous charge.
15. Silver–zinc and silver–cadmium cells are used in lightweight equipment because of their high energy density. Like the nickel–cadmium cell, they have a flat discharge characteristic.
16. Fuel cells that produce electricity by combining oxygen and hydrogen, and solar cells that directly convert light energy to electricity in a *PN* junction, are possible energy sources of the future if economically developed.

SELF-EXAMINATION

Answer T or F or indicate a, b, c, or d
8-1. In a copper–zinc cell, both the copper and zinc electrodes are used up. _____
8-2. Current inside a copper–zinc cell consists of the motion of
    a. Positive ions.   c. Both positive and negative ions.
    b. Negative ions.   d. Electrons.
8-3. The collection of hydrogen gas around the positive electrode of a cell is called
    a. Amalgamation.
    b. Polarization.
    c. Ionization.
    d. Hydrogenation.
8-4. A carbon–zinc cell is an example of a dry primary cell. _____
8-5. Local action reduces the shelf life of a cell. _____
8-6. All cells, no matter what the materials involved, produce 1.5 V. _____
8-7. The electrolyte in an alkaline–manganese cell is
    a. Ammonium chloride.
    b. Potassium hydroxide.

    c. Sulfuric acid.

    d. Manganese dioxide.

8-8. Which of the following is false? The advantages of an alkaline cell over a Leclanché cell are

    a. Longer shelf life.

    b. More energy.

    c. Lower temperature operation.

    d. Flat discharge characteristic.

8-9. Mercury and silver oxide cells are secondary cells. _____

8-10. A lead–acid cell is rechargeable because

    a. Its electrolyte is sulfuric acid.

    b. It is a wet cell.

    c. Its chemical reaction is reversible.

    d. Its electrolyte has a high specific gravity.

8-11. The cathode of a lead–acid cell is gray spongy lead. _____

8-12. As a lead–acid cell discharges, both electrodes convert to lead sulfate. _____

8-13. At 68°F a fully charged lead–acid cell should have a specific gravity close to

    a. 1.13

    b. 1.2

    c. 1.28

    d. 2.2

8-14. A 100-Ah capacity battery should deliver a current of 8 A for approximately

    a. 12 h

    b. 8 h

    c. 20 h

    d. 100 h

8-15. The nickel–cadmium cell is

    a. A wet secondary cell.

    b. A dry primary cell.

    c. A wet primary cell.

    d. A dry secondary cell.

8-16. If a nickel–cadmium cell is to be quick-charged, there is some maximum time that should not be exceeded. _____

8-17. A gelled-electrolyte lead–acid cell is similar in characteristics to a wet cell except that it is portable. _____

8-18. A flat discharge characteristic is one in which the terminal voltage of a cell is relatively constant under use but then decreases rapidly when discharged. _____

8-19. Both fuel cells and solar cells are examples of secondary cells. _____

REVIEW QUESTIONS

1. What is the basic reason for the difference between primary and secondary cells?

2. What two conditions can cause a copper–zinc cell to become fully discharged?

3. a. What do you understand by the words polarization and amalgamation?

    b. What steps are taken in a cell to overcome the first and what condition does the second present?

4. a. How are the words anode and cathode applied to a cell?
  b. Which is which in a copper–zinc cell?
5. How does a depolarizer work?
6. a. Give three advantages that an alkaline–manganese cell has over a carbon–zinc.
  b. What is one disadvantage?
7. a. What are the main applications for mercury and silver oxide cells?
  b. Why?
8. a. Where does the alkaline–manganese cell get its name?
  b. Are the mercury and silver oxide cells also alkaline cells?
9. Why do you think the lead–acid cell is the cheapest of all the secondary cells?
10. Why should a lead–acid cell be fully charged before being stored unused for any length of time?
11. What do you understand by the terms dry-charge and wet-charge, as applied to the construction of lead–acid cells?
12. How is it possible to manufacture a completely sealed lead–acid battery?
13. Why does the specific gravity of a lead–acid cell's electrolyte indicate the state of charge of the cell?
14. What is the ampere-hour capacity an indication of in a battery?
15. What is the reason behind the ability of a nickel–cadmium cell to remain on trickle charge indefinitely?
16. a. Why is it necessary to limit the charge time of a nickel–cadmium cell when it is being recharged quickly?
  b. Where does the charging energy go after the cell becomes charged?
17. a. What other secondary cells are there?
  b. What applications do they have?
18. What do you understand by the term *memory-effect* as applied to a nickel–cadmium battery?
19. a. What are the five types of energy source (as far as producing electrical energy)?
  b. Which are not practical for producing large energy quantities for commercial use?
20. a. What is the main advantage of a solar cell?
  b. What two disadvantages does it have?

PROBLEMS

(Answers to odd-numbered problems at back of book)
8-1. If a gallon of water weighs 8.3 lb, how much will a gallon of sulfuric acid weigh if a hydrometer gives the following readings?
  a. 1130
  b. 1280
8-2. A fresh alkaline C-cell has a capacity of 1.5 Ah in a certain application. Determine
  a. The available capacity of the cell after being stored for 30 months at 20°C.
  b. The length of time the cell in part (a) could deliver 40 mA.
  c. The length of time a fresh cell could deliver 40 mA.

8-3. Repeat Example 8-3 for a 40% reduction in terminal voltage. Compare the results with those of Example 8-3. What conclusions can you draw regarding the relative lengths of service of one cell compared with another?

8-4. A battery has a 100-Ah capacity.
  a. How many coulombs of charge does this represent?
  b. How much current should this battery be able to deliver for 8 hours?
  c. If the battery is recharged at the rate of 5 C/s, how long will it take to fully charge the battery?

8-5. It is required to quick-charge a fully depleted 12.2-V, 6-Ah nickel–cadmium battery in approximately 12 min.
  a. If a 36-V charger with a 0.5-Ω internal resistance is to be used, what additional series resistor is needed and what must be its power rating? (Assume that the voltages remain constant during the charging period.)
  b. What must this resistance be increased to if the battery is to remain on trickle charge?
  c. If no resistor is used between the charger and battery, what is the maximum length of time charging should be allowed to take place?
  d. If a 36-V charger is available that has an internal resistance of 3 Ω, would it be possible to recharge the battery in approximately $\frac{1}{2}$ h? Explain.

8-6. Refer to the solar cell installation in this chapter.
  a. How much power must each cell produce?
  b. Assuming each cell develops 0.432 V on load, what current is delivered by each cell?
  c. Determine a possible series-parallel connection to produce 3.5 kW at 120 V.
  d. What is the maximum available current from this combination?
  e. If the battery system can supply full power for 10 days, what is the average daily use of energy?
  f. Assuming that the cells produce full peak power for 10 hours a day, how long will it take to fully charge the batteries from full discharge while supplying the average daily use of energy?

8-7. A No. 6, 1.5-V dry cell is short-circuited at 70°F.
  a. What is the current?
  b. What is the cell's internal resistance?

8-8. Repeat Problem 8-7 using a temperature of 20°F.

8-9. a. What would be the short circuit current of a 1.5-V D-size flashlight cell at 70°F?
  b. What is the cell's internal resistance?

8-10. Repeat Problem 8-9 using a temperature of 20°F.

CHAPTER 9

INTERNAL RESISTANCE AND MAXIMUM POWER TRANSFER

Internal resistance **is an inevitable part of any voltage source.** Although the effects of internal resistance in a power supply can be reduced by voltage regulators, *all* batteries (because of internal resistance) suffer a reduction in terminal voltage when a load is connected. When cells are interconnected to form batteries, the total internal resistance must be determined. This resistance may be found by using the methods introduced for determining resistance in series and parallel circuits. The voltage, current, and ampere-hour capacities of the batteries also must be considered.

The internal resistance of a source determines the maximum *power* that may be delivered to a load. To achieve maximum transfer of power, the load resistance must be made equal to the internal resistance of the source. In this condition—known as *maximum power transfer*—the load is said to be *matched* to the source. The *wattmeter* is introduced in this chapter as a means of power measurement.

9-1 INTERNAL RESISTANCE OF A CELL

It was shown in Chapter 8 that cells are made up of electrodes and electrolytes, all of which have some amount of electrical resistance. **A cell, while generating an emf, has a resistance distributed throughout its component parts.**

The sum of these *internal* resistances is usually lumped together. It is shown as a single equivalent resistance (r) in series with the emf, as shown in Fig. 9-1a. If a voltmeter that draws a negligible current is connected across the terminals of the cell, it will indicate the cell's emf. This is referred to as the open circuit or *no-load* voltage (V).

But when a load is connected to the cell (Fig. 9-1b), the current that flows, according to Ohm's law, is:

$$I = \frac{V}{R_L + r} \tag{9-1}$$

This current, flowing through the internal resistance of the cell, causes an *internal* voltage drop given by:

$$V_r = Ir \tag{9-2}$$

As a result, when the load current (I) flows, the terminal voltage (V_t) of the cell drops below the open circuit voltage (V):

$$V_t = V - Ir \tag{9-3}$$

Equation 9-3 shows that the greater the load current, and the greater the internal resistance, the more the terminal voltage will decrease. This, in fact, is how a battery's condition is checked—by noting its terminal voltage *under load*. When a cell deteriorates, its internal resistance *increases,* and (for the same load current) it shows a lower terminal voltage.

EXAMPLE 9-1

A battery has an open-circuit voltage of 6.5 V and an internal resistance of 2.5 Ω. A load of 10 Ω is connected across the battery. Calculate:
a. The current through the load.
b. The internal voltage drop in the battery.
c. The terminal voltage of the battery under load.
d. The terminal voltage if the battery's internal resistance increases to 5 Ω.

Solution

a. $I = \dfrac{V}{R_L + r}$ (9-1)

 $= \dfrac{6.5 \text{ V}}{10 + 2.5 \ \Omega}$

 $= \textbf{0.52 A}$

b. $V_r = Ir$ (9-2)

 $= 0.52 \text{ A} \times 2.5 \ \Omega$

 $= \textbf{1.3 V}$

c. $V_t = V - Ir$ (9-3)

 $= 6.5 \text{ V} - 1.3 \text{ V}$

 $= \textbf{5.2 V}$

d. $I = \dfrac{V}{R_L + r}$ (9-1)

 $= \dfrac{6.5 \text{ V}}{10 + 5 \ \Omega}$

 $= 0.43 \text{ A}$

 $V_t = V - Ir$ (9-3)

 $= 6.5 \text{ V} - 0.43 \text{ A} \times 5 \ \Omega$

 $= 6.5 \text{ V} - 2.15 \text{ V} = \textbf{4.35 V}$

If the load resistance (R_L) is connected so that it drops the terminal voltage to one-half the open circuit

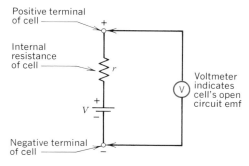

(a) Measuring the open-circuit emf, V, of a cell

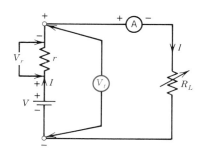

(b) Measuring the terminal voltage. V_t, of a cell under load

FIGURE 9-1

Measuring the open-circuit and terminal voltage of a cell.

voltage, then R_L must be equal to the internal resistance (r). This is because the emf is evenly divided between the two resistances. For sources with low internal resistance, a more practical way of determining r is to measure the terminal voltage at some known load current, then calculate the value of r by using Eq. 9-3.

EXAMPLE 9-2

The terminal voltage of a 1.2-V nicad cell drops to 1.1 V when supplying a full-load current of 150 mA. Calculate:
a. The cell's internal resistance.
b. The load resistance that would drop the terminal voltage to 0.6 V.

Solution

a. $V_t = V - Ir$ (9-3)

Therefore,

$$r = \frac{V - V_t}{I}$$

$$= \frac{1.2\ \text{V} - 1.1.\ \text{V}}{0.15\ \text{A}}$$

$$= \frac{0.1\ \text{V}}{0.15\ \text{A}} = \mathbf{0.67\ \Omega}$$

b. When the terminal voltage equals one-half the open-circuit voltage, the load resistance must be equal to the internal resistance of **0.67 Ω**.

Figure 9-2 shows the change in terminal voltage with load current. An *ideal* voltage source would have *no* change in terminal voltage. The change that occurs from

no-load to full load is given by the percentage *voltage regulation:*

$$\% \text{ voltage regulation} = \frac{V_{NL} - V_{FL}}{V_{FL}} \times 100\% \quad (9\text{-}4)$$

where: V_{NL} = the no-load or open-circuit voltage
V_{FL} = the terminal voltage at full-load current

Note that an ideal voltage source would have a 0% voltage regulation.

EXAMPLE 9-3

a. Determine the percentage voltage regulation of the 1.2-V cell in Example 9-2.
b. A power supply is known to have a voltage regulation of 40% and a no-load voltage of 21 V. What is the full load voltage?

Solution

a. % voltage

$$\text{regulation} = \frac{V_{NL} - V_{FL}}{V_{FL}} \times 100\% \quad (9\text{-}4)$$

$$= \frac{1.2\ \text{V} - 1.1\ \text{V}}{1.1\ \text{V}} \times 100\%$$

$$= \frac{0.1\ \text{V}}{1.1\ \text{V}} \times 100\%$$

$$= \mathbf{9.1\%}$$

b. % voltage

$$\text{regulation} = \frac{V_{NL} - V_{FL}}{V_{FL}} \times 100\% \quad (9\text{-}4)$$

$$40\% = \frac{21\ \text{V} - V_{FL}}{V_{FL}} \times 100\%$$

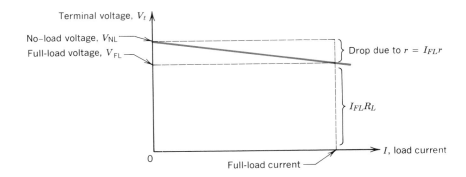

FIGURE 9-2
Effect of internal resistance and load current on terminal voltage, V_t.

$$0.4 = \frac{21\ V - V_{FL}}{V_{FL}}$$

$$0.4\ V_{FL} = 21\ V - V_{FL}$$

$$1.4\ V_{FL} = 21\ V$$

$$V_{FL} = \frac{21\ V}{1.4} = \mathbf{15\ V}$$

EXAMPLE 9-4

A 12-V, 1.5-Ah gelled-electrolyte lead–acid battery has a terminal voltage of 10.5 V when delivering a full-load current of 10 A. Calculate:

a. The battery's internal resistance.
b. The battery's voltage regulation.
c. The battery's current and terminal voltage when a 2-Ω load is connected.
d. The load resistance that would drop the terminal voltage to 6 V.
e. The terminal voltage of a dc generator to recharge the cell at 5 A.
f. The amount of current the fully charged cell can deliver for 20 h.

Solution

a. $r = \dfrac{V - V_t}{I}$ (9-3)

$$= \frac{12\ V - 10.5\ V}{10\ A} = \frac{1.5\ V}{10\ A} = \mathbf{0.15\ \Omega}$$

b. % voltage regulation

$$= \frac{V_{NL} - V_{FL}}{V_{FL}} \times 100\% \qquad (9\text{-}4)$$

$$= \frac{12\ V - 10.5\ V}{10.5\ V} \times 100\% = \mathbf{14.3\%}$$

c. $I = \dfrac{V}{R_L + r}$ (9-1)

$$= \frac{12\ V}{2 + 0.15\ \Omega}$$

$$= \frac{12\ V}{2.15\ \Omega} = \mathbf{5.58\ A}$$

$V_t = V - Ir$ (9-3)

$$= 12\ V - 5.58\ A \times 0.15\ \Omega$$

$$= 12\ V - 0.84\ V$$

$$= \mathbf{11.16\ V}$$

d. When the terminal voltage is one-half the no-load voltage, the load resistance equals the internal resistance, **0.15 Ω**.

e. To charge the battery, the generator must be con-

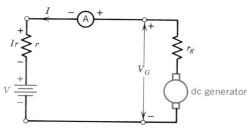

FIGURE 9-3
Charging circuit for Example 9-4.

nected across the battery with like polarities, as shown in Fig. 9-3.
Clearly, $V_G = V + Ir$

$$= 12\ V + 5\ A \times 0.15\ \Omega$$

$$= 12\ V + 0.75\ V$$

$$= \mathbf{12.75\ V}$$

f. $\text{current} = \dfrac{\text{ampere-hour capacity}}{\text{hours}}$

$$I = \frac{1.5\ Ah}{20\ h} = 0.075\ A = \mathbf{75\ mA}$$

9-2 SERIES AND PARALLEL CONNECTIONS OF CELLS TO FORM A BATTERY

9-2.1 Series Connections

As you learned in Chapter 8, the emf developed by an individual cell is generally from 1 to 2 V. To provide a higher terminal voltage, cells can be connected in *series* to form a battery, as shown in Fig. 9-4. One result of the series connection, however, is an increase in overall internal resistance. If the cells are identical (which is not necessarily the case), the total voltage and total internal resistance can be determined by:

$$V_T = n \cdot V \qquad (9\text{-}5)$$

$$r_T = n \cdot r \qquad (9\text{-}6)$$

where n is the number of cells in series.

Note that the current *capability* of the battery is not increased over that of a single cell. For example, if three equal cells are series-connected, the maximum current de-

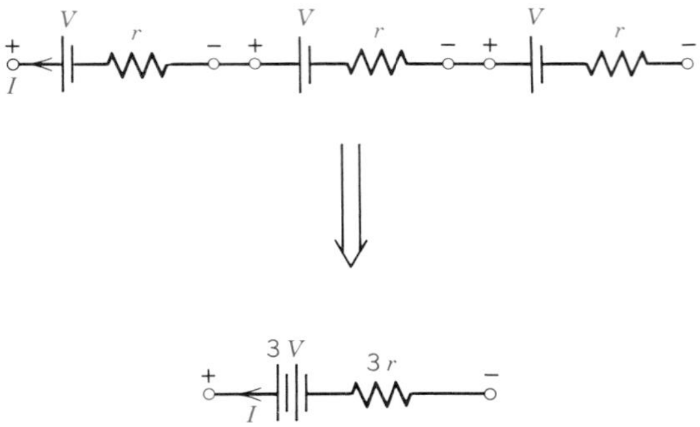

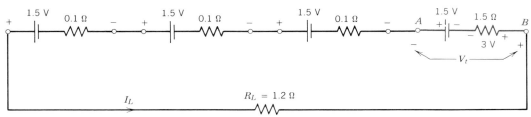

FIGURE 9-4
Series interconnection of cells.

liverable (current capability) is the same as one cell, as shown by the following:

$$I = \frac{V_T}{r_T} = \frac{3V}{3r} = \frac{V}{r}$$

If the cells are different, the current must not exceed the current capability of the cell with the lowest current capability. Otherwise, the terminal voltage of this cell may be reversed due to the high internal resistance of the cell causing a high internal voltage drop. Not only does this defeat the purpose of putting cells in series to increase

overall terminal voltage, but internal heating in the cell could cause damage to the cell. The following example illustrates these points.

EXAMPLE 9-5

a. Three series-connected 1.5-V cells, each with an internal resistance of 0.1 Ω, deliver current to a load of 1.2 Ω. Determine the load current and voltage.
b. In an attempt to increase load current and voltage, another 1.5-V cell that has an internal resistance of 1.5 Ω is connected in series with the first three cells. Determine the new load current, load voltage, the terminal voltage across the fourth cell, and its internal power dissipation.

Solution

a. See Fig. 9-5a. The open-circuit terminal voltage of the three cells is given by:

$$V_T = n \cdot V \qquad (9\text{-}5)$$
$$= 3 \times 1.5 \text{ V} = 4.5 \text{ V}$$

The total internal resistance is given by:

$$r_T = n \cdot r \qquad (9\text{-}6)$$
$$= 3 \times 0.1 \text{ Ω} = 0.3 \text{ Ω}$$

(a) Three equal series-connected cells

(b) Circuit with high internal resistance cell added

FIGURE 9-5
Circuits for Example 9-5.

The load current is

$$I_L = \frac{V_T}{r_T + R_L}$$
$$= \frac{4.5 \text{ V}}{0.3 + 1.2 \ \Omega}$$
$$= \frac{4.5 \text{ V}}{1.5 \ \Omega} = \textbf{3 A}$$

The load voltage is

$$V_L = I_L R_L$$
$$= 3 \text{ A} \times 1.2 \ \Omega = \textbf{3.6 V}$$

b. See Fig. 9-5*b*. The open-circuit terminal voltage of the four cells is

$$V_T = n \cdot V \qquad (9\text{-}5)$$
$$= 4 \times 1.5 \text{ V} = 6 \text{ V}$$

The total internal resistance is

$$r_T = 3 \times 0.1 \ \Omega + 1.5 \ \Omega$$
$$= 0.3 \ \Omega + 1.5 \ \Omega = 1.8 \ \Omega$$

The load current is

$$I_L = \frac{V_T}{r_T + R_L}$$
$$= \frac{6 \text{ V}}{1.8 + 1.2 \ \Omega}$$
$$= \frac{6 \text{ V}}{3 \ \Omega} = \textbf{2 A}$$

The load voltage is

$$V_L = I_L R_L$$
$$= 2 \text{ A} \times 1.2 \ \Omega = \textbf{2.4 V}$$

Terminal voltage of the fourth cell is

$$V_{AB} = V - I_L r$$
$$= 1.5 \text{ V} - 2 \text{ A} \times 1.5 \ \Omega$$
$$= 1.5 \text{ V} - 3 \text{ V} = \textbf{-1.5 V}$$

That is, $V_t = V_{BA} = +1.5$ V. The terminal voltage polarity under load is opposite to the open-circuit voltage polarity.
Internal power dissipation is

$$P_r = I_L^2 r$$
$$= (2 \text{ A})^2 \times 1.5 \ \Omega$$
$$= \textbf{6 W}$$

Note that the short-circuit current of the first three cells in Example 9-5 is 1.5 V/0.1 Ω or 15 A. However, the

short-circuit current of the fourth cell is only 1.5 V/1.5 Ω or 1 A. The addition of the fourth cell actually reduces load current and load voltage due to its high internal resistance and low current capability. Also, the internal power dissipation of 6 W is four times higher than the short-circuit power dissipation—a condition that will lead to certain overheating and damage to the fourth cell.

9-2.2 Parallel Connections

The advantages of connecting cells in *parallel* to form a battery (Fig. 9-6) are an *increase* in current capability and a *reduction* in the total internal resistance. If the cells are identical or closely matched (which is necessary to avoid large internal circulating currents), the result is

$$I_T = n' \cdot I \qquad (9\text{-}7)$$
$$r_T = \frac{r}{n'} \qquad (9\text{-}8)$$

where n' is the number of cells in parallel.

Note that, when cells of equal voltage are connected in parallel, there is *no* increase in terminal voltage.

9-2.3 Series-Parallel Connections

A series-parallel connection, shown in Fig. 9-7, is typical for solar cells and for 6-V lead–acid batteries. (Refer back to Fig. 8-8, which shows that each 2-V group consists of three parallel-connected cells.) With this form of connection, both voltage and current capability are increased. The internal resistances are determined by using series-parallel resistance computation methods.

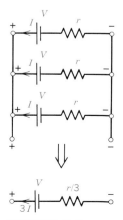

FIGURE 9-6
Parallel interconnection of cells.

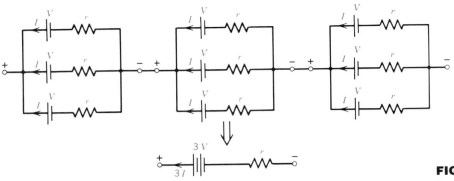

$3V$

$3I$

(c) Series-parallel

FIGURE 9-7
Series-parallel interconnection of cells.

9-2.4 Ampere-Hour (Ah) Capacities

For similar capacity cells connected in *series,* the total ampere-hour (Ah) capacity is the same as the *individual* capacity of *one* cell. That is, the total ampere-hour capacity is not increased above that of a single cell. When the cells are connected in *parallel,* however, the total ampere-hour capacity is the *sum* of the individual cell capacities. For series-parallel combinations, the total ampere-hour capacity depends on the particular form of connection, as shown in Example 9-6.

EXAMPLE 9-6

Six 2-Ah, 1.5-V cells, each with a resistance of 0.1 Ω, are to be interconnected. If each cell is capable of a current of 100 mA determine the terminal voltage, current capability, internal resistance, and ampere-hour capacity of the following arrangements:

a. All connected in series.
b. All connected in parallel.
c. Three series connections of two each in parallel.

Solution

a. $V_T = n \cdot V$ (9-5)
    $= 6 \times 1.5\ V$
    $= \mathbf{9\ V}$

The current capability is the same as one cell, **100 mA.**

$$r_T = n \cdot r \qquad (9\text{-}6)$$
$$= 6 \times 0.1\ \Omega = \mathbf{0.6\ \Omega}$$

The total ampere-hour capacity is the same as one cell = **2 Ah.**

b. The open-circuit terminal voltage is the same as one cell, **1.5 V.**

$$I_T = n' \cdot I \qquad (9\text{-}7)$$
$$= 6 \times 100\ mA$$
$$= \mathbf{600\ mA}$$

$$r_T = \frac{r}{n'} \qquad (9\text{-}8)$$
$$= \frac{0.1\ \Omega}{6} = \mathbf{0.017\ \Omega}$$

The total ampere-hour capacity is the sum of the capacities of the six cells = **12 Ah.**

c. Each of the three series connections has a voltage of 1.5 V. The total terminal voltage is given by

$$V_T = n \cdot V \qquad (9\text{-}5)$$
$$= 3 \times 1.5\ V$$
$$= \mathbf{4.5\ V}$$

The total current capability is the same as each parallel group of two cells, which has a current capability:

$$I_t = n' \cdot I \qquad (9\text{-}7)$$
$$= 2 \times 100\ mA$$
$$= \mathbf{200\ mA}$$

Each parallel group of two cells has an internal resistance given by

$$r_T = \frac{r}{n'} \qquad (9\text{-}8)$$
$$= \frac{0.1\ \Omega}{2} = 0.05\ \Omega$$

The total resistance of the three series groups is

$$r_T = n \cdot r \qquad (9\text{-}6)$$
$$= 3 \times 0.05\ \Omega$$
$$= \mathbf{0.15\ \Omega}$$

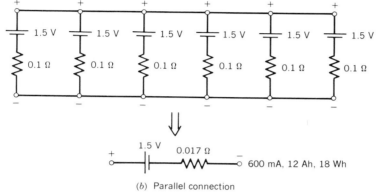

(a) Series connection

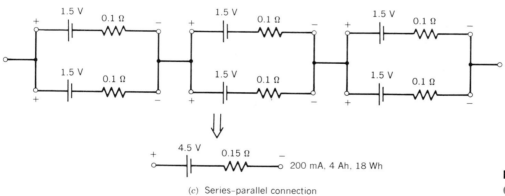

(b) Parallel connection

(c) Series–parallel connection

FIGURE 9-8
Circuits for Example 9-6.

total energy of the cells,

$$W = VIt \qquad (3-4)$$
$$= 4.5\ V \times 0.2\ A \times 20\ h = 18\ Wh$$

(Each parallel cell can deliver 0.1 A for 20 h.)

$$\text{total ampere-hour capacity} = \frac{18\ Wh}{4.5\ V}$$
$$= \textbf{4 Ah}$$

Or each of the two parallel cells (in each series combination) can deliver 0.1 A for 20 h.

Thus total ampere-hour

capacity = 0.1 A × 20 h × 2
= **4 Ah**

Note that although the ampere-hour capacities are different for each connection, the total energy is the same for each connection.

Series: 0.1 A at 9 V for 20 h = 18 Wh of energy.
Parallel: 0.6 A at 1.5 V for 20 h = 18 Wh of energy.
Series-parallel: 0.2 A at 4.5 V for 20 h
= 18 Wh of energy.

See Fig. 9-8.

9-3 MAXIMUM POWER TRANSFER

Internal resistance (r) is not a property that belongs exclusively to cells and batteries. All ac or dc power sources (such as amplifiers, microphones, or any other source of emf) have an internal resistance, more generally called *internal impedance*.

In many applications, it is desirable to achieve the greatest possible transfer of power from the source to the load. With a source of fixed open-circuit voltage and internal resistance, the problem is to determine what value of resistance the load must have to permit the maximum transfer of power to it.

Assume that a 40-V open-circuit source of emf with an internal resistance of 8 Ω is connected to a variable load, as shown in Fig. 9-9*a*. And, assume that the load resistance can be varied in 1-Ω steps. By recording the ammeter and voltmeter readings at each resistance-step increase, the product of the two readings will give the power delivered to the load. Curves for load voltage V_L and load current I are plotted in Fig. 9-9*b*. Note that the product of an increasing quantity (V_L) and a decreasing quantity (I) is shown to have a peak power value $P_{L,max}$ at $R_L = r$. In this example, a definite peak in the power curve will occur at $R_L = r = 8$ Ω.

The shape of this curve can be justified by calculating the load power at several different load resistance values:

1. $R_L = 0$ (short circuit); $P_L = I^2R_L = 0$
2. $R_L = 4$ Ω

$$P_L = I^2R_L = \left(\frac{V}{R_L + r}\right)^2 R_L$$

$$= \left(\frac{40 \text{ V}}{4 + 8 \text{ }\Omega}\right)^2 4 \text{ }\Omega = \textbf{44.4 W}$$

3. $R_L = 8$ Ω; $P_L = \textbf{50 W}$
4. $R_L = 16$ Ω; $P_L = \textbf{44.4 W}$
5. $R_L = \infty$ (open circuit); $P_L = I^2R = 0 \times \infty = \textbf{0}$

The *maximum power transfer theorem* states that **maximum power will be delivered to a load from a source when the load resistance equals the internal resistance of the source.**

This is not meant to imply that the circuit is operating at maximum *efficiency* when $R_L = r$. This is because efficiency (η) is defined as:

$$\eta = \% \text{ efficiency} = \frac{P_L}{P_T} \times 100\% \qquad (9\text{-}9)$$

$$= \frac{P_L}{P_L + P_r} \times 100\%$$

where: $P_L =$ load power
$P_r =$ power dissipated in the source
$P_T =$ total power supplied by the emf

EXAMPLE 9-7

For the above circuit in Fig. 9-9*a*, calculate the load voltage and efficiency at $R_L = 4, 8, 16,$ and 100 Ω. $V = 40$ V, $r = 8$ Ω.

Solution

$$\text{load voltage } V_L = V \times \frac{R_L}{R_L + r} \qquad (5\text{-}5)$$

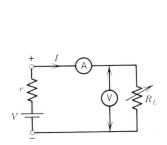

(*a*) Circuit to determine maximum load power

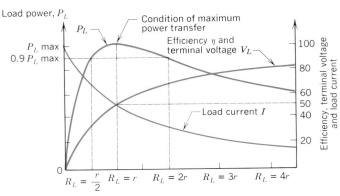

(*b*) Graphs of load power, terminal voltage, load current, and efficiency against load resistance

FIGURE 9-9

Circuit and graphs for maximum power transfer.

When $R_L = 4 \ \Omega$, $V_L = 40 \text{ V} \times \dfrac{4 \ \Omega}{4 + 8 \ \Omega} = \textbf{13.3 V,}$

$$P_L = \frac{V_L^2}{R_L} \tag{3-8}$$

$$= \frac{(13.3 \text{ V})^2}{4 \ \Omega} = 44.2 \text{ W}$$

$$P_T = \frac{V^2}{R_T} = \frac{V^2}{R_L + r} \tag{3-8}$$

$$= \frac{(40 \text{ V})^2}{4 + 8 \ \Omega} = 133.3 \text{ W}$$

efficiency, $\eta = \dfrac{P_L}{P_T} \times 100\%$ \hfill (9-9)

$$= \frac{44.2 \text{ W}}{133.3 \text{ W}} \times 100\% = \textbf{33.2\%}$$

Similarly, when

$R_L = 8 \ \Omega$, $V_L = \textbf{20 V,}$
$\qquad\qquad P_L = 50 \text{ W}$, $P_T = 100 \text{ W}$, $\eta = \textbf{50\%}$
$R_L = 16 \ \Omega$, $V_L = \textbf{26.6 V,}$
$\qquad\qquad P_L = 44.2 \text{ W}$, $P_T = 66.6 \text{ W}$, $\eta = \textbf{66.4\%}$
$R_L = 100 \ \Omega$, $V_L = \textbf{37 V,}$
$\qquad\qquad P_L = 13.7 \text{ W}$, $P_T = 14.8 \text{ W}$, $\eta = \textbf{92.6\%}$

The maximum efficiency, as defined above, occurs at loads approaching an open circuit. (See Fig. 9-9b.) This is because the power wasted in the internal resistance, in the form of heat, is less with a smaller current.

Efficiency is *always* 50% at the value of load resistance that provides *maximum power transfer* (that is, when $R_L = r$). Under these conditions, the load voltage is *one-half* the open-circuit voltage. The load is said to be *matched* to the source. This is desirable in situations such as the connection of a loudspeaker to an amplifier. Assuming a 40-V audio source with a total internal resistance of 8 Ω, an 8-Ω speaker would have a maximum power of 50 W delivered to it. If either a 4-Ω or a 16-Ω speaker were used with the same 40-V, 8-Ω source, the power delivered would be only 44.4 W. A *mismatch* between the load and the source can have other consequences besides low-power transfer. If the load (speaker) has a much lower resistance than the source (amplifier), damage can occur in the amplifier due to excessive current. Even if damage does not occur, distortion may occur in the amplifier and be heard in the speaker.

Maximum power transfer is a more important consideration than efficiency in *electronic* systems, where low-power signals are involved. In *power* systems, however, such as high-voltage transmission lines, it is impractical to match the load to the source. It is more desirable in power sources (such as batteries or a 60-Hz ac supply) to aim for maximum efficiency, rather than maximum power transfer. This will minimize any high-power loss within the source.

9-4 POWER MEASUREMENT—THE WATTMETER

A *wattmeter* can be used to verify the maximum power transfer theorem experimentally. The deflection of the wattmeter is directly proportional to the product of load

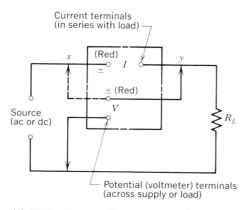

(a) Wattmeter connections to measure load power

(b) Wattmeter with a 25-W range

FIGURE 9-10
Power measurement.
[Photograph in (b) courtesy of Hickok Teaching Systems, Inc. Woburn, Mass.]

current and voltage if the circuit is purely resistive. In this chapter, only the way in which the wattmeter is *connected* will be covered. The internal operation and construction of the wattmeter will be described in Chapter 16.

The instrument has one pair of current terminals and one pair of potential (voltage) terminals. One terminal of each pair is usually red or may be marked with a ± sign. As shown in Fig. 9-10a, the current that flows from source to load passes through the current terminals. The voltmeter terminals are usually connected across the load. For proper pointer deflection and to avoid possible meter damage, the ± voltage terminal must be connected to either point x or point y, as identified on the schematic. That is, to make the meter read upscale properly, the ± terminals must have the same instantaneous polarity whether the source is ac or dc.

SUMMARY

1. The internal resistance of a cell arises from its components (electrodes and electrolyte) and causes a reduction in terminal voltage from the open-circuit voltage when a load is connected.
2. The internal resistance of a cell may be determined by noting the change in terminal voltage when a known current is drawn from the cell.
3. The voltage regulation of a source is an indication of the change in terminal voltage under load. An ideal source would have a voltage regulation of 0%.
4. If the terminal voltage is reduced to one-half of the open-circuit voltage when a load is connected, the load resistance is equal to the internal resistance. Source and load are said to be *matched*.
5. Connecting cells in *series* increases terminal voltage and internal resistance, but not current capability. A *parallel* connection increases current capability and reduces internal resistance, but does not increase terminal voltage.
6. The total ampere-hour (Ah) capacity of series-connected similar capacity cells is the same as each cell's individual capacity.
7. The total ampere-hour capacity of parallel-connected cells is the sum of the individual cell capacities.
8. For a series-parallel combination, the total ampere-hours depends upon the particular connection.
9. The total energy of a given number of interconnected cells is independent of the type of connection.
10. Maximum power transfer from source to load occurs when the load resistance and internal resistance of the source are equal. At this point, terminal voltage is one-half the open-circuit voltage, and the efficiency of the circuit is 50%.
11. A wattmeter essentially multiplies the current and voltage supplied to a load to indicate, on a direct-reading scale, the power being delivered to the load. The instrument can measure both ac and dc power.

SELF-EXAMINATION

Answer true or false or a, b, c, or d
(Answers at back of book)

  9-1. Internal resistance is inevitable in all sources of emf. _____
  9-2. If a current of 20 mA is drawn from a source of open circuit voltage, 24 V, and internal resistance, 200 Ω, the terminal voltage is

    a.   22 V
    b.   24 V
    c.   20 V
    d.   18 V

9-3. If the connection of a load to a voltage source reduces the terminal voltage to one-half of the open-circuit voltage, the load resistance is one-half of the internal resistance. _____

9-4. If the terminal voltage of a source drops from an open-circuit voltage of 30 V to a value of 20 V when a 60-Ω load is connected, the source's internal resistance is
    a.   20 Ω
    b.   30 Ω
    c.   40 Ω
    d.   60 Ω

9-5. The voltage regulation of the source in Question 9-4 is
    a.   33.3%
    b.   66.6%
    c.   50%
    d.   10%

9-6. A power supply that has a 100% voltage regulation has an effective internal resistance of 0 Ω. _____

9-7. The terminal voltage of a dc generator required to charge a 24-V battery (whose internal resistance is 0.05 Ω) with a current of 50 A is
    a.   21.5 V
    b.   24 V
    c.   25.5 V
    d.   26.5 V

9-8. The total voltage of a battery that consists solely of interconnected series cells is the sum of the emfs of all the cells. _____

9-9. The object of connecting cells in parallel is to produce a battery with increased current capability and decreased internal resistance. _____

9-10. The ampere-hour capacity of a number of interconnected cells is a constant regardless of the method of cell interconnection. _____

9-11. Four $1\frac{1}{2}$-V cells are connected in parallel with each other. This combination is then connected in series with three 2-V cells also connected in series with each other. If each cell has an internal resistance of 1 Ω, the combined voltage and internal resistance is
    a.   8 V, 7 Ω
    b.   14 V, 7 Ω
    c. $3\frac{1}{2}$ V, 3.25 Ω
    d. $7\frac{1}{2}$ V, 3.25 Ω

9-12. When a load is *matched* to a source, the load resistance equals the internal resistance of the source. _____

9-13. Maximum power transfer occurs in a circuit when the circuit efficiency is 100%.

_____

9-14. The efficiency for the conditions given in Question 9-4 is
    a.   33.3%
    b.   66.6%
    c.   50%
    d.   10%

9-15. The graph of load power versus load resistance is symmetrical on either side of the maximum power transfer condition. _____

9-16. A wattmeter in a circuit will read a maximum value if the load resistance is adjusted to one-half the value of the source's internal resistance. _____

REVIEW QUESTIONS

1. What makes up the internal resistance of a battery?
2. a. Why do the headlights of a car dim momentarily when the engine is being started?
   b. What visible indication might there be in such a situation when the battery's warranty period has just expired?
3. Is it possible to "jump" a car with a positive ground system using another car with a negative ground? Explain, using a diagram.
4. Describe how you can determine the internal resistance of a supply having:
   a. A high internal resistance.
   b. A very low internal resistance.
5. a. A power supply is known to have a 100% voltage regulation. Is this good? Explain.
   b. An electronic voltage regulator effectively maintains a constant output voltage (in a given power supply), independent of the load current. What can you say about the supply's internal resistance and voltage regulation?
6. What is the result of connecting cells in the following ways, as far as voltage, current, internal resistance, Ah capacity, and total energy are concerned?
   a. Series
   b. Parallel
   c. Series-parallel
7. What is the relation between the power transfer, efficiency, load voltage, internal resistance, and load resistance when a source and load are "matched"?
8. Explain how it is possible for there to be two different load resistances that cause only one-half of the maximum power to be delivered from a source of given internal resistance.
9. When is maximum *power transfer* most desirable and when is maximum *efficiency* most desirable? Why?
10. Describe the proper method for connecting a wattmeter in a circuit.

PROBLEMS

(Answers to odd-numbered problems at back of book)

9-1. A 100-V power supply has an internal resistance of 2.5 kΩ. A 7.5-kΩ load is connected across the power supply terminals. Calculate:
   a. The current through the load.
   b. The internal voltage drop in the power supply.
   c. The terminal voltage of the supply under load.
   d. The power dissipated in the power supply when a 5-kΩ load is connected.
   e. The current that flows when the power supply terminals are short-circuited.

9-2. Repeat Problem 9-1 using a 5-kΩ internal resistance.

9-3. A 1.5-V cell's terminal voltage drops to 1.25 V when a load is connected that draws a full-load current of 1 A.
  a.   What is the cell's internal resistance?
  b.   What is the load resistance?
  c.   What load resistance would drop the terminal voltage to 1 V?

9-4. A cell has a no-load voltage of 1.2 V. The terminal voltage becomes 1 V when a 2-Ω load is connected. What is the cell's internal resistance?

9-5. A cell has a terminal voltage of 2.2 V with a load of 10 Ω. This drops to 2 V when the load is changed to 2 Ω.
  a.   Calculate the cell's open-circuit voltage.
  b.   Calculate the cell's internal resistance.

9-6. A battery has a terminal voltage of 13.6 V when delivering 10 A to a load. When the battery is being charged with 20 A, its terminal voltage is 14.1 V.
  a.   What is the battery's open-circuit voltage?
  b.   What is the battery's internal resistance?

9-7. Determine the percent voltage regulation of the cell in Problem 9-3.

9-8. Determine the percent voltage regulation of the cell in Problem 9-4.

9-9. A battery has a no-load voltage of 13.2 V. If the battery is known to have a voltage regulation of 15%, what is its full-load voltage?

9-10. A power supply has a full-load voltage of 22 V and a voltage regulation of 30%. What is its open-circuit voltage?

9-11. Two 6-V cells and one 12-V cell are connected in series. Each of the 6-V cells has an internal resistance of 0.05 Ω, and the 12-V cell has a resistance of 0.1 Ω. How much charging current will flow from a 30-V charger with a 0.3-Ω resistance when connected to the three series-connected cells?

9-12. A 14.1-V charger with an internal resistance of 0.05 Ω is connected across a 12.6-V battery with an internal resistance of 0.1 Ω.
  a.   Determine the charging current.
  b.   Determine the terminal voltage.
  A second battery, identical to the first, is connected in parallel with the battery and charger.
  c.   Determine the total current from the charger.
  d.   Determine the current into each battery.
  e.   Determine the terminal voltage.

9-13. Four 50-Ah, 6-V cells, each with an internal resistance of 0.01 Ω and a current capability of 10 A are to be interconnected. Determine the terminal voltage, current capability, and internal resistance of the following arrangements:
  a.   All connected in series.
  b.   All connected in parallel.
  c.   Two series connections of two each in parallel.

9-14. For each of the arrangements in Problem 9-13, find:
  a.   The Ah capacity.
  b.   The total energy.

9-15. Two batteries are connected in parallel with each other. The first has an emf of 12.2 V and an internal resistance of 0.06 Ω; the second, an emf of 13 V and a resistance of 0.04 Ω.
  a.   What is the terminal voltage of the combination?
  b.   What is the internal resistance of the combination?
  If the batteries were mistakenly connected with positive-to-negative polarity calculate:

    c.   The terminal voltage.

    d.   The current flowing.

    e.   The power dissipated in each battery in the form of heat.

9-16.   A wattmeter is connected between an ac supply and a variable resistive load. The load is adjusted until at a resistance of 10 Ω the wattmeter indicates a maximum reading of 40 W. Determine:

    a.   The voltage across the load.

    b.   The internal resistance of the supply.

    c.   The open-circuit voltage of the supply.

    d.   The voltage regulation of the supply.

    e.   The efficiency at which the circuit is operating.

    f.   The load resistance to operate at an efficiency of 60%.

9-17.   a.   What should be the resistance of an automotive starter motor if it is to be matched to a 12-V battery with an internal resistance of 0.02 Ω?

    b.   What is the *initial* current that will flow for matched conditions?

    c.   How much power is delivered to the motor?

    d.   If the motor is 70% efficient, how much horsepower does the motor develop?

    e.   How much power is dissipated in heat in the battery?

    f.   What is the terminal voltage of the battery?

    g.   What is the efficiency of the circuit?

9-18.   a.   A power supply is known to have a 50% voltage regulation. If its no-load voltage is 15 V, what will be its full-load voltage?

    b.   If the current drawn from the above power supply at full load is 0.5 A, what is the internal resistance of the power supply?

    c.   If the above power supply is connected to a load of 50 Ω, how much power will a wattmeter indicate when connected in series with the load?

    d.   What value of load resistance will provide a maximum indication of the wattmeter in part (c) and what will this reading be?

    e.   What will be the load voltage in part (d) ignoring all wattmeter losses?

9-19.   An audio power amplifier has an open-circuit voltage of 20 V and an internal resistance of 8 Ω.

    a.   What two values of load resistance will give one-half of the maximum power possible?

    b.   What will be the circuit efficiency at these two values?

    c.   Which of the two loads would be preferable?

CHAPTER 10

NETWORK ANALYSIS

All the circuits you have considered thus far have been series, parallel, or series-parallel. But there are many circuits, especially those involving more than one source of emf, where components do not have these simple relationships. Such circuits are called *networks,* and analyzing them requires the use of different *techniques.* Some of the more useful methods of analysis, allowing you to solve both ac and dc networks, will be covered in this chapter.

Two of these, the *branch current* method and the *loop current* method, have similarities. They use a combination of Kirchhoff's and Ohm's laws, and involve the simultaneous solution of two or more equations to solve for the unknown currents.

The *superposition* theorem may be used to advantage in analyzing circuits that contain more than one source of emf. This theorem considers the currents resulting from each emf applied separately, and superimposes those currents to find the overall values when all emfs are in the circuit. It permits a solution using simple series-parallel combinations.

The fourth method is *Thévenin's theorem.* It permits the replacement of any complex circuit of emfs and linear resistors by an *equivalent circuit* consisting of an emf in *series* with a single resistance.

Similarly, *Norton's theorem* provides an equivalent circuit consisting of a *constant current source* in *parallel* with a single resistance. Both Norton's and Thévenin's equivalent circuits are useful in dealing with transistors and other active electronic devices.

Nodal analysis uses only Kirchhoff's current law, in which total current entering a node (or junction) is equated to total current leaving that node.

The final method considered in this chapter is the *delta-wye* conversion. Sometimes referred to as a π-T conversion, this method is used to change a complex network into a series-parallel combination.

10-1 BRANCH CURRENT METHOD

In this method, each branch current is assigned a designation and Kirchhoff's voltage law is applied to each complete circuit or loop.

Kirchhoff's voltage law can be restated as:

Around any complete circuit (loop), the sum of the voltage rises equals the sum of the voltage drops.

Rules for Branch Current Procedure

1. Draw a circuit diagram large enough to show all information clearly.
2. Determine the number of ''windows'' (openings) in the circuit. This determines *how many* different branch currents I_1, I_2, I_3, and so on should be shown. Indicate these on the diagram, using any arbitrary direction.
3. Use Kirchhoff's current law to identify the remaining currents in the circuit *in terms of I_1, I_2,* and so on.
4. Using the *assumed* directions of current I_1, I_2, and so on, show the *corresponding* polarity signs of the voltage drops across each component.

5. On the diagram, draw as many loops as there are windows. The direction of each loop, shown by an arrow, is completely arbitrary.
6. Trace (move) around each loop. Write a Kirchhoff voltage equation with voltage *rises* on one side and voltage *drops* on the other.
    a. A voltage *rise* occurs as you move through a component from − to + in the direction of the loop.
    b. A voltage *drop* occurs as you move through a component from + to − in the direction of the loop.
7. If not done in Step 6, insert in the equation the emf values and the values of R_1, R_2, and so on.
8. Collect like terms and solve for the unknown currents. There must be as many equations as there are unknown currents.
9. Find the current in each component, using the current designations from Steps 2 and 3. Interpret any negative currents as having an actual direction *opposite* to the assumed direction.

The branch current method is used in Example 10-1.

EXAMPLE 10-1

A battery of open-circuit voltage V_B and internal resistance R_B is connected in parallel with a generator of open-circuit voltage V_G and internal resistance R_G. This combination feeds a load resistance R_L. For the following values find the battery current, generator current, load current, and load voltage: $V_B = 13.2$ V, $V_G = 14.5$ V, $R_B = 0.5\ \Omega$, $R_G = 0.1\ \Omega$, and $R_L = 2\ \Omega$.

Solution

1. Draw a circuit diagram. See Fig. 10-1a.
2. It can be seen from Fig. 10-1a that there are *two* "windows" in the circuit. This means that we must show two currents, in two branches, in any *arbitrary* direction. (We shall show currents I_B and I_G in the direction we *think* the current might flow.) See Fig. 10-1b.
3. Using Kirchhoff's current law, we can identify the current through the load resistor as

$$I_L = I_B + I_G \qquad (6\text{-}1)$$

Indicate this current in Fig. 10-1b.

4. Show the polarity signs of the voltage drops across each resistor using the assumed directions of current. See Fig. 10-1b.
5. Indicate, in each window, a loop that goes around a *complete* circuit. The direction is arbitrary, but it is often convenient to use a direction that goes from − to + through an emf. See loops 1 and 2 in Fig. 10-1b. (Note that a third loop could be drawn around the *outside* of the circuit that includes V_B, R_B, R_G, and V_G. Any two of these three loops could be used to provide two independent equations.)
6. Trace around each loop, writing a Kirchhoff's voltage law equation with the voltage rises on one side and the voltage drops on the other.

For loop 1:

$$V_B = I_B R_B + (I_B + I_G)R_L \qquad (1)$$

For loop 2:

$$V_G = I_G R_G + (I_B + I_G)R_L \qquad (2)$$

7. Insert in Eqs. 1 and 2 the numerical values.

$$13.2 = 0.5I_B + 2(I_B + I_G) \qquad (3)$$
$$14.5 = 0.1I_G + 2(I_B + I_G) \qquad (4)$$

8. Collect together like terms and solve for I_B and I_G:*

$$13.2 = 2.5I_B + 2.0I_G \qquad (5)$$
$$14.5 = 2.0I_B + 2.1I_G \qquad (6)$$

Multiply Eq. 5 by four and Eq. 6 by five:

$$52.8 = 10I_B + 8I_G \qquad (7)$$
$$72.5 = 10I_B + 10.5I_G \qquad (8)$$

Subtract Eq. 7 from Eq. 8

$$19.7 = 2.5I_G$$

Therefore, $I_G = \dfrac{19.7}{2.5} = \mathbf{7.88\ A}$

Substitute $I_G = 7.88$ A in Eq. 6.

$$2I_B = 14.5 - 2.1I_G$$
$$= 14.5 - 2.1 \times 7.88 = -2.048$$

Therefore, $I_B = \dfrac{-2.048}{2} = \mathbf{-1.02A}$

9. The negative sign with I_B indicates that the *actual* direction of current is opposite to the *assumed* direction. Thus a current of 1.02 A *enters* the positive terminal of V_B and acts to recharge the battery. The load current is given by

---

*Linear equations in two or more unknowns may also be solved using the method of determinants. See Appendix D.

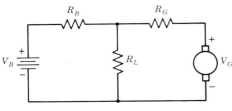

(a) Original circuit

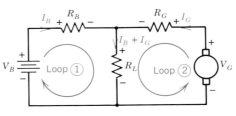

(b) Circuit with current designations, voltage drops, and loops

FIGURE 10-1
Circuits for Example 10-1.

$$I_L = I_G + I_B \tag{6-1}$$
$$= 7.88 - 1.02 = \textbf{6.86 A}$$

and the load voltage is

$$V_L = I_L R_L \tag{3-1b}$$

This is a value somewhere between the two values of the voltage sources.

EXAMPLE 10-2

Given the circuit in Fig. 10-2 use the method of branch currents to find the current in each resistor: $V_1 = 10\ V$, $V_2 = 15\ V$, $R_1 = 1\ k\Omega$, $R_2 = 2\ k\Omega$, $R_3 = 3\ k\Omega$, $R_4 = 4\ k\Omega$, and $R_5 = 5\ k\Omega$.

Solution

Since there are three windows, assign currents I_1, I_2, and I_3, and show I_{R_2} and I_{R_4} in terms of these, as shown in Fig. 10-2. Indicate the voltage drop polarity signs across the resistors and three tracing loops, with directions arbitrarily selected. We now write three equations based on Kirchhoff's voltage law.

Loop 1: $V_1 = I_1 R_1 + (I_1 + I_3)R_2$
Loop 2: $(I_2 - I_3)R_4 = I_3 R_3 + (I_1 + I_3)R_2$
Loop 3: $0 = V_2 + I_2 R_5 + (I_2 - I_3)R_4$

Substituting values and collecting terms:

$$10 = 3I_1 + 0I_2 + 2I_3 \tag{1}$$
$$0 = 2I_1 - 4I_2 + 9I_3 \tag{2}$$
$$-15 = 0I_1 + 9I_2 - 4I_3 \tag{3}$$

Multiply Eq. 1 by two and add to Eq. 3 to obtain:

$$5 = 6I_1 + 9I_2 \tag{4}$$

Multiply Eq. 1 by 9, multiply Eq. 2 by -2 and add to obtain

$$90 = 23I_1 + 8I_2 \tag{5}$$

Multiply Eq. 4 by 8, multiply Eq. 5 by -9 and add to obtain

$$770 = 159I_1$$

Therefore, $I_1 = \dfrac{770}{159} = \textbf{4.84 mA}$ (in direction assumed)

NOTE Since resistance values are in $k\Omega$ and emfs in volts, the current has units of mA.
By substitution, we obtain

$$I_2 = \textbf{-2.67 mA} \text{ (opposite to the direction assumed)}$$
$$I_3 = \textbf{-2.26 mA} \text{ (opposite to the direction assumed)}$$
$$I_{R_2} = I_1 + I_3$$
$$= 4.84 + (-2.26)\ mA$$
$$= \textbf{2.58 mA} \text{ (in the direction assumed)}$$
$$I_{R_4} = I_2 - I_3$$
$$= -2.67 - (-2.26)\ mA$$
$$= \textbf{-0.41 mA} \text{ (opposite to the direction assumed)}$$

10-2 LOOP OR MESH CURRENT METHOD

At first glance, it may seem that the loop current method is very similar to the branch current method. In this method, however, a current is indicated for each loop, rather than each branch. Only Kirchhoff's voltage law is applied to each loop.

Rules for Loop Procedure

1. Draw a circuit diagram large enough to show all information clearly.

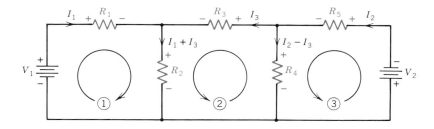

FIGURE 10-2
Circuit for Example 10-2.

2. In each window, indicate a loop current (I_1, I_2, and so on) flowing in an arbitrary direction. (A loop current may also be shown that encompasses two or more windows.) There must be as many *independent* loop currents as there are windows.

3. In *each loop*, indicate the polarity signs of voltage drops across the resistors for the assumed direction of loop current. If a resistor has more than one loop current through it, there will be two sets of voltage drops (one in each loop). They may have the same or opposing polarity signs. (See R_5 in Fig. 10-3b.)

4. Apply Kirchhoff's voltage law to each loop. Follow the assumed direction of the loop current, equating voltage rises to voltage drops.
   a. If a resistor has *opposing* current loops through it, the net voltage drop is the product of the resistance and the *difference* of the two currents. The larger of the two currents is assumed to be that of the loop you are tracing.
   b. If a resistor has current loops through it in the *same* direction, the total voltage drop is the product of the resistance and the *sum* of the two loop currents.
   c. If a loop has no emfs in it, the voltage-rise side of the equation is zero.

5. Insert in the equations the numerical values for all the components and sources of emf.

6. Collect like terms and solve for the loop currents. There must be as many equations as there are loop currents.

7. If a resistor has more than one loop current through it, find the algebraic sum of the loops. Use the signs ($+$ and $-$) as found in Step 6 and the assumed directions as assigned in Step 2. Interpret any negative currents as having an actual direction opposite to the assumed direction.

EXAMPLE 10-3

Given resistors R_1, R_2, R_3, R_4, and R_5 connected with voltage sources V_1, V_2, and V_3 (as shown in Fig. 10-3a), determine the current, and its direction, through V_2.

Solution

Indicate the two loop currents (I_1, I_2) as shown in Fig. 10-3b. Then show the polarity signs of the resistor voltage drops for each loop current. Write a Kirchhoff's voltage

law equation for each loop, with voltage rises on one side and voltage drops on the other.

Loop 1: $V_1 + V_2 = I_1R_1 + (I_1 - I_2)R_5 + I_1R_2$
Loop 2: $0 = V_2 + (I_2 - I_1)R_5 + I_2R_3 + V_3 + I_2R_4$

NOTE There are no voltage rises in Loop 2, so the left side of the equation is zero.

Substituting values:

$$15 + 20 = 10\,I_1 + 5(I_1 - I_2) + 15\,I_1$$

or
$$35 = 30\,I_1 - 5I_2 \tag{1}$$

$$0 = 20 + 5(I_2 - I_1) + 8I_2 + 10 + 12I_2$$

or
$$-30 = -5I_1 + 25I_2 \tag{2}$$

Solving simultaneously, we obtain

$$I_1 = 1\text{ A}, \qquad I_2 = -1\text{ A}$$

Current through V_2 is given by

$$I_1 - I_2 = 1\text{ A} - (-1\text{ A})$$
$$= 1\text{ A} + 1\text{ A} = \mathbf{2\ A\ \text{“downward”}}$$

EXAMPLE 10-4

Given the Wheatstone bridge circuit in Fig. 10-4a, use the loop current method to calculate:

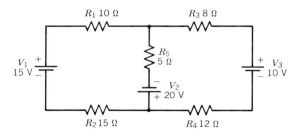

(a) Original circuit

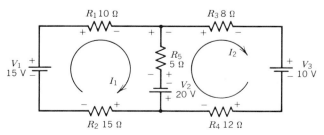

(b) Circuit with current loops and voltage drops

FIGURE 10-3

Circuits for Example 10-3.

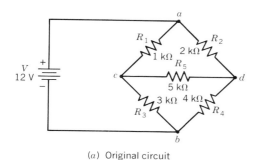

(a) Original circuit

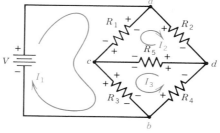

(b) Circuit with current loops and voltage drops

FIGURE 10-4

Circuits for Example 10-4.

a. The current in each resistor.
b. The total resistance of the circuit between a and b given $V = 12$ V, $R_1 = 1$ kΩ, $R_2 = 2$ kΩ, $R_3 = 3$ kΩ, $R_4 = 4$ kΩ, and $R_5 = 5$ kΩ.

NOTE It is impossible to solve for the total resistance between a and b using simple series-parallel methods, since there are no resistors either in series or in parallel with each other.

Solution

a. Indicate the three loop currents I_1, I_2, and I_3. Show the voltage drops across the resistors in each loop. See Fig. 10-4b. Write a Kirchhoff's voltage law equation for each loop:

Loop 1: $V = (I_1 - I_2)R_1 + (I_1 + I_3)R_3$
Loop 2: $0 = (I_2 + I_3)R_5 + (I_2 - I_1)R_1 + I_2R_2$
Loop 3: $0 = (I_2 + I_3)R_5 + (I_1 + I_3)R_3 + I_3R_4$

Substituting values and collecting terms, we obtain:

$$12 = 4I_1 - 1I_2 + 3I_3$$
$$0 = -1I_1 + 8I_2 + 5I_3$$
$$0 = 3I_1 + 5I_2 + 12I_3$$

By simultaneous solution we obtain

$I_1 = 5.03$ mA, $I_2 = 1.92$ mA, $I_3 = -2.07$ mA

Thus

$I_{R_1} = I_1 - I_2 = 5.03 - 1.92$ mA $=$ **3.11 mA**
$I_{R_2} = I_2 =$ **1.92 mA**
$I_{R_3} = I_1 + I_3 = 5.03 + (-2.07)$mA $=$ **2.96 mA**
$I_{R_4} = I_3 =$ **−2.07 mA**
  (opposite to the direction shown)
$I_{R_5} = I_2 + I_3 = 1.92 + (-2.07)$ mA $=$ **−0.15 mA**
  (opposite to the direction assumed)

b. The total current from the supply $= I_1 = 5.03$ mA. resistance between a and b

$$= \frac{V}{I_1} = \frac{12\ V}{5.03\ mA} = \textbf{2.39 k}\Omega$$

10-3 THE SUPERPOSITION THEOREM

As you learned in the preceding sections, circuits that contain more than one source of emf will often cause the resistive components to *not* have a simple series-parallel relation. In such cases, one method is to write an equation for each loop and solve simultaneously for each current.

The principle of the superposition theorem states that, for any given component, the currents that result from *one voltage source at a time* can be superimposed. Since the removal of all but one voltage source usually results in a series-parallel combination, you can solve for the required current or voltage using the methods you learned in Chapters 5, 6, and 7. An *algebraic addition* of each component's voltage or current will then yield the total when all voltage sources are in the circuit.

The superposition theorem may be stated as follows: **The total component voltage or current in a multi-emf circuit is the algebraic sum of the individual voltages and currents resulting from each voltage source applied one at a time, with all other voltage sources removed and replaced by their respective internal resistances.**

Note that only if a voltage source has zero internal resistance can it be removed by simply shorting it out. This is shown in Example 10-5.

EXAMPLE 10-5

A battery of open-circuit voltage 13.2 V and internal resistance 0.5 Ω is connected in parallel with a generator of open-circuit voltage 14.5 V and internal resistance 0.1 Ω. This combination feeds a load having a resistance of 2 Ω. See Fig. 10-5a. Using the superposition theorem, calculate:

a. Battery current.
b. Generator current.
c. Load current.
d. Load voltage.

Solution

a. With V_G reduced to zero as in Fig. 10-5b, the total circuit resistance is, by series-parallel techniques:

$$R_T = R_B + \frac{R_L \times R_G}{R_L + R_G}$$

$$= 0.5\ \Omega + \frac{2\ \Omega \times 0.1\ \Omega}{2\ \Omega + 0.1\ \Omega} = 0.595\ \Omega$$

$$I_T = I_{B_1} = \frac{V_B}{R_T} \tag{3-1a}$$

$$= \frac{13.2\ \text{V}}{0.595\ \Omega} = 22.2\ \text{A}$$

The component of load current due to V_B alone is

$$I_{L_1} = I_T \times \frac{R_G}{R_G + R_L} \tag{6-12}$$

$$= 22.2\ \text{A} \times \frac{0.1\ \Omega}{0.1\ \Omega + 2\ \Omega} = 1.057\ \text{A}$$

$$\text{and } I_{G_1} = I_T \times \frac{R_L}{R_L + R_G} \tag{6-13}$$

$$= 22.2\ \text{A} \times \frac{2\ \Omega}{2\ \Omega + 0.1\ \Omega} = 21.143\ \text{A}$$

Similarly, for V_B reduced to zero as in Fig. 10-5c,

$$R_T = R_G + \frac{R_L \times R_B}{R_L + R_B}$$

$$= 0.1\ \Omega + \frac{2\ \Omega \times 0.5\ \Omega}{2\ \Omega + 0.5\ \Omega} = 0.5\ \Omega$$

$$I_T = I_{G_2} = \frac{V_G}{R_T} \tag{3-1a}$$

$$= \frac{14.5\ \text{V}}{0.5\ \Omega} = 29\ \text{A}$$

$$I_{L_2} = I_T \times \frac{R_B}{R_B + R_L} \tag{6-12}$$

$$= 29\ \text{A} \times \frac{0.5\ \Omega}{0.5\ \Omega + 2\ \Omega} = 5.8\ \text{A}$$

$$I_{B_2} = 29\ \text{A} - 5.8\ \text{A} = 23.2\ \text{A}$$

The total battery current is the algebraic sum of I_{B_1} and I_{B_2}:

$$I_B = 22.2\ \text{A} \uparrow + 23.2\ \text{A} \downarrow = \mathbf{1\ A} \downarrow$$

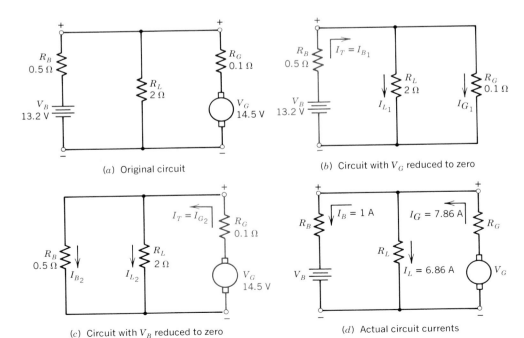

(a) Original circuit

(b) Circuit with V_G reduced to zero

(c) Circuit with V_B reduced to zero

(d) Actual circuit currents

FIGURE 10-5
Circuits for Example 10-5.

This means, because I_{B_2} is greater than I_{B_1}, the actual current through the battery is in the same direction as I_{B_2} in Fig. 10-5c. The battery is being recharged because current is flowing *into* the positive terminal.

b. The total generator current is

$$I_G = I_{G_1} + I_{G_2} \qquad (6\text{-}1)$$
$$= 21.143 \text{ A} \downarrow + 29 \text{ A} \uparrow = \textbf{7.86 A} \uparrow$$

c. The total load current is

$$I_L = I_{L_1} + I_{L_2} \qquad (6\text{-}1)$$
$$= 1.057 \text{ A} \downarrow + 5.8 \text{ A} \downarrow = \textbf{6.86 A} \downarrow$$

Note that because I_{L_1} and I_{L_2} are in the same direction, the actual current is the sum of the two.

d. The total load voltage may be found by superimposing the load voltage components due to each voltage source:

$$V_{L_1} = I_{L_1} \times R_L \qquad (3\text{-}1b)$$
$$= 1.057 \text{ A} \times 2 \text{ } \Omega = 2.114 \text{ V}$$
$$V_{L_2} = I_{L_2} \times R_L \qquad (3\text{-}1b)$$
$$= 5.8 \text{ A} \times 2 \text{ } \Omega = 11.6 \text{ V}$$
$$V_L = V_{L_1} + V_{L_2} \qquad (5\text{-}1)$$
$$= 2.114 \text{ V} + 11.6 \text{ V} = \textbf{13.7 V}$$

Actual circuit currents are shown in Fig. 10-5d.

Compare the solution of Example 10-5, and the work needed to reach it, with Example 10-1, which used the method of branch currents. It is doubtful that the superposition method reduces the amount of work needed to reach a solution, although it does allow use of basic series-parallel theory. This is especially true for more complex circuits, such as the one shown in Fig. 10-2.

Finally, note that the superposition theorem works only for circuits consisting of *linear* and *bilateral* components. A linear component is one in which there is a directly proportional relationship between voltage and current. To be bilateral, the characteristics of a device must not depend upon polarity—current must be able to flow through it equally well in either direction. For example, a diode inserted in the battery lead would prevent use of the superposition theorem, since the diode is neither linear nor bilateral. Also, the total power dissipated in the load resistor (Fig. 10-5a) is *not* the sum of the two powers dissipated in the load resistor (Figs. 10-5b and 10-5c), because

power is *not* related in a linear manner to voltage and current.

10-4 THÉVENIN'S THEOREM

This is a very powerful method of converting a complex circuit containing emfs and linear components into a simple equivalent circuit of one emf and one resistance. It is especially valuable when repeated calculations are to be made for different values of a single component. This method allows the rest of the circuit to which that component is connected to be represented by an unchanging equivalent circuit.

Thévenin's theorem can be stated as follows:

Any linear two-terminal network consisting of fixed resistances and sources of emf can be *replaced* by a single source of emf (V_{Th}) in series with a single resistance (R_{Th}), whose values are given by the following:

1. V_{Th} **is the open-circuit voltage at the two specified terminals in the original circuit.**
2. R_{Th} **is the resistance looking back into the original network from the two specified terminals, with all sources of emf shorted out and replaced by their internal resistances.**

Experimentally, the Thévenin voltage (V_{Th}) of the circuit shown in Fig. 10-6a could be measured by connecting a high-resistance voltmeter between terminals A and B. To measure the equivalent resistance with an ohmmeter between A and B, it would be necessary to remove any emf by shorting the voltage source (V).

The Thévenin equivalent circuit shown in Fig. 10-6b will then behave, as far as any load between terminals A and B is concerned, in the same manner as the original circuit (Fig. 10-6a). The values of V_{Th} and R_{Th} may also be found by calculation. For example, in Fig. 10-6a, let $V = 20$ V, and $R_1 = R_2 = R_3 = 40$ kΩ. The open-circuit voltage between A and B is the same as V_{R_3}. (R_2 has no effect since there is no voltage drop across it with an open circuit between A and B.) Thus:

$$V_{\text{Th}} = V_{R_3} = V \times \frac{R_3}{R_1 + R_3}$$
$$= 20 \text{ V} \times \frac{40 \text{ k}\Omega}{40 + 40 \text{ k}\Omega}$$
$$= 20 \text{ V} \times \frac{40}{80} = \textbf{10 V}$$

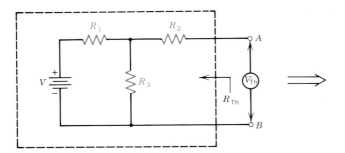

(a) Initial, linear, two-terminal circuit

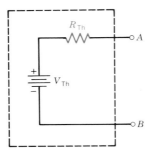

(b) Thévenin equivalent circuit

FIGURE 10-6
Circuits for
Thévenin's theorem.

To obtain R_{Th} between A and B, consider V is shorted.

$$R_{Th} = R_1 \| R_3 + R_2$$
$$= \frac{40 \text{ k}\Omega}{2} + 40 \text{ k}\Omega = \mathbf{60 \text{ k}\Omega}$$

The Thévenin equivalent circuit of a 10-V source in series with a 60-kΩ resistance will take the place of the original circuit as far as any load connected between terminals A and B is concerned.

Further examples to calculate V_{Th} and R_{Th} follow.

EXAMPLE 10-6

Use Thévenin's theorem to find the current through the load resistor R_L in Fig. 10-7 for the following values of R_L:

a. 1 Ω
b. 2 Ω
c. 3 Ω

Solution

1. Draw the circuit with the load resistor R_L removed. Identify the load terminals as A and B. See Fig. 10-8.

2. We now have a simple series circuit with a current given by

$$I = \frac{V_G - V_B}{R_B + R_G}$$
$$= \frac{14.4 - 13.2 \text{ V}}{0.5 + 0.1 \ \Omega} = \frac{1.2 \text{ V}}{0.6 \ \Omega} = 2 \text{ A}$$

3. The open-circuit voltage between A and B may be found by two methods:

$$V_{R_B} = IR_B \qquad\qquad (3\text{-}1b)$$
$$= 2 \text{ A} \times 0.5 \ \Omega = 1 \text{ V}$$

Therefore, $V_{Th} = V_B + V_{R_B} = V_{AB}$ (5-1)
$$= 13.2 + 1 = \mathbf{14.2 \text{ V}}$$

or $\qquad V_{R_G} = IR_G \qquad\qquad (3\text{-}1b)$
$$= 2 \text{ A} \times 0.1 \ \Omega = 0.2 \text{ V}$$

so that $V_{Th} = V_G - V_{R_G} = V_{AB}$ (5-1)
$$= 14.4 - 0.2 = \mathbf{14.2 \text{ V}}$$

Thus 14.2 V is the open-circuit voltage between A and B (with the load removed) and is the Thévenin equivalent circuit voltage.

4. To find the Thévenin resistance, draw the circuit, with the load resistor R_L still removed and then short-out (take out) the emfs V_B and V_G. See Fig. 10-9.

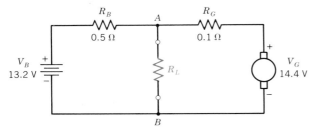

FIGURE 10-7
Circuit for Example 10-6.

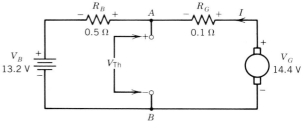

FIGURE 10-8
Open-circuit voltage determination for Example 10-6.

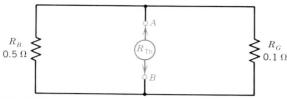

FIGURE 10-9

Resistance determination for Example 10-6.

5. Find the resistance between terminals A and B as seen by an ohmmeter connected to these terminals. (Since the voltage sources have been shorted, it would be safe to connect an ohmmeter to $A - B$.) Note that R_B and R_G are now *in parallel* with each other as far as an ohmmeter connected to terminals A and B is concerned. The resistance seen by the ohmmeter is

$$R_{Th} = \frac{R_B \times R_G}{R_B + R_G} = R_{AB} \qquad (6\text{-}10)$$

$$= \frac{0.5 \times 0.1}{0.5 + 0.1} = \frac{0.05}{0.6} = \mathbf{0.083 \ \Omega}$$

6. The Thévenin equivalent circuit can now be drawn, as shown in Fig. 10-10.
7. The load resistor can now be reconnected to terminals A and B. It will behave exactly the same as when it was connected to the original circuit in Fig. 10-7.
8. The load currents for the various loads can now be obtained:

$$I = \frac{V}{R_T} = \frac{V_{Th}}{R_{Th} + R_L} \qquad (3\text{-}1a)$$

a. For $R_L = 1 \ \Omega$, $I = \dfrac{14.2 \ \text{V}}{0.083 + 1 \ \Omega} = \mathbf{13.1 \ A}$

b. For $R_L = 2 \ \Omega$, $I = \dfrac{14.2 \ \text{V}}{0.083 + 2 \ \Omega} = \mathbf{6.8 \ A}$

c. For $R_L = 3 \ \Omega$, $I = \dfrac{14.2 \ \text{V}}{0.083 + 3 \ \Omega} = \mathbf{4.6 \ A}$

NOTE If the branch or loop current methods were used, new equations would have to be written and solved for *each* value of load resistance. Since the Thévenin equivalent circuit is found with the variable (load) component *removed*, the *same* Thévenin equivalent circuit can be used with each *different* load.

A second example of Thévenin's theorem involves finding the current in R_3 in Fig. 10-11. The problem is similar to the one solved in Example 10-2.

EXAMPLE 10-7.

Given the circuit of Fig. 10-11, find the current in R_3 using Thévenin's theorem.

Solution

1. To find the Thévenin voltage, remove R_3, label the terminals A and B, and solve for the open-circuit voltage, V_{Th}. See Fig. 10-12.
   $V_{AB} = V_{Th}$ is the difference between V_{R_2} and V_{R_4}. By the voltage divider equation:

$$V_{R_2} = V_1 \times \frac{R_2}{R_1 + R_2} \qquad (5\text{-}5)$$

$$= 10 \ \text{V} \times \frac{2 \ \text{k}\Omega}{1 \ \text{k}\Omega + 2 \ \text{k}\Omega}$$

$$= 10 \ \text{V} \times \frac{2}{3} = 6.67 \ \text{V}$$

$$V_{R_4} = V_2 \times \frac{R_4}{R_4 + R_5} \qquad (5\text{-}5)$$

$$= 20 \ \text{V} \times \frac{4 \ \text{k}\Omega}{4 \ \text{k}\Omega + 5 \ \text{k}\Omega}$$

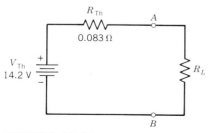

FIGURE 10-10

Thévenin equivalent circuit for Example 10-6.

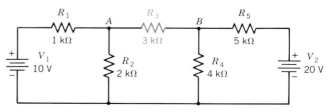

FIGURE 10-11

Circuit for Example 10-7.

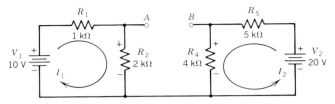

FIGURE 10-12
Open-circuit voltage determination for Example 10-7.

$$= 20 \text{ V} \times \frac{4}{9} = 8.89 \text{ V}$$

Therefore, $V_{AB} = V_{R_2} - V_{R_4}$

$$= 6.67 - 8.89 \text{ V} = -2.22 \text{ V}$$

This means that the Thévenin equivalent voltage is 2.22 V with B positive with respect to A.

2. To find the Thévenin resistance, reduce the voltages in Fig. 10-12 to zero and determine the resistance between A and B looking back into the circuit. See Fig. 10-13.

resistance $R_{AB} = R_{\text{Th}} = R_1 \| R_2 + R_4 \| R_5$

Therefore, $R_{\text{Th}} = \dfrac{R_1 R_2}{R_1 + R_2} + \dfrac{R_4 R_5}{R_4 + R_5}$

$$= \frac{1 \text{ k}\Omega \times 2 \text{ k}\Omega}{1 \text{ k}\Omega + 2 \text{ k}\Omega} + \frac{4 \text{ k}\Omega \times 5 \text{ k}\Omega}{4 \text{ k}\Omega + 5 \text{ k}\Omega}$$

$$= \frac{2}{3} \text{ k}\Omega + \frac{20}{9} \text{ k}\Omega$$

$$= 0.67 \text{ k}\Omega + 2.22 \text{ k}\Omega = \mathbf{2.89 \text{ k}\Omega}$$

3. The Thévenin equivalent circuit between terminals A and B with resistor R_3 replaced is now given by Fig. 10-14. Note how B is positive with respect to A.
4. The current in R_3 can now be determined:

$$I_3 = \frac{V}{R_T} = \frac{V_{\text{Th}}}{R_{\text{Th}} + R_3} \qquad \text{(3-1a)}$$

$$= \frac{2.22 \text{ V}}{(2.89 + 3) \text{ k}\Omega} = \mathbf{0.38 \text{ mA}}$$

flowing from B to A or right to left in the original circuit.

NOTE There is no such thing as the Thévenin equivalent for the *overall* circuit of Fig. 10-11. It depends upon which resistor or portion of the circuit is to be *removed* that determines the nature of the equivalent circuit for what remains. Thus different equivalent circuits would result for R_2, on the one hand, or R_4 on the other.

In some complex circuits, the determination of the open-circuit Thévenin voltage may involve the use of branch or loop current methods. That is, the removal of a given component to determine a Thévenin equivalent does not guarantee that the remaining circuit will be a simple series-parallel type. For example, determining the Thévenin equivalent for resistor R_4 in the circuit shown in Fig. 10-11 requires the use of superposition, branch currents, or the loop current method. (See Example 10-8.)

EXAMPLE 10-8A

Given the circuit in Fig. 10-15a, find the current in R_4, using a Thévenin equivalent circuit.

Solution

Remove R_4 and calculate the open-circuit voltage between terminals A and B, as shown in Fig. 10-15b. Since the resistors do not have a simple series-parallel relationship, let us use the loop current method for analysis. Select loop currents I_1 and I_2 as shown, and write Kirchhoff's voltage law equations.

Loop 1: $V_1 = I_1 R_1 + (I_1 + I_2) R_2$
Loop 2: $V_2 = I_2 R_5 + I_2 R_3 + (I_1 + I_2) R_2$

Substituting values and collecting terms:

$$10 = 3I_1 + 2I_2$$
$$20 = 2I_1 + 10I_2$$

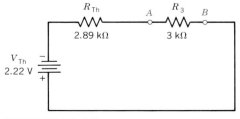

FIGURE 10-14
Thevenin equivalent for Example 10-7.

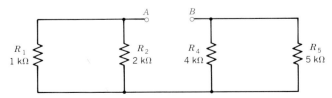

FIGURE 10-13
Resistance determination for Example 10-7.

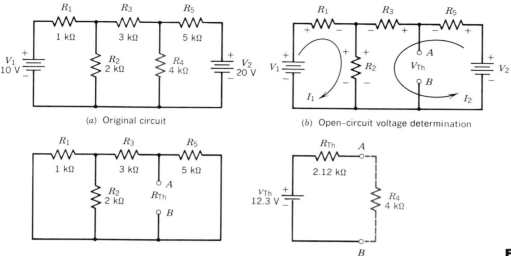

(a) Original circuit

(b) Open-circuit voltage determination

(c) Resistance determination

(d) Thévenin equivalent

FIGURE 10-15
Circuits for Example 10-8.

Solving for I_2 we obtain (since resistance is in kilohms),

$$I_2 = 1.54 \text{ mA}$$
$$V_{R_5} = I_2 R_5$$
$$= 1.54 \times 10^{-3} \text{ A} \times 5 \times 10^3 \text{ Ω} = 7.7 \text{ V}$$
$$V_{Th} = V_{AB} = V_2 - V_{R_5}$$
$$= 20 - 7.7 \text{ V} = \mathbf{12.3 \text{ V}}$$

NOTE   It is not necessary to solve for I_1.

To solve for R_{Th}, consider the circuit in Fig. 10-15c. The load R_4 is still removed, and both voltage sources are shorted. The resistance between A and B may be written as follows:

$$R_{Th} = R_{AB} = [(R_1 \| R_2) + R_3] \| R_5$$
$$= [(1\|2) + 3]\|5 \text{ kΩ}$$
$$= \left(\frac{1 \times 2}{1 + 2} + 3\right)\|5 \text{ kΩ}$$
$$= (\tfrac{2}{3} + 3)\|5 \text{ kΩ}$$
$$= \frac{3\frac{2}{3} \times 5}{3\frac{2}{3} + 5} \text{ kΩ}$$
$$= \frac{18\frac{1}{3}}{8\frac{2}{3}} \text{ kΩ} = \mathbf{2.12 \text{ kΩ}}$$

The final Thévenin equivalent circuit is shown in Fig. 10-15d. The current through R_4 is now given by:

$$I_{R_4} = \frac{V_{Th}}{R_{Th} + R_4}$$
$$= \frac{12.3 \text{ V}}{2.12 + 4 \text{ kΩ}}$$

$$= \frac{12.3 \text{ V}}{6.12 \text{ kΩ}} = \mathbf{2.01 \text{ mA}}$$

and is flowing downward in the original circuit.

Example 10-8 can also be solved by the repeated use of Thévenin's theorem. This avoids using loop currents (or one of the other methods) to find the final Thévenin voltage directly. See Example 10-8B.

EXAMPLE 10-8B

Given the circuit in Fig. 10-16a, find the current in R_4, using a Thévenin equivalent first for V_1, R_1, R_2, and R_3, then again for the rest of the circuit.

Solution

Remove R_4, and find the Thévenin equivalent for the circuit to the left of the dotted line shown in Fig. 10-16b. That is, consider the 5-kΩ resistor and 20-V source to be removed.

$$V'_{AB} = V_1 \times \frac{R_2}{R_1 + R_2}$$
$$= 10 \text{ V} \times \frac{2 \text{ kΩ}}{1 + 2 \text{ kΩ}}$$
$$= 10 \text{ V} \times \tfrac{2}{3} \approx 6.67 \text{ V}$$
$$R'_{Th} = R_1 \| R_2 + R_3$$

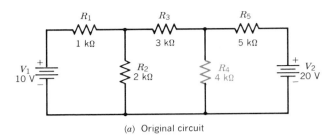

(a) Original circuit

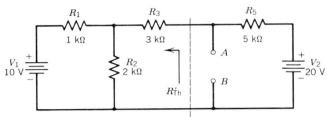

(b) Preparation to obtain first Thévenin equivalent

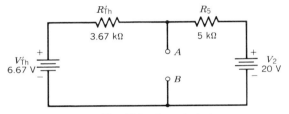

(c) First Thévenin equivalent

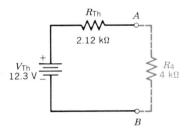

(d) Final Thévenin equivalent circuit

FIGURE 10-16
Circuits for Example 10-8B.

$$= \frac{1 \text{ k}\Omega \times 2 \text{ k}\Omega}{1 + 2 \text{ k}\Omega} + 3 \text{ k}\Omega$$
$$= \tfrac{2}{3} \text{ k}\Omega + 3 \text{ k}\Omega \approx 3.67 \text{ k}\Omega$$

Now consider the 5-kΩ resistor and 20-V source to be reconnected to this first Thévenin equivalent, as shown in Fig. 10-16c.

$$V_{R_5} = (20 - 6.67) \text{ V} \times \frac{5 \text{ k}\Omega}{5 + 3.67 \text{ k}\Omega}$$

$$= 13.33 \text{ V} \times \frac{5}{8.67} = 7.7 \text{ V}$$
$$V_{\text{Th}} = V_{AB} = V_2 - V_{R_5}$$
$$= 20 \text{ V} - 7.7 \text{ V} = \textbf{12.3 V}$$
$$R_{\text{Th}} = 3.67 \text{ k}\Omega \| 5 \text{ k}\Omega$$
$$= \frac{3.67 \text{ k}\Omega \times 5 \text{ k}\Omega}{3.67 + 5 \text{ k}\Omega} = \textbf{2.12 k}\Omega$$

The final Thévenin equivalent circuit is shown in Fig. 10-16d with the 4-kΩ load reconnected.

$$I_{R_4} = \frac{V_{\text{Th}}}{R_{\text{Th}} + R_4}$$
$$= \frac{12.3 \text{ V}}{2.12 + 4 \text{ k}\Omega}$$
$$= \frac{12.3 \text{ V}}{6.12 \text{ k}\Omega} = \textbf{2.01 mA}$$

Note that if the loop current method had been used to solve for I_{R_4} directly (instead of using Thévenin's theorem), it would have been necessary to solve *three* simultaneous equations.

10-5   CURRENT SOURCES

You have seen that a complex circuit may be represented by a constant voltage source in series with a resistance. An equivalent circuit can also be obtained in terms of a *current source*.

Consider Fig. 10-17a. This may be a 12-V battery connected in series with a 4-Ω resistor, or it may be a Thévenin equivalent of some complex circuit. In either case, an open-circuit voltage of 12 V exists between A and B with an internal resistance of 4 Ω.

If A and B are shorted together, as in Fig. 10-17b, a short-circuit current of 12 V/4 Ω = 3 A will flow. Now, assume that it is possible to make a power supply that produces a constant output of 3 A, regardless of the load resistance connected.* This *constant-current* source is shown in Fig. 10-17c, indicated by a circle enclosing an arrow that points in the direction of the conventional current. In this example, a resistance of 4 Ω is connected in

---

*Using modern IC voltage regulators and current-regulating diodes, it is possible to construct circuits with a current output that is substantially constant over very wide ranges of load resistance (from zero to several thousand ohms), depending upon the particular constant current required.

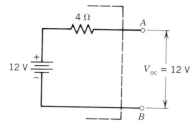

(a) Initial constant-voltage-source circuit

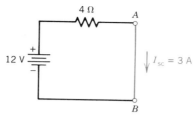

(b) Constant-voltage-source circuit provides a short-circuit current of 3 A

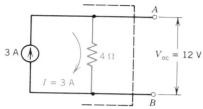

(c) Equivalent constant-current-source circuit

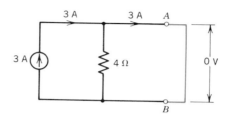

(d) Constant-current-source circuit provides a short-circuit current of 3 A

FIGURE 10-17

Constant-voltage-source and equivalent constant-current-source circuits.

parallel with the current source. This resistance is needed to provide a complete current path and to develop a voltage of 12 V. In other words, the new equivalent circuit in Fig. 10-17c must provide the same open-circuit voltage as the initial voltage source in Fig. 10-17a. The required *parallel* resistance is the same as the initial *series* resistance.

The constant-current circuit should provide the same short-circuit current as the initial circuit in Fig. 10-17a. If terminals A and B are short-circuited, as shown in Fig. 10-

17d, all of the current source's 3 A will flow between the terminals. This provides the same effect as the circuit in Fig. 10-17b.

Another test that can be performed to determine equivalency between the two circuits is to measure the circuit's internal resistance. To connect an ohmmeter to the circuit shown in Fig. 10-18a, the constant-current source must be removed. As shown in Fig. 10-18b, the circuit is left *open* when the current source is removed, in contrast to the

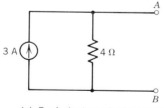

(a) Equivalent constant-current source circuit

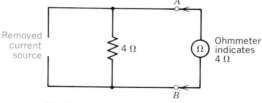

(b) Measuring internal resistance by removing the current source

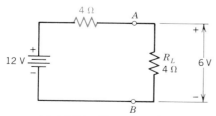

(c) A 4-Ω load is connected to the initial voltage-source circuit

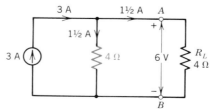

(d) A 4-Ω load is connected to the equivalent current-source circuit

FIGURE 10-18

Determining the equivalence of the constant-current-source and constant-voltage-source circuits.

shorting of the leads when a *voltage* source is removed. The ohmmeter indicates 4 Ω.

When you connect a 4-Ω load to the equivalent constant-current circuit in Fig. 10-18a, does it provide the same load voltage and current as when the load is connected to the initial constant-voltage circuit in Fig. 10-18c? Figure 10-18d shows that the 3 A from the constant-current source must divide equally between the internal resistance and the load resistance, since both are 4 Ω. As a result, the load voltage is 6 V, due to a load current of 1.5 A. These are the same values as in Fig. 10-18c,

$$\text{where } I = \frac{V}{R} = \frac{12 \text{ V}}{8 \text{ Ω}} = 1.5 \text{ A.}$$

There are two points to note.

1. While the voltage across the current source on open circuit is 12 V in Fig. 10-18a, it is only 6 V when a 4-Ω load is connected, as in Fig. 10-18d. The voltage across a constant-current source varies, according to the resistance connected across its terminals.
2. A constant-current-source circuit and a constant-voltage-source circuit can be completely equivalent, as far as any *externally connected load* is concerned. But the circuits *themselves* are not equivalent. For example, on open circuit, no power is dissipated *at all* in the initial voltage-source circuit of Fig.

10-17a, but the equivalent current-source circuit (Fig. 10-17c) dissipates 36 W on open circuit.

10-6 NORTON'S THEOREM

An understanding of the concept of a constant-current source is necessary to understanding the equivalent circuit proposed by Norton. His theorem may be stated as follows:

> Any linear two-terminal network of fixed resistances and sources of emf may be *replaced* by a single constant-current source (I_N) in parallel with a single resistance (R_N). The values of I_N and R_N are given by the following:

1. I_N is the current that flows between the designated terminals of the original circuit when short-circuited.
2. R_N is the resistance looking back into the original network from the two designated terminals, with all sources of emf shorted-out and replaced by their internal resistances (and any current sources in the network opened).

The Norton equivalent resistance is the same as the Thévenin equivalent resistance for any given circuit. The only difference is that, in the Norton equivalent, the resis-

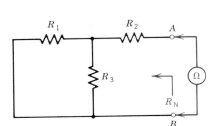

(a) Original circuit

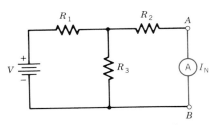

(b) Obtaining I_N. The ammeter indicates short-circuit current I_N

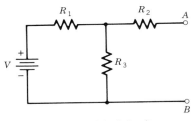

(c) Obtaining R_N. The ohmmeter indicates Norton resistance, R_N

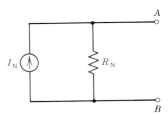

(d) Norton equivalent circuit

FIGURE 10-19
Experimental determination of a Norton equivalent circuit.

tance is placed in *parallel* with the constant-current source (I_N).

Experimentally, Norton's equivalent for the circuit shown in Fig. 10-19a would be found as follows:

1. As shown in Fig. 10-19b, connect a low-resistance ammeter between A and B to measure the short-circuit current (I_N).

2. Reduce the voltage source to zero and connect an ohmmeter between A and B to measure the resistance (R_N) looking back into the circuit. (See Fig. 10-19c.)

The Norton equivalent circuit is shown in Fig. 10-19d.

Alternatively, the Norton equivalent circuit can be obtained by calculation. The following examples illustrate this.

EXAMPLE 10-9

Given the circuit in Fig. 10-20a, calculate for loads of 55 kΩ and 110 kΩ, using Norton's theorem:

a. Load current.
b. Load voltage and polarity.

Solution

Remove the load, R_L, mark the load terminals A and B, and calculate the short-circuit current between A and B. See Fig. 10-20b. The total resistance seen by the 132-V source is given by

$$R_T = R_1 + R_2 \| R_3$$

$$= R_1 + \frac{R_2 \times R_3}{R_2 + R_3}$$

$$= 20 \text{ k}\Omega + \frac{40 \text{ k}\Omega \times 60 \text{ k}\Omega}{40 \text{ k}\Omega + 60 \text{ k}\Omega}$$

$$= 20 \text{ k}\Omega + 24 \text{ k}\Omega = 44 \text{ k}\Omega$$

The total current from the 132-V source is

$$I_T = \frac{V}{R_T} \qquad (3\text{-}1a)$$

$$= \frac{132 \text{ V}}{44 \text{ k}\Omega} = 3 \text{ mA}$$

The current, I_N, between B and A is I_{R_2}:

$$I_N = I_{R_2} = I_T \times \frac{R_3}{R_2 + R_3} \qquad (6\text{-}12)$$

$$= 3 \text{ mA} \times \frac{60 \text{ k}\Omega}{60 \text{ k}\Omega + 40 \text{ k}\Omega} = \textbf{1.8 mA}$$

The Norton resistance is found by shorting the voltage source as in Fig. 10-20c.

$$R_N = R_2 + R_1 \| R_3$$

$$= R_2 + \frac{R_1 \times R_3}{R_1 + R_3}$$

$$= 40 \text{ k}\Omega + \frac{20 \text{ k}\Omega \times 60 \text{ k}\Omega}{20 \text{ k}\Omega + 60 \text{ k}\Omega}$$

$$= 40 \text{ k}\Omega + 15 \text{ k}\Omega = \textbf{55 k}\Omega$$

The Norton equivalent circuit is shown in Fig. 10-20d. Note that the current source points "downward," causing the load current to flow from B to A.

a. For a load resistance of 55 kΩ, I_N must divide equally

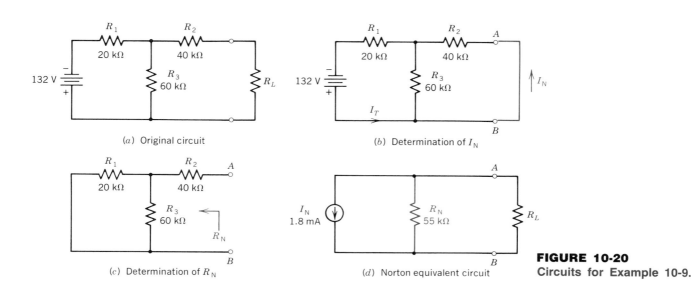

(a) Original circuit

(b) Determination of I_N

(c) Determination of R_N

(d) Norton equivalent circuit

FIGURE 10-20
Circuits for Example 10-9.

between R_N and R_L, so $I_L = $ **0.9 mA.** When $R_L = $ 110 kΩ, the load current is only one-third of I_N or **0.6 mA.** That is,

$$I_L = I_N \times \frac{R_N}{R_N + R_L} \qquad (6\text{-}12)$$

$$= 1.8 \text{ mA} \times \frac{55 \text{ k}\Omega}{55 \text{ k}\Omega + 110 \text{ k}\Omega}$$

$$= 1.8 \text{ mA} \times \frac{55 \text{ k}\Omega}{165 \text{ k}\Omega}$$

$$= 1.8 \text{ mA} \times \frac{1}{3} = \textbf{0.6 mA}$$

b. For a load resistance of 55 kΩ,

$$V_L = I_L \times R_L \qquad (3\text{-}1b)$$

$$= 0.9 \text{ mA} \times 55 \text{ k}\Omega = \textbf{49.5 V}$$

with B positive with respect to A.

For $R_L = $ 110 kΩ,

$$V_L = I_L \times R_L \qquad (3\text{-}1b)$$

$$= 0.6 \text{ mA} \times 110 \text{ k}\Omega = \textbf{66 V}$$

with B positive with respect to A.

As with Thévenin's theorem, note that Norton's theorem allows you to readily obtain the load voltage and current for *any* load, without repeating the whole problem. This is *not* the case when simple series-parallel theory is used.

The next example involves two voltage sources and cannot be solved using simple series-parallel theory. The first part of the example describes a *direct* application of Norton's theorem.

EXAMPLE 10-10A

Given the battery-generator charging circuit in Fig. 10-21a, determine the load current and voltage by obtaining a Norton equivalent for R_L directly.

Solution

Removing R_L and shorting A to B as in Fig. 10-21b, we find that

$$I_N = I_B + I_G \qquad (6\text{-}1)$$

$$= \frac{V_B}{R_B} + \frac{V_G}{R_G}$$

$$= \frac{13.2 \text{ V}}{0.3 \text{ }\Omega} + \frac{14.4 \text{ V}}{0.2 \text{ }\Omega}$$

$$= 44 \text{ A} + 72 \text{ A} = \textbf{116 A}$$

Shorting-out the voltage sources, but maintaining their internal resistances, as in Fig. 10-21c, gives

$$R_N = R_B \| R_G$$

$$= \frac{R_B \times R_G}{R_B + R_G} \qquad (6\text{-}10)$$

$$= \frac{0.3 \text{ }\Omega \times 0.2 \text{ }\Omega}{0.3 \text{ }\Omega + 0.2 \text{ }\Omega} = \textbf{0.12 }\Omega$$

The Norton equivalent circuit is shown in Fig. 10-21d.
      For $R_L = $ 0.68 Ω,

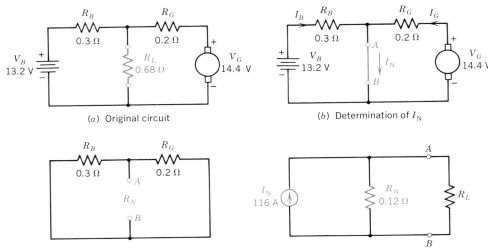

(a) Original circuit

(b) Determination of I_N

(c) Determination of R_N

(d) Norton equivalent circuit

FIGURE 10-21

Circuits for Example 10-10A.

$$I_L = I_N \times \frac{R_N}{R_N + R_L} \qquad (6\text{-}12)$$

$$= 116 \text{ A} \times \frac{0.12 \ \Omega}{0.12 \ \Omega + 0.68 \ \Omega} = \textbf{17.4 A}$$

$$\text{and } V_L = 17.4 \text{ A} \times 0.68 \ \Omega = \textbf{11.83 V}$$

Note that the arrow for I_N in the Norton equivalent circuit is shown pointing upward. This ensures that the current will flow *downward* from A to B in Fig. 10-21d, and will be consistent with the direction of I_N in Fig. 10-21b. Although the determination of I_N was made as downward in Fig. 10-21b, this direction is not used directly for I_N in the Norton constant-current generator in Fig. 10-21d.

10-6.1 Converting a Thévenin Equivalent Circuit to a Norton Equivalent

Earlier, you saw how a current source in parallel with a resistance may be equivalent to a voltage source in series with the same resistance (as far as any external load is concerned). Figure 10-22 depicts the conversion from a Thévenin equivalent to a Norton equivalent.

The value of the constant current source in the Norton equivalent is given by

$$I_N = \frac{V_{Th}}{R_{Th}} \qquad (3\text{-}1a)$$

and the parallel-connected Norton resistance equals the Thévenin resistance:

$$R_N = R_{Th}$$

This means that any voltage source in *series* with a re-sistance may be replaced by an equivalent current source in *parallel* with the *same* resistance.

This concept is useful in applying Millman's theorem.

10-6.2 Millman's Theorem—Combining Current Sources in Parallel

Millman's theorem states:

Any number of constant-current sources and their associated resistances, connected in parallel, may be combined into a single current source (I_T) and a parallel-connected resistance (R_{eq}), where:

1. I_T is the *algebraic* sum of the original current sources.
2   R_{eq} is the parallel equivalent resistance of the original parallel-connected resistors.

Although Millman's theorem refers directly to *current* sources in parallel, it may also be used to advantage with *voltage* sources that are connected in parallel. This is done by first converting each voltage source (and its series internal resistance) into its Norton equivalent current source in parallel with the same resistance. Then, Millman's theorem can be applied, as in Example 10-10B.

EXAMPLE 10-10B

Solve Example 10-10A by obtaining a Norton equivalent for each voltage source and combining the result using Millman's theorem.

Solution

The Norton equivalent for V_B and R_B is shown in Fig. 10-23b where:

$$I_{NB} = \frac{V_B}{R_B} \qquad (3\text{-}1a)$$

$$= \frac{13.2 \text{ V}}{0.3 \ \Omega} = 44 \text{ A}$$

and $\qquad R_{NB} = R_B = 0.3 \ \Omega$

Similarly, $\qquad I_{NG} = \frac{V_G}{R_G} \qquad (3\text{-}1a)$

$$= \frac{14.4 \text{ V}}{0.2 \ \Omega} = 72 \text{ A}$$

and $\qquad R_{NG} = R_G = 0.2 \ \Omega$

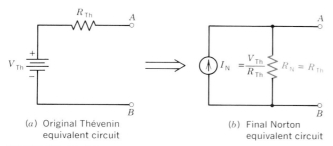

(a) Original Thévenin equivalent circuit

(b) Final Norton equivalent circuit

FIGURE 10-22
Converting a Thévenin equivalent to a Norton equivalent.

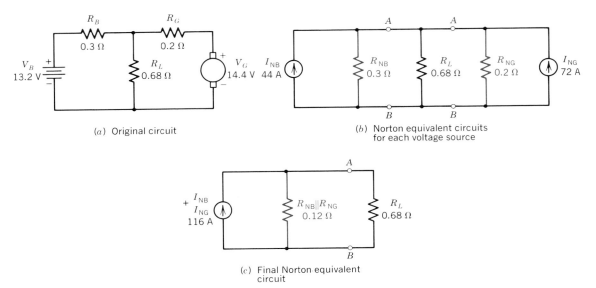

FIGURE 10-23
Circuits for Example 10-10B.

Next, using Millman's theorem, the current sources may be combined into one:

$$I_N = I_{NB} + I_{NG} \qquad (6\text{-}1)$$
$$= 44\ A + 72\ A = \textbf{116 A}$$

and the parallel resistances may also be combined:

$$R_N = \frac{R_{NB} \times R_{NG}}{R_{NB} + R_{NG}} \qquad (6\text{-}10)$$
$$= \frac{0.3\ \Omega \times 0.2\ \Omega}{0.3\ \Omega + 0.2\ \Omega} = \textbf{0.12 }\Omega$$

This yields the same Norton equivalent (see Fig. 10-23c) as in Fig. 10-21d. Then, with $R_L = 0.68\ \Omega$,

$$I_L = I_N \times \frac{R_N}{R_N + R_L} = 116\ A \times \frac{0.12\ \Omega}{0.12 + 0.68\ \Omega} = \textbf{17.4 A}$$
$$V_L = I_L R_L = 17.4\ A \times 0.68\ \Omega = \textbf{11.8 V}$$

10-6.3 Converting a Norton Equivalent Circuit to a Thévenin Equivalent

After applying Millman's theorem and obtaining a Norton equivalent, it may be desirable to return to a voltage-source circuit. **A constant-current source in *parallel* with a resistance may be converted to an equivalent voltage source in *series* with the same resistance.**

The conversion from a Norton equivalent to the Thévenin equivalent is shown in Fig. 10-24, where the constant voltage source is given by

$$V_{Th} = I_N \times R_N \qquad (3\text{-}1b)$$

and the series-connected Thévenin resistance equals the Norton resistance

$$R_{Th} = R_N$$

For example, consider the Norton equivalent in Fig. 10-25a, with $I_N = 116\ A$ and $R_N = 0.12\ \Omega$. The Thévenin equivalent voltage is

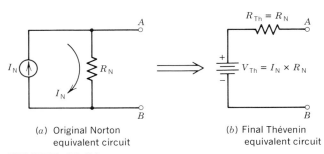

(a) Original Norton   (b) Final Thévenin
    equivalent circuit   equivalent circuit

FIGURE 10-24
Converting a Norton equivalent to a Thévenin equivalent.

$V_{Th} = I_N \times R_N = 116 \text{ A} \times 0.12 \text{ }\Omega$

$= 13.92 \text{ V} \approx \textbf{13.9 V}$

(Load R_L is removed for this step). The Thévenin equivalent resistance is

$R_{Th} = R_N = \textbf{0.12 }\Omega$

The load current in Fig. 10-25b would then be obtained as

$$I_L = \frac{V_{Th}}{R_{Th} + R_L} \qquad (3\text{-}1a)$$

$$= \frac{13.92 \text{ V}}{0.12 + 0.68 \text{ }\Omega} = \textbf{17.4 A}$$

as before.

Whether to use Norton's or Thévenin's equivalent circuit in solving a given problem depends on the circuit in question. For example, it is easier to find the Thévenin equivalent than the Norton equivalent for the circuit in Fig. 10-20a. This is because, with the load removed, the open-circuit voltage (which is the Thévenin voltage, V_{Th}) may be obtained *directly* by applying the voltage-divider equation to R_3.

On the other hand, if a circuit contains two or more parallel sources of emf (as in Fig. 10-21a), Norton's theorem may be applied most easily. In some circuits, such as one equivalent circuit for a transistor, a Thévenin equivalent is used to represent the input to the transistor and a Norton equivalent to represent the output of the transistor. (See Section 10-8.)

10-6.4 Combining Current Sources in Series

Sometimes, it is necessary to analyze a circuit in which two or more current sources are in *series* with each other. In this case, each current source and its associated parallel resistance is converted into its Thévenin equivalent. The voltage sources and resistances are then combined into a single voltage source. In turn, this could be converted back into a single current source if the conversion would be an advantage in solving the circuit. (See Example 10-11.)

EXAMPLE 10-11

Given the circuit in Fig. 10-26a, calculate the load current and its direction.

Solution

The Thévenin equivalents for the two current sources are

$$V_1 = I_1 R_1 \qquad (3\text{-}1b)$$

$$= 4 \text{ mA} \times 2 \text{ k}\Omega = \textbf{8 V}$$

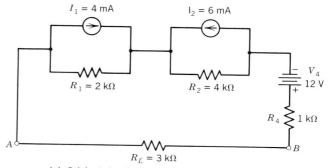

(a) Original circuit with series-current sources

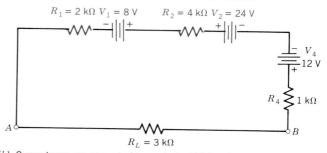

(b) Current sources converted to equivalent Thévenin voltage sources

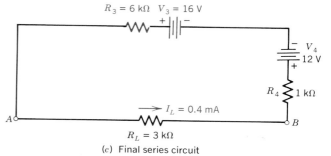

(c) Final series circuit

FIGURE 10-26

Circuits for Example 10-11.

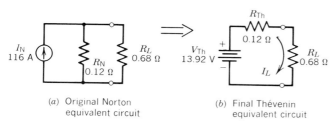

(a) Original Norton equivalent circuit

(b) Final Thévenin equivalent circuit

FIGURE 10-25

Converting a Norton equivalent to a Thévenin equivalent.

in series with a **2-kΩ** resistance

$$V_2 = I_2 R_2 \tag{3-1b}$$
$$= 6 \text{ mA} \times 4 \text{ k}\Omega = \textbf{24 V}$$

in series with a **4-kΩ** resistance. These two voltage sources are series-connected, as in Fig. 10-26b. They have a total value of 16 V in series with 6 kΩ, as shown in Fig. 10-26c. Note the opposite polarities for V_1 and V_2 resulting from the opposite directions of the original current sources. The load current is given by

$$I_L = \frac{V_T}{R_T} = \frac{V_3 - V_4}{R_3 + R_4 + R_L} \tag{3-1a}$$
$$= \frac{16 \text{ V} - 12 \text{ V}}{6 \text{ k}\Omega + 1 \text{ k}\Omega + 3 \text{ k}\Omega}$$
$$= \frac{4 \text{ V}}{10 \text{ k}\Omega} = \textbf{0.4 mA}$$

This current flows from A to B because V_3 is larger than V_4.

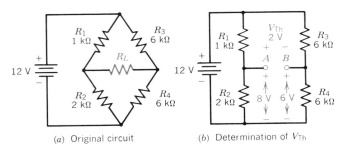

(a) Original circuit   (b) Determination of V_{Th}

FIGURE 10-27
Circuits for Example 10-12.

10-7 APPLICATION OF THÉVENIN AND NORTON EQUIVALENTS TO A WHEATSTONE BRIDGE CIRCUIT

As a final example of finding Norton and Thévenin equivalents, consider the Wheatstone Bridge circuit in Fig. 10-27. This type of circuit was solved earlier, using the loop current method. With that method, three equations had to be solved. The application of Thévenin's theorem is simpler.

EXAMPLE 10-12

Given the Wheatstone Bridge circuit in Fig. 10-27a, find
a. The Thévenin equivalent circuit for R_L.
b. The Norton equivalent circuit for R_L.
c. The value of R_L to provide maximum power transfer.
d. The power delivered to R_L for the condition in part c.

Solution

a. Remove the load (R_L) and find V_{AB} in Fig. 10-27b. Note that the circuit has been redrawn to show clearly that V_{AB} is the difference between V_{R_2} and V_{R_4}.

$$V_{R_2} = 12 \text{ V} \times \frac{R_2}{R_2 + R_1} \tag{5-5}$$
$$= 12 \text{ V} \times \frac{2 \text{ k}\Omega}{2 + 1 \text{ k}\Omega} = 8 \text{ V}$$
$$V_{R_4} = 12 \text{ V} \times \frac{R_4}{R_4 + R_3} \tag{5-5}$$
$$= 12 \text{ V} \times \frac{6 \text{ k}\Omega}{6 + 6 \text{ k}\Omega} = 6 \text{ V}$$
$$V_{\text{Th}} = V_{AB} = V_{R_2} - V_{R_4}$$
$$= 8 \text{ V} - 6 \text{ V} = \textbf{2 V}$$

To find the Thévenin resistance (R_{Th}), remove the load (R_L), and short the voltage source, as in Fig. 10-28a. When the circuit is redrawn (Fig. 10-28b), it is clear that R_1 and R_2 are in parallel with each other. Resistors R_3 and R_4 are also in parallel with each other. The two parallel combinations (R_1 and R_2, R_3 and R_4) are in series with each other, so far as terminals A and B are concerned

$$R_{\text{Th}} = R_{AB} = R_1 \| R_2 + R_3 \| R_4$$
$$= 1 \text{ k}\Omega \| 2 \text{ k}\Omega + 6 \text{ k}\Omega \| 6 \text{ k}\Omega$$
$$= \frac{1 \times 2}{1 + 2} + \frac{6 \times 6}{6 + 6} \text{ k}\Omega$$
$$= \frac{2}{3} + 3 \text{ k}\Omega = \textbf{3.67 k}\Omega$$

The final Thévenin equivalent circuit is shown in Fig. 10-29a.

b. The Norton equivalent is most easily found from the Thévenin equivalent:

$$I_N = \frac{V_{\text{Th}}}{R_{\text{Th}}}$$
$$= \frac{2 \text{ V}}{3.67 \text{ k}\Omega} = \textbf{0.545 mA}$$
$$R_N = R_{\text{Th}} = \textbf{3.67 k}\Omega$$

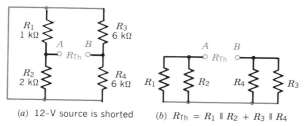

(a) 12-V source is shorted   (b) $R_{Th} = R_1 \parallel R_2 + R_3 \parallel R_4$

FIGURE 10-28

Circuits for Example 10-12.

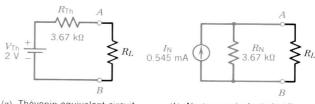

(a) Thévenin equivalent circuit   (b) Norton equivalent circuit

FIGURE 10-29

Thévenin and Norton equivalents for Example 10-12.

See Fig. 10-29b. This is simpler than trying to find the short-circuit current between A and B in the original circuit.

c. For maximum power transfer, $R_L = R_{Th} = $ **3.67 kΩ.**

d. Power delivered $= \dfrac{V_L^2}{R_L} = \dfrac{(1\ V)^2}{3.67 \times 10^3\ \Omega} = $ **0.27 mW.**

NOTE $V_L = \dfrac{V_{Th}}{2}$ since $R_L = R_{Th}$ and the open-circuit voltage (V_{Th}) divides equally between R_L and R_{Th}.

10-8* APPLICATION OF THÉVENIN AND NORTON EQUIVALENTS TO A TRANSISTOR

Semiconductor circuits that involve transistors may be analyzed using graphical or circuit analysis techniques. For *circuit analysis*, an equivalent circuit for the device (transistor) must be used. One such circuit, for a bipolar junction transistor, consists of both a Thévenin equivalent and a Norton equivalent.

Figure 10-30 shows a *PNP*-junction-type transistor connected in an amplifier circuit. A small varying voltage (v_s) is applied at the input. This causes an input current (i_b) to flow into the base (B) of the transistor. The input current, in turn, causes a larger current to flow in the transistor between the emitter (E) and the collector (C). R_C is called the collector resistor and V_{CC} is a negative supply voltage.

The transistor itself may be represented, as far as ac quantities at medium frequency are concerned, by four hybrid parameters: h_{ie}, h_{re}, h_{fe}, and h_{oe}. The input resistance is given by h_{ie} (typically 2–5 kΩ), while $h_{re}v_o$ is a small

ac voltage source (where h_{re} is typically 1×10^{-4}). Together, these two form a Thévenin equivalent for what the input to the transistor looks like.

The current flowing out of the transistor depends upon the current flowing in. This is represented by a current source ($h_{fe} \times i_b$), where h_{fe} may be 50–400. This means that output current may be 50 to 400 times greater than the input current. A resistance, h_{oe} (typically 100 kΩ), is shown in parallel with the current source. This determines how much current flows through the collector resistor (R_C), and how much voltage is developed at the output. The current source and h_{oe} (usually specified as a conductance) form a Norton equivalent circuit. Not shown in Fig. 10-30a, but necessary for proper operation as an amplifier, is a *dc bias resistor* connected from V_{CC} to the base (B).

EXAMPLE 10-13

Consider the equivalent circuit in Fig. 10-30b, with $h_{fe} = 200$, $h_{oe} = 100$ kΩ, and $R_C = 10$ kΩ. Assume an input voltage (v_s) of 0.02 V causes an input base current (i_b) of 10 μA. Calculate:

a. Output current (i_c).
b. Output voltage (v_o).
c. Voltage gain ($A_v = v_o/v_s$).

Solution

a. The value of the current source is

$$h_{fe} \times i_b = 200 \times 10\ \mu A$$
$$= 2\ mA$$
$$i_c = h_{fe}i_b \times \frac{h_{oe}}{h_{oe} + R_C}$$
$$= 2\ mA \times \frac{100\ k\Omega}{100 + 10\ k\Omega}$$
$$= \textbf{1.82 mA}$$

---

*This section may be omitted without loss of continuity.

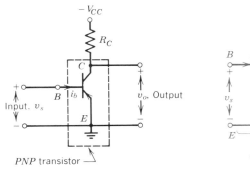

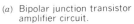

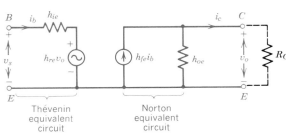

(a) Bipolar junction transistor amplifier circuit.

(b) Equivalent ac circuit for the transistor.

FIGURE 10-30

A hybrid equivalent circuit for a transistor, consisting of a Thévenin equivalent at the input and a Norton equivalent at the output.

b. Output voltage, $v_o = i_c R_C$
$$= 1.82 \text{ mA} \times 10 \text{ k}\Omega$$
$$= \textbf{18.2 V}$$

c. Voltage gain, $A_v = \dfrac{v_o}{v_s}$
$$= \dfrac{18.2 \text{ V}}{0.02 \text{ V}} = \textbf{910}$$

In nodal analysis, the currents entering or leaving a junction must be determined in terms of the nodal voltages and each component's resistance. Then, the currents leaving the junction are equated to the currents entering the junction. If the circuit contains only two nodes (counting the reference node), only a single nodal equation is needed. For each additional node in the circuit, one more nodal equation is required. Solving these equations provides the voltage at each node, allowing determination of the current in each component.

10-9 NODAL ANALYSIS

In the previous sections, you have seen how branch current analysis involves both Kirchhoff's voltage and current laws, while loop (mesh) analysis involves only Kirchhoff's voltage law. *Nodal analysis* is a method that uses only Kirchhoff's current law.

In this method, all the nodes (junctions) in the circuit are identified, and one is used as a reference. All other nodes are assumed to have a positive voltage, called a *nodal voltage,* with respect to the reference node. Each node is assigned a voltage notation, such as V_1, V_x, and so on.

A *node,* in simplest form, is a point at which two or more components have a common connection. In this section, *major* nodes will be emphasized. These are nodes where three or more components have a common connection, and are sometimes called a *junction.* Any node may be chosen as a reference, but it is usually the one with the greatest number of common connections. (See Fig. 10-31.)

EXAMPLE 10-14

Given the battery-generator charging circuit shown in Fig. 10-32a, use nodal analysis to find:

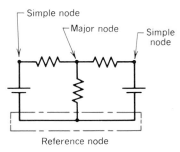

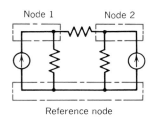

(a) Difference between simple and major nodes

(b) A three-node circuit

FIGURE 10-31

Two- and three-node circuits.

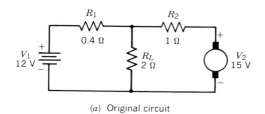

(a) Original circuit

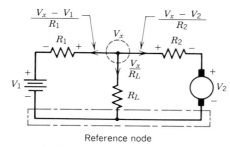

(b) Showing component currents at
node V_x

FIGURE 10-32
Circuits for Example 10-14.

a. The load voltage and current.
b. The battery current and direction.

Solution

a. In Fig. 10-32b, the node at the bottom of the circuit has been identified as the reference node, since it contains lines common to both sources and the load. Let the voltage at the other node be V_x. It is assumed to be positive with respect to the reference node. Assuming V_x is larger than V_1, current will flow through R_1 from right to left with a value given by

$$I_{R_1} = \frac{V_x - V_1}{R_1}$$

Similarly, assuming V_x is larger than V_2, we show a current flowing through R_2 from left to right (Fig. 10-32b), with a value given by

$$I_{R_2} = \frac{V_x - V_2}{R_2}$$

The current through R_L must be downward, since V_x is positive with respect to the bottom of R_L, or the reference node:

$$I_L = \frac{V_x}{R_L}$$

Since there are no currents indicated as entering the node at V_x, we have, by Kirchhoff's current law:

$$I_{R_1} + I_{R_2} + I_L = 0$$

Therefore, $\dfrac{V_x - V_1}{R_1} + \dfrac{V_x - V_2}{R_2} + \dfrac{V_x}{R_L} = 0$

$$\frac{V_x - 12}{0.4} + \frac{V_x - 15}{1} + \frac{V_x}{2} = 0$$

Multiplying through by 2:

$$5V_x - 60 + 2V_x - 30 + V_x = 0$$
$$8V_x = 90$$
$$V_x = \frac{90}{8} \text{ V} = \textbf{11.25 V}$$

Load current $I_L = \dfrac{V_L}{R_L} = \dfrac{V_x}{R_L} = \dfrac{11.25 \text{ V}}{2 \text{ }\Omega} = \textbf{5.625 A}$

b. Current through the battery, into the positive terminal, is given by

$$
\begin{aligned}
I_{R_1} &= \frac{V_x - V_1}{R_1} \\
&= \frac{11.25 - 12 \text{ V}}{0.4 \text{ }\Omega} \\
&= \frac{-0.75 \text{ V}}{0.4 \text{ }\Omega} = -1.875 \text{ A}
\end{aligned}
$$

The negative sign indicates that current is actually flowing in the *opposite* direction, so battery current is **1.875 A out of the positive terminal.**

Note that the current through R_2 is also negative, so V_2 also supplies current to the junction. Note also, that only *one* equation with one unknown had to be solved. Compare this with the loop or branch current methods, which require two equations with two unknowns. As you might expect, nodal analysis is also much less time-consuming than the superposition method.

The problem in Example 10-14 also can be solved by converting the two voltage sources and their associated internal resistances into their Norton equivalents, then using nodal analysis.

EXAMPLE 10-15

For the circuit in Fig. 10-33a, convert each voltage source and its internal resistance into a Norton equivalent, then use nodal analysis to solve for the load voltage.

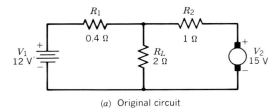

(a) Original circuit

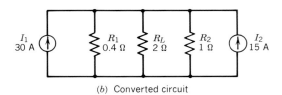

(b) Converted circuit

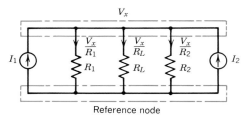

Reference node

(c) Showing currents at node V_x

FIGURE 10-33

Circuits for Example 10-15.

Solution

The current source for V_1 and R_1 is given by

$$I_1 = \frac{V_1}{R_1}$$

$$= \frac{12 \text{ V}}{0.4 \text{ }\Omega} = 30 \text{ A}$$

And for V_2 and R_2 by

$$I_2 = \frac{V_2}{R_2}$$

$$= \frac{15 \text{ V}}{1 \text{ }\Omega} = 15 \text{ A}$$

The converted circuit containing Norton equivalents is shown in Fig. 10-33b. By choosing the lower node as the reference and letting the upper node have a voltage V_x, we can equate the currents *leaving* the junction through the resistors to the current from the sources *entering* the junction. (See Fig. 10-33c.)

$$\frac{V_x}{R_1} + \frac{V_x}{R_L} + \frac{V_x}{R_2} = I_1 + I_2$$

$$V_x \left(\frac{1}{0.4} + \frac{1}{2} + \frac{1}{1} \right) = 30 + 15$$

$$V_x (2.5 + 0.5 + 1) = 45$$

$$4V_x = 45$$

$$V_x = \frac{45}{4} \text{ V} = 11.25 \text{ V}$$

Therefore, $V_L = V_x = $ **11.25 V**, as before.

Example 10-16 involves a circuit with three nodes, and requires the writing of two nodal equations.

EXAMPLE 10-16

Given the circuit in Fig. 10-34a, find the currents through the three resistors.

Solution

The circuit has been redrawn in Fig. 10-34b to show the reference node. The remaining two nodes have been assigned the nodal voltage notations V_1 and V_2. Note that the resistance values have been converted to their corresponding conductance values. This will demonstrate the nodal analysis *procedure*.

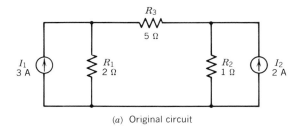

(a) Original circuit

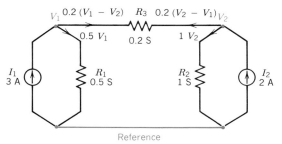

Reference

(b) Showing node voltages and branch currents.

FIGURE 10-34

Circuits for Example 10-16.

At node 1, assuming V_1 is greater than V_2, the current flowing to the right through R_3 is given by

$$I_{R_3} = 0.2(V_1 - V_2)$$

Also leaving the junction is the current through R_1:

$$I_{R_1} = 0.5 \, V_1$$

The sum of the two currents must equal the current entering due to the current source ($I_1 = 3$ A). Thus

$$I_{R_1} + I_{R_3} = I_1$$
$$0.5V_1 + 0.2(V_1 - V_2) = 3$$

or

$$0.7V_1 - 0.2V_2 = 3 \qquad (1)$$

Similarly, at node 2, assuming V_2 is greater than V_1:

$$1V_2 + 0.2(V_2 - V_1) = 2$$

or

$$-0.2V_1 + 1.2V_2 = 2 \qquad (2)$$

Solving equations (1) and (2) simultaneously:

$$V_1 = 5 \text{ V}, \qquad V_2 = 2.5 \text{ V}$$

Therefore, $I_{R_1} = 0.5V_1$

$$= 0.5 \text{ S} \times 5 \text{ V} = \textbf{2.5 A}$$
$$I_{R_2} = 1V_2$$
$$= 1 \text{ S} \times 2.5 \text{ V} = \textbf{2.5 A}$$

in the direction shown in Fig. 10-34b, downward.

$$I_{R_3} = 0.2(V_1 - V_2) \text{ to the right}$$
$$= 0.2 \text{ S}(5 \text{ V} - 2.5 \text{ V})$$
$$= 0.2 \text{ S} \times 2.5 \text{ V} = \textbf{0.5 A to the right}$$

or

$$I_{R_3} = 0.2(V_2 - V_1) \text{ to the left}$$
$$= 0.2 \text{ S}(2.5 \text{ V} - 5 \text{ V})$$
$$= 0.2 \text{ S} \times (-2.5 \text{ V}) = -0.5 \text{ A to the left}$$
$$= \textbf{0.5 A to the right}$$

or

$$I_{R_3} = I_1 - I_{R_1}$$
$$= 3 \text{ A} - 2.5 \text{ A} = \textbf{0.5 A}$$

Note that the use of conductance values avoids fractions in the nodal equations. Current is obtained by multiplying conductance in siemens by voltage in volts.

As a final demonstration of nodal analysis, consider a circuit with both current and voltage sources, as shown analyzed in Example 10-17.

EXAMPLE 10-17

Given the circuit in Fig. 10-35a, determine the current flowing through R_4 and its direction.

Solution

Assigning the nodal voltages V_x and V_y, as in Fig. 10-35b, leads to the two required equations.

At node x:

$$\frac{V_x}{R_3} + \frac{V_x - V_y}{R_4} + \frac{V_x - V_1}{R_1 + R_2} = 0 \qquad (1)$$

At node y:

$$\frac{V_y - V_x}{R_4} + \frac{V_y + V_2}{R_5} = I_1 \qquad (2)$$

Note that

$$I_{R_5} = \frac{V_y - (-V_2)}{R_5} = \frac{V_y + V_2}{R_5}$$

because of the polarity of V_2 with respect to the reference node. Substituting values in equations (1) and (2),

$$\frac{V_x}{3 \text{ k}\Omega} + \frac{V_x - V_y}{4 \text{ k}\Omega} + \frac{V_x - 8 \text{ V}}{1 + 2 \text{ k}\Omega} = 0 \qquad (3)$$
$$\frac{V_y - V_x}{4 \text{ k}\Omega} + \frac{V_y + 20 \text{ V}}{5 \text{ k}\Omega} = 2 \text{ mA} \qquad (4)$$

Since current is in milliamperes and resistance is in kilohms, the values of V_x and V_y will be in volts, so the units can be dropped for now. Multiply equation (3) by 12:

$$4V_x + 3V_x - 3V_y + 4V_x - 32 = 0$$

or

$$11V_x - 3V_y = 32 \qquad (5)$$

and multiply equation (4) by 20:

$$5V_y - 5V_x + 4V_y + 80 = 40$$

or

$$-5V_x + 9V_y = -40 \qquad (6)$$

Multiply equation (5) by 3 and add to equation (6):

$$28V_x = 56$$
$$V_x = 2 \text{ V}$$

Substitute $V_x = 2$ V in equation (5),

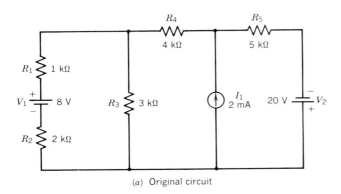

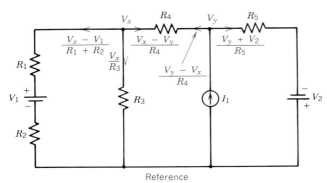

(a) Original circuit

(b) Showing node voltages and branch currents.

FIGURE 10-35
Circuits for Example 10-17.

$$3V_y = 11V_x - 32$$
$$= 22 - 32 = -10$$
$$V_y = \frac{-10}{3} V = -3.33 \text{ V}$$

Current through R_4 is

$$I_{R_4} = \frac{V_x - V_y}{R_4}$$
$$= \frac{2 - (-3.33) \text{ V}}{4 \text{ k}\Omega} = \frac{5.33 \text{ V}}{4 \text{ k}\Omega} = \textbf{1.33 mA left to right}$$

10-10   DELTA-WYE CONVERSIONS

To solve complex networks, special techniques are necessary because, very often, the circuit components do not have a simple series-parallel relationship. In some cases, a

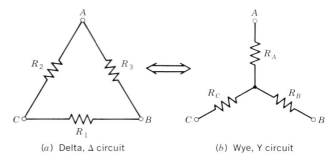

(a) Delta, Δ circuit   (b) Wye, Y circuit

FIGURE 10-36
Delta-wye and wye-delta conversions.

complex network may be reduced to a simple series-parallel combination by substituting an equivalent network. Figure 10-36 shows two networks—a "delta" (Δ) and a "wye" (Y). The names of these two networks are derived from their shapes.

By using the appropriate equations, it is possible to convert a given delta network to an equivalent wye network. These equations are

$$\Delta \rightarrow Y$$

$$R_A = \frac{R_2 R_3}{R_1 + R_2 + R_3} \quad \text{ohms} \qquad (10\text{-}1)$$

$$R_B = \frac{R_3 R_1}{R_1 + R_2 + R_3} \quad \text{ohms} \qquad (10\text{-}2)$$

$$R_C = \frac{R_1 R_2}{R_1 + R_2 + R_3} \quad \text{ohms} \qquad (10\text{-}3)$$

where R_A, R_B, and R_C are elements in the wye
R_1, R_2, and R_3 are elements in the delta.

Note that, in the equation for each resistor in the wye, the numerator is equal to the product of the two adjacent resistors in the delta.

Similarly, a wye may be converted to a delta, using the following:

$$Y \rightarrow \Delta$$

$$R_1 = \frac{R_A R_B + R_B R_C + R_C R_A}{R_A} \qquad (10\text{-}4)$$

$$R_2 = \frac{R_A R_B + R_B R_C + R_C R_A}{R_B} \qquad (10\text{-}5)$$

$$R_3 = \frac{R_A R_B + R_B R_C + R_C R_A}{R_C} \qquad (10\text{-}6)$$

where R_1, R_2, and R_3 are elements in the delta
R_A, R_B, and R_C are elements in the wye.

Note that, in the equation for each resistor in the delta, the denominator is equal to the opposite resistor in the wye.

Example 10-18 shows how a delta-wye conversion can be used to solve a problem you saw earlier, involving a Wheatstone bridge.

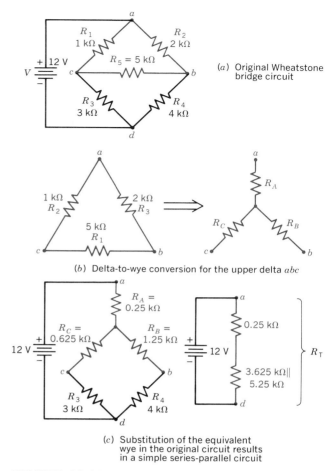

(a) Original Wheatstone bridge circuit

(b) Delta-to-wye conversion for the upper delta *abc*

(c) Substitution of the equivalent wye in the original circuit results in a simple series-parallel circuit

FIGURE 10-37
Circuits for Example 10-18.

EXAMPLE 10-18

Determine the total current supplied by the 12-V source in Fig. 10-37a by calculating the total resistance of the network between *a* and *d*.

Solution

There are four possible conversions in the circuit of Fig. 10-37a. We can identify an upper delta—*a, b, c;* a lower delta—*c, b, d;* a left wye—*ca, cb, cd;* and a right wye— *ba, bc, bd.* Let us convert the upper delta into an equivalent wye as in Fig. 10-37b. It is suggested that for reference purposes, the delta resistors be renumbered to conform to the delta-wye conversion equations.

$$R_A = \frac{R_2 R_3}{R_1 + R_2 + R_3} \tag{10-1}$$

$$= \frac{1 \text{ k}\Omega \times 2 \text{ k}\Omega}{5 \text{ k}\Omega + 1 \text{ k}\Omega + 2 \text{ k}\Omega} = 0.25 \text{ k}\Omega$$

$$R_B = \frac{R_3 R_1}{R_1 + R_2 + R_3} \tag{10-2}$$

$$= \frac{2 \text{ k}\Omega \times 5 \text{ k}\Omega}{5 \text{ k}\Omega + 1 \text{ k}\Omega + 2 \text{ k}\Omega} = 1.25 \text{ k}\Omega$$

$$R_C = \frac{R_1 R_2}{R_1 + R_2 + R_3} \tag{10-3}$$

$$= \frac{5 \text{ k}\Omega \times 1 \text{ k}\Omega}{5 \text{ k}\Omega + 1 \text{ k}\Omega + 2 \text{ k}\Omega} = 0.625 \text{ k}\Omega$$

We now see that a series-parallel circuit is formed, as in Fig. 10-37c, with a total resistance given by

$$R_T = R_A + (R_C + R_3)\|(R_B + R_4)$$

$$= 0.25 \text{ k}\Omega + \frac{3.625 \text{ k}\Omega \times 5.25 \text{ k}\Omega}{3.625 \text{ k}\Omega + 5.25 \text{ k}\Omega}$$

$$= 0.25 \text{ k}\Omega + 2.14 \text{ k}\Omega = \textbf{2.39 k}\Omega$$

$$I_T = \frac{V}{R_T} \tag{3-1a}$$

$$= \frac{12 \text{ V}}{2.39 \text{ k}\Omega} = \textbf{5.02 mA}$$

These answers and the method of solution should be compared with Example 10-4.

A wye-delta conversion could have been used, instead, to reduce the bridge circuit to a series-parallel combination. In that case, however, you would have to do more work because of the larger number of parallel combinations. The particular conversion to be used depends upon the circuit to be solved. The work may sometimes be reduced if it is possible to choose a conversion in which all three resistance values are identical. In that case, the three "arms" of the equivalent network will be the same, requiring only one calculation.

It is interesting to note that delta-wye networks are also referred to as pi-tee networks, as shown in Fig. 10-38.

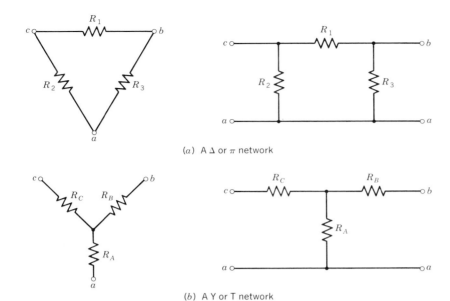

(a) A Δ or π network

(b) A Y or T network

FIGURE 10-38

Comparison of delta-pi and wye-tee networks.

A delta network can be redrawn in the shape of a (π) figure, and a wye network in the shape of a tee. The pi-tee designations are common in electronics applications, since an "input" and "output" pair of terminals can be identified (for example, in a filter circuit). The delta-wye designations are more common in electrical machinery (such as ac motors, generators, and transformers) where a three-phase, three-wire, or four-wire supply is used. This topic is discussed further in Chapter 14.

SUMMARY

1. The branch current method of analyzing a circuit involves assigning branch current designations, showing polarity signs of voltage drops across resistors, choosing arbitrary loops for tracing, and writing as many independent Kirchhoff's voltage law equations as there are windows in the circuit.

2. After simultaneous solution of the Kirchhoff's voltage law equations, any negative currents are given a direction opposite to the assumed direction.

3. The loop current method involves indicating a current loop in each window, showing the voltage drops for each loop, and writing a Kirchhoff's voltage law equation for each loop. The algebraic sum of loop currents through each resistor is used to determine the voltage drops.

4. The superposition theorem is used with a multi-emf circuit. It states that the total circuit current is the algebraic sum of the currents resulting from each emf acting independently.

5. The superposition method usually results in a series-parallel combination of components, making it unnecessary to write a set of Kirchhoff's voltage law equations.

6. Thévenin's theorem shows how a complex linear network with at least one emf may be represented by a single source of voltage in series with a single resistor.

7. The Thévenin equivalent voltage is the open-circuit voltage with any load (that may later be connected) removed.

8. The Thévenin equivalent resistance is the resistance looking back into the circuit with all emf sources removed and replaced by their internal resistances.

9. The Thévenin equivalent of a circuit is very useful in making repeated calculations for various loads to be connected to the original circuit.

10. Norton's theorem shows how a complex linear network with at least one emf may be represented by a single constant-current source in parallel with a single resistance.

11. The magnitude of the current source is the short-circuit current that flows in the original circuit with the load (that may later be connected) removed and replaced with a short.

12. The Norton equivalent resistance is the resistance looking back into the circuit with all emf sources removed and replaced by their internal resistances. It has the same value as the Thévenin resistance.

13. A Thévenin equivalent circuit can be converted to a Norton equivalent circuit, and vice-versa. Each behaves in the same way as the original circuit, as far as any externally connected load is concerned.

14. Millman's theorem applies to a circuit in which two or more emfs, and their internal resistances, are parallel-connected. Each is converted into a Norton equivalent and the respective current sources and resistances are combined. Finally, a conversion to a Thévenin equivalent is made.

15. When two or more current sources are series-connected, each is converted into an equivalent Thévenin voltage, which may then be combined into a single voltage source.

16. Nodal analysis involves identifying a reference node (junction) and assuming that all other nodes are positive with respect to the reference node.

17. At each node, the current leaving is equated to the current entering. This allows sufficient equations to be written to solve for each nodal voltage.

18. Three components may be connected in a delta or wye configuration. By using the appropriate equations, a delta (or pi) may be converted to a wye (or tee), and vice-versa. These conversions may form some series-parallel combination to simplify the solution of a complex network.

SELF-EXAMINATION

Answer true or false
(Answers at back of book)

10-1. The branch current method of analyzing a circuit involves the use of *both* Kirchhoff's current and voltage laws. _____

10-2. The number of "windows" that a network has determines how many independent Kirchhoff's voltage law equations must be written. _____

10-3. Using the branch current method, it is possible to have more than one set of voltage-drop polarity signs across a given resistor. _____

10-4. The direction of loops shown for tracing currents must always be drawn to make an emf appear as a voltage rise. _____

10-5. The interpretation of a negative current is simply to reverse the direction from the assumed direction. _____

10-6. Whether using the branch current or loop current method, there must *always* be as many independent equations as there are unknown currents. _____

10-7. The loop current method does not involve Kirchhoff's current law in order to write Kirchhoff's voltage law loop equations. _____

10-8. In the loop current method a loop with no emf will always have the voltage-rise side of the equation equal to zero. _____

10-9. The branch current and loop current methods are generally used only when the circuit is not some combination of series and parallel components. _____

10-10. The superposition theorem only applies to a multi-emf network. _____

10-11. There are no restrictions on the type of circuit to which the superposition theorem can be applied. _____

10-12. The advantage of the superposition theorem is that the removal of all emfs but one usually leaves a simple series-parallel circuit. _____

10-13. Thévenin's theorem only applies to nonseries-parallel networks. _____

10-14. The Thévenin equivalent voltage is the open-circuit voltage of the initial circuit, and the Thévenin equivalent resistance is the resistance looking back into the circuit with all voltage sources replaced by their internal resistances. _____

10-15. A Thévenin equivalent circuit acts exactly as the initial circuit, as far as the load is concerned. _____

10-16. The Thévenin equivalent for any given circuit is the same no matter which component in the initial circuit is being removed to find the equivalent. _____

10-17. The resistance in a Norton equivalent is always the same as the resistance in a corresponding Thévenin equivalent circuit. _____

10-18. The current source in a Norton equivalent is always the short-circuit current in the initial circuit. _____

10-19. The voltage across a constant current source varies depending upon the load connected to it. _____

10-20. It is impossible to have two current sources of different magnitude in series with each other. _____

10-21. A Wheatstone bridge circuit has four major nodes and requires three nodal equations for solution. _____

10-22. A delta-connected network is found only in three-phase circuits. _____

10-23. A delta-wye conversion (or vice-versa) is made if the result is a simple series-parallel combination with other components in the original circuit. _____

REVIEW QUESTIONS

1. Explain why the direction of the loops in either the branch current method or the loop current method is arbitrary.
2. In the loop current method, what is the rule for writing the Kirchhoff's voltage law equation if a resistor has more than one loop of current through it?
3. What circuit conditions are required to apply the superposition theorem?
4. How would you experimentally apply the superposition theorem?
5. Given a circuit containing a number of emfs (with negligible internal resistance) and resistors, describe how you would experimentally determine the Thévenin equivalent circuit.
6. What is one advantage of solving the current in a component (which takes on different values) using Thévenin's theorem compared with the branch current, loop current, or superposition methods?
7. a. In what way does a Norton equivalent differ from a Thévenin equivalent circuit?
   b. If both were contained in two separate black boxes, how may it be possible to determine which one contains the Norton equivalent?

8. Describe how you would experimentally determine the Norton equivalent of a circuit.

9. Draw a network that contains one voltage source, two current sources, six resistors, and four nodes.

10. Draw a network (other than a Wheatstone bridge) in which a delta-wye conversion would permit a series-parallel solution for total resistance.

PROBLEMS

(Answers to odd-numbered problems at back of book)

10-1. Use the branch current method and Fig. 10-1 with $V_B = 12.6$ V, $V_G = 14.1$ V, $R_B = 0.05$ Ω, $R_G = 0.1$ Ω, and $R_L = 1$ Ω to solve for:
a. Battery current.   c. Load current.
b. Generator current.   d. Load voltage.

10-2. Repeat Problem 10-1 using the loop current method.

10-3. Assume that the battery in Problem 10-1 was installed with reverse polarity. Recalculate the currents and load voltage using either the branch or loop current method.

10-4. Given the circuit and values as in Problem 10-1, to what would the generator open-circuit voltage have to change in order to make the battery current zero?

10-5. Given Fig. 10-11 with values as shown except that $V_2 = 15$ V, find the current in each resistor using the loop current method.

10-6. Repeat Problem 10-5 using the branch current method.

10-7. Given Fig. 10-4a with $R_1 = 5.1$ kΩ, $R_2 = 4.7$ kΩ, $R_3 = 3.3$ kΩ, $R_4 = 1$ kΩ, $R_5 = 2.2$ kΩ, and $V = 15$ V, use the branch current method to calculate:
a. The current in each resistor.
b. The voltage from c to d.
c. The total resistance between a and b.

10-8. Repeat Problem 10-7 using the loop current method.

10-9. Given Fig. 10-39 with $R = 150$ Ω, use the loop current method to find the current through R.

10-10. Use the loop current method to find the current and its direction through R_3 in Fig. 10-40.

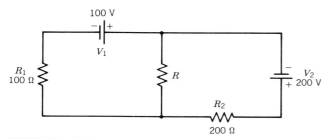

FIGURE 10-39
Circuit for Problems 10-9, 10-15, 10-31, and 10-39.

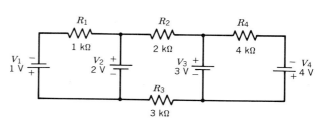

FIGURE 10-40
Circuit for Problems 10-10, 10-32, 10-33, and 10-41.

10-11. Given the circuit and values in Problem 10-7, determine:

    a. The value of resistance that R_4 must be changed to so that no current flows between c and d.

    b. The total resistance between a and b for this condition.

10-12. Repeat Problem 10-1 using the superposition theorem.

10-13. Given Fig. 10-11 with values as shown, except that $V_2 = 15$ V, find the current in each resistor using the superposition theorem.

10-14. Given Fig. 10-41, find the current through $R_L = 1$ kΩ using the superposition theorem.

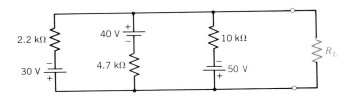

FIGURE 10-41

Circuit for Problems 10-14, 10-26, 10-29, and 10-40.

10-15. Given Fig. 10-39, find the current through $R = 100$ Ω, using the superposition theorem.

10-16. Given Fig. 10-42, a circuit of a three-wire Edison system, use superposition to find the current through R_{L_2}.

10-17. Use the mesh (loop) current method to solve for the current in line y of Fig. 10-42.

10-18. Use the branch current method to solve for the current in R_{L_1} in Fig. 10-42.

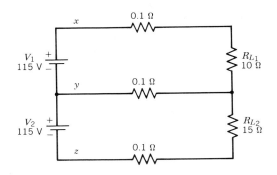

FIGURE 10-42

Circuit for Problems 10-16, 10-17, 10-18, 10-34, and 10-42.

10-19. Given Problem 10-1, find the Thévenin equivalent circuit for R_L and thus determine all the currents and load voltage.

10-20. Given Problem 10-3, find the Thévenin equivalent circuit for R_L and thus determine all the currents and load voltage.

10-21. Given Fig. 10-2 with $V_1 = 15$ V, $V_2 = 12$ V, $R_1 = 5.1$ kΩ, $R_2 = 4.7$ kΩ, $R_3 = 2.2$ kΩ, $R_4 = 4.7$ kΩ, and $R_5 = 1$ kΩ, find the current through R_3 using Thévenin's theorem.

10-22. Given Fig. 10-11 with values as shown except that $V_2 = 15$ V, use Thévenin's theorem to find the current through R_3.

10-23. Given Fig. 10-4a with $V = 15$ V, $R_1 = 5.1$ kΩ, $R_2 = 4.7$ kΩ, $R_3 = 3.3$ kΩ, $R_4 = 1$ kΩ, and $R_5 = 2.2$ kΩ, use a Thévenin equivalent circuit to find the current through R_5.

10-24. Given Fig. 10-11 and the values as shown, use a Thévenin equivalent circuit to find the current through R_4.

10-25. Given Fig. 10-11 and the values as shown, use a Thévenin equivalent circuit to find the current through R_1.

10-26. Given the circuit of Fig. 10-41, use a Thévenin equivalent circuit to find the current through $R_L = 1$ kΩ.

10-27. Given the circuit of Fig. 10-20a with $R_1 = 15$ kΩ, $R_2 = 56$ kΩ, $R_3 = 33$ kΩ, and $V = 40$ V, use Norton's theorem to find the current through R_L when
  a. $R_L = 82$ kΩ
  b. $R_L = 47$ kΩ

10-28. Repeat Problem 10-1 using a Norton equivalent for R_L.

10-29. Given the circuit of Fig. 10-41, use a Norton equivalent circuit to find the current through $R_L = 1$ kΩ.

10-30. Repeat Problem 10-29 using Millman's theorem.

10-31. Given the circuit in Fig. 10-39, calculate the value of R for maximum transfer of power, and find what this power is, if $R_1 = 200$ Ω and $R_2 = 100$ Ω.

10-32. Given the circuit in Fig. 10-40, determine the Thévenin equivalent for resistor R_2.

10-33. Repeat Problem 10-32 with $R_3 = 0$.

10-34. Find the Thévenin and Norton equivalents for R_{L_2} in Fig. 10-42.

10-35. Given the circuit in Fig. 10-43, determine the Thévenin equivalent circuit for R_L.

10-36. For the circuit shown in Fig. 10-43, find the resistance seen by the 10-V battery, when $R_L = 1$ kΩ.

10-37. Use mesh (loop) analysis to find the current through R_4 when $R_L = 10$ kΩ, in Fig. 10-43.

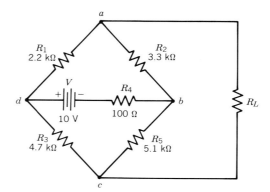

FIGURE 10-43
Circuit for Problems 10-35, 10-36, and 10-37.

10-38. Given the circuit in Fig. 10-26a with $R_1 = 6$ kΩ, and $R_2 = 1$ kΩ, find the current through $R_L = 1$ kΩ.

10-39. Use nodal analysis to solve for the current through $R = 75$ Ω in Fig. 10-39, if $R_1 = 200$ Ω and $R_2 = 100$ Ω.

10-40. Use nodal analysis to solve for the voltage across $R_L = 5$ kΩ in Fig. 10-41.

10-41. Use nodal analysis to determine the current through R_2 with $R_3 = 0$ in Fig. 10-40.

10-42. Find the current flowing through the y-line in Fig. 10-42, using nodal analysis.

10-43. Use nodal analysis to find the voltage across R_L in Fig. 10-44.

10-44. Given the delta circuit in Fig. 10-36a, with R_1 = 1 kΩ, R_2 = 2.2 kΩ, and R_3 = 3.3 kΩ, determine the equivalent wye circuit in Fig. 10-36b.

10-45. Given the wye circuit in Fig. 10-36b, with R_A = R_B = R_C = 10 kΩ, determine the equivalent delta circuit in Fig. 10-36a.

10-46. Given the circuit in Fig. 10-37a, find the current in R_2 by converting the lower delta—*cbd*—into an equivalent wye.

10-47. Given the circuit in Fig. 10-37a, with R_1 = 1.5 kΩ, find the total resistance between *a* and *d* by converting the wye on the right—*ab*, *cb*, *db*—into an equivalent delta.

10-48. Given the bridged tee attenuator circuit in Fig. 10-44, find the current through the load and the voltage across the load using a delta-wye conversion.

10-49. Solve Problem 10-48 using Thévenin's theorem.

10-50. Solve Problem 10-48 using Norton's theorem.

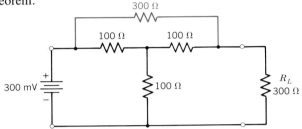

FIGURE 10-44
Circuit for Problems 10-43 and 10-48.

CHAPTER 11

MAGNETISM AND ITS APPLICATIONS

Magnetism plays a very important part in the operation of such devices as electrical measuring instruments, motors, and generators. It was first used for a practical purpose, navigation, more than 1800 years ago. The Chinese found that a natural magnet, called a *lodestone,* would align itself with the earth's magnetic field along a north–south direction. The word *magnet* is apparently derived from the name of the ancient Greek city Magnesia. Fragments of the mineral magnetite, found in the vicinity of Magnesia, display the property of aligning themselves with Earth's magnetic field.

In 1819, the Danish physicist Hans Christian Oersted discovered that magnetism could be produced by an electrical current. This led to the development of "temporary" magnets, or *electromagnets,* which display magnetic properties as long as an electrical current is supplied to them.

Electromagnetism is used in such devices as relays, solenoid valves, motors, electrical measuring instruments, and even magnetic-core memory storage in computers.

The magnetic properties of a material are revealed by its *hysteresis loop,* a presentation in graphical form of *flux density* plotted against the *magnetizing force* or *intensity*. The application of the material in ac and dc *magnetic circuits* is determined by the shape and area of the loop. The system of units used for measurement of magnetic quantities and their relationships is covered in *Ohm's law for magnetic circuits.*

Also covered in this chapter are *nonlinear* magnetic circuits, which require a graphical approach using the *B-H* curve of the material involved.

11-1 PERMANENT MAGNETS

A common observation is that a permanent magnet will attract certain metals and not others. The only three common *elements* that are attracted to a magnet at room temperature are *iron, nickel,* and *cobalt*. They are called *magnetic* or ferromagnetic materials. All other materials are considered to be *nonmagnetic*.*

On a bar-shaped magnet, the two areas that have the strongest attraction for a magnetic material are called the *poles* of the magnet. As shown in Fig. 11-1, these are distinguished as the *north pole* and *south pole,* after the earth's magnetic poles.

11-1.1 Magnetic Field

The region surrounding a magnet is called a *magnetic field*. It is useful to *represent* the direction and strength of this field by showing lines of magnetic force (or magnetic

*Some materials, such as platinum and aluminum, are *very slightly* magnetic, and are called *paramagnetic* materials. Other elements, such as silver and copper, offer a very slight *opposition* to magnetic lines of force in comparison to free space (air), and are termed *diamagnetic* materials. In both cases, however, these effects are so small (approximately 0.001% different) that all elements except iron, nickel, and cobalt can be considered nonmagnetic for practical purposes.

flux) as in Fig. 11-1. Although there is space between the lines, this does not indicate the absence of magnetic force. The spacing of the lines is used to indicate the relative *strength* of the force, or magnetic flux *density*. Thus, the more crowded together, or closer, the lines are (as at the poles), the stronger the magnetic field and its effect.

The lines of magnetic force also have *direction,* as indicated by arrows. In Fig. 11-1, they are shown leaving the north pole and entering the south pole. The lines (and thus the magnetic field) are continuous, completing a closed loop or path from south to north *inside* the magnet.

The *direction* of the magnetic field is simply the direction in which the north end of a compass needle points when the compass is placed inside a magnetic field. The lines and arrows do *not* indicate the *flow* of any electrical charge, either inside or outside the magnet. The magnetic field around a permanent magnet would continue to exist even in a perfect vacuum (such as deep space), where there is no matter to conduct electricity.

11-1.2 Attraction and Repulsion

The attraction and repulsion of two magnets placed near each other is shown in Fig. 11-2. Since lines of force cannot intersect, the repulsion exhibited by two *like* magnetic poles (north or south) is represented as shown in Fig. 11-2a. The parallel paths of the lines of force at the two like poles show the repelling force. The attraction between two

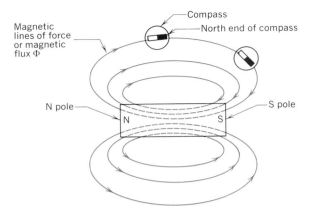

(a) Lines of force leave a north pole and enter a south pole

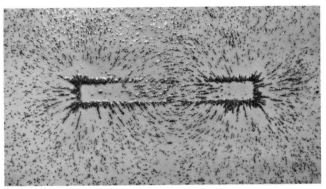

(b) Iron filings line up with the magnetic field

FIGURE 11-1
Magnetic field of a permanent magnet.

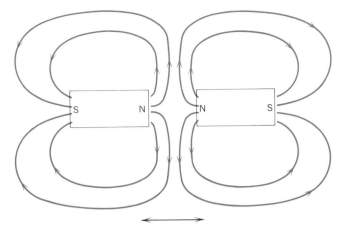

(a) Repulsion of like poles

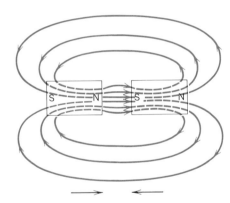

(b) Attraction of unlike poles

FIGURE 11-2
Using magnetic lines of force to show why like poles repel and unlike poles attract.

unlike poles is represented, in Fig. 11-2b, by lines of force passing from the north pole of one magnet through the south pole of the other magnet. This demonstrates an important rule of magnetism:

Like poles repel; unlike poles attract.

The attraction of magnetic (but not magnetized) material by a magnet is illustrated in Fig. 11-3. The magnetic field (as shown by the lines of force) is distorted as it passes through the magnetic material—in this case, an iron nail. Iron, as a magnetic material, is said to be more *permeable* than air. The lines of force would rather pass through iron than through air.

In effect, the nail becomes a *temporary* magnet by the process of magnetic *induction,* as shown in Fig. 11-3. Where the field enters the nail, the effect of a south pole is created. Where the field leaves the nail, a north pole is induced. As a result, the "south" end of the nail is at-

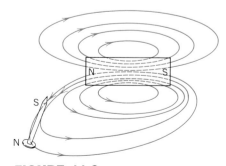

FIGURE 11-3
Attraction of an iron nail by a permanent magnet.

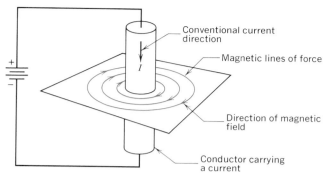

FIGURE 11-4

Magnetic field set up by a current-carrying conductor.

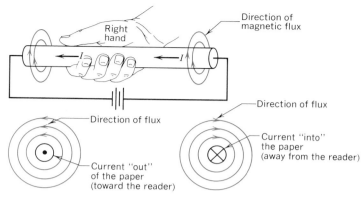

FIGURE 11-5

Right-hand rule to determine the direction of the magnetic lines of force around a straight current-carrying conductor.

tracted to the unlike (north) pole of the magnet. The same effect would take place, with the induced poles of the nail reversed, if it were near the south pole of the magnet. In other words, either pole of the magnet will attract a magnetic material that has not been magnetized.

A magnetic material (such as the iron nail) *will* become magnetized to some extent by *induction* if it remains in contact with the magnet for a period of time. When removed from the magnet, the nail will exhibit *residual magnetism*. This property of a material to retain magnetism is called *magnetic retentivity*.

11-2 OERSTED'S DISCOVERY

Hans Christian Oersted, a Danish physicist, discovered in 1819 that a current flowing through a wire generated a magnetic field that would affect a nearby compass. By using a compass to explore the region around the wire, you would find that the magnetic field takes a circular (concentric) form, as shown in Fig. 11-4. The magnetic field is strongest at the surface of the wire and weakens as distance from the wire is increased. If the current is increased, the field strength also increases. The magnetic field can be made visible by passing the wire through a piece of cardboard. Iron filings sprinkled on the cardboard will align themselves in the direction of the field, forming circular patterns.

The magnetic field has a direction given by the *right-hand rule,* as shown in Fig. 11-5. The rule is

Grasp the conductor with the right hand, so that your thumb is pointing in the direction of *conventional* current. The fingers curled around the conductor will point in the direction of the magnetic field.

The direction of the current with respect to the field around the conductor is sometimes designated with a dot or a cross, as shown in Fig. 11-5. The dot represents the current moving *toward* the reader; the cross represents the current moving *away* from the reader. By applying the right-hand rule, the direction of the *magnetic field* can be established as clockwise or counterclockwise.

11-3 THE ELECTROMAGNET

If a conductor is wrapped around a nonmagnetic core, such as a hollow cardboard tube, the result is a coil (helix). When current is passed through the coil, a magnetic field like that of a bar magnet is generated. (See Fig. 11-6a.) The current-carrying coil is an *electromagnet*. The strength of the electromagnet's magnetic field can be increased by using a soft iron core (rather than the cardboard tube or other nonmagnetic core), by increasing the number of turns of wire around the core, or by increasing the current.

The direction of the field around an electromagnet can be established by using the *right-hand rule for coils,* illustrated in Fig. 11-6b.

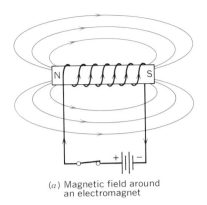

(a) Magnetic field around an electromagnet

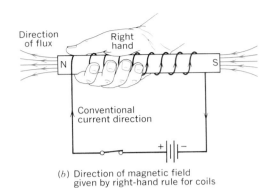

(b) Direction of magnetic field given by right-hand rule for coils

FIGURE 11-6

Magnetic effect of an electromagnet.

Grasp the coil with the *fingers* of the right hand pointing in the direction of *conventional* current. Your thumb will point in the direction of the magnetic field through the electromagnet and the coil's *north* pole.

An electromagnet is a *temporary* magnet—if current is interrupted, the magnetic field disappears. This property is used for control functions in many devices, such as relays, solenoids, bells, buzzers, and door chimes.

11-3.1 Solenoids

A *solenoid* is an electromechanical device with a *movable* iron core that acts *against* an internal spring. Figure 11-7a shows a typical use of a solenoid, such as to control a water valve. A spring acts on the plunger, holding it in the

open position. When current (either ac or dc) is applied to the coil, the plunger is drawn into the solenoid. That is, a magnetic force is exerted that produces the shortest path for magnetic flux to pass through. In this sense, the magnetic lines of force exhibit an *elastic property,* producing a force that finds the shortest path for magnetic flux. This movement of the plunger actuates the water valve, opening or closing it (depending on design).

11-3.2 Door Chimes

A solenoid used in a door chime operates in a similar manner, except that the force of gravity replaces the spring. The plunger normally rests on the bottom chime bar (Fig. 11-7b). When the circuit is completed by pressing the push button, the solenoid draws the plunger upward, so that it

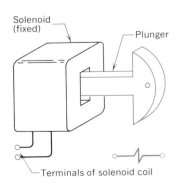

(a) Solenoid control mechanism and symbol

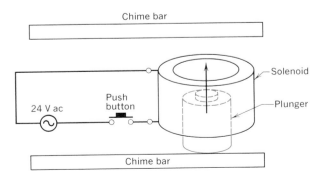

(b) Operation of a door chime

FIGURE 11-7

Applications of solenoids.

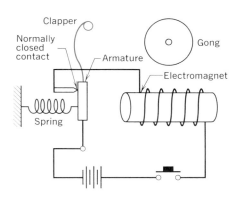

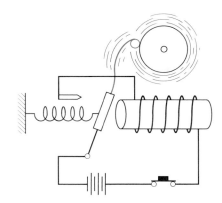

FIGURE 11-8
Operation of a doorbell.

strikes the top chime bar and makes a sound. Releasing the button breaks the circuit. The magnetic force disappears, and the plunger drops back on the lower chime bar, making another sound.

11-3.3 Bells and Buzzers

The intermittent (on-off) action of the door chime should not be confused with the *continuous* vibrating effect used in door bells, buzzers, or horns. The vibration is obtained by *interrupting* current through an electromagnet as an armature is drawn toward it. As shown in Fig. 11-8, a clapper is attached to the armature. When the door bell is pressed, the electromagnet pulls the armature toward it, and the clapper strikes the bell. As the armature moves away from the contact, current to the electromagnet is in-

terrupted. This allows the spring to pull the armature back toward the contact. When the circuit is completed, the process begins again. This make/break cycle is very rapid, and continues as long as the pushbutton is held down.

11-3.4 Application of Solenoid Valves in a Home Heating System

Many gas-fired, forced-air home heating systems involve an interesting application of the solenoid valve. (See Fig. 11-9.) Two series-connected solenoid valves (located in the same mechanical unit) are placed in the gas line feeding the furnace. A thermocouple in the pilot flame controls the *safety* valve, while the room thermostat controls the *main* gas valve.

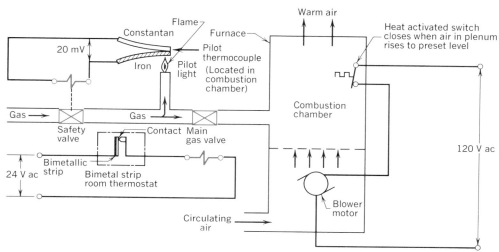

FIGURE 11-9

Application of solenoid valves in a domestic forced-air heating unit.

A thermocouple consists of two different metals (such as iron and constantan) welded together at one end and open at the other. When the closed end of the thermocouple is heated by the pilot flame, an emf of 20 to 30 mV is generated at the open end. This voltage is used to energize the solenoid in the safety valve (after the valve has been manually depressed), holding the valve open and allowing gas to flow to the pilot flame. If the pilot flame goes out, the thermocouple no longer supplies a voltage to the safety valve solenoid. The valve closes, cutting off gas flow.

The thermostat is a bimetallic switch that operates at room temperature. It consists of two different metals (often iron and brass) that are bonded together and touching a stationary contact. Since the two metals expand at different rates (the brass expands more than the iron), an increase in room temperature will cause the bimetal strip to bend. This moves the strip away from the contact, breaking a circuit and turning off the furnace. When the room cools to a predetermined temperature, the strip straightens and once again touches the contact. The furnace is turned on again.

As shown in Fig. 11-9, the circuit completed by the bimetallic strip and contact in the room thermostat applies 24-V ac to the solenoid in the main gas valve, causing it to open. If the pilot flame is lit, the safety valve will be open. This allows gas to flow to the combustion chamber and be ignited by the pilot flame. (Of course, if the pilot flame is out, the safety valve will prevent gas flow, even if the main valve opens.)

After the main burners operate for a minute or two, the temperature in the air plenum rises far enough to close a heat-activated switch, applying 120 V to the furnace blower motor. Warm air circulates through the house, until the bimetallic strip in the thermostat opens the contact. This closes the main gas valve, cutting off the supply to the combustion chamber. The blower motor will continue to operate until the air temperature in the plenum drops far enough to open the heat-activated switch. There are three separate circuits involved in this heating system—the pilot flame-safety valve circuit, the thermostat-main gas valve circuit, and the plenum switch-blower motor circuit.

11-3.5 Relays

A *relay* is a magnetically operated switch that can be used to control many high-voltage, high-power circuits from a single low-voltage input. The movement of a single ar-

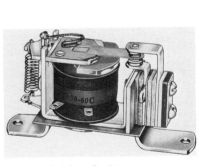

(a) A form C relay

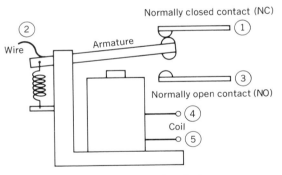

(b) Pictorial representation of relay

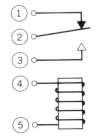

(c) Symbol for form C relay

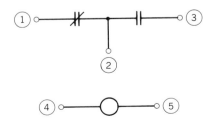

(d) Alternate symbol for form C relay

FIGURE 11-10

Magnetic relay and symbols. [Photograph in (a) courtesy of Potter and Brumfield Division AMF Incorporated.]

mature, caused by energizing one electromagnet, can simultaneously open and close many contacts. These contacts usually control loads operating at higher voltage and current than the input voltage to the electromagnet's coil. Relays used to control very large loads, such as high horsepower motors, are often referred to as *contactors* or starters. Their operating principle is the same as any other relay, however.

One common type of relay has two contacts—one *normally open*, the other *normally closed*. (See Fig. 11-10.) In this device, usually referred to as a ''Form C'' relay, a spring holds the armature against the upper (normally closed) contact, providing continuity. When the rated voltage is applied to the coil, energizing it, the armature is pulled downward. This ''breaks'' the normally closed contact and ''makes'' (provides continuity with) the normally open contact.

In a typical relay, a potential difference of 10 V at just a few milliamperes is sufficient to energize the relay and move the armature. The load contacts may be rated at 120-V ac, 1 A, or more. A wide variety of both ac and dc relays is available. The ac type can often be recognized by a copper segment inserted in the face of the electromagnet's core. This strip, called a ''shading coil,'' helps to prevent ''chattering,'' the tendency of the coil to become de-energized on each half cycle of the coil's ac input voltage.

The heating system described in Section 11-3.4 can be used to illustrate the application of an SPDT (single pole double throw) relay. Many forced-air systems are equipped to provide air conditioning through the same ducts used for heating. When used for cooling, however, the blower motor must run at a higher speed to move the heavier cold air at the proper rate. A relay can be used to automatically switch over from the low heating speed to the high cooling speed, as shown in Fig. 11-11.

In the heating mode, the relay is in its normally closed position, supplying 120-V ac to the blower motor's low-speed winding. (The relay contact is also in series with the heat-activated plenum switch and a high-temperature limit switch (not shown) that will shut down the system if temperature becomes excessive.) When the room thermostat is switched to the cooling mode, and the air conditioning compressor starts running, a contact closes in the condensing unit to energize the 24-V fan speed relay. The relay's armature is moved to the normally open contact, applying 120-V ac to the blower motor's high-speed winding. When the room is cooled to the desired temperature, the thermostat contact, not shown in Fig. 11-11, opens. This, in turn, opens the contact in the condensing unit and de-energizes the relay. The blower motor shuts off.

11-4 MAGNETIC UNITS

The magnetic properties of a material may be displayed graphically by means of a *B-H* curve. To understand this term, consider Fig. 11-12, which shows a magnetic core with a given number of turns of wire wrapped around it. A variable voltage source allows the current in the coil to be varied. This, in turn, determines the amount of magnetic flux in the core.

11-4.1 Flux Density

An increase in the magnetic lines of force in the core can be thought of as an increase in the *magnetic flux density*, *B*. Magnetic flux density can be defined as follows:

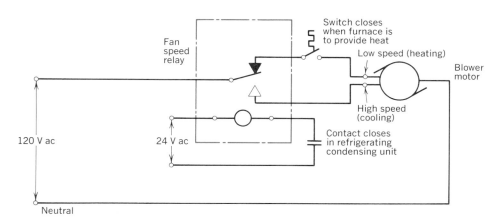

FIGURE 11-11

Relay used to change blower motor from low-speed heating to high-speed cooling.

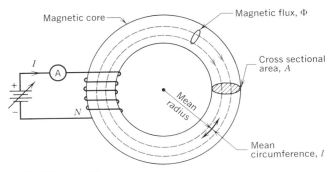

FIGURE 11-12
Quantities for determining *B* and *H* in a magnetic core.

$$B = \frac{\Phi}{A} \quad \text{webers per square meter} \quad (11\text{-}1)$$

where: *B* is the flux density, in webers per square meter (Wb/m²) or tesla (T)

Φ is the magnetic flux, in webers (Wb)

A is the core's cross-sectional area, in square meters (m²)

The weber (Wb) is the unit for the measurement of magnetic flux in the MKS (meter, kilogram, second) metric system. A flux *density* of 1 Wb/m² is also known as a flux density of 1 *tesla* (T). That is,

$$1 \text{ T} = 1 \text{ Wb/m}^2$$

11-4.2 Magnetomotive Force

Five factors determine the amount of flux density set up in the core: current, number of turns, material used in the core, length (circumference) of the core, and cross-sectional core area. The more current and the more turns of wire used, the greater will be the magnetizing effect. The product of the turns and current is called the *magnetomotive force* (mmf), and is analogous to the electromotive force (emf).

$$F_m = NI \quad \text{ampere-turns} \quad (11\text{-}2)$$

where: F_m is the magnetomotive force, mmf, in ampere-turns (At)

N is the number of turns wrapped on the core

I is the current in the coil, in amperes (A)

11-4.3 Magnetizing Intensity

It should be clear that a given number of ampere-turns on a core with a smaller circumference (length) represents a greater magnetizing *intensity* than the same number of turns on a core with a larger circumference. This *magnetizing intensity* is a useful term and is given the symbol *H*. It is the number of ampere-turns per meter of length of the magnetic circuit:

$$H = \frac{F_m}{l} = \frac{NI}{l} \quad \text{ampere-turns/meter} \quad (11\text{-}3)$$

where: *H* is the magnetizing intensity, in ampere-turns/meter (At/m)

NI is the magnetomotive force in ampere-turns (At)

l is the length of the magnetic core, in meters (m).

EXAMPLE 11-1

A cast iron toroid (doughnut) as in Fig. 11-12 has an inside radius of 5 cm and an outside radius of 7 cm. It has a circular cross section. A coil of 400 turns is wrapped around the core and carries a current of 2 A. A flux of 1.5×10^{-4} Wb is set up inside the core. Calculate:

a. The flux density in tesla.
b. The magnetomotive force in ampere-turns (At).
c. The magnetizing intensity in At/m.

Solution

a. Diameter of the circular core cross section,

$$d = 7 - 5 \text{ cm}$$
$$= 2 \times 10^{-2} \text{ m}$$

area of core's cross section,

$$A = \frac{\pi d^2}{4}$$
$$= \frac{\pi}{4} \times (2 \times 10^{-2} \text{ m})^2$$
$$= \pi \times 10^{-4} \text{ m}^2$$

flux density $B = \dfrac{\Phi}{A}$ $\qquad (11\text{-}1)$

$$= \frac{1.5 \times 10^{-4} \text{ Wb}}{\pi \times 10^{-4} \text{ m}^2}$$
$$= 0.48 \text{ Wb/m}^2 = \textbf{0.48 T}$$

b. Magnetomotive force $F_m = NI$ (11-2)

$$= 400 \text{ turns} \times 2 \text{ A}$$
$$= \textbf{800 At}$$

c. Magnetizing intensity $H = \dfrac{NI}{l}$ (11-3)

mean length of core $= 2\pi r$

$$= 2\pi \times 6 \text{ cm} = 0.377 \text{ m}$$

Therefore, $H = \dfrac{NI}{l} = \dfrac{800 \text{ At}}{0.377 \text{ m}} = \textbf{2120 At/m}$

11-4.4 Permeability

As you have seen, the flux density (B) set up in a core is directly proportional to H. (That is, B is directly proportional to N and I and inversely proportional to l.) A fourth variable in determining the amount of flux density is the material used for the core. **The ability of a material to set up a magnetic field is called its** *permeability* **(μ).** Thus, $B \, \alpha \, H$ and $B = \mu H$. This provides a method for determining a material's permeability:

$$\mu = \frac{B}{H} \quad \textbf{webers per ampere-turn meter} \quad \textbf{(11-4)}$$

where: μ is the permeability in webers per ampere-turn meter (Wb/At-m)

B is the flux density in webers per square meter (Wb/m^2)

H is the magnetizing intensity in ampere-turns per meter (At/m)

For *nonmagnetic materials,* $\mu = \mu_0 = 4\pi \times 10^{-7}$ Wb/At-m. *Magnetic* materials, however, have a much higher permeability that varies widely with the amount of flux density. To show how much more permeable a magnetic material is than air (or any nonmagnetic material) *relative permeability* (μ_r) is used.

$$\mu_r = \frac{\mu}{\mu_0} \quad\quad (11\text{-}5)$$

where: μ_r is the relative permeability (no units)

μ is the permeability of a given material, in webers per ampere-turn meter (Wb/At-m)

μ_0 is the permeability of free space (vacuum), $4\pi \times 10^{-7}$ Wb/At-m

NOTE μ_r is dimensionless (much like specific gravity), because it is a ratio of identical units. It can be as high as several thousand in some modern magnetic materials.

EXAMPLE 11-2

a. Determine the permeability of the iron core in Example 11-1.

b. What is the relative permeability of the iron core?

c. If the iron core in Example 11-1 were replaced by a sheet steel core having a relative permeability of 600, what would be the flux density in the core?

Solution

a. $\mu = \dfrac{B}{H}$ (11-4)

$$= \frac{0.48 \text{ Wb/m}^2}{2122 \text{ At/m}} = \textbf{2.26} \times \textbf{10}^{-4} \textbf{ Wb/At-m}$$

b. $\mu_r = \dfrac{\mu}{\mu_0}$ (11-5)

$$= \frac{2.26 \times 10^{-4} \text{ Wb/At-m}}{4\pi \times 10^{-7} \text{ Wb/At-m}} = \textbf{180}$$

c. $B = \mu H$

$$= \mu_r \mu_0 H$$
$$= 600 \times 4\pi \times 10^{-7} \text{ Wb/At-m} \times 2122 \text{ At/m}$$
$$= \textbf{1.6 Wb/m}^2 = \textbf{1.6 T}$$

11-5 HYSTERESIS

Now, the graphical relation between B and H for a magnetic material can be considered. Since $\mu = B/H$, the graphical relation will show how the permeability of a material *varies* with the magnetizing intensity (H).

Assume that the magnetic core in Fig. 11-12 is initially in a completely demagnetized state. As the current is increased, $H = NI/l$ increases, as does the flux density (B). Since the number of turns and length of the coil are fixed, H is directly proportional to the current or ammeter reading. The flux density can be measured by inserting the probe of a flux meter into a small hole drilled in the core.

Plotting the values of B and H gives the *normal* magnetization curve (Fig. 11-13a). There is evidently a linear portion of the curve where B is relatively proportional to H. But then, a condition of *saturation* occurs where a very large increase in H is needed to significantly increase B.

If the current is then gradually reduced toward zero, H returns to zero, but B does not. The core *retains* some residual magnetism (exhibits *retentivity*). The retentivity is represented in Fig. 11-13a by the distance *OR*.

If the connections to the coil are reversed and the cur-

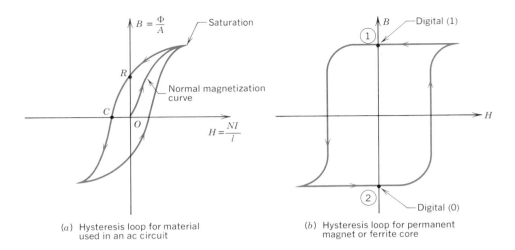

(a) Hysteresis loop for material used in an ac circuit

(b) Hysteresis loop for permanent magnet or ferrite core

FIGURE 11-13

Hysteresis loops.

rent again increased, then a certain amount of H is needed to bring the magnetism in the core down to zero again. This is called *coercivity* and is represented by the distance OC. As the current is further increased, the core is once again magnetized (in the opposite direction from before), until saturation again takes place.

Reduction of the current and subsequent reversal of direction will produce a closed figure called a *B-H* curve or *hysteresis loop*. Hysteresis comes from the Greek word, *hysteros,* meaning ''to lag behind.'' In magnetic terms, the state of the flux density is always lagging behind the efforts of the magnetizing intensity.

The shape of a *B-H* loop indicates the magnetic properties of a material, and thus, its potential applications. To understand this, the basic theory of magnetism must be considered briefly.

11-6 BASIC THEORY OF MAGNETISM

You have seen how an electrical current sets up a magnetic field. In the atomic structure of materials, each electron not only rotates around the nucleus of the atom, but also spins on its own axis. This spinning movement of a negative charge is, in effect, a very small current and produces a tiny magnetic field. The combined magnetic effect of between 10^{17} and 10^{21} molecules is called a *magnetic domain*. Each domain acts as a tiny bar magnet with its own north and south poles. Only ferromagnetic materials (iron, nickel, and cobalt) possess magnetic domains because these atoms act *collectively* in their magnetic effect, in-

stead of *singly* as in nonmagnetic materials. In an unmagnetized material, these domains have a random orientation with no *overall* magnetic effect. (See Fig. 11-14a.) But when the material is placed in a magnetic field, as shown in Fig. 11-14b, the domains become *aligned*, producing an overall magnetic effect.

In materials used to make permanent magnets, most of the domains remain in alignment after the external magnetic field is removed. In other materials, such as soft iron, the domains quickly lose their alignment after removal of the external field. Thus, the material quickly loses its magnetism.

11-6.1 Hysteresis Losses

It is now obvious that *saturation* is the effect of the domains becoming almost completely aligned. Also, if the magnetic material is subjected to an *alternating* magnetiz-

(a) Random orientation of domains in unmagnetized material

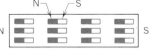

(b) Alignment of magnetic domains in magnetized material

FIGURE 11-14

Schematic representation of magnetic domains in unmagnetized and magnetized materials.

ing force (as is the case when ac current flows in a transformer), the magnetic domains in the core will undergo a reversal many times each second. If the supply frequency is 60 Hz, for example, the hysteresis loop will be repeated 60 times per second. The magnetic core will become warm (in part due to the constant reversal of the magnetic field), and as a result, lose some energy. This is referred to as "hysteresis losses." In a transformer, the result is less power available at the output compared to the input.

Hysteresis losses, in addition to being proportional to frequency, are also proportional to the *area* of the B-H curve. For this reason, materials used in motors, alternators, and other ac equipment should have low values of coercivity and retentivity. This will reduce the area of the hysteresis loop. Soft iron and alloys known as "Supermalloy" and "Permalloy" meet this need. They have a *B-H* loop more closely resembling that in Fig. 11-13a.

11-7 MATERIALS FOR PERMANENT MAGNETS

For a permanent magnet, the *B-H* loop should be as *square* as possible. (Refer to Fig. 11-13b.) Permanent magnet materials should have high values of coercive force (approximately 100,000 At/m) and residual flux density (approximately 1 T).

Alnico, an alloy of aluminum, nickel, iron, cobalt, and copper, is widely used for permanent magnets in audio speakers. However, the increasing cost of cobalt is expected to expand the use of *hard ferrites* in these applications. Hard ferrites are ceramic materials consisting of barium carbonate with particles of ferric oxides. They have good ferromagnetic properties, but are poor conductors of electricity. This lack of electrical conductivity is an advantage at high frequencies, since it significantly reduces losses due to *eddy currents* in communications coils. (Eddy currents are discussed in Section 16-1.3.)

Ferrites are not usually found in power circuits at low frequencies because they tend to saturate at fairly low *H* values. A manufacturing advantage of ferrites is that the material can be molded into any desired shape, then fired in an oxygen atmosphere to produce the finished magnet.

In addition to Alnico and hard ferrites, a third group of materials is used for permanent magnets. Referred to as the *rare earth* group, these materials are all elements that have three electrons in their valence shells. They are expected to replace Alnico in many applications.

11-8 SOME APPLICATIONS OF MAGNETIC MATERIALS

An interesting application of a ferrite core is the *digital magnetic core memory.* (See Fig. 11-15.) When the core (which has a square *B-H* loop like the one shown in Fig. 11-13b) is magnetized in one direction, it can be considered as storing a digital *one* (1). When a pulse of current is passed through a wire that runs through the center of the ferrite "bead," it is magnetized in the opposite direction, storing a digital *zero* (0). By using suitable wires and controls, the stored information can be "read" (or "sensed"), or new information can be "written" into the core. A series of such cores, assembled into flat plates like the one shown in Fig. 11-15, can be used to store "words" of information. By "addressing" the proper location in a stack of such plates, stored information can be obtained very quickly.

Another type of computer information storage is *magnetic tape*. A layer of magnetic oxide is applied to a thin mylar backing, and the tape is passed in front of a recording head, as shown in Fig. 11-16. The recording head consists of a soft iron core with a coil wrapped around it. When current passes through the coil of the recording head in one direction, a certain magnetic polarity is developed. This magnetizes the coating on the tape in a corresponding direction, representing a digital *one*. Current in the opposite direction in the coil reverses the stored magnetic po-

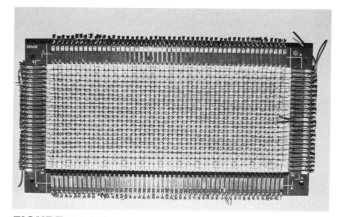

FIGURE 11-15
This 3 × 8 in. plane of magnetic core memory has 1000 ferrite "beads" to store digital information. (This is an example of an early generation of core storage.)

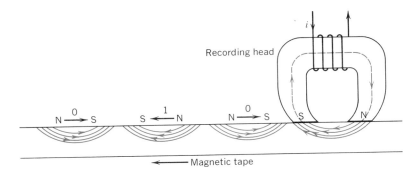

Recording head

N →O→ S   S ←1← N   N →0→ S   S   N

← Magnetic tape

FIGURE 11-16
Magnetic tape recording head principle.

larity on the tape to represent a digital *zero*. Bits of information (in a "1" state or a "0" state) are stored along the length of the tape. The material can be "read back" by passing the tape under a *read* head. Voltages of varying polarity will be *induced* in the coil of the read head as the magnetic fields on the tape pass by. (See Section 14-1.)

A similar method is used to record *audio frequency* signals on magnetic tape. In this case, the frequency of the current in the coil of the recording head (or playback head, during playback) is varied and alternated at an audio rate. Loudness of the signal is determined by the amount of current, and thus the strength of field set up on the tape. The tape must move past the head at a steady, fairly high rate of speed to record high fidelity music. Fidelity (closeness of the recording to the original performance) of the recording also depends on the quality of the tape's magnetic coating and the quality of the heads used for recording and playback.

11-9 DEMAGNETIZATION

The magnetic domain theory, as described in Section 11-6, supports methods that can be used to demagnetize a magnet. Mechanical vibrations, for example, or repeatedly striking a magnet, will tend to rearrange the aligned domains. This causes the magnet to lose its magnetic properties.

A similar, but more effective, method is to heat the magnet above the *Curie temperature*. This temperature, which varies from material to material, causes an abrupt and complete loss of ferromagnetism. The Curie temperature for iron is 770°C, for cobalt 1131°C, for nickel 358°C, and for various alloys, points between 358 and 1131°C, depending upon their composition. Apparently, the high temperature causes the atoms and molecules to reach a high state of agitation, disturbing the magnetic domain alignment. The material can be remagnetized only after cooling to below its Curie temperature.

Neither of the methods already described is suitable for demagnetizing manufactured articles. A widely used practical method is the *demagnetizing coil*, which has many turns and carries an alternating current. This sets up an alternating magnetic field in the center of the coil. If the article to be demagnetized is inserted in the center of the coil, it undergoes a cyclical magnetic reversal around the hysteresis loop. When the article is *slowly* withdrawn from the coil (or the ac current in the coil slowly reduced to zero), its magnetic state follows ever-decreasing amplitudes of hysteresis loops, as shown in Fig. 11-17a. When the article has been completely removed from the loop, the coercivity and retentivity have been essentially reduced to zero.

Demagnetization has a number of practical applications. The recording head of a tape recorder, for example, must set up a magnetic field only when current flows through its coil. But after continual use, the head retains some residual magnetism, which can cause background "noise" while recording. A demagnetizing tool, similar to the coil just described, is used to remove the unwanted magnetism from the head.

Another application involves color television sets. The metal parts around the picture tube can become permanently magnetized, affecting quality of the picture. This magnetism can be eliminated by placing a coil (operating from a 120-V, 60-Hz supply) near the picture tube and slowly moving it away. Most modern sets incorporate a demagnetizing coil around the picture tube that automatically energizes for a short time whenever the set is turned on. This eliminates the need for periodic demagnetizing, and allows the set to be moved from one location to another without being affected by variations in the earth's magnetic field.

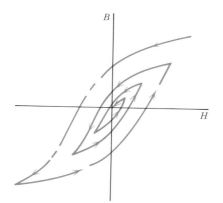

(a) Successive hysteresis loops used in demagnetizing a magnetic material

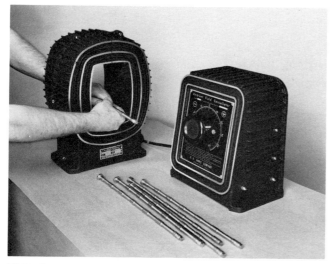

FIGURE 11-17

B-H curve and demagnetizer. [Photograph in (b) courtesy of the R. B. Annis Co.]

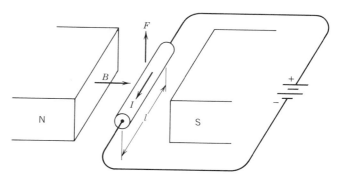

(a) Force *F* is exerted on a conductor of length *l*, carrying current *I*, in a magnetic field *B*

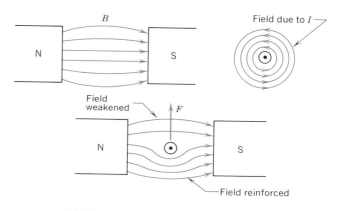

(b) The separate and combined magnetic fields

FIGURE 11-18

Force acting on a current-carrying conductor.

Demagnetizing is sometimes referred to as *degaussing* (the coil used to demagnetize TV picture tubes, in fact, is usually called a "degaussing coil"). The name comes from the older CGS (centimeter-gram-second) metric system's unit of measurement for magnetic flux density, the *gauss* (G). The gauss is a small amount of flux density, so the term kilogauss is often used. The earth's magnetic field is about one-half gauss.

Research in magnetism is often reported in CGS units, rather than MKS units. For example, the CGS unit of magnetizing intensity is the *oersted* (1 At/m = 0.0126 Oe). Thus, a *B-H* curve in the CGS system would consist of a graph of gauss against oersteds. Many of the meters

available to measure flux densities are calibrated in gauss, and are thus known as *gaussmeters*. (See Section 11-11.)

11-10 FORCE ON A CURRENT-CARRYING CONDUCTOR IN A MAGNETIC FIELD

In this section, the *interaction* of magnetic fields will be considered. In particular, you will learn about the interaction of a permanent magnetic field with that due to a current-carrying conductor placed in the field. (See Fig. 11-18a.)

As you learned early in this chapter, the field set up around a current-carrying wire is circular. If the wire, or some part of it, is perpendicular to the field of a permanent magnet, the flux lines on the lower side of the wire will be in the same direction as the flux lines of the permanent magnet. Due to the additive effect, this means that the flux lines are *concentrated* in the area beneath the wire. In the area above the wire, on the other hand, the flux lines *oppose* each other, resulting in a *weakened* field. The overall distribution of fields is shown in Fig. 11-18*b*.

The result of this concentration of flux beneath the wire is to move it *upward* into the weaker field. If either the current or the permanent magnet's field is reversed, the situation would be just the opposite—the concentration of flux would be above the wire, moving it *downward*.

The interaction of the fields produces a *force*. In fact, this is how it can be determined whether a magnetic field exists in a region:

> **A magnetic field is said to exist at a point if a force is exerted on a *moving* charge at that point.**

Thus, you could use a moving stream of charge, such as the electron beam in an oscilloscope, to explore an unknown magnetic field. If the spot of light produced by the beam on the oscilloscope's fluorescent screen remains at the same point on the screen as the tube is moved around, one of two conclusions is possible. Either no magnetic field exists, or the electron beam is parallel to the field. But if the spot on the screen is deflected upward or downward as the tube is moved, you can conclude that a magnetic field exists, and can determine its direction.

In theory, this method could be used to examine the earth's magnetic field. If the beam is placed parallel to the earth's field (roughly north–south), the spot will not be deflected. If the beam is turned 90° (to travel west to east), the spot would be deflected downward. This indicates that the earth's magnetic field has a direction from *geographic* south to *geographic* north. The lines of force leave a region near the south geographic pole and travel to a region near the north geographic pole. Since the lines of magnetic flux move from the *magnetic* north pole to the *magnetic* south pole, the magnetic and geographic poles are reversed. What we usually call the north magnetic pole is, in fact, the *south* pole—even though it is physically located near the north *geographic* pole. Thus the end of a compass needle that points north is actually a north pole, and not a "north-seeking" pole as it is often called.

The earth's distribution of magnetic flux is similar to what it would be if the earth's core were a huge bar magnet. This is illustrated in Fig. 11-19. The angle between the earth's magnetic field and the true geographic north–south direction is called *variation* or *declination*. The an-

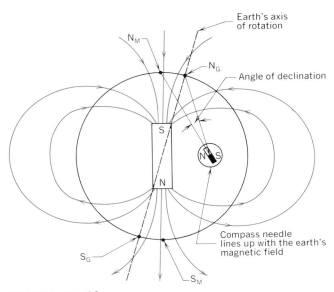

FIGURE 11-19
Simplified diagram of the earth's magnetic field.

gle of declination varies in different parts of the world and from year to year.

The amount of force acting on a moving charge is given by

$$F = qvB \text{ newtons} \qquad (11\text{-}6)$$

where: F is the force in newtons (N)
  q is the charge of the particle in coulombs (C)
  v is the velocity of the charge perpendicular to B in meters per second (m/s)
  B is the flux density of the field in teslas (T)

If the particles are moving at some angle with B other than $\theta = 90°$, the force is reduced by a factor of $\sin \theta$.

Using the preceding equation, it is possible to obtain the force acting on a conductor as in Fig. 11-18*a*.

$$F = BIl \qquad \text{newtons} \qquad (11\text{-}7)$$

where: F is the force on the conductor in newtons (N)
  B is the flux density of the field in which the conductor is placed, in teslas (T)
  I is the current in the conductor in amperes (A)
  l is the length of the conductor perpendicular to the field in meters (m)

If the conductor is at some angle with B other than $\theta = 90°$, the force is reduced by a factor of $\sin \theta$.

EXAMPLE 11-3

A 1-m length of No. 14 copper wire is located between two pole faces, each of which is 10 cm². The total flux between the two poles is 1.2×10^{-2} Wb. Calculate:

a. The force acting on the wire when it is carrying 5 A and is perpendicular to the field.

b. The force acting on the wire when it is carrying 5 A and is at an angle of 45° with respect to the field.

c. How much current must pass through the wire when it is perpendicular to the field if it is to support its own weight of 18 g and remain suspended in the magnetic field.

Solution

a. The configuration is as shown in Fig. 11-18a, with the *active* length of the conductor in the magnetic field equal to 10 cm. That is, $l = 0.1$ m.

$$B = \frac{\Phi}{A} \qquad (11\text{-}1)$$

$$= \frac{1.2 \times 10^{-2} \text{ Wb}}{0.1 \text{ m} \times 0.1 \text{ m}} = 1.2 \text{ T}$$

$$F = BIl \qquad (11\text{-}7)$$

$$= 1.2 \text{ T} \times 5 \text{ A} \times 0.1 \text{ m}$$

$$= \textbf{0.6 N} \text{ (0.13 lb)}$$

b. $F = BIl \sin \theta$

$$= 0.6 \text{ N} \times \sin 45° = 0.6 \text{ N} \times 0.707 = \textbf{0.42 N}$$

c. $w = mg$ (weight of the 1-m length of wire)

$$= 18 \times 10^{-3} \text{ kg} \times 9.8 \frac{\text{m}}{\text{s}^2} = 0.176 \text{ N}$$

This must be balanced by the magnetic force

$$F = BIl \qquad (11\text{-}7)$$

$$= 1.2 \times 0.1 \times I \qquad \text{newtons}$$

Therefore, $0.12 I = 0.176$

and $$I = \frac{0.176}{0.12} \text{ A} = \textbf{1.47 A}$$

There are many applications of a force being exerted on a current-carrying conductor in a magnetic field. Some of them are examined in the following sections.

11-11 HALL EFFECT AND THE GAUSSMETER

Figure 11-20a shows a conductor in the form of a flat strip, perpendicular to a magnetic field (B). Charges are made to flow through the strip by the external cell. The charges have a magnetic force ($F = qvB$) exerted on them, where v is the drift velocity. If the charge carriers are electrons, they are driven toward the upper edge of the strip. An excess *negative* charge accumulates at this edge, leaving an excess *positive* charge on the lower edge. This establishes a potential difference of a few microvolts, called the *Hall emf* or Hall voltage. The Hall emf is always at right angles to both the magnetic field and the current, respectively.

The Hall emf depends directly on the current *and* the magnetic field strength. Thus, if the current is held constant, the emf is a *direct indication* of the magnitude of the magnetic flux density (B). This allows it to be used in instruments to measure the magnetic field strength. The generated Hall emf is much larger if a semiconductor (such as indium arsenide) is used. Even so, amplification of several hundred times is necessary to drive a dc meter, usually calibrated in gauss. Multirange full-scale deflections from 10 to 10^4 G are available on typical portable gaussmeters (fluxmeters) like the one shown in Fig. 11-20b. The probe that contains the semiconductor strip must be rotated until a maximum reading is obtained.

The Hall effect, it is interesting to note, provides a means of distinguishing between N-type and P-type semiconductors. (See Chapter 28.) In P-type semiconductors, current is conducted by means of positive charges called *holes*. Hole flow coincides with the direction of *conventional* current. Thus, if the N-type semiconductor in Fig. 11-20a is replaced by a P-type, and all other variables remain unchanged, the Hall emf will have an *opposite polarity*. A hole flowing in one direction is *not the same* as an electron moving in the opposite direction.

11-12 HIGH-FIDELITY LOUDSPEAKER

As shown in Fig. 11-21a, a loudspeaker consists of a fixed permanent magnet and a moving coil located in an air gap. The moving coil, or "voice coil," may have a resistance of from 4 to 20 Ω. It is held in place and centered by a stiffening element called a *spider*. Attached to the spider and suspended from the basket (or main frame) is the flexible *cone*.

When current from the audio amplifier flows in the voice coil, a force acts on the coil and causes it to move in the magnetic field in the air gap. The cone vibrates back and forth, sending out air pressure waves with the same frequency as the audio input signal current.

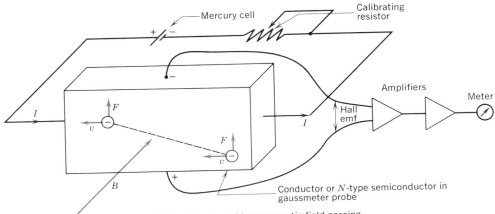

(a) Hall emf developed by a magnetic field passing through a current-carrying semiconductor

(b) Photograph of a gaussmeter

FIGURE 11-20

**Hall effect and gaussmeter.
[Photograph in (b) courtesy of
RFL Industries, Inc.]**

Modern loudspeaker systems (Fig. 11-21b) can typically reproduce frequencies ranging from 30 to 18 kHz, using three transducers ranging in size from 1 in. to 12 or 15 in. diameter. A *crossover* or dividing network directs frequencies to the proper transducer. The highest (treble) frequencies are handled by a dome radiator or horn that produces very small displacements. The bass frequencies, however, contain more energy. They cause the cone of the large "woofers" to vibrate more slowly, but with displacements of as much as 1-in. Woofers operate as low as 20 to 30 Hz.

11-13 DC MOTORS

The operation of a dc motor is the direct result of the force acting on a current-carrying conductor in a magnetic field (as described in Section 11-10 and illustrated in Fig. 11-18b). A motor takes electrical energy and converts it into *rotational* mechanical energy. Some small dc motors use a permanent magnet to set up the stationary or *stator* field. Most, however, use field windings wrapped around pole pieces located around the motor frame, as shown in Fig. 11-22.

The armature, or *rotor,* is a cylinder of soft steel mounted on a shaft so that it can rotate about its axis. A number of insulated copper wires are placed in slots running the length of the rotor. Current is led into and out of these conductors through stationary carbon *brushes* riding on a *commutator* on the end of the rotor shaft. The commutator (shown in Fig. 11-22b) consists of copper segments insulated from each other and connected to the conductors. The commutator, as it rotates, provides an automatic switching arrangement. It keeps current flowing

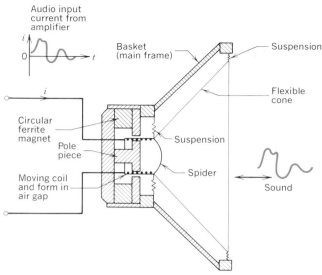

(a) Cross-sectional view of large-cone loudspeaker

(b) A three-way loudspeaker system with dividing network, 1-in. dome radiator, 5-in. midrange speaker, and a 12-in. low-frequency transducer with a 10¼ pound magnet, capable of handling 300 W rms

FIGURE 11-21

Loudspeaker details and system. [Photograph in (*b*) courteys of James B. Lansing Sound, Inc.]

in the rotor conductors in the proper direction (as shown in Fig. 11-22), no matter what the position of the rotor. As a result, all of the conductors under one field (stator) pole experience a side thrust in one direction, and the conductors under the other field pole experience a thrust in the opposite direction. The armature thus assumes a counter-clockwise rotation for the given current direction. Note the way in which the motor frame serves as a path for the magnetic field that passes through the rotor.

The way in which the field windings and the armature are connected to the dc supply determines the type and characteristics of the motor. There are four possible connections, as shown in Fig. 11-23.

A *shunt* motor has a constant field current, because the field winding is in parallel with the supply. The field winding consists of many turns of fine wire to produce a large flux. Although its speed does decrease slightly with an increase in load, the shunt motor is considered a constant-speed motor. It is commonly used to drive a fan (as in an automobile heater). The speed can be controlled by varying a rheostat wired in series with the field winding. An increase in resistance reduces the field and increases the speed of the motor.

In the *series* motor, the field winding is connected in series with the armature. Since the field must carry the full armature current, it normally consists of a few turns of comparatively large-diameter wire. High starting torque is the major characteristic of the series motor. When the motor is started, there is a large in-rush of current, setting up a very strong field in both the stator and the armature. The large torque developed makes this type of motor ideal for such applications as automobile starters. If the load is removed from a series motor, the current and the magnetic field both drop, causing the motor's speed to increase. (It is trying to generate a counter-emf equal to the applied emf with a much lower flux. It can only do this by rotating faster. See Chapter 14.) For this reason, large series motors should not be used in situations where the load might become disconnected. The speed could increase to a dangerously high rate, causing the motor to fly apart as a result of centrifugal force.

A *compound* motor has *two* field windings (shunt *and* series) that can be connected in either the long- or short-shunt configuration. As shown in Fig. 11-24, compound motors have speed and torque characteristics midway between those of the series and shunt motors.

It should be noted that a dc motor's direction of rotation is *not* changed by reversing the applied polarity (except in the case of small motors with permanent magnets used for the stator field). To obtain opposite rotation, the connec-

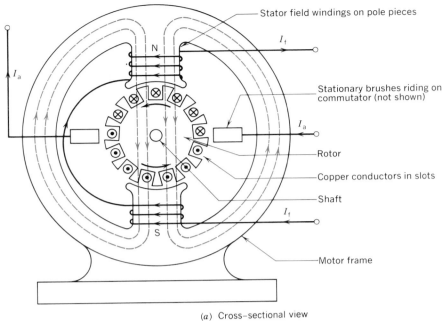

Stator field windings on pole pieces

I_f

Stationary brushes riding on commutator (not shown)

I_a

Rotor

Copper conductors in slots

Shaft

I_f

Motor frame

I_a

N

S

(a) Cross-sectional view

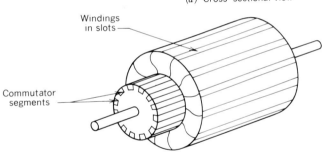

Windings in slots

Commutator segments

(b) Armature or rotor

FIGURE 11-22
Pictorial representation of a dc motor.

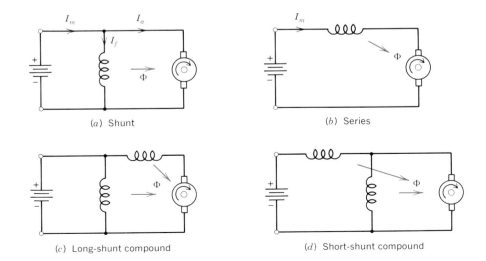

(a) Shunt

(b) Series

(c) Long-shunt compound

(d) Short-shunt compound

FIGURE 11-23
Schematic diagrams of dc motor connections.

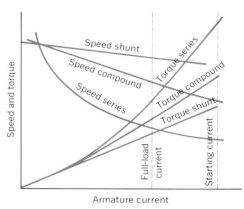

FIGURE 11-24
Comparison of speed and torque characteristics for varying loads on shunt, series, and compound motors.

tions to either the field *or* the armature must be reversed. This fact makes possible a *universal* motor that will run on either alternating current or direct current. Universal motors, commonly used for hand drills, mixers, and sewing machines, are series motors, developing high torque at low speeds. Speed control in appliance applications is accomplished by means of a solid-state device called a *triac*. It varies how much of the 120-V, 60-Hz sine wave is applied to the motor, without substantially reducing the torque (in contrast to most dc motors with simple rheostat control).

11-14 OHM'S LAW FOR A MAGNETIC CIRCUIT*

The amount of magnetic flux in a magnetic circuit can be calculated using a relationship similar to Ohm's law for a dc electric circuit. If you think of magnetic flux (Φ) corresponding to electric current (I), and magnetomotive force (F_m) corresponding to electromotive force (V), there should be some magnetic quantity corresponding to electric resistance (R). This quantity is called *magnetic reluctance* (R_m), and is a measure of a magnetic circuit's opposition to the setting up of a magnetic flux. That is,

$$\text{Result} = \frac{\text{Cause}}{\text{Opposition}}$$

---
*This section may be omitted with no loss of continuity.

For an electric circuit, $I = V/R$. For a magnetic circuit:

$$\Phi = \frac{F_m}{R_m} \quad \text{webers} \quad (11\text{-}8)$$

where: Φ is the magnetic flux in the magnetic circuit in webers (Wb)

F_m is the magnetomotive force NI in ampere-turns (At)

R_m is the reluctance of the magnetic circuit, in At/Wb

In the same way that the resistance of a conductor can be calculated using $R = \rho l/A$, the *reluctance* of a magnetic circuit can be obtained by using:

$$R_m = \frac{l}{\mu A} \quad \text{ampere-turns per weber} \quad (11\text{-}9)$$

where: R_m is the reluctance of the magnetic circuit in ampere-turns per weber

l is the length of the magnetic circuit in meters (m)

A is the cross-sectional area of the magnetic circuit in square meters (m²)

μ is the permeability of the magnetic material in webers per ampere-turn meter (Wb/At-m)

Using Eq. 11-2 ($F_m = NI$), you can now obtain an expression for Φ that shows all five factors affecting the magnitude of a magnetic field set up by a current-carrying coil:

$$\Phi = \frac{F_m}{R_m} = \frac{NI}{l/\mu A} = \frac{NI\mu A}{l} \quad (11\text{-}10)$$

where all of the terms are as previously defined.

EXAMPLE 11-4

A cast iron toroid has an inside radius of 5 cm and an outside radius of 7 cm with a circular cross section. A coil of 400 turns is wrapped around the core and carries a current of 2 A. If the permeability of the cast iron is 2.26×10^{-4}, calculate:

a. The reluctance of the cast iron toroid.
b. The flux set up in the toroid.
c. The new value of the current required to maintain the same flux level if a 5-mm air gap is cut in the toroid, as shown in Fig. 11-25.

Solution

a. The toroid has the same dimensions as in Example 11-1:

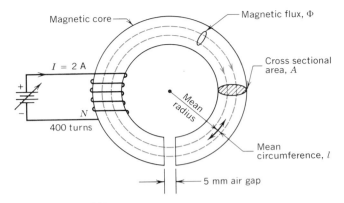

FIGURE 11-25
Magnetic circuit with airgap for Example 11-4.

Mean length of the magnetic circuit, $l = 0.377$ m
Cross-sectional area of core, $A = \pi \times 10^{-4}$ m^2

$$R_m = \frac{l}{\mu A} \qquad (11\text{-}9)$$

$$= \frac{0.377 \text{ m}}{2.26 \times 10^{-4} \text{ Wb/At-m} \times \pi \times 10^{-4} \text{ m}^2}$$

$$= \mathbf{5.31 \times 10^6 \text{ At/Wb}}$$

b. $$\Phi = \frac{F_m}{R_m} \qquad (11\text{-}8)$$

$$= \frac{400 \text{ turns} \times 2 \text{ A}}{5.31 \times 10^6 \text{ At/Wb}} = \mathbf{1.5 \times 10^{-4} \text{ Wb}}$$

c. There will be an additional mmf required to set up this flux in the air gap given by

$$F_m = \Phi R_m$$

where R_m is the reluctance of the air gap

$$R_m = \frac{l}{\mu A} \qquad (11\text{-}9)$$

$$= \frac{5 \times 10^{-3} \text{ m}}{4\pi \times 10^{-7} \text{ Wb/At-m} \times \pi \times 10^{-4} \text{ m}^2}$$

$$= 1.27 \times 10^7 \text{ At/Wb}$$

$$F_m = \Phi R_m$$

$$= 1.5 \times 10^{-4} \text{ Wb} \times 1.27 \times 10^7 \text{ At/Wb}$$

$$= 1.9 \times 10^3 \text{ At}$$

The total ampere turns required for the cast iron and the air gap is

$$F_m \text{ (total)} = F_m \text{ (cast iron)} + F_m \text{ (air gap)}$$

$$= 800 \text{ At} + 1900 \text{ At}$$

$$= 2700 \text{ At}$$

Therefore, NI total $= 2700$ At

and $$I = \frac{2700 \text{ At}}{400 \text{ t}} = \mathbf{6.75 \text{ A}}$$

Note in Example 11-4 the very large increase in current required due to the air gap. Even though the length of the air gap is very small (5 mm), its permeability is very low and its reluctance is very high. It should be observed that an air gap of this size represents the practical clearance required in such magnetic circuits as dc motors and loud-speakers. The circuit in Example 11-4c is a series magnetic circuit, whereas the dc motor is actually an example of a *series-parallel* magnetic circuit. Also, because of the nonlinear relationship between *B* and *H,* the permeability of a material is not constant. The solution of magnetic flux in such circuits involves graphical methods as shown in Section 11-15.

11-15 NONLINEAR MAGNETIC CIRCUITS*

Since the *B-H* curve for a magnetic material is not linear, the material's permeability is not constant. Thus, any problem involving a magnetic material for which the permeability is not given must include a *B-H* curve for that material. Typical curves for cast iron, cast steel, and sheet steel are shown in Fig. 11-26.

If the *B* value is known for one of these materials, the graph allows you to find the corresponding *H* value, and vice-versa. From this information, the permeability (μ) may be obtained. However, it is possible to solve the problem using *H* directly, as shown in Example 11-5.

EXAMPLE 11-5

The core of a transformer is to be built from sheet steel with the dimensions shown in Fig. 11-27. If a 600-turn coil is to provide a flux of 5.2×10^{-4} Wb in the core, calculate the current required.

Solution

Cross-sectional area of the core,

$$A = 2 \text{ cm} \times 2 \text{ cm} = 4 \text{ cm}^2 = 4 \times 10^{-4} \text{ m}^2$$

---
*This section may be omitted with no loss of continuity.

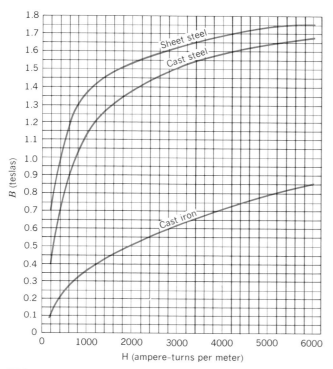

FIGURE 11-26
Typical magnetization curves.

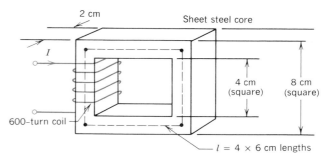

FIGURE 11-27
Magnetic core for Example 11-5.

Flux density required,

$$B = \frac{\Phi}{A} \tag{11-1}$$

$$= \frac{5.2 \times 10^{-4} \text{ Wb}}{4 \times 10^{-4} \text{ m}^2} = 1.3 \text{ T}$$

From the *B-H* curve for sheet steel in Fig. 11-26,

$$\text{for } B = 1.3 \text{ T}, \qquad H = 800 \text{ At/m}$$

Average length of magnetic core,

$$l = 4 \times 6 \text{ cm} = 24 \text{ cm} = 0.24 \text{ m}$$

$$H = \frac{NI}{l} \tag{11-3}$$

Therefore, $NI = H \times l$

$$= 800 \frac{\text{At}}{\text{m}} \times 0.24 \text{ m}$$

$$= 192 \text{ At}$$

$$I = \frac{NI}{N}$$

$$= \frac{192 \text{ At}}{600 \text{ t}}$$

$$= 0.32 \text{ A} = \textbf{320 mA}$$

Note that it would have been possible to determine the magnetic circuit's permeability after determining *H* ($\mu = B/H$).

Then, the circuit's reluctance ($R_m = l/\mu A$) could have been found, from which the magnetomotive force could have been calculated ($NI = \Phi R_m$). However, the method indicated clearly shows the meaning of magnetizing intensity (*H*). That is, multiplying the required ampere-turns per meter by the actual length in meters provides directly the necessary ampere-turns.

EXAMPLE 11-6

Assuming the core shown in Fig. 11-27 is cast iron, with a current of 1.12 A in the 600-turn coil, determine the flux in the core.

Solution

Magnetomotive force applied,

$$F_m = NI \tag{11-2}$$

$$= 600 \text{ T} \times 1.12 \text{ A} = 672 \text{ At}$$

Magnetizing intensity in the core,

$$H = \frac{NI}{l} \tag{11-3}$$

$$= \frac{672 \text{ At}}{0.24 \text{ m}} = 2800 \text{ At/m}$$

From the *B-H* curve for cast iron in Fig. 11-26,

$$\text{for } H = 2800 \text{ At/m}, \qquad B = 0.6 \text{ T}$$

But $B = \Phi/A$. Therefore,

$$\Phi = B \times A$$
$$= 0.6 \frac{\text{Wb}}{\text{m}^2} \times 4 \times 10^{-4} \text{ m}^2$$
$$= \mathbf{2.4 \times 10^{-4} \text{ Wb}}$$

As a final example of nonlinear magnetic circuits, consider a series circuit involving two different materials. The diagram in Fig. 11-28 could be a representation of a magnetic relay in the energized condition, with the upper portion representing the armature in the closed position. The electromagnet portion is made of cast steel; the rest of the magnetic circuit consists of sheet steel.

EXAMPLE 11-7

How many turns must there be in the coil in Fig. 11-28 if a flux of 0.7×10^{-4} Wb is required throughout the magnetic circuit and a current of 100 mA is necessary to hold the armature of the relay in the closed position.

Solution

Cross-sectional area for cast steel,

$$A_{\text{c.s.}} = 1 \text{ cm} \times 1 \text{ cm} = 1 \text{ cm}^2 = 1 \times 10^{-4} \text{ m}^2$$

Cross-sectional area for sheet steel,

$$A_{\text{s.s.}} = 1 \text{ cm} \times 0.5 \text{ cm} = 0.5 \text{ cm}^2 = 0.5 \times 10^{-4} \text{ m}^2$$

Flux density for cast steel,

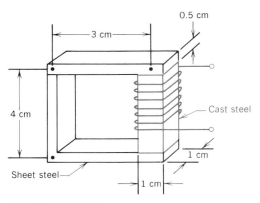

0.5 cm

3 cm

4 cm

Cast steel

1 cm

Sheet steel

1 cm

FIGURE 11-28
Series magnetic circuit for Example 11-7.

$$B_{\text{c.s.}} = \frac{\Phi}{A} \tag{11-1}$$
$$= \frac{0.7 \times 10^{-4} \text{ Wb}}{1 \times 10^{-4} \text{ m}^2} = 0.7 \text{ T}$$

Flux density for sheet steel,

$$B_{\text{s.s.}} = \frac{\Phi}{A} \tag{11-1}$$
$$= \frac{0.7 \times 10^{-4} \text{ Wb}}{0.5 \times 10^{-4} \text{ m}^2} = 1.4 \text{ T}$$

From the *B-H* curve for cast steel, Fig. 11-26,

$$H_{\text{c.s.}} = 400 \text{ At/m}$$

and for sheet steel,

$$H_{\text{s.s.}} = 1100 \text{ At/m}$$

Path length for cast steel,

$$l_{\text{c.s.}} = 3.5 \text{ cm} = 3.5 \times 10^{-2} \text{ m}$$

and for sheet steel,

$$l_{\text{s.s.}} = 10.5 \text{ cm} = 10.5 \times 10^{-2} \text{ m}$$

Therefore, since $H = NI/l$ (11-3)

$$NI_{\text{c.s.}} = H_{\text{c.s.}} \times l_{\text{c.s.}}$$
$$= 400 \frac{\text{At}}{\text{m}} \times 3.5 \times 10^{-2} \text{ m}$$
$$= 14 \text{ At}$$
$$NI_{\text{s.s.}} = H_{\text{s.s.}} \times l_{\text{s.s.}}$$
$$= 1100 \frac{\text{At}}{\text{m}} \times 10.5 \times 10^{-2} \text{ m}$$
$$= 115.5 \text{ At}$$
$$NI_{\text{total}} = NI_{\text{c.s.}} + NI_{\text{s.s.}}$$
$$= 14 + 115.5 \text{ At} = 129.5 \text{ At}$$

Therefore,

$$N = \frac{NI_{\text{total}}}{I}$$
$$= \frac{129.5 \text{ At}}{100 \times 10^{-3} \text{ A}} = \mathbf{1295 \text{ turns}}$$

Note that in a series *magnetic* circuit, the flux is the same everywhere, just as in a series *electrical* circuit, the current is the same everywhere. However, if the areas through which the magnetic flux passes are different (as in Example 11-7), there will be different flux *densities*.

Table 11-1 summarizes the equations and magnetic quantities used in this chapter. Other applications of magnetic circuits will be covered in Chapter 12.

TABLE 11-1
Summary of Magnetic Quantities

Quantity	Symbol or Equation	Unit
Flux	Φ	weber (Wb)
Flux density	$B = \dfrac{\Phi}{A}$	Wb/m^2 or tesla (T)
Magnetomotive force	$F_m = NI$	ampere-turn (At)
Magnetizing intensity	$H = \dfrac{NI}{l}$	ampere-turn per meter (At/m)
Permeability	$\mu = \dfrac{B}{H}$	weber/At-m $\left(\dfrac{\text{Wb}}{\text{At-m}}\right)$
Permeability of free space (vacuum)	$\mu_o = 4\pi \times 10^{-7}$	weber/At-m $\left(\dfrac{\text{Wb}}{\text{At-m}}\right)$
Relative permeability	$\mu_r = \dfrac{\mu}{\mu_o}$	none, dimensionless number
Reluctance	$R_m = \dfrac{l}{\mu A}$	ampere-turn per weber (At/Wb)
Ohm's law	$\Phi = \dfrac{F_m}{R_m} = \dfrac{NI\mu A}{l}$	weber (Wb)

SUMMARY

1. Common ferromagnetic elements are iron, nickel, and cobalt. Most other materials, such as wood, air, paper, copper, and lead, are nonmagnetic.
2. The attractive force that a magnet has for a magnetic material is shown graphically by lines of magnetic flux. These lines are shown leaving the magnet's north pole and entering its south pole.
3. Like magnetic poles repel each other; unlike poles attract each other.
4. A current passing through a wire sets up a magnetic field. The direction of this field around the wire can be established by using the right-hand rule: point the thumb of the right hand in the direction of conventional current and wrap the fingers of that hand around the wire. The fingers point in the direction of the magnetic field.
5. An electromagnet is a coil (helix) of wire wrapped around a core. It develops a temporary magnetic field whenever a current flows in the coil. If the fingers of the right hand are made to point in the direction of conventional current around the coil, the thumb points to the coil's north pole.
6. Solenoids, door bells, and door chimes are among the applications of electromagnets.
7. A relay is an electromagnetically operated switch. A load can be controlled from a remote location by energizing the relay's coil.
8. Permeability is an indication of a material's ability to permit the setting up of a magnetic field within it.

9. Relative permeability shows how much more permeable a magnetic material is than air or other nonmagnetic materials.

10. A hysteresis loop shows the nonlinear manner in which B (flux density) is related to H (magnetizing intensity) for a magnetic material.

11. Materials used for permanent magnets, such as Alnico and ferrites, have high values of retentivity (residual magnetism) and coercivity (demagnetizing force).

12. To minimize hysteresis losses, materials used for magnetic circuits involving an alternating magnetizing force should have a B-H curve with a small area.

13. Hysteresis losses are caused by the continuous reversal of magnetic domains in a material. A domain is a group of a large number of molecules having an overall magnetic effect like a tiny bar magnet.

14. The demagnetization (degaussing) of a magnetic material can be accomplished by continuing vibration or repeated blows, by heating above the Curie temperature, or by slow withdrawal from an alternating magnetic field.

15. A charge moving in a magnetic field experiences a force given by $F = qvB$. When the charge is moving in a wire located in a magnetic field, the force acting on the wire is given by $F = BIl$.

16. The earth's magnetic field is represented by lines of force leaving a north *magnetic* pole (located near the south *geographic* pole) and entering a south *magnetic* pole (located near the north *geographic* pole).

17. The Hall effect is the generation of an emf across the edges of a conducting strip when charges flowing through the strip are deflected by a magnetic field perpendicular to the current. This effect is used by a gaussmeter to measure flux density.

18. The force acting on a current-carrying conductor in a magnetic field provides a rotating motion in a dc motor and a back-and-forth movement in a loudspeaker.

19. Direct current motors may be of the shunt (constant speed) type, the series (high starting torque) type, or the compound type, which has both series and shunt fields.

20. B and H have a nonlinear relationship in magnetic materials. Since permeability is not constant, a graphical approach aids in solving problems that involve permeability.

SELF-EXAMINATION

Answer true or false
(Answers at back of book)

11-1. Nickel and cobalt are ferromagnetic materials with a high permeability. _____

11-2. Magnetic lines of flux indicate the paths followed by electrons from a north pole to a south pole. _____

11-3. The end of a compass that points to the earth's magnetic north pole will be attracted to a magnet's south pole. _____

11-4. Like poles attract; unlike poles repel. _____

11-5. A nail is attracted to a magnet because it becomes a temporary magnet of the opposite polarity by induction. _____

11-6. Oersted discovered that a current flowing in a wire affected a compass placed near the wire. _____

11-7. An electromagnet is a temporary magnet, most of whose magnetism can be removed simply by turning off the current through the coil. _____

11-8. There are two right-hand rules, one for determining the direction of magnetic flux around a current-carrying wire and another to determine the direction of flux set up in an electromagnet. _____

11-9. A door chime operates in essentially the same way as a doorbell. _____

11-10. A relay is a magnetically operated switch that can only be operated on dc. _____

11-11. Relays can be used to open contacts as well as close contacts on the same relay at the same time. _____

11-12. The unit of flux density is the weber. _____

11-13. A total flux of 2×10^{-3} Wb uniformly distributed over an area of 5×10^{-3} m^2 is a flux density of 0.4 T. _____

11-14. A current of 0.3 A passing through a coil of 500 turns wrapped on a magnetic core of length 10 cm provides a magnetomotive force of 150 At and a magnetizing intensity of 1500 At/m. _____

11-15. If the core in Question 11-14 has a flux density of 0.15 T, the core's permeability is 1×10^4 Wb/At m. _____

11-16. The relative permeability of the core in Question 11-15 is $250/\pi$. _____

11-17. A hysteresis curve results because of the magnetizing intensity H lagging behind the flux density B. _____

11-18. All materials, whether magnetic or nonmagnetic, have their own hysteresis loop. _____

11-19. The magnetic effect of certain materials is believed to be due to tiny clusters of molecules, called domains, that act collectively in their magnetic effect. _____

11-20. Hysteresis losses are due to a dissipation of energy in making the domains switch their alignment back and forth when under the influence of an alternating magnetizing force. _____

11-21. A material used for magnetic core storage should have as thin a B-H loop as possible whereas a material used in a transformer should ideally have a square B-H loop. _____

11-22. Raising a magnetic material above its Curie temperature is just one of the ways of degaussing a magnet. _____

11-23. A degaussing coil is used with an ac current to cycle a material through successively smaller hysteresis loops and to eventually reduce the residual magnetism to zero. _____

11-24. One gauss is the CGS unit of flux. _____

11-25. It is possible for a charge to move perpendicular to a magnetic field and not experience a force. _____

11-26. The force acting on a current-carrying conductor depends only on the field strength, current, and length of conductor. _____

11-27. The Hall effect depends upon the deflection of charge moving perpendicular to a magnetic field to develop a small emf. _____

11-28. Direct current motors use graphite brushes to conduct current through the commutator to the conductors set in slots in the armature. _____

11-29. Shunt motors have field windings in series with the armature to develop a high starting torque. _____

11-30. The flux set up by a magnetomotive force of 500 At applied to a magnetic circuit having a reluctance of 1×10^6 At/Wb is 5×10^{-4} Wb. _____

REVIEW QUESTIONS

1. Name three magnetic materials and four nonmagnetic metals.
2. What do the directions of the lines of force around a magnet indicate?
3. What is magnetic induction?
4. Explain why a nail can be picked up using *either* end of a permanent magnet.
5. What does it mean to say that iron is more permeable than copper?
6. Describe how you would determine the magnetic polarity of an unmarked magnet using a compass.
7. What effect did Oersted discover?
8. How does the right-hand rule for coils differ from the right-hand rule for a single wire?
9. Name four applications of electromagnets.
10. Why does a doorbell provide a continuous vibrating action when a door chime does not?
11. Draw two different symbols for a relay that has two sets of normally open and normally closed contacts.
12. Draw a relay circuit that uses an SPST switch to connect 6 V to a 6-V form C relay. When the switch is open, one 120-V, 100-W lamp is lit. When the switch is closed, the 100-W lamp goes off and a 120-V, 60-W lamp comes on.
13. a. If the core in Fig. 11-12 is of a very high permeability, how much magnetic flux would you expect to find in the circular region of the hole of the doughnut and also immediately outside the core?
    b. Why?
    c. Would this be different if the core were made of wood?
    d. How?
14. Distinguish between magnetomotive force and magnetizing intensity.
15. On the hysteresis loop of Fig. 11-13a, sketch a graph showing how the permeability varies with H.
16. In what units are retentivity and coercivity measured? What are typical values for a good permanent magnet?
17. a. Explain in your own words why a material having a narrow B-H loop is desirable for ac applications.
    b. Explain how the magnetic domain theory supports your answer.
18. Explain how a ferrite core with a square B-H loop can be used to store digital information.
19. a. What are the three methods of demagnetization?
    b. Describe the most practical method of demagnetization as applied to degaussing a carbon steel drill bit.
20. a. What is the operating principle behind a gaussmeter?
    b. How can this be used to distinguish between N-type and P-type semiconductors?
21. Assuming that the electron beam of an oscilloscope is sufficiently deflected by the earth's magnetic field to be visible, describe how you would determine the direction of the earth's magnetic field.
22. Describe how a high-fidelity loudspeaker converts audio current variations to sound waves.
23. Explain why a dc motor may be reversed, simply by reversing the applied emf if the motor employs permanent magnets for the field, but not if field windings are used for the stator.

24. Draw a diagram showing how a DPDT switch can be used to reverse the direction of rotation of a series motor.

PROBLEMS

(Answers to odd-numbered problems at back of book)

11-1. A flux of 1.8×10^{-3} Wb is uniformly distributed over a rectangular path of 5 cm $\times$ 6 cm. What is the flux density?

11-2. For the flux distribution in Problem 11-1, how much flux passes through a 2-cm-diameter circle?

11-3. The flux density at the pole of a horseshoe magnet is 0.8 T. How much flux passes between the two poles if each pole face is 2 cm $\times$ 1 cm?

11-4. If the flux between the two poles in Problem 11-3 is concentrated into a 1-cm $\times$ 1-cm region, what is the flux density in this region?

11-5. How much current must flow through a 350-turn coil of wire to provide a mmf of 850 At?

11-6. How many turns of wire are needed to provide a mmf of 150 At if the current to be used is 250 mA?

11-7. If the length of the magnetic circuit in Problem 11-5 is 40 cm, what is the magnetizing intensity?

11-8. How many At are necessary to provide a magnetizing intensity of 2000 At/m in a magnetic circuit of length 60 cm?

11-9. A cast steel toroid (as in Fig. 11-12) has an inside radius of 6 cm and an outside radius of 8 cm with a square cross section. A coil of 300 turns carrying 2.5 A is wrapped around the core. A flux of 5.4×10^{-4} Wb is set up inside the core. Calculate:

    a. The flux density.   d. The permeability of the core.

    b. The magnetomotive force.   e. The relative permeability of the core.

    c. The magnetizing intensity.

11-10. Repeat Problem 11-9 assuming a circular cross-sectional core.

11-11. If a 2-mm air gap is cut into the toroid of Problem 11-9, calculate the new current required to maintain the flux in the core at its initial level.

11-12. Repeat Problem 11-11 using a 3-mm air gap assuming a circular cross-sectional core.

11-13. A cast steel toroid (as in Fig. 11-12) has an inside diameter of 10 cm and an outside diameter of 15 cm with a circular cross section. A coil of 200 turns carrying a current of 750 mA is wrapped around the core. If the permeability of the cast steel is 7.5×10^{-4} Wb/At-m, calculate:

    a. The reluctance of the cast steel toroid.

    b. The flux set up in the toroid.

11-14. Repeat Problem 11-13 assuming a square cross-sectional core.

11-15. A dc motor has five turns of wire located in a single slot directly under a pole whose field strength is 0.5 T. The length of the armature is 30 cm and the current in each wire is 5 A.

    a. Calculate the total force acting on the conductors in this one slot.

    b. If current flows in the opposite direction in five turns of wire, located in a single slot on the opposite side of the armature (under the opposite field pole), calculate the torque if the diameter of the armature is 15 cm.

11-16.   The electrons in an oscilloscope are accelerated through a potential difference of 2 kV reaching a velocity of 3×10^7 m/s (one-tenth the speed of light). Assuming that the beam is placed in an east-west direction perpendicular to the earth's magnetic field, which has a horizontal component of 0.17 G and a downward vertical component of 0.55 G, calculate the magnitude and direction of the two forces acting on a single electron. Indicate the resultant direction in which the beam should be deflected if it is moving from east to west.

11-17.   A dc shunt motor is operating from a 120-V source. The resistance of the field winding is 240 Ω and the resistance of the armature is 2 Ω. When the motor is running, the total current drawn is 3.5 A. Calculate:
   a.   The field current.
   b.   The armature current.
   c.   The voltage drop due to the armature resistance.
   d.   The amount of back emf generated within the armature.
   e.   The total I^2R power lost in the field windings and armature.
   f.   The total power input to the motor.
   g.   The output power from the motor in watts and horsepower if 60 W are lost in friction and windage.
   h.   The efficiency of the motor.

11-18.   Given the magnetic core in Fig. 11-29, determine the number of turns required on the coil if a current of 0.5 A is to provide a flux in the core of 4×10^{-4} Wb.

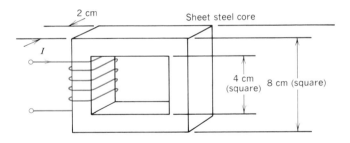

FIGURE 11-29
Magnetic core for Problems 11-18 through 11-24.

11-19.   Repeat Problem 11-18, assuming that cast steel is used instead of sheet steel.

11-20.   Find the permeability and relative permeability of the sheet steel in Problem 11-18.

11-21.   Given the magnetic core in Fig. 11-29 with a 500-turn coil carrying 1.5 A, determine the flux in the core.

11-22.   A 3-mm air gap is cut through the magnetic core in Fig. 11-29. Determine the number of turns required on the coil if a current of 0.5 A is to provide a flux of 4×10^{-4} Wb. Compare your answer with the answer for Problem 11-18 and comment.

11-23.   Repeat Problem 11-22, assuming a cast steel core. Compare with Problem 11-19.

11-24.   A 4-mm air gap is cut through the magnetic core in Fig. 11-29. If the coil has 500 turns and carries 1.5 A, determine the flux in the core. (*Hint:* Assume a flux and see if the total mmf required for the sheet steel and air gap equals the amount available.)

11-25. Refer to Example 11-7 and determine how many more turns will be needed on the coil if a 1-mm air gap exists between the cast steel and sheet steel at both ends of the cast steel core.

11-26. Given the magnetic circuit in Fig. 11-30, determine how many turns are required if a current of 50 mA is needed to operate the relay. Assume that a flux of 0.4×10^{-4} Wb is necessary and that cast iron is used in place of cast steel.

11-27. What current is required through a 1000-turn coil wrapped around the cast steel core in Fig. 11-30 to produce a flux of 0.4×10^{-4} Wb if the sheet steel is replaced by cast iron?

11-28. The series circuit in Fig. 11-30 has a coil of 1600 turns carrying a current of 100 mA. Determine the flux in the core. (*Hint:* Use a trial and error method by assuming a flux and checking to see if the required mmf is available.)

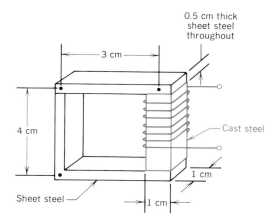

FIGURE 11-30

Series magnetic circuit for Problems 11-26 through 11-28.

CHAPTER 12

ANALOGUE DC AMMETERS AND VOLTMETERS

Most *analogue* (nondigital) electrical instruments, such as ammeters and voltmeters, use a basic *meter movement* to provide deflection of a pointer. This meter movement depends upon the force exerted on a current-carrying conductor located in a magnetic field. Such a meter movement (in an ammeter) would cause full-scale deflection with only a few microamperes of current through it. However, the range can be extended to allow measurement of much higher values (amperes) by the use of *parallel-connected* resistors, called *shunts.* If resistors (called *voltmeter multipliers*) are connected in *series* with the movement, the result is a *voltmeter* capable of any desired range.

An ideal ammeter would have *zero* internal resistance; an ideal voltmeter, *infinite* internal resistance. Practical meters, because of the way they are constructed, have finite resistance. Inevitably, they have *some* effect on the circuit they are measuring. If this effect is noticeable, it is called *loading.* In this chapter, you will learn about ammeter and voltmeter loading, and the percentage error of these meters due to what are called full-scale deflection (FSD) accuracies.

12-1 THE D'ARSONVAL OR PMMC METER MOVEMENT

The most widely used instrument for dc measurement makes use of a permanent-magnet, moving-coil (PMMC) movement. It is sometimes referred to as a D'Arsonval movement, after the man who first patented a device based on the moving-coil principle. The movement used in modern PMMC meters is also referred to as a Weston movement, since Edward Weston was primarily responsible for converting the delicate D'Arsonval meter into a rugged portable instrument.

As the name implies, the PMMC movement consists of a rectangular movable coil suspended in a strong magnetic field provided by a permanent magnet. (See Fig. 12-1.) Spiral hairsprings of phosphor-bronze lead the current to and from the coil, as well as providing a *restoring torque*. When current flows through the coil, a force acts on its two vertical sides to cause a turning motion (torque). As the pointer attached to the coil is deflected upscale, the hairsprings tighten and resist the motion. The pointer comes to rest when the applied torque (caused by current through the coil) is balanced by the tightened hairsprings. When the current is removed, the springs return *(restore)* the coil and its attached pointer to the zero position. Provision is made for mechanical adjustment *(zeroing)* of the pointer from the outside of the meter by turning a small

screw. This screw is indirectly connected to the hairspring of the meter movement.

It can be shown from the previous chapter that the torque acting on the movable coil is given by:

$$T = NBIA \sin\alpha \qquad \text{newton-meters} \qquad (12\text{-}1)$$

where: T is the torque acting on the coil in newton-meters (N·m)

N is the number of turns in the coil

B is the flux density of the permanent magnet in webers/m^2 (Wb/m^2)

I is the current in the movable coil in amperes (A)

A is the area of the coil in square meters (m^2)

α is the angle between the pointer and the magnetic field (B)

This equation (not to be used here for calculations) shows that torque is related in a linear manner to the coil current if N, B, A, and $\sin\alpha$ are all constant. B and α are kept constant by using curved pole pieces and the soft iron cylindrical core. These ensure a constant *radial* magnetic field in the air gap where the coil moves. (See Fig. 12-2.) Thus $\alpha = 90°$ and $\sin\alpha = 1$, no matter where the coil is located as it rotates. For this reason, dc ammeters and voltmeters have *linear* scales—showing equal divisions for equal current increments. Doubling the current doubles the deflection, and so on.

The PMMC meter movement is *polarity-sensitive*. If

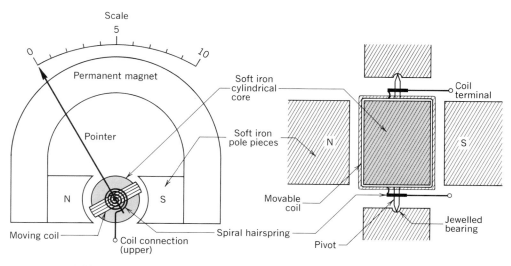

(a) PMMC movement   (b) Cross-section through magnet, core and coil

FIGURE 12-1

The D'Arsonval or permanent-magnet, moving-coil movement.

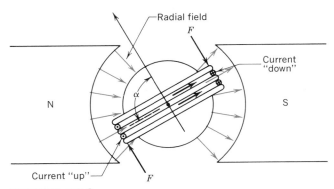

FIGURE 12-2

Detail of radial field in air gap for movable coil of a D'Arsonval or Weston meter movement.

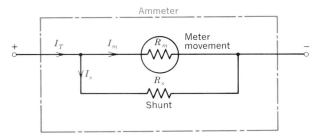

FIGURE 12-3

Basic ammeter shunt circuit.

current is allowed to flow through the meter in the opposite direction, deflection of the pointer is reversed (down scale) and the meter may be damaged.

A modification of the horseshoe magnet shown in Fig. 12-1 is the *concentric* magnet. This type has a complete outer ring of soft steel that makes possible an expanded scale covering 240° in a very compact space. Even more important, the outer ring provides *magnetic shielding* from external fields (such as a large current-carrying conductor) that could affect meter readings.

12-1.1 Current Sensitivities

The *current sensitivity* of a meter movement is the amount of current through the coil that will cause full-scale deflection (FSD). To make a very sensitive movement (for example, $I_{FSD} = 50$ μA), a very strong magnetic field B, and a large number of turns N, on the coil are desirable (See Eq. 12-1). A large number of turns increases the resistance and the weight of the coil, however, so a design compromise is usually necessary. Table 12-1 shows the typical resistance of one manufacturer's meter movements and ammeters. Note that meter resistance *decreases* as full-scale deflection current *increases*. Other movements may be slightly lower in resistance or (in some cases) as much as seven times higher.

12-2 AMMETER SHUNTS

To extend the range of a basic meter movement and make it possible to measure currents larger than the

full-scale deflection current, parallel resistors called *shunts* are used to direct most of the current being measured away from the meter movement. Design of the ammeter is based on full-scale deflection currents for the meter movement and the ammeter as a whole.

As shown in Fig. 12-3, I_T is the total current into the ammeter that causes full-scale deflection. This is the *range* of the ammeter. Current I_T divides up, with current I_m going through the meter movement (R_m) and current I_s going through the shunt resistor (R_s). By Kirchhoff's current law:

$$I_T = I_m + I_s$$

And by Ohm's law, the voltage across the meter is given by

$$V_m = I_m R_m = I_s R_s$$

Thus $$R_s = R_m \times \frac{I_m}{I_s} \quad \text{ohms} \qquad (12\text{-}1)$$

where: R_s is the shunt resistance in ohms (Ω)

TABLE 12-1

Typical Resistances of Meter Movements and Ammeters

Range, I_{FSD}	Approximate Movement Resistance, R_m, Ohms
50 μA	3000
100 μA	2000
200 μA	600
1 mA	20
10 mA	5
50 mA	1
1 A	0.05
10 A	0.005

R_m is the meter movement resistance in ohms (Ω)

I_m is the FSD current of the meter movement in amperes (A)

I_s is the shunt current, equal to $I_T - I_m$, in amperes (A)

Note that this may also be obtained by determining the parallel equivalence of 3 kΩ and 158 Ω. Note also that if only 0.5 mA is flowing through the meter, only 25 μA will pass through the movement so that it will deflect only to half-scale. This point would be marked as 0.5 mA on the *meter scale,* as shown in Fig. 12-4b.

EXAMPLE 12-1

Given a 50-μA meter movement of resistance $R_m = 3$ kΩ, calculate:

a. The value of shunt resistance to make a 1-mA range meter.

b. The voltage drop across the meter at full-scale deflection.

c. The total resistance of the meter.

Solution

Refer to Fig. 12-4a.

a. $I_s = I_T - I_m$

$= 1 \text{ mA} - 50 \ \mu\text{A}$

$= 1000 \ \mu\text{A} - 50 \ \mu\text{A} = 950 \ \mu\text{A}$

$$R_s = R_m \times \frac{I_m}{I_s} \qquad (12\text{-}1)$$

$$= 3 \text{ k}\Omega \times \frac{50 \ \mu\text{A}}{950 \ \mu\text{A}}$$

$$= 3000 \ \Omega \times \frac{1}{19} = \textbf{158} \ \boldsymbol{\Omega}$$

b. $V_m = I_m R_m \qquad (3\text{-}1b)$

At FSD, $V_m = 50 \ \mu\text{A} \times 3 \text{ k}\Omega = \textbf{150 mV}$

c. Total meter resistance,

$$R_T = \frac{V_m}{I_T} \qquad (3\text{-}1)$$

$$= \frac{150 \times 10^{-3} \text{ V}}{1 \times 10^{-3} \text{ A}} = \textbf{150} \ \boldsymbol{\Omega}$$

12-3 MULTIRANGE AMMETERS

Many ammeters used in the electronics laboratory are *multirange* meters. That is, the current causing full-scale deflection can be selected by a switch to be close to the value being measured. This allows larger deflections and more accurate measurements. One method commonly used in multirange meters consists simply of a *range selector switch* that can be rotated to select the appropriate shunt resistor. This is illustrated in Example 12-2.

EXAMPLE 12-2

Given a 50μA, 3-kΩ meter movement, it is desired to construct a multirange milliammeter with ranges of 1 mA, 10 mA, 100 mA, and 1 A. Calculate:

a. Appropriate shunt resistor values.

b. The voltage drop across the meter on each range at FSD.

c. The total resistance of the meter on each range.

Draw a circuit diagram of the switching arrangement with all values.

Solution

a. We have already calculated R_s for a 1-mA range in Example 12-1.

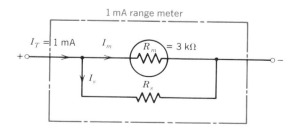

(a) Shunt resistor used to convert meter movement to 1-mA range meter

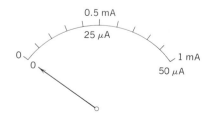

(b) 1-mA range meter scale marking corresponding to 50-μA meter movement

FIGURE 12-4

Ammeter shunt circuit and scale for Example 12-1.

$$R_s = 158 \ \Omega$$

For 10 mA, $I_s = 10 \text{ mA} - 0.05 \text{ mA} = 9.95 \text{ mA}$.

$$R_s = R_m \times \frac{I_m}{I_s} \qquad (12\text{-}1)$$

$$= 3000 \ \Omega \times \frac{0.05 \text{ mA}}{9.95 \text{ mA}} = \mathbf{15.1 \ \Omega}$$

Similarly, for 100 mA, $R_s = \mathbf{1.5 \ \Omega}$
           for 1 A, $\quad R_s = \mathbf{0.15 \ \Omega}$

b. Since, at FSD on each range, there will be 50 μA through the 3-kΩ movement, the voltage drop on each range is given by

$$V_m = I_m R_m \qquad (3\text{-}1b)$$
$$= 50 \ \mu\text{A} \times 3 \text{ k}\Omega = \mathbf{150 \ mV}$$

c. Total meter resistance, $R_T = \dfrac{V_m}{I_T} \qquad (3\text{-}1)$

for 1-mA range, $\quad R_T = \mathbf{150 \ \Omega}$

for 10-mA range, $\quad R_T = \dfrac{150 \times 10^{-3} \text{ V}}{10 \times 10^{-3} \text{ A}} = \mathbf{15 \ \Omega}$

for 100-mA range, $R_T = \dfrac{150 \times 10^{-3} \text{ V}}{100 \times 10^{-3} \text{ A}} = \mathbf{1.5 \ \Omega}$

for 1-A range, $\quad R_T = \dfrac{150 \times 10^{-3} \text{ V}}{1 \text{ A}} = \mathbf{0.15 \ \Omega}$

See Fig. 12-5.

A multirange ammeter of the type designed in Example 12-2 is shown in Fig. 12-6. Note that it is necessary to use a *make-before-break* rotary switch for current range selection. This type of switch ensures that a shunt resistor is always part of the circuit, even while switching between ranges. If an ordinary switch were used, contact with one shunt resistor would be broken before contact was made with the next resistor. This would allow *all* the current to briefly flow through the meter movement, causing damage.*

12-4 AYRTON OR UNIVERSAL SHUNT

In many multirange meters, an *Ayrton shunt* is used instead of a make-before-break switch. This type of shunt

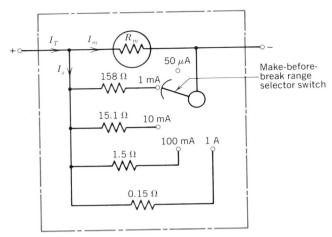

FIGURE 12-5
Multirange ammeter shunt circuit for Example 12-2.

(also known as a *universal* shunt) has the advantage of using resistors that are closer to standard values. This was not the case in Example 12-2, where accurate values of 158 Ω and 15.1 Ω were required.

Example 12-3 illustrates the method of selecting resistors for an Ayrton shunt. All that is necessary is the application of series-parallel circuit theory. No special equations are involved.

FIGURE 12-6
Typical analogue multirange dc milliammeter. (Courtesy of Hickok Teaching Systems, Inc., Woburn, Mass.)

---

*Many milliammeters protect the meter movement by using a pair of *breakdown diodes* connected directly across it. Then, if the meter is set on too low a range for the circuit being checked, much of the current is diverted through one of the diodes. When the range setting is correct, however, the diodes have no effect because there is not enough voltage across the movement to make them conduct.

EXAMPLE 12-3

Given a 50-μA, 3-kΩ meter movement, determine the values of R_1, R_2, and R_3 in Figure 12-7 to design an Ayrton shunt with ranges of 100 μA, 1 mA, and 10 mA.

Solution

100 μA range; Fig. 12-7a:
The total shunt resistance equals $R_1 + R_2 + R_3$ and must carry 50 μA. In order for the current to divide equally, the resistance of the two paths must be equal, so

$$R_1 + R_2 + R_3 = R_m = 3 \text{ k}\Omega$$

1-mA range; Fig. 12-7b:
R_1 is now in series with R_m, with $R_2 + R_3$ shunting these two.
The voltage across the meter movement and R_1 must equal the voltage across the shunt ($R_2 + R_3$).

Therefore, $I_m (R_1 + R_m) = I_s (R_2 + R_3)$

$$0.05 \text{ mA } (R_1 + R_m) = 0.95 \text{ mA } (R_2 + R_3)$$

But $\quad R_m = 3 \text{ k}\Omega$ and $R_2 + R_3 = 3 \text{ k}\Omega - R_1$

Thus $0.05 \text{ mA } (R_1 + 3 \text{ k}\Omega) = 0.95 \text{ mA } (3 \text{ k}\Omega - R_1)$

$$R_1 + 3 \text{ k}\Omega = \frac{0.95 \text{ mA}}{0.05 \text{ mA}} (3 \text{ k}\Omega - R_1)$$

Therefore, $\quad R_1 + 3 \text{ k}\Omega = 19 (3 \text{ k}\Omega - R_1)$

$$20 \quad R_1 = 54 \text{ k}\Omega$$
$$R_1 = \textbf{2.7 k}\Omega$$

10-mA range; Fig. 12-7c:
$R_1 + R_2$ are now in series with R_m, with R_3 shunting all three.

$$0.05 \text{ mA } (R_1 + R_2 + R_m) = 9.95 \text{ mA} \times R_3$$

But $R_1 = 2.7 \text{ k}\Omega$, $R_m = 3 \text{ k}\Omega$, and $R_2 + R_3 = 300 \ \Omega = 0.3 \text{ k}\Omega$, so $R_3 = 0.3 \text{ k}\Omega - R_2$.

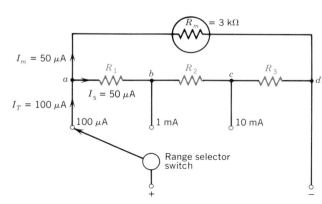

(a) 100-μA range

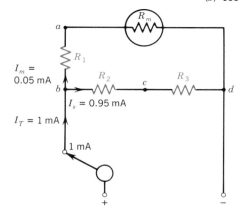

(b) 1-mA range

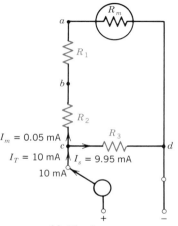

(c) 10-mA range

FIGURE 12-7
Ayrton shunt multirange milliammeter for Example 12-3.

Thus $0.05 \text{ mA} (2.7 \text{ k}\Omega + R_2 + 3 \text{ k}\Omega)$

$\qquad\qquad = 9.95 \text{ mA}(0.3 \text{ k}\Omega - R_2)$

solving, $R_2 = 0.27 \text{ k}\Omega = \mathbf{270\ \Omega}$

and $R_3 = 300\ \Omega - 270\ \Omega = \mathbf{30\ \Omega}$

It can be shown that, in general, the total resistance of a meter using an Ayrton shunt will be slightly higher (on any given range) than that of a multirange ammeter using the multishunt arrangement. Another disadvantage of the Ayrton shunt is that if *one* of the shunt resistors changes value or opens up, *all* ranges are affected. This is not the case for the multishunt ammeter.

12-5 ACCURACY AND ERRORS

If a multirange ammeter is being used, which range should be selected for maximum accuracy? This can be answered by realizing that some errors are inherent in the meter and some are due to the operator. Some of the operator errors are as follows:

1. Imprecise interpolation between scale markings when the pointer falls between two main divisions.
2. Line-of-sight errors caused by the operator's eye not being perpendicular to the pointer and scale (parallax). Some meters have a mirrored scale, making it possible to line up the pointer with its reflected image to avoid this problem.
3. Failure to check the mechanical zero before using the instrument.

Other errors that have a similar effect on accuracy are due to the manufacture or use of the instrument:

1. A scale error, due to markings not being in exactly the correct place. Mass-produced scales may not fit the actual characteristics of the meter movement. (The instrument needs individual calibration.)
2. An error caused by frictional effects (due to constant use or misuse of the instrument) that prevents the pivot from swinging freely to its proper location. (This error can be minimized by making it a practice to gently tap the meter case before taking a reading.)

All the above errors are *constant amounts,* independent of the deflection of the pointer across the scale. There are other errors that are *proportional* to the deflection of the pointer. They include:

1. Inexact resistances for shunts or multipliers in ammeters or voltmeters.

2. Temperature effects on the resistance values in (1) and on the movement coil resistance, as well as the flexibility of springs.

The errors due to the last two effects are relatively small compared to the errors that are *constant.* Thus, the total error (due to all effects combined) is a relatively constant *amount* at different points on the scale, rather than a *percentage of the reading* that is being taken. This error is usually expressed by the manufacturer as a percentage of the range or full-scale reading. For example, if a meter is quoted as having an accuracy of $\pm 3\%$, it means that the error at *any* point of the scale will not exceed 3% of the *full-scale* reading. Such a percentage accuracy may be very poor for a reading taken on the lower part of the scale. This is illustrated in Example 12-4.

EXAMPLE 12-4

A 10-mA range ammeter is guaranteed to have an accuracy of $\pm 3\%$, FSD. For readings on the scale of 1 mA, 5 mA, and 10 mA determine:
a. The possible error in mA.
b. The range of values for the true current.
c. The percentage error in each reading.

Solution

a. For all three readings
    the error $= \pm 3\%$ of 10 mA

$$= \frac{\pm 3}{100} \times 10 \text{ mA} = \pm\mathbf{0.3\ mA}$$

b. The actual reading may be:
    For 1-mA indication:

$$1 \pm 0.3 \text{ mA} = \mathbf{0.7\ to\ 1.3\ mA}$$

For 5-mA indication:

$$5 \pm 0.3 \text{ mA} = \mathbf{4.7\ to\ 5.3\ mA}$$

For 10-mA indication:

$$10 \pm 0.3 \text{ mA} = \mathbf{9.7\ to\ 10.3\ mA}$$

c. Percentage error $= \dfrac{\text{error}}{\text{reading}} \times 100\%$
    For 1-mA reading, % error

$$= \frac{0.3 \text{ mA}}{1 \text{ mA}} \times 100\% = \mathbf{30\%}$$

For 5-mA reading, % error

$$= \frac{0.3 \text{ mA}}{5 \text{ mA}} \times 100\% = \mathbf{6\%}$$

For 10-mA reading, % error

$$= \frac{0.3 \text{ mA}}{10 \text{ mA}} \times 100\% = \textbf{3\%}$$

Thus percentage error due to FSD errors can be minimized by selecting a range that gives as close to full-scale deflection as possible.

12-6 AMMETER LOADING

When an instrument is used for measurement, it is expected that application of the instrument to the circuit will *not* change conditions in that circuit.

Since an ammeter is connected in series in a circuit, the perfect or *ideal* instrument would have *zero* resistance. This would prevent its presence from affecting the circuit under any conditions. But in practice, an ammeter *does* have some internal resistance. If the ammeter's resistance is in any way comparable to the circuit resistance, there is a problem of *loading,* as shown by Example 12-5.

EXAMPLE 12-5

A technician, attempting to determine current in a circuit using Ohm's law, measures the voltage across a 1.2-kΩ resistor and finds it to be exactly 1 V. Not being sure that the resistor is actually 1.2 kΩ, he decides to measure the current by inserting a VOM on the 1-mA range in series with the resistor. If the VOM's resistance is 1 kΩ (typical for many VOMs), calculate:
a. The actual current in the circuit assuming that the resistor is exactly 1.2 kΩ.
b. The indication of the meter on the 1-mA range.
c. The indication of the meter if it is switched to the 10-mA range where its resistance is 100 Ω.
d. The effect of FSD errors on the 10-mA range reading, assuming an accuracy of ±3%.

Solution

a. $I = \dfrac{V}{R}$ (3-1a)

$\quad = \dfrac{1 \text{ V}}{1.2 \text{ k}\Omega} = \textbf{0.83 mA}$

b. $I = \dfrac{V}{R_T}$

$\quad = \dfrac{1 \text{ V}}{1.2 \text{ k}\Omega + 1 \text{ k}\Omega} = \textbf{0.45 mA}$

c. $I = \dfrac{V}{R_T}$

$\quad = \dfrac{1 \text{ V}}{1.2 \text{ k}\Omega + 0.1 \text{ k}\Omega} = \textbf{0.77 mA}$

d. It is evident that the higher range (10 mA) with its much lower resistance (100 Ω) has reduced loading error compared with the 1-kΩ, 1-mA range meter. However, the 0.77 mA will cause such a small deflection on the 10-mA range that it will be difficult to read accurately.

$$\text{FSD error} = \frac{\pm 3}{100} \times 10 \text{ mA} = \pm 0.3 \text{ mA}$$

Thus the 10-mA ammeter could indicate anything between 0.47 and 1.07 mA. On the 1-mA range,

$$\text{FSD error} = \frac{\pm 3}{100} \times 1 \text{ mA} = \pm 0.03 \text{ mA}$$

Thus the 1-mA ammeter could indicate anything between 0.42 and 0.48 mA. The range of values on the 10-mA scale at least includes the values of 0.77 mA and 0.83 mA whereas the 1-mA scale could never read higher than 0.48 mA. Put another way, on the 10-mA range the readings could be from 43% low to 29% high compared with the true value of 0.83 mA. On the 1-mA range, the readings could be from 42% low to 49% low compared with the true value of 0.83 mA.

Conclusion: Where loading is a problem, switching to the next higher range will reduce the effect of the loading. However, the increase in FSD errors may remove much of the increased accuracy.

12-6.1 Clamp-On Ammeter

A meter that causes no loading because it is not connected in series with the circuit is the *clamp-around probe* or clamp-on ammeter shown in Fig. 12-8.

There is no need to break the circuit, since the jaws open to clamp around the insulated wire in which the current is flowing. The current in the wire (ac or dc) sets up a magnetic field with a strength that is proportional to the current. The instrument accurately concentrates this field in a magnetic core surrounding the jaws. A Hall-effect generator (Section 11-11) is mounted in the air gap in the

FIGURE 12-8
Clamp-around ammeter probe or "current gun" that measures alternating and direct current (Courtesy F. W. Bell Inc.)

core. The Hall emf generated is directly proportional to the current in the conductor.

A dc-coupled linear operational amplifier (powered by four AA cells) drives a three-and-a-half digit LCD readout located in the instrument's handle. The "current gun," as this meter is sometimes called, has a range of ± 200 A from dc to 1 kHz with a resolution of 0.1 A.

Currents of less than 1 A can be read with reasonable accuracy by looping several turns of the conductor through the jaws. This effectively "multiplies" the current.

12-7 VOLTMETER MULTIPLIERS

As you have seen, the 50-μA, 3-kΩ meter movement requires a voltage of only 150 mV across it to cause full-scale deflection. In other words, the movement may only be used to measure voltages up to a maximum of 150 mV. To extend the range to higher voltages, it is necessary

to use a resistor in *series* with the meter movement. Such a resistor is called a voltmeter *multiplier*.

Whenever the voltmeter is connected across its rated voltage, the multiplier *limits* the current to the full-scale deflection value of the meter movement. The method used to determine the required series resistor (multiplier resistor) for a given voltage range involves simple series circuit theory. (See Fig. 12-9.)

As was the case with the ammeter, the design is based on full-scale deflection values.

Total voltmeter resistance,

$$R_T = \frac{\text{voltmeter range}}{\text{FSD current of movement}}$$
$$= \frac{V_T}{I_{\text{FSD}}}$$

But $R_T = R_s + R_m$
and $R_s = R_T - R_m$

Therefore, $R_s = \dfrac{V_T}{I_{\text{FSD}}} - R_m$ ohms   (12-2)

where: R_s is the resistance of the voltmeter multiplier in ohms (Ω)

R_m is the resistance of the meter movement in ohms (Ω)

V_T is the range of the voltmeter in volts (V)

I_{FSD} is the meter movement's full-scale deflection current in amperes (A)

If we let

$$\frac{1}{I_{\text{FSD}}} = S \qquad (12\text{-}3)$$

Equation 12-2 can be rewritten as

$$R_s = S \times \text{Range} - R_m \qquad \text{ohms} \qquad (12\text{-}4)$$

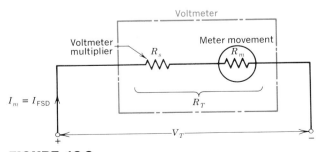

FIGURE 12-9
Basic voltmeter multiplier circuit.

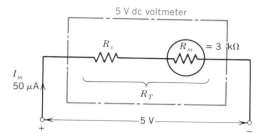

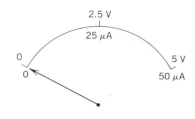

(a) Connection of voltmeter multiplier to convert meter movement to 5-V range voltmeter

(b) Voltmeter markings corresponding to meter movement's current

FIGURE 12-10

Voltmeter multiplier circuit and scale for Example 12-6.

Since S is the reciprocal of current, it has units of ohms per volt. This is called the voltmeter's sensitivity, as further explained in Section 12-10.

Note that if the voltmeter is connected across only 2.5-V dc, only 25 μA will flow through the meter. The pointer will deflect halfway toward full scale. As shown in Fig. 12-10b, this point will be marked 2.5 V on the voltmeter scale.

EXAMPLE 12-6

It is required to convert a 50μA, 3-kΩ meter movement to a voltmeter having a range of 5 V. Calculate:
a. The total resistance of the meter.
b. The necessary resistance of the voltmeter multiplier using Eqs. 12-2 and 12-4.

Solution

Refer to Fig. 12-10a
a. Total resistance of voltmeter,

$$R_T = \frac{V}{I_m} \tag{3-1}$$

$$= \frac{5\ V}{50\ \mu A} = \textbf{100 k}\Omega$$

b. $R_s = \frac{V_T}{I_{FSD}} - R_m \tag{12-2}$

$$= R_T - R_m$$

$$= 100\ k\Omega - 3\ k\Omega = \textbf{97 k}\Omega$$

or $\quad S = \frac{1}{I_{FSD}} \tag{12-3}$

$$= \frac{1}{50 \times 10^{-6}A} = 20\ k\Omega/V$$

$R_s = S \times$ Range $- R_m \tag{12-4}$

$$= 20\ k\Omega/V \times 5\ V - 3\ k\Omega$$

$$= 100\ k\Omega - 3\ k\Omega = \textbf{97 k}\Omega$$

12-8 MULTIRANGE VOLTMETERS

One method of making a multirange voltmeter is to use a *range selector switch* to series-connect the appropriate multiplier. (See Example 12-7 and Fig. 12-11.)

EXAMPLE 12-7

Convert a 50-μA, 3-kΩ meter movement into a multirange voltmeter having ranges of 5 V, 15 V, and 50 V.

Solution

For 5-V range, $R_T = 100$ kΩ, $R_s = \textbf{97 k}\Omega$

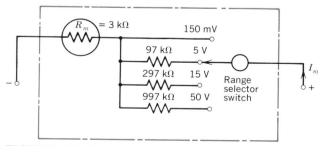

FIGURE 12-11

Multirange voltmeter circuit using individual multipliers for Example 12-7.

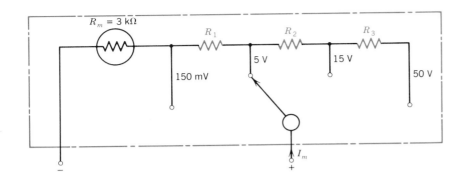

FIGURE 12-12

Series-connected multiplier circuit for Example 12-8.

For 15-V range, $R_T = \dfrac{V}{I_m}$ (3-1)

$$= \frac{15 \text{ V}}{50 \text{ μA}} = 300 \text{ k}\Omega$$

$R_s = R_T - R_m$ (12-2)

$= 300 \text{ k}\Omega - 3 \text{ k}\Omega = \textbf{297 k}\Omega$

For 50-V range,

$R_s = S \times \text{Range} - R_m$ (12-4)

$= 20 \text{ k}\Omega/\text{V} \times 50 \text{ V} - 3 \text{ k}\Omega$

$= 1 \text{ M}\Omega - 3 \text{ k}\Omega = \textbf{997 k}\Omega$

12-9. SERIES-CONNECTED MULTIRANGE VOLTMETER

Instead of using individual multipliers, each with its own nonstandard resistance value, the multipliers can be connected in series, as in Fig. 12-12.

EXAMPLE 12-8

Determine the values of R_1, R_2, and R_3 in Fig. 12-12 to convert the 50-μA, 3-kΩ movement into a multirange voltmeter having ranges of 150 mV, 5 V, 15 V, and 50 V.

Solution

$$S = \frac{1}{I_{FSD}}$$ (12-3)

$$= \frac{1}{50 \times 10^{-6} \text{ A}} = 20 \text{ k}\Omega/\text{V}$$

On the 5-V range,

$R_1 = S \times \text{range} - R_m$ (12-4)

$= 20 \text{ k}\Omega/\text{V} \times 5 \text{ V} - 3 \text{ k}\Omega$

$= 100 \text{ k}\Omega - 3 \text{ k}\Omega = \textbf{97 k}\Omega$

On the 15-V range,

$R_2 = S \times \text{range} - (R_m + R_1)$ (12-4)

$= 20 \text{ k}\Omega/\text{V} \times 15 \text{ V} - (3 \text{ k}\Omega + 97 \text{ k}\Omega)$

$= 300 \text{ k}\Omega - 100 \text{ k}\Omega = \textbf{200 k}\Omega$

On the 50-V range,

$R_3 = S \times \text{range} - (R_m + R_1 + R_2)$ (12-4)

$= 20 \text{ k}\Omega/\text{V} \times 50 \text{ V} - (3 \text{ k}\Omega + 97 \text{ k}\Omega + 200 \text{ k}\Omega)$

$= 1000 \text{ k}\Omega - 300 \text{ k}\Omega = \textbf{700 k}\Omega$

Note that, apart from the first multiplier (R_1), the resistors are different from those used in the individual multiplier multirange voltmeter of Example 12-7. These resistors can be obtained in standard resistance values. A disadvantage of this circuit is the effect that a change in value of one resistor will have on all the other ranges.

The series-connected type of multiplier circuit in Example 12-8 is used as the range selector switch circuit of the VTVM* in Fig. 12-13.

12-10 VOLTMETER SENSITIVITY

The sensitivity of a voltmeter may be defined as follows:

$$\text{Sensitivity} = \frac{\text{total resistance of voltmeter}}{\text{range of voltmeter}}$$

$$S = \frac{R_T}{V_T} \quad \text{ohms per volt}$$ (12-5)

*VTVM is an acronym for vacuum tube voltmeter. Although most modern voltmeters no longer contain vacuum tubes, the abbreviation is a commonly used term.

FIGURE 12-13
Typical electronic voltmeter or VTVM. (Courtesy of Heath Company.)

For example, determine the sensitivity of the multirange voltmeter from Example 12-7.

5-V range: sensitivity $= \dfrac{R_T}{V_T} = \dfrac{100 \text{ k}\Omega}{5 \text{ V}} = 20 \text{ k}\Omega/\text{V}$

15-V range: sensitivity $= \dfrac{R_T}{V_T} = \dfrac{300 \text{ k}\Omega}{15 \text{ V}} = 20 \text{ k}\Omega/\text{V}$

50-V range: sensitivity $= \dfrac{R_T}{V_T} = \dfrac{1 \text{ M}\Omega}{50 \text{ V}} = 20 \text{ k}\Omega/\text{V}$

Evidently, the sensitivity (S) is a constant for a given meter movement. The voltmeter sensitivity, in fact, may also be expressed as the reciprocal of the movement's current sensitivity, as shown in Section 12-7.

$$S = \frac{1}{I_{\text{FSD}}} \qquad (12\text{-}3)$$

In this case,

$$S = \frac{1}{50 \text{ }\mu\text{A}} = \frac{1}{50 \times 10^{-6} \text{ V}/\Omega} = 20 \text{ k}\Omega/\text{V}$$

The value of knowing a voltmeter's sensitivity is the ability to determine the total resistance of the voltmeter for any given *range*. For example, if the meter has a sensitivity of 20 kΩ/V (typical of many VOMs), the total resistance of the voltmeter on its 150-V range is

$$R_T = S \times V_T$$
$$= 20 \text{ k}\Omega/\text{V} \times 150 \text{ V} = 3000 \text{ k}\Omega = 3 \text{ M}\Omega$$

Note that, if this voltmeter is measuring a voltage of 75 V on this range, the total resistance of the meter is *still* 3 MΩ (not 1.5 MΩ). If used on a 1.5-V range, this voltmeter will present a resistance of only 30 kΩ (20 kΩ/V $\times$ 1.5 V).

12-11   VOLTMETER LOADING

The effect of a voltmeter when it is connected in a circuit is called voltmeter *loading*. Since a voltmeter is connected *across* a device (in *parallel* with it), the instrument repre-

FIGURE 12-14
Typical VOM or multimeter. (Courtesy of Triplett Corporation.)

sents a shunting resistance. If this resistance is low in comparison with the circuit resistance, the voltmeter loading may be extreme, depending on circuit conditions. Ideally, a voltmeter would have an *infinite* resistance. This condition can best be approached by using a voltmeter with as *high* a sensitivity or as high an ohms-per-volt rating as possible. (Portable voltmeters found in VOMs may have sensitivities ranging from 1 kΩ/V up to 100 kΩ/V. The VOM or *multimeter* in Fig. 12-14 has an ohms-per-volt rating of 20 kΩ/V on dc and 5 kΩ/V on ac.)

Thus, the resistance of a given meter can vary widely, according to the range being used. For many electronic and digital voltmeters (EVMs and DVMs), the meter resistance is a constant. For example, the electronic voltmeter shown in Fig. 12-13 has an input (total) resistance of 11 MΩ on all ranges from 1.5 to 1500 V dc.

EXAMPLE 12-9

Two 1-MΩ resistors are series-connected across a 10-V source. Calculate and compare with initial conditions:

a. The reading of a 20 kΩ/V VOM on the 5-V range connected across one of the resistors.
b. The new reading if the range is increased to 50 V.
c. The reading of an EVM on the 5-V range connected in place of the VOM.
d. Each reading taking into account a ±3% FSD accuracy.

See Fig. 12-15.

Solution

The initial circuit conditions consist of a current of 5 μA causing a voltage drop of 5 V across each resistor. See Fig. 12-15a.

a. The resistance of the 5-V range meter = 20 kΩ/V × 5 V = 100 kΩ. Refer to Fig. 12-15b.
Total circuit resistance

$$= 1\ M\Omega + 1\ M\Omega \| 100\ k\Omega \approx 1.1\ M\Omega$$

New circuit current, $I = \dfrac{V}{R_T}$

$$= \frac{10\ V}{1.1\ M\Omega} \approx 9.1\ \mu A$$

Voltage drop across $R_1 = V_1 = I R_1$
$$= 9.1\ \mu A \times 1\ M\Omega = 9.1\ V$$
Voltage across R_2 and VOM reading

$$= 10\ V - 9.1\ V = \mathbf{0.9\ V}$$

Note that the low resistance of the meter has almost doubled the circuit current causing a much larger voltage drop across R_1. This has drastically changed the voltage across R_2 from 5 to 0.9 V.

b. The resistance of the 50-V range meter = 20 kΩ/V × 50 V = 1 MΩ. See Fig. 12-15c.
Total circuit resistance
$$= 1\ M\Omega + 1\ M\Omega \| 1\ M\Omega = 1.5\ M\Omega$$
New circuit current, $I = \dfrac{V}{R_T}$
$$= \frac{10\ V}{1.5\ M\Omega} = 6.7\ \mu A$$
Voltage drop across $R_1 = V_1 = IR_1$
$$= 6.7\ \mu A \times 1\ M\Omega = 6.7\ V$$
Voltage across R_2 and VOM reading
$$= 10\ V - 6.7\ V = \mathbf{3.3\ V}$$

Although perhaps difficult to read on a 50-V range this reading is much closer to the desired 5 V. It results from not as large a voltage drop across R_1 because the current has not increased as much.

c. Refer to Fig. 12-15d. Repeating the calculation for

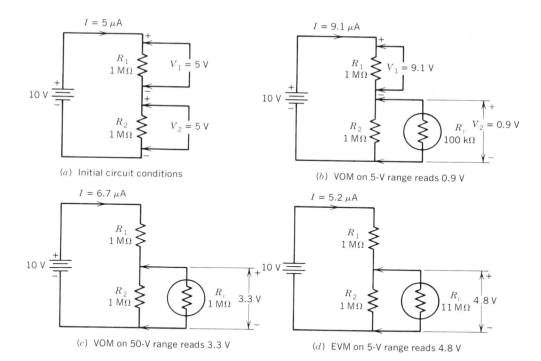

an 11-MΩ resistance (the EVM) in parallel with R_2, we obtain:

$$\text{Total circuit resistance} = 1.92 \text{ M}\Omega$$
$$\text{New circuit current} = 5.2 \text{ } \mu\text{A}$$
$$\text{Voltage drop across } R_1 = 5.2 \text{ V}$$
$$\text{VTVM reading across } R_2 = \mathbf{4.8 \text{ V}}$$

d. VOM on 5-V range: 0.9 V $\pm$3% of 5 V
   $$= 0.9 \text{ V} \pm 0.15 \text{ V} = \mathbf{0.75 \text{ to } 1.05 \text{ V}}$$
   VOM on 50-V range: 3.3 V $\pm$3% of 50 V
   $$= 3.3 \text{ V} \pm 1.5 \text{ V} = \mathbf{1.8 \text{ to } 4.8 \text{ V}}$$
   VTVM on 5-V range: 4.8 V $\pm$3% of 5 V
   $$= 4.8 \text{ V} \pm 0.15 \text{ V} = \mathbf{4.65 \text{ to } 4.95 \text{ V}}$$

EXAMPLE 12-10*

Repeat Examples 12-9A–12-9C using a Thévenin equivalent circuit to represent the 10-V source and the two 1-MΩ resistors.

Solution

For the initial conditions, the voltage across R_2 can be obtained from the Thévenin equivalent circuit looking

back between terminals A and B, as shown in Fig. 12-16b.

$$V_{\text{Th}} = 10 \text{ V} \times \frac{R_2}{R_2 + R_1}$$
$$= 10 \text{ V} \times \frac{1 \text{ M}\Omega}{1 + 1 \text{ M}\Omega}$$
$$= 10 \text{ V} \times \tfrac{1}{2} = \mathbf{5 \text{ V}}$$
$$R_{\text{Th}} = R_1 \| R_2$$
$$= \frac{R_1 \times R_2}{R_1 + R_2}$$
$$= \frac{1 \text{ M}\Omega \times 1 \text{ M}\Omega}{1 + 1 \text{ M}\Omega} = \mathbf{0.5 \text{ M}\Omega}$$

The 0.5-MΩ resistor can be thought of as the internal resistance of the 5-V source. With no voltmeter (load) connected, the voltage V_2 across R_2 is the open-circuit voltage $V_{\text{Th}} = 5$ V.

a. With the 100-kΩ 5-V range VOM connected across R_2, the equivalent circuit appears as in Fig. 12-16c.

$$V_2 = 5 \text{ V} \times \frac{R_v}{R_v + R_{\text{Th}}}$$
$$= 5 \text{ V} \times \frac{100 \text{ k}\Omega}{100 + 500 \text{ k}\Omega}$$
$$= 5 \text{ V} \times \tfrac{1}{6} = \mathbf{0.83 \text{ V}}$$

---

*This example may be omitted with no loss of continuity.

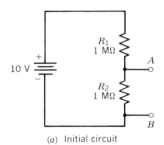

(a) Initial circuit

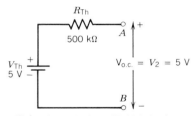

(b) Thévenin equivalent of initial circuit

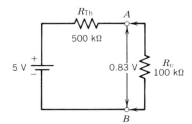

(c) VOM on 5-V range reads 0.83 V

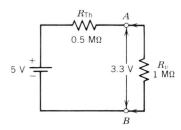

(d) VOM on 50-V range reads 3.3 V

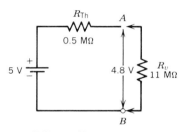

(e) EVM on 5-V range reads 4.8 V

FIGURE 12-16

Loading effects of voltmeters on various ranges for Example 12-10 using a Thévenin equivalent circuit approach.

Note how a low-resistance voltmeter load of 100 kΩ causes a high voltage drop across the high "internal resistance" of 500 kΩ.

b. With the 1-MΩ 50-V range VOM connected across R_2, the equivalent circuit appears as in Fig. 12-16d.

$$V_2 = 5 \text{ V} \times \frac{R_v}{R_v + R_{Th}}$$

$$= 5 \text{ V} \times \frac{1 \text{ M}\Omega}{1 + 0.5 \text{ M}\Omega}$$

$$= 5 \text{ V} \times \frac{1}{1.5} = \textbf{3.3 V}$$

c. With the 11-MΩ 5-V range EVM connected across R_2, the equivalent circuit is as shown in Fig. 12-16e.

$$V_2 = 5 \text{ V} \times \frac{R_v}{R_v + R_{Th}}$$

$$= 5 \text{ V} \times \frac{11 \text{ M}\Omega}{11 + 0.5 \text{ M}\Omega}$$

$$= 5 \text{ V} \times \frac{11}{11.5} = \textbf{4.8 V}$$

The "internal resistance" of 0.5 MΩ is now relatively small compared with the load (11-MΩ voltmeter) so very little internal voltage drop takes place. The voltmeter reads closer to the open-circuit voltage of 5 V. Note how obvious it is (with the Thévenin equivalent circuit) that an ideal voltmeter of infinite resistance would indicate the true voltage of 5 V.

It is clear that, despite increased FSD errors, the 50-V range on the VOM causes less loading and a reading closer to the initial condition.

Whenever loading is a problem, switching to a *higher* range will generally improve the meter's accuracy. However, if voltage measurements are to be made in high-resistance circuits, use of an electronic voltmeter (with its constant high resistance) is essential to obtain accurate readings.

It should be noted that loading occurred in the circuit of Fig. 12-15 because of the high resistance of R_1. If the value of R_1 is changed from 1 MΩ to 10 kΩ, the *additional* current that flows when the VOM is connected (though it *is* appreciable) will cause very little *additional* voltage drop across R_1. Thus the voltage across R_2 will be relatively unchanged.

Voltmeter loading is not caused merely by low-resistance meters connected in parallel with high-resistance devices. It also is dependent upon other high resistances in *series* with the component whose voltage is being measured.

SUMMARY

1. A D'Arsonval or PMMC meter movement consists of a movable coil pivoted in the strong magnetic field of a permanent magnet.
2. The PMMC movement has a deflection directly proportional to the dc current in the coil. This produces a linear scale.
3. *Current sensitivity* is the current required in the movable coil to cause full-scale deflection (FSD) of the pointer.
4. Ammeters use a basic meter movement paralleled by a shunting resistor to extend the movement's current range.
5. In multirange ammeters the movement's current range is extended by individual shunts selected by a make-before-break switch or by a series arrangement of resistors called an Ayrton shunt.
6. The *higher* the current range of an ammeter, the *lower* the value of the shunt resistance and the *lower* the total resistance of the meter itself.
7. An *ideal* ammeter would have zero resistance, regardless of current range.
8. Ammeter loading occurs when the meter's resistance is comparable to the circuit's resistance, thus upsetting the circuit's original state.
9. Ammeter loading can be decreased by moving to a higher-current range.
10. The accuracy of a meter is quoted as a percentage of its *full-scale* reading.
11. Where ammeter loading is not a problem, for maximum percentage accuracy, readings should be taken on a range that most nearly gives full-scale deflection.
12. In a voltmeter, a multiplier resistor is connected in series with the basic meter movement.
13. Multirange voltmeters may use individual or series-connected multipliers.
14. A voltmeter's sensitivity is rated in ohms/volt and is the reciprocal of the meter movement's current sensitivity.
15. A high-sensitivity voltmeter causes less loading in high-resistance circuits because the meter's large resistance draws negligible additional current.
16. An ideal voltmeter has *infinite* resistance. Many voltmeters (VOMs) exhibit higher resistance on higher ranges, but electronic voltmeters (EVMs) have a constant high resistance through all ranges.
17. Where voltmeter loading is not a problem, readings should be made on a range that most nearly gives full-scale deflection.

SELF-EXAMINATION

Answer true or false or a, b, c, or d
(Answers at back of book)

12-1.   A PMMC movement responds only to dc current. _____

12-2. The hairsprings serve solely to provide a restoring torque. _____

12-3. A radial magnetic field is produced in a PMMC movement to ensure a linear scale. _____

12-4. Current sensitivity is the current required through a movement to produce half of full-scale deflection. _____

12-5. A higher-current sensitivity is usually accompanied by an increase in resistance of the moving coil. _____

12-6. An ammeter shunt has the same voltage drop across it as the meter movement.

_____

12-7. Higher-range ammeters have higher shunt resistances. _____

12-8. A 100-μA, 2-kΩ meter movement has a voltage drop across it at half of full-scale deflection of:

   a.   100 μV
   b.   200 mV
   c.   100 mV
   d.   200 μV

12-9. A multirange ammeter has the same voltage drop across the meter on each range at FSD. _____

12-10. A multirange ammeter has lower resistance on the higher ranges. _____

12-11. An Ayrton shunt consists of a number of series-connected resistors connected in parallel with the meter movement. _____

12-12. A make-before-break selector switch is required in both the Ayrton shunt and the multishunt multirange ammeter. _____

12-13. A disadvantage of the universal shunt is the change in calibration of all ranges should the resistance of just one resistor change. _____

12-14. The value of shunt resistance required to double the range of a 100-μA, 2-kΩ meter movement is:

   a.   1 kΩ
   b.   2 kΩ
   c.   3 kΩ
   d.   4 kΩ

12-15. Ammeter loading occurs when the insertion of the ammeter presents a significant resistance that reduces the current flowing to a value much less than the initial value. _____

12-16. Loading can be minimized by using a higher range. This applies to both voltmeters and ammeters (VOMs). _____

12-17. A voltmeter multiplier is a resistor connected in series with a meter movement to extend the voltage range. _____

12-18. A 50-μA, 2-kΩ movement requires a multiplier to produce a 10-V range voltmeter of:

   a.   198 kΩ
   b.   19.8 kΩ
   c.   18 kΩ
   d.   1.98 MΩ

12-19. The lowest range that a multirange voltmeter can have that uses a 50-μA, 2-kΩ movement is:

   a.   100 μV
   b.   50 mV
   c.   100 mV
   d.   50 μV

12-20. A multirange voltmeter may use individual multipliers or a number of series-connected multipliers. _____

12-21. A voltmeter that has a total resistance of 1 MΩ on a 25-V range has a sensitivity of:
   a. 4 kΩ/V
   b. 25 kΩ/V
   c. 40 kΩ/V
   d. 2.5 kΩ/V

12-22. The higher the sensitivity of a voltmeter, the higher the total resistance of the voltmeter and loading. _____

12-23. Voltmeter loading is due to relatively low-resistance voltmeters connected in high-resistance circuits. _____

12-24. A 50-kΩ/V voltmeter used on a 10-V range indicates 5 V. The total resistance of the voltmeter is
   a. 250 kΩ
   b. 500 kΩ
   c. 10 kΩ
   d. 50 kΩ

12-25. A 10-V range voltmeter has an accuracy of $\pm 2\%$ of FSD. The percentage error of a reading of 4 V is
   a. 2%
   b. 3%
   c. 4%
   d. 5%

REVIEW QUESTIONS

1. What two functions do the phosphor-bronze hairsprings serve in a D'Arsonval meter movement?

2. What is the principle of operation of a PMMC movement?

3. What construction features ensure that the PMMC scale is linear?

4. a. Why would you expect a more-sensitive meter movement to have higher resistance?
   b. What would be necessary to achieve higher sensitivity without an increase in resistance?

5. a. What is the primary purpose of a shunt?
   b. Why are shunt values selected on the basis of FSD currents in the meter movement?

6. a. Why is it necessary to use a make-before-break selector switch in a multishunt ammeter?
   b. Why is it not necessary in an Ayrton shunt?

7. a. Why does the total resistance decrease for higher-range ammeters?
   b. Why would an ideal ammeter have zero resistance whereas an ideal voltmeter would have infinite resistance?

8. What factors contribute to manufacturers quoting the accuracy of a meter as a percentage of FSD rather than percentage of reading?

9. Why is the percentage error higher when making a very low reading on a scale than when the reading is near full scale?

10. a. What do you understand by "parallax"?
    b. How is this problem reduced in some meters?
11. a. When is ammeter loading likely to occur?
    b. If you were using a multirange ammeter, how could you determine if loading is taking place?
12. What is the design principle on which a voltmeter multiplier is selected?
13. Is it necessary to use a make-before-break switch in multirange voltmeters?
14. a. What is an advantage of using series-connected multipliers in multirange voltmeters?
    b. What is a disadvantage?
    c. Which do you think is most expensive: series-connected or individual multipliers?
15. a. What purpose does knowing a voltmeter's sensitivity serve?
    b. Why is it meaningless to refer to an EVM's voltmeter sensitivity?
16. Justify why a voltmeter's sensitivity should also be given by the reciprocal of the meter movement's current sensitivity.
17. Why is voltmeter loading more likely to occur with low-sensitivity voltmeters and high-resistance circuits?
18. a. How can you determine if a VOM is loading a circuit in voltmeter measurements?
    b. How can you reduce loading?
    c. What penalty do you pay?
19. Will voltmeter loading *always* occur across a high resistance with a low-sensitivity voltmeter? Explain.

PROBLEMS

12-1. Given a 100-μA, 2-kΩ meter movement, calculate:
   a. The shunt resistance needed to make a 5-mA range meter.
   b. The voltage drop across the meter at full-scale deflection.
   c. The total resistance of the meter.
12.2. Repeat Problem 12-1 for a 10-mA range.
12-3. Given a 100-μA, 2-kΩ meter movement, it is required to construct a multirange milliammeter with ranges of 1 mA, 5 mA, 25 mA, 100 mA, and 500 mA. Calculate:
   a. Appropriate shunt values.
   b. Voltage drop across the meter on each range at FSD.
   c. Total resistance of the meter on each range.
   Draw a circuit diagram showing switch arrangements and all values.
12-4. Repeat Problem 12-3, using a 200-μA, 600-Ω movement.
12-5. A 20-mA, 5-Ω meter movement must be converted into a 10-A range ammeter.
   a. What shunt resistance is needed?
   b. If the only shunt available is 0.02 Ω, what additional resistance is needed in *series* with the meter movement itself connected as shown in Fig. 12-17?

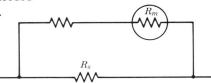

FIGURE 12-17
Circuit for Problem 12-5.

c. How much power does the 0.02-Ω shunt dissipate when the meter indicates 5 A?

12-6. A certain manufacturer's meter movement has a current sensitivity of 100 μA and a resistance of 5 kΩ. In an effort to increase sensitivity, the designer intends to double the number of turns on the coil (using wire of half the diameter), use a permanent magnet with a 50% stronger field, and increase the area of the movable coil by one-third. Calculate:
a. The sensitivity of the new meter.
b. The resistance of the new meter, assuming that the new coil has the same perimeter as the old one but increased area.

12-7. Given a 100-μA, 2-kΩ meter movement, design an Ayrton shunt with ranges of 200 μA, 1 mA, and 5 mA.

12-8. Repeat Problem 12-7 for ranges of 0.5 mA, 1 mA, and 10 mA.

12-9. a. Given a 50-μA, 3-kΩ meter movement, design an Ayrton shunt with ranges of 1 and 10 mA.
b. Determine the total resistance of the meter on the 10-mA range.
c. Using the above meter movement, determine the single-shunt resistance to make a 10-mA range meter.
d. Determine the total resistance of the 10-mA range meter in part (c) and compare with the answer in part (b).

12-10. Repeat Problem 12-7 for ranges of 1 mA, 5 mA, 10 mA, and 25 mA.

12-11. A 25-mA range ammeter has an accuracy of ±4% of FSD. For readings on the scale of 5 mA, 15 mA, and 25 mA determine:
a. The possible error in mA.
b. The range of values of the true current.
c. The percentage error in each reading.

12-12. Repeat Problem 12-11 for a 50-mA range meter.

12-13. A 2.2-kΩ resistor is connected across a 1.5-V dry cell. Assuming the resistance to be exactly 2.2 kΩ, calculate:
a. The actual current and the power dissipated in the resistor. (Use $P = I^2R$.)
b. The reading of a 1-mA range, 1-kΩ meter connected in series with the resistor, and the power in the resistor. (Use $P = I^2R$.)
c. The range of readings possible in part (b) assuming a ±3% of FSD accuracy.
d. The reading of a 10-mA range, 100-Ω meter in series with the resistor.
e. The range of readings possible in part (d) assuming a ±3% of FSD accuracy.

12-14. Repeat Problem 12-13 using a 1.6-kΩ resistor.

12-15. Given a 100-μA, 2-kΩ meter movement, it is required to produce a multirange voltmeter having ranges of 1 V, 10 V, and 50 V using individual multipliers. Calculate:
a. The total resistance of the meter on each range.
b. The voltmeter multiplier resistors.
c. The voltmeter sensitivity.
d. The resistance of the meter when indicating 20 V on the 50-V range.
Draw a circuit diagram of the switching arrangement with values.

12-16. Repeat Problem 12-15 using a 50-μA, 5-kΩ movement.

12-17. Repeat Problem 12-15 using series-connected multipliers.

12-18. a. What must be the current sensitivity of a meter movement if it is to be used to produce a 50-kΩ/V voltmeter?

    b.   Assume that the meter movement in part (a) has a resistance of 10 kΩ and is shunted by another 10-kΩ resistor. What series resistor is required to produce a 2-V range voltmeter?

    c.   What is the voltmeter sensitivity of the voltmeter in part (b)?

    d.   If the 10-kΩ shunt resistor is removed, what is the new range of the voltmeter?

12-19. A 10-V source of emf with an internal resistance of 100 kΩ is checked by a 10-kΩ/V voltmeter on its 15-V range. Calculate:

    a.   The current drawn by the voltmeter.

    b.   The reading of the voltmeter.

    c.   The reading of the voltmeter if its range is increased to 50 V.

    d.   The reading of an 11-MΩ EVM on the 15-V range in place of the original meter.

12-20. Repeat Problem 12-19 using a 50-kΩ/V voltmeter.

12-21. A 500-kΩ resistor is series-connected with a 1-MΩ resistor across a 12-V source of negligible internal resistance. Calculate:

    a.   The voltage across the 0.5-MΩ resistor with no meters connected in the circuit.

    b.   The reading of an 11-MΩ EVM on the 5-V range connected across the 0.5-MΩ resistor.

    c.   The reading of a 20-kΩ/V VOM on the 5-V range connected in parallel with the EVM.

    d.   The new reading of the EVM for the condition in part (c).

    e.   The reading of both instruments when the VOM is switched to the 50-V range.

12-22. Repeat Problem 12-21 using a 2.2-MΩ resistor instead of a 1-MΩ, and a 25-V source instead of 12 V.

12-23. Repeat Problem 12-21 using a 10-kΩ resistor instead of a 1-MΩ, and a 5-V source instead of 12 V.

12-24. In Problem 12-21, assume that both meters have a $\pm 3\%$ of FSD accuracy. Calculate the possible reading and the percentage error (due to FSD errors—not loading) of the meters in parts (b), (c), and (e).

12-25. A neon lamp that fires (conducts) at 60 V is series-connected with a 100-kΩ resistor (to limit current after firing) across an 80-V source. A milliammeter on the 1-mA range is inserted in the circuit and indicates 200 μA. When a VOM on the 25-V range is connected across the resistor, the current reading increases to 216 μA.

    a.   What is the sensitivity of the VOM?

    b.   If the VOM has a FSD accuracy of $\pm 5\%$, what range of readings could it indicate across the resistor?

    c.   What would be the milliammeter reading if the VOM were connected across the neon lamp on the 100-V range?

    d.   What percentage error would the VOM indication in part (c) represent?

CHAPTER 13

OTHER DC MEASURING INSTRUMENTS

In Chapter 12, you learned that voltmeter loading is caused by current drawn from the circuit by the voltmeter. A *potentiometer* is an instrument that is used to make voltage measurements with no load at all. This is done by comparing an accurately known and variable voltage with the unknown voltage. When a galvanometer connected between the two voltages shows a "balance," you know that they are equal. A potentiometer is very useful in calibrating voltmeters and ammeters.

The measurement of resistance is important, but when done by simply using a voltmeter and ammeter, the result is not very accurate. An ohmmeter is convenient for general use in measuring resistances, but where high accuracy is required, a Wheatstone bridge is employed. This is a very versatile circuit with four resistance arms, one of which is the unknown. Accuracy is determined only by the resistance values.

Also in this chapter, digital multimeter specifications will be considered and compared to analogue-type volt-ohm meters (VOMs) and electronic voltmeters (EVMs).

13-1 PRINCIPLE OF A NO-LOAD VOLTMETER

All of the voltmeters you learned about in Chapter 12 drew some current from the circuit for their operation. Because of this current draw, connection of such a voltmeter *disturbs* the circuit. Even when the current drawn by the voltmeter is very small, it will cause a meter reading that is less than the *true* voltage in any multibranch circuit.

When a source voltage or open-circuit voltage is being measured, the voltmeter reading is also reduced if the internal resistance of the source is significant. (See Fig. 13-1*a*.) This is especially true when measuring the small emf (only a few millivolts) of a thermocouple, where the voltage drop across connecting leads can be comparable to the *generated* emf.

Figure 13-1*b* shows a measurement method in which *no* current is drawn from the source. If voltages are equal on both sides of a *galvanometer* (a sensitive current-indicating device with a zero-center pointer), there is no deflection of the pointer. This is called a *null* condition, and no current is drawn from the source. When a null condition is indicated, the voltage of the "unknown" emf (V) is *precisely* the same as the accurately known voltage of the source used for comparison. Since no current flows in the circuit, the measurement is *independent* of the resistances of the galvanometer (r_G), the unknown source (r_v), and the known, or comparison, source (r_k). Calibration of the galvanometer movement itself is unimportant, so long as it is sensitive enough to clearly show a null condition (zero deflection) when the voltages are balanced. The greater the sensitivity of the galvanometer, of course, the more accurate the measurement. When the galvanometer is combined with a device that can provide a *variable* accurate voltage, the resulting instrument is called a *potentiometer*.

13-2 THE POTENTIOMETER

The commercial potentiometer is essentially a *high-resolution, variable voltage divider*. As shown in Fig. 13-2, it consists of a *standard cell*, one or more voltage-dividing networks and a galvanometer (which may or may not be built into the instrument).

The standard cell is a source of accurately known voltage used only for *calibration* of the potentiometer. Measurement of an unknown emf is done by comparing that emf to *known* voltage drops caused by current flowing through the voltage-dividing networks (R_1 and R_2).

There are two types of standard cells available for use in potentiometers. The *saturated* or *normal* cell, uses $CdSO_4$ crystals and is manufactured to a high degree of uniformity. It develops a known emf of 1.0183 V. For many years, the National Bureau of Standards (NBS) used this type of cell for calibrating purposes, defining the *international volt* as 1/1.01830 of the emf produced by this cell at 20°C.

A portable version of standard cell, shown in Fig. 13-3, uses a cadmium sulfate solution and is called an *unsaturated* cadmium cell. New cells of this type range between 1.0190 and 1.0194 *absolute* volts. The actual value is determined by comparison with a normal *(saturated)* cell. Each cell of the unsaturated type is accompanied by a certificate from the NBS showing the actual voltage (1.01830 international volts equals 1.01864 absolute volts). The voltage of a standard cell should be checked yearly with a potentiometer (*not* an ordinary voltmeter). If a cell is inverted, it should be allowed to settle for 24 hours before use.

The following steps are used to operate a potentiometer like the one shown in Fig. 13-2*a*:

1. Set the dials of the instrument to the indicated emf of the standard cell (R_1 to 1.0 V, R_2 to 19 mV).
2. Move the DPDT (double-pole, double-throw) switch to the *calibrate* position. This puts the standard cell in the circuit on one side of the galvanometer.
3. Close K_1 and adjust the rheostat until the galvanometer shows *near-zero* deflection.

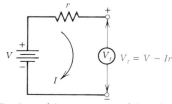

(*a*) Voltmeter resistance causes an internal voltage drop in the source

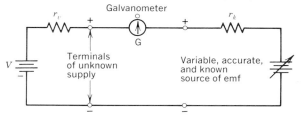

(*b*) Potentiometer principle. Galvanometer indicates "null" (zero) when the two voltages are equal

FIGURE 13-1

Principle of a no-load voltmeter.

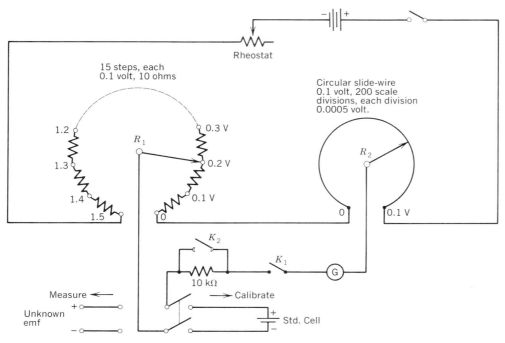

15 steps, each
0.1 volt, 10 ohms

Circular slide-wire
0.1 volt, 200 scale
divisions, each division
0.0005 volt.

(a) Schematic diagram of a student-type potentiometer

(b) Photograph of a student-type potentiometer

FIGURE 13-2
Circuit diagram and
photograph of a
potentiometer. (The
latter courtesy of the
Leeds and Northrup
Co.)

4. Close K_2 as well, then vary the rheostat to obtain a null reading.

The potentiometer is now calibrated or *standardized*. This means that R_1 is calibrated in steps of 0.1 V up to a maximum of 1.5 V, and R_2 is calibrated up to a maximum of 0.1 V. Now, the current delivered by the potentiometer's *working* battery will cause voltage drops across R_1 and R_2 that will correspond *exactly* to the marked values on the dials. The *rheostat* should not be varied unless a

(a) External view of mounted cell

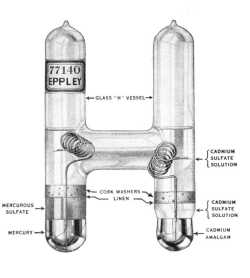

(b) View showing construction of a cell

FIGURE 13-3
Standard cell made by the Eppley Laboratory, Inc. (Courtesy of Eppley Laboratory, Inc.)

later calibration check indicates that adjustment is needed. Note that no current is drawn from the standard cell at balance. This prevents it from being consumed or damaged when checking calibration.

Once the potentiometer has been calibrated, measurement of unknown voltages can be carried out:

5. Move the DPDT switch to *measure,* and connect the source of unknown emf to the potentiometer.
6. Adjust the *dials* (not the *rheostat*) until the galvanometer shows a null condition with both K_1 and K_2 closed.
7. Read the unknown voltage from the dials ($R_1 + R_2$).

A maximum voltage of 1.6 V can be measured directly by the potentiometer. To read higher voltages, a separate voltage-divider ("volt-box") is connected between the potentiometer and source of the unknown voltage. The volt-box reduces the voltage by some known factor to a maximum of 1.5 V, so that it can be measured by the potentiometer. Multiplying the potentiometer reading by the volt-box factor will give the actual unknown voltage. (The presence of the volt-box in the circuit, however, puts a load on the voltage to be measured and may degrade the reading.)

13-3 CALIBRATION OF VOLTMETERS AND AMMETERS

Since the potentiometer is capable of making very accurate voltage measurements, it is commonly used to calibrate digital voltmeters.

A *calibration curve* for a given voltmeter may be obtained by using the circuit shown in Fig. 13-4a. Note that the potentiometer and the voltmeter are connected in parallel. The *correction* that must be made (added to or subtracted from the voltmeter reading) can be obtained at various points on the voltmeter scale and plotted on a graph like the one shown in Fig. 13-4c. On a 1.5-V range voltmeter, the correction would normally be small (usually in millivolts).

Note that the correction curve consists of measured points joined by straight lines. A good meter should be correct at three points along the scale, as shown in Fig. 13-4c. Many meters, however, will get progressively worse (less accurate) as the reading increases toward full-scale deflection. The *FSD accuracy* of the meter is established by expressing the largest correction (no matter *where* it appears on the curve) as a *percentage* of the full-scale range.

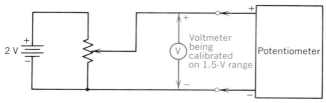

(a) Circuit for voltmeter calibration

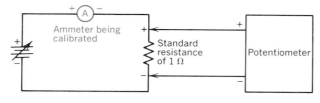

(b) Circuit for ammeter calibration

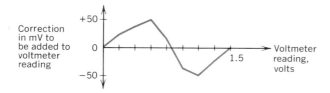

(c) Typical correction curve for a voltmeter

FIGURE 13-4

Calibration circuits for the voltmeter and ammeter and a typical calibration curve for a voltmeter.

$$= \frac{0.054 \text{ V}}{1.5 \text{ V}} \times 100\%$$

$$= \frac{5.4}{1.5} \% = \textbf{3.6\%}$$

The potentiometer can also be used to calibrate an ammeter. As shown in Fig. 13-4b, the ammeter is simply connected in series with a highly accurate *precision* resistor and the voltage across the resistance is measured by the potentiometer.

Then, by Ohm's law, the true current can be obtained (the potentiometer reading is divided by the resistance value). If the precision resistor is exactly 1.0000 Ω, the voltage readings of the potentiometer are numerically the same as the true ammeter current. A correction curve can be drawn in the same manner used for the voltmeter.

EXAMPLE 13-2

A potentiometer is being used to calibrate a 250-mA range milliammeter. The largest error occurs when the milliammeter indicates 235 mA. At this value, the two dials of the potentiometer indicate 0.2 V and 30 mV across a precision 1.0-Ω resistor connected in series with the milliammeter. Calculate:
a. The true milliammeter current.
b. The milliammeter correction at this point.
c. The percent FSD accuracy of the meter.

Solution:

a. True current $I = \dfrac{V}{R}$

$$= \frac{0.2 \text{ V} + 30 \text{ mV}}{1.0 \text{ } \Omega}$$

$$= 0.230 \text{ A} = \textbf{230 mA}$$

b. Correction = Potentiometer − Ammeter

$$= 230 \text{ mA} - 235 \text{ mA}$$

$$= \textbf{−5 mA}$$

(The negative sign indicates that the meter reading is high and must be reduced by adding a negative quantity.)

c. FSD accuracy $= \dfrac{\text{maximum correction}}{\text{meter range}} \times 100\%$

$$= \frac{5 \text{ mA}}{250 \text{ mA}} \times 100\%$$

$$= \frac{500}{250} \% = \textbf{2\%}$$

EXAMPLE 13-1

A potentiometer is being used to calibrate a dc voltmeter set to its 1.5-V range. The largest correction occurs when the voltmeter indicates 0.960 V. At this value, the potentiometer dials indicate 1.0 V and 14 mV. Calculate:
a. The true reading that the voltmeter should indicate.
b. The voltmeter correction in millivolts.
c. The percent FSD accuracy of the meter.

Solution

a. True reading = 1.0 V + 14 mV
$$= 1.0 \text{ V} + 0.014 \text{ V}$$
$$= \textbf{1.014 V}$$

b. Correction = Potentiometer − Voltmeter
$$= 1.014 \text{ V} - 0.960 \text{ V}$$
$$= 0.054 \text{ V} = \textbf{54 mV}$$

c. FSD accuracy $= \dfrac{\text{maximum correction}}{\text{meter range}} \times 100\%$

13-4 AMMETER–VOLTMETER METHOD OF MEASURING RESISTANCE

An apparently simple method of measuring a circuit's resistance is to divide the voltage across the circuit by the current through the circuit, using Ohm's law:

$$R_x = \frac{V_x}{I_x}$$

This simply requires a voltmeter reading and an ammeter reading. But how accurate is such a measurement? There are many possible sources of error.

First, ignore the resistances of the meters themselves and consider only their FSD accuracies.

Assume that your meters both give you *full-scale* readings—the 5-V range voltmeter indicates 5 V, and the 10-μA range ammeter indicates a circuit current of 10 μA. Now, assume that *both* meters are accurate to ±3% FSD. Then

$$R_x = \frac{V_x}{I_x} = \frac{5\ V}{10\ \mu A} = 500\ k\Omega$$

The circuit's resistance is thus 500 kΩ with an accuracy of ±6% That is, the voltmeter *could* be reading 3% high and the ammeter 3% low (or vice versa), for a maximum error of 6%. The *actual* resistance could be as low as 470 kΩ or as high as 530 kΩ. The other possibility is that the errors could cancel each other, but without calibration curves for the two instruments, there is no way to be sure. You must consider the worst case—the *sum* of the two errors.

If the meter readings were not at or near *full-scale,* the percentage error could be much worse. If both readings were near half-scale, for example, the maximum error could be as high as 12% (and readings lower on the scales could produce even *larger* percentage errors).

In the next section, the effect on accuracy of the voltmeter's location in the circuit will be considered.

13-4.1 Measurement of High Resistance

If the circuit resistance is considered high (compared to the voltmeter's resistance), the voltmeter should be connected across the series combination of R_x and the ammeter to avoid loading effects. (See Fig. 13-5a.) With this arrangement, the ammeter indicates the true resistor current, but the voltmeter reading will be a bit high. The following

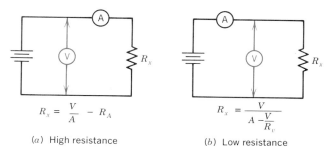

$$R_x = \frac{V}{A} - R_A \qquad\qquad R_x = \frac{V}{A - \frac{V}{R_v}}$$

(a) High resistance (b) Low resistance

FIGURE 13-5
Showing the voltmeter location to measure an unknown resistance in high- and low-resistance circuits.

equation will correct for the resistance effect of the ammeter.

$$R_x = \frac{V}{A} - R_A \qquad \text{ohms} \qquad (13\text{-}1)$$

where: R_x is the resistance of the unknown in ohms (Ω)
   V is the voltmeter reading in volts (V)
   A is the ammeter reading in amperes (A)
   R_A is the resistance of the ammeter in ohms (Ω)

EXAMPLE 13-3

A resistor is being measured using the circuit shown in Fig. 13-5a. A 10-V range, 3%-FSD accuracy voltmeter indicates 5 V; a 50-μA, 20-kΩ, 2%-FSD accuracy microammeter indicates 10 μA. Calculate:
a. The resistance.
b. The maximum possible error.

Solution

a. $R_x = \dfrac{V}{A} - R_A$ (13-1)

$$= \frac{5\ V}{10 \times 10^{-6}\ A} - 20\ k\Omega$$

$$= 500\ k\Omega - 20\ k\Omega = \mathbf{480\ k\Omega}$$

b. Percentage error in voltmeter reading

$$= \frac{\dfrac{3}{100} \times 10\ V}{5\ V} \times 100\% = 6\%$$

Percentage error in ammeter reading

$$= \frac{\dfrac{2}{100} \times 50\ \mu A}{10\ \mu A} \times 100\% = 10\%$$

Maximum possible error in measurement of R_x = 6% + 10% = **16%**

13-4.2 Measurement of Low Resistance

For measurements of circuits where resistance is considered *low* (compared to the voltmeter's resistance), the arrangement in Fig. 13-5b is used. In this case, the voltmeter reading will be correct, but the ammeter will read high by an amount equal to the current drawn by the voltmeter (V/R_V). That is,

$$R_x = \frac{V}{A - V/R_v} \qquad \text{ohms} \qquad (13\text{-}2)$$

where: R_x is the resistance of the unknown in ohms (Ω)
  V is the voltmeter reading in volts (V)
  A is the ammeter reading in amperes (A)
  R_v is the resistance of the voltmeter in ohms (Ω)

EXAMPLE 13-4

The resistance of a circuit is being measured as in Fig. 13-5b. The 10-mA range milliammeter indicates 10 mA and the 10-kΩ/V voltmeter indicates 15 V on the 15-V range. If both meters have a 2% FSD accuracy, determine:
a. The resistance.
b. The maximum possible error.

Solution

a. $R_v = \Omega/V \times \text{range} = 10\ \text{k}\Omega/V \times 15\ V = 150\ \text{k}\Omega$

$$R_x = \frac{V}{A - V/R_v} \qquad\qquad (13\text{-}2)$$

$$= \frac{15\ V}{10 \times 10^{-3}\ A - \dfrac{15\ V}{150 \times 10^3\ \Omega}}$$

$$= \frac{15\ V}{9.9 \times 10^{-3}\ A} = \mathbf{1.52\ k\Omega}$$

b. Maximum possible error = 2% + 2% = **4%**

Note that, if an ideal ammeter ($R_A = 0$) and an ideal voltmeter ($R_v = \infty$) are used, both Eq. 13-1 and Eq. 13-2 reduce to $R_x = V/A$, and the location of the voltmeter in the circuit is immaterial.

13-5 SERIES TYPE OF OHMMETER

An *ohmmeter* provides a direct reading of resistance in ohms, eliminating the need for two readings and calculating a correction. The ohmmeter is not an instrument of high accuracy, but it is adequate for many commercial applications. (In electronics work, many resistors are only accurate to 5% or 10%, anyway.)

The simplest ohmmeter is the series type shown in Fig. 13-6. **Unlike the ammeter or voltmeter, it requires a battery (1.5 V or larger) as a source of emf, and is used in circuits with no other power applied. To avoid damage to the meter, power to the circuit must always be turned OFF before the ohmmeter is connected.** The ohmmeter uses a basic meter movement and includes provision for *zeroing* the pointer when the two leads are shorted (connected together).

For the basic circuit shown in Fig. 13-6a, you can see that (for any value of R_x),

$$V = IR_T \qquad\qquad (3\text{-}1b)$$

and $\qquad\qquad V = I(R_x + R_m + R_z) \qquad (13\text{-}3)$

where: V is the emf of the cell in volts (V)
  I is the current through the resistor and meter movement in amperes (A)

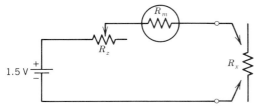

(a) Circuit of ohmmeter

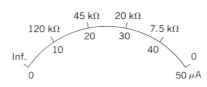

(b) Calibration of ohmmeter scale

FIGURE 13-6
Series type of ohmmeter.

R_x is the external resistance to be measured in ohms (Ω)

R_m is the meter movement resistance in ohms (Ω)

R_z is the resistance of the zeroing control in ohms (Ω)

When the leads are shorted together, R_x is zero, and

$$R_m + R_z = \frac{V}{I_{FSD}} \tag{13-4}$$

Equation 13-4 determines the value of R_z to give full-scale deflection of the meter. This point on the scale is marked zero ohms. The markings on the remainder of the scale are determined as in Example 13-5. In general, the procedure for marking the ohmmeter scale involves assuming a convenient value of current and determining the external resistance (R_x) that permits this current to flow.

EXAMPLE 13-5

Given a 0–50-μA, 3-kΩ meter movement, determine the resistance markings at 0, 10, 20, 30, 40, and 50 μA when used in a series-type ohmmeter (Fig. 13-6) with a 1.5-V cell. Also determine a suitable value for the zero-ohms control.

Solution

When the leads are shorted together, $R_x = 0$, and

$$R_m + R_z = \frac{V}{I_{FSD}} \tag{13-4}$$

$$= \frac{1.5 \text{ V}}{50 \times 10^{-6} \text{ A}} = 30 \text{ k}\Omega$$

therefore, $R_z = 30 \text{ k}\Omega - R_m$

$$= 30 \text{ k}\Omega - 3 \text{ k}\Omega = \textbf{27 k}\boldsymbol{\Omega}$$

This is the value to which the zero-ohms control (maximum value **30 kΩ**) must be adjusted for a *full-scale deflection* of 50 μA, corresponding to 0 Ω. (This mechanical process is often called *calibrating* the ohmmeter.)

When $I = 40 \mu$A, corresponding to $\frac{4}{5}$ of full-scale deflection,

$$R_x + R_m + R_z = \frac{V}{I} \tag{3-1}$$

$$= \frac{1.5 \text{ V}}{40 \ \mu\text{A}} = 37.5 \text{ k}\Omega$$

Thus $R_x = 37.5 \text{ k}\Omega - (R_m + R_z)$

$$= 37.5 \text{ k}\Omega - (3 \text{ k}\Omega + 27 \text{ k}\Omega)$$

$$= 37.5 \text{ k}\Omega - 30 \text{ k}\Omega = \textbf{7.5 k}\boldsymbol{\Omega}$$

This is the ohmmeter calibration at $\frac{4}{5}$ full-scale deflection. For $I = 30 \mu$A, corresponding to $\frac{3}{5}$ of full-scale deflection,

$$R_x + R_m + R_z = \frac{V}{I} \tag{3-1}$$

$$= \frac{1.5 \text{ V}}{30 \ \mu\text{A}} = 50 \text{ k}\Omega$$

Thus $R_x = 50 \text{ k}\Omega - (R_m + R_z)$

$$= 50 \text{ k}\Omega - 30 \text{ k}\Omega = \textbf{20 k}\boldsymbol{\Omega}$$

This is the ohmmeter calibration at $\frac{3}{5}$ full-scale deflection. Similarly, for $I = 20 \mu$A, $R_x = \textbf{45 k}\boldsymbol{\Omega}$,

calibration at $\frac{2}{5} I_{FSD}$

$I = 10 \mu$A, $R_x = \textbf{120 k}\boldsymbol{\Omega}$,

calibration at $\frac{1}{5} I_{FSD}$

When the test leads are apart, R_x is infinite, the ohmmeter is on open circuit, and no current flows. **Thus $I = 0$ corresponds to infinity, ∞, on the left side of the scale.** The markings are shown in Fig. 13-6b.

The type of scale described in Example 13-5 and shown in Fig. 13-6b is often called a *back-off* scale, since the numbers run in the opposite direction from those on the scale of a voltmeter or ammeter.

From the calibration values shown, it is evident that the ohmmeter scale is *nonlinear,* being very crowded between 120 kΩ and infinity. This is one disadvantage of the series-type ohmmeter—it is difficult to accurately read resistance values above 120 kΩ.

Another disadvantage of this type of ohmmeter circuit is caused by aging of the battery. As the battery gets older, its voltage drops, causing a large change in the resistance reading. The decreased voltage means that a given current (and thus a given meter deflection) corresponds to a *lower* external resistance. Although the error can be offset to some extent by adjusting the zeroing control (R_z), the result is a *reading* that is *higher* than the true external resistance (R_x). This is demonstrated in Example 13-6.

EXAMPLE 13-6

When the ohmmeter battery voltage in Example 13-5 drops to 1.4 V, what value of R_x corresponds to 30 μA on the meter (marked 20 kΩ)?

Solution

When the battery voltage drops, R_z must be readjusted and reduced to be able to zero the ohmmeter. Substituting the reduced battery voltage in Eq. 13-4, we find that

$$R_m + R_z = \frac{V}{I_{FSD}} \qquad (13\text{-}4)$$

$$= \frac{1.4 \text{ V}}{50 \text{ } \mu\text{A}} = 28 \text{ k}\Omega$$

Thus $R_z = 28 \text{ k}\Omega - 3 \text{ k}\Omega = 25 \text{ k}\Omega$. (Note that R_z has been reduced from 27 to 25 kΩ because of the reduction in battery voltage.)

For $I = 30 \text{ } \mu\text{A}$,

$$R_x + R_m + R_z = \frac{V}{I} \qquad (13\text{-}3)$$

$$= \frac{1.4 \text{ V}}{30 \text{ } \mu\text{A}} = 46.7 \text{ k}\Omega$$

Thus $R_x = 46.7 \text{ k}\Omega - (25 \text{ k}\Omega + 3 \text{ k}\Omega) = \mathbf{18.7 \text{ k}\Omega}$.

Since the calibration scale value is 20 kΩ, the ohmmeter is indicating a value that is 1.3 kΩ too high, an *additional* error of 7%.

The error due to battery aging can be reduced to less than 1% by connecting the zeroing control in parallel with the meter movement, rather than in series with it. (See Problem 13-12.) It should be apparent, from what you have read so far, that the series-type ohmmeter is *not* capable of measuring either very low or very high resistances accurately. The *shunted-series* type of ohmmeter overcomes this problem (in part) and also lends itself to multirange capability. A shunted-series ohmmeter can provide several scales, ranging from $R \times 1$ for low resistance up to $R \times 1000$ for higher resistances.

13-6 THE SHUNTED-SERIES TYPE OF OHMMETER

A point of comparison between ohmmeters is the center scale (or *half-scale*) deflection resistance. The lower this half-scale resistance, the greater the meter's ability to measure low resistance values. A value of 10 to 12 Ω is considered an acceptable half-scale calibration. (For comparison, the half-scale calibration in the series-type ohmmeter shown in Fig. 13-6b is 30 kΩ.)

As shown in Fig. 13-7, a resistor equal to the half-scale resistance ($R_{s1} = 10 \text{ } \Omega$) is used to shunt the movement and the zero control on the $R \times 1$ range.

On any range, the total resistance (R_T) seen by the 1.5-V cell is given by

$$R_T = R_x + R_s \| (R_m + R_z) \qquad (13\text{-}5)$$

where: R_s is the shunt resistance selected by the range selector switch in ohms (Ω). All other symbols are as in Eq. 13-3.

When the instrument is being zeroed, $R_x = 0$. This connects the cell in parallel with the shunt resistance (R_s) and the series combination (R_m and R_z). The total resistance seen by the cell is now:

$$R_T = R_s \| (R_m + R_z) \qquad (13\text{-}6)$$

Since the current through $R_m + R_z$ is typically only 50 μA, the resistance of this series combination is much larger than that of R_s, which may be only 10 or a few hundred ohms. Thus, to a close approximation: $R_T \approx R_s$ when $R_x = 0$. This causes a certain current to flow through R_s, and full-scale deflection current (I_{FSD}) through the movement.

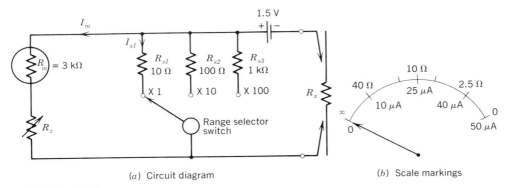

(a) Circuit diagram

(b) Scale markings

FIGURE 13-7
Shunted-series type of multirange ohmmeter.

If $R_x = R_s$, the total resistance seen by the cell doubles, so that $R_T \approx 2R_s$. The current through both R_s and the meter movement is thus cut in half, and the meter will deflect only to half scale. This half-scale deflection is calibrated with a value equal to R_{s1}, as shown in Example 13-7.

EXAMPLE 13-7

With the range selector switch in Fig. 13-7a set on the $R \times 1$ position, calculate:

a. The current through R_{s1} and the movement when $R_x = \infty$.

b. The value of R_z to zero the ohmmeter.

c. The current through R_{s1} when $R_x = 0$.

d. The current through R_{s1} and the movement when $R_x = 10 \ \Omega$.

e. The current through R_{s2} and the movement when the range is changed to $R \times 10$ and $R_x = 100 \ \Omega$. Given $I_m = 50 \ \mu A$, $R_m = 3 \ k\Omega$, $V = 1.5 \ V$ cell.

Solution

a. When $R_x = \infty$, there is no complete path for the current, so $I_m = $ **0**, $I_{s1} = $ **0**, and the pointer indicates infinity at the extreme left of the scale.

b. To zero the ohmmeter, $R_x = 0$.
This connects the 1.5-V cell in parallel with R_{s1} and the series combination of R_m and R_z. I_{FSD} must flow through the movement for full-scale deflection.

$$R_m + R_z = \frac{V}{I_{FSD}} \quad (3\text{-}1)$$

$$= \frac{1.5 \ V}{50 \ \mu A} = 30 \ k\Omega$$

Thus $R_z = 30 \ k\Omega - 3 \ k\Omega = $ **27 kΩ**.

c. $I_{s1} = \dfrac{V}{R_{s1}} \quad (3\text{-}1a)$

$$= \frac{1.5 \ V}{10 \ \Omega} = 0.15 \ A = \textbf{150 mA}$$

d. When $R_x = 10 \ \Omega$, the total resistance seen by the 1.5-V cell is

$$R_T = R_x + R_{s1}\|(R_m + R_z) \quad (13\text{-}5)$$

$$= 10 \ \Omega + 10 \ \Omega\|(3 \ k\Omega + 27 \ k\Omega)$$

$$= 10 \ \Omega + 10 \ \Omega\|30{,}000 \ \Omega$$

$$= 10 \ \Omega + 9.997 \ \Omega$$

$$\approx 20 \ \Omega$$

It is to be noted that this is double the resistance

seen by the 1.5-V cell when R_x was 0 Ω. Consequently, the battery current will be one-half of what it was when $R_x = 0$.

Thus the current through both R_{s1} and R_m will be cut in half.

$$I_{s1} = \frac{150 \ mA}{2} = \textbf{75 mA}$$

$$I_m = \frac{50 \ \mu A}{2} = \textbf{25 } \boldsymbol{\mu}\textbf{A}$$

Since the movement current is 25 μA, which is half of the full-scale deflection, the center scale reading of the ohmmeter is 10 Ω.

e. On the $R \times 10$ range, and $R_x = 0 \ \Omega$,

$$I_{s2} = \frac{V}{R_{s2}} = \frac{1.5 \ V}{100 \ \Omega} = 15 \ mA$$

and $\qquad I_m = 50 \ \mu A$

with $R_T = R_{s2}\|(R_m + R_z) \approx R_{s2} = 100 \ \Omega$. When $R_x = 100 \ \Omega$, $R_T \approx 200 \ \Omega$, so $I_{s2} = $ **7.5 mA** and $I_m = $ **25 μA**. The ohmmeter will indicate 10 Ω, which must be multiplied by 10 to obtain the deflection scale value of $R_x = 100 \ \Omega$.

NOTE On the $R \times 1$ range a large amount of current is required from the cell compared with the $R \times 10$ range. It is because of this that an aging cell may not be able to zero a commercial ohmmeter on the $R \times 1$ range but can on higher ranges. Also, because of the increased internal resistance of the aging cell, the terminal voltage will *not* be constant. This means that the zero-ohms control will have to be readjusted, each time the range is changed, to zero the meter.

On very-high-resistance ranges (such as $R \times 10$ k), higher voltages are needed to provide enough current to operate the meter movement. In commercial ohmmeters, this is accomplished by switching in additional cells (for a total of as much as 10.5 V). An ohmmeter with the capability of switching in an additional 9-V cell is shown in Fig. 13-8.

Many commercial ohmmeters do not use independent shunts (such as R_{s1}, R_{s2}, and R_{s3} in Fig. 13-7). Instead, they use series-parallel resistor combinations based on the Ayrton shunt principle described in Section 12-4.

13-7 THE WHEATSTONE BRIDGE

Of all resistance-measuring devices, the ohmmeter is the *least* accurate. To make *precision* resistance measure-

FIGURE 13-8
Model 60A VOM with back removed to show the 1.5-V cell for most ohmmeter ranges and the additional 9-V cell for the high-resistance range. (Courtesy of Triplett Corporation.)

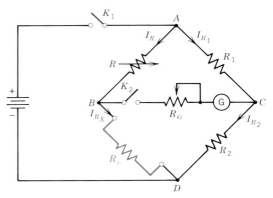

(a) Basic Wheatstone bridge circuit

(b) Commercial Wheatstone bridge

FIGURE 13-9
Wheatstone bridge for measuring resistance. [Photograph in (b) courtesy of Leeds and Northrup Co.]

ments, the instrument most often used is the Wheatstone bridge. Named for the English physicist Charles Wheatstone, the basic instrument circuit consists of an emf source, a galvanometer, two keys (switches), and a diamond-shaped arrangement of four resistors. A dc voltage is connected across one opposite set of corners, and the galvanometer across the other set, as shown in Fig. 13-9a.

Resistors R_1 and R_2 are usually decade resistors with values in some accurately known ratio to one another. R is a variable decade resistance with finely divided markings to indicate its value with a high degree of accuracy; R_x is the unknown value to be measured.

Keys K_1 and K_2 are closed and R is varied until (with the galvanometer resistance R_g at its minimum value), the galvanometer shows zero deflection. This zero deflection

(null) means that there is no potential difference between B and C. The bridge is said to be "balanced."

A null reading can only be obtained if

$$V_{BD} = V_{CD}$$

This also implies that

$$V_{AB} = V_{AC}$$

Thus $I_{R_x}R_x = I_{R_2}R_2$ and $I_R R = I_{R_1}R_1$

Dividing these two equations:

$$\frac{I_{R_x}R_x}{I_R R} = \frac{I_{R_2}R_2}{I_{R_1}R_1}$$

But if there is no current through the galvanometer,

$$I_R = I_{R_x} \quad \text{and} \quad I_{R_1} = I_{R_2}$$

$$\frac{R_x}{R} = \frac{R_2}{R_1} \qquad (13\text{-}7)$$

and

$$R_x = \left(\frac{R_2}{R_1}\right)R \qquad (13\text{-}7a)$$

or

$$R_x R_1 = R R_2 \qquad (13\text{-}7b)$$

Equation 13-7a shows that R_x can be obtained if the "ratio arms" (the ratio of R_2 to R_1) and R are accurately known.

Equation 13-7b shows a convenient way of remembering the balanced bridge resistance equation: *The cross products of the opposite arms are equal.*

You should note that the accuracy with which R_x can be determined in Eq. 13-7a depends on the *sum* of the precision errors of the known resistors (R_2, R_1, and R). Thus, if all three are known to an accuracy of 0.1%, the value of R_x is known to a maximum possible error of 0.3%. This measurement is independent of the applied voltage, the galvanometer accuracy, and the value of R_G in Fig. 13-9a.

EXAMPLE 13-8

Using the notation in Fig. 13-10a, a student-built Wheatstone bridge consists of 1% ratio arms with $R_1 = 3.3$ kΩ

and $R_2 = 1.2$ kΩ. Bridge balance occurs when R is adjusted to 310 ohms, with an accuracy of 0.1%. Calculate:
a. The unknown resistance and the maximum possible error.
b. The current through R_x if a 9-V battery is used.
c. The power dissipated in R.

Solution

a. $R_x = \left(\dfrac{R_2}{R_1}\right)R \qquad (13\text{-}7a)$

$= \left(\dfrac{1.2 \text{ kΩ}}{3.3 \text{ kΩ}}\right) 310 \text{ Ω} = \mathbf{112.7 \text{ Ω}}$

Maximum possible error $= 1\% + 1\% + 0.1\%$
$= \mathbf{2.1\%}$

b. $I_{R_x} = \dfrac{V}{R + R_x} \qquad (3\text{-}1a)$

$= \dfrac{9 \text{ V}}{310 + 112.7 \text{ Ω}} = \mathbf{21.3 \text{ mA}}$

c. $P_R = I_R^2 R \qquad (3\text{-}7)$

$= (21.3 \times 10^{-3} \text{ A})^2 \times 310 \text{ Ω}$
$= 0.14 \text{ W} = \mathbf{140 \text{ mW}}$

In the commercial Wheatstone bridge shown in Fig. 13-9b, R_2/R_1 is adjustable to provide ratio factors of 1, 10, 100, and so on, and is referred to as the *multiplier*. It multiplies the value of R indicated on the right dial.

The range of values that may be accurately measured by a Wheatstone bridge extends from 1 to 5 Ω at the lower limit (due to contact resistance) to as high as 1 MΩ. At high resistance the sensitivity of the galvanometer to imbalance in the bridge is severely restricted. This is due to the very small currents in the circuit.

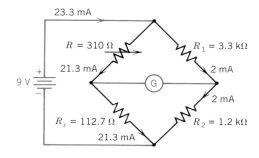

(a) Original bridge circuit

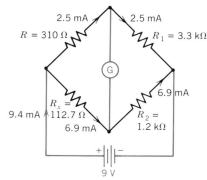

(b) Bridge circuit with galvanometer and battery interchanged

FIGURE 13-10
Wheatstone bridge circuits for Example 13-8.

The following example illustrates the effect of interchanging the galvanometer and battery locations.

EXAMPLE 13-9

Repeat Example 13-8 with the positions of the galvanometer and battery interchanged as in Fig. 13-10b.

Solution

a. $R_x = 112.7\ \Omega$
    Maximum error $= 2.1\%$

b. $I_{R_x} = \dfrac{V}{R_x + R_2}$ (3-1a)

    $= \dfrac{9\ V}{112.7 + 1200\ \Omega} = 6.9\ mA$

c. $P_R = I_R^2 R$ (3-7)

    $I_R = \dfrac{V}{R + R_1}$ (3-1a)

    $= \dfrac{9\ V}{310 + 3300\ \Omega} = 2.5\ mA$

    $P_R = I_R^2 R$ (3-7)

    $= (2.5 \times 10^{-3}\ A)^2 \times 310\ \Omega = 1.9\ mW$

Although the same value for R_x is obtained with either connection, the one illustrated in Fig. 13-10b draws much less current and the power dissipation is greatly reduced. (In R_x, the power is reduced from 51 mW to 5 mW.) If the device being measured is either temperature-sensitive or has a high-temperature coefficient of resistance (resistance that varies greatly with temperature changes), the connection shown in Fig 13-10b is usually desirable.

Maximum sensitivity of the galvanometer is claimed to occur with the connection shown in Fig. 13-10a. That is, the battery should be connected with the two low resistances in series. However, the difference in sensitivity is slight, and when using a *commercial* bridge instrument, there is no choice in the type of connection.

A very versatile circuit, the Wheatstone bridge is used in many applications beyond measuring resistance. For example, if a *thermistor* (which changes resistance with temperature) is substituted for one of the resistance arms, the bridge will become unbalanced as the temperature changes. The varying current flowing through the meter then reflects temperature changes. Through the substitution

of the appropriate *transducer* (any device whose resistance changes in response to a change in a physical quantity), the circuit can be used as a pressure gauge, a flow meter, a vacuum gauge, an anemometer (wind-speed gauge), and so on.

13-8 DIGITAL MULTIMETERS

All the meters discussed so far have been of the analogue type—they used a pointer to indicate a reading, much as the hands of a conventional watch indicate the time. A digital multimeter (like a digital watch) may use light-emitting diodes (LEDs) or a liquid-crystal display (LCD) to show the digital value of the electrical quantity being measured. Instead of interpreting the position of the pointer on the scale as a numeric value, you read the numbers directly from the digital display.

Figure 13-11 shows a $4\frac{1}{2}$-digit multimeter (the farthest left position displays a "1" when necessary; the remaining four positions can display numerals from 0 to 9). Both function and range buttons are provided on this meter, although some available meters automatically select the required range. Polarity is automatic, with a minus sign displayed when the quantity is negative with respect to common.

A digital multimeter (DMM) usually performs five measuring functions: ac and dc current, ac and dc volts, and ohms. The meter's characteristics are compared, in Table 13-1, to the analogue type of VOM and EVM. Unlike the VOM and EVM, the DMM has no need for an ohms ad-

FIGURE 13-11

A $4\frac{1}{2}$-digit five-function multimeter. (Courtesy of Hewlett-Packard.)

TABLE 13-1

Comparison of Multimeter Characteristics

Characteristic	VOM	EVM	DMM
Volts dc	Yes	Yes	Yes
Volts ac	Yes	Yes	Yes
Milliamperes dc	Yes	No	Yes
Milliamperes ac	No	No	Yes
Resistance ranges	$R \times 1$ to $R \times 100$ k	$R \times 1$ to $R \times 1$ M	100 Ω–10 MΩ
Adjustments	$R = 0$	$R = 0; R = \infty$	None
Dc volts input resistance	20 kΩ/V	10 MΩ	10 MΩ to 10^{10} Ω
Ac volts input resistance	5–20 kΩ/V	1 MΩ	1 MΩ
Ac volts frequency range	20 Hz–100 kHz	25 Hz–1 MHz	40 Hz–20 kHz
Power	Battery (ohms)	Ac or battery	Ac or battery
Display	Pointer	Pointer	LED or LCD
Accuracy	±2-3% of FSD	±3-5% of FSD	± (% reading + % range) (See Table 13-2)

justment. Some have a zeroing adjustment, however, for use when making measurements in the low (10-mV) dc volts range.

In addition to ease of reading, the primary features of a digital multimeter are its resolution and accuracy. For example, one $3\frac{1}{2}$-digit meter has seven ranges, measuring current from 1 pA to 2 mA, with less than 0.2-mV drop caused by the instrument. Some of the accuracies attaina-ble with the multimeter shown in Fig. 13-11 are indicated in Table 13-2. The stated accuracies remain valid for a period of 30 days following calibration of the instrument, provided a temperature of ±5°C from 23°C is maintained. Note that accuracy depends on both the reading and the range in use. Some DMM manufacturers may express the accuracy in terms of a percentage of the reading and plus or minus one digit.

TABLE 13-2

Typical Digital Multimeter Accuracies

Function	Accuracy
Dc volts (100 mV–100 V)	± (0.02% of reading + 0.01% of range)
Ac volts (100 mV–100 V, 40–10 kHz)	± (0.15% of reading + 0.05% of range)
Dc amperes (100 μA–10 mA)	± (0.1% of reading + 0.01% of range)
Ac amperes (100 μA–10 mA, 1–10 kHz)	± (0.25% of reading + 0.75% of range)
ohms (1 kΩ–1 MΩ)	± (0.02% of reading + 0.01% of range)

EXAMPLE 13-10

A digital multimeter having an accuracy given by Table 13-2 is being used on the 10-V dc range.

a. Determine the possible reading and the percentage error when measuring 5-V dc.

b. Repeat for a measurement of 10-V dc.

Solution

a. From Table 13-2:

Possible reading = 5 V ± (0.02% of reading
                          + 0.01% of range)

$$= 5\ V \pm \left(\frac{0.02}{100} \times 5\ V\right.$$

$$\left. + \frac{0.01}{100} \times 10\ V\right)$$

$$= 5\ V \pm (0.001\ V + 0.001\ V)$$

$$= \textbf{5 V} \pm \textbf{0.002 V}$$

Percentage error $= \dfrac{0.002\ V}{5\ V} \times 100\% = \textbf{0.04\%}$

b. Possible reading = 10 V ± (0.02% of 10 V
                              + 0.01% of 10 V)

$$= 10\ V \pm \left(\frac{0.02}{100} \times 10\ V\right.$$

$$\left. + \frac{0.01}{100} \times 10\ V\right)$$

$$= 10\ V \pm (0.002\ V + 0.001\ V)$$

$$= \textbf{10 V} \pm \textbf{0.003 V}$$

Percentage error $= \dfrac{0.003\ V}{10\ V} \times 100\% = \textbf{0.03\%}$

Example 13-10 illustrates, once again, the advantage of making readings that are as close to full scale as possible to reduce percentage errors.

SUMMARY

1. A potentiometer is an accurate, high-resolution, variable-voltage divider used to measure unknown voltages without drawing any current.

2. A standard cell is a source of accurately known voltage used for calibrating potentiometers.

3. A potentiometer can be used to calibrate voltmeters and ammeters.

4. A calibration curve consists of a graph of corrections that must be added to or subtracted from the reading on an ammeter or voltmeter scale to obtain a true reading.

5. Resistance may be measured by applying Ohm's law to actual ammeter and voltmeter measurements.

6. Equations 13-1 and 13-2 are used to account for meter resistance when measuring circuit resistance.

7. For calculations involving products and quotients, the total possible error is the *sum* of the errors of the individual variables.

8. A simple type of ohmmeter consists of a series-connected cell, meter movement, and zeroing control. Such an ohmmeter has a high half-scale resistance.

9. An ohmmeter scale is nonlinear, and reads from zero ohms on the right to infinity on the left. This scale is called a ''back-off'' type.

10. When the battery voltage drops in a simple series-type ohmmeter, the indicated readings will be too high.

11. The shunted-series type of ohmmeter is adaptable to multirange operation and has a low half-scale resistance.

12. To measure resistance with a high degree of accuracy, a Wheatstone bridge is used.

13. A bridge is balanced (in a ''null'' condition) when the cross products of the opposite arms are equal.

14. Interchanging the location of the battery and galvanometer in a Wheatstone bridge does not change the resistance values for balance, but may have a large effect on the power dissipation in the circuit.
15. Digital multimeters are easy to read, more accurate, and have higher resolution than analogue multimeters.

SELF-EXAMINATION

Answer true or false or, in the case of multiple choice, a, b, c, or d
(Answers at back of book)

13-1. A no-load voltmeter measures the open-circuit voltage of a source of emf. _____

13-2. A potentiometer is a source of variable voltage known to a high degree of accuracy for measuring unknown voltages. _____

13-3. A potentiometer compares the voltage from a standard cell with an unknown voltage to determine the unknown voltage. _____

13-4. The calibration of a potentiometer requires setting the dials to the standard cell voltage and adjusting the current from the working battery until a null is indicated on the galvanometer. _____

13-5. A potentiometer can directly measure voltages up to 150 V. _____

13-6. A potentiometer, because it can only measure voltage, can be used to calibrate a voltmeter but not an ammeter. _____

13-7. In calibrating a voltmeter, the correction may be either positive or negative. _____

13-8. The ammeter–voltmeter method of measuring resistance is one of the most accurate methods available. _____

13-9. If a 15-V 3%-accurate voltmeter indicates 7.5 V and a series-connected 10-mA range 4% accurate milliammeter indicates 10 mA, the resistance and its maximum percentage error are
   a. 750 Ω, 7%
   b. 75 Ω, 7%
   c. 750 Ω, 10%
   d. 7.5 kΩ, 10%

13-10. When using the ammeter–voltmeter method to measure high resistance, the voltmeter should be connected directly across the resistor. _____

13-11. A series-type of ohmmeter, like all analogue ohmmeters, has a zero-ohms control. _____

13-12. A series type of ohmmeter has a large change in indicated resistance as the battery voltage drops. _____

13-13. A series type of ohmmeter is especially suited for making low-resistance measurements. _____

13-14. A shunted-series type of ohmmeter is used for multirange ohmmeters and has a low center-scale resistance. _____

13-15. Failure to be able to zero an ohmmeter (especially on the X-1 range) is an indication that the main battery needs replacing. _____

13-16. An ohmmeter in a VOM cannot be zeroed on the highest range, but it can on the lowest. This may be an indication that the additional cells were not installed. _____

13-17. In a balanced Wheatstone bridge, the products of the adjacent resistor arms are equal.

13-18. At balance, no current flows through the galvanometer in a Wheatstone bridge, so meter accuracy is unimportant. _____

13-19. In measuring an unknown resistance, the choice of location of the battery and galvanometer gives the same result. _____

13-20. In a commercial Wheatstone bridge the ratio arms are usually adjusted as multiples or submultiples of 10 and become the multiplier factor. _____

13-21. A Wheatstone bridge can be used only for measuring resistance. _____

13-22. A digital multimeter can also measure ac current while most analogue multimeters do not. _____

REVIEW QUESTIONS

1. Describe the principle of a no-load voltmeter.

2. Describe briefly the complete procedure for using a potentiometer to measure an unknown emf.

3. a. Why doesn't the accuracy of the galvanometer affect the accuracy of voltage determination in a potentiometer or resistance determination in a Wheatstone bridge?

  b. What is a very desirable characteristic for a galvanometer in these applications?

4. A standard cell has a high internal resistance of approximately 500 Ω. Why doesn't this affect the accuracy of voltage measurement?

5. a. What is the maximum voltage measurable by a potentiometer?

  b. How can this be extended?

  c. What disadvantage does this have?

6. Describe the process of calibrating an ammeter.

7. Given an accurately known resistance, suggest how you could use a potentiometer to make an accurate measurement of another resistor similar in value to the first.

8. If you were using the ammeter–voltmeter method for measuring circuit resistance, what *visual* indications would you have on the meters as you tried to decide where the voltmeter should be placed for the following circuits?

  a. A high-resistance circuit.

  b. A low-resistance circuit.

9. When determining resistance by applying Ohm's law to voltmeter and ammeter readings, under what conditions will the maximum possible error be given by the following sums?

  a. The sum of the FSD accuracies of the two meters.

  b. Twice the sum of the FSD accuracies of the two meters.

10. Give a simple explanation of why an ohmmeter scale must be nonlinear.

11. Why are most VOM ohmmeter scales of the ''back-off'' type?

12. Why does a series type of ohmmeter inherently have a high center-scale resistance?

13. Why does a reduction in voltage in a series-type ohmmeter cause the indicated resistances to be high?

14. In a shunted-series type of ohmmeter, why is it more likely to be unable to zero the instrument on the $R \times 1$ range than on a higher range?

15. a. What is often necessary in many VOM ohmmeter circuits to be able to make measurements on the $R \times 10$ k range?

    b. Why is this so?

16. a. Why does no current flow through the galvanometer of a balanced Wheatstone bridge?

    b. If, in Fig. 13-9a, at balance, a connection were made from B directly to C, would this upset the balanced conditions? Explain.

17. Why are two of the resistors in a Wheatstone bridge referred to as the "ratio arms"?

18. Why should interchanging the locations of the dc supply and galvanometer have any effect upon the power dissipations in the Wheatstone bridge circuit?

19. What are four advantages that digital multimeters have over most analogue multimeters?

20. a. How does the error on a digital multimeter measurement differ from that on an analogue instrument?

    b. Does minimum percent error occur for readings near full scale on a digital multimeter as it does for an analogue multimeter?

PROBLEMS

(Answers to odd-numbered problems at back of book)

13-1. A potentiometer is being used to calibrate a dc voltmeter on the 1.5-V range. The largest correction occurs when the voltmeter indicates 1.3 V. At this value, the two dials of the potentiometer indicate 1.2 V and 53 mV. Calculate:

    a. The true reading of the voltmeter.

    b. The correction in mV in magnitude and sign.

    c. The percent FSD accuracy of the meter.

13-2. A potentiometer is being used to calibrate a 100-mA range milliammeter. The largest correction occurs when the milliammeter indicates 60 mA. At this value, the two dials of the potentiometer indicate 0.5 V and 76 mV, when connected across a standard 10-Ω resistor connected in series with the ammeter. Calculate:

    a. The true reading of the milliammeter.

    b. The correction in magnitude and sign.

    c. The percent FSD accuracy of the meter.

13-3. A potentiometer is being used to measure resistance. A standard 10-Ω 0.1% resistor is connected in series with the unknown resistor, and the combination is connected to a dc supply. When the potentiometer is connected across the 10-Ω resistor, it indicates 0.105 V at balance; across the unknown resistor it indicates 1.189 V. If the potentiometer is accurate to 0.5% calculate:

    a. The unknown resistance.

    b. The accuracy of the measured resistance.

13-4. A voltmeter is being calibrated on the 300-V range using a volt-box with a maximum 300-V input and maximum 1.5-V output. When the voltmeter indicates 180 V, the maximum correction occurs for dial readings on the potentiometer of 0.9 V and 35 mV. If the volt-box is 0.1% accurate and the potentiometer 0.2%, calculate:

    a.   The true reading of the voltmeter.

    b.   The correction in magnitude and sign.

    c.   The percent FSD accuracy of the voltmeter.

    d.   The accuracy to which the determination in part (c) is made.

13-5. A resistor is being measured using the circuit of Fig. 13-5a. A 15-V range, 3%-FSD accuracy voltmeter indicates 10 V; and a 1-mA, 100-Ω, 4%-FSD accuracy milliammeter indicates 0.5 mA. Calculate:

    a.   The resistance.

    b.   The maximum possible error.

13-6. Repeat Problem 13-5 assuming that both meters indicate FSD.

13-7. A resistor is being measured using the circuit of Fig. 13-5b. The 100-mA range milliammeter indicates 60 mA, and the 5 kΩ/V voltmeter indicates 3 V on the 5-V range. If both meters have a 3% FSD accuracy calculate:

    a.   The resistance.

    b.   The maximum possible error.

13-8. Repeat Problem 13-7 assuming that both meters indicate 50% of FSD.

13-9. Given a 0–100-μA, 2-kΩ meter movement, determine the resistance markings every 20 μA when used in a series-type ohmmeter with two 1.5-V cells.

13-10. Repeat Problem 13-9 using a 0–200-μA, 5-kΩ meter movement, every 40 μA.

13-11. a.   Assuming that each cell in Problem 13-9 decreases by 0.1 V, calculate the true resistance being measured when the meter indicates 60 μA.

    b.   Using the resistance marking for 60 μA from Problem 13-9, determine the additional error introduced by the aging batteries.

    c.   What could the total battery voltage drop to before being unable to zero the ohmmeter?

13-12. A series-type ohmmeter consists of a variable zeroing rheostat in parallel with the meter movement and this combination in series with a fixed resistor R_1 and a 1.5-V cell. If the meter movement is a 3-kΩ, 50-μA FSD, and the center-scale resistance marking is to be 25 kΩ, calculate:

    a.   The value of R_1 and the zero-ohms control value.

    b.   The new value of the zero-ohms control when the voltage drops to 1.4 V.

    c.   The true resistance being measured when the ohmmeter indicates its center-scale resistance reading.

    d.   The percentage change in center-scale resistance due to a drop of 0.1 V.

    e.   Compare the results with the percentage change of center-scale resistance in Examples 13-5 and 13-6.

13-13. Given a shunted-series type of multirange ohmmeter as in Fig. 13-7 with R_{s1} = 12 Ω, R_{s2} = 120 Ω, R_{s3} = 1.2 kΩ, and I_m = 50 μA, R_m = 3 kΩ, V = 3 V, calculate, with the range switch on $R \times 1$:

    a.   The current through R_{s1} and the movement when R_x = ∞.

    b.   The value of R_z to zero the ohmmeter.

    c.   The current through R_{s1} when R_x = 0.

    d.   The current through R_{s1} and the movement when R_x = 12 Ω.

    e.   The current through R_{s2} and the movement when the range is changed to $R \times 10$ and R_x = 120 Ω.

    f.   The resistance markings at 25% and 75% of FSD.

13-14. Repeat Problem 13-13 using a 100-μA, 5-kΩ meter movement and a 1.5-V cell.

13-15. If the range selector switch in Example 13-7 has a fourth position, where no shunt is connected into the circuit, determine the multiplier for this position.

13-16. Repeat Problem 13-15 as applied to Problem 13-13.

13-17. A Wheatstone bridge has the following values at balance, using the notation in Fig. 13-9: $R_2/R_1 = 10$ to an accuracy of 0.5%, $R = 4.52\ \Omega$, $\pm 0.1\%$.

 If the dc supply is 9 V, calculate:

 a. The value of R_x.

 b. The accuracy of the determination of R_x.

 c. The current through R_x at balance.

13-18. a. If three of the resistors in a Wheatstone bridge are each 5 kΩ, $\pm 1\%$, what must be the value of the fourth at balance?

 b. What range of values could the fourth resistor have?

 c. What would be the total resistance seen by the battery at balance?

13-19. Using the notation in Fig. 13-10a, a student-built Wheatstone bridge consists of $R_1 = 10$ kΩ, $\pm 2\%$, $R_2 = 220$ kΩ, $\pm 1\%$. Bridge balance occurs when R is adjusted to 4.75 kΩ, $\pm 0.1\%$. Calculate:

 a. The unknown resistance.

 b. The maximum possible error.

 c. The current through R_x if a 6-V battery is used.

 d. The power dissipated in R_x.

13-20. Repeat Problem 13-19 with $R_1 = 2.2$ kΩ, $\pm 1\%$.

13-21. Repeat Problem 13-19 with the positions of the galvanometer and battery interchanged.

13-22. Repeat Problem 13-20 with the positions of the galvanometer and battery interchanged.

13-23. A digital multimeter has accuracies given by the data in Table 13-2. It is being used as a dc voltmeter. Determine the possible reading and the percentage error when making a measurement of 1 V on:

 a. The 1-V range.

 b. The 10-V range.

13-24. Repeat Problem 13-23 for the multimeter being used to make ac voltage measurements at 1 kHz.

13-25. A digital multimeter has accuracies given by the data in Table 13-2. It is being used as a dc milliammeter on the 10-mA range. Determine the possible reading and the percentage error when making a measurement of

 a. 2 mA.

 b. 9 mA.

13-26. Repeat Problem 13-25 for the multimeter being used to make ac current measurements at 10 kHz.

13-27. a. By how many ohms could the multimeter of Table 13-2 be in error when it indicates 820 kΩ on the 1-MΩ range?

 b. What percentage accuracy does this represent?

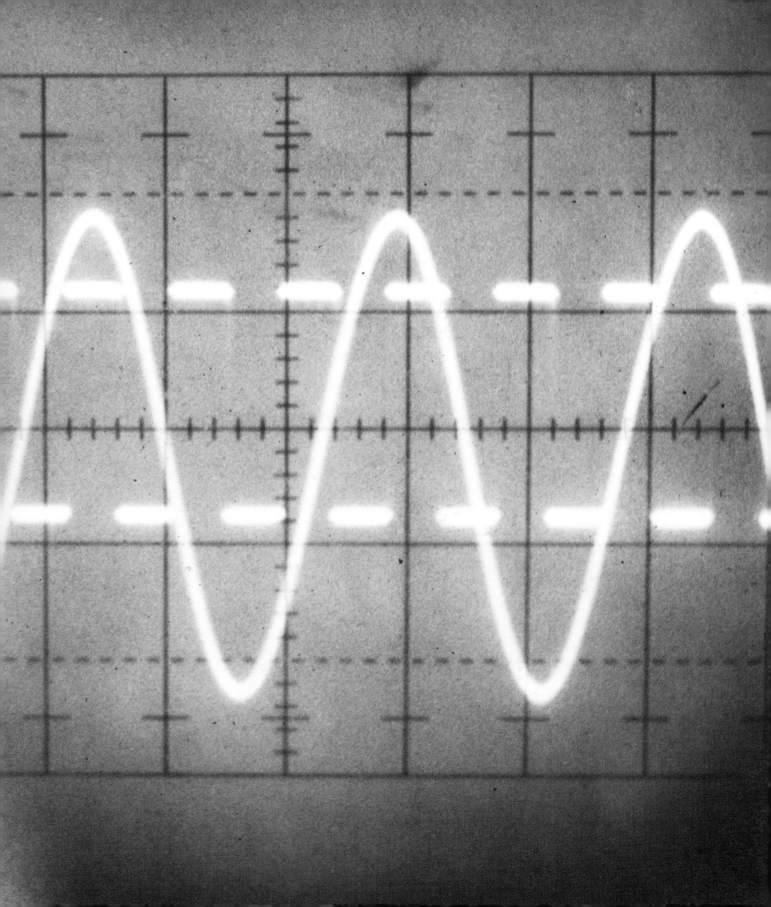

PART II

AC CIRCUITS

CHAPTER 14

GENERATING AC AND DC

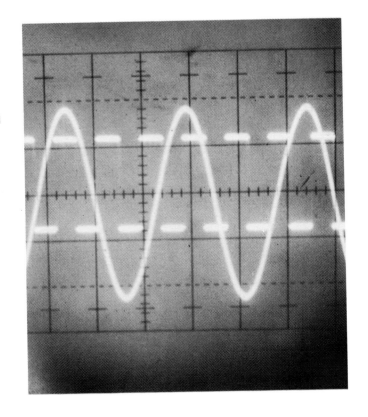

Earlier, you learned about the magnetic field that is created around a conductor by current flowing through that conductor. In this chapter, you will investigate how a current can be *induced* in a conductor by placing it in a *changing* magnetic field. This process, called *electromagnetic induction,* is used by ac alternators and dc generators. The *amount* of voltage induced in a conductor is given by Faraday's law (Section 14-3), and the *polarity* of that voltage by Lenz's law (Section 14-2).

If a coil is rotated in a *fixed* magnetic field, a voltage is produced that alternates in polarity (ac or alternating current). The waveform of this voltage can be made to have a *sinusoidal* variation. That is, its instantaneous value of induced emf depends upon the sine of the angle that the coil makes with the magnetic field. The factors that determine the amplitude and frequency of the induced sine wave of voltage will also be considered. Practical dc generators and three-phase ac alternators will be discussed in this chapter as well.

14-1 ELECTROMAGNETIC INDUCTION

In the process of electromagnetic induction, a voltage is induced in a conductor whenever there is relative motion between that conductor and a magnetic field.

One of the ways of producing such a relative motion is shown in Fig. 14-1. In this case, the conductor is being moved through a stationary *(fixed)* magnetic field. The direction and strength of the magnetic field is represented by lines of force. In Fig. 14-1*a,* the conductor is said to be "cutting" the magnetic field (moving in a line across the lines of force). Note the polarity of the voltage induced in the conductor, and compare it to the polarity of the induced voltage in Fig. 14-1*b,* where the conductor is moved in the opposite direction. Reversing the direction of motion reverses the polarity of the induced voltage. Moving the conductor *parallel* to the lines of magnetic force

produces *no* induced voltage, since it is not cutting any of the lines in the field.

The same effects can be produced by holding the conductor stationary and moving the magnetic field. The basic requirements for electromagnetic induction are *relative* motion and the cutting of the lines of force. (Section 14-3 will discuss modification of the cutting requirement.)

What is the *source* of the emf produced in the conductor? In Section 11-10, you learned that a magnetic field exists at a point if a force is exerted on a moving charge at that point ($F = qvB$). By moving the conductor, the free electrons within that conductor are made to move through the magnetic field. As a result, a force is exerted on those free electrons, making them move from one end of the conductor toward the other. This causes a concentration of electrons at one end and a deficiency at the other. In other words, a *potential difference* has been gen-

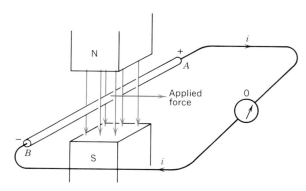

(*a*) Polarities of induced voltage and current are due to moving the conductor to the right

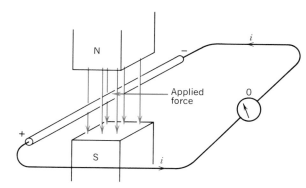

(*b*) Polarities of induced voltage and current are reversed because of the conductor's motion in the opposite direction

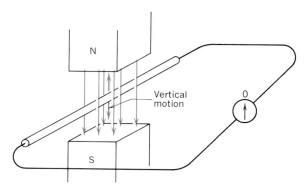

(*c*) No voltage is induced when the conductor is moved parallel to the magnetic field

FIGURE 14-1

Electromagnetic induction.

erated that can cause current to flow around an externally connected circuit. It is important to note that the polarity signs in Fig. 14-1 indicate a *source* of emf, not a voltage *drop* caused by the current.

The force needed to move the electrons is supplied by the external force applied to the conductor *(applied force)*. If the applied force is increased, moving the conductor through the field more quickly, a *higher* voltage is induced. The polarity of this voltage is determined by the application of Lenz's law.

14-2   LENZ'S LAW

The polarity of the induced emf is such that any current resulting from it produces a magnetic flux that *opposes* the motion or change producing the emf.

Basically, Lenz's law states that the polarity of the induced emf opposes the cause that produced the emf. This is shown in Fig. 14-2a, where the conductor is being moved to the right. The polarity of the induced emf causes current to flow ''into'' the paper (the direction is indicated by the cross symbol in the conductor). The current sets up a magnetic field around the conductor in a *clockwise* direction. (This can be established by applying the ''right-hand rule'' described in Chapter 11.) The result is a strengthening of the overall field to the *right* of the conductor and a weakening to the *left* of the conductor. A ''motor action'' force, or magnetic repelling effect, is developed on the conductor in a direction opposed to the applied force. Thus, *work* must be done by the externally applied force to develop the emf, as stated by Lenz's law.*

In Fig. 14-2b, the applied force is in the opposite direction, reversing the polarity of the induced emf, so the magnetic repelling effect (opposing force) is developed on the opposite side of the conductor. Later in this chapter, you will have other opportunities to apply Lenz's law.

14-3   FARADAY'S LAW

The more rapidly a conductor is moved through a magnetic field, the larger the voltage that will be induced. The volt-

---

*The polarity of the induced emf may also be obtained by the Right-Hand Generator Rule: Hold the thumb and first two fingers of the right hand mutually perpendicular to each other. With the thumb pointing in the direction of conductor motion (applied force), and the first finger pointing in the direction of the magnetic flux, the second finger points in the direction of induced current or from − to + of the induced emf.

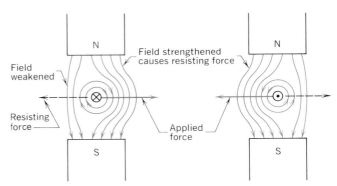

(a) Induced current into the paper due to an applied force to the right

(b) Induced current out of the paper due to an applied force to the left

FIGURE 14-2
Using Lenz's law to determine the polarity of induced emf.

age may also be increased by using a stronger magnetic field. These two effects are combined in Faraday's law:

$$v_{ind} = \frac{\Delta\Phi}{\Delta t} \qquad \text{volts} \qquad (14-1)$$

where: v_{ind} is the emf induced in a conductor, in volts (V)

$\Delta\Phi$ is the magnetic flux cut by the conductor, in webers (Wb)

Δt is the time required for the conductor to move, in seconds (s)

$\Delta\Phi/\Delta t$ is the rate at which magnetic flux lines are cut by the conductor, in webers per second (Wb/s)

The symbol Δ means ''change,'' so $\Delta\Phi$ means a change in the magnetic flux, and Δt means a change in time. Δt is used here to refer to an *interval* of time equal to $t_2 - t_1$, where t_2 is the final time and t_1 is the initial time. Similarly, $\Delta\Phi = \Phi_2 - \Phi_1$ is a *quantity* of flux cut by the conductor in the time interval Δt.

Thus Faraday's law states that the voltage induced in a conductor is *directly proportional* to the *rate* at which the magnetic flux is cut by the conductor.

EXAMPLE 14-1

The magnetic field strength between the pole faces of a horseshoe magnet is 0.6 T (Wb/m²). The pole faces are 2 cm square. Calculate the voltage induced in a wire if it is moved perpendicularly through the field at a uniform rate in

a. 0.1 s
b. 0.01 s

Solution

a. The flux cut by the wire can be calculated from

$$\Phi = BA \qquad (11\text{-}1)$$

or $\quad \Delta\Phi = B \times \Delta A$

$$= 0.6 \frac{Wb}{m^2} \times (2 \times 10^{-2}\ m)^2$$

$$= 2.4 \times 10^{-4}\ Wb$$

$$v_{ind} = \frac{\Delta\Phi}{\Delta t} \qquad (14\text{-}1)$$

$$= \frac{2.4 \times 10^{-4}\ Wb}{0.1\ s}$$

$$= 2.4 \times 10^{-3}\ Wb/s = \mathbf{2.4\ mV}$$

b. $v_{ind} = \dfrac{\Delta\Phi}{\Delta t} \qquad (14\text{-}1)$

$$= \frac{2.4 \times 10^{-4}\ Wb}{0.01\ s}$$

$$= 2.4 \times 10^{-2}\ Wb/s = \mathbf{24\ mV}$$

Note that the length of the conductor is not important, so long as it cuts through *all* of the flux. The voltage depends only on the *rate* at which the flux is cut.

Now, the unit of magnetic flux, the *weber*, can be formally defined:

The **weber** is the amount of magnetic flux that, when cut at a constant rate by a conductor in 1 s, will induce an emf of 1 V in the conductor. (1 volt = 1 weber/second)

So far, you have considered the emf generated when a single wire is moved through a magnetic field. If the conductor consists of a number of wires, series-connected as in the turns of a coil, the induced emf is increased accordingly:

$$v_{ind} = N\!\left(\frac{\Delta\Phi}{\Delta t}\right) \qquad \text{volts} \qquad (14\text{-}2)$$

where the symbols are as in Eq. 14-1 and N equals the number of turns in the coil.

EXAMPLE 14-2

The north end of a permanent magnet is moved into a solenoid as in Fig. 14-3. If the coil is cut by magnetic flux lines at the rate of 0.03 Wb/s, determine:

a. The voltage induced in the coil of 100 turns.
b. The polarity of induced voltage in the coil.
c. The effect of withdrawing the magnet at twice the speed.

Solution

a. $v_{ind} = N\!\left(\dfrac{\Delta\Phi}{\Delta t}\right) \qquad (14\text{-}2)$

$$= 100 \times 0.03\ Wb/s = \mathbf{3\ V}$$

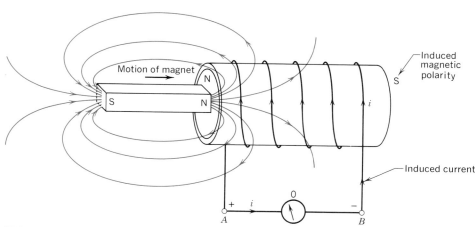

FIGURE 14-3
Using Lenz's law to determine the polarity of induced emf when a permanent magnet is moved toward a solenoid.

b. By Lenz's law, the current must flow in the coil in such a direction as to set up an opposing force. This means that the left side of the solenoid must become a north pole while the permanent magnet is approaching. Applying the right-hand rule for a coil, the current must flow upward on the front of the coil. **This means that point _A_ is positive with respect to _B_.**

c. If the magnet is removed at twice the speed it entered, the rate of change of flux in the coil, $\frac{\Delta\Phi}{\Delta t}$, is doubled to 0.06 Wb/s, and the induced voltage is doubled to **6 V.**

The polarity of the induced voltage is **reversed** to set up a south pole at the **left** side of the solenoid. This attempts to keep the north pole of the permanent magnet from being removed.

A galvanometer connected between _A_ and _B_ will deflect alternately to the left and to the right as the magnet is moved in and out of the coil. As this indicates, an _alternating current_ is produced by the movement of the magnet. Whenever the magnet stops moving, as at the end of travel in one direction before moving in the opposite direction, the induced voltage drops to zero (as required by Faraday's law).

This is the principle used in the dynamic microphone shown in Fig. 14-4. The sound-pressure waves cause a flexible cone to move in and out. A coil attached to the cone, as a result, moves back and forth in the field of a permanent magnet. The voltage induced in the coil varies

in frequency with the sound waves that caused the motion of the cone and coil.

This variable voltage, when properly amplified, can be used to drive a loudspeaker (as described in Section 11-12). The action of the microphone is the reverse of the speaker's action. This allows the use of the same device (for example in a walkie-talkie or intercom) for both functions. Depending on whether the unit is being used to _transmit_ or _receive_, the device converts sound into electric current or electric current into sound.

Faraday's law is also useful for applications where no _cutting_ of the magnetic field appears to take place. This occurs when two coils are wrapped on a common iron core, as in Fig. 14-5. A _toroid_ (''doughnut-shaped'') core restricts all the magnetic flux to within itself. If the current in the first coil is made to change by varying a rheostat, a _changing magnetic field_ is set up in the core. The magnetic flux _enclosed_ by the second coil changes, inducing a voltage. (This is the principle of the transformer.) The polarity of the induced voltage can be obtained by applying Lenz's law, as in Example 14-3. In effect, a voltage is induced whenever a change in flux around a conductor takes place. This is true whether the conductor moves _through_ a magnetic field or the field moves (changes) _about_ a stationary conductor.

EXAMPLE 14-3

Refer to Fig. 14-5. The rheostat is decreased in value in such a way as to cause the magnetic flux in the core to

_(a)_ Undirectional dynamic microphone

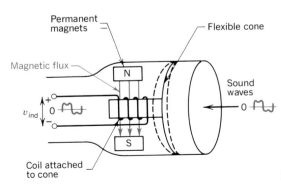

_(b)_ Cross-sectional view through a dynamic microphone

FIGURE 14-4
Dynamic microphone. (Photograph courtesy of Shure Brothers Inc.)

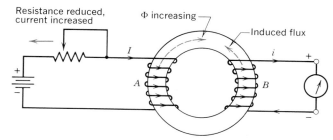

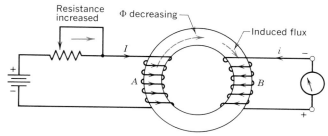

(a) An increasing current in coil A sets up an increasing field in the core. The induced current in coil B sets up a flux that opposes the increase

(b) A decreasing current in coil A reduces the field in the core. The induced current in coil B sets up a flux that opposes the decrease

FIGURE 14-5
Application of Faraday's and Lenz's laws to coils on a common core.

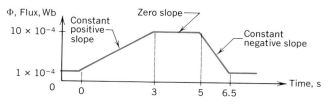

(a) Variation of flux in iron core of Fig. 14-5

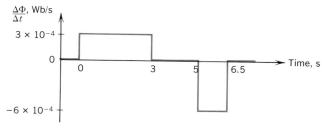

(b) Rate of change of flux in iron core

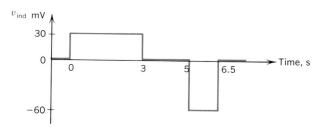

(c) Induced voltage in coil B

FIGURE 14-6
Waveforms for Example 14-3.

increase uniformly from 1×10^{-4} Wb to 10×10^{-4} Wb in 3 s. After a delay of 2 s, the rheostat is then increased, causing the flux to drop uniformly to its initial value in 1.5 s. Coil B has 100 turns.

a. Draw a graph showing how magnetic flux changes in the core.

b. Underneath the above graph, draw a second graph showing the quantity $\dfrac{\Delta \Phi}{\Delta t}$.

c. Draw a third graph indicating the voltage induced in coil B.

d. Show polarity signs of induced voltage on the second coil for when the flux is increasing and decreasing.

Solution

a. The uniform increase and decrease of the flux in the iron core is shown in Fig. 14-6a.

b. The rate of change of flux during the first 3 s is given by

$$\frac{\Delta \Phi}{\Delta t} = \frac{10 \times 10^{-4} - 1 \times 10^{-4} \text{ Wb}}{3 - 0 \text{ s}}$$

$$= \frac{9 \times 10^{-4} \text{ Wb}}{3 \text{ s}}$$

$$= \mathbf{3 \times 10^{-4} \text{ Wb/s}}$$

The flux decreases at a rate given by

$$\frac{\Delta \Phi}{\Delta t} = \frac{1 \times 10^{-4} - 10 \times 10^{-4} \text{ Wb}}{6.5 - 5 \text{ s}}$$

$$= \frac{-9 \times 10^{-4} \text{ Wb}}{1.5 \text{ s}}$$

$$= \mathbf{-6 \times 10^{-4} \text{ Wb/s}}$$

These quantities are actually the slopes of the flux variation graph and are plotted as positive and negative quantities, as in Fig. 14-6b.

c. $v_{\text{ind}} = N\left(\dfrac{\Delta \Phi}{\Delta t}\right)$ (14-2)

When the flux is increasing,

$$v_{ind} = 100 \times 3 \times 10^{-4} \text{ V} = \textbf{30 mV}$$

When the flux is decreasing,

$$v_{ind} = 100 \times (-6 \times 10^{-4}) \text{ V} = \textbf{-60 mV}$$

These voltages are plotted in Fig. 14-6c.

d. The polarities of the induced voltages are indicated in Fig. 14-5. When the current increases in coil *A*, the flux enclosed by coil *B* also increases. The induced current in coil *B* flows in such a direction as to set up a flux that opposes the flux *increase* due to coil *A*. When the current in coil *A* decreases, as in Fig. 14-5b, the induced voltage has the opposite polarity. The current in coil *B* now tends to *maintain* the decreasing flux. That is, it is opposing the *change* in core flux, not the flux itself.

14-4 INDUCED EMF DUE TO CONDUCTOR MOTION IN A STATIONARY MAGNETIC FIELD

Faraday's law is uniquely suited to the situation described in Example 14-3, where the flux is changing within a stationary coil. But in the case of a conductor moving through a stationary magnetic field, it is convenient to express the induced voltage in terms of more readily measured quantities.

Assume that the conductor in Fig. 14-7 is moved through a distance Δs in time Δt. If the magnetic field is uniform, the conductor cuts through a total flux ($\Delta\Phi$) given by

$$\Delta\Phi = B \times A \qquad (11\text{-}1)$$

where A is the area $= l \times \Delta s$. Then

$$v_{ind} = \frac{\Delta\Phi}{\Delta t} \qquad (14\text{-}1)$$

$$= \frac{B \times l \times \Delta s}{\Delta t}$$

$$= B \times l \times \frac{\Delta s}{\Delta t}$$

But $\dfrac{\Delta s}{\Delta t}$ is the velocity (v) of the conductor. Thus

$$v_{ind} = Blv \qquad \text{volts} \qquad (14\text{-}3)$$

where: v_{ind} is the emf induced, in volts (V)
$\qquad\quad$ B is the flux density of the field, in teslas (T)

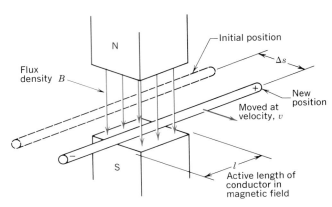

FIGURE 14-7
Quantities involved in $v_{ind} = Blv$.

l is the active length of the conductor, in meters (m)
v is the velocity of the conductor, in meters per second (m/s)

If N conductors are series-connected, the total voltage induced is

$$v_{ind} = NBlv \qquad \text{volts} \qquad (14\text{-}4)$$

EXAMPLE 14-4

A car with a 0.8-m vertical radio antenna is traveling due west at 90 km/h (approximately 56 mph). If the horizontal component of the earth's magnetic field is 1.7×10^{-5} T at this point, calculate:

a. The voltage induced in the antenna.
b. The voltage if the car travels due north.

Solution

a. 90 km/h $= 90 \times 10^3$ m/h $\times \dfrac{1 \text{ h}}{3600 \text{ s}} = 25$ m/s

$$v_{ind} = Blv \qquad (14\text{-}3)$$
$$= 1.7 \times 10^{-5} \text{ Wb/m}^2 \times 0.8 \text{ m} \times 25 \text{ m/s}$$
$$= 34 \times 10^{-5} \text{ V} = \textbf{340 μV}$$

b. No voltage is induced when traveling north, since the antenna is traveling *with* the field and *not cutting across* it.

NOTE The voltage induced in part (a) is dc, and while relatively large, it has no effect on the radio, since the radio reacts to RF (radio frequency) or *ac* signals from a transmitting station.

14-4.1 Effect of Angle on Induced Electromotive Force

Equation 14-3 gives the maximum induced voltage (V_m) when the motion of the conductor is at a *right angle* to the field. (See Fig. 14-8*a*.) As you have seen in Example 14-4, no voltage is produced when the motion of the conductor is *parallel* to the field. (See Fig. 14-8*c*.) The voltage produced for an angle *between* these two extremes, as in Fig. 14-8*b*, is given by:

$$v_{ind} = V_m \sin \theta \qquad \text{volts} \qquad (14\text{-}5)$$

where: v_{ind} is the emf induced, in volts (V)

B is the flux density of the field, in teslas (T)

l is the active length of the conductor, in meters (m)

v is the velocity of the conductor, in meters per second (m/s)

θ is the angle between the magnetic field and the direction of motion of the conductor

$\sin \theta$ is a trigonometric ratio, between 0 and 1 (see Appendix F)

$V_m = Blv$ is the maximum voltage that can be induced in volts (V)

Appendix E provides additional explanation of this topic.

EXAMPLE 14-5

a. What is the maximum voltage that can be induced in a 10-cm-length conductor moving at 20 m/s through a magnetic field of flux density 0.7 T?

b. What voltages are induced in this conductor when moving at angles of 60°, 30°, and 10° through the field?

Solution

a. $V_m = Blv$ (14-3)

   $= 0.7 \text{ Wb/m}^2 \times 0.1 \text{ m} \times 20 \text{ m/s}$

   $= \textbf{1.4 V}$

b. $v_{ind} = V_m \sin \theta$ (14-5)

   $\theta = 60°, v_{ind} = 1.4 \text{ V} \times \sin 60°$

   $= 1.4 \text{ V} \times 0.866 = \textbf{1.21 V}$

   $\theta = 30°, v_{ind} = 1.4 \text{ V} \times \sin 30°$

   $= 1.4 \text{ V} \times 0.5 = \textbf{0.7 V}$

   $\theta = 10°, v_{ind} = 1.4 \text{ V} \times \sin 10°$

   $= 1.4 \text{ V} \times 0.174 = \textbf{0.24 V}$

The variation of induced voltage as the angle changes in Example 14-5 occurs continually in alternating current

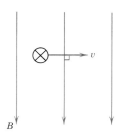

(*a*) Conductor moves at 90° through the field. Maximum voltage induced, V_m

(*b*) Conductor moves at angle Θ through the field. Less than maximum voltage induced, $V_m \sin \Theta$

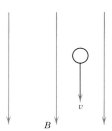

(*c*) Conductor moves at 0° through the field. Zero voltage induced

FIGURE 14-8

Effect of direction angle on induced voltage.

generators. That is, the constantly *changing* angle θ (in rotary motion) is responsible for generating a sinusoidal alternating current.

14-5 ELEMENTARY AC GENERATOR

To generate an emf continually, it is simpler to *rotate* a conductor in a magnetic field than to move it back and forth. The arrangement of the rotating conductor and the fixed magnetic field is illustrated in Fig. 14-9.

Mechanical power is used to rotate a coil in the magnetic field, inducing an emf in the series conductors A and B. This voltage is made available for external use by two brushes riding on slip rings that rotate with the coil. The brushes are connected to an external circuit and the load.

Some of the angular positions assumed by conductors A and B as the coil makes one complete revolution are shown in Fig. 14-10. The induced voltage in a given conductor varies instantaneously with the changing angle between the direction of motion and the magnetic field, as described in Section 14-4.1. Thus, when θ = 0° in Fig. 14-10a, no voltage is induced, since the conductors are not cutting any lines in the magnetic field. When θ = 90°, the induced voltage is at its maximum, since the maximum *rate* of cutting is taking place. (See Fig. 14-10c.)

After half a revolution (Fig. 14-10e), the induced voltage again drops to zero. Voltage is again being induced

when θ = 225°, as shown in Fig. 14-10f. Note, however, that the direction of induced current has *reversed*, compared to the direction during the first half of the revolution. The voltage generated between the two brushes (shown in Fig. 14-9) thus *alternates* and follows a *sinusoidal* variation given by $v_{ind} = V_m \sin \theta$.

14-6 THE SINE WAVE

The shape of the voltage waveform produced by the ac generator may be obtained by graphing the values of v_{ind} against selected values of θ in the equation

$$v_{ind} = V_m \sin \theta \qquad (14-5)$$

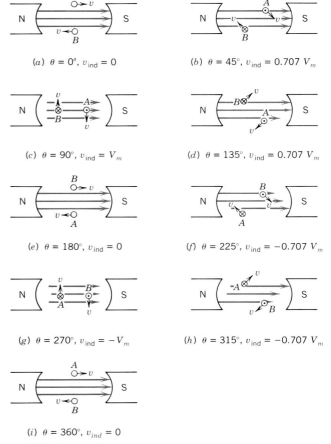

(a) θ = 0°, v_{ind} = 0

(b) θ = 45°, v_{ind} = 0.707 V_m

(c) θ = 90°, v_{ind} = V_m

(d) θ = 135°, v_{ind} = 0.707 V_m

(e) θ = 180°, v_{ind} = 0

(f) θ = 225°, v_{ind} = −0.707 V_m

(g) θ = 270°, v_{ind} = −V_m

(h) θ = 315°, v_{ind} = −0.707 V_m

(i) θ = 360°, v_{ind} = 0

FIGURE 14-10

Various angles and corresponding induced voltages during one revolution of a simple ac generator.

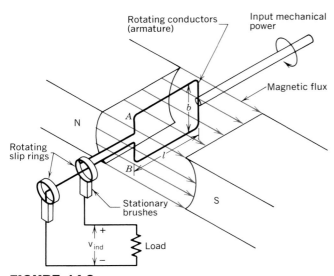

FIGURE 14-9

A simple ac generator.

The result is shown in Fig. 14-11 for a complete revolution from $\theta = 0°$ to $\theta = 360°$. This is called a *cycle,* since the variation in voltage (for every 360° revolution) will repeat itself. Note that the sine wave consists of four symmetrical segments, each like the first 90° interval.

For example, note that when $\theta = 135°$,

$$v_{ind} = V_m \sin 135° = 0.707 V_m$$

the same as when $\theta = 45°$. Also, when $\theta = 225°$,

$$v_{ind} = V_m \sin 225° = -0.707 V_m$$

or the *negative* of the value when $\theta = 45°$.

The maximum value of a sine wave (V_m) is known as the *amplitude* of the wave, and is often referred to as the *peak* value. In oscilloscope measurements, it is often more convenient to find the *peak-to-peak* value. As shown in Fig. 14-11, this is the *total voltage* from the positive peak to the negative peak. That is, for symmetrical waves such as a pure sine wave,

$$V_{p-p} = 2V_m \qquad \text{(14-6)}$$

EXAMPLE 14-6

The peak-to-peak value of a sine wave of voltage is 50 V. Determine:
a. The peak voltage or amplitude.
b. The instantaneous voltages at one-third, three-fifths, and one-and-a-quarter cycles.

Solution

a. $V_{p-p} = 2\,V_m \qquad \text{(14-6)}$

Therefore, $V_m = \dfrac{V_{p-p}}{2}$

$$= \frac{50 \text{ V}}{2} = 25 \text{ V}$$

b. At $\frac{1}{3}$ of a cycle, $\theta = \dfrac{360°}{3} = 120°$

$$v = V_m \sin \theta \qquad \text{(14-5)}$$
$$= 25 \text{ V} \times \sin 120°$$
$$= 25 \text{ V} \times 0.866 = 21.7 \text{ V}$$

At $\frac{3}{5}$ of a cycle, $\theta = 360° \times \dfrac{3}{5} = 216°$.

$$v = V_m \sin \theta \qquad \text{(14-5)}$$
$$= 25 \text{ V} \times \sin 216°$$
$$= 25 \text{ V} \times (-0.588) = -14.7 \text{ V}$$

At $1\frac{1}{4}$ cycles, $\theta = 360° \times 1\frac{1}{4} = 450°$.

$$v = V_m \sin \theta \qquad \text{(14-5)}$$
$$= 25 \text{ V} \times \sin 450°$$
$$= 25 \text{ V} \times 1 = 25 \text{ V}$$

NOTE One-and-a-quarter cycles corresponds to a quarter of a cycle or 90° where $v = V_m$. Note also, that the use of a scientific calculator permits the sine of any angle to be obtained directly. It is not necessary to convert the angle to one that lies in the first quadrant (0 − 90°) unless you are using a set of trigonometric tables that are so limited.

14-7 FACTORS AFFECTING THE AMPLITUDE OF A GENERATOR'S SINE WAVE

The faster the generator's coil rotates in the magnetic field, the higher will be the induced voltage. Other factors af-

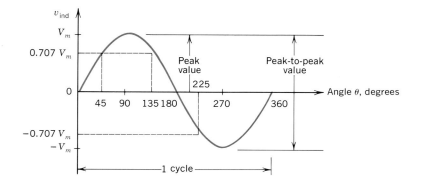

FIGURE 14-11
A sine wave of voltage.

fecting amplitude (peak value) of the wave are considered in Appendix G. The following equation results:

$$V_m = 2\pi nBAN \qquad \text{volts} \qquad (14\text{-}7)$$

where: V_m is the peak voltage generated, in volts (V)

n is the speed of rotation of the coil, in revolutions per second (rps)

B is the flux density of the magnetic field, in teslas (T)

A is the area of the rotating coil in square meters (m^2)

N is the number of turns in the rotating coil

For any given machine, A and N are fixed. B and n, however, may be varied to control the frequency and output voltage.

EXAMPLE 14-7

An ac generator runs at 1200 rpm and uses a flux density of 0.6 T. The length of the rotating coil is 40 cm with a radius of 10 cm. If the coil has 50 turns, calculate the amplitude of the voltage produced.

Solution

$$n = 1200 \text{ rpm} = \frac{1200}{60} \text{ rps} = 20 \text{ rps}$$

$B = 0.6$ T
$A = \text{length} \times \text{diameter}$
$\quad = 40 \times 10^{-2} \times 2 \times 10 \times 10^{-2} \text{m}^2 = 8 \times 10^{-2} \text{ m}^2$
$N = 50$
$V_m = 2\pi nBAN \qquad\qquad\qquad\qquad\qquad (14\text{-}7)$
$\quad = 2\pi \times 20 \text{ rps} \times 0.6 \text{ T} \times 8 \times 10^{-2} \text{ m}^2 \times 50$
$\quad = \textbf{301.6 V}$

14-8 FACTORS AFFECTING THE FREQUENCY OF A GENERATOR'S SINE WAVE

The simple ac generator shown in Fig. 14-9 had one pair of poles, and required one full mechanical revolution of the coil to induce one cycle of voltage. (See Fig. 14-12a.) Thus, if the coil is rotated at a steady n revolutions per second, that same number of complete cycles of voltage will be produced every second. The number of cycles of voltage produced in one second is called the *frequency* (*f*). If the coil completes 60 revolutions in a second, a voltage with a frequency of 60 cycles per second is generated. The unit of 1 cycle per second (1 cps) is called 1 hertz (Hz). Thus, 60 cps = 60 Hz.

Figure 14-12b shows the addition of a second pair of poles to the generator. Assuming that the coil rotates at the

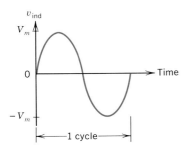

(a) One pair of poles produces 1 cycle in 1 mechanical revolution

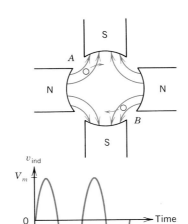

(b) Two pairs of poles produce 2 cycles in 1 mechanical revolution

FIGURE 14-12
Effect of the number of pairs of poles on the frequency of a generator running at a fixed speed.

same speed as before, it will generate a complete cycle of voltage every *half* revolution. That is, a conductor cuts the lines of force between a north and a south pole during that half revolution, producing a complete cycle of voltage. At the end of one full *mechanical* revolution (360°), *two* complete cycles of voltage have been produced, amounting to 720° *electrical* degrees. If the coil rotates at 60 rps, this generates a voltage with a frequency of 120 Hz. (See Fig. 14-12.)

Clearly then, combining the effects of speed and magnetic poles,

$$f = pn \quad \text{hertz} \quad (14\text{-}8)$$

where: f is the frequency of the induced emf, in hertz (Hz)
p is the number of *pairs* of poles
n is the speed of rotation, in revolutions per second (rps)

EXAMPLE 14-8

At what speed in rpm must the rotor of a water turbine turn if it is to produce voltage at a frequency of 60 Hz in a machine having a total number of 30 poles?

Solution

Number of *pairs* of poles, $p = \dfrac{30}{2} = 15$.

$$n = \frac{f}{p} \quad (14\text{-}8)$$

$$= \frac{60 \text{ Hz}}{15} = 4 \text{ rps} = \textbf{240 rpm}$$

As the frequency of a voltage increases, the time needed to generate one complete cycle decreases. The time needed to generate one cycle is called the *period* of the wave (T). The period is the reciprocal of the frequency (f), or

$$T = \frac{1}{f} \quad \text{seconds} \quad (14\text{-}9)$$

and

$$f = \frac{1}{T} \quad \text{hertz (Hz or s}^{-1}) \quad (14\text{-}9a)$$

where: T is the period of the wave, in seconds (s)
f is the frequency of the wave, in hertz (Hz)

EXAMPLE 14-9

a. What is the period of a 60-Hz voltage?
b. What is the frequency of a waveform that has a period of 1 ms?

Solution

a. $T = \dfrac{1}{f}$ $\quad (14\text{-}9)$

$= \dfrac{1}{60 \text{ Hz}} = 0.01667 \text{ s} = \textbf{16.7 ms}$

b. $f = \dfrac{1}{T}$ $\quad (14\text{-}9a)$

$= \dfrac{1}{1 \times 10^{-3} \text{ s}} = 1 \times 10^3 \text{ Hz} = \textbf{1 kHz}$

14-9 PRACTICAL ALTERNATORS

The voltage developed by the simple ac generator is called a *single-phase* output. Most practical ac generators, or alternators (including those in modern automobiles), produce *three-phase* ac. They use a rotating magnetic field *(rotor)* to induce voltages in three sets of windings on a stationary *stator*.

Direct current is used to excite the rotor's field windings (turning it into an electromagnet). This current enters the rotor through slip rings on its shaft that ride against two stationary brushes. (See Fig. 14-13a.)

The armature windings are spaced around the stator's surface so that the emfs generated in them reach their peak value 120° apart, as shown in Fig. 14-13b. Large alternators can produce voltages as high as 18,000 V.

As you have learned, the speed of coil (or rotor) rotation determines both the amplitude and the frequency of the generated voltage. Commercial power stations maintain a frequency of 60 Hz within ±0.02 Hz by carefully controlling rotational speed. How much voltage is produced is controlled by varying the direct current to the rotor's field windings, and thus, the field strength (B). When power output must be reduced because the electrical demand cannot be met, rotational speed is held constant to maintain the proper frequency, but the field strength is reduced to produce a lower voltage. This is often referred to as a ''brown-out.''

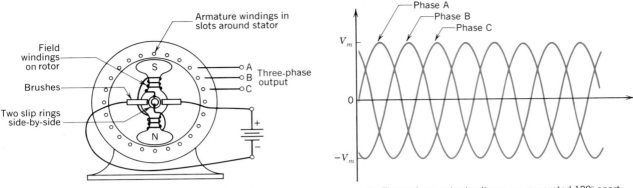

(a) Cross-sectional sketch of a three-phase alternator

(b) Three-phase output voltages are generated 120° apart

FIGURE 14-13

Practical three-phase alternator and its output waveforms.

14-9.1 Voltage Regulator for an Automobile Alternator

The alternator in an automobile experiences wide variations of speed, from a few hundred to several thousand rpm. In this application, the frequency is not important, since the alternator's output is rectified by diodes to produce direct current. But a large variation in output *voltage* cannot be tolerated. The alternator's voltage output is therefore controlled by a *voltage regulator* that varies the amount of field current. A voltage regulator that uses relays (solid-state versions are also used) is shown in Fig. 14-14.

When the ignition switch is turned on, current flows from the battery, through the red warning light and the regulator relay's normally closed contact, to the field

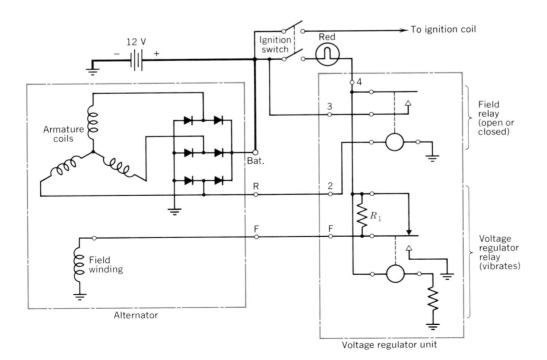

FIGURE 14-14

Schematic of a voltage regulator circuit for an automotive alternator system.

winding of the alternator. Once the engine starts, the alternator turns and generates a voltage at terminal R. This energizes the field relay of the voltage regulator, closing the normally open contact, and providing an alternate electrical path for current to the field winding. The red warning light goes out. At idle speeds, approximately 14.1 V is generated at the battery terminal.

If the alternator's output increases due to a higher speed, the *voltage regulator* relay is partially energized, opening its normally closed contact. Current to the field winding of the alternator must now pass through resistor R_1. This causes a drop in field strength and output voltage, and the relay contact will tend to close again. The regulator relay, in effect, *vibrates* open and closed as it attempts to maintain the *average* output of the alternator at close to 14.1 V.

If the speed and output voltage of the alternator increase *far* above the required value, the relay's arm is pulled all the way down to the ground contact. This shunts the field current to ground, causing a sudden drop in output voltage. (Other features of the voltage regulator circuit are considered in the Review Questions at the end of this chapter.)

14-9.2 Delta-Wye Connections

The advantage of a three-phase system is its ability to transmit the *same* power as a single-phase system while using less copper in the conductors. Three-phase *motors* run more smoothly than single-phase motors, and are smaller and more efficient for the same capacity. Three-phase *supplies* may use either a delta or a wye connection. (See Fig. 14-15.)

The delta connection is a three-wire system often used in power applications such as three-phase motors. Line-to-

line voltages of 220 V, 440 V, 550 V, and so on are typical.

The wye connection is a four-wire, three-phase system with a neutral that is usually grounded. This allows the operation of single-phase equipment (such as receptacles, lights, or single-phase motors) between any phase and the neutral, where 120 V exists. Three-phase motors connected directly to the three lines operate at 208 V. Note that the line-to-line voltage equals $\sqrt{3}$ × line-to-neutral voltage.

14-10 ELEMENTARY DC GENERATOR

By modifying the brush and slip-ring arrangement of the simple one-coil alternator, the negative half-cycles can be "inverted." This produces a unidirectional but pulsating dc output, as shown in Fig. 14-16b.

The *split* ring functions as a simple *commutator*. (See Fig. 14-16a.) The commutator "switches" the armature coil connections to the external circuit at the instant the induced emf in the coil reverses. This maintains the polarity of one brush at a positive potential with respect to the other (even though the instantaneous value still varies between 0 and V_m).

A dc voltmeter connected across the output of the generator indicates the *average* value of the wave. Mathematically, this can be shown to be

$$V_{dc} = \frac{2V_m}{\pi} \approx 0.637 V_m \qquad \text{volts} \qquad (14\text{-}9)$$

where: V_{dc} is the dc or average emf, in volts (V)
  V_m is the peak induced emf, in volts (V)

For example, if a one-coil dc generator produced a peak voltage of 20 V, its dc output would be 12.74 V.

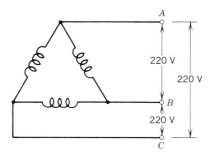

(a) 220 V, three-phase, three-wire delta (Δ)

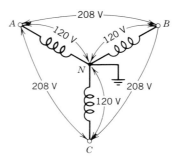

(b) 120/208 V, three-phase, four-wire wye (Y)

FIGURE 14-15
Coil connections in three-phase delta and wye systems.

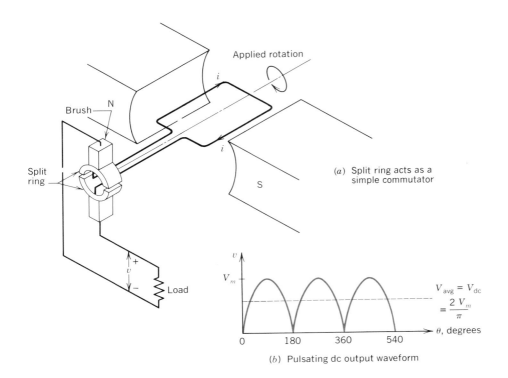

(a) Split ring acts as a simple commutator

(b) Pulsating dc output waveform

FIGURE 14-16
Elementary dc generator and output voltage waveform.

14-11 PRACTICAL DC GENERATORS

In a practical dc generator, there are a large number of coils evenly distributed around the surface of the armature and connected to an equally large number of commutator segments. This arrangement ensures that the output voltage developed across the brushes is nearly always at some peak value. The result is a small "ripple" voltage riding on a dc level, as shown in Fig. 14-17.

It is interesting to note that the dc generator produces an output that is practically the same *shape* as the full-wave rectified output of a three-phase alternator. The three-phase alternator is widely used in automobile electri-

cal systems. It is superior to the dc generator in that application because it can deliver a higher voltage at lower speeds. Even at idle speeds, most automobile alternators can supply the electrical load while continuing to recharge the battery used for starting.

Most dc generators use field windings, rather than a permanent magnet, to create the magnetic field. These electromagnets may be *separately* excited from an outside source, or *self*-excited. Self-excited field windings use the output of the generator itself to supply the exciting current. Residual magnetism in the windings is relied upon to induce a small voltage, which increases the field current and thus the voltage. The field current "builds up" until the generator is operating at normal values. For this cumulative action to take place, the field must be properly polarized so that magnetic flux resulting from the small initial voltage *adds to* the residual flux, rather than *opposing* it.

Like the dc motors discussed in Section 11-13, dc generators may be of the series, shunt, or compound type. In fact, a dc machine can be run as either a motor *or* a generator. However, they are usually optimized to run as one or the other. The output of a dc generator, like that of an alternator, may be controlled easily by means of a rheostat in the field winding circuit.

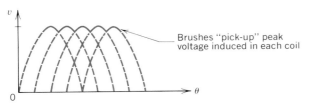

FIGURE 14-17
Output voltage of a practical dc generator.

SUMMARY

1. Electromagnetic induction is the process of generating a voltage in a wire or coil located in a changing magnetic field.

2. Lenz's law: The polarity of the induced emf is such that it opposes the *change* that produced it.

3. Faraday's law: When magnetic flux is changing at a rate given by $\dfrac{\Delta\Phi}{\Delta t}$ Wb/s, the emf induced in N turns is

$$v_{\text{ind}} = N\left(\frac{\Delta\Phi}{\Delta t}\right)$$

4. When flux is increasing with time, the quantity $\dfrac{\Delta\Phi}{\Delta t}$ is positive. When the flux decreases, $\dfrac{\Delta\Phi}{\Delta t}$ is negative, inducing a voltage of opposite polarity.

5. When a coil of N turns, each having an effective length l is moved through a magnetic field B at a velocity v, the total induced emf is given by $v_{\text{ind}} = NBlv$.

6. If a conductor is moved through a magnetic field at an angle θ with the field, the induced emf is $v_{\text{ind}} = V_m \sin \theta$.

7. The sine of an angle is a trigonometric ratio that has a fixed value for every angle. In a right-triangle, for example, $\sin \theta$ is given by:

$$\sin \theta = \frac{\text{opposite side}}{\text{hypotenuse}}$$

8. Maximum voltage (V_m) is induced when a conductor cuts the field at a right angle, since $\theta = 90°$ and $\sin 90° = 1$.

9. A simple ac generator consists of a single coil rotating in a fixed magnetic field. The sinuosoidal output voltage from the generator is available at two stationary brushes rubbing against rotating slip rings on the coil shaft.

10. The sine wave output of an alternator is

$$v = V_m \sin \theta$$

where the amplitude V_m is referred to as the peak value and $2V_m = V_{\text{p-p}}$ is the peak-to-peak value.

11. The amplitude of an induced sine wave is determined by the speed of rotation (n), flux density (B), area of coil (A), and the number of turns (N) in the coil, as given by

$$V_m = 2\pi nBAN$$

12. The frequency of an induced sine wave is determined by speed of rotation (n) and the number of *pairs* of poles (p),

$$f = pn$$

13. A waveform goes through a complete cycle when it generates all of its possible values that are periodically repeated.

14. The *frequency* of a waveform is the number of cycles produced in one second. One cycle per second (cps) = 1 hertz (Hz).

15. The period (T) of a wave of frequency f is the time required for one complete cycle:

$$T = \frac{1}{f}$$

16. A practical alternator consists of a rotating magnetic field that induces emf in armature windings arranged around the stator. The frequency is determined by rotational speed and the amount of emf is determined in part by the field current.

17. Either delta- or wye-connected windings may be used to develop three-phase ac. In a wye connection, the line-to-line voltage is $\sqrt{3}$ times the line-to-neutral voltage.

18. In an elementary dc generator, the split ring acts as a commutator to invert the negative half cycle and produce a waveform that has an average or dc value given by

$$V_{dc} = \frac{2V_m}{\pi}$$

19. A practical dc generator has many coils on the armature connected to a like number of segments on the commutator. It produces an almost-constant voltage with a small "ac ripple."

SELF-EXAMINATION

Answer T or F or, in the case of multiple choice, a, b, c, or d
(Answers at back of book)

14-1. Electromagnetic induction always requires a conductor to be "cut" by a moving magnetic field. _____

14-2. A potential difference is produced in a wire if there is relative motion between the wire and the field such that a conductor cuts across the magnetic field. _____

14-3. Lenz's law determines the amount of induced voltage. _____

14-4. The induced current always sets up a flux that opposes the change that produced it. _____

14-5. If a coil is moved back and forth across the end of a permanent magnet, an alternating voltage is induced in the coil. _____

14-6. Faraday's law states that the induced emf is inversely proportional to the rate of change of flux. _____

14-7. A dynamic microphone is an application of Faraday's law. _____

14-8. When magnetic flux changes from 3 to 8 mWb in 5 ms, the rate of change of flux, $\frac{\Delta\Phi}{\Delta t}$, is
    a. 1 mWb/s
    b. 1 μWb/s
    c. 1 Wb/s
    d. 1×10^3 Wb/s

14-9. If a coil of 10 turns is subject to the flux change in Question 14-8, the emf induced in the coil is

    a.   10 mV
    b.   10 μV
    c.   10 V
    d.   0.1 V

14-10. The maximum voltage that can be induced in a 50-cm-length wire moving at 10 m/s through a field of 0.5 T is
    a.   25 V
    b.   2.5 V
    c.   250 V
    d.   250 mV

14-11. If the conductor in Question 14-10 is moving at an angle of 30° with the field, the induced voltage is
    a.   12.5 V
    b.   2.165 V
    c.   1.25 V
    d.   125 mV

14-12. The sine of 30° is one-half the sine of 60°. _____

14-13. In a simple ac generator with two poles, the induced current drops to zero every half a revolution. _____

14-14. The output voltage of an ac generator is a sine wave because the conductor moves in a circle in a uniform horizontal magnetic field. _____

14-15. If the peak-to-peak voltage of a sine wave is 15 V, the amplitude is 30 V. _____

14-16. If the amplitude of a sine wave is 10 V, the emf after one-sixth of a cycle is
    a.   8.66 V
    b.   5 V
    c.   10 V
    d.   7.07 V

14-17. All other factors remaining the same, the output voltage from an alternator coil of length 50 cm and diameter 20 cm is the same as a coil of length 40 cm and diameter 25 cm. _____

14-18. If the speed of an alternator is doubled, the output voltage will double, and so will its frequency. _____

14-19. If the flux density of an alternator is doubled, but the speed is cut in half, the frequency will be half of its original value and the output voltage doubled. _____

14-20. Increasing the number of pairs of poles in an alternator means that the speed can be reduced to produce the same frequency. _____

14-21. Five cycles are counted on an oscilloscope over a time span of 100 μs. The frequency of this waveform is
    a.   5 kHz
    b.   5 MHz
    c.   20 kHz
    d.   50 kHz

14-22. The period of the wave in Question 14-21 is
    a.   20 ms
    b.   20 μs
    c.   500 μs
    d.   500 ms

14-23. The period of a wave is 0.04 s. Its frequency is
  a. 2.5 Hz
  b. 25 Hz
  c. 2.5 kHz
  d. 40 Hz

14-24. In a practical alternator the voltage is always induced in the armature. _____

14-25. The output voltage of a commercial alternator is maintained at its correct value by making slight adjustments in the speed of rotation. _____

14-26. Where three-phase power is to supply both three-phase motors and lighting circuits a three-phase, four-wire wye connection is used. _____

14-27. In an elementary, single-coil dc generator with a peak value of 80 V, the dc output voltage is approximately:
  a. 25.5 V
  b. 51 V
  c. 80 V
  d. 56.6 V

14-28. A practical, self-excited dc generator relies upon residual magnetism to build up. _____

REVIEW QUESTIONS

  1. What are the required conditions for electromagnetic induction to take place?
  2. Give three practical applications where electromagnetic induction is put to use.
  3. What is the origin of the emf induced in a wire that is being moved across a magnetic field?
  4. State Lenz's law in your own words.
  5. a. Give two examples of where Faraday's law is at work in an automobile.
    b. What are the fundamental units of $\frac{\Delta\Phi}{\Delta t}$?
    c. How is the weber defined?
  6. a. If the voltage induced in a conductor is constant, how is the magnetic field changing?
    b. If the voltage induced in a conductor is uniformly increasing, how is the magnetic field changing?
  7. How can an alternating current be produced using only a solenoid and a permanent magnet?
  8. Under what conditions does a magnetic field not have to "cut" through a coil or wire to induce an emf?
  9. Under what conditions does a conductor moving through a magnetic field induce:
    a. Maximum emf?
    b. Zero emf?
    c. 0.707 of maximum emf?
  10. In Example 14-4 the earth's magnetic field induced 340 μV in a single antenna. Would it be possible to use a number of vertical wires, series interconnected, to increase the total voltage to a higher value? Explain.

11. Why is the sine of 30° not one-half the sine of 60° or one-third the sine of 90°?

12. a. Sketch a right-angled triangle and show why the tangent of a 45° angle is 1. See Appendix F.
    b. What is the name of this triangle?
    c. Why is the tangent of 90° infinite?

13. a. Sketch a right triangle with a 30° angle. Why is the sine of 30° equal to the cosine of 60°?
    b. What is the *exact* value of the sine of 45°?

14. Refer to Fig. 14-9. If conductors *A* and *B* in the armature coil were held stationary, would the same emf be induced between the brushes if the external permanent magnets were rotated around the coil instead?

15. Refer to Fig. 14-9. If the coil were rotated in the opposite direction from that shown how, if at all, would the output waveform of voltage be different? Explain.

16. a. Why is the output of an alternator sinusoidal?
    b. Could we make it have any other shape if we wanted to? How?
    c. Why do we choose a sine wave? Why not a triangular shape or semicircular shape?

17. Once an alternator is built, what factors affect the following?
    a. The output voltage's amplitude.
    b. The output voltage's frequency.
    How could you double the output voltage without affecting the frequency?

18. a. What is the relationship between the frequency and period for any alternating wave?
    b. Does a change in amplitude change the period?

19. What are the major differences between a practical alternator and a simple single-coil ac generator?

20. a. What is meant by a "brownout"?
    b. How is this brought about?
    c. How does this affect electrical appliances?
    d. Why must the frequency be maintained at 60 Hz, even in a brownout?

21. Refer to Fig. 14-14.
    a. If, while the car is running, the alternator fails to develop an output (due to a broken belt perhaps), what visual indication will be noticed?
    b. Describe the electrical sequence that takes place following this failure.
    c. Does current continue to flow into the field winding?
    d. Can the battery discharge through the alternator armature coils?

22. If, after the car engine has been switched off, the field relay remains closed due to a malfunction, what possible effect would be noticed if the car is not used for several days? Explain.

23. What is an advantage of a three-phase, four-wire system over a single-phase system?

24. What is the purpose of a commutator in a dc generator?

25. How does a practical dc generator prevent the output voltage from falling to zero periodically as it does in a single-coil elementary generator?

26. Older automobiles that use a dc generator often had to have the field of the generator "polarized" before it would build up properly.
    a. What is meant by this?
    b. Why was it necessary?

PROBLEMS

(Answers to odd-numbered problems at back of book)

14-1. Refer to Fig. 14-1*a*. What is the polarity of the induced emf at *A* with respect to *B* when
   a. The field is moved to the right?
   b. The field is moved to the left?

14-2. Repeat Problem 14-1 with the magnetic polarity reversed.

14-3. Refer to Fig. 14-3. What is the polarity of the induced emf at *A* with respect to *B* when
   a. The south pole is moved into the coil?
   b. The north pole is moved into the coil?

14-4. Repeat Problem 14-3 with the windings wrapped around the solenoid in the opposite direction.

14-5. Refer to Fig. 14-18. What is the polarity of the induced emf at *A* with respect to *B* when
   a. The switch is first closed?
   b. The switch is just opened?

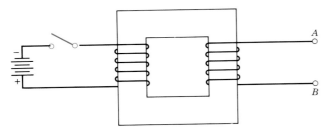

FIGURE 14-18
Circuit for Problems 14-5 and 14-6.

14-6. Repeat Problem 14-5 with the winding connected to the battery wound around the core in the opposite direction.

14-7. What is the rate of change of flux in Wb/s for the following?
   a. 2 Wb in 10 min
   b. 5 mWb in 10 ms
   c. 4.5 μWb in 5 μs
   d. 8 mWb in 2 μs

14-8. What emf is induced in a 10-turn coil subjected to the rate of change of flux in Problem 14-7?

14-9. At what rate is the flux changing in a 100-turn coil if 0.5 V is induced in the coil?

14-10. How many turns are there in a coil that has an emf of 80 mV induced when the flux changes from 28 to 30 mWb in one-tenth of a second?

14-11. A wire is moved across the face of a permanent magnet that is 3 cm square, in 0.1 s. If an emf of 9 mV is induced in the wire, what is the flux density of the magnetic field?

14-12. A wire is moved across the face of a permanent magnet that is 4 cm square and 0.8 T in strength. Calculate the voltage induced in the wire if it is moved perpendicular to the field in:

a. 150 ms

b. 5 ms

14-13. Refer to Fig. 14-5. The magnetic flux in the core is increased uniformly from 2 to 8 mWb in 4 ms, held constant for 3 ms, increased again to 10 mWb in 1 ms, then suddenly reduced uniformly to zero in 2 ms. Coil *B* has 50 turns. Draw graphs showing:

a. How the flux changes with time.

b. The rate of change of flux with time.

c. The voltage induced in coil *B*.

14-14. Refer to Fig. 14-5. The flux density in the core is increased uniformly from 0.7 to 1 T in 4 ms, held constant for 2 ms, reversed to 0.6 T in the opposite direction in 8 ms, then reduced to zero in 3 ms. If coil *B* has 40 turns and the toroid has a circular cross section of radius 2 cm, draw graphs showing:

a. How the flux changes with time.

b. The rate of change of flux with time.

c. The emf induced in coil *B*.

14-15. Five series-connected conductors, each 25-cm long, are moved at 8 m/s perpendicular to a magnetic field of 1.5 T. What is the total induced emf?

14-16. At what velocity must a 40 cm-long conductor be moved perpendicular to a magnetic field of 1.2 T to induce an emf of 5 V?

14-17. What emf will be induced in the five conductors in Problem 14-15 if they move through the field at the following angles with the field?

a. 30°

b. 45°

c. 80°

d. 90°

14-18. At what angle is the conductor in Problem 14-16 moving through the field if the induced emf is as follows?

a. 2.5 V

b. 3.535 V

c. 4.0 V

14-19. The peak-to-peak value of a sine wave is 25 V. Determine:

a. The amplitude.

b. The instantaneous voltages at one-quarter, four-fifths, and two-and-one-third cycles.

14-20. If the peak value of a sine wave is 20 V, what are the instantaneous voltages at the following angles:

a. 60°

b. 110°

c. 270°

d. 345°

14-21. The armature of an ac generator, running at 3600 rpm, has a length of 50 cm, a diameter of 25 cm, and has 20 turns. If the field strength is 0.5 T, calculate the amplitude of the induced voltage.

14-22. What must be the area of an armature coil rotating at 1500 rpm in a field of 0.8 T if the coil has 30 turns and the induced voltage has a peak-to-peak value of 150 V?

14-23. a. If the generator in Problem 14-21 has one pair of poles, what is the frequency of the generated voltage?

b. What is its period?

14-24. a. What is the total number of poles required by the generator in Problem 14-22 if it is to generate a voltage having a frequency of 50 Hz?

b. What is the period of this waveform?

14-25. Consider a three-phase, four-wire supply.

a. If the line-to-line voltage is 10 V, what is the line-to-neutral voltage?

b. What must be the line-to-line voltage of a system if the line-to-neutral voltage is 115 V?

14-26. a. An elementary single-coil dc generator produces a peak voltage of 40 V. What is its dc value?

b. If a single-coil dc generator is used to produce an output voltage of 14.5 V dc, what is the peak value?

CHAPTER 15

ALTERNATING VOLTAGE AND CURRENT

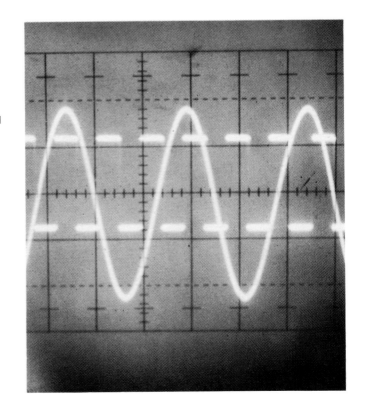

You have seen how the frequency of a sine wave of voltage may be expressed in cycles per second, or *hertz.* In this chapter, you will be introduced to the *radian,* rather than the degree, as a measure of an angle. This allows the frequency to be given in radians per second, and is known as the *angular frequency,* ω. A sinusoidal voltage (or current) may then be expressed in terms of the amplitude, frequency, and time.

One of the questions that will be resolved in this chapter is the relationship between an oscilloscope display of a sinusoidal voltage and the single reading of a voltmeter. This requires consideration of how current varies in a resistor with a sinusoidal applied voltage and the corresponding variation of power. The *average* power dissipated in the resistor is equated to the *same* amount of power produced by a direct current flowing through the same resistance. The amount of dc that produces the same heating effect as the ac is called the *effective* value of the ac. Because of the steps followed in determining the effective value, it is often called the *rms* (root-mean-square) value.

Finally you will see that Ohm's law, and the power equations developed for dc circuits, can be used for an ac circuit *if* rms values of current and voltage are used, and only *pure* resistance is considered in any calculation.

15-1 INDUCED VOLTAGE IN TERMS OF FREQUENCY AND TIME

As you have seen, a sine wave of voltage can be expressed by the equation

$$v = V_m \sin \theta \qquad (15\text{-}1)$$

The angle θ can be expressed in terms of the frequency (f) and the instant of time (t) for which you wish to know the voltage. (See Appendix H.) The result is

$$\theta = 2\pi ft \qquad \text{radians} \qquad (15\text{-}2)$$

where: θ is the angle, in radians (rad)

f is the frequency of the voltage, in hertz (Hz)

t is the time since the voltage was zero, in seconds (s)

The 2π term is a result of expressing the angle in radians, rather than degrees. The radian is commonly used in engineering work. As shown in Fig. 15-1, a radian is the angle between two radii of a circle that cut off (on the circumference of the circle) an arc equal in length to the circle's radius.

The number of radians in a complete circle is therefore equal to the number of times the *radius* is contained in the *circumference* of the circle. The circumference equals $2\pi r$. This means that 2π radians equals 360°.

$$2\pi \text{ radians } = 360°$$
$$\pi \text{ radians } = 180°$$
$$\frac{\pi}{2} \text{ radians } = 90°$$
$$1 \text{ radian } = \frac{180°}{\pi} \approx 57.3°$$

The way in which a sine wave of voltage varies with the angle given in radians is shown in Fig. 15-2.

An advantage of using radians is that they are *dimensionless*, being the ratio of two lengths (arc length to radius).

Thus the induced voltage can be written in the following form:

$$v = V_m \sin 2\pi ft \qquad \text{volts} \qquad (15\text{-}3)$$

or

$$v = V_m \sin \omega t \qquad \text{volts} \qquad (15\text{-}4)$$

and

$$\omega = 2\pi f \qquad \text{radians per second} \qquad (15\text{-}5)$$

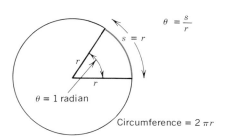

FIGURE 15-1

One radian (approximately 57.3°) is the angle subtended at the center of a circle by an arc equal in length to the radius.

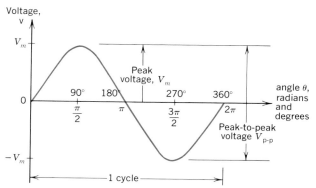

FIGURE 15-2

A sine wave of voltage with the angle given in radians and degrees.

where: v is the induced emf, in volts (V)

V_m is the maximum induced emf, in volts (V)

f is the frequency of the emf, in hertz (Hz)

t is the instant of time under consideration, in seconds (s)

$\omega = 2\pi f$ is the angular frequency, in radians per seconds (rad/s)

15-1.1 Angular Frequency

The Greek letter omega (ω) is referred to, in some texts, as the angular *velocity*. This term is appropriate to describe the speed of a rotating coil or conductor. But, as pointed out in Section 14-8, the frequency of the generated voltage depends not only on the speed of rotation, but on the number of pairs of magnetic poles the conductor passes under.

For example, a coil that makes a full 360° revolution in 1 s has an angular *velocity* of 2π rad/s. If the generator has one pair of poles, the generated voltage will have a frequency of 1 cycle per second (cps). Assuming the same velocity of 2π rad/s, a generator with *two* pairs of poles will generate a voltage with *frequency* of 2 cps. The angular *velocity* of the coil is 2π rad/s in both cases. But since a cycle of voltage is equivalent to 2π rad, the angular *frequency* of the generated voltage is 2π rad/s in the first case and 4π rad/s in the second case. That is, we are more concerned with the *frequency* of the *voltage* or *current* than we are with the speed or *velocity* of the conductor which generated that voltage or current.

To summarize: if f is the frequency of a voltage in cycles per second, $2\pi f$ is the frequency in radians per second. The term *angular* frequency (ω) is used for $2\pi f$, since an *angle* of 2π rad corresponds to one *cycle*.

Note that the dimension of ω is per second, or s^{-1}. This is because the radian is a *dimensionless* quantity. Thus ω has the same dimension as f, and it is consistent to refer to ω as frequency.

EXAMPLE 15-1

A sine wave of voltage has a peak value of 5 V and a frequency of 1 kHz. Determine:
a. The angular frequency.
b. The instantaneous voltage when $t = 400$ μs.

Solution

a. $\omega = 2\pi f$ (15-5)

$= 2\pi \times 1 \times 10^3$ rad/s

$= \mathbf{6.28 \times 10^3 \ rad/s}$

b. $v = V_m \sin \omega t$ (15-4)

$= 5\ \text{V} \sin 6.28 \times 10^3\ \dfrac{\text{rad}}{\text{s}} \times 400 \times 10^{-6}\ \text{s}$

$= 5\ \text{V} \times \sin 2.51\ \text{rad}$

$= 5\ \text{V} \times 0.59 = \mathbf{2.95\ V}$

Note that in Example 15-1, you must find the sine of an angle where the angle is given in radians. When doing this with a calculator, be careful to enter the angle in radians, not degrees. This can be done by means of a slide switch or by pushing a button, depending on the calculator.

EXAMPLE 15-2

The voltage applied to a circuit is given by $v = 170 \sin 377\ t$ volts. Determine:
a. The amplitude of the voltage.
b. The peak-to-peak voltage.
c. The angular frequency.
d. The frequency in hertz.
e. The period.

Solution

a. By comparison with Eq. 15-4:

$$v = V_m \sin \omega t$$
$$= 170 \sin 377t$$

The amplitude $V_m = \mathbf{170\ V}$.

b. $V_{p\text{-}p} = 2V_m$ (14-6)

$= 2 \times 170\ \text{V}$

$= \mathbf{340\ V}$

c. The angular frequency (ω) = $\mathbf{377\ rad/s}$

d. $\omega = 2\pi f$ (15-5)

$f = \dfrac{\omega}{2\pi}$

$= \dfrac{377}{2\pi} = \mathbf{60\ Hz}$

e. $T = \dfrac{1}{f}$ (14-9)

$= \dfrac{1}{60\ \text{Hz}} = 0.01667\ \text{s} = \mathbf{16.7\ ms}$

15-2 CURRENT AND VOLTAGE WAVEFORMS WITH A RESISTIVE LOAD

Consider the application of a sinusoidally alternating voltage to a resistor, as shown in Fig. 15-3a. What is the nature of the current variation in the circuit?

Given:
$$v = V_m \sin \omega t \qquad (15\text{-}4)$$

By Ohm's law,
$$i = \frac{v}{R} \qquad (3\text{-}1a)$$

That is, the instantaneous value of current is determined by the instantaneous value of the applied voltage and R. Note the use of *small* letters to represent quantities *varying with time*.

$$i = \frac{v}{R} = \frac{V_m}{R} \sin \omega t$$

Therefore, $i = I_m \sin \omega t$ **amperes**   (15-6)

and
$$I_m = \frac{V_m}{R} \qquad (15\text{-}7)$$

where: i is the instantaneous current, in amperes (A)
I_m is the peak current, in amperes (A)
V_m is the peak voltage, in volts (V)
R is the resistive load, in ohms (Ω)
$\omega = 2\pi f$ is the angular frequency, in radians/second (rad/s)
t is the time, in seconds (s)

Equation 15-6 shows that the current (i) varies at the same frequency as the applied voltage (v) and *in-phase*

with v. In other words, as shown in Fig. 15-3b, the current and voltage waveforms pass through their minimum and maximum values at the same instant. These waveforms may be observed on an oscilloscope to identify V_m and I_m. (See Chapter 16.)

EXAMPLE 15-3

A sinusoidal voltage of peak-to-peak value, 8 V, is observed on an oscilloscope when connected across a 2.2-kΩ resistor. The period of the voltage waveform is 2 ms. Determine:

a. The peak value of the current.
b. The frequency of the current.
c. The equation representing the current.

Solution

a. $V_m = \dfrac{V_{\text{p-p}}}{2}$ (14-6)

     $= \dfrac{8\text{ V}}{2} = 4\text{ V}$

     $I_m = \dfrac{V_m}{R}$ (15-7)

     $= \dfrac{4\text{ V}}{2.2\text{ k}\Omega} =$ **1.82 mA**

b. $f = \dfrac{1}{T}$ (14-9a)

     $= \dfrac{1}{2 \times 10^{-3}\text{ s}} =$ **500 Hz**

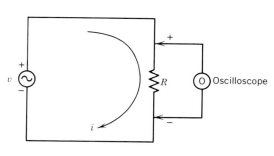

(a) Sinusoidal voltage applied to the resistor

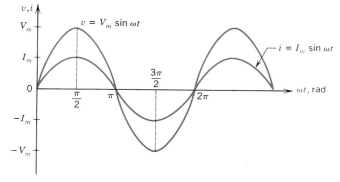

(b) Waveforms of voltage and current

FIGURE 15-3
Circuit and waveforms for a resistive load.

c. $\omega = 2\pi f$ (15-5)
$= 2\pi \times 500$
$= 1000\pi$ rad/s
Therefore, $i = I_m \sin \omega t$ (15-6)
$= 1.82 \sin 1000\pi t$ mA

NOTE The polarity signs shown in Fig. 15-3 establish a *reference* to interpret the positive and negative values in the graphs for v and i. Also, the graph for i has been drawn arbitrarily smaller than for v.

As you can see, the current variation in the circuit is the *same* as the voltage variation across a resistor that has the (same) current flowing through it. This is a *very important* principle. Although an oscilloscope reacts only to voltage, it can be used to show how current varies by displaying the voltage across a resistor in the circuit. And, if the *resistance* is known, the *current* is also known.

15-3 POWER IN A RESISTIVE LOAD

You know that in a dc circuit,

the average power $P = I^2 R$ (3-7)

In an ac circuit, where the instantaneous current (i) varies, the instantaneous power (p) must also vary. Thus

$$p = i^2 R$$

and since $i = I_m \sin \omega t$

$$p = (I_m \sin \omega t)^2 R$$
$$= I_m^2 R \sin^2 \omega t$$

Therefore, $p = P_m \sin^2 \omega t$ watts   (15-8)
and $P_m = I_m^2 R$ (15-9)

where: p is the instantaneous value of the power in the resistor, in watts (W)
P_m is the peak power, in watts (W)
I_m is the peak current, in amperes (A)
R is the resistive load, in ohms (Ω)
ω is the angular frequency, in radians/second (rad/s)
t is the time, in seconds (s)

The power waveform is shown in Fig. 15-4b. It evidently has a frequency twice that of the current waveform, varying from zero to a peak value (P_m) twice as often as the current.

However, even though the power pulsates, in a purely resistive circuit it is always *positive*. This simply means that an alternating current in a resistor dissipates power in the resistor (in the form of heat), no matter in which direction the current flows.

As you can see, by inspection of the symmetrical power curve in Fig. 15-4b, the *average* power (P_{av}) is one-half of the peak power. Thus

$$P_{av} = \frac{P_m}{2} = \frac{I_m^2 R}{2} \quad \text{watts} \quad (15\text{-}10)$$

where the symbols are as in Eqs. 15-8 and 15-9.

Consider a line drawn through the power curve at one-half its peak value. The *average* value of a curve is the point at which a horizontal line can be drawn so that it

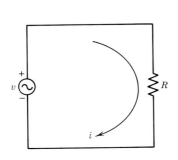

(a) Sinusoidal voltage applied to the resistor

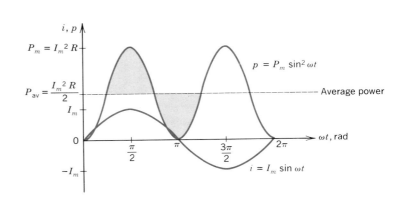

(b) Waveforms of current and power

FIGURE 15-4
Circuit and waveforms for power in a resistive load.

encloses the same area below it as enclosed by the original curve. In Fig. 15-4b, the shaded portion of the waveform *above* the average power line may be used to ''fill in'' the shaded portion *below* the line. This is shown in Example 15-4.

EXAMPLE 15-4

An ac voltage given by the equation $v = 50 \sin 800\pi t$ is applied to a 25-Ω resistor. Determine:
a. The peak power dissipated in the resistor.
b. The frequency of the power variation.
c. The average power dissipated in the resistor.

Solution

a. $I_m = \dfrac{V_m}{R}$ (15-7)

$= \dfrac{50\text{ V}}{25\ \Omega} = 2\text{ A}$

$P_m = I_m^2 R$ (15-9)

$= (2\text{ A})^2 \times 25\ \Omega = \textbf{100 W}$

b. $v = V_m \sin 2\pi f t$ (15-3)

$= 50 \sin 800\pi t$

$2\pi f = 800\ \pi$ rad/s

Therefore, $f = 400$ Hz

Power varies at a frequency of *2f* or **800 Hz.**

c. $P_{av} = \dfrac{P_m}{2}$ (15-10)

$= \dfrac{100\text{ W}}{2} = \textbf{50 W}$

Example 15-4 shows that a constant power level of 50 W provides the same heating effect as a power waveform varying sinusoidally between 0 and 100 W, no matter what the frequency.

That is, the *frequency* of an alternating current flowing through a resistor does not affect the *power* dissipated in that resistor. The power is determined only by the resistance value and the peak value of the current (or voltage).

15-3.1 Equations for Average Power

Consider the expression for the average power dissipated in a resistor by a symmetrical sine wave of current:

$$P_{av} = \frac{I_m^2 R}{2} \qquad (15\text{-}10)$$

The average power determines the amount of heat developed in a resistor, and is more important than the peak power. It is convenient, therefore, to calculate the average power in terms of the current indicated by an ammeter, rather than from an oscilloscope measurement. This is done as follows:

$$P_{av} = \frac{I_m^2 R}{2}$$

$$= \left(\frac{I_m}{\sqrt{2}}\right)^2 R$$

and $\qquad P_{av} = I^2 R \qquad$ (15-11)

where $I = \dfrac{I_m}{\sqrt{2}}$ is called the rms value of the current. This is the value that an ac ammeter, calibrated for sine waves, will indicate in a sinusoidal circuit. A more detailed explanation of the term *rms* will be provided in Section 15-4. It should be noted that the rms value is *not* an average value. (The rms value is sometimes called the *effective* value. The *average* value of a sine wave over a complete cycle is zero.)

It is important to realize that Eq. 15-11 can be used to calculate power in an ac circuit *only* if the whole circuit is purely resistive, or if the current through just the resistive portion is considered. To emphasize this, the notation I_R should be used. **Also, since you can obtain the rms value of voltage indicated by a voltmeter using** $V = \dfrac{V_m}{\sqrt{2}}$ **you can state the following equations for calculating power in a resistor:**

$$P = I_R^2 R \qquad \text{watts} \qquad (15\text{-}12)$$

$$P = \frac{V_R^2}{R} \qquad \text{watts} \qquad (15\text{-}13)$$

$$P = V_R I_R \qquad \text{watts} \qquad (15\text{-}14)$$

where: P is the average or rms* power, in watts (W)
I_R is the rms current in the resistor, in amperes (A)
V_R is the rms voltage across the resistor, in volts (V)
R is the resistance, in ohms (Ω)

---

*The term ''rms power'' is often used when referring to the output power that an audio amplifier can deliver to a loudspeaker. This indicates the *continuous* power that can be delivered, using rms values of current and voltage. Audio power ratings, in the past, were sometimes given in watts of ''music power'' and were generally much higher than the rms power ratings. This is because music has momentary peaks that an amplifier may be able to handle, but not on a sustained basis.

These power equations may be used with waveforms of any shape, regardless of frequency, including dc, so long as rms values are used.

Equations 15-12 through 15-14 imply that Ohm's law can be applied to an ac circuit. This is true only if you consider the current in the resistor and the voltage across the resistor. That is,

$$R = \frac{V_R}{I_R} \quad \text{ohms} \tag{15-15}$$

$$I_R = \frac{V_R}{R} \quad \text{amperes} \tag{15-15a}$$

$$V_R = I_R R \quad \text{volts} \tag{15-15b}$$

with the symbols as above.

EXAMPLE 15-5

A resistor R is connected in a complex ac circuit as shown in Fig. 15-5. An ac voltmeter across the resistor indicates 15 V and an ac ammeter in series with the resistor indicates 50 mA. Calculate:
a. The resistance of R.
b. The average power dissipated in R using Eqs. 15-12 through 15-14.

Solution

a. $R = \dfrac{V_R}{I_R}$ $\hspace{3cm}$ (15-15)

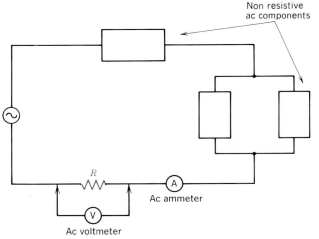

FIGURE 15-5
Circuit for Example 15-5.

$$= \frac{15 \text{ V}}{50 \times 10^{-3} \text{ A}} = \mathbf{300\ \Omega}$$

b. $P = I_R^2 R$ $\hspace{3cm}$ (15-12)

$$= (50 \times 10^{-3} \text{ A})^2 \times 300 \text{ }\Omega$$

$$= 0.75 \text{ W} = \mathbf{750 \text{ mW}}$$

$$P = \frac{V_R^2}{R} \tag{15-13}$$

$$= \frac{(15 \text{ V})^2}{300 \text{ }\Omega}$$

$$= 0.75 \text{ W} = \mathbf{750 \text{ mW}}$$

$$P = V_R I_R \tag{15-14}$$

$$= 15 \text{ V} \times 50 \times 10^{-3} \text{ A}$$

$$= 0.75 \text{ W} = \mathbf{750 \text{ mW}}$$

15-4   RMS VALUE OF A SINE WAVE

To understand the term *rms*, you have to return to the relationship between the readings of an ac voltmeter or ammeter and the instantaneous values of voltage and current in a complete cycle of a sine wave.

If an ac voltmeter is connected across a sine wave of voltage, what should its reading be? The meter can only indicate one value. Which of the values between $-V_m$ and $+V_m$ should this be? Similarly, which value should an ac ammeter indicate? (See Fig. 15-6.)

The answer lies in comparing the *average* heating effect of the ac with that produced by some steady dc value. More specifically, **the effective value of an alternating current or voltage is that value of *dc* which is *as effective* in producing *heat* in a given resistance as the given quantity of *ac* in the same resistance.**

Since heating is dependent upon *power*, the *average power* produced by the ac should be the same as that produced by the dc.

The power dissipated by a nonvarying current (I_{dc}) flowing through a resistance (R) is given by

$$P_{av} = P_{dc} = I_{dc}^2 R \tag{3-7}$$

The *average* power dissipated by an alternating current of peak value (I_m) flowing through the same resistance (R) is given by

$$P_{av} = \frac{I_m^2 R}{2} \tag{15-10}$$

Since these two must be equal,

$$I_{dc}^2 R = \frac{I_m^2 R}{2}$$

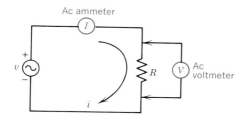

(a) Ac circuit with ammeter and voltmeter

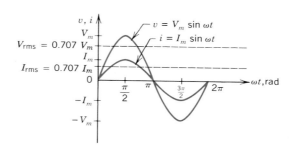

(b) Waveforms of current and voltage
with rms values

FIGURE 15-6
Readings of an ac ammeter and ac voltmeter.

Therefore, $I_{dc}^2 = \dfrac{I_m^2}{2}$

and $\qquad I_{dc} = \sqrt{\dfrac{I_m^2}{2}} = \dfrac{I_m}{\sqrt{2}} = 0.707I_m$

Thus an alternating current of peak value (I_m) is as effective in producing heat in a pure resistance as a dc current equal to $0.707I_m$ or $\dfrac{I_m}{\sqrt{2}}$. (See Fig. 15-7.)

Thus, $\qquad I_{eff} = \dfrac{I_m}{\sqrt{2}} = 0.707I_m$

Similarly, $V_{eff} = \dfrac{V_m}{\sqrt{2}} = 0.707V_m$

If you examine the steps taken to obtain this result, you will see that you first *squared* the current, then found the average or *mean* value of the power, and finally took the square *root*. This is summarized by referring to the effective value of the ac as I_{rms} (root-mean-square) or V_{rms}. The subscripts are used only for clarification.

Unless otherwise stated, all ac quantities are understood to be *rms* values. Thus:

$$I = I_{rms} = \dfrac{I_m}{\sqrt{2}} = 0.707I_m \qquad (15\text{-}16)$$

$$V = V_{rms} = \dfrac{V_m}{\sqrt{2}} = 0.707V_m \qquad (15\text{-}17)$$

where symbols are as previously described.

15-4.1 Nominal Line Voltages and Frequencies

The nominal ac line voltage for domestic use (lighting and wall receptacles) in the United States (including Puerto Rico) and Canada is 120 V at a frequency of 60 Hz. The actual voltage maintained by the local power company may be lower, such as 115 V or 117 V. Also, a 240-V service is usually available for appliances such as electric stoves, ovens, and air conditioners, which require higher power. (This will be covered in detail in Section 18-3.)

Other countries have voltages ranging from 100 to 240 V at frequencies of 50 or 60 Hz. Table 15-1 lists the nominal line voltages (rms) and frequencies for a number of countries.

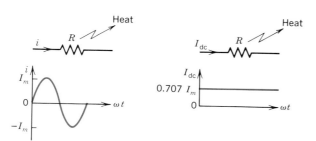

(a) Heat produced by alternating current of peak value I_m

(b) Same heat produced by direct current equal to $0.707\,I_m$

FIGURE 15-7
Effective or rms value of an alternating current.

EXAMPLE 15-6

The domestic ac line voltage, given by the equation $v = 170 \sin 377t$, is applied to a 33-Ω resistor as in Fig. 15-6a. Calculate:

a. The reading of an ac voltmeter connected across the resistor.

b. The reading of an ac ammeter connected in series with the resistor.

TABLE 15-1
Nominal AC Line Voltages and Frequencies of Selected Countries

Country	Volt.	Freq.	Country	Volt.	Freq.
Argentina	220	50	Japan	100	50/60
Australia	240	50	Mexico	127	60
Austria	220	50	Netherlands	220	50
Belgium	220	50	Norway	230	50
Brazil	127	60	South Africa	220	50
Canada	120	60	Spain	127/220	50
China	220	50	Sweden	220	50
Shanghai	110	50	Switzerland	220	50
Denmark	220	50	Taiwan	110	60
Finland	220	50	United Kingdom	240	50
France	127/220	50	U.S.A.	120	60
Germany	220	50	U.S.S.R.	127	50
Ireland	220	50	Venezuela	120	60
Italy	220	50			

Solution

a. $V_m = 170$ V
Voltmeter indicates

$$V = \frac{V_m}{\sqrt{2}} \qquad (15\text{-}17)$$

$$= \frac{170 \text{ V}}{\sqrt{2}} = \mathbf{120\ V}$$

b. $I_m = \dfrac{V_m}{R} \qquad (15\text{-}15a)$

$$= \frac{170 \text{ V}}{33 \ \Omega} = 5.2 \text{ A}$$

Ammeter indicates

$$I = \frac{I_m}{\sqrt{2}} \qquad (15\text{-}16)$$

$$= \frac{5.2 \text{ A}}{\sqrt{2}} = \mathbf{3.6\ A}$$

Alternatively, after obtaining the rms value of voltage, the rms value of current can be obtained directly from Ohm's law:

$$I_R = \frac{V_R}{R} \qquad (15\text{-}15a)$$

$$= \frac{120 \text{ V}}{33 \ \Omega} = \mathbf{3.6\ A}$$

Thus Ohm's law may be applied to an ac circuit containing *only* resistance, using oscilloscope *or* meter values for I and V.

As shown in Example 15-6, the 120-V, 60-Hz supply has a peak value of 170 V and a peak-to-peak value of 340 V. However, this ac waveform is *just as effective* in heating the filament of an incandescent lamp (and thus producing as much light) as 120-V dc.

15-4.2 True rms Values

It should be noted that ac ammeters and voltmeters (unless labeled "true rms") are usually calibrated to indicate rms values only when used to measure *sine* waves. If the meters are connected in a circuit where the waveform is *not* sinusoidal (for example, a square wave or a sawtooth wave), the readings will be erroneous, since the 0.707 factor applies *only* to sine waves. An rms value can be obtained for any waveform, but a different factor must be used for each type of wave. This is the reason why an oscilloscope is so important in electronic measurements. It displays much more information about a voltage or a current than does a single-reading meter. The oscilloscope is essential in dealing with nonsinusoidal circuits, such as pulse and digital circuits.

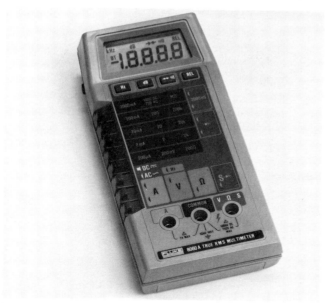

FIGURE 15-8

Hand-held $4\frac{1}{2}$-digit true rms multimeter. (Courtesy of the John Fluke Mfg. Co., Inc.)

Only a few instruments will show the true rms value regardless of waveform. They are called true-reading rms meters. (See Fig. 15-8.)

This hand-held $4\frac{1}{2}$-digit multimeter will read true rms ac voltages having frequencies up to 100 kHz and true rms ac currents up to 30 kHz, depending on the range in use. Accuracy is 0.5% of reading for line voltage frequencies and 0.04% for dc voltages, with an input resistance of 10 MΩ on all ranges. The meter can measure frequency up to 20 kHz with an accuracy of 0.005% and conductance from 2 mS to 2000 nS. Also included is a continuity checker that provides either a visual display or an audible tone when continuity exists between the two probes.

Equations 15-16 and 15-17 can be rearranged so that peak-to-peak oscilloscope readings can be obtained from sinusoidal rms meter values:

$$V_m = \sqrt{2} \times V_{rms} = 1.414V_{rms} = 1.414V \quad (15\text{-}18)$$

$$\text{and } V_{p-p} = 2V_m = 2\sqrt{2}V_{rms} = 2.828V \quad (15\text{-}19)$$

Similarly, $\quad\quad\quad I_m = 1.414I \quad\quad (15\text{-}20)$

and $\quad\quad\quad I_{p-p} = 2.828I \quad\quad (15\text{-}21)$

EXAMPLE 15-7

An ac voltmeter across a resistor indicates 3.5 V and an ammeter in series with the resistor reads 15 mA. Calculate:

  a. The peak-to-peak voltage across the resistor.
  b. The peak current through the resistor.
  c. The resistance of the resistor.

Solution

  a. $V_{p-p} = 2\sqrt{2}\,V$ (15-19)
       $= 2 \times \sqrt{2} \times 3.5\text{ V}$
       $= \textbf{9.9 V}$

  b. $I_m = \sqrt{2}\,I$ (15-20)
       $= \sqrt{2} \times 15\text{ mA}$
       $= \textbf{21.2 mA}$

  c. $R = \dfrac{V_R}{I_R}$ (15-15)

       $= \dfrac{3.5\text{ V}}{15 \times 10^{-3}\text{ A}} = \textbf{233 Ω}$

15-5 SERIES AND PARALLEL AC CIRCUITS WITH PURE RESISTANCE

The solution of series and parallel ac circuits that involve only resistance follows the same principles used for dc circuits. As noted previously, the same equations for power and Ohm's law can be applied if rms quantities are used throughout. In problems that contain data for peak or peak-to-peak values, care must be taken to make the appropriate conversions, so that a consistent set of units results.

The laws of voltage division and current division hold for ac as they do for dc, provided that only pure resistances are involved. Problems 15-19 and 15-20 at the end of the chapter are designed to give you practice in solving series- and parallel-resistive ac circuits with a mixed set of units. Examples 15-8 and 15-9 illustrate some of the principles.

EXAMPLE 15-8

Two series-connected resistors, 3.3 kΩ and 4.7 kΩ, have a 36-V peak-to-peak sinusoidal voltage connected to them as shown in Fig. 15-9. Determine:

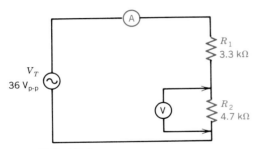

FIGURE 15-9
Circuit for Example 15-8.

a.   The reading of a series-connected ammeter.
b.   The voltage across the 4.7-kΩ resistor.
c.   The power dissipated in the 3.3-kΩ resistor.

Solution

a.   The rms applied voltage,

$$V_T = \frac{V_{p-p}}{2\sqrt{2}} \qquad (15\text{-}17)$$

$$= \frac{36}{2\sqrt{2}} V = 12.73\ V$$

Total circuit resistance,

$$R_T = R_1 + R_2 \qquad (5\text{-}2)$$
$$= 3.3\ k\Omega + 4.7\ k\Omega = 8\ k\Omega$$

The ammeter indicates

$$I = \frac{V_T}{R_T} \qquad (15\text{-}15a)$$

$$= \frac{12.73\ V}{8\ k\Omega} = \textbf{1.59 mA}$$

b.   Voltage across the 4.7-kΩ resistor,

$$V_2 = V \times \frac{R_2}{R_2 + R_1} \qquad (5\text{-}5)$$

$$= 12.73\ V \times \frac{4.7\ k\Omega}{4.7\ k\Omega + 3.3\ k\Omega}$$

$$= \textbf{7.48 V}$$

c.   Power dissipated in the 3.3-kΩ resistor,

$$P = I_R^2 R \qquad (15\text{-}12)$$
$$= (1.59 \times 10^{-3}\ A)^2 \times 3.3 \times 10^3\ \Omega$$
$$= \textbf{8.34 mW}$$

EXAMPLE 15-9

Consider $R_1 = 4.7\ k\Omega$ and $R_2 = 3.3\ k\Omega$ connected in parallel across a sinusoidal voltage source, as shown in Fig. 15-10. The resistors draw a total current of 50 mA. Calculate:

a.   The current in the 3.3-kΩ resistor.
b.   The peak-to-peak voltage indicated by an oscilloscope across the source.
c.   The total power dissipated in the two resistors.

Solution

a.   Current in the 3.3-kΩ resistor,

$$I_2 = I_T \times \frac{R_1}{R_1 + R_2} \qquad (6\text{-}13)$$

$$= 50\ mA \times \frac{4.7\ k\Omega}{4.7\ k\Omega + 3.3\ k\Omega}$$

$$= 50\ mA \times \frac{4.7}{8} = \textbf{29.4 mA}$$

b.   Voltage across the circuit,

$$V_R = I_R R \qquad (15\text{-}15b)$$
$$= 29.4 \times 10^{-3}\ A \times 3.3 \times 10^3\ \Omega$$
$$= 97\ V$$

Oscilloscope indication,

$$V_{p-p} = 2\sqrt{2}\ V \qquad (15\text{-}19)$$
$$= 2\sqrt{2} \times 97\ V$$
$$= \textbf{274 } V_{p-p}$$

(Note that it is not necessary to find the total parallel resistance.)

c.   Total power,

$$P = V_R I_T \qquad (15\text{-}14)$$
$$= 97\ V \times 50 \times 10^{-3}\ A$$
$$= \textbf{4.85 W}$$

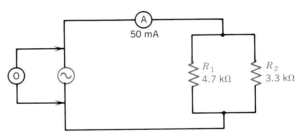

FIGURE 15-10
Circuit for Example 15-9.

SUMMARY

1. The equation of a sinusoidally varying voltage may be given in terms of frequency and time by $v = V_m \sin \omega t$, where $\omega = 2\pi f$ is the angular frequency in rad/s.
2. A sine wave of voltage applied to a resistor causes a sinusoidal current that is in phase with the voltage.
3. The power dissipated in a resistor varies between 0 and a peak value (P_m) at a frequency twice that of the applied voltage.
4. The average ac power dissipated in a resistor (with a sinusoidal voltage applied) is half the peak power, and is given by:

$$P_{av} = \frac{I_m^2 R}{2}$$

5. Ohm's law and the power equations apply equally well to a dc circuit and to an ac resistive circuit (if rms values are used and care is taken to ensure that voltages and currents are limited to the resistive portion of the circuit).
6. The rms value of an alternating current is that value of dc that is *as effective* in producing heat in a given resistance as the ac.
7. For sine waves only,

$$I_{rms} = \frac{I_m}{\sqrt{2}}$$

$$V_{rms} = \frac{V_m}{\sqrt{2}}$$

and

$$I_m = \sqrt{2}\, I_{rms}$$
$$V_m = \sqrt{2}\, V_{rms}$$

8. Unless otherwise specified, all ac quantities indicated without subscripts are understood to be rms values.

SELF-EXAMINATION

Answer T or F or, in the case of multiple choice, a, b, c, or d
(Answers at back of book)

15-1. The angular frequency of a 60-Hz waveform is
    a.   120 rad/s
    b.   240 rad/s
    c.   377 rad/s
    d.   754 rad/s

15-2. Given $v = 50 \sin 4000\, \pi t$, the amplitude and frequency are
    a.   50 V, 4000 Hz
    b.   100 V, 4 kHz
    c.   100 V, 2000 Hz
    d.   50 V, 2 kHz

15-3. A sinusoidal voltage causes a sinusoidal current through a resistor of twice the frequency of the applied voltage. _____

15-4. The peak current through a resistor is given by V_m/R and is in phase with the peak of the applied voltage. _____

15-5. When a 60-Hz ac voltage is applied to a resistor, the power varies from 0 to a peak at a rate of 120 Hz. ⎯⎯⎯⎯⎯

15-6. In a 10-Ω resistor carrying a peak current of 3 A, the peak and average power are
   a.   90 W, 45 W
   b.   9 W, 4.5 W
   c.   30 W, 15 W
   d.   45 W, 22.5 W

15-7. The rms value of a sine wave of voltage of peak-to-peak value 10 V is
   a.   7.07 V
   b.   6.36 V
   c.   3.18 V
   d.   3.535 V

15-8. The alternating current through a 5-Ω resistor is 2 A. The rms voltage across the resistor is
   a.   10 V
   b.   5 V
   c.   7.07 V
   d.   3.535 V

15-9. An ac ammeter with a sinusoidal current indicates 4 mA. The peak-to-peak current is
   a.   2.828 mA
   b.   11.312 mA
   c.   5.656 mA
   d.   1.414 mA

15-10. The alternating current flowing through a 1-kΩ resistor is 2 A. A suitably connected wattmeter will indicate:
   a.   4 W
   b.   400 W
   c.   4 kW
   d.   4 mW

REVIEW QUESTIONS

1.   a.   What units are used to measure angular frequency?
         b.   How is angular frequency related to frequency in Hz?
         c.   What do you understand by the term angular frequency?
2.   a.   How many degrees in $\pi/3$ radians?
         b.   How many radians in 225°?
3.   a.   If an alternating triangular voltage is applied to a resistor, what is the shape of the current waveform?
         b.   How can you support this?
         c.   Is this also true of a square wave?
4.   Can you explain qualitatively why power dissipation in a resistor occurs at a rate twice that of the alternating current in the resistor? Given

$$\sin^2 \omega t = \tfrac{1}{2}(1 - \cos 2\omega t)$$

use Eq. 15-8 to show that the power dissipation has an average value of $\dfrac{P_m}{2}$ and quantitatively why the overall power varies at twice the current frequency.

5. a. Explain why the effective value of a sine wave is *not* the average value of the sine wave.
    b. Why is it the average value of the *square* of the sine wave?

6. a. What would be the rms value of a full-wave rectified sine wave as in Fig. 14-16b, compared with a pure sine wave?
    b. What is the rms value of a square wave that alternates from $+10$ V to -10 V?
    c. What is the rms value of a direct current of 3 A?

7. Under what special conditions can Ohm's law and the power equations be used in ac circuit calculations?

8. What measurement and calculation would you have to make to determine the peak current through a resistor, connected in a sinusoidal circuit, without the aid of an oscilloscope?

9. a. If the power and light in an incandescent lamp are pulsating at a rate of 120 Hz, why is there no visible flicker?
    b. What peak ac voltage is required for a 120-V (sinusoidal) lamp?
    c. In which circuit do you think a 120-V lamp will last the longer, in a 120-V ac or a 120-V dc circuit? Why?

10. If we state ac voltages and current in terms of effective dc values, why don't we generate, transmit, and use dc instead of ac?

PROBLEMS

(Answers to odd-numbered problems at back of book)

15-1. A sinusoidal voltage has a peak-to-peak value of 10 V and a frequency of 20 kHz.
    a. Express this voltage in the form $v = V_m \sin \omega t$.
    b. Determine the instantaneous value at $t = 105$ μs.

15-2. Repeat Problem 15-1 for a current waveform with a peak-to-peak value of 50 mA and a frequency of 30 kHz.

15-3. Given the equation $v = 70 \sin 50t$, determine:
    a. The amplitude of the voltage.
    b. The peak-to-peak value.
    c. The angular frequency.
    d. The frequency in Hz.
    e. The period.
    f. The voltage when $t = 1$ s.

15-4. A sinusoidal voltage has a period of 20 μs and has a value of 2 V at $t = 1$ μs. Express this voltage in the form $v = V_m \sin \omega t$.

15-5. A sinusoidal voltage with a peak-to-peak value of 120 V and a period of 2.5 ms is connected across a 33-Ω resistor. Calculate:
    a. The peak value of the current.
    b. The frequency of the current.
    c. The equation representing the voltage.
    d. The equation representing the current.

15-6. Repeat Problem 15-5 with a 47-kΩ resistor.

15-7. For the information in Problem 15-5 calculate:
    a. The peak power in the resistor.
    b. The average power in the resistor.
    c. The frequency of the power variation.

15-8. Repeat Problem 15-7 with a 47-kΩ resistor.

15-9. Determine the rms value of the following:
    a. 7.5 V peak
    b. 14 mA peak-to-peak
    c. 20 mV peak-to-peak
    d. 17 A peak

15-10. Determine the peak values of the following:
    a. 8 V
    b. 20 μA rms
    c. 120 V effective
    d. 10 mA p–p

15-11. Determine the peak-to-peak values of the following:
    a. 120 V
    b. 240 V rms
    c. 15 A
    d. 1.6 μV

15-12. Determine the rms values of the following:
    a. 340 V p–p
    b. 50 mV peak
    c. 325 kV peak
    d. 100 A p–p

15-13. An ac ammeter connected in series with a 220-Ω resistor indicates 580 mA. Calculate:
    a. The applied voltage.
    b. The power dissipated in the resistor.

15-14. An electric heater is rated at 1.5 kW at 120 V ac. Calculate:
    a. The hot resistance of the heater.
    b. The current drawn by the heater.

15-15. What is the maximum permissible reading of an ac ammeter in series with a 33-Ω 2-W resistor?

15-16. What is the minimum resistance that a $\frac{1}{2}$-W resistor may have if it is to be safely connected across an alternating voltage of 6.3 V?

15-17. If an ac voltmeter connected across a resistor indicates 5 V and a series-connected milliammeter reads 2.2 mA, calculate:
    a. The power dissipated in the resistor.
    b. The resistance of the resistor.

15-18. A voltage having the equation $v = 30 \sin 3000\pi t$ V is applied to a 6.8-kΩ resistor. Calculate:
    a. The reading of an ac voltmeter across the resistor.
    b. The reading of an ac ammeter connected in series with the resistor.
    c. The power dissipated in the resistor.

15-19. Three series-connected resistors have a sinusoidal voltage applied to them. An ac voltmeter across the first resistor indicates 12 V; an ac ammeter in series with the second resistor indicates 2.4 A; and an oscilloscope across the third resistor reads 10 V peak-to-peak. If a wattmeter, connected to read the total power delivered to the series circuit, indicates 55 W, calculate:

a. The resistance of each resistor.

b. The power dissipated in each resistor.

c. The applied voltage.

15-20. Three resistors R_1, R_2, and R_3 are connected in parallel across a sinusoidal voltage. An ammeter in series with the first reads 1.5 A while an oscilloscope across the second indicates a peak voltage of 35 V. If $R_3 = 2R_2$ and a watt-meter indicates a total power of 200 W in the whole circuit, calculate:

a. The resistance of each resistor.

b. The power dissipated in each resistor.

15-21. Given the circuit in Fig. 15-11, find:

a. The rms value of the applied voltage (V_T).

b. The values of R_4, R_5, and R_6, if they are equal in value.

c. The total resistance of the circuit.

15-22. Given the circuit in Fig. 15-12, find:

a. The ammeter reading.

b. The reading of the voltmeter across the supply.

c. The wattmeter reading.

d. The values of R_1 and R_2 if $R_1 = 2R_2$.

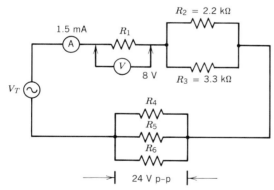

FIGURE 15-11
Circuit for Problem 15-21.

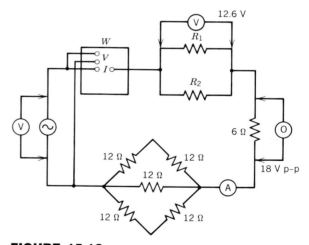

FIGURE 15-12
Circuit for Problem 15-22.

CHAPTER 16

AC MEASURING INSTRUMENTS

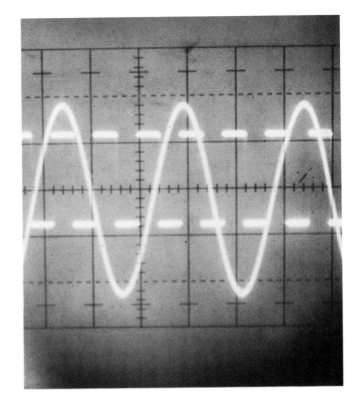

One of the most important instruments used to measure alternating voltage is the cathode-ray oscilloscope (CRO). This instrument measures and displays peak-to-peak voltage, period, and phase relationships. Since the oscilloscope is widely used in industry and in the laboratory, understanding its operation is essential. In this chapter, you will learn the construction and operation features that are common to most oscilloscopes. The different probes that are used will also be covered.

Most often, alternating current voltmeters and ammeters are calibrated to indicate rms values. Many of these meters, however, only indicate rms values when used in sine wave applications. They are generally of the permanent-magnet moving-coil (PMMC) type, as described in Chapter 12, but use additional diodes to rectify or convert the ac to a proportional dc. Other meters show the *true-reading* rms value *regardless of the waveform*. Finally, this chapter will consider precautions arising from common-ground connections when using equipment operated from a 120-V, 60-Hz supply.

16-1 THE CATHODE-RAY OSCILLOSCOPE

The manner in which any voltage (or current) varies with time can be displayed on an instrument called an *oscilloscope*. This device relies upon a beam of electrons that strikes a fluorescent screen, providing a "picture" of the input waveform variation. In the case of a sine wave, for example, it will show all the instantaneous values over a full cycle by providing a continuous display of one or more cycles.

16-1.1 Construction

The "heart" of the oscilloscope is the cathode-ray tube, or CRT. In a modern solid-state instrument, this is usually the only tube to be found (older instruments used a number of vacuum tubes that have been replaced by solid-state components). The CRT consists of a glass envelope that has been highly evacuated (i.e., a vacuum has been drawn inside the tube). Figure 16-1 is a schematic representation of the CRT.

A *heater* or *filament* is used to bring the *cathode* to a high temperature, which causes the cathode to emit electrons. (Before this process of electron emission was fully understood, these electrons were referred to as "cathode rays.") The *control grid* regulates the flow of electrons from the cathode, and thus the brightness or intensity of the image on the CRT screen. The *focusing anode,* which has a hole in its center, shapes the stream of electrons into

a tight stream, or *beam*. The beam of electrons next passes through the *accelerating anode,* which is held positive with respect to the cathode by 2 kV or more. Electrons accelerate toward this anode, pass through the hole in the center of the anode, and travel toward the *screen*. Taken together, the cathode, grid, focusing anode, and accelerating anode form the *electron gun,* which has the sole function of providing a *controlled beam* of electrons.

After emerging from the accelerating anode, the beam passes between two sets of *deflecting* plates. Voltages applied to these plates "bend" the beam of electrons up, down, left, or right to control the point where it strikes the inside face of the screen. This is called an *electrostatic* deflection system, and is quite suitable for the small deflections needed for oscilloscope operation. In television set CRTs, larger deflections are needed and are provided by magnetic fields set up by coils arranged in a *yoke* around the outside of the tube.

The inside of the CRT's screen is *fluorescent*—it is coated with *phosphors* that are excited (made to glow) when struck by the beam of electrons. The result is a bright spot or line visible on the screen. Different phosphors are used to provide green or yellow displays, depending upon the manufacturer. Phosphors also have differing degrees of *persistence,* causing a spot to continue to glow for a shorter or longer time after the electron beam has moved on. The electrons arrive at the screen moving at a speed of approximately one-tenth the speed of light. If the beam is allowed to remain in one spot for many hours, it can "burn in" that spot on the phosphor screen.

The inside of the tube is coated with a conductive film

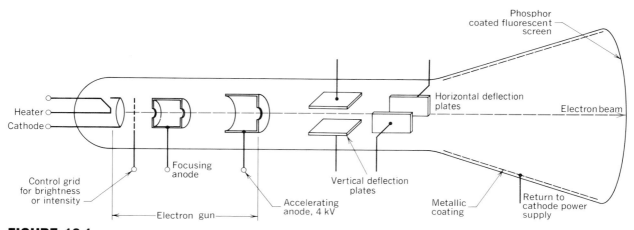

FIGURE 16-1
Schematic cross section through a cathode-ray tube (CRT).

of graphite material called *aquadag*. This coating serves as an electrostatic shield against outside fields and, in some oscilloscopes, as an additional accelerating anode. The aquadag coating also serves as a means of avoiding electron *accumulation* on the screen. When the electron beam arrives at the screen, secondary electrons are emitted, which must be returned to the power supply. These electrons are collected by the aquadag coating and make their way back to the cathode through the power supply.

16-1.2   Operation

If no voltage is applied to the deflecting plates, the beam of electrons will continue in a straight line from the accelerating anode, causing a bright spot in the center of the screen. (See Fig. 16-2*a*.)

If a potential difference is applied to the horizontal (*X*-axis) deflecting plates, the negatively charged electron beam will move toward the *positive* potential. This causes the spot on the screen to change its position *horizontally*. If a varying (sawtooth) voltage is used to provide this horizontal deflection, at a minimum frequency of about 16 Hz, the movement of the spot of light on the screen will appear as a straight, continuous line. (See Fig. 16-2*b*.) This is usually referred to as the *trace*.

If a sine wave of voltage is applied to the vertical (*Y*-axis) set of deflecting plates, in *addition* to the voltage applied to the horizontal plates (as previously described), the pattern shown in Fig. 16-2*c* will appear on the screen. As the electron beam moves from left to right, it is also deflected *vertically* by the sine wave—upward when the upper plate is positive, downward when the upper plate is negative. The frequency of the voltage on the horizontal plates can be adjusted (usually from less than 1 Hz to many megahertz) so that one or two full cycles of the voltage on the vertical plates will be displayed on the screen. (Provision is made electronically to *blank* the electron

beam while it is being returned from the right side of the screen to the left side. If the oscilloscope's STABILITY control is improperly adjusted, retrace or multiple trace problems will occur.)

16-1.3   Function of Controls

Figure 16-3 shows a representative example of the common *single-beam* oscilloscope. The controls typically found on such instruments include:

BRILLIANCE   Adjusted to give convenient image intensity.

FOCUS AND ASTIG(MATISM)   Adjusted for sharpest definition.

TRIG(GER) LEVEL   Controls the point at which the applied input waveform is initially displayed. Normally left at the counterclockwise **AUTO** setting, so that the display begins at its average position.

STABILITY   Adjusted to provide a stationary, stable (nonoverlapping) waveform.

Y SHIFT, VERNIER   Used to make coarse and fine vertical adjustment of the trace position on the screen.

X SHIFT   Used to adjust the position of the trace horizontally on the screen.

TRIG SELECTOR
   TV FIELD, TV LINE   Used in conjunction with **TRIG LEVEL** to provide correct triggering for video pulses. Normally set to the **OUT** position.
   HF   Used in conjunction with **TRIG LEVEL** for high frequencies (1 to 10 MHz).
   +/−   Used to select triggering of the observed input waveform on either the positive or negative going *slope* of the waveform. "Inverts" the waveform.
   INT/EXT TRIG   When the button in pushed **IN**, the triggering sweep of the beam is controlled by an

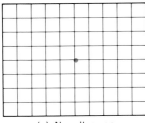

(*a*) No voltage on
         V or H plates

(*b*) Voltage applied
         to H plates only

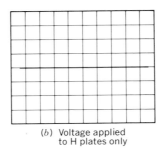

(*c*) Combined voltages
         on H and V plates

FIGURE 16-2

Screen display for different voltage conditions on deflecting plates.

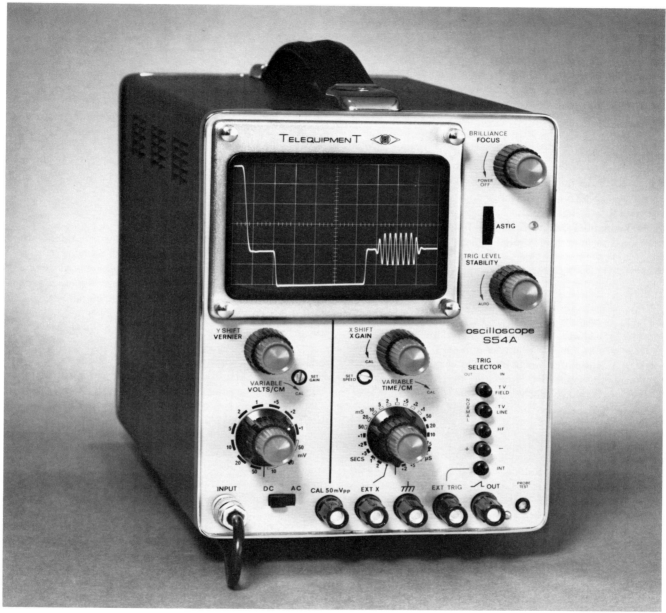

FIGURE 16-3
A single-beam oscilloscope showing the controls. (Courtesy of Tektronix, Inc.)

internally generated voltage. When the button is **OUT,** the **EXTERNAL TRIGGER** control is active.

EXT(ERNAL) TRIG(GER)   Allows an external voltage to be connected for triggering purposes. This provides more stable displays for nonperiodic waveforms.

VOLTS/CM   Determines what voltage connected at the **INPUT** will cause a vertical deflection of 1 cm. The **VOLTS/CM** knob controls the gain of a vertical amplifier so that a 1-cm deflection on the screen may represent an input from 10 mV to 50 V. This is only the case when the VARIABLE knob in the center of the control is set to its fully clockwise or **CALIBRATED** position.

TIME/CM   Determines the rate at which the beam sweeps across the screen. When the inner **VARIABLE** knob is in its **CALIBRATED** position (and the X GAIN knob is rotated fully counterclockwise to *its* **CALIBRATED** position), each centimeter of horizontal movement corresponds to the time selected with the **TIME/CM** knob. This control actually varies the frequency of the internally generated voltage that is applied to the horizontal deflection plates. It is adjusted to give a convenient number of cycles on the screen.

EXT X   If the **TIME/CM** knob is rotated fully counterclockwise an external voltage may be applied directly to the horizontal plates. (See Section 16-1.6.)

INPUT   This is the connection point for a *probe* used to apply to the oscilloscope the voltage to be measured.

DC/AC   In the DC (direct coupling) position, a voltage to be observed at the INPUT is coupled *directly* to the amplifier. This allows both *ac* and *dc* voltages to be observed. In the AC (alternate coupling) position, only the *ac* portion of an input can be observed. If the input happens to contain a *dc* level, the dc is blocked by a capacitor and does not appear on the screen. The DC setting gives a trace of the *full information* contained in the input; the AC setting gives a trace of only the *ac portion* of the input.

CAL 50 mV pp   Provides an accurate, internally generated reference voltage used to check and calibrate the oscilloscope's vertical amplifier by adjusting the SET GAIN control.

PROBE TEST   A positive-going, fast-rise pulse used to adjust (compensate) 10X probes. (See Section 16-1.4.)

16-1.4   Oscilloscope Probes

Measuring voltage between two points with an oscilloscope is initially very similar to measuring voltage with a voltmeter. An oscilloscope probe consists of two leads. The input *tip* is connected through the body of the probe via a coaxial cable to the input connection of the oscilloscope. The *ground lead* of the probe is connected to the metal sheath of the coaxial cable. The tip is considered the *positive* lead; the ground lead, the negative side. The ground lead is usually connected to the ground side of the 120-V ac supply that powers the oscilloscope. As described in Section 16-6, care must be taken with the

ground connection to avoid erroneous readings and possible damage to the instrument.

There are two basic types of oscilloscope probes. The most common type is referred to as a *10X* or 10:1 probe. It incorporates a voltage-dividing circuit that reduces the input voltage to the oscilloscope to *one-tenth* of the value at the tip. The voltage divider may be either in the body of the probe or in the form of an *attenuator* at the end of the cable. The advantage of the 10X probe is that it increases the usual input resistance to the oscilloscope from 1 MΩ to 10 MΩ, reducing loading problems. It also extends, by a factor of 10, the maximum voltage that the oscilloscope can be used to measure. (This factor must be taken into account, however, when reading the oscilloscope.)

To provide the proper flat response to a square wave input, the voltage-dividing circuit in the probe requires adjustment from time to time. This *probe compensation* is done by connecting the probe to a suitable square wave (as at the **PROBE TEST** terminal) and making an adjustment in the probe. Each probe must normally be compensated for the oscilloscope with which it is to be used.

Direct probes do not include a voltage-dividing circuit, and are generally used for measuring voltages in the millivolt range. Such low voltages require the maximum sensitivity of the oscilloscope.

Direct probes are usually marked 1X; if a probe is *unmarked*, it is generally a 10X probe.

For direct measurement and display of current waveforms, *current* probes are available. These probes provide a slot in which the current-carrying wire is placed, allowing measurement without breaking the circuit under test. With a suitable amplifier system (either separate from, or included in, the oscilloscope), either dc or ac currents from 1 mA to more than 100 A can be measured.

16-1.5   Measurement of Voltage and Frequency

Assume that an oscilloscope probe has been connected across a sinusoidal voltage, and that the controls have been adjusted to display the waveform shown in Fig. 16-4.

In this example, the peaks of the sine wave coincide with the *graticule* lines 2 cm *above* and 2 cm *below* the 0-V level. Thus, the peak-to-peak displacement occupies 4 cm. To convert this to a voltage, you must use the volts/cm calibration (in this case, 0.2 V/cm). If you take into account the 10X factor, the full peak-to-peak voltage is given by

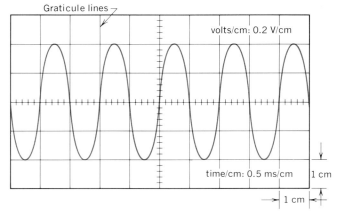

FIGURE 16-4
Typical waveform for measurement of voltage and frequency.

$$V_{p-p} = \text{number of cm} \times \text{volts/cm} \times 10 \quad \text{(16-1)}$$

Therefore, $V_{p-p} = 4 \text{ cm} \times 0.2 \text{ V/cm} \times 10$
$$= \textbf{8 V}$$

When a *direct* probe is used, the 10X factor is omitted.

The peak voltage that has been measured (V_p) equals 4 V, although only 0.4 V was actually connected to the oscilloscope itself (since the probe's voltage-dividing circuit reduced the voltage by a factor of 10).

After some practice, you should be able to read voltage directly from the waveform on the oscilloscope by noting how many volts each cm represents. For example, in the above, combining the 0.2 V/cm with the 10X probe factor means that each centimeter is "worth" 2 V. Starting from the bottom of the waveform and moving centimeter by centimeter you read 2 V, 4 V, 6 V, 8 V for the 4-cm peak-to-peak waveform. Each small subdivision would represent 0.4 V for any waveform that does not occupy a whole number of centimeters.

Frequency is measured in a similar way, but the 10X factor is *not* used. First, you have to obtain the number of cm for one cycle, or *period* (T). In Fig. 16-4, the period is represented by 2 cm, at 0.5 ms/cm.

$$T = \text{number of cm for one cycle} \times \text{time/cm} \quad \text{(16-2)}$$

Therefore, $T = 2 \text{ cm} \times 0.5 \text{ ms/cm}$
$$= 1 \text{ ms}$$
$$f = \frac{1}{T} \quad \text{(14-12a)}$$
$$= \frac{1}{1 \times 10^{-3} \text{ s}} = \textbf{1 kHz}$$

NOTE When making horizontal and vertical measurements, both the horizontal and vertical *variable* controls must be in the CALIBRATED position. Also, when measuring dc levels, the location of the trace for 0 V (zero volt) must be identified.

When measuring peak-to-peak voltage, the waveform does not have to be centered on the screen. Where the peak-to-peak value is not a whole number of centimeters, the Y-SHIFT is used to line up either the positive or negative peak with a graticule line. Then, the X-SHIFT is used to center one of the peaks (positive or negative) on the finely divided vertical graticule line. This allows you to determine the overall height of the waveform. A similar method is used to measure the *period* when a waveform does not occupy a whole number of centimeters horizontally. This is demonstrated in Example 16-1.

EXAMPLE 16-1

Given the waveform and oscilloscope settings shown in Fig. 16-5, determine, assuming a 10X probe, the following:

a. The peak voltage, V_p.
b. The period, T.
c. The frequency, f.

Solution

a. V_{p-p} = number of cm × volts/cm × 10   (16-1)
$$= 4.6 \text{ cm} \times 50 \times 10^{-3} \text{ V/cm} \times 10$$
$$= 2.3 \text{ V}$$
$$V_p = \frac{V_{p-p}}{2} = \frac{2.3}{2} \text{ V} = \textbf{1.15 V}$$

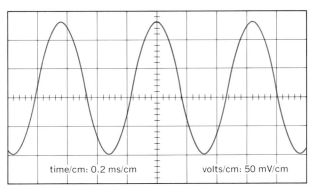

FIGURE 16-5
Waveform for Example 16-1.

b. T = number of cm × time/cm   (16-2)
      = 3.2 × 0.2 ms/cm
      = **0.64 ms**

c. $f = \dfrac{1}{T}$ (14-12a)

    $= \dfrac{1}{0.64 \times 10^{-3} \text{ s}}$

    = **1.56 kHz**

The accuracy of the vertical and horizontal controls ranges from ±3% to ±5%, depending upon cost of the instrument. Errors arising from reading the vertical and horizontal deflections will negatively affect the accuracy level.

16-1.6 Lissajous Figures

Another method of measuring frequency eliminates the inaccuracy of the **TIME/CM** sweep control. The frequency to be measured is applied to the vertical input, as before, but a variable frequency sine wave is connected to the *horizontal* input through the **EXT X** connection. (See Fig. 16-6.) The electron beam is now swept from left to right by the external sine wave, instead of the internally generated wave.

The frequency of the horizontal input is varied until a stationary pattern is displayed on the screen. This pattern, called a *Lissajous figure,* takes on different shapes according to the ratio of the two frequencies. Lissajous figures for various frequency ratios are shown in Fig. 16-7.

The Lissajous figures actually displayed on the oscilloscope might not be as symmetrical as those shown in Fig.

16-7. For example, when the horizontal and vertical frequencies are the same, an ellipse or a straight line can be seen gradually changing to a circle. This is caused by slight differences in time or phase relationships between the two waves. Actually, this is useful information, and is covered under "phase angle measurements" in Section 24-8.

Similarly, the other patterns may "drift," but at the ratios shown, they will become stationary. The *unknown* frequency is then known in terms of the *variable* frequency. In general, the frequency ratio is given by the ratio of the number of *tangent points* on the horizontal to the number of such points on the vertical. That is,

$$\frac{f_v}{f_h} = \frac{\text{number of horizontal tangent points}}{\text{number of vertical tangent points}} \quad (16\text{-}3)$$

For example, see Fig. 16-7d. A line drawn horizontally on the top of the figure touches the figure at two points. A line drawn vertically at the side of the figure touches it at three points. Thus $f_v/f_h = \frac{2}{3}$. If $f_h = 120$ Hz,

$$f_v = \frac{2}{3}f_h = \frac{2}{3} \times 120 \text{ Hz} = \mathbf{80 \text{ Hz.}}$$

The frequency of the unknown is determined only as accurately as the variable frequency is known. This frequency may be obtained from a *signal generator,* of which there are two common types.

An *audio* signal or function generator, as shown in Fig. 16-8, provides sine and square (and sometimes triangular) output voltages. The frequency is variable, usually from approximately 1 Hz to 1 MHz, and the amplitude is also variable, with a maximum value from 10 to 20 V peak-to-peak. Signal generators like this one are used to inject signals into circuits (such as amplifiers) so that their gain and

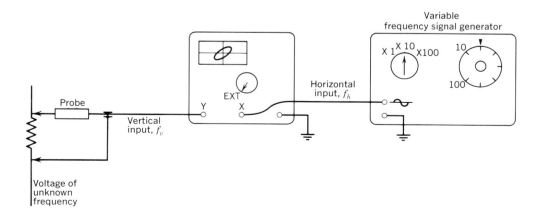

FIGURE 16-6
Measuring frequency using Lissajous figures.

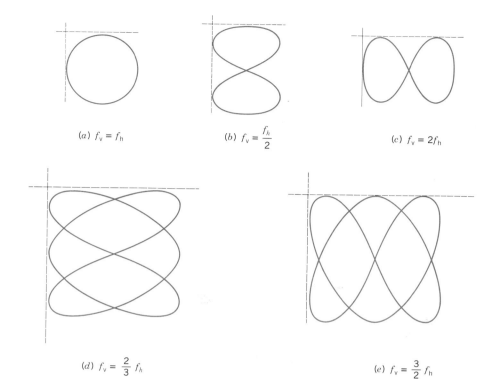

(a) $f_v = f_h$ (b) $f_v = \dfrac{f_h}{2}$ (c) $f_v = 2f_h$

(d) $f_v = \dfrac{2}{3} f_h$ (e) $f_v = \dfrac{3}{2} f_h$

FIGURE 16-7

Lissajous figures for various frequency ratios.

FIGURE 16-8

Typical function generator. (Courtesy of Hewlett-Packard Company.)

frequency characteristics can be measured with an oscilloscope. *Radio frequency* (RF) generators provide higher frequencies (up to hundreds of megahertz) for use in servicing radio, TV, and communications equipment.

Where high accuracy is required, the use of Lissajous figures for frequency measurement is very limited. A more accurate and convenient method, using a frequency meter, is covered in Section 16-8. The Lissajous method, however, can be used as a quick check on the calibration of an audio signal generator at low frequencies. This can be done by comparing the line frequency (60 ± 0.02 Hz) with the 60-Hz setting on the generator. Also, as mentioned earlier, Lissajous figures can be used to make phase angle measurements in circuits at a given (constant) frequency.

16-1.7 Other Oscilloscopes

A common type of oscilloscope on which two waveforms can be displayed simultaneously is the *dual beam* or *dual trace* oscilloscope shown in Fig. 16-9. Channels *A* and *B* control the input signal from two independent probes. This

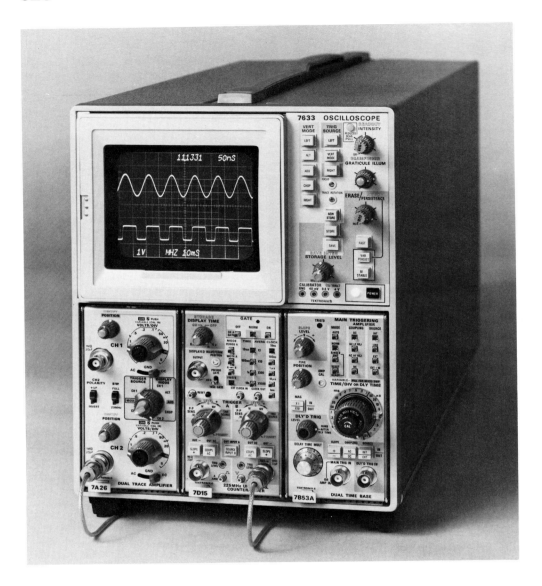

FIGURE 16-9
Dual-beam storage
oscilloscope. (Courtesy
of Tektronix, Inc.)

allows the time and phase relationships of the two wave-forms to be immediately visible. Modern oscilloscopes of the dual-beam type can display waveforms with frequencies of several GHz.

With a ''develop while you wait'' camera, a photograph *(hard copy)* of the waveforms displayed on the screen can be made for reference. Alternatively, oscilloscopes with a *storage* feature can ''freeze'' or store away a given waveform, displaying it later upon demand.

Other modifications of oscilloscopes and their circuitry include *curve tracers* that display the characteristics of solid-state devices, *logic state analyzers* that display digi-

tal events in a data stream and automotive *ignition analyzers*.

In the next several sections, you will examine some of the meters designed to measure ac quantities.

16-2 AC METERS USING THE PMMC MOVEMENT

As you will recall from Chapter 12, the permanent-magnet moving-coil (PMMC) meter movement is polarity-sensitive and is used for dc measurements. Consider the appli-

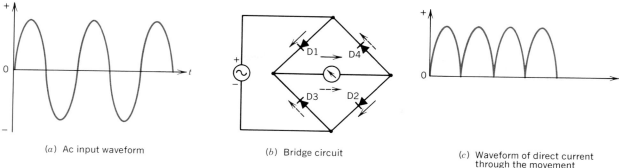

(a) Ac input waveform (b) Bridge circuit (c) Waveform of direct current through the movement

FIGURE 16-10

Operation of an ac meter using a PMMC movement.

cation of a sinusoidal voltage to a PMMC movement. During the first half of the cycle, the meter movement experiences a torque in one direction, followed by a torque in the opposite direction during the remaining half of the cycle. At a frequency of 60 Hz, the reversals are so rapid that the pointer on the meter scale will be unable to follow the alternations—it will remain at zero and exhibit a small vibrating motion. The meter actually indicates the *average* value of the sine wave, which is zero.

To make the pointer deflect in only one direction, the current must be made to flow through the movement in one direction (i.e, not alternate). In ac milliammeters and some other meters, this is done by using diodes connected in a bridge circuit, as shown in Fig. 16-10b.* (Ac voltmeters usually convert the ac to dc by using one or two diodes in a shunt rectifier or voltage doubler circuit, followed by a filter.)

A *diode* is a device that allows current to flow through in only one direction (indicated on schematics by an arrow) when the polarity across the diode is favorable. (See Section 28-5 for more detailed information on diodes.) In the bridge circuit shown in Fig. 16-10b, diodes D1 and D2 will conduct on the positive half-cycle; during the negative half-cycle, diodes D3 and D4 will conduct instead. The waveform of the current through the meter movement is shown in Fig. 16-10c. This is a *full-wave rectified waveform,* which you should recognize as the output of a simple dc generator. Its average value is 0.636 $(2/\pi)$ of the peak value. The pointer moves through an angle proportional to this average value. The point at which it comes to rest on the scale is marked with the *effective* value of

the applied sine wave. The ac scale of a meter using a PMMC movement is *linear,* although some nonlinearity occurs on very low ranges, due to the characteristics of the diodes. Separate scales are often used for the 1.5-V and 5-V ac ranges to take this nonlinearity into account.

Because of the need for rectification, the ac sensitivity (or ohms per volt rating) of ac meters using a PMMC movement is lower than the dc sensitivity. A typical VOM with a dc sensitivity of 20 kΩ/V may have an *ac* sensitivity of only 5 kΩ/V. In some electronic voltmeters (VTVM or solid state) with 11-MΩ resistance on direct current, the input resistance is only 1 MΩ for alternating current. Despite these shortcomings, PMMC movements are widely used for analogue multimeters capable of measuring both alternating and direct current. Such meters are capable of indicating ac voltages at frequencies of 1 MHz and higher, which the ac meters described below are unable to do.

16-3 THE ELECTRODYNAMOMETER MOVEMENT

The inability of the PMMC movement to handle ac current without rectification is due to the fixed direction of the permanent magnet's field. If the magnetic field could be reversed each time the current in the movable coil reversed, the result would be pointer movement in one direction. This is achieved by replacing the poles of the permanent magnet with two *stationary coils,* as shown in Fig. 16-11a. The movable coil is free to turn in the space between the two stationary coils, which act as solenoids. This type of movement is called an *electrodynamometer.*

If the fixed and movable coils are series-connected, as

---

*In addition, a current transformer is used to provide a low resistance when the meter is inserted in a circuit.

shown in Fig. 16-11a, the result is an ac ammeter. This basic meter movement can then be used with shunts and multipliers to make any range of ammeter and voltmeter. Because the deflection of the pointer is proportional to the *square* of the current, scales used for these meters are nonlinear. The markings are crowded together toward the low end and more widely spaced toward the high end.

This type of movement will also operate on dc, but is too expensive to compete with dc ammeters using the PMMC movement. The electrodynamometer movement allows calibration on ac by noting the degree of deflection for given values of dc current. In fact, this movement indicates *true-reading* rms values for *all* waveforms.

The sensitivity of the electrodynamometer movement is much lower than that of the PMMC movement, having a value of about 50 Ω/V for an ac voltmeter. Also, the frequency range of this movement extends from dc up to only about 125 Hz, so meters using it are limited to making measurements in power circuits.

The real value of the electrodynamometer movement is in its application to measuring average power (either ac or dc) in a wattmeter, as shown in Fig. 16-11b.

You have already considered, in Chapter 9, the external connection of the wattmeter terminals. You can now see that the current terminals make the load current pass through the heavy stationary coils, setting up a field whose strength depends on the amount of current. (There is no iron in these coils, so there is a linear relationship between the field and the current.)

The potential terminals connect the movable coil across either the source *or* the load. The force acting to turn the coil thus depends on *both* the current and the voltage. The resulting pointer deflection is proportional to the *product* of the two—$P = VI$ for a pure resistor.

However, if the circuit contains components other than resistance, the wattmeter reading *still* gives the true power in watts, although the result is no longer *solely* given by the product $P = VI$. (See Section 25-6.)

It should be noted that if the potential (movable) coil is connected across the load, a slight error is introduced by the additional current drawn by the potential coil. If connected across the supply, the potential coil includes an additional voltage drop across the load-carrying current coils. Consequently, the potential coil is usually connected directly across the *load,* as shown in Fig. 16-11b. This produces a *constant* error. The error can be compensated for and corrected by opening the load circuit and subtracting the no-load wattmeter reading from all future readings. Note that the scale of the wattmeter is *linear,* unlike the scales of ammeters and voltmeters using the electrodynamometer movement.

It should be clear that, with direct current, polarity must be observed when making wattmeter connections to avoid backward deflection of the pointer. This also applies to

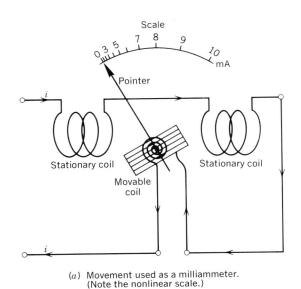

(a) Movement used as a milliammeter.
(Note the nonlinear scale.)

(b) Movement used as a wattmeter

FIGURE 16-11

The **electrodynamometer movement.**

alternating current, because each ac half-cycle has its own *instantaneous polarity*. The ± (or red) terminal of the potential coil must be connected to the side of the line where the current coil is located. (See Fig. 16-11*b*.) As is the case with any electrodynamometer movement, most wattmeters are *not* suitable for frequencies above 100 Hz. Special meters must be used for higher frequencies.

16-4 MOVING-IRON INSTRUMENTS

A group of devices called *moving-iron instruments* has been developed primarily for making ac measurements. These consist of iron vanes attached to a pointer and free to move in a magnetic field. The vanes carry no current (they move because of induced magnetic effects), so the movement is rugged, simple and less expensive than the electrodynamometer movement. This movement is also an inherent rms indicator in all waveforms. However, an instrument using this movement is less sensitive than the PMMC movement, very limited in frequency, and has a nonlinear scale. Moving-iron ammeters and voltmeters, like those using the electrodynamometer movement, are limited to low-frequency power circuit applications.

16-5 FREQUENCY METER

In Chapter 13, you learned how a digital multimeter can be used to measure voltage, current, and resistance with a high degree of accuracy and resolution. Although an oscilloscope (which is an analogue device) can be used to measure period and frequency, it seldom is more accurate than 2%. An instrument that provides direct digital measurement of these quantities is shown in Fig. 16-12.

Typical of many such devices, this instrument is a multiple-function counter. It presents a 6-digit display of frequency and period for waveforms ranging from 5 Hz to 80 MHz, giving a resolution of 0.1 Hz and 100 ns. The instrument may also be used to display a running total of input events, in the form of pulses, up to a maximum of 999999. The counter requires only 25-mV input for operation of all functions.

16-6 PRECAUTIONS WITH GROUNDED EQUIPMENT

The neutral side of the 120-V, 60-Hz domestic supply is grounded, usually to a cold water pipe where the supply

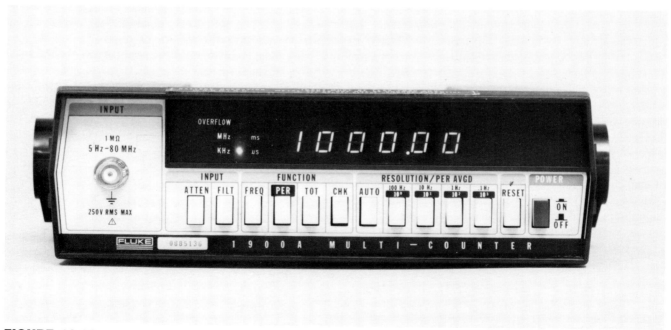

FIGURE 16-12
Multiple-function counter to measure frequency and period. (Courtesy of John Fluke Mfg. Co., Inc.)

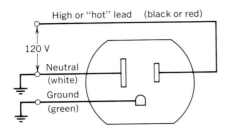

High or "hot" lead   (black or red)

120 V

Neutral
(white)

Ground
(green)

(a) Wiring of a 120-V receptacle

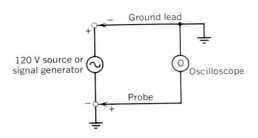

Ground lead

120 V source or
signal generator

Oscilloscope

Probe

(b) Improper connection of grounded
oscilloscope causes a short circuit

FIGURE 16-13

Ground connections in a receptacle outlet and oscilloscope.

enters the building. The neutral must be a *continuous, unbroken* line throughout the *whole* wiring system. As shown in Fig. 16-13a, the larger opening on a 120-V receptacle is the grounded neutral, while the "hot" or high side is connected to the smaller opening.* (In 120-V, single-phase polarized systems, the *third* opening of the receptacle is also connected to ground. The safety feature provided by this third connection is discussed in Chapter 18.)

Most electronic equipment that operates from the 120-V, 60-Hz supply has one side of its output (in the case of a signal generator) or one side of the input leads (in the case of meters and oscilloscopes) connected to the power line ground. This side is usually designated as the negative (−) or low side.

Consider a situation in which an oscilloscope, with its negative side grounded, is connected across a source that also has one side grounded. If the grounds are not *matched*, the result is a direct short across the source, as shown in Fig. 16-13b. Since a signal generator has a rather high internal resistance, from 50 to 600 Ω, the result is merely a 0-V indication. But if the source is a 120-V receptacle, a high current will flow at the *instant* of connecting the ground lead, whether or not the probe has already been connected. In some cases, this can cause expensive-to-repair damage to the measuring instrument.

The path that is provided by unmatched grounds between circuit ground and instrument ground is called a *ground loop*. This is usually a shorter path for current to flow through and generally results in larger currents flowing, as well as erroneous readings.

Obviously, grounds should be connected to *each other* to avoid such a problem. If the identity of the ground is unknown, only the *probe* should be connected. If no voltage is indicated on the measuring instrument, the probe has been connected to ground, and should be moved to the other terminal. In many cases, connection of the ground lead is unnecessary, since all voltages being measured will be with respect to ground.

Figure 16-14 shows another problem arising from ground loops. If the oscilloscope is connected across R_1 (Fig. 16-14a), R_2 is shorted out. The oscilloscope indicates the full voltage from the source, *not* the true voltage that was across R_1. If it is possible to interchange R_1 and R_2, as in Fig. 16-14b, the true value can be obtained if grounds are observed. Otherwise, the voltage across R_2, subtracted from the source voltage, will give the voltage across R_1. This will work, however, only where all the components are resistors, or capacitors, or pure inductors.

In situations where measurements are being made across mixtures of resistors, inductors, and capacitors, the problem of grounds can be eliminated by using an *isolation transformer* to supply power to either the source or the measuring instrument. This is discussed further in Section 18-4.

*For safe and proper operation of some equipment, two-prong plugs *must* be inserted in the receptacle with the proper orientation to maintain correct polarity. In a TV set, for example, proper orientation of the plug ensures that the metal cabinet is connected to the grounded neutral rather than the "hot" side of the circuit. To ensure correct polarity, the plugs provided with such appliances have blades of differing sizes. The wider blade on the neutral side can be inserted in the receptacle only with proper orientation.

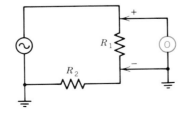

(a) Oscilloscope ground shorts out R_2

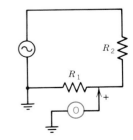

(b) Oscilloscope measures true voltage across R_1

FIGURE 16-14

Oscilloscope measurements taking grounds into account.

SUMMARY

1. A cathode-ray oscilloscope generates a beam of electrons that is deflected vertically and horizontally to display a "picture" of a voltage waveform on a fluorescent screen.
2. An oscilloscope allows voltage and period (time) to be measured using calibrated vertical and horizontal controls (VOLTS/CM and TIME/CM) as follows:

$$V_{p-p} = \text{number of cm} \times \text{volts/cm}$$
$$T = \text{number of cm} \times \text{time/cm}$$

3. A 10X probe reduces the input voltage by a factor of 10. This requires the *voltage* equation given above to be multiplied by 10.
4. A Lissajous figure results from applying sine waves to both the horizontal and vertical deflection plates of the oscilloscope. The resulting pattern on the screen allows the ratio of the two frequencies to be determined by the method of tangent points:

$$\frac{f_v}{f_h} = \frac{\text{number of horizontal tangent points}}{\text{number of vertical tangent points}}$$

5. A PMMC movement can be used in an ac meter if diodes are used to convert the ac to a proportional dc.
6. An electrodynamometer movement replaces the permanent magnet poles of a PMMC movement with fixed solenoid coils for ac applications.
7. The fixed and movable coils of an electrodynamometer are used to measure power in a wattmeter.
8. Moving-iron instruments use iron vanes moving in a current-carrying coil to cause true-reading rms deflections, primarily in ac power circuits.
9. A frequency meter can measure period and frequency of ac waveforms with a high degree of accuracy and resolution.
10. Most electronic equipment operated from the 120-V, 60-Hz supply has one side of its input or output grounded. Ground connections must be matched to avoid possible damage to measuring equipment.

SELF-EXAMINATION

Answer T or F, or, in the case of multiple choice, a, b, c, or d
(Answers at back of book)

16-1.   The electron gun is that part of a CRT that produces the beam of electrons.
        _____

16-2.   An electrostatic deflection method is used in an oscilloscope whereas a magnetic deflection method is used in TV tubes. _____

16-3.   The voltage waveform to be examined is connected to the X-axis or horizontal deflection plates. _____

16-4.   The Y-SHIFT control can be used to move the display to one side or the other.
        _____

16-5.   The VOLTS/CM control is adjusted for the largest, complete vertical display on the screen. _____

16-6.   The TIME/CM control is adjusted for the desired number of cycles or portion of a cycle. _____

16-7. The dc input switch can be used only with pure dc voltages. _____

16-8. When using a 10X probe, both the horizontal and vertical deflections must be multiplied by 10. _____

16-9. A direct probe may be used where small input voltages require the full vertical sensitivity of the oscilloscope. _____

16-10. An oscilloscope displays a peak-to-peak value of 3.2 cm with a 10X probe and a VOLTS/CM setting of 20 V/cm. The peak voltage is
   a. 32 V   c. 640 V
   b. 320 V   d. 64 V

16-11. An oscilloscope displays 4 cycles over an 8-cm distance with a 10X probe and a TIME/CM setting of 50 μs/cm. The period and frequency are
   a. 10 μs, 100 kHz   c. 100 μs, 10 kHz
   b. 100 μs, 100 kHz   d. 400 μs, 2.5 kHz

16-12. The frequency of an unknown sinusoidal frequency may be determined to an accuracy of 0.1%, even if the oscilloscope is only 3% accurate, by the use of Lissajous figures. _____

16-13. A Lissajous figure is a stationary pattern only when the two frequencies are exactly equal. _____

16-14. A dual-beam (or dual trace) oscilloscope may simultaneously display two waveforms of the same or related frequency on the same screen. _____

16-15. If a symmetrical ac voltage, at power line frequencies, is applied directly to a PMMC movement, the resultant deflection will be approximately 0.707 of the applied peak voltage. _____

16-16. PMMC movements are used in ac meters because of their high sensitivity. _____

16-17. The electrodynamometer movement may only be used to measure power. _____

16-18. When measuring power line voltages with an ac-operated voltmeter or oscilloscope, it is best not to make any ground connection at all. _____

REVIEW QUESTIONS

1. a. What is an "electron gun"?
   b. What polarity voltage must be applied to the control grid to reduce intensity in the CRT trace?
   c. What type of deflection system does an oscilloscope use?

2. a. What produces the visible trace on a CRT screen?
   b. What might happen if electrons were allowed to accumulate on the screen?
   c. How are these excess electrons removed?

3. a. Where does the voltage come from to normally sweep the electron beam across the CRT screen?
   b. What would be the visual effect on a displayed waveform if the sweep frequency were reduced to a very low value?
   c. When would you want the sweep frequency to be very high?
   d. What display would occur for a sine wave input if the TIME/CM control were turned to external?

4. a. What type of probes are commonly available?
   b. What advantages does a 10X probe provide?
   c. How does it affect the determination of voltage and frequency?

5. a. Describe in your own words the oscilloscope technique to make a peak-to-peak voltage measurement that does not happen to be a whole number of centimeters in height.

b. Repeat part (a) for a frequency determination.

c. Why is it generally more convenient to obtain the peak voltage by first determining the peak-to-peak value?

6. a. Assume that the *same* sine wave is applied to both the vertical and horizontal plates of an oscilloscope. What kind of Lissajous figure will result?

b. Draw two sine waves, one under the other and in phase with each other. Assume that one is connected to the vertical, the other to the horizontal deflection plates. Also assume that positive voltages deflect the electron beam upwards or to the right; negative voltages deflect the beam downward or to the left. Determine what pattern results in a Lissajous figure.

c. Repeat part (b), assuming that one is displaced by 90° or a quarter of a cycle from the other.

7. a. Why does a PMMC movement indicate zero when a pure ac voltage is applied to it?

b. How may a PMMC movement be used to indicate ac values?

c. What effect does this have on the sensitivity of an ac voltmeter in a VOM?

8. a. Explain how an electrodynamometer can be used to measure power.

b. Why is the wattmeter's scale linear but an ammeter's scale nonlinear?

9. Why is it possible for a wattmeter to read backward when measuring ac power?

10. Why is there some slight error in a wattmeter reading according to the way the potential terminals are connected?

11. a. What is a common drawback that moving-iron and electrodynamometer voltmeters have compared with a PMMC voltmeter?

b. What features do moving-iron and electrodynamometer meters have in common?

12. a. What problems can arise when making ac measurements with power-line-operated equipment in circuits that are also supplied from the 120-V, 60-Hz line?

b. If ac has "no polarity," why does it matter how a voltmeter is connected in a circuit?

c. Does this apply to portable meters such as a VOM?

d. If a circuit is battery-operated, is there any difficulty in using an oscilloscope without regard to grounds?

e. What problems may occur if a line-operated voltmeter and an oscilloscope are used simultaneously? Explain.

PROBLEMS

(Answers to odd-numbered problems at back of book)

16-1. Given a 5.3 cm peak-to-peak oscilloscope deflection, determine the peak voltages for the following VOLTS/CM settings assuming a 1X probe:

a. 10 mV/cm

b. 0.2 V/cm

c. 5 V/cm

16-2. Repeat Problem 16-1 using a 10X probe.

16-3. An oscilloscope displays five complete cycles in 8 cm. Determine the frequency for the following TIME/CM settings assuming a 10X probe:
   a. 0.5 s/cm
   b. 2 ms/cm
   c. 0.1 ms/cm
   d. 5 μs/cm

16-4. Repeat Problem 16-3 assuming three complete cycles in 7 cm.

16-5. The frequency of a voltage is being determined using the method of Lissajous figures as in Fig. 16-6. A stationary figure is obtained with five horizontal tangent points and three vertical tangent points when the horizontal frequency is adjusted to 360 Hz. What is the frequency of the unknown?

16-6. Repeat Problem 16-5 with the vertical and horizontal inputs interchanged.

16-7. A sinusoidal voltage having a peak-to-peak deflection of 4.6 cm at 0.2 V/cm is observed on an oscilloscope using a 10X probe across a 4.7-kΩ resistor. If one cycle occupies 2 cm at 0.2 ms/cm, determine:
   a. The peak value of the current.
   b. The frequency of the current.
   c. The equation representing the voltage.
   d. The equation representing the current.

16-8. A voltage having the equation $v = 20 \sin 2000\pi t$ volts is applied to a 5.1-kΩ resistor. Calculate:
   a. The reading of an ac voltmeter connected across the resistor.
   b. The reading of an oscilloscope connected across the resistor in volts.
   c. The reading of an ac ammeter connected in series with the resistor.
   d. The reading of a frequency meter, in the period mode, connected across the resistor.

16-9. An oscilloscope connected across a resistor indicates 18 V_{p-p} while a series-connected ammeter indicates 2.4 mA. What is the resistance of the resistor?

16-10. What peak-to-peak voltage does an oscilloscope indicate when connected across a 1.2-kΩ resistor when an ac ammeter indicates a resistor current of 10 mA?

16-11. Repeat Problem 16-8 with $v = 60 \sin 10,000\pi t$ and $R = 22$ kΩ.

16-12. An oscilloscope indicates 100 Vp–p when connected across a resistor that dissipates 100 mW. What is the resistance of the resistor?

16-13. Refer to Fig. 16-15. Given: the oscilloscope across R_2 indicates 28.28 V_{p-p}; the ac ammeter indicates 3 mA; and the ac voltmeter across R_4 indicates 9 V, $R_1 = R_4$. Also, $R_3 = 10$ kΩ. Calculate:
   a. The reading of an ac voltmeter connected across V_T.
   b. The total resistance of the circuit.
   c. The resistance of R_2.
   d. The resistance of R_4.

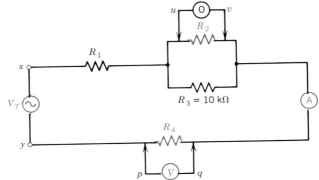

FIGURE 16-15
Circuit for Problems 16-13 through 16-16.

16-14. Repeat Problem 16-13 for $R_3 = 20$ kΩ and a voltmeter reading of 5 V.

16-15. Refer to Fig. 16-15. V_T is a 24-V source with point y grounded. Also, $R_1 = 2$ kΩ, $R_2 = 10$ kΩ, $R_3 = 10$ kΩ, and $R_4 = 5$ kΩ.

    a. With neither oscilloscope nor voltmeter connected, what is the reading of the ac ammeter?

    b. What are the readings of the ac ammeter and voltmeter when the voltmeter is connected across R_4 with side q grounded?

    c. What are the readings of the ammeter, voltmeter, and oscilloscope when the oscilloscope is connected across R_2 with side v grounded?

16-16. Repeat Problem 16-15 with V_T as a 36-V ac source with point x grounded.

CHAPTER 17

INDUCTANCE

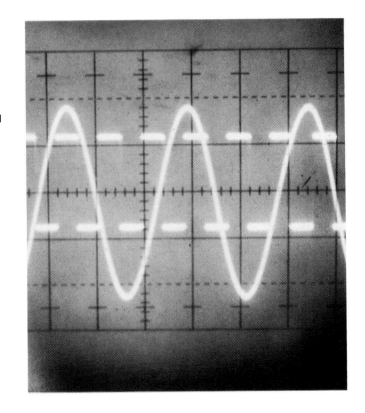

So far, you have considered only one property of an electric circuit: *resistance.* Resistance opposes the flow of current no matter whether the current is steady or changing. In this chapter, you will learn about inductance. Inductance is the property of a circuit or component that opposes *only* a *change* in current through the circuit or component.

This property relies on electromagnetic induction to induce an *opposing* emf whenever the current is *changing.* If this emf is induced in the *same* coil as the changing current, the effect is called *self-inductance.* When the emf is induced in a *nearby* coil that is magnetically coupled to the coil with the changing current, the effect is called *mutual inductance.* Discussed in this chapter will be the factors that affect inductance and the result of making series and parallel connections of inductors.

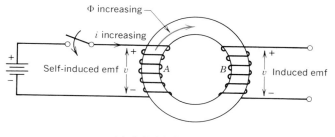

(a) Switch being closed

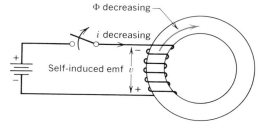

(b) Switch being opened

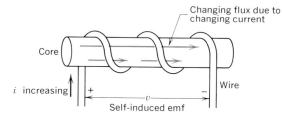

(c) Changing flux from one turn "cuts" adjacent turns in the coil

FIGURE 17-1

Self-induced emf due to inductance and changing current.

17-1 SELF-INDUCTANCE

Inductance is a property of a circuit or coil that arises from electromagnetic induction in the coil itself or in a neighboring coil. (See Fig. 17-1a.)

In Section 14-3, you saw how an emf was induced in a coil, at the instant a switch was closed, by a changing field *within* that coil. This is the situation shown in coil *B* of Fig. 17-1a. But the conditions that induce the emf in coil *B* are also present in coil *A*. Thus, a *self-induced* emf is developed in coil *A* (within the coil itself), opposing the *applied* emf.

Similarly, when the switch is opened, as in Fig. 17-1b, the decreasing current (and flux) in coil *A* self-induces an emf with a polarity that attempts to keep the current flowing. This is another example of Lenz's law at work, as explained in Chapter 14.

The way in which an expanding (or collapsing) magnetic field in one turn of a coil links and "cuts" adjacent turns is shown in Fig. 17-1c. This is the source of the self-induced emf in a coil with a changing current.

The ability of a coil to self-induce an emf when current through it is changing is called the coil's *self-inductance*, or simply *inductance*. The letter symbol for this property is *L*; its basic unit is the henry (H).

Inductance (by Lenz's law) always creates an induced voltage that is *opposed* to the effect that created it. But it only does this when the current is *changing* and thus, causing a change in magnetic flux. Also, it is found that the self-induced emf is higher if the current is made to change at a faster rate. The following equation states these two effects:

$$v = L\left(\frac{\Delta i}{\Delta t}\right) \qquad \text{volts} \qquad (17\text{-}1)$$

where: v is the emf induced, in volts (V)

L is the inductance, in henrys (H)

Δi is the change in current, in amperes (A)

Δt is the change in time, in seconds (s)

$\left(\dfrac{\Delta i}{\Delta t}\right)$ is the rate of change of current, in amperes/second (A/s)

The polarity of the induced emf is best determined from an application of Lenz's law.*

---

*Some texts use a negative sign in Eq. 17-1:

$$v = -L\frac{\Delta i}{\Delta t} \qquad \text{volts} \qquad (17\text{-}1a)$$

to signify that the voltage induced in the coil is a *counter* emf. We shall use Eq. 17-1 to give the voltage *across* the coil (either as an emf or voltage drop) and Lenz's law to obtain the polarity from the conditions that caused it.

EXAMPLE 17-1

The current in a 200-mH coil increases from 2 to 5 A in 0.1 s. Calculate:

a. The rate of change of current.
b. The self-induced emf.

Solution

a. $\dfrac{\Delta i}{\Delta t} = \dfrac{5\text{ A} - 2\text{ A}}{0.1\text{ s}} =$ **30 A/s**

b. $v = L\left(\dfrac{\Delta i}{\Delta t}\right)$ (17-1)

$= 200 \times 10^{-3}\text{ H} \times 30\text{ A/s} =$ **6 V**

Equation 17-1 can be rearranged to obtain the definition for inductance in terms of the induced voltage and rate of change of current:

$$L = \frac{v}{\left(\dfrac{\Delta i}{\Delta t}\right)} \qquad (17\text{-}1b)$$

From Eq. 17-1b, you can see that, **an inductance of 1 H may be defined as that inductance which, when current through it is changing at the rate of 1 A/s, induces an emf of 1 V.** (See Fig. 17-2.)

Note the schematic symbol for inductance used in Fig. 17-2. The parallel lines next to the coil indicate a magnetic (iron) core. Such a core is needed for large values of inductance (1 H or higher). However, as you will learn in the next section, the magnetic core causes the inductance to vary with the current through the coil.

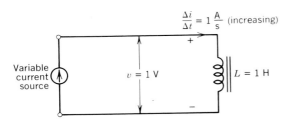

FIGURE 17-2

An inductance of 1 H induces an emf of 1 V due to a changing current of 1 A/s.

EXAMPLE 17-2

The current in a 5-H coil varies as shown in Fig. 17-3a. Draw the waveform of voltage induced in the coil.

Solution

During the first 10 ms:

$$v = L\frac{\Delta i}{\Delta t} \qquad (17\text{-}1)$$

$$= 5\text{ H} \times \frac{(1.0 - 0.4)\text{ A}}{10 \times 10^{-3}\text{ s}}$$

$$= 5\text{ H} \times 60\text{ A/s} = \mathbf{300\ V}$$

Note that $\dfrac{\Delta i}{\Delta t}$ is constant at 60 A/s during this interval (0 to 10 ms), so the induced voltage is constant as shown in Fig. 17-3b.

From 10 to 20 ms there is no change in current, so $\dfrac{\Delta i}{\Delta t} = 0$. As a result, the induced voltage is zero from 10 to 20 ms.

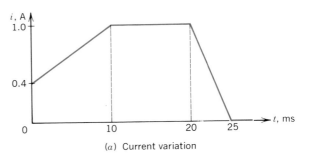

(a) Current variation

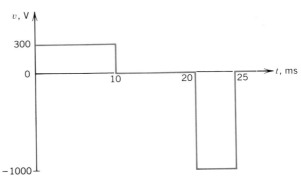

(b) Induced voltage

FIGURE 17-3

Waveforms for Example 17-2.

During the interval from 20 to 25 ms, the current is *decreasing* at a constant rate, where:

$$v = L \frac{\Delta i}{\Delta t} \tag{17-1}$$

$$= 5 \text{ H} \times \frac{(0 - 1) \text{ A}}{(25 - 20) \times 10^{-3} \text{ s}}$$

$$= 5 \text{ H} \times (-200 \text{ A/s}) = -\textbf{1000 V}$$

Since the current is *decreasing,* the induced voltage has the *opposite,* (negative) polarity, as shown in Fig. 17-3b, and is constant at -1000 volts.

17-2 FACTORS AFFECTING INDUCTANCE

The self-induced voltage in a coil with a changing current and flux was given earlier by Faraday's law:

$$v = N \left(\frac{\Delta \Phi}{\Delta t} \right) \tag{14-2}$$

But

$$\Phi = \frac{Ni}{R_m} \tag{11-8}$$

where Φ is the instantaneous value of flux due to the instantaneous value of the current (i), and

$$R_m = \frac{l}{\mu A} \tag{11-9}$$

is the reluctance of the coil's magnetic circuit.

Thus

$$v = N \frac{\Delta}{\Delta t} \left(\frac{N \mu A i}{l} \right)$$

But $\frac{N \mu A}{l}$ are constant for a given coil, and so:

$$v = \frac{N^2 \mu A}{l} \left(\frac{\Delta i}{\Delta t} \right)$$

But

$$v = L \left(\frac{\Delta i}{\Delta t} \right) \tag{17-1}$$

and therefore

$$L = \frac{N^2 \mu A}{l} \quad \text{henrys} \tag{17-2}$$

where: L is the inductance of the coil, in henrys H (or Wb/At)

N is the number of turns in the coil (dimensionless)

μ is the permeability of the magnetic core, in Wb/At-m

A is the cross-sectional area of the core, in square meters (m^2)

l is the length of the core, in meters (m) (length of the coil for air-core coils)

Note that Eq. 17-1 is merely a restatement of Faraday's law, using current variation instead of flux change. Equation 17-2, therefore, assumes that *all the flux* set up by the coil links *all the turns* of the coil. For iron-cored coils, this is reasonably accurate. But for *long* air-core coils, you must use empirical equations found in handbooks to take *leakage of flux* into account. In such a coil, some turns do not have the same flux linkage as other turns. If the air-core coil is relatively short in length, however, it may be treated as shown in Example 17-3.

EXAMPLE 17-3

A coil of 50 turns is wrapped on a hollow nonmagnetic core of cross-sectional area 1 cm^2 and length 2 cm. Calculate:

a. The inductance of the air-core coil $\left(\mu_0 = 4\pi \times 10^{-7} \frac{\text{Wb}}{\text{At-m}} \right)$.

b. The new inductance when an iron core of *relative* permeability 200 is inserted in the hollow core.

c. The inductance if the number of turns on the iron core is doubled, with no other changes.

Solution

a. $L = \dfrac{N^2 \mu_0 A}{l} \tag{17-2}$

$$= \frac{50^2 \times 4\pi \times 10^{-7} \text{ Wb/At-m} \times 1 \times 10^{-4} \text{ m}^2}{2 \times 10^{-2} \text{ m}}$$

$$= \textbf{15.7 } \mu\textbf{H}$$

b. $\mu = \mu_r \mu_0 \tag{11-5}$

$$= 200 \, \mu_0$$

$$L = 200 \times 15.7 \, \mu\text{H} = \textbf{3.1 mH}$$

c. If the number of turns is doubled, L will quadruple

$$L = 4 \times 3.1 \text{ mH} = \textbf{12.4 mH}$$

In developing Eq. 17-2, the permeability (μ) was assumed to be constant. This is true for coils with cores made of *nonmagnetic* materials (Example 17-3A). But as shown in Section 11-4.4, the permeability of magnetic materials (such as iron) varies widely with the amount of flux density and thus the current in the coil. This means that the inductance of any coil with a magnetic core is *never* constant; it varies with the amount of current (and flux) in the coil. Many coils have their inductance quoted at some specified value of current. For example, a coil may have 8 H of inductance at 85 mA, but 14 H if operated with only 1 mA of current.

17-3   PRACTICAL INDUCTORS

Coils wound intentionally to produce a specified amount of inductance are known as *inductors* or *chokes*. They range in value from a few microhenrys in high-frequency communication circuits through millihenrys and up to several henrys in power-supply filter circuits. Figure 17-4 shows a typical filter coil (other inductors are shown in Fig. 1-4b).

In some applications, a *variable* inductance is required. Variability is achieved by using a movable core or *slug* that can be screwed into or out of the coil. (See Appendix A for variable inductor schematic symbols.)

It should be noted that *every* conductor, even a straight wire (one turn), has *some* inductance. When the wire is wrapped into a coil, the flux set up by one turn can more readily link the other turns, increasing the self-induced voltage and inductance.

17-4   INDUCTANCES IN SERIES AND PARALLEL

Series or parallel combinations of inductors can be used to obtain a desired inductance value.

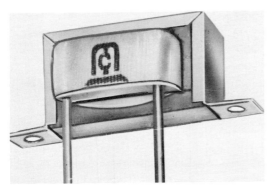

FIGURE 17-4
Typical power frequency filter coil. (Courtesy of Microtran Company, Inc.)

If the inductances are connected in series, as shown in Fig. 17-5a, their self-induced voltages are *additive*, and so are their inductive effects. Thus,

$$L_T = L_1 + L_2 + L_3 + \cdots + L_n \qquad \text{henrys} \quad (17\text{-}3)$$

where L_T is the total inductance due to series-connected inductances having individual values of L_1, L_2, and so on.

When connected in parallel, as in Fig. 17-5b, the total inductance is given by

$$\frac{1}{L_T} = \frac{1}{L_1} + \frac{1}{L_2} + \frac{1}{L_3} + \cdots + \frac{1}{L_n} \qquad \text{henrys} \quad (17\text{-}4)$$

For two inductors in parallel:

$$L_T = \frac{L_1 \times L_2}{L_1 + L_2} \qquad \text{henrys} \qquad (17\text{-}5)$$

Note that the above equations for inductors have the *same* form as those used for series- and parallel-connected *resistors*. They apply, however, *only* if the coils are separated far enough to prevent interaction of their magnetic fields (known as *coupling*). The equations assume that there is *no* mutual inductance. This occurs when coils are widely separated or are at right angles to one another.

EXAMPLE 17-4

Two inductors placed at right angles have values of 8 mH and 12 mH. Calculate the total inductance when they are connected in

a.   Series.
b.   Parallel.

Solution

a. $L_T = L_1 + L_2$ (17-3)
        $= 8 \text{ mH} + 12 \text{ mH}$

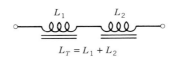

(a) Series connection of iron-core coils

(b) Parallel connection of iron-core coils

FIGURE 17-5
Series- and parallel-connected inductances.

$$= 20 \text{ mH}$$

b. $\quad L_T = \dfrac{L_1 \times L_2}{L_1 + L_2}$ \hfill (17-5)

$$= \frac{8 \text{ mH} \times 12 \text{ mH}}{8 \text{ mH} + 12 \text{ mH}} = \frac{96}{20} \text{ mH} = \textbf{4.8 mH}$$

17-5 MUTUAL INDUCTANCE

When two coils are parallel to each other (or on the same core) and sufficiently close to each other that their magnetic fields interact, an additional inductive effect occurs. Figure 17-6 illustrates such a situation: two coils that are not connected electrically but *are* magnetically coupled.

As you have seen, a changing current (i) not only self-induces an emf in L_1, but also causes a voltage to be induced in L_2. This is due to the changing flux (Φ_1). But the voltage induced in L_2 causes a current i_2 that sets up its own changing flux (Φ_2). This, in turn, not only self-induces an *additional* voltage in L_2, but also induces an additional voltage in L_1.

In other words, a changing current in *one* coil can induce an emf in the *other* coil due to a mutual flux. This effect is called *mutual induction,* and the two coils are said to have a mutual inductance (M), apart from their own self-inductances.

Mutual inductance, like self-inductance, is measured in henrys.

Two coils have a mutual inductance of 1 H when a current changing at the rate of 1 A/s in one coil induces an emf of 1 V in the other coil.

Thus the mutually induced emfs in the circuit of Fig. 17-6 are given by

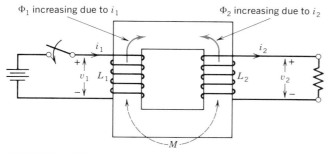

FIGURE 17-6
Mutual inductance M between two coils of self-inductance L_1 and L_2.

$$v_2 = M \frac{\Delta i_1}{\Delta t} \tag{17-6}$$

and $\qquad\qquad v_1 = M \dfrac{\Delta i_2}{\Delta t}$ \hfill (17-7)

These are in addition to the self-induced emfs

$$v_1 = L_1 \frac{\Delta i_1}{\Delta t} \tag{17-1}$$

$$v_2 = L_2 \frac{\Delta i_2}{\Delta t} \tag{17-1}$$

17-5.1 Series-Connected Coils with Mutual Inductance

Now consider the total inductance when two coils are series-connected, as in Fig. 17-7a. The total emf induced in coil 1 is due to L_1 and M:

$$v_1 = L_1 \frac{\Delta i}{\Delta t} + M \frac{\Delta i}{\Delta t}$$

The total emf induced in coil 2 is due to L_2 and M:

$$v_2 = L_2 \frac{\Delta i}{\Delta t} + M \frac{\Delta i}{\Delta t}$$

The total emf in both coils $= v_1 + v_2$:

$$v_T = \frac{\Delta i}{\Delta t} (L_1 + M + L_2 + M)$$

$$= \frac{\Delta i}{\Delta t} (L_1 + L_2 + 2M)$$

$$= L_T \frac{\Delta i}{\Delta t}$$

where $L_T = L_1 + L_2 + 2M \quad$ henrys \hfill (17-8)

Thus, the total inductance is increased above the self-inductances ($L_1 + L_2$) by an amount ($2M$) because the coils set up a *series-aiding* flux.

When coils are connected *series-opposing,* as in Fig. 17-7b, the total inductance is

$$L_T = L_1 + L_2 - 2M \quad \text{henrys} \tag{17-9}$$

In general, the total inductance of two series-connected coils, with mutual inductance M is

$$L_T = L_1 + L_2 \pm 2M \quad \text{henrys} \tag{17-10}$$

where the *positive* sign is used for *series-aiding* coils and the *negative* sign for *series-opposing* coils.

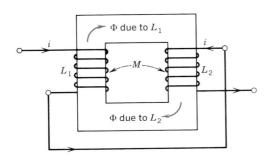

(a) Series-aiding coils set up flux in same direction
$L_T = L_1 + L_2 + 2M$

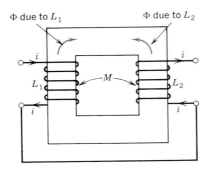

(b) Series-opposing coils set up flux in opposite directions
$L_T = L_1 + L_2 - 2M$

FIGURE 17-7
Two series-connected coils with mutual coupling.

EXAMPLE 17-5

Two coils of inductance 5 and 15 H have a mutual inductance of 4 H. Determine the following inductances available from series-connecting the two coils.
a. The maximum inductance.
b. The minimum inductance.

Solution

a. $L_{Tmax} = L_1 + L_2 + 2M$ (17-8)
$= 5\ H + 15\ H + 2 \times 4\ H$
$= 20\ H + 8\ H = \mathbf{28\ H}$

b. $L_{Tmin} = L_1 + L_2 - 2M$ (17-9)
$= 5\ H + 15\ H - 2 \times 4\ H$
$= 20\ H - 8\ H = \mathbf{12\ H}$

17-5.2 Dot Notation

Whether two coils are to be connected series-*aiding* or series-*opposing* is often indicated on schematics by using a *dot notation*. When current enters both dots (or leaves both dots), the effect is *additive* or series-aiding, as shown in Fig. 17-8*a*. But when the current enters one dot and leaves the other (Fig. 17-8*b*), the mutual inductance is subtractive, or series-opposing.

This dot notation is also used in both transformers and loudspeakers to show correct *phasing*. For example, when paralleling two loudspeakers, you want the two cones to move in the same direction at the same time. This reinforces the sound. Making a common connection to the two dotted inputs will ensure that the two speakers work in phase with each other.

17-5.3 Coefficient of Coupling

The amount of mutual inductance between two coils depends on two factors: the self-inductance of each coil and the amount of *mutual flux* between the two. The physical placement of the two coils determines the amount of mutual flux—the flux that links both coils. Mutual flux is indicated by the coefficient of coupling (*k*).

$$k = \frac{\Phi_m}{\Phi} \quad \text{(dimensionless)} \quad (17\text{-}11)$$

where: k is the coefficient of coupling between the two coils, $0 \le k \le 1$
Φ_m is the mutual flux between the two coils, in webers (Wb)
Φ is the total flux set up by one coil, in webers (Wb)

For example, when the coils are wrapped on the same iron toroid (as in Fig. 17-7), there is very little leakage of flux and k is practically equal to 1.

In Fig. 17-9*a*, if only 30% of the flux set up by coil 1 links coil 2, the coefficient of coupling is only 0.3. If the two coils are perpendicular to each other, as in Fig. 17-

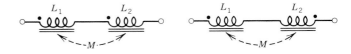

(a) Series-aiding coils, M positive
$L_T = L_1 + L_2 + 2M$

(b) Series-opposing coils, M negative
$L_T = L_1 + L_2 - 2M$

FIGURE 17-8
Dot notation for coils with mutual inductance.

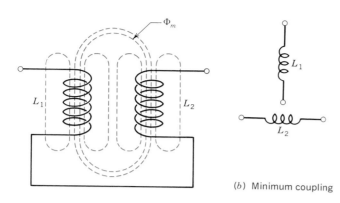

(a) Mutual flux, Φ_m between two coils

(b) Minimum coupling

FIGURE 17-9
Coefficient of coupling between coils.

9b, the coupling is minimum and the coefficient is close to zero.

It can be shown that the mutual inductance is given by

$$M = k\sqrt{L_1 \cdot L_2} \qquad \text{henrys} \qquad (17\text{-}12)$$

where: M is the mutual inductance, in henrys (H)
L_1 is the self-inductance of coil 1, in henrys (H)
L_2 is the self-inductance of coil 2, in henrys (H)
k is the coefficient of coupling (dimensionless)

EXAMPLE 17-6

Two coils having inductances of 5 and 8 H are wound on the same core giving a coefficient of coupling of 0.8. Determine:
a. The mutual inductance between the two coils.
b. The maximum inductance if series-connected.
c. The minimum inductance if series-connected.

Solution

a. $M = k\sqrt{L_1 \cdot L_2}$ (17-12)
    $= 0.8 \times \sqrt{5\,H \times 8\,H}$
    $= 0.8 \times \sqrt{40}\,H$
    $= \mathbf{5.1\ H}$

b. $L_{Tmax} = L_1 + L_2 + 2M$ (17-8)
    $= 5\,H + 8\,H + 2 \times 5.1\,H$
    $= \mathbf{23.2\ H}$

c. $L_{Tmin} = L_1 + L_2 - 2M$ (17-9)
    $= 5\,H + 8\,H - 2 \times 5.1\,H$
    $= \mathbf{2.8\ H}$

17-5.4 Measurement of Mutual Inductance

The mutual inductance and coefficient of coupling between two coils can be measured using a technique suggested by Eq. 17-10. If the total inductance is measured with the fields *aiding:*

$$L_{T_a} = L_1 + L_2 + 2M \qquad (17\text{-}8)$$

and when opposing:

$$L_{T_o} = L_1 + L_2 - 2M \qquad (17\text{-}9)$$

Subtracting Eq. 17-9 from Eq. 17-8:

$$L_{T_a} - L_{T_o} = 4M$$

thus $M = \dfrac{L_{T_a} - L_{T_o}}{4}$ henrys   (17-13)

and $k = \dfrac{M}{\sqrt{L_1 \cdot L_2}}$ (dimensionless)   (17-14)

where L_{T_a} is the total inductance with the coils connected series-aiding, in henrys (H)
L_{T_o} is the total inductance with the coils connected series-opposing, in henrys (H)

An instrument used to measure inductance is described in Chapter 20. Example 17-7 shows how such an instrument can be used, with Eqs. 17-13 and 17-14, to obtain the mutual inductance and coefficient of coupling between two coils.

EXAMPLE 17-7

A coil of 40 mH is placed parallel to a second coil of unknown inductance. When series-connected one way, a total inductance of 100 mH is measured; when the connections are interchanged a total inductance of 36 mH is measured. Calculate:
a. The mutual inductance.
b. The inductance of the second coil.
c. The coefficient of coupling.

Solution

a. $M = \dfrac{L_{T_a} - L_{T_o}}{4}$ (17-13)

    $= \dfrac{100\ mH - 36\ mH}{4} = \mathbf{16\ mH}$

b. $L_{T_a} = L_1 + L_2 + 2M$ (17-8)

Therefore,

$$L_2 = L_{T_a} - L_1 - 2M$$
$$= 100\ \text{mH} - 40\ \text{mH} - 2 \times 16\ \text{mH}$$
$$= 60\ \text{mH} - 32\ \text{mH} = \textbf{28 mH}$$

c. $k = \dfrac{M}{\sqrt{L_1 \cdot L_2}}$ (17-14)

$$= \dfrac{16\ \text{mH}}{\sqrt{40\ \text{mH} \times 28\ \text{mH}}}$$

$$= \dfrac{16\ \text{mH}}{33.5\ \text{mH}} = \textbf{0.48}$$

EXAMPLE 17-8

Two *equal* coils are wrapped on the same core. When connected in series one way, a total inductance of 500 mH is measured; when the connections are interchanged, the total inductance is 50 mH. Calculate:
a. The mutual inductance.
b. The self-inductance of each coil.
c. The coefficient of coupling.

Solution

a. $M = \dfrac{L_{T_a} - L_{T_o}}{4}$ (17-13)

$$= \dfrac{500\ \text{mH} - 50\ \text{mH}}{4} = \textbf{112.5 mH}$$

b. $L_{T_a} = L_1 + L_2 + 2M$ (17-8)

Therefore, $2L = L_{T_a} - 2M$

$$= 500\ \text{mH} - 2 \times 112.5\ \text{mH}$$
$$= 275\ \text{mH}$$

and $L = \dfrac{275\ \text{mH}}{2} = \textbf{137.5 mH}$

c. $k = \dfrac{M}{\sqrt{L_1 \cdot L_2}}$ (17-14)

$$= \dfrac{112.5\ \text{mH}}{\sqrt{137.5\ \text{mH} \times 137.5\ \text{mH}}}$$

$$= \dfrac{112.5\ \text{mH}}{137.5\ \text{mH}} = \textbf{0.82}$$

Comparing Examples 17-7 and 17-8, note how two coils wrapped on the same core have a higher coefficient of coupling than when merely placed next to each other.

SUMMARY

1. Self-inductance is the property of a coil to induce an emf within itself due to the presence of a changing current.
2. A coil's self-induced emf depends upon the inductance (L) in henrys and the rate of change of current $\dfrac{\Delta i}{\Delta t}$ in amperes per second as given by

$$v = L\left(\dfrac{\Delta i}{\Delta t}\right)$$

3. The polarity of an *induced* emf is always such that it opposes any change in the applied current.
4. A coil has a self-inductance of 1 H when current changing at the rate of 1 A/s causes a self-induced emf of 1 V.
5. The inductance of a given coil depends on the square of the number of turns in the coil, the permeability and area of the core, and the length of the magnetic circuit as given by:

$$L = \dfrac{N^2 \mu A}{l}$$

6. Variable inductors use a movable iron core to vary the permeability, and thus the inductance of the coil.
7. The equations for inductances in series and parallel are the same as those for resistances if there is no mutual inductance (magnetic interaction) between the inductors.

8. Two coils have a mutual inductance (M) of 1 H if the current changing at the rate of 1 A/s in one coil induces an emf of 1 V in the second coil.

9. When two coils of self-inductance L_1 and L_2 are series-connected with a mutual inductance M, the total inductance is

$$L_T = L_1 + L_2 \pm 2M$$

depending on whether the mutual inductance of the coils is aiding or opposing.

10. The coefficient of coupling (k) between two coils determines the amount of mutual inductance (M) as given by

$$M = k\sqrt{L_1 L_2}$$

SELF-EXAMINATION

Answer T or F, or, for multiple choice, a, b, c, or d
(Answers at back of book)

17-1. Self-inductance always induces an emf that opposes the change in current that produced the emf. _____

17-2. A conductor must be wrapped into a coil before it can display the property of inductance. _____

17-3. If the current in a 50-mH coil changes from 10 to 20 mA in 5 ms, the induced emf is
    a. 500 mV
    b. 0.1 V
    c. 10 mV
    d. 100 V

17-4. If the number of turns in a coil is tripled, with all other factors constant, the inductance is
    a. doubled
    b. tripled
    c. quadrupled
    d. increased by a factor of nine.

17-5. The inductance of a coil is a constant, regardless of its application. _____

17-6. A variable inductor has a movable core that changes the permeability and therefore the inductance. _____

17-7. The same equations can be used for inductances in series and parallel as for resistors, in all applications. _____

17-8. Mutual inductance between coils arises because of an interaction of their magnetic fields. _____

17-9. Mutual inductance always causes an increase in the total inductance when two coils are series-connected. _____

17-10. Two 8-H coils with a mutual inductance of 8 H are connected in series. Their total maximum and minimum possible values are
    a. 16 H and 0 H
    b. 16 H and 8 H
    c. 32 H and 0 H
    d. 24 H and 0 H

17-11. The two coils in Question 17–10 must have a coefficient of coupling of:
    a. 0

b. 0.5
c. 0.9
d. 1

REVIEW QUESTIONS

1. a. Why should a higher rate of change of current in a coil cause a higher emf to be induced?
    b. What other factors will determine how much voltage is induced?
    c. How would you determine the polarity of the induced emf?
2. Refer to Fig. 17-1a.
    a. How will the voltages induced in coils A and B vary as the switch is closed and then opened?
    b. What would have to be done to coil B to make its induced voltage have the opposite polarity when the switch is closed and opened compared with part (a)?
3. Define 1 H of self-inductance and 1 H of mutual inductance.
4. a. What causes the inductance of a coil to vary with the current?
    b. What would you expect to happen to the inductance at saturation? Why?
    c. Why should Eq. 17-2 not hold for long air-core coils?
5. Under what conditions do series- and parallel-connected inductors obey the same equations as those for resistors?
6. a. When two inductors that have mutual inductance are series-connected, how many sources of induced emf are there when the current through the coils changes?
    b. What determines whether all these emfs are additive?
    c. What is meant by the "dot notation"?
7. a. What is the highest coefficient of coupling that is possible between two coils?
    b. How can this be achieved?
    c. If the coefficient of coupling is unity, and the two coils have equal self-inductances, what can you say about the value of the mutual inductance?
    d. What does this mean about the total inductance when the two coils are connected series-opposing?

PROBLEMS

(Answers to odd numbers at back of book)

17-1. The current in a 7-H coil increased from 0 to 1.4 A in 100 ms. Calculate:
    a. The rate of change of current.
    b. The self-induced emf.
17-2. The current in a 200-μH coil decreased from 12 to 7 mA in 10 μs. Calculate:
    a. The rate of change of the current.
    b. The self-induced emf.
17-3. A 50-mH coil of negligible resistance has a voltage across its terminals of 30 V. Calculate the rate at which the current is changing in the coil.
17-4. What is the inductance of a coil if an emf of 300 V is induced when the current changes from 100 to 101 A in 30 ms?

17-5. A coil of 100 turns is wrapped on a hollow nonmagnetic core of diameter 2 cm and length 3 cm. Calculate:
  a. The inductance of the coil.
  b. The new inductance when an iron core having a relative permeability of 180 is inserted in the hollow core.

17-6. An 8-H inductor is to be made by wrapping wire around a circular cross section iron toroid having inside and outside diameters of 5 and 8 cm, respectively. How many turns are required if the iron's relative permeability is 250?

17-7. The current in a 5-mH coil increases from 0.5 to 1.2 mA in 10 μs, remains constant at this value for 20 μs, and falls to 0 in 2 μs. Draw waveforms of current variation and induced emf in the coil showing all the numerical values.

17-8. A square wave of current of peak-to-peak value 20 mA passes through a 2-H coil. If the current takes 100 μs to change from one level to another and has a frequency of 1 kHz, draw the waveform of induced voltage across the coil as seen by an oscilloscope. Show representative time intervals on the horizontal axis as well as peak voltages on the vertical axis.

17-9. Three inductors have values of 20, 40, and 60 mH. Determine:
  a. The total inductance when connected in series.
  b. The total inductance when connected in parallel.
  c. How they should be connected to give a total inductance of 44 mH.

17-10. Two inductors, with no mutual inductance, have a total inductance of 50 H when connected in series and 5 H when connected in parallel. What are the values of the two inductors?

17-11. a. What are the maximum and minimum possible values of total inductance when two coils of self-inductance, 7 H and 8 H, are series-connected with a mutual inductance of 3 H between them?
  b. What is the coefficient of coupling between the two coils?

17-12. Two *equal* coils wound on the same core have a total inductance of 10 H when connected series-aiding and 5 H when connected series-opposing. Calculate:
  a. The self-inductance of each coil.
  b. The mutual-inductance between the coils.
  c. The coefficient of coupling between the two coils.

17-13. If the coefficient of coupling between two coils of self-inductance, 150 μH and 200 μH, is 0.2, find the *maximum* inductance when these two coils are series-connected.

17-14. A coil is known to have twice the self-inductance of a second coil. When connected series-aiding, their total inductance is 100 mH; when connected series-opposing, the total inductance is 10 mH. Calculate:
  a. The self-inductance of each coil.
  b. The mutual inductance between the coils.
  c. The coefficient of coupling between the two coils.

CHAPTER 18

TRANSFORMERS

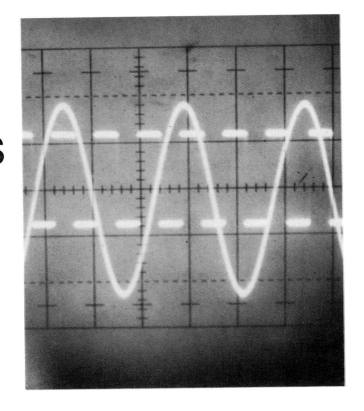

In Chapter 17, you saw that two coils wound on the same iron core have a property called *mutual inductance*—a changing current in one coil induces an emf in the other coil. An important application of this property is in iron-core *power transformers.* These devices have the ability to step-up or step-down alternating voltages, making the distribution of electrical power between the generator and the customer practical and economical. They can also transform load resistances to effect *matching* for maximum power transfer.

A special type of transformer, with the output voltage the same as the input voltage, is used for *isolation* where voltage supply grounds can cause problems while working with grounded equipment. Another special type is the *auto-transformer,* which is particularly adaptable to variable voltage operation. *Instrument* transformers, in both *potential* and *current* types, are designed to transform the high voltage and current in distribution systems, allowing the safe measurement of power. Finally, this chapter will cover briefly a residential wiring system, the safety features of a ground system, and the operating principles of a ground-fault circuit-interrupter.

18-1 THE IRON-CORE TRANSFORMER

As you have seen, the mutual inductance between two coils can cause an emf to be induced in one coil as a result of a changing current in the other coil. A *transformer* is a direct application of the property of mutual inductance. Two coils are wound on the same magnetic core to provide a coefficient of coupling close to unity (1.0). An alternating current is applied to one coil (the *primary* winding), while the other coil (the *secondary* winding) is connected to the load. Generally speaking, the two windings have different numbers of turns. The ratio of the number of turns on the primary (N_p) to the number of turns on the secondary (N_s) is called the *turns ratio* (*a*).

$$a = \frac{N_p}{N_s} \quad \text{(dimensionless)} \quad (18\text{-}1)$$

For example, if there are 200 primary turns and 50 secondary turns, $a = 4$.

18-1.1 Voltage Ratio

Consider the application of a sinusoidal voltage to the primary of a transformer such as the one shown in Fig. 18-1. Since the input voltage is sinusoidal, the flux set up in the magnetic core is *also* sinusoidal. This means that the flux is constantly changing, setting up a sinusoidal voltage in the secondary by mutual induction.

The primary and secondary voltages are given by Faraday's law:

$$V_p = N_p \frac{\Delta \Phi_p}{\Delta t} \quad \text{volts} \quad (18\text{-}2)$$

$$V_s = N_s \frac{\Delta \Phi_s}{\Delta t} \quad \text{volts} \quad (18\text{-}3)$$

In a practical transformer, the primary and secondary coils are wound on top of each other on the same iron core. (See Fig. 18-1c.) This means that practically all the flux produced by the primary links the secondary. Thus

$$\frac{\Delta \Phi_p}{\Delta t} = \frac{\Delta \Phi_s}{\Delta t}$$

in the above equations for V_p and V_s. If, for example, $N_s > N_p$, then $V_s > V_p$. Thus, a transformer can transform one voltage level to another.

Dividing Eq. 18-2 by Eq. 18-3 gives

$$\frac{V_p}{V_s} = \frac{N_p}{N_s} = a \quad (18\text{-}4)$$

where all terms are as previously defined.

Equation 18-4 is often called the *transformation ratio*. It states that the ratio of primary to secondary *voltage* is the same as the ratio of primary to secondary *turns*.

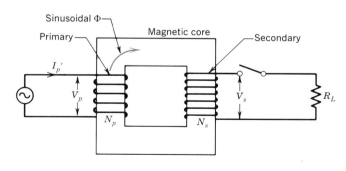

(a) Pictorial diagram showing common flux between the primary and secondary

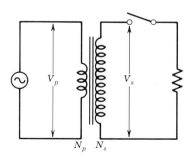

(b) Schematic symbol of a step-up transformer

(c) Power line transformer for use in a low voltage dc power supply suitable for mounting on a printed circuit board

FIGURE 18-1

Iron core transformer. [Photograph in (c) courtesy of Microtran Company, Inc.]

Clearly, from Eq. 18-4,

$$V_s = V_p \times \frac{N_s}{N_p}$$

Thus, if $N_s > N_p$, then $V_s > V_p$, and the result is a *step-up* transformer—one that increases the output voltage compared to the input voltage. It has a transformation ratio (*a*) *less* than unity.

If $N_s < N_p$, then $V_s < V_p$, and a *step-down* transformer results. In the step-down transformer, the output voltage is lower than the applied input voltage. The transformation ratio (*a*) is *greater* than unity.

Some transformers are designed to develop an output voltage the *same* as the input voltage. These *isolation* transformers are discussed in Section 18-4. (The *auto-transformer*, which is primarily used to step-up or step-down voltage, *may* also develop an output voltage the same as the input voltage. But since the auto-transformer includes a common electrical connection between input and output, it does not provide *isolation*. See Section 18-5.)

EXAMPLE 18-1

A transformer with 50 primary turns and 300 secondary turns has a primary voltage of 120 V. Determine:
a. The transformation ratio.
b. The secondary voltage.
c. The type of transformer.

Solution

a. $a = \dfrac{N_p}{N_s}$ (18-1)

   $= \dfrac{50 \text{ turns}}{300 \text{ turns}} = \mathbf{0.167}$

b. $\dfrac{V_p}{V_s} = \dfrac{N_p}{N_s}$ (18-4)

   $V_s = V_p \times \dfrac{N_s}{N_p}$

   $= 120 \text{ V} \times \dfrac{300 \text{ turns}}{50 \text{ turns}}$

   $= 120 \text{ V} \times 6 = \mathbf{720 \text{ V}}$

c. Since *a* is less than unity or since $V_s > V_p$, this is a step-up transformer.

It should be noted that the ability of a transformer to function depends on the continual changing of the field $\dfrac{\Delta \Phi}{\Delta t}$ set up by the alternating current. For this reason, a transformer cannot operate with constant dc.

18-1.2 Current Ratio

So far, transformers have been considered without a *load* connected to the secondary. Without a load, the only current flowing in the primary (I_p') is called the *exciting* or *magnetizing* current. This current is quite small, but is sufficient to set up the changing flux in the core. Now, consider what happens when a load is connected, as shown in Fig. 18-2.

A secondary current (I_s) flows, determined by V_s and R_L. But I_s must flow through N_s in such a way that (by Lenz's law) it sets up a flux opposing the changing flux that induced it.

This *demagnetizing* force (mmf) produced by the load current is given by $N_s I_s$. To restore and maintain the flux in the core at its initial level and supply the load, an *additional* current (I_p) must flow in the primary. This *magnetizing* force ($N_p I_p$) must equal the demagnetizing force ($N_s I_s$):

$$N_p I_p = N_s I_s$$

Thus $\dfrac{I_s}{I_p} = \dfrac{N_p}{N_s} = a$ (dimensionless)   (18-5)

where all the terms are as already defined.

Equating the value of *a* in Eqs. 18-4 and 18-5, you can see that the current ratio is the *inverse* of the voltage ratio:

Thus $\dfrac{I_s}{I_p} = \dfrac{V_p}{V_s} = a$ (dimensionless)   (18-5a)

This implies, for example, that if a transformer steps-*up* a voltage by a given ratio, the current will be stepped-*down* by the same ratio.

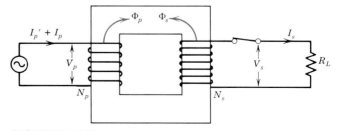

FIGURE 18-2

Transformer with a load sets up a demagnetizing force in the secondary winding.

EXAMPLE 18-2

A 120-V primary, 12.6-V secondary transformer is connected to a 10-Ω load. Calculate:

a. The turns ratio.
b. The secondary current.
c. The primary current.

Solution

a. $\dfrac{N_p}{N_s} = \dfrac{V_p}{V_s}$ \hfill (18-4)

$\quad = \dfrac{120 \text{ V}}{12.6 \text{ V}} = \mathbf{9.5}$

b. $I_s = \dfrac{V_s}{R_L}$ \hfill (3-1a)

$\quad = \dfrac{12.6 \text{ V}}{10 \ \Omega} = \mathbf{1.26 \ A}$

c. $I_p = I_s \times \dfrac{V_s}{V_p}$ \hfill (18-5a)

$\quad = 1.26 \text{ A} \times \dfrac{12.6 \text{ V}}{120 \text{ V}} = \mathbf{0.13A}$

Note that the primary current calculated in Example 18-2C is actually the *additional* current over and above the exciting current (I_p'). And, since I_p' is generally only 5% of the full-load current, it is small enough to be ignored, except in miniature transformers.

18-1.3 Power, Volt-Ampere Relations, and Transformer Efficiency

Equation 18-5a leads directly to consideration of the input and output power of the transformer.

Since $\quad \dfrac{I_s}{I_p} = \dfrac{V_p}{V_s}$ \hfill (18-5a)

then $\quad V_p I_p = V_s I_s \quad$ volt-amperes \hfill (18-5b)

This states that, assuming a 100% efficient transformer, the product of the primary volts and amperes equals the product of the secondary volts and amperes. The output *rating* of a transformer is given in *volt-amperes* (VA), rather than watts. This is because the load may not be purely resistive (in which case, the current and voltage are *not* in phase with each other). Recall, from Chapter 15, that $P = V_R I_R$ watts. That is, the power in watts is given

by the product of volts and amperes *only* when the voltage and current are in phase with each other.*

Unless otherwise noted, assume that all loads are purely resistive. In such a case, you can say that the input power $V_p I_p$ equals the output power $V_s I_s$, for a 100% efficient transformer. That is,

$$P_{\text{in}} = V_p I_p \quad \text{watts} \qquad (18\text{-}6)$$
$$P_{\text{out}} = V_s I_s \quad \text{watts} \qquad (18\text{-}7)$$

It is important to realize that a transformer, even a (voltage) step-up transformer, does not and cannot step-up *power*. An increase in *voltage* is accompanied by a similar decrease in *current* (and vice-versa). At best, only as much power is delivered to the load by the transformer secondary as is drawn from the supply by the transformer primary.

The transformer is merely an *energy transfer* device, using magnetic coupling to transfer energy from the primary to the secondary circuit. There is *no* direct electrical connection between the primary and the secondary.

In a practical transformer, the output power is always slightly less than the input power, due to losses in the transformer itself.

There are three sources of losses in a transformer:

1. *Winding I^2R losses*. The resistance of the primary and secondary coil windings means that some of the input power is converted internally to heat, as given by the equation $P = I^2R$. (These are commonly called copper losses.)
2. *Hysteresis losses*. The magnetic field of the iron core undergoes a complete reversal 60 times each second (for power line frequencies). Reversing magnetism of the iron core expends energy in the form of heat. The heat loss is proportional to the area of the magnetic material's *B-H* loop. (See Chapter 11.)
3. *Eddy current losses*. The changing magnetic field in the iron core induces an emf within the core, which is itself a conductor. This emf sets up currents that circulate in the core perpendicular to the flux through the core.

These circulating eddy currents generate heat due to the resistance of the core ($P = I^2R$), resulting in less power

---

*In general, the product *VI* is called the *apparent* power in volt-amperes. *True* power in watts is obtained by multiplying the apparent power by the *power factor*. This is the cosine of the angle between voltage and current ($\cos\theta$). That is, $P = VI\cos\theta$. For a pure resistor, $\theta = 0°$, so $\cos\theta = 1$ and the true power in watts is the same as the apparent power in volt-amperes. (See Section 25-6.)

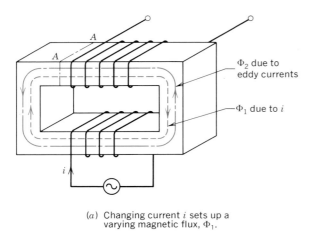

(a) Changing current i sets up a varying magnetic flux, Φ_1.

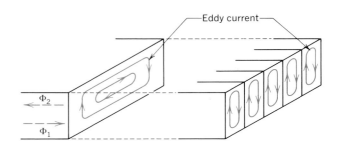

(b) Section A-A of a solid core showing circulating eddy currents that set up an opposing flux Φ_2 in the core.

(c) Section A-A of a laminated core showing how longer paths increase the resistance and reduce the eddy currents.

FIGURE 18-3
Eddy currents in solid and laminated cores.

available at the output of the transformer. The induced eddy currents also set up an opposing flux (Φ_2) in the core. (See Fig. 18-3a.) This results in more current flowing in the primary to try to maintain the core's magnetic field, further increasing losses.

Eddy currents can be induced in a core that is *not* a magnetic material—any good conductor, such as brass, will become very hot due to eddy currents. (This is the principle behind an industrial heating method called *induction heating*.)

To reduce eddy current losses, the core is often made from laminated sheets of steel, separated by thin layers of iron oxide and varnish. (See Fig. 18-3c.) Because the varnish has a high resistance, the eddy currents must travel longer paths, resulting in a greatly increased resistance. The induced emf in the core is not changed, but the greater resistance reduces the eddy currents considerably. As a result, the power loss is greatly reduced.

Losses increase with frequency, so transformers that operate at high frequencies (*radio frequency,* or RF, transformers) use a *ferrite* core. As explained in Chapter 11, ferrites are ferromagnetic but also relatively good insulators. This reduces eddy currents and their heating effects to a minimum.

A nonmagnetic *copper* or *aluminum* cover is often used for RF transformers. This cover provides a *shield* to prevent any nearby external circuits from being affected by the varying magnetic flux set up in the transformer windings by the RF current. The shielding is provided by the opposing magnetic field set up in the cover by the eddy currents. The cover also shields the transformer from ef-

fects that might be caused by *external* varying magnetic fields.

Power transformers, on the other hand, usually have a cover that is a good *magnetic* material. This is because the shielding effect resulting from eddy currents is only effective at high frequencies. The magnetic material shields the power transformer from external low-frequency and *static* magnetic fields (as from wires carrying direct current) that would pass through an aluminum or copper cover. (See Fig. 18-4.) The iron cover provides mechanical protection and also protects external circuits from the transformer's leakage flux.

Large power transformers can be as much as 98 to 99.5% efficient. Even at such high efficiencies, the heat generated within the transformer is still appreciable and can be a problem if not removed. Very large transformers, such as the one shown in Fig. 18-5, circulate an oil that is

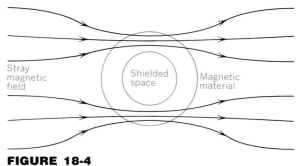

FIGURE 18-4
Magnetic shielding.

not electrically conductive through the windings and core to remove heat. Note that

$$P_{in} = P_{out} + \text{losses} \tag{18-8}$$

and efficiency $\eta = \dfrac{P_{out}}{P_{in}} = \dfrac{P_{out}}{P_{out} + \text{losses}} \tag{18-9}$

EXAMPLE 18-3

A 4160-V/230-V step-down transformer has a load of 4 Ω pure resistance connected to the secondary. Calculate the primary current assuming:
a. 100% efficiency.
b. 98% efficiency.
c. Calculate the losses in the case of part (b).
d. Determine the necessary output volt-ampere (VA) rating of the transformer.
e. On a circuit diagram show the conditions in part (b).

Solution

a. Secondary current, $I_s = \dfrac{V_s}{R_L}$ (15-15a)

FIGURE 18-5
A 15-kV, 3000-kVA, multiple-tapped oil-filled transformer used in an electrostatic precipitator system. (Courtesy of NWL Transformers.)

$$= \frac{230\ V}{4\ \Omega} = 57.5\ A$$

For 100% efficiency, $V_p I_p = V_s I_s$ (18-5b)

The primary current,

$$I_p = \frac{V_s}{V_p} \times I_s$$

$$= \frac{230\ V}{4160\ V} \times 57.5\ A = \mathbf{3.18\ A}$$

b. $P_{out} = \dfrac{V_s^2}{R_L}$ (15-13)

$$= \frac{(230\ V)^2}{4\ \Omega} = 13.23\ kW$$

efficiency $\eta = \dfrac{P_{out}}{P_{in}}$ (3-6)

$$P_{in} = \frac{P_{out}}{\eta}$$

$$= \frac{13.23\ kW}{0.98} = 13.5\ kW$$

Since the load is pure resistance:

$$P_{in} = V_p I_p \tag{18-6}$$

and $\quad I_p = \dfrac{P_{in}}{V_p}$

$$= \frac{13.5 \times 10^3\ W}{4160\ V} = \mathbf{3.25\ A}$$

c. Losses $= P_{in} - P_{out}$ (18-8)

$$= 13.5\ kW - 13.23\ kW$$

$$= 0.27\ kW = \mathbf{270\ W}$$

d. Transformer output rating (minimum)

$$= V_s I_s$$

$$= 230\ V \times 57.5\ A$$

$$= 13,225\ VA$$

$$= \mathbf{13.2\ kVA}$$

e. The conditions in part (b) are shown in Fig. 18-6.

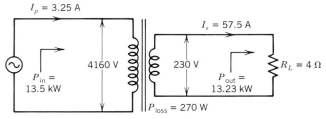

FIGURE 18-6
Circuit for Example 18-3.

18-1.4 Matching Resistances for Maximum Power Transfer

As you learned in Chapter 9, maximum dc power is transferred from a dc source to a load if the load resistance (R_L) is equal to the internal resistance (r) of the source. This requirement must also be met to achieve maximum ac power transfer in the special case where the source has *only* pure internal resistance and the load is also purely resistive. (The requirement for the more general case, including inductive and capacitive effects, is considered in Chapter 25.)

In this section, the use of a transformer to effect maximum power transfer if the source and load resistances are *not* equal will be considered. (This is only practical for electronic circuits, *not* power circuits.)

When a load (R_L) is connected to the secondary of a transformer, an additional current (I_p) flows in the primary, as shown in Fig. 18-7a.

As far as the input voltage (V_p) is concerned, it appears that a load (R_p) has been connected directly in parallel with the input. This load has a value of

$$R_p = \frac{V_p}{I_p} \qquad (15\text{-}15)$$

since, for pure resistance, the primary voltage and current are in phase with each other.

But
$$V_p = aV_s \qquad (18\text{-}4)$$

and
$$I_p = \frac{I_s}{a} \qquad (18\text{-}5)$$

Thus
$$R_p = \frac{V_p}{I_p} = a^2\frac{V_s}{I_s}$$

But, ignoring the resistance of the secondary winding itself,

$$\frac{V_s}{I_s} = R_L$$

Thus,
$$R_p = a^2 R_L \qquad \text{ohms} \qquad (18\text{-}10)$$

where: R_L is the load resistance on the secondary, in ohms (Ω)

R_p is the resistance "reflected" into the primary, in ohms (Ω)

a is the turns ratio of the transformer, $\frac{N_p}{N_s}$ (dimensionless)

Equation 18-10 shows that the load (R_L) on the secondary has been "reflected" into the primary side as an equivalent load equal to $a^2 R_L$. This means that a transformer can transform not only voltage and current, but also resistance.

A transformer with the proper turns ratio (a) can make a given load resistance (R_L) appear as a resistance equal to the source's internal resistance (r), as shown in Fig. 18-7b. Whenever $r = a^2 R_L$, it will provide maximum power transfer into the transformer and ultimately to the load connected across the secondary. This is shown in Example 18-4.

EXAMPLE 18-4

An audio amplifier has an open-circuit voltage of 20 V and an internal resistance of 125 Ω. A transformer is to be selected to match a 4-Ω loudspeaker to the amplifier. Calculate, assuming 100% efficiency:
a. The necessary turns ratio of the transformer.
b. The primary input voltage and current.
c. The secondary output voltage and current.
d. The output power transferred to the loudspeaker by the matching transformer.
e. The power transferred to the loudspeaker if connected directly to the amplifier.

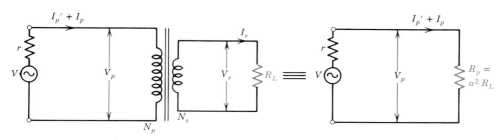

(a) Load R_L connected to the secondary

(b) Equivalent load a^2R_L connected to the primary

FIGURE 18-7

A load R_L connected to the secondary is reflected into the primary as an equivalent load of a^2R_L.

Solution

a. For matching, $R_p = 125\ \Omega$

$$a^2 = \frac{R_p}{R_L} \qquad (18\text{-}10)$$

$$a^2 = \frac{125\ \Omega}{4\ \Omega} = 31.25$$

Turns ratio $a = \sqrt{31.25} = \mathbf{5.6}$

b. Refer to Fig. 18-7b. When

$$R_p = r$$

$$V_p = \frac{V}{2}$$

$$= \frac{20\ \text{V}}{2} = \mathbf{10\ V}$$

Note that this equal division of voltage only holds for pure resistances. It does *not* apply to the general case where reactances are involved.

$$I_p = \frac{V_p}{R_p} \qquad (15\text{-}15a)$$

$$= \frac{10\ \text{V}}{125\ \Omega} = 0.08\ \text{A} = \mathbf{80\ mA}$$

c. $$\frac{V_p}{V_s} = a \qquad (18\text{-}4)$$

$$V_s = \frac{V_p}{a}$$

$$= \frac{10\ \text{V}}{5.6} = \mathbf{1.8\ V}$$

$$\frac{I_s}{I_p} = a \qquad (18\text{-}5)$$

$$I_s = aI_p$$

$$= 5.6 \times 0.08\ \text{A} = \mathbf{0.45\ A}$$

d. $$P_L = I_s{}^2 R_L \qquad (15\text{-}12)$$

$$= (0.45\ \text{A})^2 \times 4\ \Omega$$

$$= \mathbf{0.81\ W}$$

e. $$I = \frac{V}{r + R_L} \qquad (9\text{-}1)$$

$$= \frac{20\ \text{V}}{125\ \Omega + 4\ \Omega} = 0.16\ \text{A}$$

$$P_L = I^2 R_L \qquad (15\text{-}12)$$

$$= (0.16\ \text{A})^2 \times 4\ \Omega = \mathbf{0.1\ W}$$

Note that eight times as much power is delivered to the load when the matching transformer is used, compared to connecting the load directly across the source. Matching transformers are generally specified in terms of the resistances they are to match, rather than their turns ratio. An audio output transformer, suitable for mounting on a printed circuit board, is shown in Fig. 18-8. Note the multiple taps available to allow matching to a wide variety of loads.

18-2 POWER DISTRIBUTION SYSTEM

Figure 18-9 shows the three parts of a power system: the generating system, the transmission system, and the distribution system. Three-phase transformers, with ratings as high as 1000 MVA, are used to feed the generated power to the transmission line. At the substations, three single-phase transformers can be interconnected for three-phase operation. The step-up and step-down process is usually carried out in stages, rather than in a single step. High voltages are used directly by some industries, but residences are fed from a 120/240-V secondary, single-phase transformer.

The main reason for generating alternating (rather than direct) current is the ability of a transformer to step-up the voltage and distribute the power economically. Although some transmission systems use dc, the process of stepping up direct voltage is not simple.

If an alternator's output voltage of approximately 20 kV is increased by a factor of 30, the current is *reduced* by a factor of 30. This allows the use of smaller diameter transmission lines and reduces the I^2R power losses that occur.

FIGURE 18-8
Audio output transformer suitable for printed circuit board mounting. (Courtesy of Microtran Company, Inc.)

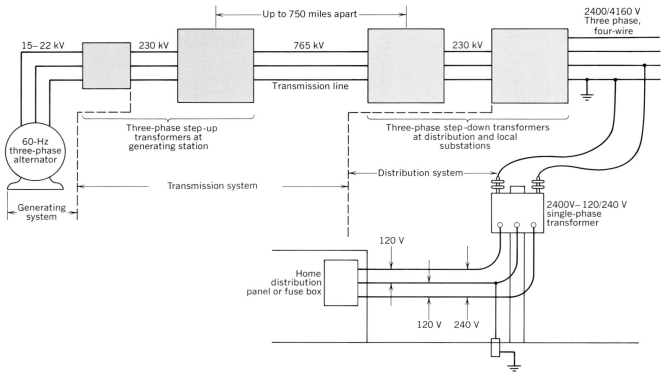

FIGURE 18-9

Use of step-up and step-down transformers in an electric power transmission and distribution system.

For long-distance transmission lines, the rule of thumb is that a potential of 1 kV is needed for each mile. The highest voltage in use today is 765 kV, but experiments are being conducted in the 1.5 to 2 million volt range.

18-3 RESIDENTIAL WIRING SYSTEM

A typical modern residential wiring system is shown in Fig. 18-10. It is important to note that the local transformer (located either above or below ground) is single-phase with a *center-tapped* secondary that is grounded. This produces a 120/240-V, three-wire single-phase (*not* three-phase) system.

After passing through a watthour meter that records the amount of energy used, the three wires feed the fuse box or circuit-breaker panel for the branch circuits. The grounded center tap is connected to the neutral and ground bus in the panel. To ensure a good earth ground, the ground bus is connected to a metal cold water pipe.

As shown in Fig. 18-10, some of the circuit breakers

are *ganged* together in pairs to provide the 240-V circuit needed by ovens and other large electrical appliances. Lights need only a 120-V "live" wire and a neutral. National and local electrical codes specify that the switch controlling a circuit be placed on the live side.

A light can be controlled from two different locations using *three-way* switches as shown in Fig. 18-11. The switches are actually of the single-pole, double-throw type. In the situation shown in Fig. 18-11, the circuit will be completed (lighting the lamp) if S1 is thrown to position *A* or S2 is thrown to position *B*. By adding *four-way* switches between S1 and S2, it is possible to control the light from many locations (such as a stairway serving several floors).

Receptacles require a third (grounding) wire that runs separately to the neutral or ground bus. The reason for this is explained in Section 18-3.1. Some local codes may also specify that a ground line be run to both lighting fixtures and wall receptacles.

The size of the wire to be used in a residential system is determined by the load and the fuse or circuit-breaker rating. A minimum of No. 14 gauge wire should be used

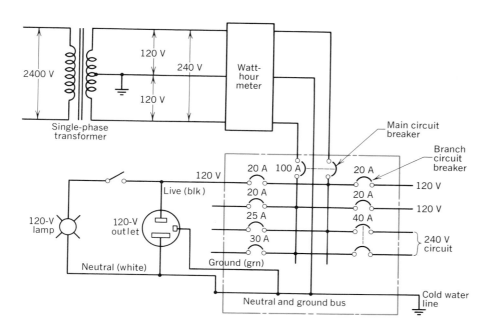

FIGURE 18-10

Residential wiring system.

with a 15-A circuit breaker, No. 12 gauge for 20-A, No. 10 gauge for 30-A, and No. 8 gauge for 40-A. (These are the commonly available types of wire for residential use.) If wire and breaker or fuse sizes are properly matched, the breaker will "trip" or the fuse "blow" before the safe current-carrying capacity of the wire is exceeded. This avoids an overheating that may cause a fire.

Where the use of aluminum wire is permitted by local codes, special care must be taken to ensure that connections are tight. Loose connections create a high resistance and can develop enough heat to ignite the aluminum.

18-3.1 Safety Features of a Ground Connection

Figure 18-12 shows an appliance with a metal case fed from a 120-V receptacle. Assume that the appliance cord

is of the two-wire type, with only a live and a neutral wire, and that a short has developed on the live side due to the wire fraying at the point where it passes through the metal case. This causes the case to become live or "hot"—it is part of the 120-V circuit.

If a person touches both the case and a ground at the same time, his or her body completes a path for current to flow. This can happen easily in the damp conditions around a clothes washer or dishwasher. The result would be a severe (possibly fatal) electrical shock.

Now, consider the same situation, but with the addition of a third *ground* wire connecting the metal case to the grounded receptacle. This is shown by the dotted line in Fig. 18-12. If a short to the metal case develops, the ground wire provides a path for the current to flow. In most cases, this is sufficient to cause the circuit breaker to open, disconnecting the power. Even if the current flow

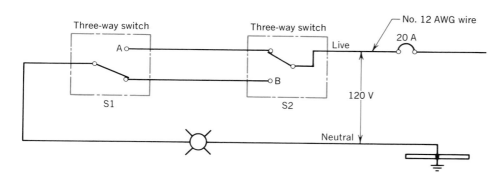

FIGURE 18-11

Three-way switched circuit that controls a light from two locations.

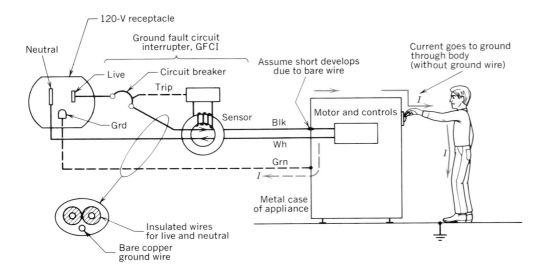

FIGURE 18-12

Current paths with and without ground-return wire.

isn't sufficient to trip the breaker, the case will no longer have a dangerous 120-V potential.

18-3.2 Ground-Fault Circuit-Interrupters (GFCIs)

Ground protection using circuit breakers is not sufficient in some situations, since circuit breakers are thermally operated and may not open quickly enough to prevent injury or damage. An example would be a swimming pool, where short-circuit current could be instantly fatal. In such situations, and where portable equipment is operated, the National Electrical Code requires the use of ground-fault circuit-interrupters (GFCIs). These devices, as shown in Fig. 18-12, use a toroidal sensing transformer with both the live and neutral wires from the power source passing through the toroid. When operation is normal, the two equal currents in opposite directions through the wires set up equal opposing magnetic fields. But if a fault occurs, the current takes an alternate path to ground, causing the currents in the wires through the toroid to be unbalanced. This develops a voltage in the coils of the sensing transformer, opening contacts to disconnect the equipment from the power source. The GFCI is very sensitive and fast-reacting—a difference of only 5 mA will cause the current to be interrupted in just 25 ms.

18-4 ISOLATION TRANSFORMERS

In Section 16-6, you learned about the precautions that must be taken when using grounded equipment. This is particularly important when measurements must be made on a circuit operated from the 120-V, 60-Hz electrical supply.

Figure 18-13a shows what will occur if the equipment ground of the measuring instrument is connected to the "live" side of the supply. This connection provides a low-resistance path (short circuit) that allows a large current to flow, damaging the instrument and the circuit.

That type of problem is avoided by using an *isolation transformer*, which accepts a 120-V input and produces a 120-V output *without* any electrical connection between input and output. (See Fig. 18-13b.) The secondary side does not have a supply ground. This means that the equipment ground of an instrument connected to the transformer secondary side cannot provide a low resistance return path to the ungrounded side of the primary. Note, however, that if two grounded instruments are used simultaneously on the secondary side, their unmatched grounds can cause a short circuit in the transformer secondary. (See Fig. 18-13c.) To avoid this problem, the instrument grounds must be connected to each other.

Figure 18-14 shows a low-powered isolation transformer, suitable for mounting on a printed circuit board.

It should be noted that the isolation transformer can be used to operate either the circuit under test or the measuring instrument. (In either case, the volt-ampere rating of the transformer must be checked to determine if it can handle the load.)

The use of an ordinary isolation transformer, however, permanently removes the normal ground protection for the circuit. To provide both protection *and* isolation, a device called a *ground isolation monitor* has been developed. Figure 18-15 shows such a monitor.

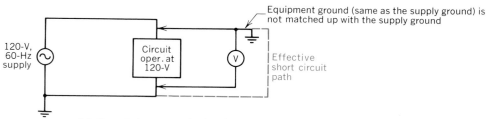

(a) Grounded ac measuring instrument shorts out the 120-V, 60-Hz supply.

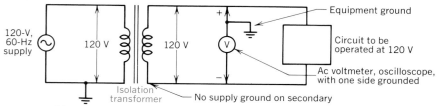

(b) Isolation transformer isolates the grounded supply from the load circuit and the ac measuring instrument. The instrument may be connected either way.

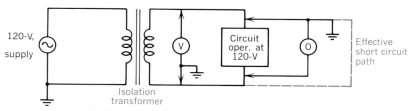

(c) Two grounded measuring instruments used simultaneously with their grounds not matched cause the secondary side to be shorted.

FIGURE 18-13

Use of an isolation transformer and precautions with grounded equipment.

FIGURE 18-14

Isolation transformer suitable for printed circuit board mounting. (Courtesy of Microtran Company, Inc.)

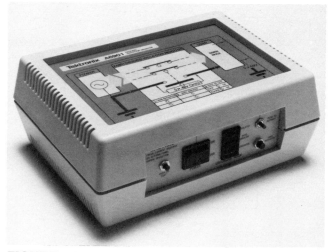

FIGURE 18-15

A ground isolation monitor provides protection and isolation. (Courtesy of Tektronix, Inc.)

The monitor provides ac power to the test instrument while temporarily interrupting the protective ground to the instrument. This eliminates the common path for unwanted ground current and allows the test instrument (specifically, an oscilloscope) to "float." However, if predetermined voltage and current limits are exceeded (indicating a possible danger to the operator), the power is removed from the instrument and the ground connection is re-established. In a sense, this device operates much like a GFCI.

18-5 AUTOTRANSFORMERS

Where isolation between input and output is not important, considerable savings can be realized by using an autotransformer in preference to a conventional transformer. As shown in Fig. 18-16, the autotransformer has a common electrical connection between input and output. It also has a sliding contact like that of a potentiometer, making it

adaptable to variable voltage operation. However, the autotransformer does *not* function as a simple voltage divider.

Assume that the autotransformer shown in Fig. 18-16b has a total of 240 turns from point a to point c. With a primary voltage of 120 V, 80 V will be across the 160 turns from point a to point b, and 40 V across the 80-turn section from point b to point c. A secondary output voltage of 80 V across a 10-Ω pure resistance load causes a secondary current of 8 A. The power delivered to the load is 80 V × 8 A = 640 W.

If transformer losses are ignored, the input power from the 120-V source must also be 640 W, requiring an input current of 5.3 A (640 W/120 V = 5.3 A). Applying Kirchhoff's current law at point a shows that 2.7 A must flow from a to b (8 A − 5.3 A = 2.7 A).

The actual power transformed by the 160-turn secondary is 80 V × 2.7 A or 216 W, compared with a total load of 640 W, or only 33.8%. The largest portion of the load power (424 W) is conducted directly from the input through the 80-turn section of the primary winding. As a

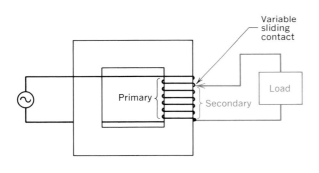

(a) Arrangement of windings on core.

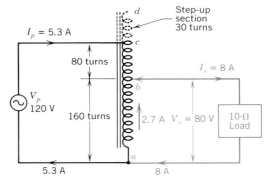

(b) Current distribution for a step-down condition

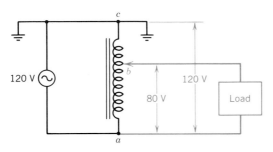

(c) Reversing the input voltage connections makes output terminal a 120 V above ground at all times.

(d) A variable autotransformer.

FIGURE 18-16

Variable autotransformer. [Photograph in (d) courtesy of Staco Energy Products Co.]

result, an autotransformer is smaller and uses less iron than a conventional two-winding transformer with the same rating.

Some transformers used for variable voltage operations (often referred to by the trade name Variac) incorporate a small step-up section. This is shown by the dotted section in Fig. 18-16*b*. The step-up section allows the transformer to develop a variable voltage output from 0 to 135 V when connected to a 120-V input.

Note that if a two-prong unpolarized plug is used to connect the autotransformer to the 120-V supply, the grounded neutral may be connected to point *c*, rather than point *a*. As shown in Fig. 18-16*c*, the output voltage between *a* and *b* may be only 80 V, but the output terminal will be at 120 V with respect to ground (point *c*).

A variable autotransformer is shown in Fig. 18-16*d*. This transformer, with dimensions of 14 × 15 × 8 in., can deliver a maximum of 8.4 kVA and 60 A at 0–140 V from a 120-V input.

It should be noted that *variable* transformers are now available with separate primary and secondary windings that are *not* electrically connected, and thus provide isolation. However, the current and volt-ampere ratings of these transformers are approximately one-half of the values that are in effect when operating in a nonisolated mode.

18-6   INSTRUMENT TRANSFORMERS

Voltage, current, and power in high-voltage ac power circuits are not generally measured by direct connection of the measuring instruments. Instead, *instrument* transformers are used to drop the voltage and current by known ratios to a range allowing the connection of standard ac voltmeters and ammeters. This arrangement also lessens the shock hazard to operators and the potential for instrument damage. Instrument transformers are available as *potential* transformers or *current* transformers.

18-6.1   Potential Transformers

These are small (50–600 VA) transformers that operate on the principle of power transformers. They step-down the voltage connected across the primary by a definite ratio, and are designed to produce 120 V on the output side when the rated voltage is applied to the primary. Multiplying the voltmeter reading by the known ratio provides the primary side high voltage.

18-6.2   Current Transformers

Current transformers are designed to be connected in *series* with the line carrying the high current to be measured. To keep the resistance inserted in the line as low as possible, only a few turns are used in the primary. (See Section 18-7 for a discussion of pass-through, clamp-on meters that operate as current transformers.) Since the current must be stepped-down, the number of turns in the secondary must be greater than the number in the primary. The turns ratio is usually selected so that 5 A will flow in the secondary when rated current flows in the primary. In fact, the *current transformer ratio* is given as a ratio of primary current to secondary current, where secondary current is usually 5 A. Thus, for example, a current ratio may appear as 300 A : 5 A or a 60 : 1 ratio.

Since the connection of an ammeter on the secondary provides a very low resistance, very little voltage appears across the secondary (despite the transformer being a voltage step-up type). However, if the current transformer is operated on open circuit, there is no secondary demagnetizing mmf set up, so the flux density in the core becomes very high. This can cause dangerously high voltages to be induced in the secondary on open circuit. The secondary should always be short-circuited before removing the ammeter or making any other changes in the secondary circuit.

EXAMPLE 18-5

Determine the load current, voltage, and power for the given instrument readings and the instrument transformer ratios shown in Fig. 18-17.

Solution

load current = ammeter reading × ratio
= 4 A × 15 = **60 A**
load voltage = voltmeter reading × ratio
= 115 V × 20 = **2300 V**
load power = wattmeter reading × ammeter ratio
× voltmeter ratio
= 350 W × 15 × 20
= 105,000 W = **105 kW**

Note how the load power is *not* the product of the load voltage and load current. This means that the load is not purely resistive, so the current and voltage are not in phase with each other (i.e., the power factor, cos θ, is less than unity).

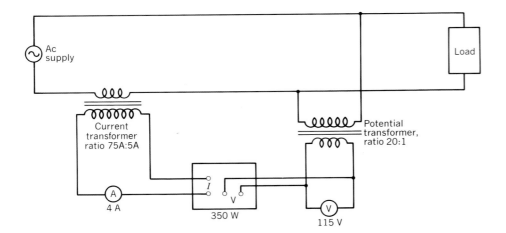

FIGURE 18-17

Instrument transformers to measure the line current, voltage, and power for Example 18-5.

A current transformer is also used in ac milliammeters that have the PMMC type of meter movement. As explained in Chapter 16, diodes are used to convert the alternating current to be measured to a proportional direct current that will operate the dc meter movement. The voltage drop across the diodes, however, would cause an excessive voltage drop across the meter. The current transformer, using a primary with just a few turns, introduces a very small voltage drop. As shown in Fig. 18-18, taps on the primary winding can be used to give a multirange ac ammeter. Diodes connected to the secondary in a bridge configuration rectify the alternating current to operate the PMMC meter movement.

Note how fewer turns are used on the primary for the 1000-mA range than for any other current range. This makes the turns ratio, *a, smaller* and causes a *greater* step-down of current.

18-7 CLAMP-ON AC AMMETER

The ac clamp-on ammeter shown in Fig. 18-19 operates on the principle of a current transformer. A wire carrying an alternating current, when placed between the clamp jaws, functions as a single-turn primary. The current in the wire sets up an alternating magnetic field in the iron core (the jaws) and induces current in a secondary winding. This secondary current (which is proportional to the primary current) is used, after suitable processing, to operate the LED display. Because of the *very* small turns ratio, the instrument can be used to measure currents up to 1000 A.

With suitable probes, the instrument shown can also measure ac volts and ohms.

This type of instrument is not suitable, however, for measuring direct current. Although this meter is similar in appearance to the clamp-on dc ammeter shown in Fig. 12-7, the two instruments employ different operating principles. The dc ammeter uses a Hall effect device and can measure both dc and ac current; the ac ammeter is essentially a transformer and will operate on (and thus be able to measure) only ac current.

18-8 CHECKING A TRANSFORMER

If a transformer fails to develop an output voltage, an *open* in either the primary or secondary could be responsible. If the primary is open, no current will flow in either the primary or the secondary windings. If the secondary is open, only a very small magnetizing current will flow in the primary. In a transformer with multiple secondary windings (like the one in Fig. 18-20), an open in one secondary will not affect the others. An ohmmeter will show an infinity reading when connected across the open winding.

If windings in the secondary short to each other or the core, the additional load will draw an increased primary current, often burning out the primary winding. It is usually necessary to replace the whole transformer when windings are open or have been shorted, since repair is seldom practical.

A word of caution: Before deciding to replace a transformer because of a very low ohmmeter reading across a

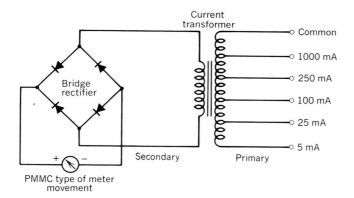

(a) Current transformer and bridge rectifier

(b) Ac milliammeter

FIGURE 18-18
Use of a current transformer to operate a PMMC meter movement in an ac milliammeter. [Photograph in (b) courtesy of Hickok Teaching Systems Inc., Woburn, Mass.]

FIGURE 18-19
Digital clamp-on ac ammeter, voltmeter, and ohmmeter. (Courtesy of TIF Instruments, Inc.)

winding, remember that the dc resistance of most windings *is* very low. This is especially true in power transformers, which use windings of wire with a large cross-sectional area to carry high currents. Such wires produce very little dc resistance.

18-9 MULTIPLE SECONDARY TRANSFORMERS

As shown in Fig. 18-20b, a transformer may have more than one secondary winding, and in fact, may have more than one primary winding. The phase relationships of the windings (shown by the dot notation) determine what voltages will result when the windings are interconnected.

The primary of the transformer shown in Fig. 18-21a is commonly found in equipment that may be operated from either 120 or 240 V. (In this way, the equipment is suitable for either domestic or foreign use.) When the two primary windings are connected in parallel, a supply of 120 V is required; when connected in series, a supply of 240 V is necessary. Example 18-6 shows how to make these connections, and the voltages available on the secondary side.

EXAMPLE 18-6

Given the transformer in Fig. 18-21a, show the wiring diagram for

a. An output of 6 V with an input of 240 V.
b. An output of 42 V with an input of 120 V.

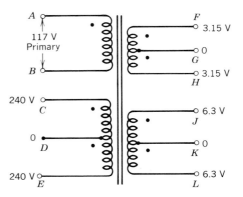

(a) Transformer has a center–tapped high–voltage secondary on one side (next to the 117–V primary terminals), and low–voltage secondary outputs on the other side. (Note the iron cover as discussed in Sect. 18–1.3.)

(b) Schematic of the transformer shown in (a) (note the dot notation on the windings to show instantaneous polarity)

FIGURE 18-20

A small power transformer, with schematic.

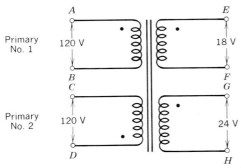

(a) Transformer with two primary and two secondary windings

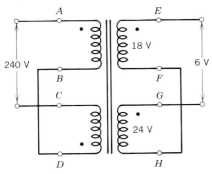

(b) Connections for 240-V input, 6-V output

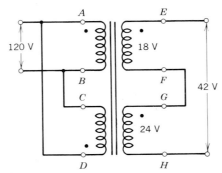

(c) Connections for 120-V input, 42-V output

FIGURE 18-21

Circuits for Example 18-6.

Solution

a. For a 240-V input, the two primary windings must be series-connected with their magnetic fluxes aiding. This is accomplished by connecting together terminals B and D. See Fig. 18-21b. In this way, if current enters primary winding number 1 at the dot (terminal A) current will also enter primary winding number 2 at the dot (terminal D). This is the requirement for fluxes produced by the two coils to be additive. (Note: If terminals B and C were connected together, the fluxes would subtract, and no voltage would be developed on the secondary side.)

On the secondary side, an output of 6 V will result if the magnetic fluxes of the 24-V and 18-V coils oppose each other. Connecting together terminals F

and H will make current *entering* the dot at terminal E *exit* the dot at terminal G. This causes the magnetic fluxes on the secondary side to be subtractive.

b. For 120-V operation, the two primary windings must be parallel-connected, again with their fluxes additive. This requires that A and D be connected together, and terminals B and C connected together, as in Fig. 18-21c. Then input current will always either flow into both dots or away from both dots. (Note again that if the dot notation had not been observed, the fluxes would be subtractive, and the output voltage would be zero.)

On the secondary side, connecting terminals F and G will cause the fluxes to be additive, producing an output of 18 V + 24 V = 42 V.

SUMMARY

1. The primary winding of a transformer is connected to the source; the secondary to the load. The two windings are magnetically coupled to each other on the same core.

2. The turns ratio, voltage ratio, and current ratio are related by

$$a = \frac{N_p}{N_s} = \frac{V_p}{V_s} = \frac{I_s}{I_p}$$

3. A step-up transformer increases output voltage compared with the input, but decreases output current by the same ratio. The output power can never be greater than the input power.

4. A transformer can be used to match resistances for maximum power transfer because a load resistance (R_L) is reflected into the primary as a load resistance (R_p) given by

$$R_p = a^2 R_L$$

5. Transformers depend upon the continual change in magnetic field set up by an alternating current in the primary winding. They cannot operate from pure direct current.

6. Generator outputs may be stepped-up to as much as 765 kV in a power distribution system for transmission to distant customers.

7. A residential wiring system is fed from a single-phase, center-tapped secondary transformer to provide a 120/240-V, three-wire system with a grounded neutral.

8. A third-wire grounded receptacle allows the metal frames of appliances and power tools to be grounded, preventing shock hazards in the event of a short between the "live" side of the circuit and the frame.

9. A ground-fault circuit-interrupter (GFCI) is used with a 120-V outlet to sense a current imbalance between "live" and neutral wires. The circuit is quickly disconnected in the event of an imbalance indicating a short circuit.

10. An isolation transformer has two separate windings. The output voltage is equal to the input voltage, but is electrically isolated from any ground on the primary.

11. An autotransformer shares a single-tapped coil between the input and the output to transform voltages using a much smaller core than a conventional two-winding transformer.

12. Instrument transformers step-down voltage or current by a known ratio, so that standard meters can be used to make ac measurements.

SELF-EXAMINATION

Answer true or false, or, for multiple choice, a, b, c, or d
(Answers at back of book)

18-1. An iron-core transformer is an application of mutual inductance in which the coefficient of coupling is close to unity. _____

18-2. A transformer will not operate with pure direct current applied because there is no continual changing flux set up to induce a secondary voltage. _____

18-3. The voltage and current ratios have the same value as the turns ratio. _____

18-4. A 1:5 step-up transformer with a primary voltage of 120 V has a pure resistance load of 300 Ω. The primary current and secondary voltage are
a. 10 A and 600 V
b. 0.4 A and 600 V
c. 0.4 A and 24 V
d. 10 A and 24 V

18-5. Assuming an 80% efficiency, the input power to the transformer in Question 18-4 is
a. 6 kW
b. 7 kW
c. 7.5 kW
d. 8 kW

18-6. The volt-ampere rating of a transformer is the same as the power rating in watts. _____

18-7. A transformer is used to match an 8-Ω load to a 288-Ω source. The necessary turns ratio is
a. 36
b. 6
c. 8
d. 288

18-8. A power distribution system needs step-up transformers to allow smaller diameter conductors to be used for the transmission line. _____

18-9. A residential 120/240-V, three-wire system is an example of a three-phase system. _____

18-10. A variable autotransformer can be used as an isolation transformer simply by adjusting the output voltage to be the same as the input voltage. _____

18-11. When an isolation transformer is used, there is never any problem with using grounded measuring equipment on the secondary side. _____

18-12. A GFCI relies upon equal but opposite currents flowing through a toroidal transformer to indicate when a fault occurs. _____

18-13. A current instrument transformer is connected across the line like a potential transformer. _____

REVIEW QUESTIONS

1. a. If a transformer has a 120-V primary and a 6.3-V secondary, could it be operated in reverse to produce 120 V with a 6.3-V input?
    b. What would determine the maximum input voltage with this mode of operation? (Could you apply 120 V to the 6.3-V winding?)
2. What general principle can you use to justify that in a step-up transformer the current must be reduced by at least the same factor?
3. If a transformer cannot operate on pure direct current, how is it possible for an automotive coil (transformer) to induce 25 kV on the secondary from a 12-V dc supply?
4. a. What are the factors contributing to less than 100% operating efficiency in a transformer?
    b. How are these effects minimized?
5. a. What part does a transformer play in providing maximum power transfer?
    b. How does it do this?
6. a. Under what special condition is the volt-ampere rating of a transformer the same as the power rating?
    b. Why is this not *generally* the case?
7. a. What maximum distance could power be economically transmitted if transmission lines were operated at 2 million V?
    b. Why is the step-down process in substations accomplished in stages rather than by dropping immediately to 120/240 V?
8. a. How is it possible for the 120/240 V residential system to be single phase with three wires?
    b. Why is the neutral grounded?
    c. In a two-prong receptacle with no separate ground why, if the neutral is grounded, is the neutral not tied to the metal frame of the appliance?
9. a. Describe the operation of a ground-fault circuit interrupter.
    b. Where is a GFCI likely to be used? Why?
10. a. Explain how an isolation transformer can be used to avoid dangerous short circuits when using grounded test equipment on line-operated circuits.
    b. What precautions must still be observed if more than one grounded instrument is being used?
11. a. Why is an autotransformer smaller than a conventional two-winding transformer for a given load?
    b. What is a Variac?
    c. Can it be used to step-up the voltage?
    d. What is the disadvantage of an autotransformer?
12. a. How does a current transformer differ from a potential transformer?
    b. What precaution should be taken when removing an ammeter from the secondary of a current transformer?
    c. Why is this necessary?

PROBLEMS

(Answers to odd-numbered problems at back of book)
(Unless otherwise stated, neglect transformer losses.)

18-1. A transformer with 200 primary turns and 65 secondary turns has a primary voltage of 50 V. What is the secondary voltage?

18-2.   A transformer with a turns ratio of 0.4 has a secondary voltage of 230 V. What is the primary voltage?

18-3.   A 4160/230-V single-phase distribution transformer has an output rating of 7.5 kVA. Calculate:
   a.   The maximum current that can safely be drawn from the secondary.
   b.   The minimum resistance that can be connected to the secondary.

18-4.   A 2300/120-V single-phase transformer is connected to an effective load resistance of 2 Ω. Calculate:
   a.   The minimum output rating of the transformer.
   b.   The input power if the transformer is 92% efficient.

18-5.   A 230-V primary, 24-V secondary transformer is connected to a 15-Ω resistive load. Calculate:
   a.   The turns ratio.
   b.   The secondary current.
   c.   The primary current.
   d.   The volt-ampere rating of the transformer.

18-6.   A 115-V primary transformer has a turns ratio of 6.5 and a secondary current of 2 A. Calculate:
   a.   The secondary voltage.
   b.   The primary current.
   c.   The load resistance.
   d.   The volt-ampere rating of the transformer.

18-7.   A 2400 V/240-V step-down transformer has a load of 5 Ω pure resistance. Calculate the primary current assuming:
   a.   100% efficiency.
   b.   90% efficiency.

18-8.   A 4160 V/240-V step-down transformer has a load of 2.5 Ω pure resistance. If the core losses amount to 2 kW, determine:
   a.   The efficiency of the transformer.
   b.   The primary current.
   c.   The volt-ampere rating of the transformer.

18-9.   An audio amplifier has an open-circuit voltage of 100 V p-p and an output resistance of 64 Ω. A transformer is to be selected to match an 8-Ω loudspeaker to the amplifier. Calculate, assuming 100% efficiency:
   a.   The necessary turns ratio of the transformer.
   b.   The primary input voltage and current.
   c.   The secondary output voltage and current.
   d.   The output power in the loudspeaker.
   e.   The power in the loudspeaker if connected directly to the amplifier.

18-10.   A voltage source with an internal resistance of 80 Ω is connected to the primary of a step-down transformer that has a turns ratio of 5. Calculate:
   a.   The load to be connected to the secondary for maximum power transfer.
   b.   The open-circuit voltage of the source if 10 W of power are developed in the load.

18-11.   The total power in a three-phase system is given by $P = \sqrt{3}\, VI \cos \theta$, where V is the line-to-line voltage and I is the line current. Consider a 230/765-kV 1000-MW step-up transformer. Calculate, assuming a purely resistive load:
   a.   The secondary current when operating at its rated value.
   b.   The input current if 99.5% efficient.
   c.   The losses in the transformer in the form of heat.

18-12. The variable autotransformer in Fig. 18-16*b* is adjusted to provide an output voltage of 90 V with an 8-Ω load. Neglecting transformer losses, determine:

  a. The power delivered to the load.
  b. The load current.
  c. The source current.
  d. The current in the secondary winding.
  e. The percentage of power transformed by the secondary compared with the total circuit power.
  f. The instantaneous directions of current in the circuit.

18-13. Repeat Problem 18-12 with the output adjusted to 110 V and a 10-Ω load.

18-14. Repeat Problem 18-12 with the output adjusted to 135 V utilizing the 30-turn step-up section, and a 12-Ω load.

18-15. Instrument transformers are connected as in Fig. 18-17 to measure line current, voltage, and power. If the current transformer ratio is 10:1 and the potential transformer ratio is 15:1, determine:

  a. The line current.
  b. The line voltage.
  c. The line power.
  d. The apparent line power (volt-amperes).

18-16. For the instrument transformer ratios given in Fig. 18-17 determine the wattmeter reading if the ammeter indicates 5 A, the voltmeter indicates 120 V, and

  a. The load is purely resistive.
  b. The power factor is 0.9.
  c. The power factor is zero.

18-17. Refer to Fig. 18-20. With 117 V applied to the primary, determine:

  a. The rms voltage between *C* and *E*.
  b. The peak voltage at *F* with respect to *G*.
  c. The peak-to-peak voltage between *J* and *L*.
  d. The rms voltage between *F* and *L* if *H* is connected to *J*.

18-18. Refer to Fig. 18-20. With 117 V applied to the primary, determine:

  a. The rms voltage between *J* and *L*.
  b. The peak voltage at *L* with respect to *K*.
  c. The peak-to-peak voltage between *C* and *E*.
  d. The rms voltage between *F* and *J* if *H* is connected to *L*.
  e. The rms voltage between *F* and *K* if *H* is connected to *L*.

18-19. Given the transformer in Fig. 18-22, determine the connection necessary to provide the following secondary voltages:

  a. 5 V
  b. 8 V
  c. 15 V
  d. 26 V

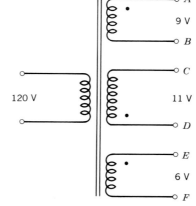

FIGURE 18-22
Transformer for Problem 18-19.

CHAPTER 19

INDUCTANCE IN DC CIRCUITS

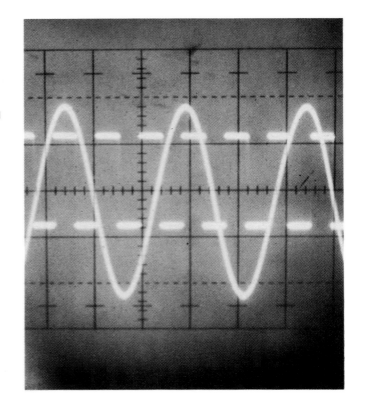

As you have seen, the property of inductance is to oppose change of current. In this chapter, you will examine in detail how current increases and decreases when a dc voltage is applied to a series circuit involving resistance and inductance. This involves defining the inductive *time constant* (L/R) of the circuit, which depends *only* upon the inductance and resistance. The time constant is not affected by current, voltage, or any other factors. This time constant allows examination of the *transient responses* in the *L-R* circuit, including those temporary changes in current and voltage that take place before *steady-state* conditions are established. These transients may be analyzed using a graphical method of universal time-constant curves.

If mechanical switching is used to interrupt current in an inductive circuit, energy stored in the coil's magnetic field is capable of inducing high voltages. This phenomenon is used to advantage in a fluorescent lamp circuit and in an automotive ignition system (with the help of a step-up autotransformer). Where high voltages could cause damage, however, means of reducing inductive effects are necessary.

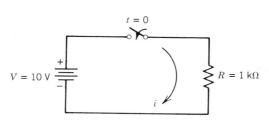

(a) Dc voltage applied to a resistor

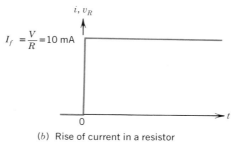

(b) Rise of current in a resistor

FIGURE 19-1

Rise of current in a purely resistive dc circuit.

19-1 RISE OF CURRENT IN AN INDUCTIVE CIRCUIT

In a *purely resistive* circuit, the rise of current is instantaneous, as shown in Fig. 19-1. Upon the closing of the switch, the current rises instantly to its final Ohm's law value. For a given applied voltage, resistance (by itself) determines only *how much* current flows, not the rate at which the current *changes*. The final steady current is determined by $I_f = V/R$, as shown in Fig. 19-1b.

Now, consider the circuit in Fig. 19-2a. It consists of a coil of negligible resistance in series with a 1-kΩ resistor. You could also think of the resistance of the coil being "lumped" into a separate series-connected resistor of 1 kΩ. In either case, the effect on the rise of current is the same, as shown in Fig. 19-2b. Note that the current rises

gradually, rather than *instantly* (as in a pure resistive circuit).*

When the switch is closed, current tries to rise instantaneously (as in a purely resistive circuit). But, as you know from Lenz's law, a voltage (v_L) will be induced across the coil, opposing any change in current in the circuit. In this case, the change would be an *increase* in current. This results in the polarity of v_L shown in Fig. 19-2a. This voltage can be thought of as an *opposing* emf or a counter*voltage* v_L. The applied voltage must balance v_L and v_R as shown in Fig. 19-2a. By Kirchhoff's voltage law, therefore,

$$V = v_R + v_L = iR + v_L$$

*In section 19-4 it is shown that the current follows a mathematical equation that involves an *exponential* relationship with time.

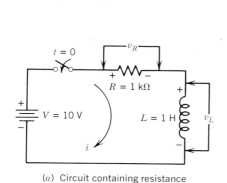

(a) Circuit containing resistance and inductance

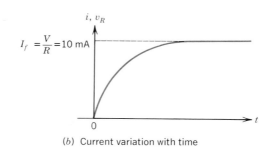

(b) Current variation with time

(c) Coil voltage variation with time

FIGURE 19-2

Current and voltage waveforms in a dc circuit containing resistance and inductance.

Now, at $t = 0^+$ (an instant after closing the switch), the instantaneous current (i) in the above equation must continue to be zero. This is because inductance is the property of a coil to oppose any change in current, and (for an initial instant) the coil is completely successful in doing this. This means that there is no voltage drop across the resistor, so:

$$v_R = iR = 0$$

and since

$$V = v_R + v_L$$

we have

$$V = 0 + v_L$$

This means that, at the instant of closing the switch, the rapid rate of change of current in the circuit is sufficient to induce a voltage across the inductor that is *equal to* the applied voltage ($v_L = V$). This is shown in Fig. 19-2c.

However, the current in the circuit *is* changing. Thus, a short time later, when the current has increased above zero, there must be a voltage drop (v_R) across the resistor, as shown in Fig. 19-2b. For Kirchhoff's voltage law to be obeyed, the voltage across the coil at any instant is now given by

$$v_L = V - v_R = V - iR$$

The voltage across the coil is now less than before as a result of the iR drop across the resistor. But the voltage across the coil is

$$v_L = L\left(\frac{\Delta i}{\Delta t}\right) \qquad (17\text{-}1)$$

so that the rate of change of current with time $\left(\dfrac{\Delta i}{\Delta t}\right)$ must be less than the initial value and must decrease as time goes on and the current rises.

This implies that, although the current is *increasing,* it is doing so at a slower *rate.* This is shown in Fig. 19-2b, where the slope of the curve at any given instant represents $\dfrac{\Delta i}{\Delta t}$. The rate continues to decrease until eventually, as shown in Fig. 19-2b, the current reaches a final steady-state value (I_f), where the slope is horizontal and no change occurs. Thus, when $i = I_f$, no voltage is induced across the inductance (see Fig. 19-2c) and all the applied voltage appears across the resistance.

The final steady-state current is now given by Ohm's law:

$$I_f = \frac{V}{R} \qquad (3\text{-}1a)$$

Using the values given in Fig. 19-2a, $I_f = \dfrac{10\ \text{V}}{1\ \text{k}\Omega} = 10$ mA. At this point, the inductance has no effect in the circuit unless a change takes place (such as opening the switch, changing the resistance, or changing the applied voltage). The time interval required for the current to rise to the steady-state value after closing the switch is referred to as a *transient*. Changes occurring in the current (Fig. 19-2b) and voltage (Fig. 19-2c) are called *transient responses*.

To determine the length of time needed to reach the steady-state value, or the *duration* of the transient, you must first determine the *time constant*. This is symbolized by the lowercase Greek letter tau (τ).

19-2   TIME CONSTANT (τ)

The rate of rise of current is highest (and its slope is greatest) when the switch is first closed, because at this instant,

$$v_L = V$$

and

$$v_L = L\left(\frac{\Delta i}{\Delta t_0}\right) \qquad (17\text{-}1)$$

Therefore

$$L\left(\frac{\Delta i}{\Delta t_0}\right) = V$$

and the initial rate of rise of current is

$$\frac{\Delta i}{\Delta t_0} = \frac{V}{L} \qquad (19\text{-}1)$$

where: $\dfrac{\Delta i}{\Delta t_0}$ is the initial rate of rise of current, in amperes per second (A/s)

V is the applied dc voltage, in volts (V)

L is the inductance of the coil, in henrys (H)

For the numerical values given in Fig. 19-2a, with $V = 10$ V and $L = 1$ H, the initial rate of rise of current is given by

$$\frac{\Delta i}{\Delta t_0} = \frac{V}{L}$$

$$= \frac{10\ \text{V}}{1\ \text{H}} = \mathbf{10\ A/s}$$

Now, consider how long it would take the current to reach its final value *if* it continued to increase at its *initial* rate of change of $\dfrac{V}{L}$ amperes per second. This length of time can be called the *time constant* (τ). As shown in Fig. 19-3, however, the current does *not* increase at this rate. In fact, when $t = \tau$, the actual current is only 0.63 of its final value. This is also used as a definition of the time constant. That is, one time constant is the time required for the current to change 63% from its present value toward its final value. (See Section 19-3.)

19-2.1 Derivation of $\tau = {}^{L}/_{R}$

Let the dotted line that is tangent to the current curve reach the final value of current in a time interval τ. This line represents the initial rate of rise of current. Its slope is given by

$$\text{initial slope} = \frac{\text{rise}}{\text{run}} = \frac{\text{vertical intercept}}{\text{horizontal intercept}}$$
$$= \frac{V/R}{\tau} \text{ amperes per second}$$

but the initial slope $= \frac{\Delta i}{\Delta t_0} = \frac{V}{L}$ amperes per second.

Therefore $\dfrac{V/R}{\tau} = \dfrac{V}{L}$

and $\tau = \dfrac{L}{R}$ seconds   (19-2)

where: τ is the time constant of the circuit, in seconds (s)
L is the inductance of the circuit, in henrys (H)
R is the resistance of the circuit, in ohms (Ω)

Note that τ depends only on the values of L and R. If the circuit is extremely inductive (L is high), the time needed to reach steady state is *long*. If the circuit is extremely resistive (R is high), the time needed to reach a steady state is *short*. This is shown in Examples 19-1 and 19-2.

EXAMPLE 19-1

If a circuit's resistance is 1 kΩ and its inductance is 1 H, what is the circuit's time constant?

Solution

Time constant $\tau = \dfrac{L}{R}$ (19-2)

$$= \frac{1 \text{ H}}{1 \times 10^3 \text{ Ω}} = \textbf{1 ms}$$

EXAMPLE 19-2

If the circuit of Example 19-1 has its inductance doubled and its resistance cut in half, what is the new time constant?

Solution

Time constant $\tau = \dfrac{L}{R}$ (19-2)

$$= \frac{2 \text{ H}}{500 \text{ Ω}} = \textbf{4 ms}$$

EXAMPLE 19-3

A coil has an inductance of 8 H and a resistance of 400 Ω. A 20-V dc supply is connected across the coil by means of a switch. Calculate:
a. The time constant of the coil.
b. The final value of current through the coil.
c. The initial rate of rise of current in the coil after the switch is closed.
d. The time it would take for the current to reach its final value *if* the current continued to increase at its initial rate of rise calculated in part (c).

Solution

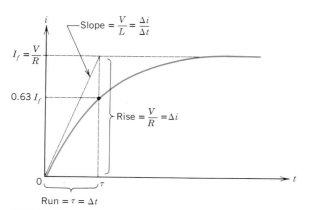

FIGURE 19-3
Determination of the time constant, τ.

a. $\tau = \dfrac{L}{R}$ (19-2)

$$= \frac{8\ H}{400\ \Omega} = 0.02\ s = \textbf{20 ms}$$

b. $I_f = \dfrac{V}{R}$ (3-1a)

$$= \frac{20\ V}{400\ \Omega} = 0.05\ A = \textbf{50 mA}$$

c. $\dfrac{\Delta i}{\Delta t_0} = \dfrac{V}{L}$ (19-1)

$$= \frac{20\ V}{8\ H} = \textbf{2.5 A/s}$$

d. If the current is changing at 2500 mA/s, it will change 50 mA in

$$\frac{50\ mA}{2500\ mA/s} = \frac{1}{50}\ s = \textbf{20 ms}$$

Note that this is the same time as the time constant in part (a).

19-3   UNIVERSAL TIME-CONSTANT CURVES

It can be shown that the rate of increase of current at *any instant* on the curve is such that if the current continued to

increase at *that* rate, the current would reach its final value of V/R amperes in L/R seconds. Thus, the time constant applies to the *whole* curve, not just initial conditions. This allows the manner in which current changes in an inductive circuit to be graphed in terms of time constants. Such a graph is called a *universal* time-constant curve. Two such curves are shown in Fig. 19-4.

Curve *a* gives the percentage of the final steady-state current when the switch is closed and the current is rising. Since $v_R = iR$, curve *a* also represents the change in the voltage across the resistance when the current is rising. Curve *b* gives the voltage across the inductance as a percentage of its initial value whenever the switch is closed and the current is rising.

The curves also may be used when the switch is being opened. In this case, i, v_R, and v_L all *decay*, and are represented by curve *b*. (The *polarity* of v_L is opposite when the current is falling, but the *magnitude* of the voltage v_L is given by curve *b* whether the current is rising or falling.)

It can be seen from Fig. 19-4 that the current reaches 63% of its steady-state value in one time constant (L/R s).

In fact, the time constant of an inductive circuit is more usually defined as the time required for the current (or coil voltage) to change 63% from its original value to its final value.

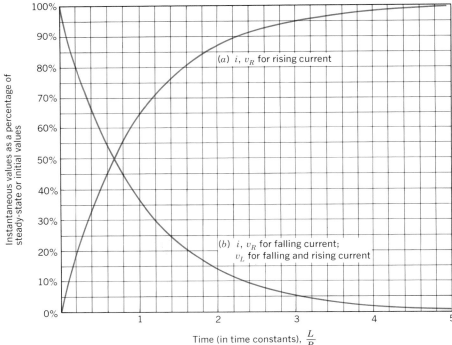

Instantaneous values as a percentage of steady-state or initial values

(a) i, v_R for rising current

(b) i, v_R for falling current; v_L for falling and rising current

Time (in time constants), $\dfrac{L}{R}$

FIGURE 19-4

Universal time-constant curves for the rise and fall of current and voltage in an inductive circuit with direct current.

Specifically, assume that the time constant is 1 ms and the circuit's final (steady-state) current is 10 mA. In this case, the current reaches 63% of 10 mA, or 6.3 mA, in 1 ms. At the end of two time constants (2 ms), the current is:

$$\text{new value} = \text{original value} + 63\% \text{ of}$$
$$(\text{final value} - \text{previous value}) \quad (19\text{-}3)$$
$$= 6.3 \text{ mA} + 0.63 \times (10 \text{ mA} - 6.3 \text{ mA})$$
$$= 6.3 \text{ mA} + 0.63 \times 3.7 \text{ mA}$$
$$= 6.3 \text{ mA} + 2.33 \text{ mA}$$
$$= 8.63 \text{ mA at the end of 2 ms } (2\tau)$$

This current corresponds to 86.3% of the steady-state current and could have been read directly from the universal time-constant curve (Fig. 19-4, curve a) after 2τ. Similarly, at 3τ, $i = 95\%$ of the final value, and at 4τ, $i = 98\%$ of the final steady-state current. After five time constants (5τ), $i = 99.3\%$ of its final value. This time of 5 τ is usually considered to be the end of the transient.

For all practical purposes, an inductive circuit reaches its steady state after 5 τ (equal to 5 L/R) has elapsed.

Thus, the steady-state value of 10 mA would be reached in approximately 5 ms, since $1\tau = 1$ ms.

EXAMPLE 19-4

Consider the coil in Example 19-3; $L = 8$ H, $R = 400$ Ω, $V = 20$ V, $\tau = 20$ ms, and $I_f = 50$ mA. Use the universal time constant curves (Fig. 19-4) to find:
a. The current 46 ms after closing the switch.
b. How long it takes for the current to reach 27.5 mA.

Solution

a. $t = 46$ ms
$$= \frac{46 \text{ ms}}{20 \text{ ms/}\tau} = 2.3 \tau$$
From Fig. 19-4, when $t = 2.3\tau$, $i = 90\%$ of I_f (see Fig. 19-5). Therefore,
$$i = 90\% \times I_f$$
$$= \frac{90}{100} \times 50 \text{ mA} = \textbf{45 mA}$$

b. $i = 27.5$ mA
$$= \frac{27.5 \text{ mA}}{50 \text{ mA}} \times 100\%$$
$$= 55\% \text{ of final current, } I_f$$

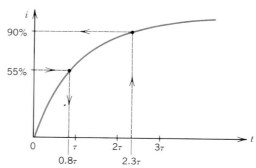

FIGURE 19-5
Sketch of i versus t for Example 19-4.

From Fig. 19-4, when $i = 55\%$, $t = 0.8\tau$ (see Fig. 19-5). Therefore,
$$t = 0.8 \tau$$
$$= 0.8 \times 20 \text{ ms} = \textbf{16 ms}$$

EXAMPLE 19-5

A circuit consists of a 30-mH coil (of negligible resistance) in series with a 2-kΩ resistor. A switch is closed to apply 50 V dc. Using the universal time constant curves (Fig. 19-4), find:
a. The time constant of the circuit.
b. The initial voltage across the coil when the switch is closed.
c. The voltage across the coil 7.5 μs after closing the switch.
d. The steady-state current in the circuit.
e. The current 45 μs after closing the switch.
f. How long it takes for the voltage across the resistor to reach 37.5 V.
g. Approximately how long it takes for the voltage across the coil to drop to zero.

Solution

a. $\tau = \dfrac{L}{R}$ \quad (19-2)
$$= \frac{30 \times 10^{-3} \text{ H}}{2 \times 10^3 \text{ }\Omega} = 15 \times 10^{-6} \text{ s} = \textbf{15 }\boldsymbol{\mu}\textbf{s}$$

b. Initially, $v_L = V = \textbf{50 V}$

c. $t = 7.5 \text{ }\mu\text{s} = \dfrac{7.5 \text{ }\mu\text{s}}{15 \text{ }\mu\text{s/}\tau} = 0.5 \tau$
From Fig. 19-4, after 0.5 time constant, the decreasing v_L curve is at 60% of its initial value.

Therefore, v_L = 60% of 50 V

                              = **30 V**

d. Steady-state current,

$$I_f = \frac{V}{R} \qquad\qquad (3\text{-}1a)$$

$$= \frac{50 \text{ V}}{2 \times 10^3 \ \Omega} = \textbf{25 mA}$$

e. $t = 45 \ \mu s = \dfrac{45 \ \mu s}{15 \ \mu s/\tau} = 3 \ \tau$

   From Fig. 19-4, after three time constants, the increasing i curve is at 95% of its final value.

   Therefore, i = 95% of 25 mA

                         = 0.95 × 25 mA

                         = **23.8 mA**

f. The resistor voltage $v_R = iR$ has the same percent variation with time as the current, with a final value of 50 V.

$$v_R = 37.5 \text{ V} = \frac{37.5 \text{ V}}{50 \text{ V}} \times 100\% = 75\%$$

   of the final value. From the increasing i, v_R curve in Fig. 19-4, the time required for the curve to reach 75% is 1.4 τ. Therefore,

$$t = 1.4 \ \tau$$
$$= 1.4 \times 15 \ \mu s$$
$$= \textbf{21} \ \boldsymbol{\mu s}$$

g. The coil voltage is approximately zero after 5 time constants. This time is

$$t = 5 \ \tau$$
$$= 5 \times 15 \ \mu s$$
$$= \textbf{75} \ \boldsymbol{\mu s}$$

19-4 ALGEBRAIC SOLUTION USING EXPONENTIAL EQUATIONS*

Instead of using a *graphical* method to obtain the circuit current and coil voltage, you can employ an *algebraic* solution. This method requires the use of the following equation, which applies to a changing current or voltage in a dc series circuit containing inductance and resistance. (It also can apply to a series circuit containing *capacitance* and resistance, as will be shown later.)

---

*This section may be omitted with no loss of continuity.

Instantaneous value
$$= \text{Final value} + (\text{Initial} - \text{Final})e^{-t/\tau} \qquad (19\text{-}4)$$

where: Instantaneous value is the value of the current or voltage at any instant of time

Final value is the steady-state value of current or voltage after the transient has ended

Initial value is the value of current or voltage at $t = 0^+$, at the beginning of the transient

e is the irrational number 2.71828 . . ., the base of natural logarithms

t is the instant of time at which the value of current or voltage is to be found

τ is the time constant of the circuit, in seconds (s)—equal to L/R for an inductive circuit

To apply Eq. 19-4, make a sketch of the response of voltage or current, and indicate initial and final values. For an inductive circuit with a dc voltage being applied, the current has an initial value of 0 and a final value of V/R. (See Fig. 19-6b.)

Instantaneous value
$$= \text{Final value} + (\text{Initial} - \text{Final})e^{-t/\tau} \qquad (19\text{-}4)$$
$$i = \frac{V}{R} + \left(0 - \frac{V}{R}\right)e^{-t/\tau}$$
$$= \frac{V}{R} - \frac{V}{R}e^{-t/\tau}$$
$$i = \frac{V}{R}(1 - e^{-t/\tau}) \qquad\qquad (19\text{-}5)$$

where: i is the value of current, in amperes (A)

V is the applied dc voltage, in volts (V)

R is the circuit resistance, in ohms (Ω)

t is the instant of time after closing the switch, in seconds (s)

$\tau = \dfrac{L}{R}$ is the time constant, in seconds (s)

For the voltage across the coil, at $t = 0^+$ the initial coil voltage is V and the final value is 0. (See Fig. 19-6c.)

Instantaneous value
$$= \text{Final value} + (\text{Initial} - \text{Final})e^{-t/\tau} \qquad (19\text{-}4)$$
$$v_L = 0 + (V - 0)e^{-t/\tau}$$
$$v_L = V \, e^{-t/\tau} \qquad\qquad (19\text{-}6)$$

where: v_L is the inductor voltage, in volts (V)

V is the applied dc voltage, in volts (V)

t is the time after closing the switch, in seconds (s)

$\tau = L/R$ is the time constant, in seconds (s)

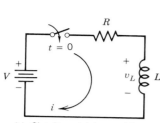

(a) Circuit containing resistance and inductance

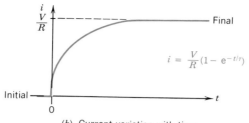

$$i = \frac{V}{R}(1 - e^{-t/\tau})$$

(b) Current variation with time

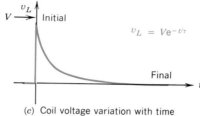

$$v_L = Ve^{-t/\tau}$$

(c) Coil voltage variation with time

FIGURE 19-6

Series *RL* circuit with dc voltage applied, showing current and inductor voltage responses with exponential equations.

EXAMPLE 19-6

Given the circuit in Fig. 19-6a, with $V = 20$ V, $L = 8$ H, and $R = 400 \ \Omega$, find:

a. The current 46 ms after closing the switch.
b. How long it takes for the current to reach 27.5 mA.

Solution

a. $\tau = \dfrac{L}{R}$ (19-2)

$\quad = \dfrac{8 \text{ H}}{400 \ \Omega} = 0.02 \text{ s} = 20 \text{ ms}$

$i = \dfrac{V}{R}(1 - e^{-t/\tau})$ (19-5)

$\quad = \dfrac{20 \text{ V}}{400 \ \Omega} (1 - e^{-46 \times 10^{-3}/20 \times 10^{-3}})$

$\quad = 50 (1 - e^{-2.3}) \text{ mA}$

The evaluation of $e^{-2.3}$ uses the e^x button on some calculators. (Enter 2.3; obtain the negative, -2.3, press e^x. The result is 0.100.)

On other scientific calculators, it may be necessary to use the ln x (natural log) button. (Enter 2.3; obtain the negative, -2.3; press the INV (inverse) button, then press ln x. The result is 0.100.)

Therefore,

$\quad\quad i = 50 (1 - 0.1) \text{ mA}$

$\quad\quad = 50 \times 0.9 \text{ mA} = \textbf{45 mA}$

b. $i = \dfrac{V}{R}(1 - e^{-t/\tau})$ (19-5)

$27.5 \text{ mA} = 50 (1 - e^{-t/20 \times 10^{-3}}) \text{ mA}$

$\dfrac{27.5}{50} = 1 - e^{-50t}$

$e^{-50t} = 1 - \dfrac{27.5}{50} = 1 - 0.55$

$e^{-50t} = 0.45$

Take the natural log ($\log_e = \ln$) of both sides:

$\quad\quad \ln e^{-50t} = \ln 0.45$

$\quad\quad -50t \ln e = \ln 0.45$

The natural log can be readily obtained with a scientific calculator. (Enter 0.45 and press the ln x button. The result is $-0.7985 \approx -0.8$.) Similarly, ln $e = 1$.

$\quad\quad -50t = -0.8$

$\quad\quad t = \dfrac{-0.8}{-50} \text{ s}$

$\quad\quad = 0.016 \text{ s} = \textbf{16 ms}$

These answers should be compared to those obtained graphically in Example 19-4.

EXAMPLE 19-7

A circuit consists of a 30-mH coil (of negligible resistance) in series with a 2-kΩ resistor. A switch is closed to apply 50-V dc to the series combination, as in Fig. 19-7a. Calculate:

a. The initial voltage across the coil when the switch is closed.

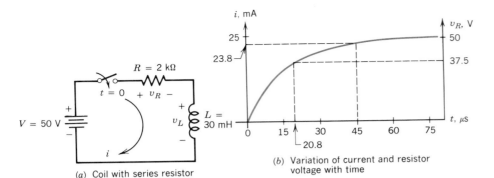

(a) Coil with series resistor

(b) Variation of current and resistor voltage with time

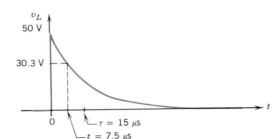

(c) Variation of coil voltage with time

FIGURE 19-7

Circuit and waveforms for Example 19-7.

b.   The voltage across the coil 7.5 μs after closing the switch.
c.   The current 45 μs after closing the switch.
d.   How long it takes for the voltage across the resistor to reach 37.5 V.

Solution

a. $v_L = V e^{-t/\tau}$ (19-6)

Initially, at $t = 0^+$,

$$v_L = 50e^{-0} \text{ V}$$
$$= 50 \times 1 \text{ V} = \textbf{50 V}$$

b. $\tau = \dfrac{L}{R}$ (19-2)

$$= \frac{30 \times 10^{-3} \text{ H}}{2 \times 10^3 \text{ } \Omega} = 15 \times 10^{-6} \text{ s} = 15 \text{ μs}$$

$$v_L = Ve^{-t/\tau}$$
$$= 50e^{-7.5 \times 10^{-6}/15 \times 10^{-6}} \text{ V}$$
$$= 50e^{-0.5} \text{ V}$$
$$= 50 \times 0.607 \text{ V} = \textbf{30.3 V}$$

c. $i = \dfrac{V}{R}(1 - e^{-t/\tau})$ (19-5)

$$= \frac{50 \text{ V}}{2 \times 10^3 \text{ } \Omega}(1 - e^{-45 \times 10^{-6}/15 \times 10^{-6}}) \text{ A}$$
$$= 25(1 - e^{-3}) \text{ mA}$$

$$= 25(1 - 0.05) \text{ mA}$$
$$= 25 \times 0.95 \text{ mA} = \textbf{23.8 mA}$$

d. $v_R = iR$

$$= \frac{V}{R}(1 - e^{-t/\tau})R$$
$$= V(1 - e^{-t/\tau})$$
$$37.5 \text{ V} = 50(1 - e^{-t/\tau}) \text{ V}$$
$$\frac{37.5}{50} = 1 - e^{-t/\tau}$$
$$e^{-t/\tau} = 1 - 0.75 = 0.25$$
$$\frac{-t}{\tau} = \ln 0.25 = -1.39$$
$$t = 1.39 \times \tau = 1.39 \times 15 \text{ μs} = \textbf{20.8 μs}$$

See Fig. 19-7 for current and voltage waveforms.

19-5   ENERGY STORED IN AN INDUCTOR

As current rises in the coil during the transient period, energy is being stored in the growing magnetic field around the coil. The total amount of stored energy when the transient ceases may be determined from the following:

$$W = \tfrac{1}{2} LI_f^2 \qquad \text{joules} \qquad (19\text{-}7)$$

where: W is the stored energy, in joules (J)
$\quad\quad$ L is the inductance, in henrys (H)
$\quad\quad$ I_f is the final current, in amperes (A)

This energy is stored in the coil as long as the current in the coil is maintained at its final, steady-state value.

EXAMPLE 19-8

How much energy is finally stored in a 1-H coil that is connected in series with a 1-kΩ resistor and a 10-V dc source?

Solution

Steady-state current through coil,

$$I_f = \frac{V}{R} \tag{3-1a}$$

$$= \frac{10\ V}{1\ k\Omega} = 10\ mA$$

$$W = \tfrac{1}{2} LI_f^2 \tag{19-7}$$
$$= \tfrac{1}{2} \times 1\ H \times (10 \times 10^{-3}\ A)^2$$
$$= 50 \times 10^{-6}\ \text{joules}$$
$$= \mathbf{50\ \mu J}$$

19-6 FALL OF CURRENT IN AN INDUCTIVE CIRCUIT (WITH ELECTRONIC SWITCHING)

Instead of applying a voltage to an inductive circuit by closing a mechanical switch, you can use the output of a square wave signal generator. This will apply the voltage for a given time (depending on the frequency), then reduce the voltage to zero for an equal amount of time. Since the switching from ON to OFF can be made repetitively at a suitably high frequency, the variation in current and voltage in the circuit may be examined on an oscilloscope. The circuit arrangement and resulting waveforms are shown in Fig. 19-8.

When the applied input voltage goes positive (ON), the current builds up gradually (as in Fig. 19-4, curve a), and the inductive voltage (v_L) decays with time, following curve b in Fig. 19-4.

When the input voltage drops to zero (OFF), the current

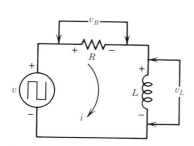

(a) Square wave generator connected to an inductive circuit

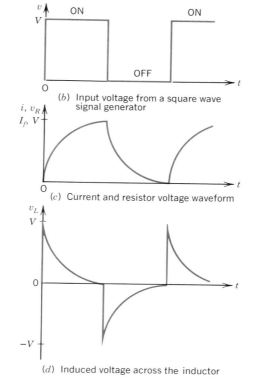

(b) Input voltage from a square wave signal generator

(c) Current and resistor voltage waveform

(d) Induced voltage across the inductor

FIGURE 19-8

Current and voltage waveforms in an inductive circuit with electronic switching of the applied voltage.

must also drop. The collapsing magnetic field around the coil induces a voltage (v_L) that, by Lenz's law, has a polarity *opposite* to that induced when the current was rising. This voltage, shown *negative* in Fig. 19-8d, is responsible for the gradual decay toward zero of current in the circuit.

That is, the voltage induced across the coil (by the decreasing current) acts as a *source* of voltage that attempts to keep the current flowing in the circuit. All the energy stored in the magnetic field is eventually dissipated in the resistor in the form of heat, however, and the current finally drops to zero.

Note that the polarity signs ($+$, $-$) shown in Fig. 19-8a across the coil and the square wave generator are for *reference* purposes only. Thus, when v_L is shown positive in Fig. 19-8d, this means that v_L has an instantaneous polarity equal to that shown in Fig. 19-8a. In the same way, when v_L goes negative in Fig. 19-8d, the opposite polarity is induced across the coil in Fig. 19-8a. Without these reference polarity signs, you would be unable to interpret a positive or negative voltage in a graph. Thus polarity signs *are* used, even where a voltage alternates, to provide a reference for interpreting graphical data.

With electronic switching, the *falling* time constant is the same as the *rising* time constant. You can use the universal time-constant curves in Fig. 19-4 to solve for instantaneous current and voltage values, as in Example 19-9.

EXAMPLE 19-9

A square wave signal generator develops a 12-V peak output at a frequency of 2.5 kHz. It is connected across a series circuit consisting of a 20-mH coil of negligible resistance and a 500-Ω resistor. Find:

a. The current 0.1 ms after the input voltage goes to zero.
b. The voltage across the coil 20 μs after the input voltage goes to zero.
c. From a consideration of the generator's frequency and the time constant of the circuit, the circuit current and inductor voltage 0.2 ms after the input voltage goes to zero.

Solution

a. When the input voltage goes to zero, the current decays following curve b in Fig. 19-4. The initial value of current was

$$I_f = \frac{V}{R} \quad (3\text{-}1a)$$
$$= \frac{12 \text{ V}}{500 \text{ }\Omega} = 24 \text{ mA}$$

The time constant of the circuit

$$\tau = \frac{L}{R} \quad (19\text{-}2)$$
$$= \frac{20 \times 10^{-3} \text{ H}}{500 \text{ }\Omega} = 40 \text{ }\mu\text{s}$$

$$t = 0.1 \text{ ms} = 100 \text{ }\mu\text{s} = \frac{100 \text{ }\mu\text{s}}{40 \text{ }\mu\text{s/}\tau} = 2.5 \text{ }\tau$$

when $t = 2.5 \tau$, $i = 8\%$ of I_f. Therefore,

$$i = 8\% \times 24 \text{ mA}$$
$$= 0.08 \times 24 \text{ mA} = \textbf{1.92 mA}$$

b. The initial coil voltage is 12 V; it also follows the decaying curve b in Fig. 19-4.

$$t = 20 \text{ }\mu\text{s} = \frac{20 \text{ }\mu\text{s}}{40 \text{ }\mu\text{s/}\tau} = 0.5 \text{ }\tau$$

when $t = 0.5 \tau$, $v_L = 60\%$ of the initial voltage. Therefore, $\quad v_L = 60\% \times 12 \text{ V} = \textbf{7.2 V}$

c. $$T = \frac{1}{f} \quad (14\text{-}9)$$
$$= \frac{1}{2.5 \times 10^3 \text{ Hz}} = 0.4 \text{ ms}$$

The duration of time that the input voltage is zero = T/2 = 0.2 ms. This corresponds to $\frac{200 \text{ }\mu\text{s}}{40 \text{ }\mu\text{s/}\tau} = 5 \text{ }\tau$. Thus both i and v_L are approximately zero after 0.2 ms.

19-6.1 Algebraic Solutions*

Equation 19-4 may also be used to calculate circuit current and inductor voltage when the current is *decreasing*. In Fig. 19-9a, it is assumed that the switch moves from position A to position B without introducing additional switch resistance. (Mechanical switching that takes into account the effect of switch arcing is covered in Section 19-7.)

With the switch in position A, the current in the circuit is V/R and the coil voltage is zero. At $t = 0^+$, after the switch has been moved to position B, the initial value of current is still V/R, since current cannot change instanta-

---

*This section may be omitted with no loss of continuity.

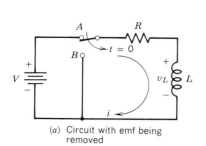

(a) Circuit with emf being removed

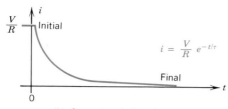

$$i = \frac{V}{R} e^{-t/\tau}$$

(b) Current variation with time

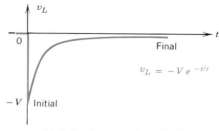

$$v_L = -V e^{-t/\tau}$$

(c) Coil voltage variation with time

FIGURE 19-9

Series *RL* circuit with dc voltage being removed, showing current and inductor voltage responses with exponential equations.

neously in an inductive circuit. The final value of current is zero. (See Fig. 19-9b.)

Instantaneous value
$$= \text{Final value} + (\text{Initial} - \text{Final}) \, e^{-t/\tau} \qquad (19\text{-}4)$$

$$i = 0 + \left(\frac{V}{R} - 0\right) e^{-t/\tau}$$

$$i = \frac{V}{R} e^{-t/\tau} \qquad (19\text{-}8)$$

where: i is the value of current, in amperes (A)
V is the source emf, in volts (V)
R is the circuit resistance, in ohms (Ω)
t is the time after opening the switch, in seconds (s)
$\tau = L/R$ is the time constant, in seconds (s)

The voltage across the coil has an initial value (at $t = 0^+$) of $-V$ and a final value of zero, as shown in Fig. 19-9c.

Instantaneous value
$$= \text{Final value} + (\text{Initial} - \text{Final}) \, e^{-t/\tau} \qquad (19\text{-}4)$$
$$v_L = 0 + (-V - 0) \, e^{-t/\tau}$$
$$v_L = -V e^{-t/\tau} \qquad (19\text{-}9)$$

where: v_L is the inductor voltage, in volts (V)
V is the source emf, in volts (V)
t is the time after opening the switch, in seconds (s)
$\tau = L/R$ is the time constant, in seconds (s)

EXAMPLE 19-10

In the circuit shown in Fig. 19-9a, $V = 12$ V, $R = 500$ Ω, and $L = 200$ mH. Assuming the switch has been in position A for a long time, calculate:

a. The current 0.1 ms after the switch has been moved to position B.
b. The voltage across the inductor 20 μs after the switch has arrived at position B.
c. The circuit current and inductor voltage 0.2 ms after the switch has been moved to position B.

Solution

a. $\tau = L/R$ (19-2)

$$= \frac{20 \times 10^{-3} \text{ H}}{500 \, \Omega} = 40 \times 10^{-6} \text{ s} = 40 \ \mu\text{s}$$

$$i = \frac{V}{R} e^{-t/\tau} \qquad\qquad (19\text{-}8)$$

$$= \frac{12 \text{ V}}{500 \, \Omega} e^{-\frac{100 \ \mu s}{40 \ \mu s}}$$

$$= 24 \times 10^{-3} \times e^{-2.5} \text{ A}$$

$$= 24 \times 0.082 \text{ mA} = \textbf{1.97 mA}$$

b. $v_L = -V e^{-t/\tau}$ (19-9)

$$= -12 e^{-\frac{20 \ \mu s}{40 \ \mu s}} \text{ V}$$

$$= -12 e^{-0.5} \text{ V}$$

$$= -12 \times 0.607 \text{ V} = \textbf{-7.28 V}$$

c. $i = \dfrac{V}{R} e^{-t/\tau}$ (19-8)

$= 24 e^{-\frac{200\ \mu s}{40\ \mu s}}$ mA

$= 24 e^{-5}$ mA

$= 24 \times 0.0067$ mA $=$ **0.16 mA**

$v_L = -V e^{-t/\tau}$ (19-9)

$= -12 e^{-5}$ V

$= -12 \times 0.0067$ V $=$ **−0.0804 V**

Compare these answers with those for Example 19-9.

Note that the exponential equations 19-5, 19-6, 19-8, and 19-9 are special forms of the general equation, Eq. 19-4. All the above examples may be solved by the direct application of Eq. 19-4 by simply sketching the response involved, and identifying the initial (at $t = 0^+$) and final values.

19-6.2 Measurement of Coil Inductance

The waveforms of Fig. 19-8 suggest that you can determine the inductance of a coil by measuring the time constant of a circuit in which the coil is series-connected with a known resistance. Figure 19-10a shows an oscilloscope connected across the resistance, and a square wave generator used to apply a voltage that varies between 0 and V_m. The signal generator's frequency is adjusted until the oscilloscope displays the transient response of current with at least five time constants. (See Fig. 19-10c.) The horizontal distance required for the voltage waveform across R to reach 63% of its final value is the time constant. From this, the inductance may be determined, as shown in Example 19-11.

EXAMPLE 19-11

A coil that has a dc resistance of 50 Ω is connected in series with a 10-Ω resistor and a square wave signal generator with a 10-V p-p output. An oscilloscope across the resistor displays a waveform as in Fig. 19-10c, reaching 63% of its peak value in a horizontal distance of 0.8 cm. If the horizontal calibration is 0.5 ms/cm, determine:

a. The time constant of the whole circuit.
b. The inductance of the coil.

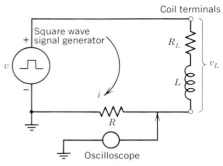

(a) Connection of an oscilloscope to measure v_R

(b) Square wave input voltage from a signal generator

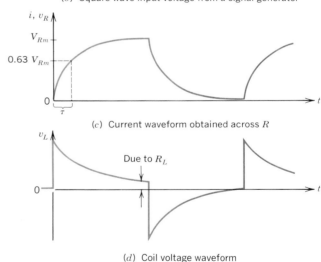

(c) Current waveform obtained across R

(d) Coil voltage waveform

FIGURE 19-10
Using the method of time-constant measurement to determine the inductance L.

Solution

a. The time constant of the whole circuit is the time required for the current to reach 63% of the final value
$= 0.8$ cm $\times$ 0.5 ms/cm
$=$ **0.4 ms**

b. $\tau = \dfrac{L}{R_T}$ (19-2)

Therefore, $L = \tau R_T$

$$= 0.4 \times 10^{-3} \text{ s} \times (50 \ \Omega + 10 \ \Omega)$$
$$= 24 \times 10^{-3} \text{ H} = \textbf{24 mH}$$

Note that the voltage waveform across the coil in Fig. 19-7d does not drop to *zero* on the positive half-cycle because of the voltage drop across the resistive portion (R_L). This also means that the final value of the voltage across R in Fig. 19-10c is not V_m, but V_{Rm}, where:

$$V_{Rm} = V_m \times \frac{R}{R + R_L}$$

Thus the sum of the voltage drops across the coil (v_L) and across the resistor (v_R) is equal to the input voltage (V_m) *at all times,* as required by Kirchhoff's voltage law.

19-7 FALL OF CURRENT IN AN INDUCTIVE CIRCUIT (WITH MECHANICAL SWITCHING)

As you have seen, current builds up gradually when a dc voltage is applied to a series circuit of resistance and inductance, with the coil's induced voltage opposing the increase in current through the coil. Also, energy is stored in the magnetic field around the coil.

Now, consider the mechanical opening of a switch to interrupt the current through the coil shown in Fig. 19-11a.

The immediate effect of the coil's inductance is to induce an emf whose polarity is such that the emf tries to *maintain* the current at its initial level. That is, v_L in Fig. 19-11a acts in series with the applied voltage (10 V, in this case) to try to keep a current of 10 mA flowing, even though the blade of the switch is being opened. The induced voltage is high enough to make the 10-mA current "jump the gap" across the opening switch contacts. This effect, called *arcing,* results from the breaking down of the insulating property of the air between the contacts of the switch. The air ionizes, conducting current for a brief instant. A blue spark is usually visible under such conditions. Why is such a high voltage induced?

Assume that the resistance of the ionized air between the opening switch contacts (R_{sw}) is 9 kΩ. If a current of 10 mA is flowing through the switch, there must be a *voltage drop* across the switch, given by

$$V_{sw} = IR_{sw}$$
$$= 10 \text{ mA} \times 9 \text{ k}\Omega$$
$$= \textbf{90 V}$$

Since at this instant *all* of V (10 V) still appears across R, this voltage (across the switch) must have been developed across the coil by the collapsing magnetic field when the current *started* to decrease. (See Fig. 19-11c.) You can determine the initial rate at which the current decreases as follows:

$$v_L = L\left(\frac{\Delta i}{\Delta t_0}\right) \qquad (17\text{-}1)$$

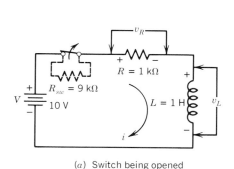

(a) Switch being opened

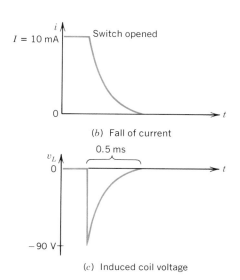

(b) Fall of current

(c) Induced coil voltage

FIGURE 19-11
Current and voltage waveforms resulting from interruption of current in an inductive circuit with dc.

and
$$\frac{\Delta i}{\Delta t_0} = \frac{v_L}{L}$$

$$= \frac{90 \text{ V}}{1 \text{ H}} = \textbf{90 A/s}$$

Note that this rate of change of decreasing current is nine times greater than when the switch was initially closed and current was rising. (For current rising, $\frac{\Delta i}{\Delta t_0} = \frac{V}{L} = \frac{10 \text{ V}}{1 \text{ H}} = 10$ A/s.) Why is this so?

Calculate the time constant for the circuit when the switch is being opened. Note that you now must use the *total* series resistance in the circuit:

$$\tau = \frac{L}{R_T} = \frac{L}{R + R_{sw}} \tag{19-2}$$

$$= \frac{1 \text{ H}}{1 \text{ k}\Omega + 9 \text{ k}\Omega} = \textbf{0.1 ms}$$

This is only one-tenth the time constant for the interval when the switch was being closed (for closing, $\tau = \frac{L}{R} = \frac{1 \text{ H}}{1 \text{ k}\Omega} = 1$ ms). This means that the current decreases much more quickly than it increases. (Actually, the time "constant" is not *constant,* because the switch resistance increases as it opens. The current will be essentially zero in less than 0.5 ms.) This very rapid rate of change of current is responsible for the high induced voltage across the coil. And, it is the high resistance introduced by the switch that makes the circuit more resistive and causes this rapid change in current.

The ability of a coil to induce this high voltage is readily seen by observing a neon lamp connected across the coil as in Fig. 19-12a. Such a lamp, sometimes used as an ON/OFF indicator in amplifiers, requires approximately 70 V to "fire" or light up.*

Closing the switch in Fig. 19-12a will not light the lamp, since a peak of only 10 V exists across the coil and lamp. But when the switch is *opened,* the lamp will flash because of at least 90 V induced. Current and voltage waveforms resulting from a repeated opening and closing of the switch are shown in Figs. 19-12b–19-12d. The large

---

*Only the electrode that is at a negative potential will glow. For this reason, a neon lamp can be used to determine the polarity of direct current. It also serves to distinguish direct from alternating current, since ac will cause *both* electrodes to glow.

negative voltage "spike" that causes the lamp to flash can also be observed on an oscilloscope.

A major application of the high voltage induced by interrupting the current through a coil is found in a fluorescent lamp circuit. (See Section 20-5.) The momentary self-induced voltage of several hundred volts is sufficient to "fire" or *start* the fluorescent lamp.

19-8 CONVENTIONAL AUTOMOTIVE IGNITION SYSTEM

Another application of interrupting current through a coil is found in the conventional automotive ignition system that uses a distributor and points. In this case, it is not the self-induced voltage that is of value, but the much higher *mutual* induced voltage.

Figure 19-13 shows that the automotive ignition coil is actually a step-up autotransformer. The primary has several hundred turns of wire and is capable of carrying several amperes. The secondary has thousands of turns of very fine wire.

If the ignition switch is ON and the engine is cranked by the starter motor, a shaft driven by the engine turns a cam in the distributor to alternately open and close the breaker points. When the points (contacts) are closed, a circuit is completed, allowing direct current to flow through the coil's primary winding, setting up a strong magnetic field. When the rotating cam causes the points to open, the current to the primary winding is interrupted. The magnetic field collapses, inducing a very high (as much as 25 kV) voltage in the secondary winding. This voltage is directed to the proper spark plug of the engine by the distributor *rotor* (a rotating switch driven by the same shaft as the breaker cam).

To produce the very high secondary voltage required, the magnetic field must collapse very quickly. To achieve this rapid collapse, the primary current must be interrupted abruptly. But as you have seen, the self-inductance of the primary induces a high voltage that tends to keep the current flowing. This results in arcing across the contacts, but even worse, drags out the decay of the current. To eliminate this problem, a capacitor (formerly called a *condenser*) is connected across the points. (See Section 21-1.)

When the points open, the current flows into the capacitor, rather than jumping the gap (arcing) at the opening points. The capacitor very quickly becomes charged, acting like an open circuit with a resistance that is effectively infinite. The time constant for current decay on the pri-

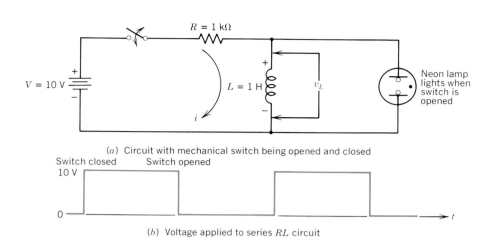

(a) Circuit with mechanical switch being opened and closed

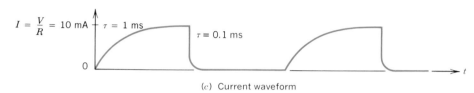

(b) Voltage applied to series RL circuit

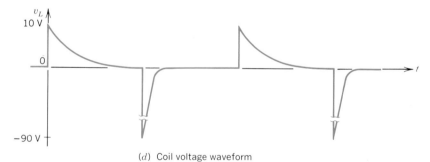

(c) Current waveform

$I = \dfrac{V}{R} = 10$ mA $\tau = 1$ ms $\tau = 0.1$ ms

(d) Coil voltage waveform

FIGURE 19-12

Waveforms resulting from mechanical switching of current in an inductive circuit (switch resistance assumed to be 9 kΩ).

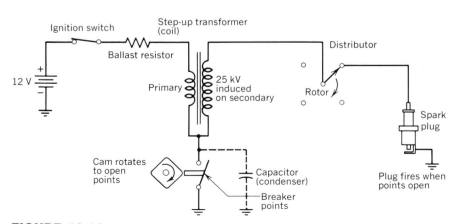

FIGURE 19-13

Conventional ignition system.

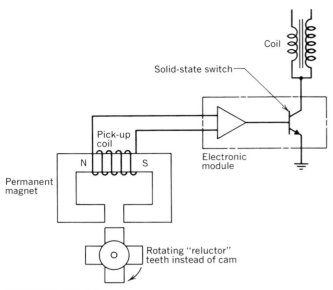

FIGURE 19-14
One type of electronic ignition system.

19-9 ELECTRONIC AUTOMOTIVE IGNITION SYSTEM

The induced voltage in a secondary coil can be increased to as much as 35 or 40 kV by using electronic switching. As shown in Fig. 19-14, this involves using a transistor as a solid-state switch in place of the conventional breaker points. The electronic switch also eliminates arcing, and thus, the need for a capacitor.

The switch is turned on and off by a signal from a pickup coil on a stationary permanent magnet core. Every time one of the rotating teeth of the *reluctor* passes through the magnet's gap, changing the field strength, an emf is induced in the pickup coil.

The signal (emf) is suitably amplified in the electronic module and "conditioned" by other information such as dwell time, engine speed, atmospheric pressure, and so on, to achieve maximum performance consistent with pollution emission controls.

mary becomes extremely small ($\tau = L/R_T$), so the current decreases rapidly and the field collapses just as quickly.

There is also a voltage induced on the secondary at the instant the points close. But since the time constant for the buildup of current is relatively long, this voltage is quite small and is not used. It does, however, provide a useful indication of point closure on an ignition analyzer oscilloscope connected for engine testing. This instrument displays a primary or secondary "parade" pattern in which the firing characteristics of all engine cylinders are shown. The pattern shows cylinder firing from left to right on the screen in normal firing order.

19-10 METHODS OF REDUCING INDUCTIVE EFFECTS

Even though the interruption of current through an inductance has many useful applications (such as those discussed above), it is often necessary to prevent dangerously induced high voltages, which can break down insulation. The problem occurs in both ac and dc circuits, but can be minimized in a number of ways.

One method used with dc circuits is to provide a relatively low resistance as an alternate path for the current. The current will flow through this path, rather than the opening switch. (See Fig. 19-15a.)

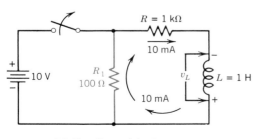

(a) Shunting resistor R_1

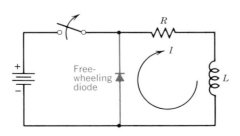

(b) Free-wheeling diode

FIGURE 19-15
Methods of reducing high inductive voltage.

For the values shown, the steady-state current through the coil with the switch closed is $\frac{10 \text{ V}}{1 \text{ k}\Omega}$ or 10 mA. When the switch is opened, this current continues to flow, but follows a path through R_1, assumed here to be 100 Ω. Thus the voltage across the coil is

$$v_L = IR_1 + IR$$
$$= 10 \text{ mA} \times 100 \text{ }\Omega + 10 \text{ mA} \times 1 \text{ k}\Omega$$
$$= 1 \text{ V} + 10 \text{ V}$$
$$= \mathbf{11 \text{ V}}$$

This compares to 90 V without R_1 and an assumed switch resistance of 9 kΩ.

The disadvantage of this method is the high current (100 mA) drawn by R all the time the switch is closed. This can be overcome by using a diode, as shown in Fig. 19-15b. When the switch is closed, no current can flow through the diode because it is reverse-biased and effectively *open*. Current flows normally through the coil.

When the switch opens, the polarity of the induced emf causes the diode to conduct, shunting the coil with a very low resistance. The time constant $\left(\frac{L}{R_T}\right)$ is large, and the current drops off gradually. Very little voltage is induced across the coil.

A diode used in this manner is often referred to as a "free-wheeling" diode. Such arrangements are often used in dc motors and relay coils to suppress high induced voltages when the equipment is turned off.

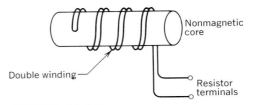

FIGURE 19-16
Bifilar noninductive resistor.

19-10.1 Noninductive Resistors

Wire-wound resistors, which are used for higher-power applications, inevitably possess some inductance. Where this cannot be tolerated, *special* resistors of the wire-wound type are available. As shown in Fig. 19-16, a *bifilar* (double) winding is used. The magnetic fields set up by the side-by-side currents flowing in opposite directions cancel each other. The result is minimized inductance and, thus, very little induced voltage.

Conversely, a bifilar winding can be used to provide a coefficient of coupling between two coils that is very close to unity (1.0). This is necessary where iron cores are not practical, as in radio frequency transformers. In that case, two terminals are provided at each end of the bifilar winding.

SUMMARY

1. When a dc voltage is applied to an inductive circuit, the opposing induced emf causes the current to build up gradually.
2. Current has the greatest rate of change when the voltage is first applied, and is given by

$$\frac{\Delta i}{\Delta t_0} = \frac{V}{L}$$

3. The final value of current is determined only by the resistance and the applied voltage, as given by Ohm's law.
4. The time constant is the length of time that would be required for the current to reach its final value if it continued to increase at its initial rate.
5. Alternatively, the time constant is the time required for the current to change 63% from its initial value toward its final value.
6. The time constant is given by

$$\tau = \frac{L}{R}$$

7. Universal time-constant curves give the current and voltage values as a percentage of maximum in terms of the time constant.

8. The instantaneous value of current or voltage in a series *RL* circuit in which a change is taking place may be calculated by using the exponential equation:

Instantaneous value = Final value + (Initial − Final) $e^{-t/\tau}$

9. The final energy stored in an inductor's magnetic field is

$$W = \tfrac{1}{2} L I_f^2$$

10. Mechanical switching in an inductive circuit causes current to fall at a much faster rate than it rises, due to the resistance introduced by the switch, which causes a much smaller time constant.

11. The rapid rate of current decay in a coil causes high induced voltages.

12. The high voltage produced on the secondary side of an automotive ignition coil results from the use of a step-up transformer in combination with the sudden interruption of current in the primary (by breaker points or an electronic switch).

13. Inductive effects can be minimized by shunting the induced voltage with a low resistance or a free-wheeling diode.

SELF-EXAMINATION

Answer T or F, or, for multiple choice, a, b, c, or d
(Answers at back of book)

19-1. In a dc circuit, inductance determines how the current changes and not the final value of current. _____

19-2. In a dc circuit with inductance and resistance, the initial voltage across the inductance is equal to the applied voltage. _____

19-3. Interrupting the current in an inductive circuit always induces the same voltage across the inductance as when the voltage is first applied. _____

19-4. The current builds up in an inductive circuit containing resistance at the same rate until the final steady state is reached. _____

19-5. The time constant determines how long it takes for steady-state conditions to be reached. _____

19-6. The time constant applies only to the first 63% change in the current. _____

19-7. If a 12-V dc supply is connected to a coil having an inductance of 100 mH and a resistance of 1 kΩ, the initial rate of rise of the current and the final current are
a.   12 A/s and 12 mA
b.   1.2 A/s and 120 mA
c.   120 A/s and 12 mA
d.   12 mA/s and 120 A

19-8. The time constant in Question 19-7 is
a.   100 ms
b.   100 μs
c.   12 ms
d.   10 s

19-9. Energy stored in an inductor is in the form of an electric field. _____

19-10. Arcing across the contacts of an opening switch is due to dirty contacts. _____

19-11. High voltage is induced across an inductor when the current through it is interrupted because of a reduction in the time constant because of switch resistance. _____

19-12. An automotive ignition system induces a high voltage as a result of mutual inductance. _____

19-13. A capacitor is used across the breaker points solely to reduce arcing and avoid the constant replacement of the points. _____

19-14. A higher voltage is induced in an electronic ignition system because the current can be interrupted much more suddenly. _____

19-15. The high voltage induced across a dc relay coil when it is shut off can be minimized by using a diode across the coil. _____

REVIEW QUESTIONS

1. a. What is the basic reason for the gradual buildup of current in an inductive circuit with an applied dc voltage?
   b. Why doesn't this happen in a pure resistance?
   c. If the voltage induced across an inductance is initially equal to the applied dc voltage, why does any current flow at any time?
   d. Why does the current eventually reach a final value?

2. a. What do you understand by the term *time constant?*
   b. Give two definitions.
   c. How long does a transient last?

3. Prove that the time constant, given by L/R, has units of seconds.

4. a. Under what conditions does the voltage across a coil not drop to zero at the end of a transient with a dc voltage?
   b. Under what conditions does the voltage across a coil exceed the input dc voltage?
   c. Why doesn't this happen when a square wave signal generator is used?

5. a. What factors determine the energy stored by an inductor?
   b. What prevents an inductor being used as a portable energy storage device like a battery?
   c. In an inductive circuit with a mechanical switch, what evidence is there of the energy stored in the coil?

6. a. What causes arcing?
   b. Why does it not occur, as badly, when closing a switch?
   c. Describe a demonstration to illustrate the high-voltage-inducing capability of an inductor.

7. a. Describe how it is possible for a 12-V dc source to produce 25 kV if a transformer cannot be operated from dc.
   b. If the capacitor were removed from a conventional ignition system, why would the engine be unlikely to start?
   c. Why is high voltage produced by the *opening* of the breaker points but not by the *closing?*

8. a. What basic difference exists between a standard and an electronic ignition system?
   b. What causes a higher secondary voltage to be induced in the latter?
   c. Why is no capacitor required?

9. a. Describe two methods of reducing inductive effects in a dc circuit.
   b. What is the advantage of each?

c.   Which system can be used for alternating current and direct current?

d.   How does a bifilar winding reduce the inductance in a wire-wound resistor?

10.   Describe how you would measure the inductance of a coil.

PROBLEMS

(Answers to odd-numbered problems at back of book)

19-1.   A 200-μH coil with a resistance of 150 Ω is connected across a 300-mV dc supply. Calculate:
a.   The initial rate of rise of current.
b.   The final value of current.
c.   The time constant.
d.   The time required for the current to reach its final value.

19-2.   Repeat Problem 19-1 with a 7-H choke, a resistance of 15 Ω, and a dc supply of 120 V.

19-3.   The current through a coil is observed to increase from 0 to a final value of 0.25 A in 0.15 s when a 12-V battery is connected across it. Calculate:
a.   The resistance of the coil.
b.   The time constant.
c.   The inductance of the coil.

19-4.   The current in a coil of 20-Ω resistance increases from 0 to 0.315 A in 2 s when connected across 10-V dc.
a.   What is the inductance of the coil?
b.   How long will it take the current to reach its final value?

19-5.   An 8-H inductance is connected in series with a 320-Ω resistance. A switch is closed to apply 120-V dc. Use the universal time constant curves to calculate:
a.   The current after 50 ms.
b.   The voltage across the inductance after 0.1 s.
c.   How long it takes for the current to reach 0.3 A.
d.   The time for the voltage across the inductance to become 60 V.

19-6.   Repeat Problem 19-5 using $L = 12$ H, $R = 200$ Ω, and $V = 150$ V.

19-7.   Solve Problem 19-5 using the exponential equations.

19-8.   Solve Problem 19-6 using the exponential equations.

19-9.   Determine the energy stored in the coil of Problem 19-5 after five time constants.

19-10.   Determine the energy stored in the coil of Problem 19-6 after seven time constants.

19-11.   A square wave signal generator develops a peak output of 20 V at a frequency of 50 kHz when connected across a 200-μH coil of negligible resistance in series with a 100-Ω resistor.
a.   Draw, in the proper time relationship, the input voltage, the current waveform, and the voltage waveform across the coil. Show peak values of voltage and current on all three waveforms.
b.   Use Fig. 19-4 to find the current 3 μs after the input voltage goes to zero.
c.   Use Fig. 19-4 to find the voltage across the coil 1 μs after the input voltage goes to zero.
d.   Use Fig. 19-4 to find how long it takes for the current to fall to 50 mA after the input voltage goes to zero.

19-12.   Repeat Problem 19-11 assuming a peak voltage of 10 V.

19-13.   Solve Problem 19-11 using the exponential equations.

19-14. Solve Problem 19-12 using the exponential equations.

19-15. A coil of dc resistance 150 Ω is connected in series with a 22-Ω resistor and a square wave signal generator. An oscilloscope across the resistor indicates that the current reaches 63% of its final value in a horizontal distance of 3.2 cm. If the horizontal calibration is 20 μs/cm, determine:

    a.   The time constant of the whole circuit.

    b.   The inductance of the coil.

    c.   What the coil voltage drops to if the signal generator has an internal resistance of 600 Ω and an open circuit voltage of 10-V peak.

19-16. What is the maximum frequency the signal generator can be set to in Problem 19-15 and still maintain the ability to measure the coil's inductance?

19-17. Refer to Fig. 19-12a: $R = 3.3$ kΩ, $L = 8$ H, and $V = 20$ V. When the switch is closed calculate:

    a.   The initial rate of rise of current.

    b.   The time constant.

    c.   The final current.

    d.   The time for the current to reach its final steady-state value.

    e.   The peak voltage across the inductance.

When the switch is opened, assume an initial switch resistance of 20 kΩ and calculate:

    f.   The peak voltage across the inductance.

    g.   The initial rate of fall of the current.

    h.   The time constant.

    i.   The approximate time for the current to drop to zero.

19-18. The circuit in Problem 19-17 has a 2.2-kΩ resistor connected across the 3.3-kΩ resistor and 8-H inductance as in Fig. 19-15a to reduce the inductive effect.

    a.   What voltage will be induced across the coil when the switch is opened?

Will the lamp light?

    b.   What is the total current drawn from the supply when the switch is closed?

19-19. Refer to Fig. 19-17. At $t = 0$, after the switch has been in position B for a long time the switch is thrown to position A, stays there for 3 s, then is returned to position B. Calculate:

    a.   The circuit current just before the switch is returned to position B.

    b.   The voltage across the inductor just before the switch is returned to position B.

    c.   The inductor voltage just after the switch is returned to position B.

    d.   The circuit current and inductor voltage 0.5 s after the switch returns to position B. (*Hint:* Use Eq. 19-4 directly.) Draw waveforms of i and v_L from $t = 0$ to $t = 4$ s.

19-20. Repeat Problem 19-19, assuming the 60-V source has an internal resistance of 5 Ω.

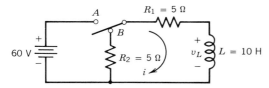

FIGURE 19-17
Circuit for Problems 19-19 and 19-20.

CHAPTER 20

INDUCTANCE IN AC CIRCUITS

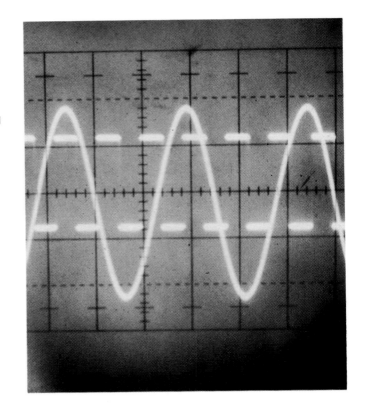

In this chapter, you will consider the current and voltage relationships of an inductance in a sinusoidal ac circuit. Because the current is always changing in an ac circuit, there is some effective opposition to the current drawn by a coil at a given frequency. This is called the *inductive reactance* (X_L), which is determined both by inductance and frequency.

Because of inductive reactance, a coil has a greater opposition to alternating current than it does to direct current. This property is taken advantage of in a dc *filter* circuit to reduce the ac ripple component. An inductor is also used in a fluorescent lamp circuit, where (as a *ballast*) it is used to induce a momentary high voltage to light the lamp, then limit the current (after the lamp is lit).

The series and parallel connection of inductive reactances can be treated similarly to resistance, except when *mutual reactance* (X_m) is involved. In this case, the coefficient of coupling permits a determination of mutual inductance and the total inductive reactance.

The *quality of a coil (Q)* is expressed in terms of inductive reactance, and is a "figure of merit" to show how inductive a coil is compared with its *ac resistance.* The ac resistance of a coil is shown to be different from its dc ohmic resistance, due partially to the *skin effect.*

20-1 EFFECT OF INDUCTANCE IN AN AC CIRCUIT

You have seen how an inductance will induce an emf whenever the current through it changes. If a sinusoidal (ac) current is made to flow through an inductance, you should expect an emf to be induced almost continually. Since this emf opposes any change in current, the inductance must provide some effective opposition to the flow of alternating current. This opposition is over and above any resistance that the coil may have.

20-1.1 Phase Angle Relationship Between Current and Voltage

Assume that an inductance of negligible resistance has a sinusoidal current flowing through it, as shown in Fig. 20-1. The current waveform is given by

$$i = I_m \sin \omega t$$

The voltage across the coil, which must also be the voltage applied from the source, is

$$v = L\left(\frac{\Delta i}{\Delta t}\right) \tag{17-1}$$

Recall that $\dfrac{\Delta i}{\Delta t}$ is the *slope* of the current waveform.

Only when the current is at its peak values of I_m and $-I_m$ is the slope zero. Only at these instants, then, is the voltage across the coil zero. The current waveform has its greatest slope where it crosses the time axis. This causes the maximum voltage to be induced across the coil. The voltage is either positive or negative, depending on whether the current is increasing or decreasing. (See Fig. 20-1b.)

It is not difficult to justify what waveform of voltage must appear across the coil, although it requires some mathematics to *prove* that the waveform is sinusoidal in shape. In fact, the voltage waveform is actually a *cosine curve* given by

$$v = V_m \cos \omega t$$

This is actually a sine curve that has been shifted to the left (advanced a quarter-cycle, or 90°). Since the voltage waveform reaches its peak value one-quarter cycle *before* the current waveform reaches its own peak, you could say that the voltage *leads* the current by 90°.

Conversely, you could say that the current in the coil *lags* behind the applied coil voltage by 90° in a *purely* inductive circuit.

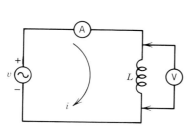

(a) Pure inductance in an ac circuit

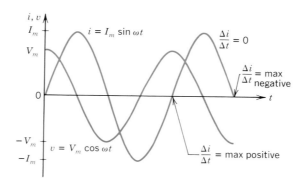

(b) Waveforms of current and voltage to show current lagging voltage by 90°

FIGURE 20-1

Effect of inductance in an ac circuit.

This angle is often referred to as the *phase angle*, because it shows how much (by what angle) the current and voltage waveforms are *out of phase*. You will recall that in a purely resistive circuit, the current and voltage are in phase with each other; the phase angle is 0°.

It is easy to remember that current *lags* behind voltage if you recall what happens in a dc circuit. When voltage is applied to a coil, the current builds up slowly to its final value. (Refer to Fig. 19-2.) Even in a dc circuit, there is a delay between the application of voltage across the coil and current flowing in the coil.

20-2 INDUCTIVE REACTANCE

Since both the voltage and current waveforms in Fig. 20-1 are sinusoidal, their rms values can be obtained as shown below:

$$V = \frac{V_m}{\sqrt{2}} \qquad (15\text{-}17)$$

and

$$I = \frac{I_m}{\sqrt{2}} \qquad (15\text{-}16)$$

These are the values of the voltmeter and ammeter readings in Fig. 20-1*a*. Now, recall that the opposition to current in a resistive circuit (either dc or ac) is given by the ratio:

$$\frac{V_R}{I_R} = R \qquad (15\text{-}15)$$

Similarly, the opposition to current provided by a *pure* inductance is given by the ratio V_L/I_L. Although this ratio of voltage to current must be expressed in units of ohms, it cannot be called resistance. (A pure inductance has no [zero] resistance.) Instead, the term *inductive reactance* (X_L) is used. Thus, for a pure inductor,

$$X_L = \frac{V_L}{I_L} \qquad \text{ohms} \qquad (20\text{-}1)$$

where: X_L is the inductive reactance, in ohms (Ω)
V_L is the voltage across the pure inductance, in volts (V)
I_L is the current through the inductance, in amperes (A)

EXAMPLE 20-1

In the circuit of Fig. 20-2, a 28.28-V p–p sinusoidal voltage causes an ammeter reading of 10 mA. What is the opposition to current caused by the inductance?

Solution

The rms value of the applied voltage is given by

$$V = \frac{V_{p\text{-}p}}{2\sqrt{2}} \qquad (15\text{-}17)$$

$$= \frac{28.28 \text{ V}}{2\sqrt{2}} = 10 \text{ V}$$

$$X_L = \frac{V_L}{I_L} \qquad (20\text{-}1)$$

$$= \frac{10 \text{ V}}{10 \times 10^{-3} \text{ A}}$$

$$= 1000 \ \Omega = \mathbf{1 \ k\Omega}$$

EXAMPLE 20-2

How much current will flow through a coil of negligible resistance whose inductive reactance is 500 Ω when 120 V are applied?

Solution

$$X_L = \frac{V_L}{I_L} \qquad (20\text{-}1)$$

Therefore, $\qquad I_L = \dfrac{V_L}{X_L}$

$$= \frac{120 \text{ V}}{500 \ \Omega} = \mathbf{0.24 \ A}$$

EXAMPLE 20-3

What is the voltage drop across a coil of inductive reactance 1.5 kΩ and negligible resistance when a current of 70 mA flows through the coil?

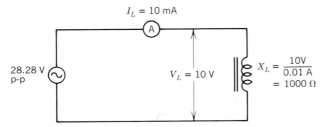

FIGURE 20-2
Circuit for Example 20-1, showing $X_L = V_L/I_L$.

Solution

$$X_L = \frac{V_L}{I_L} \qquad (20\text{-}1)$$

Therefore,
$$\begin{aligned} V_L &= I_L X_L \\ &= 70 \times 10^{-3}\,\text{A} \times 1.5 \times 10^3\,\Omega \\ &= \textbf{105 V} \end{aligned}$$

20-3 FACTORS AFFECTING INDUCTIVE REACTANCE

Inductive reactance (opposition to current) is due to the induced emf across the inductor produced by the changing current through the inductor. Anything that increases this emf must also increase the inductive reactance.

We know that

$$v_L = L\frac{\Delta i}{\Delta t} \qquad (17\text{-}1)$$

Obviously, an increase in L must increase the inductive reactance. Further, $\dfrac{\Delta i}{\Delta t}$, the slope of the current waveform, must depend upon the frequency. A higher frequency means that the peak value of the current is reached in a shorter time, so $\dfrac{\Delta i}{\Delta t}$ must be higher. Combining these two effects, and using the angular frequency ($\omega = 2\pi f$), we obtain

$$X_L = 2\pi f L = \omega L \quad \text{ohms} \qquad (20\text{-}2)$$

where: X_L is the inductive reactance, in ohms (Ω)
f is the frequency, in hertz (Hz)
L is the inductance, in henrys (H)
ω is the angular frequency, in radians per second (rad/s)

Note that this equation is valid only for sinusoidal ac circuits. (A full derivation, using the principles previously covered, is given in Appendix I.)

Equation 20-2 states that, as the frequency increases, there is a linear increase in the inductive reactance of a coil. This is represented in the graph of Fig. 20-3, which is drawn for a constant inductance of $L = 1.6$ H. Note that when $f = 0$ (direct current), the inductive reactance is zero. **This means that inductance has no steady-state effect or opposition to current in a** *direct current* **circuit.**

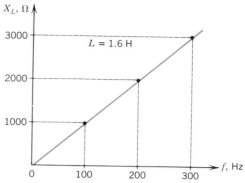

FIGURE 20-3
Linear increase of X_L with frequency, f, for a constant inductance of 1.6 H.

EXAMPLE 20-4

A 120-V, 60-Hz source is applied to a 5-H coil of negligible resistance.
a. How much current flows?
b. What is the equation for current?

Solution

a. $X_L = 2\pi f L \qquad (20\text{-}2)$

$\qquad = 2\pi\ 60\ \text{Hz} \times 5\ \text{H}$

$\qquad = 1885\ \Omega$

Therefore, $I_L = \dfrac{V_L}{X_L} \qquad (20\text{-}1)$

$\qquad = \dfrac{120\ \text{V}}{1885\ \Omega} = 0.064\ \text{A} = \textbf{64 mA}$

b. $I_m = I_L \times \sqrt{2} \qquad (15\text{-}20)$

$\qquad = 0.064\ \text{A} \times \sqrt{2}$

$\qquad = 0.091\ \text{A}$

$i = I_m \sin 2\pi f t$

$\qquad = 0.091 \sin 2\pi 60 t$

$\qquad = \textbf{0.091 sin 377}t\ \textbf{A}$

EXAMPLE 20-5

A sine wave signal generator, set to 1-kHz and 10-V output, causes a current of 50 mA through a coil of negligible resistance. What is the inductance of the coil?

Solution

$$X_L = \frac{V_L}{I_L} \qquad \text{(20-1)}$$

$$= \frac{10 \text{ V}}{50 \times 10^{-3} \text{ A}} = 200 \text{ }\Omega$$

$$X_L = 2\pi f L \qquad \text{(20-2)}$$

Therefore,

$$L = \frac{X_L}{2\pi f}$$

$$= \frac{200 \text{ }\Omega}{2 \text{ }\pi \times 1 \times 10^3 \text{ Hz}}$$

$$= 0.032 \text{ H} = \textbf{32 mH}$$

The total opposition to current arising from the combined resistance and inductance of the coil is called *impedance*. It is considered in Chapter 24.

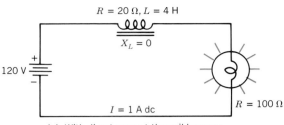

(a) With direct current, the coil has zero inductive reactance. The lamp lights.

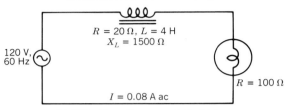

(b) With alternating current, the high inductive reactance limits the current to 80 mA. The lamp does not light.

FIGURE 20-4
Comparison of inductance in dc and ac circuits.

20-4 APPLICATION OF INDUCTORS IN FILTER CIRCUITS

It should be clear from the preceding section that a coil with negligible resistance provides a much higher opposition to alternating current than to direct current. This is illustrated in the two circuits shown in Fig. 20-4. With direct current, as in Fig. 20-4a, the inductive reactance of the coil is zero. The only opposition to current in this circuit is provided by 20 Ω of coil resistance and 100 Ω of lamp resistance. When an emf of 120-V dc is applied, a current of 1-A dc results, and the lamp lights.

With alternating current (Fig. 20-4b), an inductance of 4 H at a frequency of 60 Hz results in an inductive reactance of 1500 Ω. In comparison with X_L (the inductive reactance), the resistance of the circuit is negligible, so the current is determined by V_L/X_L. The current of 80 mA is not sufficient to light the lamp.

The different opposition to alternating and direct current provided by a coil is used in a dc power-supply *filter* circuit. After the alternating current has been converted to direct current by diodes, an ac component remains (Fig. 20-5) and must be filtered out to leave pure direct current. This may be done by using a coil, often called a *choke*, to allow passage of direct current while blocking the alternating current.

A simple filter circuit is shown in Fig. 20-5. Assume that the coil's resistance is negligible (compared with R) and that the inductive reactance of the coil (X_L) is much larger than R.

A waveform that contains both direct and alternating current is applied to the input. The dc portion appears across R without being affected by the coil. The ac portion, however, undergoes a form of voltage division be-

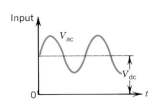

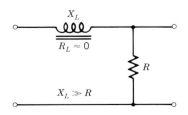

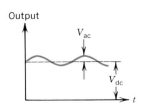

FIGURE 20-5
Inductor used in filter circuit to reduce ac ripple component.

tween X_L and R. With the assumptions made above, the ac output voltage across R is given by

$$V_{out(ac)} = V_{in(ac)} \times \frac{R}{X_L} \qquad (20\text{-}3)$$

EXAMPLE 20-6

A waveform containing 10-V dc and 10-V ac is applied to the filter of Fig. 20-5 with $L = 10$ H and $R = 100\ \Omega$. If the ac component has a frequency of 60 Hz, determine:

a. The dc output voltage.
b. The ac output voltage.

Solution

a. The dc output = the dc input = 10 V (assuming that the resistance of the coil is zero).

b. $X_L = 2\pi f L$ (20-2)

       $= 2\pi \times 60$ Hz $\times 10$ H

       $= 3770\ \Omega$

$$V_{out(ac)} = V_{in(ac)} \times \frac{R}{X_L} \qquad (20\text{-}3)$$

$$= 10\text{ V} \times \frac{100\ \Omega}{3770\ \Omega} = \textbf{0.27 V}$$

Note that Example 20-6 shows how the ac *ripple* component is reduced by an inductor, as shown in Fig. 20-5.

20-5 APPLICATION OF INDUCTANCE IN A FLUORESCENT LAMP CIRCUIT

Figure 20-6*a* shows a preheat fluorescent lamp circuit. It is typical of many desk-type fluorescent lamps, but the same principle of operation is used in larger lamps, as well.

The *ballast* (inductor) is used to induce a momentary high voltage to start the lamp. Then, after the lamp is lit, the inductive reactance of the ballast coil *limits* current through the lamp, as explained below.

A fluorescent lamp is a glass tube with a tungsten filament sealed in each end, and an inner surface coated with a phosphor material (different phosphors are used to produce different colors of light). During manufacture, most of the air is removed from the lamp and a small amount of argon gas and mercury sealed in the tube.

When the momentary contact switch is pushed CLOSED and held in that position for a few seconds, a complete series path for current is created through the two filaments and the ballast. The filaments heat up, emitting electrons, and a dull glow can be observed at either end of the tube.

When the momentary contact switch is released, the current through the ballast is interrupted, inducing a momentary high voltage. This voltage, along with the 120-V input, causes the lamp to "fire"—current is conducted through the ionized gas in the tube from filament to filament. (This does *not*, however, *by itself* produce visible light.)

As the electrons move through the tube, they collide with the now-gaseous ions of mercury. This causes the valence electrons of the mercury atoms to be dislodged from their shell and raised to a higher energy level. As these electrons fall back into their stable orbits, they give up the energy they have absorbed, in the form of invisible ultraviolet "light." When the ultraviolet radiation strikes the phosphors lining the tube, the phosphors *fluoresce*, emitting visible light.

Compared to an incandescent bulb, light produced by this process wastes much less energy in the form of heat. As a result, a fluorescent lamp produces approximately three times as much light (measured in *lumens*) as an incandescent lamp of the same wattage. Fluorescent lamps also last longer—between 10,000 and 20,000 hours, compared to 1000 to 1500 hours for most incandescent lamps.

The ballast's second function is to *limit* current. A typical 14-W fluorescent lamp, once it is lit, requires only 55 V to maintain the proper current through the lamp. The opposition to alternating current caused by inductance drops the applied 120 V to the required value across the lamp.

Some fluorescent lamps use a single ON/OFF switch, rather than a momentary contact switch. These lamps make use of a *starter*, shown in Fig. 20-6*b*, that performs the same function as a momentary contact switch.

The starter resembles a neon glow lamp and includes a bimetallic strip. Initially, the starter is an open circuit. When power is applied to the lamp circuit, 120 V appears across the open starter electrodes. Ionization of the neon gas makes the starter glow, heating the bimetallic strip until it bends and makes contact with the other electrode. This completes a path through the filaments and ballast,

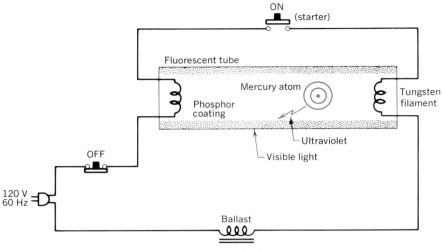

(*a*) Schematic diagram

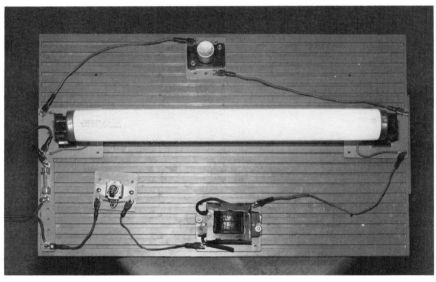

(*b*) Circuit that uses a starter. (The starter is located at the top of the board.)

FIGURE 20-6

Use of an inductor in a preheat fluorescent lamp circuit.

preheating the lamp. A few seconds after the starter switch has closed, no further heat is developed, so the bimetal strip cools and breaks the contact, interrupting the current and making the lamp "fire." Once the lamp fires, there is no longer sufficient voltage across the lamp (and starter) to initiate ionization of the neon gas in the starter, so the starter contacts remain open.

A recent development is a fluorescent lamp designed for use in incandescent-lamp sockets. It consists of a replaceable fluorescent tube that is folded in such a way that it occupies a space only $7\frac{1}{2}$ in. (178 mm) in length and $4\frac{1}{2}$ in. (114 mm) in diameter, including the reflector and lens. (See Fig. 20-7.) A ballast and starter switch are located in the screw-in base of the fixture.

With a life of 10,000 hours, the 7-W lamp produces as much light (approximately 650 lumens) as a 40-W incandescent lamp. A highly polished reflector magnifies and concentrates the light, so that the lamp can replace a 60-W incandescent bulb. The 7-W lamp and 3-W ballast can thus provide the same light as a 60-W incandescent lamp,

FIGURE 20-7

Compact fluorescent lamp designed to replace incandescent lamps. The cylindrical-shaped starter is seen on the top in the left view, and the rectangular-shaped ballast is seen on the front in the right view. (Courtesy of Lumatech Corporation.)

while consuming only *one-sixth* the energy. In addition, the useful life of the fluorescent lamp is 10 times greater than the incandescent lamp.

Other types of fluorescent lamps are the *rapid-start,* which needs no starter, and the *instant-start,* which has a single-pin base and a step-up transformer to produce the high ionization voltage needed to fire the lamp.

EXAMPLE 20-7

A 60-W incandescent lamp is used 8 hours per day through a full year. It is replaced by a fluorescent lamp that consumes 9 W (including the ballast), and costs $20.00. If electrical energy costs 8 cents per kWh, calculate:

a. How much it costs to operate the incandescent lamp for a year.
b. How much it costs to operate the fluorescent lamp for a year.
c. The payback period—the time necessary for the energy savings to pay for the fluorescent lamp—and compare with the remaining lamp life. (Total life: 10,000 hours.)

Solution

a. Hours per year = 365 days × 8 hours/day
$$= 2920 \text{ hours}$$

$$\text{Energy per year} = \frac{\text{watts}}{1000} \times \text{hours}$$

$$= \frac{60 \text{ W}}{1000} \times 2920 \text{ hours}$$

$$= 175.2 \text{ kWh}$$

$$\text{Cost per year} = \text{kWh} \times \text{cost/kWh}$$

$$= 175.2 \text{ kWh} \times .08/\text{kWh}$$

$$= 14.016 = \textbf{\$14.02}$$

b. $$\text{Energy per year} = \frac{\text{watts}}{1000} \times \text{hours}$$

$$= \frac{9 \text{ W}}{1000} \times 2920 \text{ hours}$$

$$= 26.28 \text{ kWh}$$

$$\text{Cost per year} = \text{kWh} \times \text{cost/kWh}$$

$$= 26.28 \text{ kWh} \times .08/\text{kWh}$$

$$= 2.1024 = \textbf{\$2.10}$$

c. From the above, the energy savings that occur every 12 months are

$$\$14.02 - \$2.10 = \$11.92$$

Time required for energy savings of $20.00:

$$\frac{\$20.00}{\$11.92} \times 12 \text{ months} = \textbf{20 months}$$

After this length of time, the fluorescent lamp would have a remaining life of:

$$10,000 - \frac{2920}{12} \times 20 \text{ hours}$$

$$= 10,000 - 4867 = \textbf{5133 hours}$$

NOTE During the 4867 operating hours, it would have been necessary to replace between three and four incandescent lamps. The replacement cost for these lamps would reduce the payback period for the fluorescent lamp to about 18 months. After the payback period, the fluorescent lamp will provide energy savings alone of $23.00 before needing replacement. (This does not include the cost of at least four more incandescent lamps.)

In commercial and industrial applications, where "relamping" represents a considerable labor cost, fluorescent lamps provide even greater savings than in the above example.

20-6 SERIES INDUCTIVE REACTANCES

When two or more coils are connected in series, the total inductive reactance is given by

$$X_{L_T} = X_{L_1} + X_{L_2} + X_{L_3} + \cdots + X_{L_n} \quad (20\text{-}4)$$

This equation assumes that there is no mutual inductance between the coils. The current through the coils may be obtained from

$$I = \frac{V}{X_{L_T}} \quad (20\text{-}5)$$

and the voltage across each coil from $V_L = IX_L$.

Example 20-8, with Fig. 20-8, shows the solution to a problem in which two independent inductors are connected in series.

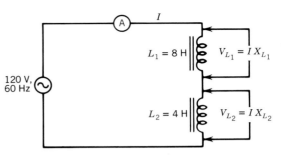

FIGURE 20-8

Circuit for Example 20-8 with series-inductive reactances.

$$V_{L_2} = IX_{L_2} \quad (20\text{-}1)$$
$$= 26.67 \times 10^{-3} \text{ A} \times 1.5 \times 10^3 \text{ }\Omega$$
$$= \mathbf{40\ V}$$

EXAMPLE 20-8

Two coils of 8 and 4 H with negligible resistance are series-connected across a 120-V, 60-Hz source, as in Fig. 20-8. Calculate:

a. The total inductive reactance.
b. The reading of an ammeter in series with the coils.
c. The voltage drop across each coil.

Solution

a. $X_{L_1} = 2\pi f L_1 \quad (20\text{-}2)$
  $= 2\pi \times 60 \text{ Hz} \times 8 \text{ H}$
  $= 3016 \text{ }\Omega \approx 3 \text{ k}\Omega$
  $X_{L_2} = 2\pi f L_2 \quad (20\text{-}2)$
  $= 2\pi \times 60 \text{ Hz} \times 4 \text{ H}$
  $= 1508 \text{ }\Omega \approx 1.5 \text{ k}\Omega$
  $X_{L_T} = X_{L_1} + X_{L_2} \quad (20\text{-}4)$
  $= 3 \text{ k}\Omega + 1.5 \text{ k}\Omega = \mathbf{4.5\ k\Omega}$

b. $I = \dfrac{V}{X_{L_T}} \quad (20\text{-}5)$

  $= \dfrac{120 \text{ V}}{4.5 \text{ k}\Omega} = \mathbf{26.67\ mA}$

c. $V_{L_1} = IX_{L_1} \quad (20\text{-}1)$
  $= 26.67 \times 10^{-3} \text{ A} \times 3 \times 10^3 \text{ }\Omega$
  $= \mathbf{80\ V}$

Note that $V_{L_1} + V_{L_2} = 120$ V, the supply voltage. Also, note that $V_{L_1} = 2V_{L_2}$ because $L_1 = 2L_2$.

20-6.1 Series Inductive Reactances with Mutual Reactance

As you learned in Chapter 17, two series-connected coils that have mutual inductance between them have a total inductance given by

$$L_T = L_1 + L_2 \pm 2M \quad (17\text{-}10)$$

When an alternating current flows through these coils, the total inductive reactance is

$$X_{L_T} = X_{L_1} + X_{L_2} \pm 2X_M \quad \text{ohms} \quad (20\text{-}6)$$

where: X_{L_1} is the inductive reactance of L_1, in ohms (Ω)
  X_{L_2} is the inductive reactance of L_2, in ohms (Ω)
  $X_M = 2\pi f M$ is the mutual reactance, in ohms (Ω)
  M is the mutual inductance between the coils, in henrys (H)

The mutual reactance is positive if the magnetic fields are aiding; negative if the fields are opposing.

As before, the current through the coils is given by $I = V/X_{L_T}$. The voltage across each coil is

$$V_L = I(X_L \pm X_M) \quad (20\text{-}7)$$

EXAMPLE 20-9

Assume that the two coils in Example 20-8 now have a coefficient of coupling of 0.6. The coils are connected with their fields aiding, as shown by the dot notation in Fig. 20-9. Calculate:
a. The total inductive reactance.
b. The current through the coils.
c. The voltage across each coil.

Solution

a. $M = k \sqrt{L_1 L_2}$ (17-12)

    $= 0.6 \sqrt{8 \text{ H} \times 4 \text{ H}}$

    $= 3.394 \text{ H} \approx 3.4 \text{ H}$

  $X_M = 2\pi f M$

    $= 2\pi \times 60 \text{ Hz} \times 3.4 \text{ H}$

    $= 1282 \ \Omega \approx 1.3 \text{ k}\Omega$

  $X_{L_T} = X_{L_1} + X_{L_2} + 2 X_M$ (20-6)

    $= 3 \text{ k}\Omega + 1.5 \text{ k}\Omega + 2 \times 1.3 \text{ k}\Omega$

    $= \mathbf{7.1 \text{ k}\Omega}$

b. $I = \dfrac{V}{X_{L_T}}$ (20-5)

  $= \dfrac{120 \text{ V}}{7.1 \text{ k}\Omega} = \mathbf{16.9 \text{ mA}}$

c. $V_{L_1} = I (X_{L_1} + X_M)$ (20-7)

    $= 16.9 \times 10^{-3} \text{ A } (3 \times 10^3 \ \Omega + 1.3 \times 10^3 \ \Omega)$

    $= 16.9 \times 10^{-3} \text{ A} \times 4.3 \times 10^3 \ \Omega$

    $= \mathbf{72.7 \text{ V}}$

  $V_{L_2} = I (X_{L_2} + X_M)$ (20-7)

    $= 16.9 \times 10^{-3} \text{ A } (1.5 \times 10^3 \ \Omega + 1.3 \times 10^3 \ \Omega)$

    $= 16.9 \times 10^{-3} \text{ A} \times 2.8 \times 10^3 \ \Omega$

    $= \mathbf{47.3 \text{ V}}$

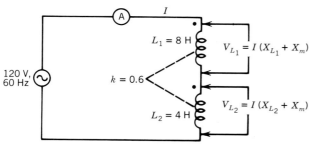

FIGURE 20-9
Circuit for Example 20-9 with series-inductive reactances and positive mutual reactance.

Note once again that $V_{L_1} + V_{L_2} = 120$ V. However, V_{L_1} is *not* twice V_{L_2} because of the effect of the mutual reactance (X_M).

20-7 PARALLEL INDUCTIVE REACTANCES

Connecting two or more inductive reactances in parallel provides a reduction in the overall reactance, compared with the branch having the lowest reactance. Thus, similar to resistances in parallel, the total inductive reactance is given by

$$\frac{1}{X_{L_T}} = \frac{1}{X_{L_1}} + \frac{1}{X_{L_2}} + \frac{1}{X_{L_3}} + \cdots + \frac{1}{X_{L_n}} \quad (20\text{-}8)$$

EXAMPLE 20-10

Two coils of inductance 8 and 4 H, with negligible resistance, are connected in parallel across a 120-V, 60-Hz supply, as in Fig. 20-10. Determine the total current drawn from the supply.

Solution

As earlier, $X_{L_1} = 3 \text{ k}\Omega$

                $X_{L_2} = 1.5 \text{ k}\Omega$

        $I_{L_1} = \dfrac{V}{X_{L_1}}$ (20-1)

            $= \dfrac{120 \text{ V}}{3 \text{ k}\Omega} = 40 \text{ mA}$

        $I_{L_2} = \dfrac{V}{X_{L_2}}$ (20-1)

            $= \dfrac{120 \text{ V}}{1.5 \text{ k}\Omega} = 80 \text{ mA}$

Total current $I_T = I_{L_1} + I_{L_2}$ (6-1)

        $= 40 \text{ mA} + 80 \text{ mA} = \mathbf{120 \text{ mA}}$

or $\dfrac{1}{X_{L_T}} = \dfrac{1}{X_{L_1}} + \dfrac{1}{X_{L_2}}$ (20-8)

     $X_{L_T} = \dfrac{X_{L_1} \times X_{L_2}}{X_{L_1} + X_{L_2}}$

         $= \dfrac{3 \text{ k}\Omega \times 1.5 \text{ k}\Omega}{3 \text{ k}\Omega + 1.5 \text{ k}\Omega} = 1 \text{ k}\Omega$

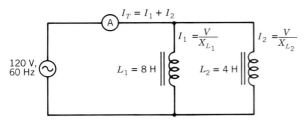

FIGURE 20-10
Circuit for Example 20-10 with parallel inductive reactances.

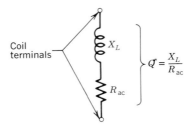

FIGURE 20-11
Equivalent circuit of a coil.

Therefore, $I = \dfrac{V}{X_{L_T}}$ (20-5)

$= \dfrac{120\text{ V}}{1\text{ k}\Omega} = \textbf{120 mA}$

or $L_T = \dfrac{L_1 \times L_2}{L_1 + L_2}$ (17-5)

$= \dfrac{8\text{ H} \times 4\text{ H}}{8\text{ H} + 4\text{ H}}$

$= \dfrac{32}{12}\text{ H} = 2\tfrac{2}{3}\text{ H}$

$X_{L_T} = 2\pi f L_T$ (20-2)

$= 2\,\pi \times 60\text{ Hz} \times 2\tfrac{2}{3}\text{ H}$

$= 1000\ \Omega$

$I = \dfrac{V}{X_{L_T}}$ (20-5)

$= \dfrac{120\text{ V}}{1\text{ k}\Omega} = \textbf{120 mA}$

X_L is the reactance of the coil, in ohms (Ω)
R_{ac} is the ac resistance of the coil, in ohms (Ω)

The Q of a coil is also known as the *storage factor*, since it is proportional to the energy storage capability of the coil.

EXAMPLE 20-11

A 200-μH coil, when operated at a frequency of 1.5 MHz, has a total ac resistance of 100 Ω. What is the quality of the coil?

Solution

$Q = \dfrac{X_L}{R_{ac}} = \dfrac{2\pi f L}{R_{ac}}$ (20-9)

$= \dfrac{2\pi \times 1.5 \times 10^6 \text{ Hz} \times 200 \times 10^{-6}\text{ H}}{100\ \Omega}$

$= \textbf{18.8}$

20-8   QUALITY OF A COIL (Q)

At high frequencies, the usefulness of a coil is judged not only by its inductance, but also by the *ratio* of its inductive reactance to the *ac* or *effective* resistance of the coil. (See Fig. 20-11.) This ratio is called the *quality* of the coil and is given the symbol Q:

$$Q = \frac{X_L}{R_{ac}} \quad (20\text{-}9)$$

where: Q is the quality of the coil (a dimensionless number)

20-9   EFFECTIVE RESISTANCE (R_{ac})

Note the use of *effective* (ac) resistance in Eq. 20-9. This is different from, and higher than, the dc (ohmic) resistance measured with an ohmmeter, for the following reason.

Consider the changing magnetic field set up by the alternating current in and around the wire conductors that make up the coil. This field has the greatest rate of change $\left(\dfrac{\Delta\Phi}{\Delta t}\right)$ at the *center* of the conductor. Thus, the induced voltage $\left(v = \dfrac{\Delta \Phi}{\Delta t} \right)$, and the opposition to the current, has

the greatest value at the center of the conductor. As a result, at radio frequencies (hundreds of kHz and above), most of the current flows along the *surface* of the conductor, and very little, if any, at the center. (The current becomes more dense toward the surface.) This effectively decreases the cross-sectional area of the conductor and increases its resistance.

This is called the *skin effect*. At microwave frequencies, the effect is so pronounced that *hollow* conductors are used, since current flows only on the surface. At radio frequencies, an additional loss occurs due to the *radiation* of energy. The combined higher resistance is called the ac resistance (R_{ac}).

The Q of a coil (at radio frequencies) is relatively constant with frequency. This is because R_{ac} increases with the frequency at approximately the same rate that X_L increases, maintaining an almost constant ratio. Thus, in Example 20-11, at a frequency of 3 MHz, both X_L and R_{ac} would approximately double so that Q remains approximately 19.

At power line frequencies of 60 Hz, the skin effect is negligible. A significant difference between dc and ac resistance can occur, however. If the coil is wrapped on a *magnetic core,* there will be both hysteresis and eddy current losses, as discussed in Section 17-6.3. Both these losses convert electrical energy to heat, so their effect is a dissipation of electrical power (as if an equivalent resistance was connected in the circuit).

The total ac resistance of a circuit can be obtained from a wattmeter reading of total power. That is, $R_{ac} = P/I^2$, which includes the additional effects of hysteresis, eddy currents, skin effect, and so on, over and above the ohmic (dc) resistance.

EXAMPLE 20-12

An inductor of 0.5 H is connected in series with an ammeter, a wattmeter, and a 60-Hz supply, as in Fig. 20-12. If the ammeter indicates 250 mA and the wattmeter indicates 5 W, calculate:

a. The ac resistance of the coil.
b. The Q of the coil.

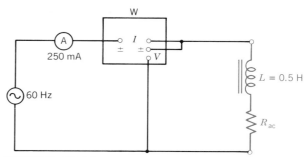

FIGURE 20-12
Circuit for Example 20-12.

Solution

a. $R_{ac} = \dfrac{P}{I^2}$ (15-12)

$= \dfrac{5 \text{ W}}{(0.25 \text{ A})^2} = \mathbf{80 \ \Omega}$

b. $Q = \dfrac{X_L}{R_{ac}}$ (20-9)

$= \dfrac{2\pi \times 60 \text{ Hz} \times 0.5 \text{ H}}{80 \ \Omega} = \mathbf{2.4}$

20-10 MEASURING INDUCTANCE

An instrument that operates on the principle of the Wheatstone bridge (as described in Chapter 13) may be used to measure the inductance and quality (Q) of a coil.

The impedance bridge, shown in Fig. 20-13, can also measure the dc and ac resistance of the coil. Both controls are adjusted (first one, then the other) until minimum deflection of the galvanometer is obtained. The Q is read directly off the left dial; the inductance off the right using the appropriate multiplying factor. An internally generated 1-kHz signal is used for the measurement, but an external signal generator may be connected to measure the Q at any other desired frequency. Inductances from 1 μH to 1100 H may be measured, with Q values from 0.02 to 1000. A digital instrument that measures inductance from 1 μH to 10 H is shown in Fig. 21-9.

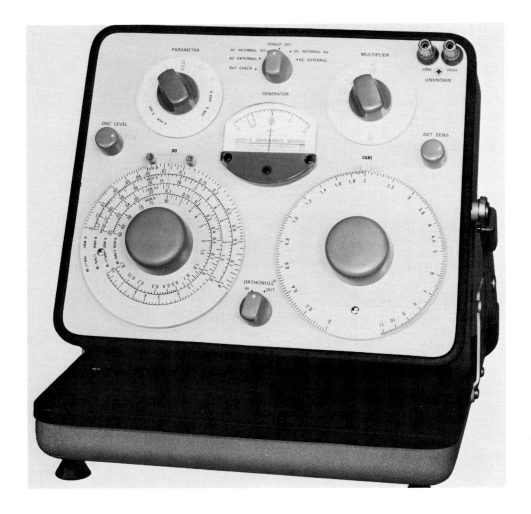

FIGURE 20-13

Impedance bridge suitable for measuring inductance and Q of a coil. (Courtesy of GenRad Inc., Concord, Mass.)

SUMMARY

1. A sinusoidal voltage across a pure inductance causes a sinusoidal current that lags behind the voltage by one-quarter cycle (90°).

2. The opposition to alternating current provided by a pure inductance is called inductive reactance and is given by $X_L = 2\pi fL$.

3. The current through an inductor that has negligible resistance is given by $I_L = V_L/X_L$.

4. A coil with low resistance provides a much higher opposition to alternating current than to direct current. This property is used in filter circuits that remove unwanted alternating current.

5. The ballast in a fluorescent lamp circuit has two functions. It induces a momentary high voltage (when current through the coil is interrupted) to fire the lamp, then limits current through the lamp after it is lit (as a result of inductive reactance).

6. Invisible ultraviolet energy provides visible light from a fluorescent lamp by causing a phosphor coating inside the tube to fluoresce. This makes the lamp much more efficient than an incandescent lamp.

7. The equations for combining inductive reactances connected in series and parallel are the same as those for resistances if no mutual inductances are involved.

8. The total inductive reactance of two series-connected coils that have mutual inductance is given by $X_{L_T} = X_{L_1} + X_{L_2} \pm 2X_M$.

9. The voltage across each of the series-connected coils that have mutual inductance is given by $V_L = I(X_L \pm X_M)$.

10. The quality of a coil is given by $Q = X_L/R_{ac}$.

11. The effective (ac) resistance is higher than the dc resistance and includes the effects of eddy currents, skin effect, radiation loss, and hysteresis losses.

12. Inductance can be measured directly on an impedance bridge. Both ac and dc resistance of a coil can be measured, as can the quality (Q) of the coil.

SELF-EXAMINATION

Answer T or F, or, for multiple choice, a, b, c, or d
(Answers at back of book)

20-1. In an ac circuit, inductance determines both the waveform of the current and the rms value of the current. _____

20-2. The phase angle between voltage across a pure inductance and the current through it is always 90° even if connected in series with a pure resistance.

_____

20-3. In a purely inductive circuit it is possible for there to be zero current through the inductance at the instant when the voltage is not zero. _____

20-4. Inductive reactance may be thought of as an effective ac resistance to the flow of alternating current. _____

20-5. A coil of 2-H inductance and negligible resistance is used at a frequency of 400 Hz. Its inductive reactance is
  a. 5000 Ω
  b. 5.1 kΩ
  c. 1600 Ω
  d. 1600π Ω

20-6. If the coil in Question 20-5 is connected to a 240-V 400-Hz supply, the current is
  a. 47.7 mA
  b. 4.8 mA
  c. 15 mA
  d. 0.15 A

20-7. The use of an inductor to filter out unwanted ac voltage relies upon a high inductive reactance compared with its dc resistance. _____

20-8. One of the functions of a ballast in a fluorescent lamp circuit is the production of a momentary high voltage to fire the lamp. _____

20-9. A fluorescent lamp has a high efficiency because of the intense white heat produced by the phosphor coating. _____

20-10. The mercury in a fluorescent lamp is necessary for the production of ultraviolet energy. _____

20-11. The inductive reactance of the ballast in a fluorescent lamp circuit is used to limit current only at the instant the lamp is being turned on. _____

20-12. The same equations can be used to find total inductive reactance for series- and parallel-connected reactances as are used for resistances, with no restrictions. _____

20-13. The dot notation used to show when mutual reactance is additive is the same as that used for mutual inductance. _____

20-14. The quality of a coil is the comparison of the coil's resistance to the coil's inductive reactance. _____

20-15. At high frequencies the Q of a coil is relatively constant. _____

20-16. Alternating current resistance is the effective resistance responsible for ac power dissipation in a circuit and can be much larger than the dc resistance. _____

20-17. An impedance bridge works on the general principle of a Wheatstone bridge to measure inductance. _____

REVIEW QUESTIONS

1. a. Why does inductance in an ac circuit determine both *how* the current changes and how *much* current flows?
   b. What is the phase relationship between current and voltage in a pure inductance?
   c. How do you think this would be different in a coil that has as much resistance as inductive reactance?

2. a. Why is it not correct to characterize inductive reactance as the effective ac resistance due to a coil's inductance?
   b. Why should an *increase* in inductance or frequency *reduce* the current in an inductive circuit if the voltage is unchanged?

3. a. In what way can direct current be thought of as zero frequency alternating current when applied to an inductor?
   b. Does this mean that inductance has no effect at all when the frequency of the applied voltage is zero?

4. If an alternating voltage of 120 V rms is as effective in producing heat as 120 V dc, explain in your own words why the lamp lights in Fig. 20-4a but does not light in Fig. 20-4b.

5. a. Describe how an inductor can be used to filter out unwanted ac voltage from an input containing both alternating current and direct current.
   b. Under what conditions can Eq. 20-3 be used?

6. Explain in detail how a fluorescent lamp circuit works, including the two functions of the ballast and the production of visible light.

7. Under what conditions can the equations for combining resistances in series and parallel be applied to inductive reactances in series and parallel?

8. Describe how you would use an ac source of known frequency, an ac ammeter, and an ac voltmeter to determine the mutual inductance between two series-connected coils whose interconnections can be interchanged. Assume that the coils have negligible resistance.

9. a. Explain why the ac resistance of a circuit can be considerably higher than the dc resistance measured with an ohmmeter.

b.   What resistance value is used in determining the Q of a coil?
c.   Why doesn't Q increase with frequency?

10.   Describe two ways of measuring the inductance of a coil.

PROBLEMS

(Answers to odd-numbered problems at back of book)

20-1.   What inductive reactance does a coil have if an applied emf of 120 V causes a current of 50 mA? The resistance of the coil is negligible.

20-2.   A coil with an inductive reactance of 2.5 kΩ and negligible resistance carries a current of 20 mA. What peak-to-peak voltage does an oscilloscope indicate when connected across the coil?

20-3.   What inductance is necessary to cause a voltage drop of 15 V due to 5 mA of current having a frequency of 400 Hz?

20-4.   What voltage drop appears across a 30-mH inductance with a 10-mA p–p current having a frequency of 1 kHz?

20-5.   What frequency of current will cause a voltage drop of 12 V p–p across a 200-μH inductance when a current of 1 mA flows?

20-6.   A 12-V p–p sinusoidal voltage is applied to an 8-H coil of negligible resistance. How much current flows at the following frequencies?
a.   100 Hz.
b.   10 kHz.
c.   1 MHz.

20-7.   A 5-V, 10-kHz signal generator is applied to a 0.1-H coil of negligible resistance. Determine:
a.   How much current flows.
b.   The equation for current assuming a sinusoidal input.

20-8.   The output of a full-wave rectifier has a 12-V dc component and an 8-V p–p ac component having a frequency of 120 Hz. It is applied to a filter circuit like that in Fig. 20-5, with $L = 4$ H and $R = 47\ \Omega$. Determine:
a.   The dc output voltage.
b.   The ac output voltage.

20-9.   Repeat Problem 20-8 using $L = 8$ H and $R = 33\ \Omega$.

20-10.   Assuming that the ballast in Fig. 20-6 has negligible resistance and a voltage drop of 75 V, calculate:
a.   The inductive reactance of the ballast when the lamp is lit and drawing 0.3 A.
b.   The inductance of the ballast.

20-11.   Three coils of 4, 8, and 12 H are series-connected across a 60-V, 400-Hz supply. Assuming negligible resistance, calculate:
a.   The current through the coils.
b.   The voltage across each coil.

20-12.   Two series-connected coils, one having twice the inductance of the other, draw a current of 200 mA when connected to a 120-V, 60-Hz source. Determine the inductance of each coil assuming negligible resistance and no mutual inductance.

20-13.   Two coils of 30 and 50 mH are series-connected with a coefficient of coupling of 0.8. They are connected across a 10-V, 1-kHz sinusoidal source with their fields aiding. Assuming negligible resistance, calculate:

   a. The total inductive reactance.

   b. The current through the coils.

   c. The voltage across each coil.

20-14. Repeat Problem 20-13 assuming opposing fields.

20-15. Repeat Problem 20-12 assuming a coefficient of coupling of 0.5 and aiding fields.

20-16. Repeat Problem 20-12 assuming a coefficient of coupling of 0.5 and opposing fields.

20-17. Two equal coils with negligible resistance are connected in series with a 20-V, 5-kHz source. With their fields aiding the current is 1 mA; with their fields opposing the current is 5 mA. Calculate:

   a. The mutual inductance between the coils.

   b. The self-inductance of the coils.

   c. The coefficient of coupling between the coils.

20-18. Three inductors of negligible resistance are connected in parallel across a 400-Hz supply. They draw a total current of 500 mA. The first coil carries a current of 100 mA; the second coil has 40 V across it; and the third coil has an inductance of 50 mH. Calculate:

   a. The total inductive reactance of the circuit.

   b. The inductances of the first two coils.

20-19. The three coils in Problem 20-11 are connected in parallel across a 60-V, 400-Hz supply. Calculate:

   a. The total inductive reactance of the circuit.

   b. The total current drawn from the supply.

20-20. A 47-μH coil, when operated at a frequency of 10 MHz, has an ac resistance of 70 Ω.

   a. What is the quality of the coil?

   b. What would you expect the Q to be at 20 MHz?

20-21. A wattmeter in an ac circuit indicates 40 W when a series ac ammeter shows 0.15 A.

   a. Determine the ac resistance of the circuit.

   b. If an ohmmeter measurement shows 1200 Ω of resistance, what resistance is due to hysteresis and eddy current losses?

20-22. Repeat *Example* 20-7 for a 40-W incandescent lamp, used 12 hours daily, which is being replaced by a 7-W fluorescent lamp costing $18.00. Assume electrical energy costs to be 10 cents per kilowatt-hour.

CHAPTER 21

CAPACITANCE

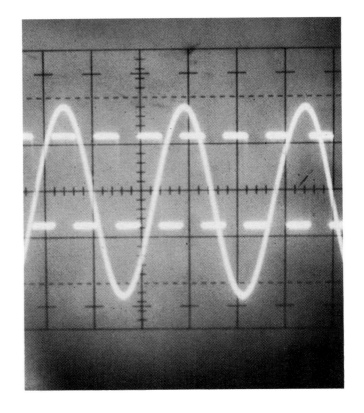

In this chapter, you will consider the third property of electrical circuits, *capacitance.* A capacitor consists of two metal plates separated by an insulator *(dielectric).* When the capacitor is connected to a voltage source, a momentary charging current deposits a charge on the plates, establishing an *electric field.* The energy stored in this field may be returned by *discharging* the capacitor through a *load.* The various types of capacitors and the factors that affect capacitance are considered in this chapter, along with the series and parallel connections of capacitors and the maximum working voltage of each combination.

21-1 BASIC CAPACITOR ACTION

A capacitor consists of two conductors separated by an insulating medium, or *dielectric*. In simplest form, a capacitor can be thought of as two parallel metal plates separated by an air space. The air space serves as the dielectric (insulator). In many practical capacitors, the conductive "plates" may be aluminum foil with a paper or mica dielectric separating them. This allows the capacitor to be rolled into a cylindrical form to conserve space. To provide characteristics for specific applications, many different dielectrics and methods of fabrication are used. (See Section 21-5.)

Figure 21-1a shows a basic parallel-plate capacitor connected to a dc source through a resistor and switch.

When the switch is closed, there is a momentary current (flow of charge) that deposits electrons on one plate and removes an equal number of electrons from the other. The capacitor is now said to be *charged,* with one plate positive and the other negative. If the dc source is removed and a high-resistance voltmeter is connected across the capacitor, the meter will indicate a potential difference equal to the previously applied emf. (See Fig. 21-1b.)

When the capacitor is *discharged* across a load (Fig. 21-1c), electrons flow in the *opposite* direction to redis-

tribute the original charge. The electrons flow from the plate with a surplus of negative charge to the plate with a deficiency of negative charge. (Note the similarity of this action to the two rods charged by friction, discussed in Section 2-4.) Current flows until the excess charge on one plate has been transferred to the other. Once the charges are equalized, there is zero potential difference between the plates.

This momentary flow of charging and discharging currents can be observed on an oscilloscope connected across the resistor (R). The shape of these current pulses is considered in Chapter 22.

21-2 CAPACITANCE

Assume that you can measure the charge deposited on either of the capacitor's plates. (This can be done with a suitable "measuring amplifier.") Figure 21-2a shows that a 5-V source deposits a charge (Q) on each plate: one is positive ($+Q$); the other negative ($-Q$). If the voltage source is *doubled,* to 10 V, the charge is also doubled to $2Q$. If the source is *tripled,* to 15 V, the accumulated charge becomes $3Q$ (where Q is the initial charge due to 5 V).

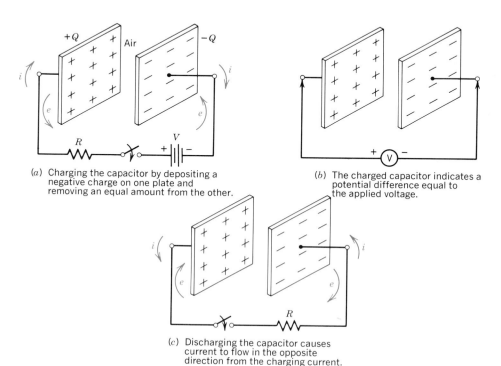

(a) Charging the capacitor by depositing a negative charge on one plate and removing an equal amount from the other.

(b) The charged capacitor indicates a potential difference equal to the applied voltage.

(c) Discharging the capacitor causes current to flow in the opposite direction from the charging current.

FIGURE 21-1
Basic capacitor charging and discharging action.

For any *given capacitor,* the ratio of the charge on one plate to the potential difference across the plates is a constant. This constant is a property of the capacitor called its *capacitance.*

Thus capacitance is defined as:

$$C = \frac{Q}{V} \quad \text{farads} \quad (21\text{-}1)$$

where: C is the capacitance of the capacitor, in farads (F)
Q is the charge per plate, in coulombs (C)
V is the potential difference across the capacitor, in volts (V)

Thus 1 farad = 1 coulomb/volt (1 F = 1 C/V).

A capacitor has a *capacitance* of 1 F if 1 C of charge must be deposited to raise the potential difference by 1 V across the capacitor.

A farad is a very *large* amount of capacitance. (A capacitor with the plates separated by a 1-mm air space, for example, would require plates approximately 10.5 km—$6\frac{1}{2}$ *miles*—on a side to produce a capacitance of 1 F.)

The microfarad and picofarad are much more practical units of capacitance.

$$1 \text{ microfarad} = 1 \ \mu\text{F} = 1 \times 10^{-6} \text{ F}$$
$$1 \text{ picofarad} = 1 \text{ pF} = 1 \times 10^{-12} \text{ pF}$$

Practical capacitors are available with capacitances ranging from 1 pF to 500,000 μF.

EXAMPLE 21-1

A certain capacitor requires 50 μC of charge to be deposited on the plates to raise the potential difference to 2 V.

a. What is the capacitance of the capacitor?
b. How much charge is required to raise the voltage across the capacitor from 0 to 10 V?

c. If the charge on the plates is reduced to 10 μC, what voltage now exists across the capacitor?

Solution

a. $\quad C = \dfrac{Q}{V}$ $\qquad\qquad\qquad$ (21-1)

$\qquad = \dfrac{50 \times 10^{-6} \text{ C}}{2 \text{ V}}$

$\qquad = 25 \times 10^{-6} \text{ F} = \textbf{25 } \boldsymbol{\mu}\textbf{F}$

b. $\quad Q = VC$

$\qquad = 10 \text{ V} \times 25 \times 10^{-6} \text{ F}$

$\qquad = 250 \times 10^{-6} \text{ C} = \textbf{250 } \boldsymbol{\mu}\textbf{C}$

c. $\quad V = \dfrac{Q}{C}$

$\qquad = \dfrac{10 \times 10^{-6} \text{ C}}{25 \times 10^{-6} \text{ F}} = \textbf{0.4 V}$

21-3 FACTORS AFFECTING CAPACITANCE

It can be shown (see Appendix J) that capacitance is determined by the physical factors of *plate area, plate separation,* and *dielectric material,* as given by

$$C = K\frac{\epsilon_0 A}{d} \quad \text{farads} \quad (21\text{-}2)$$

where: C is the capacitance, in farads (F)
A is the area of one side of one plate, in square meters (m^2)
d is the separation of the two plates, in meters (m)
ϵ_0 is the permittivity of free space (air) = $8.85 \times 10^{-12} \dfrac{\text{C}^2}{\text{Nm}^2}$
K is the dielectric constant of the particular material used between the plates. (See Table 21-1.)

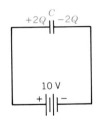

(a) 5 V deposits charge Q

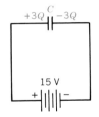

(b) 10 V deposits charge $2Q$

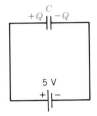

(c) 15 V deposits charge $3Q$

FIGURE 21-2

Relation between the voltage and charge for a given capacitor C.

TABLE 21-1

Material	K
Vacuum (and air)	1
Polyethylene	2
Teflon	2
Polystyrene	2.5
Paper	3
Rubber	3
Vinylite	3
Polyesters	4
Glass	5
Mica	5
Bakelite (plastic)	6
Neoprene	7
Ceramics—Porcelain	7
Titanium dioxide	14–110
Strontium titanate	7500

See Fig. 21-3 for some of the quantities involved and two alternate symbols for a capacitor.

EXAMPLE 21-2

Two metal plates, each 5 × 6 cm, are separated from each other by 0.5 mm. Calculate the capacitance for the following conditions:
a. The dielectric is air.
b. The dielectric is mica.
c. The dielectric is strontium titanate (ceramic material).

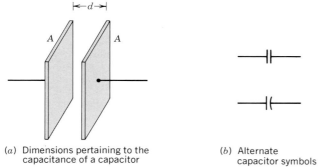

(a) Dimensions pertaining to the capacitance of a capacitor

(b) Alternate capacitor symbols

FIGURE 21-3

Physical factors that determine capacitance and alternate symbols.

---

d. The separation is reduced to 0.25 mm with the same ceramic dielectric.

Solution

a. $K_{air} = 1$

$$C = K\frac{\epsilon_0 A}{d} \qquad (21\text{-}2)$$

$$= 1 \times 8.85 \times 10^{-12} \frac{C^2}{Nm^2}$$

$$\times \frac{5 \times 10^{-2}\,m \times 6 \times 10^{-2}\,m}{0.5 \times 10^{-3}\,m}$$

$$= 53.1 \times 10^{-12} \frac{C^2}{Nm}$$

$$= 53.1 \times 10^{-12}\,F = \mathbf{53.1\ pF}$$

b. $K_{mica} = 5$

$$C = K\frac{\epsilon_0 A}{d} \qquad (21\text{-}2)$$

$$= 5 \times 53.1\ pF = \mathbf{265.5\ pF}$$

c. $K_{ceramic} = 7500$

$$C = K\frac{\epsilon_0 A}{d} \qquad (21\text{-}2)$$

$$= 7500 \times 53.1\ pF = 398{,}250\ pF = \mathbf{0.4\ \mu F}$$

d. If the separation is cut in half, the capacitance is doubled.

$$C = 2 \times 0.4\ \mu F = \mathbf{0.8\ \mu F}$$

21-4 EFFECT OF A DIELECTRIC

In Section 21-3, you saw how a larger plate area increases capacitance. This seems reasonable, since a larger area will allow a greater charge to accumulate. But why should a smaller separation between plates also increase capacitance? And, why should a dielectric material other than air increase capacitance? To answer these and other questions, it is necessary to consider the electric field between the plates of a charged capacitor, and the effect on that field of inserting a dielectric.

21-4.1 Electric Field

First, consider a single atom of a dielectric in an uncharged capacitor, as in Fig. 21-4a. The electrons of the atom follow a roughly circular path as they orbit the nucleus. After the capacitor is charged, however, the electron paths become increasingly elliptical. (See Fig. 21-4b.) This distortion of the electron orbits takes place because

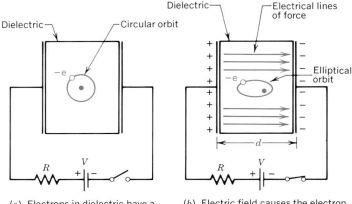

(a) Electrons in dielectric have a more circular orbit in an uncharged capacitor.

(b) Electric field causes the electron orbit to become more elliptical

FIGURE 21-4

Effect of electric field upon electron orbit in an atom of a dielectric in a capacitor.

the negatively charged electrons tend to move closer to the positive plate and away from the negative plate. This "stressing" of the electrons from generally circular orbits to more elliptical orbits accounts for the *momentary* charging current when voltage is first applied. There is no current, because there are no *free* electrons in the dielectric material. The charging current is often referred to as a "displacement" current, resulting from the shifting of the electrons in the dielectric material's atoms in their orbital paths. The atoms are then said to be polarized. When the capacitor is discharged, the electrons return to their roughly circular orbits.

The force acting on the electrons can be represented by *electric lines of force* drawn between the two charged plates (as in Fig. 21-4b). It is conventional to use an arrow to indicate the direction of force on a *positive* charge. The force on the electron is thus opposite the arrow.

The strength, or *intensity,* of the electrical field between the two plates determines the force on (and the resulting distortion of) the electrons orbiting the dielectric material's atoms. If the electric field is intensive enough to "tear" the electrons from their orbits, a continuous flow of current through the dielectric will result, shorting the capacitor. The dielectric is said to have "ruptured" or broken down. To remain functional, a capacitor must be designed to withstand a high electric field intensity.

The intensity of the electric field is given by

$$E = \frac{V}{d} \qquad \text{volts/meter} \qquad (21\text{-}3)$$

where: E is the electrical field intensity, in volts per meter (V/m)

V is the potential difference across the plates, in volts (V)

d is the plate separation, in meters (m)

EXAMPLE 21-3

The capacitor in Example 21-2 is charged to a potential difference of 20 V. With a ceramic dielectric determine the electric field intensity for the following separations:
a.   0.5 mm
b.   0.25 mm

Solution

a. $E = \dfrac{V}{d}$ $\qquad\qquad\qquad\qquad\qquad$ (21-3)

$\qquad = \dfrac{20\ V}{0.5 \times 10^{-3}\ m}$

$\qquad = 40 \times 10^{3}\ V/m = \mathbf{40\ kV/m}$

b. $E = \dfrac{V}{d}$ $\qquad\qquad\qquad\qquad\qquad$ (21-3)

$\qquad = \dfrac{20\ V}{0.25 \times 10^{-3}\ m} = \mathbf{80\ kV/m}$

For a given voltage, a *smaller* separation causes a *higher* electric field intensity. This means that more electric lines of force would have to be drawn to represent the field between the two plates. Electric lines of force originate on a positive charge and terminate on a negative charge. This implies that there must be a greater charge accumulation on the two plates if the electric field intensity is higher. A greater charge for the same voltage means greater capacitance. Hence a *smaller* separation of plates (all other factors remaining unchanged) causes an *increase* in capacitance.

21-4.2 Dielectric Strength

The field *intensities* calculated in Example 21-3 are independent of the type of material used for the dielectric. The dielectric *strength* of the material, however, determines whether the capacitor can withstand the electric field intensity in a given situation. The permissible dc working voltage (WVDC) of a capacitor is determined by the separation of the plates *and* the dielectric material. Average dielectric strengths for some typical materials are shown in Table 21-2.

EXAMPLE 21-4

A 600-WVDC, 0.1-μF paper capacitor is to be constructed in a tubular form. Calculate:

a. The minimum separation of the plates using a safety factor of 1.5.
b. The area of one of the metal foil plates.
c. The length of metal foil for one plate if the capacitor is to be only 4 cm wide.

Solution

a. Maximum field intensity $E_m = \dfrac{V}{d}$ (21-3)

TABLE 21-2
Average Dielectric Strengths

Dielectric	Dielectric Strength, kV/mm
Air	4
Porcelain	8
Ceramics	10
Plastic	16
Paper	20
Glass	28
Teflon	60
Mica	100

Therefore, $d = \dfrac{V}{E_m}$

$$= \frac{600 \text{ V}}{20 \times 10^3 \text{ V/mm}}$$

$$= 3 \times 10^{-2} \text{ mm}$$

Applying a safety factor of 1.5, $d = 1.5 \times 3 \times 10^{-2}$ mm = **4.5×10^{-2} mm**

b. $C = K\dfrac{\epsilon_0 A}{d}$ (21-2)

Therefore, $A = \dfrac{Cd}{K\epsilon_0}$

$$= \frac{0.1 \times 10^{-6} \text{ F} \times 4.5 \times 10^{-5} \text{ m}}{3 \times 8.85 \times 10^{-12} \text{ C}^2/\text{Nm}^2}$$

$$= 0.17 \frac{\text{F} \times \text{m}}{\text{C}^2/\text{Nm}^2}$$

But $1\text{F} = 1 \text{ C}^2/\text{Nm}$; therefore,

$$A = 0.17 \frac{\text{C}^2/\text{Nm} \times \text{m}}{\text{C}^2/\text{Nm}^2} = \mathbf{0.17 \text{ m}^2}$$

c. Area = length × width

Therefore, length $= \dfrac{\text{area}}{\text{width}}$

$$= \frac{0.17 \text{ m}^2}{4 \times 10^{-2} \text{ m}} = \mathbf{4.25 \text{ m}}$$

NOTE When the capacitor is rolled into a tubular form, one plate will make contact with the other. A second sheet of paper is used for insulating purposes and has the effect of doubling the capacitance because of a parallel combination. (See Section 21-6.) Thus the required area in Example 21-4 need only be half as much, with a foil length of only **2.12 m.**

Note also that paper has five times the dielectric strength of air. This means that for a given voltage rating, the plates are five times as close as when air is used. This requires only one-fifth the area needed by an air dielectric and contributes to a smaller-sized capacitor.

21-4.3 Induced Electric Field*

The effect of a dielectric (other than air) on a capacitor can now be examined. Consider a pair of charged plates, separated by an air space, with a voltmeter connected as shown in Fig. 21-5a. An electric field (E_{air}) exists between the plates.

Now assume that a glass dielectric is inserted between the plates. (See Fig. 21-5b.) The voltmeter reading im-

---

*This section may be omitted with no loss of continuity.

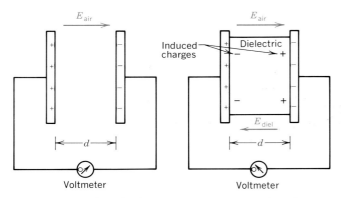

(a) Charged capacitor with plates separated by air.

(b) Insertion of a dielectric sets up an opposing electric field due to induced charges.

FIGURE 21-5
Effect of a dielectric upon the electric field of a capacitor

mediately drops to one-fifth its original value ($K_{glass} = 5$). Why? Although there are no free electrons in the dielectric, an *induced* charge is set up where the dielectric comes in contact with the metal plates. This is caused by the *polarization* of the dielectric's atoms when placed between the plates. The induced electrical charges set up an electric field (E_{diel}) in the dielectric that opposes the original field (E_{air}). This accounts for the term *di*electric, since the prefix ''di'' means *opposing*.

The resulting electric field is now given by

$$E_{total} = E_{air} - E_{diel} \qquad (21\text{-}4)$$

and E_{total} is evidently weaker than before. But as you know, electric field intensity is

$$E_{total} = \frac{V}{d} \qquad (21\text{-}3)$$

But since d is unchanged, V must have decreased, proving the results of Fig. 21-5.

But how does this insertion of a dielectric cause an *increase* in capacitance? Consider that a dc source is connected across the plates. As the dielectric is being inserted, the voltage across the plates will try to drop. But this will allow the dc source to deposit *more* charge on the plates to restore the voltage to the applied voltage again. After the dielectric is fully inserted, a new, *larger* charge will exist on the plates for the same potential *across* the plates. Since

$$C = \frac{Q}{V} \qquad (21\text{-}1)$$

the result is an increase in capacitance.

The ratio of the capacitance *with* the dielectric to the capacitance with *air* is called the relative permittivity or dielectric constant (K).

$$K = \frac{C_{diel}}{C_{air}} \qquad (21\text{-}5)$$

The *dielectric constant* is a measure of how readily the dielectric *permits* an opposing electric field to be set up in a given material. The *absolute* permittivity of a dielectric is $K\epsilon_0$. Thus the capacitance of a capacitor is

$$C = K\frac{\epsilon_0 A}{d} \qquad (21\text{-}2)$$

The dielectric has three functions in a capacitor:

1. It solves the mechanical problem of keeping the two metal plates separated by a very small distance.
2. It increases the maximum voltage that can be applied before causing breakdown (compared with air).
3. It increases the amount of capacitance, compared with air, for a given set of dimensions.

21-5   TYPES OF CAPACITORS

The properties and applications of a capacitor are determined by the type of dielectric used. Table 21-3 shows the major types of capacitors, their typical values, working voltages, and common applications.

As described at the beginning of this chapter, a capacitor consists of two conductors (metal plates) separated by an insulator (dielectric). The ''plates'' may be in the form of metal films in a *ceramic disc,* silver coatings in a *tubular ceramic,* or aluminum foil strips in a *paper capacitor.* (See Figs. 21-6a–21-6c.)

A paper capacitor often has a black band to indicate which lead is connected to the outer foil strip. In some applications, it is desirable to connect this lead to ground to provide shielding by the outer foil. The capacitance is the same, however, regardless of the way the capacitor is connected in the circuit.

To increase the effective area of the capacitor, plates may be interleaved, as in Fig. 21-6d. This construction is typical of *mica* capacitors and of the *variable* type of *air* capacitor. In the variable air capacitor, one set of plates is fixed and the other can be rotated in or out to increase or decrease capacitance. This type of capacitor is often used in low-capacitance tuning circuits for radios.

A type of capacitor that *must* be connected in a circuit

TABLE 21-3
Major Types of Capacitors

Capacitor table (handwritten note)

Type	Capacitance		Applications
Monolithic ceramics			UHF, RF coupling
Disc and tube ceramics			General, VHF
Paper			Motors, power supplies
Film—polypropylene			TV vertical circuits, RF
Polyester			Entertainment electronics
Polystyrene			General, high stability
Polycarbonate			General
Metallized polypropylene			AC motors
Metallized polyester			Coupling, RF filtering
Electrolytic-aluminum			Power supplies, filters
Tantalum	0.1–1000 μF	3–125	Small space requirement
			High reliability low leakage
Nonpolarized (either Al or Ta)	0.47–1000 μF	10–200	Loudspeaker crossovers
Mica	330 pF–0.05 μF	50–100	High frequency
Silver—mica	5–820 pF	50–500	High frequency
Variable—ceramic	1–5 to 16–100 pF	200	Radio, TV, communications
Film	0.8–5 to 1.2–30 pF	50	Oscillators, Antenna, RF circuits
Air	10–365 pF	50	Broadcast receiver
Teflon	0.25–1.5 pF	2000	VHF, UHF

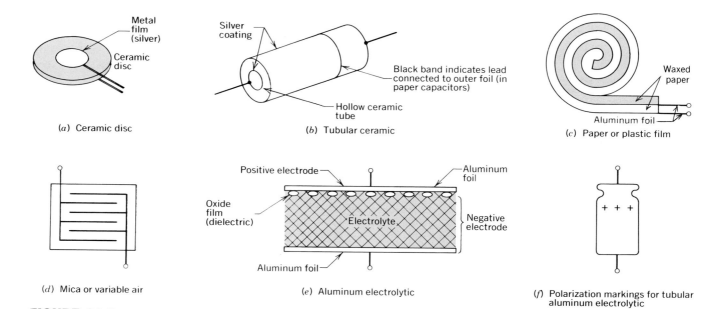

(a) Ceramic disc
Metal film (silver)
Ceramic disc

(b) Tubular ceramic
Silver coating
Black band indicates lead connected to outer foil (in paper capacitors)
Hollow ceramic tube

(c) Paper or plastic film
Waxed paper
Aluminum foil

(d) Mica or variable air

(e) Aluminum electrolytic
Positive electrode
Aluminum foil
Oxide film (dielectric)
Electrolyte
Negative electrode
Aluminum foil

(f) Polarization markings for tubular aluminum electrolytic
+ + +

FIGURE 21-6
Construction of various types of capacitors.

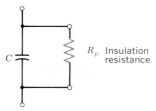

FIGURE 21-7
Equivalent circuit of a capacitor.

with proper polarity is the *polarized electrolytic* capacitor. The most common type is the *aluminum* electrolytic, shown in Fig. 21-6e. A paste electrolyte is placed between two sheets of aluminum foil and a dc *forming* voltage is applied. A current flows and (by a process of electrolysis) builds up a molecular-thin layer of aluminum oxide bubbles on the *positive* electrode. This serves as the dielectric. The rest of the electrolyte and the other sheet of aluminum foil make up the negative electrode. The very thin dielectric makes possible very large values (thousands of microfarads) of capacitance in a fairly small space.

If an electrolytic capacitor is connected with reverse polarity, the insulating layer of bubbles is removed from the positive electrode. This allows a high current to flow and the capacitor to become hot (it may even *explode*). To ensure proper connection, the polarization of the capacitor is clearly marked, as shown in Fig. 21-6f. It is considered good practice to use an electrolytic capacitor at a voltage close to its design value. If, for example, a 20-μF 400-V capacitor is used in a circuit at only 10 V, there may not

be enough voltage to maintain the aluminum oxide bubble layer in good condition, and the capacitance value may change. Also, the 400-V capacitor is bulkier and more expensive than the 10-V version.

21-5.1 Other Properties of Capacitors

One disadvantage of the aluminum electrolytic capacitor is its relatively large *leakage current*. This is a dc current that flows through the capacitor due to imperfections in the dielectric or to surface paths from one plate to the other. The effect can be represented by an insulation resistance (R_p) connected in parallel with the capacitor, as shown in Fig. 21-7. The heating effect produced within the capacitor due to current flowing through this resistance is sometimes represented by an equivalent series resistance (ESR) connected in *series* with the capacitor.

Similar to dry cells, electrolytic capacitors have a limited shelf life due to drying out of the electrolyte paste. This process, which may take 3 or 4 years, causes the leakage current to increase beyond acceptable levels. It is sometimes possible to ''repair'' such dried-out electrolytics by applying a dc voltage of the correct polarity that is near the capacitor's rated value. A resistor must be used in series with the capacitor to limit current until the aluminum oxide layer reforms.

Insulation resistances vary from 3000 MΩ for mica capacitors to more than 100,000 MΩ for plastic film capacitors. For electrolytics, ESR values and leakage current are specified. For standard aluminum electrolytic capacitors,

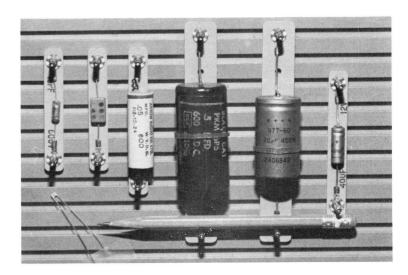

FIGURE 21-8

Typical capacitors. From left to right: 33-pF, 60-V; 15-pF, 600-V; 100-pF, mica; 0.05-μF, 600-V; 0.5-μF, 600-V, paper; 20-μF, 450-V; 40-μF, 12-V aluminum electrolytic.

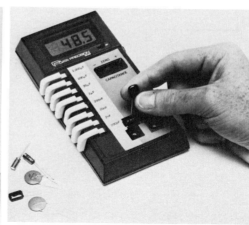

FIGURE 21-9

Capacitance meters. [Photograph in (a) courtesy of Sencore, Inc., in (b), courtesy of Data Precision Corp.]

this current is in the range of 0.01 to 0.04 mA per microfarad of capacitance, for working voltages of 100 to 500 V. Any capacitor showing leakage current in the range of 10 to 15 mA should normally be replaced, because heat produced inside such a faulty capacitor forms gases that could cause it to explode. *Computer grade* aluminum electrolytic capacitors have much lower leakage current values, with ESR values as low as a few milliohms (depending upon capacitance and voltage ratings).

In critical applications requiring *very* low leakage currents, a *tantalum* capacitor can be used. This type of capacitor is polarized and must be connected with proper polarity (although some can withstand a 3-V reverse voltage). The tantalum capacitor has a lower maximum capacitance and voltage rating than the aluminum capacitor, but occupies less space and provides high reliability. Some typical capacitors are shown in Fig. 21-8.

Many capacitors have an *operating temperature* range from −55 to +125°C. A new capacitor designed for hostile environments operates from 25 to 300°C. It has a silicon nitride dielectric and uses tungsten for the metal plates.

Like resistors, capacitors have various *temperature coefficients* and *tolerances*. Tolerances may ranges from ±1% up to ±20%. Some capacitors (because the minimum value is important) may have a specified tolerance of −20%, +80%. Ceramic capacitors are available with positive, negative, or virtually zero temperature coeffi-

cients. They can be used to compensate against temperature variations because a capacitance can be made to automatically increase or decrease with a change in temperature. A typical temperature coefficient might be +750 ppm (parts per million) per degree Celsius.

Capacitors are marked in various ways to show values. Some use a color-coded stripe method similar to that on resistors; others, a series of colored dots. In very small capacitors, it is understood that the number printed on the capacitor refers to picofarads. In some cases, the marking MF is used to show values in microfarads (millionths of a farad).

A number of instruments are available to measure the capacitance and other properties of a capacitor. Figure 21-9 shows two modern instruments that, between them, can measure capacitance from 0.1 pF to 200,000 μF, with an accuracy of 0.1%.

Capacitors can also be checked for open or short circuits with an ohmmeter (described in Section 22-3.1).

21-6 CAPACITORS IN PARALLEL

Figure 21-10a shows two capacitors connected in parallel with a dc voltage source (*V*). We wish to find the value of the single capacitor that is equivalent to the two parallel-connected capacitors.

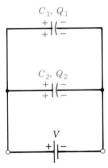
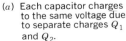

(a) Each capacitor charges to the same voltage due to separate charges Q_1 and Q_2.

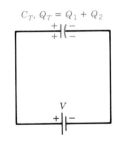

(b) The single equivalent capacitor charges to a total charge of $Q_1 + Q_2$.

FIGURE 21-10
Equivalent capacitance of two parallel-connected capacitors.

The charge that accumulates on each capacitor is given by

$$Q_1 = VC_1 \tag{21-1}$$

and

$$Q_2 = VC_2$$

If one capacitor (C_T) is to take the place of C_1 and C_2, it must store a total charge

$$Q_T = VC_T \tag{21-1}$$

where

$$Q_T = Q_1 + Q_2.$$

Therefore, $VC_T = VC_1 + VC_2$

$$= V(C_1 + C_2)$$

or

$$C_T = C_1 + C_2$$

In general, $C_T = C_1 + C_2 + C_3 + \cdots + C_N$ **(21-6)**

This equation gives the total combined capacitance for any number of *parallel*-connected capacitors. Note how the equation for *parallel*-connected capacitors is similar to a *series* connection of resistors. That is, an increase in capacitance results, due to an effective increase in plate area (A) in Eq. 21-2.

Capacitors are connected in parallel, of course, to increase the capacitance. However, if capacitors of different voltage ratings are parallel-connected, the combination is limited to the working voltage of the *lowest* voltage-rating capacitor in the group, as shown in Example 21-5.

EXAMPLE 21-5

The following three capacitors are connected in parallel with each other: 5 μF, 25 V; 10 μF, 20 V; and 1 μF, 50 V. Determine:

a. The total capacitance.
b. The maximum working voltage of the group.
c. The maximum charge that the parallel-connected group can store without damage to any capacitor.

Solution

a. $C_T = C_1 + C_2 + C_3$ (21-6)
    $= 5 + 10 + 1$ μF
    $= \mathbf{16 \ \mu F}$

b. Maximum working voltage $= \mathbf{20 \ V}$

c. $Q_T = VC_T$ (21-1)
    $= 20 \ V \times 16 \ \mu F$
    $= \mathbf{320 \ \mu C}$

21-7 CAPACITORS IN SERIES

When capacitors are connected in series with a voltage source (V), the momentary charging current is the same through the circuit, as shown in Fig. 21-11a. This means that the same charge (Q) is deposited on each capacitor.

The *voltage* to which each capacitor charges is then given by:

$$V_1 = \frac{Q}{C_1} \tag{21-1}$$

$$V_2 = \frac{Q}{C_2}$$

But Kirchhoff's voltage law requires that

$$V = V_1 + V_2 \tag{5-1}$$

Therefore,

$$V = \frac{Q}{C_1} + \frac{Q}{C_2}$$

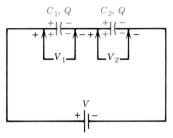

(a) Each capacitor charges to a different voltage due to the same charge.

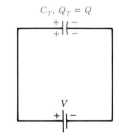

(b) The single equivalent capacitor charges to Q due to total voltage V.

FIGURE 21-11
Equivalent capacitance of two series-connected capacitors.

and
$$V = Q\left(\frac{1}{C_1} + \frac{1}{C_2}\right)$$

For one equivalent capacitor (C_T) to store a charge (Q) due to a total voltage (V), it is required that

$$V = \frac{Q}{C_T} \qquad (21\text{-}1)$$

Therefore,
$$\frac{Q}{C_T} = Q\left(\frac{1}{C_1} + \frac{1}{C_2}\right)$$

and
$$\frac{1}{C_T} = \frac{1}{C_1} + \frac{1}{C_2}$$

In general, $\dfrac{1}{C_T} = \dfrac{1}{C_1} + \dfrac{1}{C_2} + \dfrac{1}{C_3} + \cdots + \dfrac{1}{C_N}$ **(21-7)**

This equation gives the total combined capacitance for *any* number of *series*-connected capacitors. Note how the equation is similar to a *parallel* connection of resistors. This means that a *decrease* in capacitance results with a total capacitance that is always *less* than the smallest of the individual capacitances. This may be thought of as resulting from an effective *increase* in the separation (d) of the capacitor plates in Eq. 21-2.

Other equations to find the total capacitance for series-connected capacitors follow. For two series-connected capacitors only:

$$C_T = \frac{C_1 \times C_2}{C_1 + C_2} \qquad (21\text{-}8)$$

For N equal series-connected capacitors, each of capacitance C:

$$C_T = \frac{C}{N} \qquad (21\text{-}9)$$

EXAMPLE 21-6

Two capacitors—5 µF, 20 V and 10 µF, 20 V—are connected in series across a 30-V dc supply. Calculate:
a. The total capacitance of the series combination.
b. The charge on each capacitor.
c. The voltage across each capacitor.

Solution

a. $C_T = \dfrac{C_1 \times C_2}{C_1 + C_2}$ (21-8)

$\qquad = \dfrac{5\ \mu F \times 10\ \mu F}{5\ \mu F + 10\ \mu F}$

$\qquad = \dfrac{50}{15}\ \mu F = 3\tfrac{1}{3}\ \mu F$

b. $Q = VC_T$ (21-1)

$\qquad = 30\ V \times 3\tfrac{1}{3}\ \mu F$

$\qquad = \mathbf{100\ \mu C}$

c. $V_1 = \dfrac{Q}{C_1}$ (21-1)

$\qquad = \dfrac{100\ \mu C}{5\ \mu F} = \mathbf{20\ V}$

$\quad V_2 = \dfrac{Q}{C_2}$

$\qquad = \dfrac{100\ \mu C}{10\ \mu F} = \mathbf{10\ V}$

21-7.1 Voltage Division Across Capacitors in Series

The main reason for connecting capacitors in series is to *increase* the working voltage. Example 21-6, however, shows an important restriction. The total working voltage of the combination is *not* simply the sum of the individual working voltages (unless all capacitors are identical in capacitance and voltage rating). In Example 21-6, the 5-µF capacitor is working at its limit of 20 V when only 30 V are applied to the series combination.

For two capacitors in series,

$$Q = C_1V_1 = C_2V_2$$

and
$$\frac{V_2}{V_1} = \frac{C_1}{C_2} \qquad (21\text{-}10)$$

This equation implies that the *smaller* capacitor has the *larger* voltage across it, and vice-versa.

For the special case of only two capacitors in series, it can be shown that

$$V_1 = V \times \frac{C_2}{C_1 + C_2} \qquad \text{volts} \qquad (21\text{-}11a)$$

$$V_2 = V \times \frac{C_1}{C_1 + C_2} \qquad \text{volts} \qquad (21\text{-}11b)$$

where: V_1 is the voltage across capacitor C_1, in volts (V)
$\qquad\quad V_2$ is the voltage across capacitor C_2, in volts (V)
$\qquad\quad V$ is the total voltage across the series connection
$\qquad\qquad$ of capacitors C_1 and C_2, in volts (V)

These voltage-divider equations for capacitors are similar to the current-division equations for two parallel-connected resistors.

For more than two capacitors in series, the voltage

across each capacitor can be obtained by using the method shown in Example 21-6; that is, by finding the charge on each capacitor and applying Eq. 21-1.

EXAMPLE 21-7

Two capacitors—C_1 = 50 μF, 16 V and C_2 = 40 μF, 10 V—are connected in series. Calculate:
a. The maximum working voltage of this combination.
b. The voltage across each capacitor when working at the maximum total voltage.
c. The total capacitance.

Solution

a. Since C_2 is the smaller capacitor, it will have the higher voltage across it, which must be limited to 10 V.

$$V_2 = V \times \frac{C_1}{C_1 + C_2} \qquad (21\text{-}11b)$$

Maximum voltage $V = V_2 \times \dfrac{C_1 + C_2}{C_1}$

$$= 10 \text{ V} \times \frac{50 \text{ μF} + 40 \text{ μF}}{50 \text{ μF}}$$

$$= 10 \text{ V} \times \frac{9}{5} = \textbf{18 V}$$

b. Since C_2 is working at its limit of 10 V, C_1 must have a voltage across it of **8 V**.

$$\text{Or} \qquad V_1 = V \times \frac{C_2}{C_1 + C_2} \qquad (21\text{-}11a)$$

$$= 18 \text{ V} \times \frac{40 \text{ μF}}{50 \text{ μF} + 40 \text{ μF}}$$

$$= 18 \text{ V} \times \frac{4}{9} = \textbf{8 V}$$

$$V_2 = V \times \frac{C_1}{C_1 + C_2} \qquad (21\text{-}11b)$$

$$= 18 \text{ V} \times \frac{50 \text{ μF}}{50 \text{ μF} + 40 \text{ μF}}$$

$$= 18 \text{ V} \times \frac{5}{9} = \textbf{10 V}$$

c. $\quad C_T = \dfrac{C_1 \times C_2}{C_1 + C_2} \qquad (21\text{-}8)$

$$= \frac{50 \text{ μF} \times 40 \text{ μF}}{50 \text{ μF} + 40 \text{ μF}}$$

$$= \frac{2000}{90} \text{ μF} = \textbf{22.2 μF}$$

21-8 ENERGY STORED IN A CAPACITOR

As you have seen in this chapter, after a capacitor has been charged and the charging voltage removed, a potential difference remains across the capacitor because of the charge on the plates. If a photographic flashbulb (a load) is connected across the capacitor, it will "flash." This shows that energy is stored in the capacitor and can be transferred upon discharge through a load. In fact, the capacitor is the only device, apart from an electric voltaic cell, that can store electrical energy. (The charge can sometimes be stored for months, as in the application of indicating measurements of temperature or pressure from remote locations.)

An *inductor* also stores energy, but only so long as current is flowing through it. The energy in a capacitor can be thought of as being stored in its electric field, similar to the storage of energy in an inductor's magnetic field. The expression for stored energy in a capacitor is similar to that for an inductor ($\frac{1}{2} L I_f^2$). The energy stored in a capacitor is given by

$$W = \tfrac{1}{2} C V_f^2 \qquad \text{joules} \qquad (21\text{-}12)$$

where: W is the energy stored in the capacitor, in joules (J)
$\quad C$ is the capacitance, in farads (F)
$\quad V_f$ is the final voltage across the capacitor, in volts (V)

EXAMPLE 21-8

A 40-μF, 450-V capacitor and a 20-μF, 500-V capacitor are connected in parallel across a 400-V dc supply. Calculate:
a. The energy stored by each capacitor.
b. The total energy stored.

Solution

a. For the 40 μF,

$$W_1 = \tfrac{1}{2} C V_f^2 \qquad (21\text{-}12)$$

$$= \tfrac{1}{2} \times 40 \times 10^{-6} \text{ F} \times (400 \text{ V})^2$$

$$= \textbf{3.2 J}$$

For the 20 μF,

$$W_2 = \tfrac{1}{2} C V_f^2$$

$$= \tfrac{1}{2} \times 20 \times 10^{-6} \text{ F} \times (400 \text{ V})^2$$

$$= \textbf{1.6 J}$$

b. Total energy $= W_1 + W_2$

$\qquad = 3.2 \text{ J} + 1.6 \text{ J} = \textbf{4.8 J}$

or $\quad W_T = \frac{1}{2} C_T V_f^2$ $\qquad\qquad$ (21-12)

$\qquad = \frac{1}{2} \times 60 \times 10^{-6} \text{ F} \times (400 \text{ V})^2$

$\qquad = \textbf{4.8 J}$

Capacitors charged to the energy levels in Example 21-8 are considered *dangerous*. To avoid shock and to prevent damage to such instruments as ohmmeters, *always* discharge capacitors (with the power disconnected) before working on them.

SUMMARY

1. A capacitor consists of two conductors (metal plates) separated with an insulator called a dielectric.
2. When connected to a source of voltage, a capacitor charges to the applied emf, with a positive charge on one plate and an equal, but negative, charge on the other.
3. Capacitance (C) is defined as the ratio of the charge (Q) in coulombs per plate to the potential difference (V) between the plates: $C = Q/V$.
4. The basic unit of capacitance is the farad (F), with $1 \ \mu\text{F} = 1 \times 10^{-6}$ F and 1 pF $= 1 \times 10^{-12}$ F used for practical capacitor values.
5. Capacitance increases with the area of the capacitor's plates (A) and with a decrease in the plate separation (d) as given by $C = K \epsilon_0 A/d$.
6. ϵ_0 is the permittivity of free space and K is the dielectric constant, given in Table 21-1.
7. The momentary charging current, or displacement current, is due to the stressing of the valence electrons in the atoms of the dielectric when an electric field is set up. The atoms are said to be polarized.
8. The dielectric strength in kilovolts per millimeter is the electric field intensity that determines a capacitor's working voltage. It must not be exceeded or the dielectric will break down.
9. A dielectric in a capacitor has an induced electric field that opposes the electric field that arises from the charges on the plates. This increases the capacitance, permitting more charge to be deposited for a given voltage (compared with air).
10. The type of capacitor is determined by the dielectric material. Electrolytic capacitors make use of a chemical process to establish a very thin dielectric layer that is polarity sensitive but produces a large capacitance in a small volume.
11. When capacitors are connected in parallel, the total capacitance is the sum of the individual capacitances. Working voltage is equal to the lowest individual voltage rating in the group.
12. When capacitors are series-connected, capacitance decreases (similar to parallel-connected resistors). The smallest capacitor has the largest voltage across it.
13. The energy stored in a capacitor (in the electric field of the dielectric) is given by $W = \frac{1}{2} C V_f^2$

SELF-EXAMINATION

Answer T or F or a, b, c, d, or e for multiple choice
(Answers at back of book)
 21-1. Capacitance is the property of a capacitor to store electrical charge and energy.
 _____

21-2. The charge stored on one plate of a capacitor is the same as the charge on the other plate. _____

21-3. Capacitance is defined as the ratio of the potential difference across the capacitor to the charge stored per plate. _____

21-4. If a charge of 6 μC raises the potential difference of a capacitor from 0 to 3 V, the capacitance is
a. 2 pF   d. 0.5 μF
b. 18 μF   e. none of the above
c. 2 μF

21-5. A capacitance of 0.005 μF is the same as
a. 500 pF   d. 5000 pF
b. 50 nF   e. none of the above
c. 5×10^{-8} F

21-6. If the area of a given capacitor's plates is doubled and the plate separation is reduced by one-half, the new capacitance compared with the initial value is
a. not changed   d. eight times larger
b. quadrupled   e. one-eighth as large
c. doubled

21-7. If a dielectric with $K = 6$ is inserted between the plates of an air capacitor, the new capacitance compared with the initial value is
a. 6 times smaller   d. not changed
b. 36 times larger   e. none of the above
c. 6 times larger

21-8. The momentary charging current into a capacitor is due to the dielectric becoming polarized. _____

21-9. If a voltage of 500 V is applied to a capacitor whose plates are separated by 0.5 mm, the electric field intensity is
a. 250 kV/m   d. 250 V/mm
b. 1 kV/mm   e. either part (a) or (d)
c. 1000 kV/mm

21-10. The dielectric strength of paper is higher than that of air. _____

21-11. A smaller plate separation increases the capacitance because the electric field intensity is higher. This means that there is a greater charge on the plates for the same applied voltage. _____

21-12. The insertion of a dielectric between the plates of a charged capacitor (with no voltage applied) causes an increase in the electric field intensity and potential difference. This accounts for the increase in capacitance with a dielectric other than air. _____

21-13. The dielectric constant of a material is the ratio of the capacitance with the dielectric to the capacitance with air. _____

21-14. Large values of capacitance are obtainable only in electrolytic types of capacitors. _____

21-15. A tantalum capacitor, although an electrolytic type, is not polarity-sensitive. _____

21-16. A high insulation resistance indicates a large leakage current in a capacitor. _____

21-17. The high capacitance of an electrolytic capacitor results from a molecular-thin oxide layer acting as the dielectric between the plates. _____

21-18. The equations for calculating the total capacitance of parallel-connected capacitors are similar to those for parallel-connected resistors. _____

21-19. The maximum working voltage of parallel-connected capacitors is determined by the lowest individual working voltage. _____

21-20. The working voltage of a series-connected combination of capacitors is the sum of the individual working voltages. _____

21-21. A parallel combination of 0.5 μF, 100 V and 1 μF, 50 V has an equivalent value of

    a. 1.5 μF, 100 V   d. 1.5 μF, 150 V

    b. 3 μF, 50 V   e. none of the above

    c. 1.5 μF, 50 V

21-22. A series connection of the two capacitors in Question 21-21 has a total capacitance of

    a. $\frac{1}{3}$ μF   d. 0.3 μF

    b. 3 μF   e. none of the above

    c. 1.5 μF

21-23. When capacitors are series-connected, the largest capacitor has the largest voltage across it. _____

21-24. Voltage division across series-connected capacitors can be calculated using the same equations as those used for voltage division across series-connected resistors. _____

21-25. The energy stored by a capacitor can be doubled by either doubling the capacitance or the voltage. _____

REVIEW QUESTIONS

1. a. What is the mathematical definition of *capacitance?*

    b. What is 1 C/V called?

    c. Why are microfarads and not farads used to identify a capacitor's capacitance?

2. a. What are the factors that determine the capacitance of a capacitor?

    b. Which of these factors also determines the working voltage?

3. a. If a perfect insulator were used for the dielectric in a capacitor, would there still be a momentary charging current?

    b. If so, how can you explain this?

4. What is the difference between the electric field intensity of a charged capacitor and the dielectric field strength?

5. Explain why a decrease in plate separation causes an increase in capacitance.

6. a. Why does the insertion of a dielectric (in place of air) cause a decrease in voltage across the plates of a charged capacitor?

    b. What is the origin of the word *dielectric?*

    c. Explain why a dielectric increases the capacitance compared with air.

7. What are three functions provided by a dielectric?

8. a. What is the basic construction difference between an electrolytic capacitor and the other types?

    b. What accounts for the very large values of capacitance?

    c. What restriction applies to an electrolytic?

9. a. What does a capacitor's insulation resistance represent?

    b. What initial and final readings do you think an ohmmeter would indicate when connected across an uncharged capacitor?

    c. How can the final reading of the ohmmeter differ if an electrolytic is being checked compared with a plastic film capacitor? Would the reading change if the ohmmeter connections were reversed?

10. a. What is the point of connecting capacitors in parallel?
    b. How is the maximum working voltage of a parallel combination determined?
    c. If electrolytic capacitors are being parallel-connected, what precaution must be observed?

11. a. Why, when capacitors are series-connected, does the smallest capacitor have the highest voltage across it?
    b. Under what condition is the maximum working voltage of series-connected capacitors equal to the sum of the individual capacitor working voltages?

12. How is the energy stored in a capacitor different from the energy stored by an inductor?

PROBLEMS

(Answers to odd-numbered problems at back of book)

21-1. How much capacitance is required to store 100 µC of charge on each plate of a capacitor when connected to a 20-V dc supply?

21-2. A 0.015-µF capacitor is connected to a 6-V battery. How much charge is deposited on each plate?

21-3. What voltage must a 1500-pF capacitor be connected to for a charge of 0.03 µC per plate?

21-4. A charge of 2.5 µC raises the potential difference of a capacitor to 100 V.
    a. What is the capacitance of the capacitor?
    b. How much charge must be removed to drop the voltage to 20 V?
    c. What is the potential difference across the capacitor when the charge is increased to 4 µC?

21-5. A sheet of paper is 0.008 cm thick. Calculate:
    a. The capacitance of two metal plates each 21.6 × 28 cm (approximately 8½ by 11 in.) when separated by one sheet of paper.
    b. The maximum working voltage of the capacitor in part (a).

21-6. Repeat Problem 21-5 for the same thickness of mica.

21-7. A 1-µF capacitor is to be constructed with a strontium titanate (ceramic) dielectric that is 0.1 mm thick. Calculate:
    a. The necessary plate area.
    b. The maximum working voltage assuming a safety factor of three.

21-8. Repeat Problem 21-7 using a bakelite (plastic) dielectric.

21-9. A 0.0015-µF paper capacitor must have a maximum working voltage of 600 V, with a safety factor of three. Determine the necessary area of each plate.

21-10. Repeat Problem 21-9 using a bakelite (plastic) dielectric.

21-11. Determine the electric field intensity in the capacitor of Problem 21-9 when operating at its maximum working voltage.

21-12. Two metal plates of 20-cm diameter are separated by 1 cm of air. A charge of 0.003 µC is deposited on the plates. Calculate:
    a. The capacitance of the two plates.
    b. The potential difference across the plates.
    c. The electric field intensity.
    d. The new potential difference when a 1-cm-thick sheet of glass is inserted between the plates.

e.   The new electric field intensity.

f.   The new capacitance of the plates.

g.   The *additional* charge necessary to restore the potential difference to the initial value in part (b).

21-13. The following capacitors are parallel-connected: 47 μF, 25 V; 25 μF, 35 V; and 50 μF, 16 V. Determine:

a.   The total capacitance.

b.   The maximum working voltage of the group.

c.   The maximum charge that the group can store.

21-14. A 40-μF capacitor is connected across a 12-V dc supply. What size of capacitor must be connected in parallel with the first so that a total charge of 840 μC is stored?

21-15. Four 25-μF, 25-V capacitors are connected in series to a 50-V source. Calculate:

a.   The total capacitance of the series combination.

b.   The charge on each capacitor.

c.   The voltage across each capacitor.

d.   The maximum working voltage of the combination.

21-16. Repeat Problem 21-15 using two 25-μF, 25-V capacitors and two 10-μF, 30-V capacitors.

21-17. A 0.015-μF capacitor is connected in series with a 0.02-μF capacitor and a 200-V dc supply. Calculate:

a.   The total capacitance.

b.   The minimum voltage rating of each capacitor.

21-18. What is the largest value of capacitance that can safely be connected in series with a 500-μF, 12-V capacitor and a 15-V dc supply?

21-19. What size of capacitor must be connected in series with a 40-μF capacitor to give a total capacitance of 24 μF?

21-20. If the two capacitors in Problem 21-19 are each rated at 15 V, what is the maximum working voltage of the series combination?

21-21. Find the maximum energy that the capacitors in Problem 21-13 can store.

21-22. Find the maximum energy that the capacitors in Problem 21-15 can store.

21-23. What size of capacitor, charged to 250 V, is necessary to store an energy of 1 J?

21-24. A capacitor is to be charged with 0.01 C to store 2 J of energy. Calculate:

a.   The necessary voltage.

b.   The amount of capacitance.

21-25. Calculate the total capacitance of the series-parallel-connected capacitors in Figs. 21-12a and 21-12b.

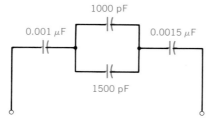

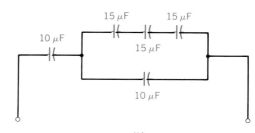

FIGURE 21-12
Circuits for Problem 21-25.   (a)   (b)

CHAPTER 22

CAPACITANCE IN DC CIRCUITS

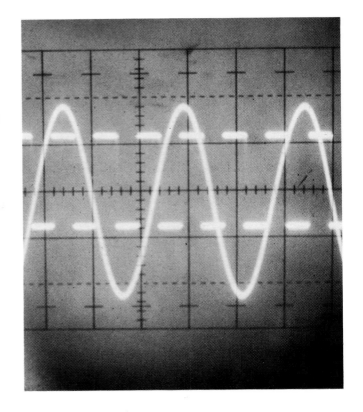

Although no *steady* current flows in a dc capacitive circuit, there is a *momentary* charging or discharging current that flows whenever the voltage across the capacitor is *changing.* (This can be compared with the induced *voltage* across a coil, which appears only when the *current* through the coil is changing.) To determine the amount of time needed to charge or discharge a capacitor, you must determine the *time constant (RC)* of the circuit. This time constant allows you to examine the *transient responses* in a series *RC* circuit and determine how *steady-state* conditions are reached. In a way similar to that used for inductive circuits, a graphical method of universal time-constant circuits can be employed to solve for instantaneous capacitor voltage and current.

An electronic *photoflash* unit is a common application of a capacitor in a dc circuit. This application involves *slowly charging* a capacitor to limit the current drawn from a battery or dc supply, then *suddenly discharging* all or some of that stored energy to provide a bright light for a very short time. Other capacitor applications include a short time-constant *differentiating* circuit, and long time-constant-*filtering* and *delay* circuits.

Finally, you will learn how to use an ohmmeter to check a capacitor and an oscilloscope to measure capacitance.

22-1 CURRENT IN A CAPACITIVE CIRCUIT

You have learned that when a dc voltage is applied to a capacitor, the charge stored is given by

$$Q = CV \qquad (21\text{-}1)$$

Assume that depositing an additional small charge (ΔQ) raises the potential difference across the capacitor by a small amount (ΔV). Therefore,

$$\Delta Q = C \Delta V*$$

If these changes take place in a short time (Δt), then:

$$\frac{\Delta Q}{\Delta t} = C \frac{\Delta V}{\Delta t}$$

But $\dfrac{\Delta Q}{\Delta t}$ is the average current (i) during this short interval (Δt). Thus

$$i = C \frac{\Delta v}{\Delta t}$$

To emphasize that this is the current into (or out of) the capacitor, and that you must consider the *changing* voltage across the capacitor, consider the following:

$$i_C = C\left(\frac{\Delta v_C}{\Delta t}\right) \qquad \text{amperes} \qquad (22\text{-}1)$$

where: i_C is the current into or out of the capacitor, in amperes (A)
  C is the capacitance, in farads (F)
  Δv_C is the change in voltage across the capacitor, in volts (V)
  Δt is the time interval, in seconds (s)
  $\dfrac{\Delta v_C}{\Delta t}$ is the rate of change of voltage, in volts per second (V/s)

Equation 22-1 states that current flows into (or out of) a capacitor *only* when the voltage across the capacitor is *changing*. This should be compared to the corresponding equation for an inductor:

$$v_L = L\left(\frac{\Delta i}{\Delta t}\right) \qquad (17\text{-}1)$$

where the *voltage* is induced *only* when the *current* is changing.

EXAMPLE 22-1

The voltage across a 4-μF capacitor is raised from 16 to 24 V in 2 ms. Calculate:
a. The average rate of change of voltage across the capacitor.
b. The average current into the capacitor during the 2-ms interval.
c. The increase in charge on the capacitor.

Solution

a. $\dfrac{\Delta v_C}{\Delta t} = \dfrac{24\text{ V} - 16\text{ V}}{2 \times 10^{-3}\text{ s}}$

$= \dfrac{8\text{ V}}{2 \times 10^{-3}\text{ s}} = $ **4000 V/s**

b. $i_C = C \times \left(\dfrac{\Delta v_C}{\Delta t}\right)$ $\qquad (22\text{-}1)$

$= 4 \times 10^{-6}\text{ F} \times 4000\ \dfrac{\text{V}}{\text{s}}$

$= 16 \times 10^{-3}\text{ A} = $ **16 mA**

c. $\Delta Q = C\ \Delta V$

$= 4 \times 10^{-6}\text{ F} \times 8\text{ V}$

$= 32 \times 10^{-6}\text{ C} = $ **32 μC**

EXAMPLE 22-2

The voltage across a 50-μF capacitor varies as shown in Fig. 22-1a. Draw the waveform of the capacitor current.

Solution

During the first 5 ms:

$i_C = C \times \dfrac{\Delta v_C}{\Delta t}$ $\qquad (22\text{-}1)$

$= 50 \times 10^{-6}\text{ F} \times \dfrac{14\text{ V} - 4\text{ V}}{5 \times 10^{-3}\text{ s}}$

$= 50 \times 10^{-6}\text{ F} \times 2 \times 10^{3}\ \dfrac{\text{V}}{\text{s}} = $ **100 mA**

Note that the rate of change of voltage, $\dfrac{\Delta v_C}{\Delta t}$, is constant at 2 kV/s during this interval, so the current into the capacitor (causing it to charge) is constant at 100 mA. See Fig. 22-1b.

From 5 to 10 ms there is no change in voltage. This

---

*If $Q = CV$, then increasing V by a small amount ΔV to $V + \Delta V$ results in an increase in Q to a *new* value $Q + \Delta Q$, such that $Q + \Delta Q = C(V + \Delta V)$. Subtracting the equation $Q = CV$, we obtain $\Delta Q = C\Delta V$.

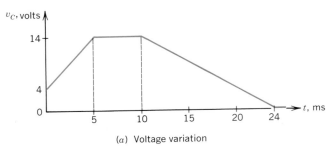

(a) Voltage variation

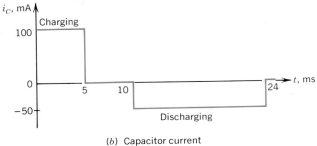

(b) Capacitor current

FIGURE 22-1

Waveforms for Example 22-2.

means that $\dfrac{\Delta v_C}{\Delta t}$ is zero, and the capacitor current is also zero.

During the interval from 10 to 24 ms the voltage is *decreasing* at a constant rate. The capacitor current is given by

$$i_C = C \times \frac{\Delta v_C}{\Delta t} \tag{22-1}$$

$$= 50 \times 10^{-6} \text{ F} \times \frac{0 \text{ V} - 14 \text{ V}}{24 - 10 \text{ ms}}$$

$$= 50 \times 10^{-6} \text{ F} \times \left(-1 \times 10^3 \frac{\text{V}}{\text{s}} \right) = \textbf{-50 mA}$$

The negative sign means that the current is flowing in the opposite direction, indicating that the capacitor is *discharging*. See Fig. 22-1*b*.

22-2 CHARGING A CAPACITOR

Example 22-2 suggests that, if a capacitor is charged with a *constant* current, the capacitor voltage increases in a *linear* manner. Next, we examine the rate at which a capacitor charges when connected *suddenly* across a dc voltage, with a series resistance, as in Fig. 22-2*a*.

Assume that the capacitor is initially discharged, so that v_C is zero. This means that, when the switch is first closed, all the applied voltage must appear across the resistor. (The capacitor appears, momentarily, as a short circuit.) Thus, the *initial* charging current is determined only by the applied voltage and the resistance:

$$I_{\text{initial}} = \frac{V}{R} = \frac{10 \text{ V}}{1 \text{ k}\Omega} = 10 \text{ mA}$$

At the instant the switch is closed, the current immediately increases from 0 to *V/R*, as shown in Fig. 22-2*c*.

This current causes a charge to be deposited on the ca-

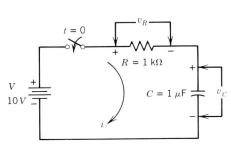

(a) Circuit containing resistance and capacitance

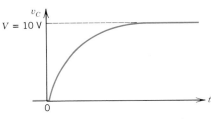

(b) Capacitor voltage variation with time

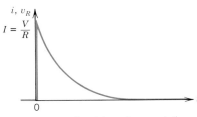

(c) Current and resistor voltage variation with time

FIGURE 22-2

Voltage and current waveforms for a charging capacitor.

pacitor, so that some short time later, the capacitor voltage has increased above zero, as shown in Fig. 22-2b.

But Kirchhoff's voltage law, as shown in Fig. 22-2a, requires that

$$V = v_R + v_C$$

This means that with some charge on the capacitor, the resistor voltage, v_R, is less than when the switch was first closed:

$$v_R = V - v_C$$

And since $v_R = iR$, the current must now have some *lower* value. But you also know that

$$i = C\frac{\Delta v_C}{\Delta t} \qquad (22\text{-}1)$$

If i is lower than its initial value, then the rate of change of the capacitor voltage $\left(\dfrac{\Delta v_C}{\Delta t}\right)$ is also lower. That is, the capacitor voltage is still increasing (as shown in Fig. 22-2b), but at a *decreasing* rate. The charging current also continues, but at a decreasing rate, as shown in Fig. 22-2c.

Eventually, the capacitor voltage reaches the applied voltage (V). There is no further change in voltage, and the charging current drops to zero. (See Fig. 22-2c.) If the switch is now opened, the capacitor value remains at its *steady-state* value of V. The increasing portion of v_C (Fig. 22-2b) and the momentary decreasing charging current (Fig. 22-2c) are called *transient responses*. They are similar to the transients that occur in an inductive circuit, but with the roles of current and voltage interchanged.

To determine how long it takes for the capacitor to fully charge to its steady-state value, the *time-constant* (τ) of the circuit must be considered.

22-3 TIME CONSTANT (τ)

Since the largest charging current occurs at the instant the switch is closed, the highest rate of capacitor voltage *change* must occur at that instant, as well. That is,

$$i = I_{\text{initial}} = \frac{V}{R}$$

and

$$i = C\left(\frac{\Delta v_C}{\Delta t_0}\right) \qquad (22\text{-}1)$$

Thus

$$C\left(\frac{\Delta v_C}{\Delta t_0}\right) = \frac{V}{R}$$

and the *initial* rate of rise of the capacitor voltage is

$$\frac{\Delta v_C}{\Delta t_0} = \frac{V}{RC} \qquad \text{volts per second} \qquad (22\text{-}2)$$

where: $\dfrac{\Delta v_C}{\Delta t_0}$ is the *initial* rate of rise of the capacitor
voltage, in volts per second (V/s)
V is the applied dc voltage, in volts (V)
R is the series resistance, in ohms (Ω)
C is the capacitance, in farads (F)

EXAMPLE 22-3

For the circuit and values given in Fig. 22-2a, determine:
a. The initial rate of rise of the capacitor voltage.
b. The initial charging current.
c. The final capacitor voltage.

Solution

a. $\dfrac{\Delta v_C}{\Delta t_0} = \dfrac{V}{RC}$ $\qquad\qquad\qquad\qquad$ (22-2)

$$= \frac{10 \text{ V}}{1 \times 10^3\ \Omega \times 1 \times 10^{-6}\ \text{F}}$$

$$= 10 \times 10^3 \text{ V/s}$$

b. Initial charging current $= \dfrac{V}{R}$

$$= \frac{10 \text{ V}}{1 \times 10^3\ \Omega} = 10 \text{ mA}$$

c. Final capacitor voltage $= V = 10 \text{ V}$

Now, consider how long it would take the capacitor voltage to reach its final value *if* it continued to increase at its initial rate of change, or $\dfrac{V}{RC}$ volts per second. This length of time can be referred to as the time constant (τ). (See Fig. 22-3.)

Figure 22-3 shows that the voltage does *not* increase at this rate. In fact, when $t = \tau$, the actual capacitor voltage is only 0.63 of its final value. This is the more usual definition of the time constant. That is, the time constant is the time required for a change of 63% to occur from the initial value to the final value. (See Section 22-5.)

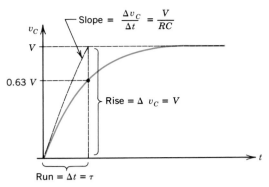

FIGURE 22-3
Determination of time constant, τ.

Let the dotted line that is tangent to the voltage curve reach the final value of the capacitor voltage in a time interval τ. This dotted line represents the initial rate of change of the capacitor voltage. It has a slope given by

$$\text{initial slope} = \frac{\text{rise}}{\text{run}} = \frac{\text{vertical intercept}}{\text{horizontal intercept}}$$
$$= \frac{V}{\tau} \text{ volts per second}$$

But initial slope $= \dfrac{\Delta v_C}{\Delta t_0} = \dfrac{V}{RC}$ volts per second   (22-2)

Therefore, $\dfrac{V}{\tau} = \dfrac{V}{RC}$

and $\tau = RC$ seconds   (22-3)

where: τ is the time constant of the circuit, in seconds (s)
R is the resistance, in ohms (Ω)
C is the capacitance of the circuit, in farads (F)

Equation 22-3 gives the time required for a capacitor to reach 63% of its final steady-state voltage. This time is equal to the product of the circuit capacitance and resistance.

It can also be shown (see Section 22-5) that it takes approximately five time constants for the capacitor to charge to its final steady-state voltage. This time is not dependent upon the voltage applied, as shown by the following examples.

EXAMPLE 22-4

A 5-μF capacitor and a 10-kΩ resistor are series-connected across a 12-V dc supply. (See Fig. 22-4.) Calculate:

a. The time constant of the circuit.
b. The initial rate of rise of the capacitor voltage.
c. The capacitor voltage after one time constant.
d. The time required for the capacitor to charge to 12 V.

Solution

a. $\tau = RC$ (22-3)
$= 10 \times 10^3 \ \Omega \times 5 \times 10^{-6} \text{ F}$
$= 50 \times 10^{-3} \text{ s} = \textbf{50 ms}$

b. $\dfrac{\Delta v_C}{\Delta t_0} = \dfrac{V}{RC}$ (22-2)
$= \dfrac{12 \text{ V}}{50 \times 10^{-3} \text{ s}} = \textbf{240 V/s}$

c.   After one time constant
$v_C = 0.63 \times V$
$= 0.63 \times 12 \text{ V} = \textbf{7.56 V}$

d.   Time to reach 12 V $= 5\tau$
$= 5 \times 50 \text{ ms} = \textbf{250 ms}$

EXAMPLE 22-5

The capacitance in the previous example is doubled to 10 μF; the resistance is cut in half to 5 kΩ; and the voltage is doubled to 24-V dc. (See Fig. 22-4.) Calculate:

a. The time constant of the circuit.
b. The initial rate of rise of the capacitor voltage.
c. The capacitor voltage after one time constant.
d. The time required for the capacitor to charge to 24 V.

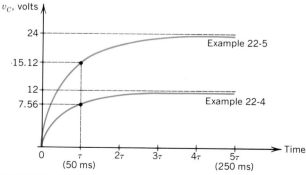

FIGURE 22-4
Capacitor charging curves for Examples 22-4 and 22-5.

Solution

a. $\tau = RC$ (22-3)

$\qquad = 5 \times 10^3 \, \Omega \times 10 \times 10^{-6} \, F$

$\qquad = 50 \times 10^{-3} \, s = \textbf{50 ms}$

b. $\dfrac{\Delta v_C}{\Delta t_0} = \dfrac{V}{RC}$ (22-2)

$\qquad = \dfrac{24 \, V}{50 \times 10^{-3} \, s} = \textbf{480 V/s}$

c. After one time constant

$$v_C = 0.63 \times V$$
$$= 0.63 \times 24 \, V = \textbf{15.12 V}$$

d. Time to reach 24 V $= 5 \, \tau$

$$= 5 \times 50 \, \text{ms} = \textbf{250 ms}$$

Both capacitors reach 63% of their final steady-state value (7.65 V for the first capacitor, 15.12 V for the second) in the *same* time of 50 ms. Also, both charge to their *final* values of 12 V and 24 V in the *same* time of 5 RC or 250 ms. This occurs because the *rate of rise* of voltage is *twice* as great in Example 22-5 as in Example 22-4. (See Fig. 22-4.)

22-3.1 Checking Capacitors with an Ohmmeter

When an ohmmeter is connected across an uncharged capacitor, there is an initial charging current determined by the resistance of the ohmmeter and its cell voltage. (An ohmmeter is essentially a series circuit consisting of resistance, dry cell, and meter movement. The ohmmeter eventually charges the capacitor to the voltage of its internal dry cell or cells.)

The effect on the ohmmeter is a sudden deflection of the pointer from infinity toward zero ohms, as shown in Fig. 22-5a, because of the relatively large initial charging current. The capacitor initially acts like a short circuit (exhibits very low resistance).

As the capacitor charges to the ohmmeter cell voltage (see Fig. 22-5b), the ohmmeter's pointer gradually moves back toward infinity. The rate at which it moves is determined by the time constant of the capacitor and the ohmmeter resistance. The final reading of the ohmmeter is an indication of the insulation resistance of the capacitor. (Section 21-5.1.) For most capacitors, this resistance is practically infinite on the average ohmmeter. For electrolytic capacitors, however, the final ohmmeter reading will be noticeably less than infinite and may depend upon the polarity applied to the capacitor.

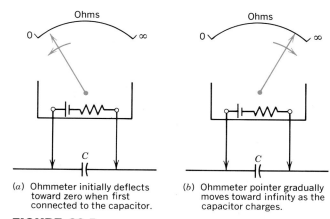

(a) Ohmmeter initially deflects toward zero when first connected to the capacitor.

(b) Ohmmeter pointer gradually moves toward infinity as the capacitor charges.

FIGURE 22-5

Checking a capacitor with an ohmmeter.

If the pointer goes to zero and stays there, it is evident that the capacitor is shorted. If there is no initial deflection from infinity toward zero, the capacitor is open. However, the capacitor must be checked on a suitable ohmmeter range to give a visible charging deflection. On the very low resistance ranges, charging can be so instantaneous that no initial deflection toward zero can be observed. This is especially true for very small capacitors of a few picofarads, where the time constant is very small. A higher resistance range (such as $R \times 1 \, M\Omega$) must be used for these capacitors. The higher resistance, however, limits the initial charging current, so little (if any) pointer deflection from infinity will result on *any* ohmmeter range for very small capacitors. Also, if the capacitor is connected in a complete circuit, the ohmmeter readings may be affected by other components.

Obviously, the use of an ohmmeter to check a capacitor must be done carefully. It provides only a simple check to see if the capacitor has any obvious problems.

22-4 DISCHARGING A CAPACITOR

You have seen how a charged capacitor stores energy ($W = \frac{1}{2} CV^2$). Now, we will examine how a capacitor *discharges,* returning this energy to a load.

Consider the circuit in Fig. 22-6a. If the switch has remained in position 1 for five or more time constants, the capacitor is fully charged (to 10 V, in this example). When the single-pole, double-throw switch is thrown to position 2, the capacitor immediately begins to discharge. The initial discharge current is determined by the initial capacitor voltage and the resistance:

$$I_{initial} = \frac{V_{initial}}{R}$$

$$= \frac{10\text{ V}}{1 \times 10^3 \ \Omega} = 10\text{ mA}$$

This occurs because, by Kirchhoff's voltage law, the initial capacitor voltage must be equal to the voltage across the resistor. Since the direction of the discharging current is opposite to that of the charging current, the waveform of the capacitor current in Fig. 22-6c is shown as negative. (Note that Fig. 22-6a shows the instantaneous polarity of v_R across R and the actual direction of current (i) when the switch is in position 2. In this sense, v_R and i in Fig. 22-6c are not negative. They *are* negative, however, with respect to the reference polarity and direction used in Fig. 22-2a, where the capacitor was charging. Showing the current waveform in Fig. 22-6c as negative is meant to emphasize the opposite direction that current flows upon capacitor discharge.)

Since charge is being removed from the capacitor, the potential difference across the capacitor drops from its initial value, as shown in Fig. 22-6b. The initial rate of decrease of capacitor voltage is again given by

$$\frac{\Delta v_C}{\Delta t_0} = \frac{V_{initial}}{RC} \qquad (22\text{-}2)$$

After one time constant ($\tau = RC$), the capacitor voltage has *changed* 63%. This means that the voltage across the capacitor is now only 0.37 of its initial value (3.7 V). Similarly, the current has dropped to 0.37 of its initial value (3.7 mA) after one time constant.

After five time constants ($5\tau = 5RC$), both the decreasing capacitor voltage and the discharging current are approximately zero. The transient is ended and the capacitor is completely discharged. All of the energy stored in the capacitor has been dissipated in the resistor.

NOTE It takes the same time for the capacitor to *discharge* as it does to *charge*, if the same series resistance is used. For Fig. 22-6a, $\tau = RC = 1$ ms for both charging and discharging. Thus it takes 5 ms to completely charge or discharge. Example 22-6 shows a circuit that has different charge and discharge time constants.

22-4.1 Application to a Capacitor Photoflash Unit

The circuit shown in Fig. 22-7 can produce short-duration, high-current pulses without drawing a large current from the supply. The large resistance (R_1) limits the peak charging current (I_1) and allows a gradual charging of the capacitor due to the large time constant ($\tau_1 = R_1C$). When

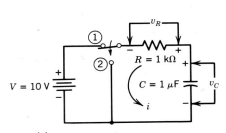

(a) Capacitor is discharged by throwing switch to position 2.

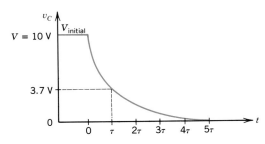

(b) Capacitor discharging voltage

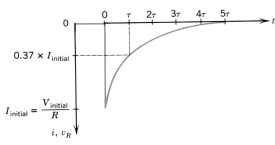

(c) Capacitor discharging current

FIGURE 22-6

Voltage and current waveforms for a discharging capacitor.

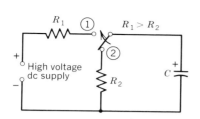

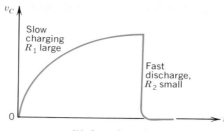

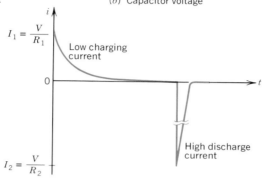

(a) Circuit provides slow charge in position 1 and fast discharge in position 2.

(b) Capacitor voltage

(c) Capacitor current

FIGURE 22-7

Typical *RC* circuit for short-duration, high-current pulses.

the switch is thrown to position 2, the low resistance of R_2 permits a high discharge current. The duration of this current is determined by the small time constant ($\tau_2 = R_2C$). In addition to its use in photoflash units, this system provides the high surge current needed in such applications as electric spot welding and radar transmitter tubes. For a given capacitance, the amount of energy delivered by each discharge pulse can be controlled (for a given capacitance) by the voltage used to charge the capacitor.

EXAMPLE 22-6

A capacitor photoflash unit uses a circuit similar to that in Fig. 22-7 with a 300-V supply, a 500-μF capacitor, and a current limiting resistor of 4 kΩ. Assume that the resistance of the lamp, R_2, stays constant at 10 Ω when the shutter contact discharges the capacitor in position 2. Calculate:

a.   The time required for the capacitor to fully charge.
b.   The peak charging current.
c.   The time required for the capacitor to fully discharge.
d.   The peak discharging current.
e.   The energy stored by the capacitor.

f.   The average power produced by the lamp, in watts, during the discharge period.

Solution

a.   Charging time = $5 R_1C$
$$= 5 \times 4 \times 10^3 \ \Omega \times 500 \times 10^{-6} \ \text{F}$$
$$= \textbf{10 s}$$

b.   Peak charging current, $I_1 = \dfrac{V}{R_1}$
$$= \dfrac{300 \ \text{V}}{4 \times 10^3 \ \Omega} = \textbf{75 mA}$$

c.   Discharging time = $5 R_2C$
$$= 5 \times 10 \ \Omega \times 500 \times 10^{-6} \ \text{F}$$
$$= \textbf{25 ms}$$

d.   Peak discharging current,
$$I_2 = \dfrac{V}{R_2}$$
$$= \dfrac{300 \ \text{V}}{10 \ \Omega} = \textbf{30 A}$$

e.   Energy stored,
$$W = \tfrac{1}{2} CV^2 \qquad (21\text{-}12)$$
$$= \tfrac{1}{2} \times 500 \times 10^{-6} \ \text{F} \times (300 \ \text{V})^2$$
$$= \textbf{22.5 J}$$

Power, $\quad P = \dfrac{W}{t}$ (3-4)

$\qquad = \dfrac{22.5\ J}{0.025\ s}$

$\qquad = 900\ J/s = \textbf{900 W}$

22-5 UNIVERSAL TIME-CONSTANT CURVES

If you wish to determine capacitor voltage and current at times other than one and five time constants, you can use universal time-constant curves. These curves give the instantaneous values as a percentage of the initial or steady-state values, with the time given in time constants. Universal time-constant curves for charge and discharge voltage and current in a dc capacitive circuit are shown in Fig. 22-8. (These are the same curves shown for series RL circuits in Fig. 19-4.) Note that, at one time constant (τ), the capacitor has charged to 63% of its final steady-state voltage, and the charging current has dropped to only 37% of its initial value. In either case, a *change* of 63% occurs in just one time constant.

These curves also confirm that the charging or discharging of a capacitor is essentially complete after *five* time constants.

EXAMPLE 22-7

Refer to the circuit of Fig. 22-9a. After the switch has remained in position 2 for a long time (until the capacitor is completely discharged), the switch is thrown to position 1. Use the universal time constant curves to determine:

a. The capacitor voltage 3.5 s later.
b. How long it takes for the capacitor voltage to discharge to 24 V if the switch is returned to position 2 after being in position 1 for 3.5 s.
c. The capacitor voltage waveform for parts (a) and (b).

Solution

a. $\quad \tau_1 = R_1 C$ (22-3)

$\qquad = 220 \times 10^3\ \Omega \times 10 \times 10^{-6}\ F$

$\qquad = 2.2\ s$

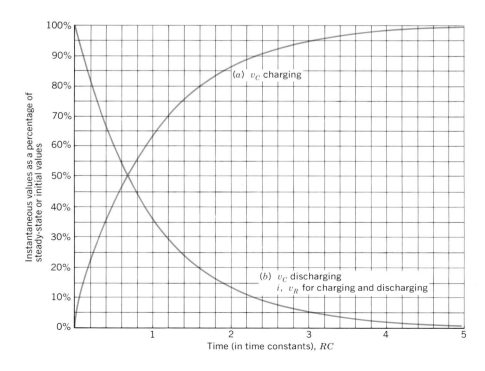

FIGURE 22-8
Universal time-constant curves for charge and discharge voltage and current in a capacitive circuit with direct current.

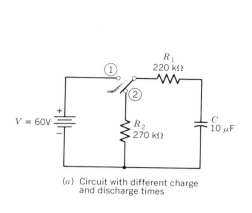

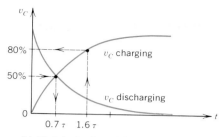

(b) Sketch of universal time-constant curves

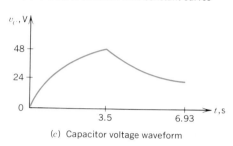

(c) Capacitor voltage waveform

(a) Circuit with different charge and discharge times

FIGURE 22-9
Circuit and waveforms for Example 22-7.

Charge time $t = 3.5$ s

$$= \frac{3.5 \text{ s}}{2.2 \text{ s/}\tau} = 1.59 \ \tau \approx 1.6 \ \tau$$

From Fig. 22-8a, when $t = 1.6 \ \tau$, $v_C = 80\%$ of V (see Fig. 22-9b).
Therefore, $v_C = 80\%$ of V

$$= 0.8 \times 60 \text{ V} = \textbf{48 V}$$

b. Initial capacitor voltage $= 48$ V
Capacitor voltage

$$= 24 \text{ V}$$

$$= \frac{24 \text{ V}}{48 \text{ V}} \times 100\% \text{ of initial value}$$

$$= 50\% \text{ of initial value}$$

From Fig. 22-8b, when $v_C = 50\%$ of initial value,

$$t = 0.7 \ \tau \quad \text{(see Fig. 22-9b)}$$

Discharge time constant
$\tau_2 = (R_1 + R_2)C$

$$= (220 \times 10^3 + 270 \times 10^3) \ \Omega \times 10 \times 10^{-6} \text{ F}$$

$$= 4.9 \text{ s}$$

Discharge time to reach 24 V $= 0.7 \ \tau$

$$= 0.7 \times 4.9 \text{ s}$$

$$= \textbf{3.43 s}$$

c. See Fig. 22-9c.

Note that the *discharge* time constant depends upon the *total* series resistance ($R_1 + R_2$).

22-6 ALGEBRAIC SOLUTIONS USING EXPONENTIAL EQUATIONS*

Instead of using a graphical approach, you can calculate the capacitor voltage and current at any instant (for charging or discharging), using the exponential equation introduced in Section 19-4:

Instantaneous value
$$= \text{Final value} + (\text{Initial} - \text{Final}) \ e^{-t/\tau} \qquad (19\text{-}4)$$

where: Instantaneous value is the value of the current or voltage at any instant of time
Final value is the steady-state value of current or voltage after the transient has ended
Initial value is the value of current or voltage at $t = 0^+$ (at the beginning of the transient)
e is the irrational number 2.71828 . . ., the base of natural logarithms

_____

*This section may be omitted without loss of continuity.

t is the instant of time at which the value of current or voltage is to be found

τ is the time constant of the circuit, in seconds (s), which is equal to L/R for an inductive circuit and RC for a capacitive circuit

22-6.1 Equations for Capacitors Charging

Specific equations for capacitor voltage and current can be developed by sketching the responses and identifying initial and final values. (See Fig. 22-10.)

For a charging capacitor, the initial capacitor voltage is zero, and its final value is V. (Remember that the voltage across a capacitor cannot change instantaneously.) See Fig. 22-10b.

Instantaneous value
$$= \text{Final value} + (\text{Initial} - \text{Final})\, e^{-t/\tau} \qquad (19\text{-}4)$$

$$v_C = V + (0 - V)e^{-t/\tau}$$
$$= V - Ve^{-t/\tau}$$
$$v_C = V(1 - e^{-t/\tau}) \qquad (22\text{-}4)$$

where: v_C is the capacitor voltage, in volts (V)
V is the applied dc voltage, in volts (V)
t is the instant of time after closing the switch, in seconds (s)
$\tau = RC$ is the time constant, in seconds (s)

To obtain the expression for capacitor charging current, observe (from Fig. 22-10c) that the initial current (at $t = 0^+$) is V/R and that the final current is 0.

Instantaneous current
$$= \text{Final current} + (\text{Initial} - \text{Final})\, e^{-t/\tau} \qquad (19\text{-}4)$$

$$i = 0 + \left(\frac{V}{R} - 0\right) e^{-t/\tau}$$
$$i = \frac{V}{R}e^{-t/\tau} \qquad (22\text{-}5)$$

where: i is the value of current, in amperes (A)
V is the applied dc voltage, in volts (V)
R is the circuit resistance, in ohms (Ω)
t is the time after closing the switch, in seconds (s)
$\tau = RC$ is the time constant, in seconds (s)
Note the result of substituting $t = RC$ in Eq. 22-4:

$$v_C = V(1 - e^{-t/\tau})$$
$$= V(1 - e^{-RC/RC})$$
$$= V(1 - e^{-1})$$
$$= V(1 - 0.368)$$
$$= 0.632V = \textbf{63.2\% of } V$$

This confirms that in one time constant, the capacitor voltage has reached approximately 63% of the final steady-state voltage.

If $t = 5RC$ is substituted in Eq. 22-4,

$$v_C = V(1 - e^{-5})$$

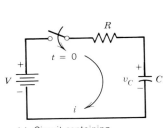

(a) Circuit containing resistance and capacitance

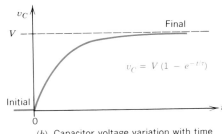

(b) Capacitor voltage variation with time

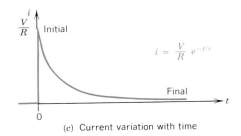

(c) Current variation with time

FIGURE 22-10

A series RC circuit with dc voltage applied, showing current and capacitor voltage responses with exponential equations.

$$= V(1 - 0.0067)$$
$$= 0.9933V = \textbf{99.3\% of } V$$

This also confirms that the transient is essentially complete in a period of five time constants.

Note that the equation for capacitor voltage (Eq. 22-4) is of the same form as the equation for inductor current (Eq. 19-5), and that the equation for capacitor current (Eq. 22-5) is of the same form as the equation for inductor voltage (Eq. 19-6).

This relationship is no accident—it arises from the fundamental relations $v_L = L\dfrac{\Delta i}{\Delta t}$ and $i_C = C\dfrac{\Delta v}{\Delta t}$, where it is seen that the roles of voltage and current are interchanged. Because of this, circuits that contain both inductance and capacitance can produce some interesting results. This interaction of inductance and capacitance will be examined in detail under the topic of *resonance* in Chapter 26.

22-6.2   Equations for Capacitors Discharging

For a discharging capacitor, with an initial voltage of V and a final value of 0 (as in Fig. 22-11b), Eq. 19-4 can again be applied.

Instantaneous voltage
$$= \text{Final voltage} + (\text{Initial} - \text{Final}) \, e^{-t/\tau} \qquad (19\text{-}4)$$

$$v_C = 0 + (V - 0)\, e^{-t/\tau}$$
$$v_C = V e^{-t/\tau} \qquad (22\text{-}6)$$

The capacitor current has an initial value (at $t = 0^+$) of $-V/R$ and a final value of 0. (See Fig. 22-11c.)

Therefore
$$i = 0 + \left(-\frac{V}{R} - 0\right) e^{-t/\tau}$$
$$i = -\frac{V}{R} e^{-t/\tau} \qquad (22\text{-}7)$$

where: v_C is the capacitor voltage, in volts (V)
i is the current, in amperes (A)
V is the initial capacitor voltage at the beginning of discharge, in volts (V)
R is the circuit resistance, in ohms (Ω)
t is the time after discharge begins, in seconds (s)
$\tau = RC$ is the time constant, in seconds (s)

The negative sign in Eq. 22-7 indicates discharging current with a direction *opposite* to that of the charging current. (This corresponds to the reversal of induced voltage across an inductor when current through the inductor is decreasing.)

Note that the V in Eqs. 22-6 and 22-7 is not necessarily the source of voltage in the circuit. That is, the capacitor may have charged to some value of voltage *less* than V when it is suddenly made to discharge, as shown in Example 22-8.

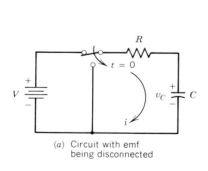

(a) Circuit with emf being disconnected

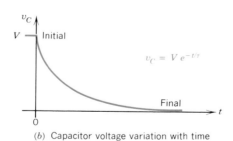

(b) Capacitor voltage variation with time

$$v_C = V e^{-t/\tau}$$

$$i = -\frac{V}{R} e^{-t/\tau}$$

(c) Current variation with time

FIGURE 22-11

A series *RC* circuit with dc voltage being removed, showing current and capacitor voltage responses wtih exponential equations.

EXAMPLE 22-8

Refer to the circuit shown in Fig. 22-12. Assume that the capacitor is initially discharged, and that the switch is thrown to position 1, remains there for 3.5 s, then is returned to position 2. Calculate, and compare (where possible) with the graphical results shown in Example 22-7:

a. The capacitor voltage 3.5 s after the switch is thrown to position 1.
b. The capacitor current at this instant.
c. The time required for the capacitor to discharge to 24 V after the switch is returned to position 2.
d. The capacitor current at this instant.

Solution

(Refer to Example 19-6 for the evaluation of exponential quantities and natural logs, using a scientific calculator.)

a. Charging time constant

$$\tau_1 = R_1 C \tag{22-3}$$
$$= 220 \times 10^3 \ \Omega \times 10 \times 10^{-6} \ F = 2.2 \ s$$

$$v_C = V(1 - e^{-t/\tau_1}) \tag{22-4}$$
$$= 60 \ V(1 - e^{-3.5s/2.2s})$$
$$= 60 \ V(1 - e^{-1.591})$$
$$= 60 \ V(1 - 0.204)$$
$$= 60 \ V \times 0.796 = \textbf{47.8 V}$$

This compares favorably with 48 V using the graphical method. (See the solution to Example 22-7A.)

b. $$i_C = \frac{V}{R} e^{-t/\tau_1} \tag{22-5}$$

$$= \frac{60 \ V}{220 \times 10^3 \ \Omega} \times e^{-3.5 \ s/2.2 \ s}$$
$$= 0.273 \ mA \times e^{-1.591}$$
$$= 0.273 \ mA \times 0.204 = \textbf{0.056 mA}$$

c. Discharging time constant

$$\tau_2 = (R_1 + R_2) \ C \tag{22-3}$$

$$= (220 \times 10^3 + 270 \times 10^3) \ \Omega \times 10 \times 10^{-6} \ F$$
$$= 4.9 \ s$$

$$v_C = V_{initial} \times e^{-t/\tau_2} \tag{22-6}$$
$$24 \ V = 47.8 \ V \times e^{-t/4.9 \ s}$$
$$e^{-t/4.9 \ s} = \frac{24 \ V}{47.8 \ V} = 0.502$$

Take the natural log of both sides:

$$ln \ e^{-t/4.9 \ s} = ln \ 0.502$$
$$-\frac{t}{4.9 \ s} = -0.689$$

Therefore, $t = 4.9 \ s \times 0.689 = \textbf{3.38 s}$

This compares favorably with 3.43 s using the graphical method. (See the solution of Example 22-7B.)

d. $$i_C = -\frac{V_{initial}}{R} \times e^{-t/\tau_2} \tag{22-7}$$

$$= -\frac{47.8 \ V}{(220 + 270) \times 10^3 \ \Omega} \times e^{-3.38 \ s/4.9 \ s}$$
$$= -0.098 \ mA \times 0.502 = \textbf{-0.049 mA}$$

The negative sign indicates a discharging current.

Note again that Eq. 19-4 could have been used directly in place of Eqs. 22-4 through 22-7, simply by knowing the initial and final values of each response and the time constant involved.

22-7 OSCILLOSCOPE MEASUREMENT OF CAPACITANCE

Although it is *possible* to observe the charging and discharging of a capacitor on an oscilloscope when using a mechanical switching circuit, this is only *practical* for relatively large time constants. With small time constants, repetitive switching at a sufficiently high frequency can be done only by connecting the output of a square wave signal generator to the capacitive circuit. The frequency of the signal is adjusted until the capacitor voltage waveform shows that a steady state has been reached, as shown in Fig. 22-13. To achieve such a steady state, the half-period of the square wave must be equal to, or greater than, five time constants ($T/2 \geq 5 \ RC$).

With the oscilloscope connected across the capacitor, the time required to reach 63% of the final voltage is the time constant (τ). (See Fig. 22-13c.) If the total resistance

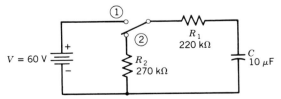

FIGURE 22-12
Circuit for Example 22-8.

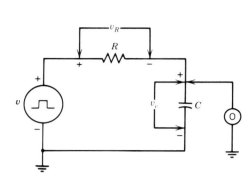

(a) Square wave generator connected to a capacitive circuit

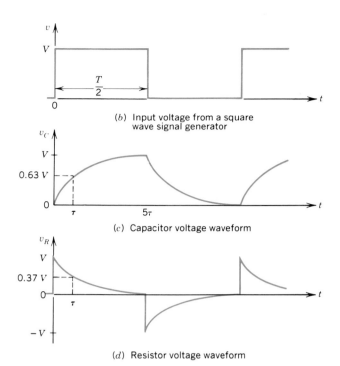

(b) Input voltage from a square wave signal generator

(c) Capacitor voltage waveform

(d) Resistor voltage waveform

FIGURE 22-13

Voltage waveforms in a capacitive circuit with a square wave signal generator.

of the circuit is known, the capacitance can be obtained, as demonstrated in Example 22-9.

With the oscilloscope connected across the resistor (interchanging R and C to avoid common ground problems), the time constant is obtained by observing the time required for v_R to fall to 37% of its initial value. (See Fig. 22-13d.) This method is particularly useful if the circuit's capacitance is not clearly defined. That is, *stray* capacitance in the circuit (due to the method of wiring or other causes) may make it impossible to connect the oscilloscope across the capacitance itself. (See Problem 22-18.)

EXAMPLE 22-9

An oscilloscope is connected across a capacitor as in Fig. 22-13a. The display reaches a final vertical value of 5 cm. It requires a horizontal distance of 1.5 cm for the capacitor voltage display to reach 3.15 cm. If the oscilloscope horizontal calibration is 2 ms/cm, determine:

a. The time constant of the circuit.
b. The capacitance if $R = 4.7$ kΩ.

c. The maximum frequency allowable for the signal generator.

Solution

a. 63% of 5 cm = 0.63 × 5 cm
                = 3.15 cm (vertical distance)
The *horizontal* distance of 1.5 cm is thus proportional to the time constant:

$$\tau = 1.5 \text{ cm} \times 2 \text{ ms/cm}$$
$$= \textbf{3 ms}$$

b. $\tau = RC$ (22-3)

Therefore, $C = \dfrac{\tau}{R}$

$$= \dfrac{3 \times 10^{-3} \text{ s}}{4.7 \times 10^{3} \text{ }\Omega}$$
$$= 0.64 \times 10^{-6} \text{ F} = \textbf{0.64 }\boldsymbol{\mu}\textbf{F}$$

c. $\dfrac{T}{2} \geq 5\,RC$

Therefore, $T_{\text{min}} = 10\,RC$
$$= 10 \times 3 \text{ ms} = 30 \text{ ms}$$

$$f_{max} = \frac{1}{T} \qquad (14\text{-}9a)$$

$$= \frac{1}{30 \times 10^{-3} \text{ s}} = \textbf{33.3 Hz}$$

22-8 SHORT TIME-CONSTANT DIFFERENTIATING CIRCUIT*

The circuit shown in Figure 22-14a has a very small time constant, compared with the period of applied voltage (e.g., $5RC \leq T/10$). This means that the capacitor charges very rapidly, producing a waveform that is very similar to the input voltage waveform. (Compare Figs. 22-14b and 22-14c.)

You know that the current in the circuit is given by

$$i_C = C\frac{\Delta v_C}{\Delta t} \qquad (22\text{-}1)$$

---

*This section may be omitted with no loss of continuity.

and $v_R = iR = RC\frac{\Delta v_C}{\Delta t}$. But since $v_C \approx v$ (the input voltage),

$$v_R \approx RC\frac{\Delta v}{\Delta t} \qquad (22\text{-}4)$$

Thus the resistor voltage (v_R) is proportional to the *slope* of the input voltage $\left(\frac{\Delta v}{\Delta t}\right)$.

The mathematical process of finding the slope of a curve is called *differentiation*. The term $\frac{\Delta v}{\Delta t}$ is an approximation to the derivative $\frac{dv}{dt}$, which is found by *differentiating* a mathematical equation, so v_R is said to be the output of a *differentiating circuit*.

Thus, the resistor voltage gives a sharp, narrow positive pulse when the *leading edge* of the input voltage goes positive, due to the very steep slope of the input waveform at this instant.

The resistor voltage then drops to zero, corresponding to the zero slope of the input voltage. A sharp *negative* pulse is then produced at the *trailing edge* of the input

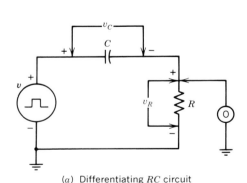

(a) Differentiating *RC* circuit

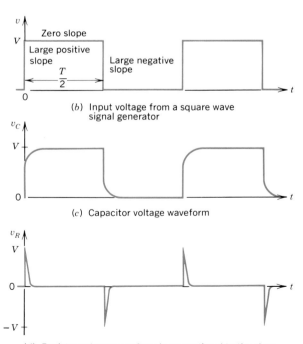

(b) Input voltage from a square wave signal generator

(c) Capacitor voltage waveform

(d) Resistor voltage waveform is proportional to the slope of the input voltage waveform.

FIGURE 22-14
Voltage waveforms in a differentiating circuit.

voltage, showing the negative slope of the input wave-form. These fast-rising and fast-falling voltage pulses are useful in digital logic circuits, where they are used for triggering purposes.

22-9 LONG TIME-CONSTANT CIRCUITS

The circuit in Fig. 22-15a shows an application in which a very long time constant is desirable. When the switch is open, the diode produces a half-wave-rectified voltage variation across the load, as shown in Fig. 22-15b. Although this is a *direct current* output, it is not very smooth because it contains a large *ac* ripple voltage.

When the switch is closed, the capacitor is placed in parallel with the load. Now the capacitor can charge to the peak voltage (V_m) when the input voltage increases. But when the input voltage *decreases*, the capacitor discharges through the load (following the curve shown in Fig. 22-8), attempting to keep the voltage near V_m. The basic capacitor property of *opposing* any change in voltage is being used to smooth the load voltage. The larger the time constant ($R_L C$), the higher will be the average output load

voltage, as shown in Fig. 22-15c. The energy delivered to the load during the discharge period is replaced during the very short charging period, when the input voltage again nears V_m. Note that the ac ripple voltage content is very small, due to the action of the capacitor.

EXAMPLE 22-10

A 1000-μF capacitor is used to provide filtering for a 100-Ω load, as in Fig. 22-15a, with a peak voltage input of 9 V and a frequency of 60 Hz. Assuming that the capacitor discharges for a full period of the input waveform, calculate:

a. The lowest value of load voltage.
b. The peak-to-peak ripple voltage across the load as indicated in Fig. 22-15c.
c. The effect on load and ripple voltage of using a 2000-μF capacitor.

Solution

a. $\tau = RC$ (22-3)
    $= 100 \ \Omega \times 1000 \times 10^{-6} \ F$

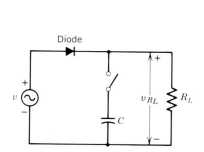

(a) Capacitor filters voltage fluctuations when the switch is closed.

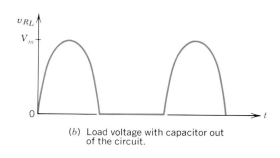

(b) Load voltage with capacitor out of the circuit.

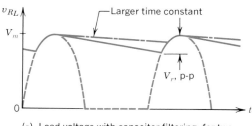

(c) Load voltage with capacitor filtering, for two different time constants.

FIGURE 22-15
Use of a capacitor to filter voltage fluctuations.

$$= 0.1 \text{ s}$$

$$T = \frac{1}{f} \tag{14-9}$$

Therefore, discharge time (approximately equal to T) is given by

$$t = \frac{1}{60 \text{ Hz}}$$

$$= 16.6 \text{ ms}$$

$$= \frac{16.6 \text{ ms}}{0.1 \text{ s}/\tau} = 0.166 \ \tau$$

From curve (b) in Fig. 22-8, when $t = 0.17 \ \tau$,

$$v_C = 85\% \text{ of } V_{initial}$$

$$= 0.85 \times 9 \text{ V}$$

$$= \textbf{7.65 V}$$

b. Peak-to-peak ripple voltage $= V_{r,p-p}$

$$= v_{C_{max}} - v_{C_{min}}$$

$$= 9 \text{ V} - 7.65 \text{ V}$$

$$= \textbf{1.35 V}$$

c. When $C = 2000 \ \mu\text{F}$, $\tau = 0.2$ s and $t = 0.083 \ \tau$

$$v_C = 92\% \text{ of } V_{initial} = \textbf{8.28 V}$$

and $\hspace{3em} V_{r,p-p} = \textbf{0.72 V}$

Example 22-10 shows that use of a larger capacitor reduces ripple voltage.

22-9.1   RC Delay Circuits

An RC circuit with a long time constant can be used to introduce a *delay,* as in the timing circuit shown in Fig. 22-16.

The neon lamp in this circuit acts as an *open* until a firing voltage of between 60 and 80 V is reached. The voltage across the capacitor charges toward 100 V at a rate

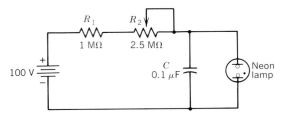

FIGURE 22-16

RC relaxation oscillator timing circuit.

determined by the time constant $(R_1 + R_2)C$. But when the firing voltage of the neon lamp is reached, the capacitor discharges through the lamp, lighting it briefly. Because of the low resistance of the conducting lamp, the capacitor voltage drops quickly and the lamp goes out. Since the lamp is once again an open circuit, the capacitor recharges. This provides a controlled delay before the lamp fires again. The rate of flashing for the lamp can be varied by adjusting R_2. (See Problem 22-20.)

The delay introduced by the capacitor in the circuit shown in Fig. 22-16 illustrates a useful principle. For example, if the operation of a dc relay must be delayed following application of voltage to the coil, the delay can be achieved by connecting a capacitor across the coil. For longer delays, a resistor in series with the parallel-connected coil and capacitor can be used. This arrangement will affect the final voltage applied to the coil, however. (See Problem 22-22.)

Finally, the voltage across the capacitor in an RC circuit with a medium or long time constant is proportional to the *integral* of the input voltage. Since it ''sums up'' or ''averages out'' the input voltage, such a circuit is referred to as an *integrating circuit.* See Appendix K for an application of an integrating circuit that displays a *B-H* hysteresis curve on an oscilloscope.

SUMMARY

1. Capacitor current flows only when the voltage across the capacitor changes, as given by the equation:

$$i_C = C\frac{\Delta v_C}{\Delta t}$$

2. The initial charging current of a capacitor is determined only by the applied voltage and series resistance, as given by:

$$I_{\text{initial}} = \frac{V}{R}$$

3. The gradual rise of capacitor voltage during charging is determined by the time constant:

$$\tau = RC$$

4. The *time constant* is the length of time needed for the capacitor voltage to rise to 63% of its final steady-state value, or for the charging current to fall to 37% of its initial value.

5. After five time constants, the transient response is essentially complete, and the capacitor is considered fully charged or discharged.

6. The initial rate of change of capacitor voltage, during charging or discharging, is given by

$$\frac{\Delta v_C}{\Delta t} = \frac{V}{RC}$$

7. If a capacitor continued to charge (or discharge) at its initial rate of change, the capacitor voltage and current would reach steady-state values in one time constant.

8. A quick check of a capacitor can be made with an ohmmeter by watching for an initial deflection toward zero followed by a gradual charging effect toward infinity.

9. The high discharge current from a capacitor is used to provide a brief, very bright light in an electronic photoflash unit.

10. Universal time-constant curves can be used to graphically determine instantaneous capacitor voltage and current with time given in terms of time constants.

11. Exponential equations, used to algebraically determine the values of capacitor voltage and current at any instant, may be obtained from:

$$\text{Instantaneous value} = \text{Final} + (\text{Initial} - \text{Final})\, e^{-t/\tau}$$

12. An oscilloscope and square-wave signal generator can be used to measure a circuit's time constant, and thus, its capacitance.

13. If a series RC circuit has a short time constant ($5RC \leq T/10$), the voltage waveform across the resistor is an approximation to the derivative or the slope of the input voltage, and the circuit is called a differentiating circuit.

14. Long time-constant circuits are used to provide filtering in dc power supplies and to provide a delay in timing and relay circuits.

SELF-EXAMINATION

Answer T or F or a, b, c, d, or e for multiple choice
(Answers at back of book)

22-1. When a dc voltage is applied to a capacitor, the voltage across the capacitor cannot change instantaneously. _____

22-2. Capacitor current can only flow if the voltage across the capacitor is changing. _____

22-3. Capacitor charging current has its greatest rate of change at one time constant. _____

22-4. A larger capacitor has a higher initial charging current when a dc voltage is applied than a smaller capacitor. _____

22-5. If a capacitor voltage increased from 25 to 50 V in 5 ms, the rate of change of voltage is
   a. 5 V/s
   b. 5 kV/s
   c. 5 mV/s
   d. 125 mV/s
   e. none of the above

22-6. Given the rate of change in Question 22-5, what capacitor current will a capacitance of 10 μF cause?
   a. 5 mA
   b. 50 A
   c. 50 mA
   d. 1.25 A
   e. 5 μA

22-7. If 10-V dc is applied to a series combination of a 2-kΩ resistor and a 5-μF capacitor, the initial charging current is
   a. 20 mA
   b. 0.2 mA
   c. 10 mA
   d. 5 μA
   e. 5 mA

22.8 A series circuit of 2.2-MΩ resistance and 0.1-μF capacitance has a time constant of
   a. 0.22 s
   b. 0.22 ms
   c. 0.22 μs
   d. 22 s
   e. 22 ms

22-9. Checking a capacitor with an ohmmeter determines only whether the capacitor is good or bad, not the actual capacitance. _____

22-10. A 0.1-μF capacitor, initially charged to 20 V, is discharged through a 10-kΩ resistor. After 1 ms the capacitor voltage is
   a. 6.3 V
   b. 3.7 V
   c. 7.4 V
   d. 12.6 V
   e. none of the above

22-11. The charge and discharge times of a capacitor are always equal. _____

22-12. The universal time constant curves of Fig. 22-8 can be used either for the charging or discharging of a capacitor. _____

22-13. A capacitor's voltage after discharging for five time constants is exactly zero. _____

22-14. If a series RC circuit with R = 20 kΩ shows a time constant on an oscilloscope of 50 ms, the circuit capacitance is
   a. 20 μF
   b. 40 pF
   c. 20 pF
   d. 2.5 μF
   e. 2.5 pF

22-15. The output of a differentiating circuit is taken across the resistor in a series RC circuit. _____

22-16. When a capacitor is used in a filter circuit, a higher load contributes to a longer time constant and therefore a smoother output voltage. _____

22-17. A larger capacitor used across the coil of a relay will increase the delay in the operation of the relay. _____

REVIEW QUESTIONS

1. Which two equations can you cite to show that capacitors and inductors have opposite effects as far as voltage and current are concerned?

2. a. How would you interpret a negative $\dfrac{\Delta v_C}{\Delta t}$?
   b. What effect does it have on the capacitor current?

3. Explain in your own words why the capacitor voltage increases *gradually* (and not instantaneously) in a series RC circuit with direct current applied.

4. Give two definitions of a time constant as applied to a series RC circuit.

5. Why is the rate of rise of the capacitor voltage highest when direct current is first applied to a series RC circuit?

6. a. Why doesn't it take longer (for a given RC circuit) to charge to a final steady-state value of 20 V than it does for a final value of 10 V?
   b. What makes this possible?

7. Describe the effect on an ohmmeter of checking:
   a. A shorted capacitor.
   b. An open capacitor.
   c. A very small capacitor on a very low range.
   d. A very large capacitor on a medium range.

8. a. If the capacitor in a photoflash unit charges to the applied voltage, why not use the voltage source (battery) to light the flash lamp directly?
   b. What other advantages does the use of a capacitor provide?
   c. If you had the choice of increasing the lamp brightness by increasing the size of the capacitor or the applied voltage, which would be more effective and why?

9. Describe how you would measure the capacitance in a series RC circuit using a square wave generator and an oscilloscope connected across:
   a. The capacitor.
   b. The resistor.

10. If a sinusoidal voltage is applied to a differentiating circuit, draw the output waveform, in proper time relationship with the input waveform.

11. a. What is the result of a long time constant in a capacitor filtering circuit?
    b. How would you expect the peak-to-peak ripple output voltage to change as the load resistance is increased?

12. Draw the capacitor voltage waveform for the neon lamp oscillating circuit in Fig. 22-16. Indicate when the lamp is on and off. What will happen to the flashing frequency if R_2 is increased?

PROBLEMS

(Answers to odd-numbered problems at back of book)

22-1. The voltage across a 0.2-μF capacitor is increased uniformly from 50 to 100 V in 10 ms. Calculate:
   a. The rate of change of the capacitor voltage.
   b. The capacitor current.
   c. The increase in stored charge.

22-2. The voltage across a 1500-pF capacitor is decreased uniformly from 60 to 20 V in 100 μs. Calculate:
   a. The rate of change of the capacitor voltage.
   b. The capacitor current.
   c. The decrease in stored charge.

22-3. The voltage across a 15-μF capacitor increases uniformly from 3 to 15 V in 4 ms, remains constant for 6 ms, and then decreases uniformly to zero in 6 ms. Draw the waveforms of the capacitor voltage and current showing all values.

22-4. The voltage across a 200-pF capacitor varies as shown in Fig. 22-17. Draw the waveform of capacitor current.

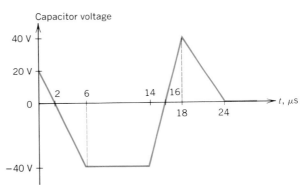

FIGURE 22-17
Capacitor voltage waveform for Problem 22-4.

22-5. A 6-V battery is connected in series with a 22-kΩ resistor and a 0.15-μF capacitor. Calculate:
   a. The initial rate of increase of the capacitor voltage.
   b. The initial charging current.
   c. The final capacitor voltage.

22-6. When a 20-V dc supply is connected to a series RC circuit, the initial rate of rise of the capacitor voltage is determined to be 500 V/s. Calculate:
   a. The time constant of the circuit.
   b. The time for the capacitor to reach its steady-state value if it continues to charge at the initial rate of rise.
   c. The approximate time required for the capacitor to charge to 20 V.

22-7. A 250,000-μF capacitor and a 100-Ω resistor are series-connected across a 5-V dc supply. Calculate:
   a. The time constant of the circuit.
   b. The initial rate of rise of the capacitor voltage.
   c. The capacitor voltage after one time constant.
   d. The approximate time required for the capacitor to charge to 5 V.
   e. The initial charging current.

22-8. It is required that the initial charging current in a series RC circuit be limited to 50 mA when a 200-V dc supply is connected. It is also required that the charging transient be complete in 100 ms. Determine:

    a.   The resistance of the circuit.

    b.   The capacitance of the circuit.

    c.   The charging current after one time constant.

    d.   The capacitor voltage after one time constant.

22-9. After a 40-μF capacitor was fully charged to 450 V, a 2.2-MΩ resistor was connected across the capacitor. Calculate:

    a.   The initial discharge current.

    b.   The current after one time constant.

    c.   The capacitor voltage after one time constant.

    d.   The time to completely discharge the capacitor.

22-10. A 500,000-μF capacitor is charged to 6 V. Calculate:

    a.   The resistance load that will fully discharge the capacitor in one minute.

    b.   The initial discharge current.

    c.   The capacitor current after one time constant.

    d.   The capacitor voltage after one time constant.

    e.   The initial rate of decrease of capacitor voltage upon discharge.

22-11. Repeat Example 22-6 assuming that $C = 300$ μF and $V = 400$ V.

22-12. Consider the circuit of Fig. 22-7. Determine:

    a.   The voltage necessary if a 1000-μF capacitor is to store 30 J of energy.

    b.   The value of R_1 if the initial charging current is to be limited to 100 mA.

    c.   The time required for the capacitor to fully charge.

    d.   The discharge time if $R_2 = 5$ Ω.

    e.   The peak discharging current.

    f.   The average power dissipated in R_2.

22-13. Repeat Example 22-7 with $R_1 = 150$ kΩ, $R_2 = 68$ kΩ, $C = 40$ μF, and $V = 80$ V.

22-14. Refer to Fig. 22-9. After the switch has been left in position 1 for more than five time constants, it is thrown to position 2. Determine, using Fig. 22-8,

    a.   How long it takes for the capacitor voltage to drop to 30 V.

    b.   What time is required for the voltage to increase to 45 V if the switch is then returned to position 1.

22-15. Refer to the circuit of Fig. 22-9a. $R_1 = 4.7$ kΩ, $R_2 = 3.3$ kΩ, $C = 0.15$ μF, and $V = 150$ V. The capacitor is initially uncharged. Calculate:

    a.   The capacitor voltage 1.5 ms after the switch is thrown to position 1.

    b.   The capacitor current at this instant.

    c.   The time required for the capacitor to discharge to 20 V if the switch is returned to position 2 after being in position 1 for 1.5 ms.

    d.   The capacitor current at this instant.

22-16. Repeat Problem 22-15 with $C = 0.3$ μF.

22-17. An oscilloscope is connected across a capacitor as in Fig. 22-13a. The oscilloscope display reaches a final value of 6 cm. It requires a horizontal distance of 1.2 cm for the capacitor voltage display to reach 3.8 cm. If the oscilloscope time axis has a calibration of 100 μs/cm, determine:

    a.   The time constant of the circuit.

    b.   The capacitance if $R = 100$ Ω.

    c.   The maximum permitted frequency of the signal generator.

22-18. A series RC circuit of unknown resistance and capacitance has a square wave signal generator of 10-V amplitude applied to it. A 10-Ω resistor is connected in series with the generator, and an oscilloscope across the resistor displays a peak value of 50 mV. The waveform drops to 18.5 mV in a horizontal distance of 2.2 cm. If the time axis of the oscilloscope has a calibration of 5 μs/cm, calculate:

    a.   The peak capacitor current.
    b.   The resistance of the original circuit.
    c.   The time constant of the circuit.
    d.   The capacitance of the circuit.

22-19. A differentiating circuit consists of a 10-Ω resistor in series with a 0.15-μF capacitor. Determine:

    a.   The maximum frequency up to which this circuit should operate satisfactorily.
    b.   The approximate slope of the input square wave voltage if the peak resistor voltage is 0.5 V.

22-20. Refer to Fig. 22-16. Assuming that it requires one time constant for the capacitor to charge to the firing voltage of the neon lamp, and the discharge of the capacitor takes only one-tenth of this time, determine the maximum and minimum flashing rate possible with the given circuit.

22-21. Repeat Example 22-10 for a load resistance of 1 kΩ.

22-22. A 100-Ω, 12-V dc relay will operate satisfactorily at 10 V. It is required to delay the operation of the relay by 10 ms when the voltage is applied by connecting a 20-Ω resistor in series with the relay and a capacitor in parallel with the relay. Determine the necessary size of capacitor. (*Hint:* With the capacitor removed, obtain the Thévenin equivalent circuit [as seen by the capacitor] for the 12-V source in series with the 20-Ω resistor and 100-Ω relay coil.)

22-23. Solve Problem 22-13 using exponential equations.

22-24. Solve Problem 22-14 using exponential equations.

22-25. Solve Problem 22-15 using exponential equations.

22-26. Repeat Problem 22-15 with $C = 0.3$ μF using exponential equations.

22-27. Given the circuit in Fig. 22-18 and a 10-V peak square wave input applied as shown, with a frequency of 500 Hz. Calculate:

    a.   The capacitor voltage at the end of 2.5 periods.
    b.   The capacitor current at the end of 2.5 periods.

    Sketch the waveforms, indicating starting and ending values at each half-period.

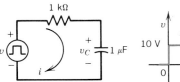 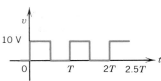

FIGURE 22-18
Circuit and waveform for Problems 22-27 and 22-28.

22-28. Repeat Problem 22-27, using $f = 1$ kHz.

22-29. Study the circuit in Fig. 22-19, considering both capacitors initially uncharged. If the switch is thrown to position 2, then after 3 s returned to position 1, calculate:

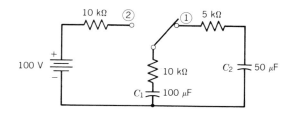

FIGURE 22-19
Circuit for Problems 22-29 and 22-30.

a. The initial capacitor current when the switch reaches position 2.
b. The voltage that C_1 charges to with the switch in position 2.
c. The capacitor current at the instant just before the switch is returned to position 1.
d. The charge stored on C_1 while the switch was in position 2.
e. The initial discharge current of C_1 when the switch is returned to position 1.
f. The discharging time constant.
g. The final voltage of both capacitors. (*Hint:* The initial charge on C_1 must equal the sum of the final charges on C_1 and C_2.)

22-30. Repeat Problem 22-29 with $C_1 = 50 \ \mu F$ and $C_2 = 100 \ \mu F$.

CHAPTER 23

CAPACITANCE IN AC CIRCUITS

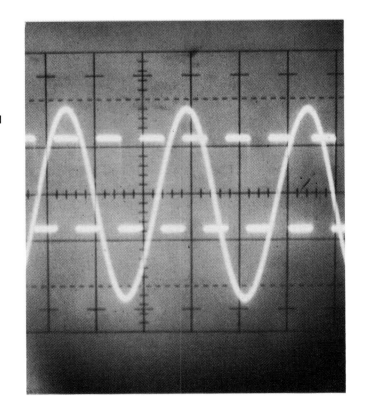

In this chapter, you will consider the current and voltage relationships of a capacitor in a sinusoidal ac circuit. Because the voltage changes *continuously,* there is some rms capacitor current at all times. The ratio of capacitor voltage to current is called *capacitive reactance* (X_C) and is the opposition in ohms provided by the capacitor. The reactance is inversely proportional to both the frequency and the capacitance. Thus, for direct current, the capacitor is an *open circuit.* This allows it to be used to "block" direct current but "pass" alternating current.

The series and parallel connections of capacitive reactances can be treated in a way similar to resistances and inductances. But (as in dc circuits) it is found that a higher voltage occurs across the *smaller* capacitor in a series circuit.

Finally, the *dissipation factor* (*D*) of a capacitor will be defined. This is an indication of the total losses that take place in a capacitor at a given frequency, and includes the *dielectric hysteresis* effect produced by the alternating electric field.

23-1 EFFECT OF A CAPACITOR IN AN AC CIRCUIT

When a capacitor is connected across a sinusoidally varying voltage, it will show a pattern of continuous charging and discharging. Consequently, an ac ammeter connected in series with the capacitor will indicate an effective value of current. Even though it may appear that current is *flowing through* the capacitor, the charge is actually being transferred from one capacitor plate to the other, back and forth, *through the ac supply*. In this section, you will determine the relationship between the capacitor voltage and current, and the opposition to current that the capacitor provides.

23-1.1 Phase Angle Relationship Between Voltage and Current

Assume that a sinusoidal voltage is applied to a capacitor, as in Fig. 23-1. The voltage waveform is given by

$$v = V_m \sin \omega t$$

You can deduce the waveform of the resulting current by using the relationship

$$i = C \left(\frac{\Delta v}{\Delta t} \right) \qquad (22\text{-}1)$$

Note, in Fig. 23-1b, that at instants 2 and 4 of the voltage waveform, the slope $\frac{\Delta v}{\Delta t}$ is zero. Hence, the current at these instants must also be zero. At instants 1 and 5 the voltage curve has its maximum *positive* slope $\left(\frac{\Delta v}{\Delta t} \right)$, so the current must be at its positive peak (I_m). Similarly, when the voltage is *decreasing* at its maximum rate (at instant 3), the capacitor current must have its maximum *negative* value ($-I_m$).

The manner in which the current varies between these instants can be shown to follow a cosine curve given by

$$i = I_m \cos \omega t$$

The waveform of the current is of the same general shape (and frequency) as the sine curve of voltage, but is displaced one-quarter-cycle to the left. Note that the current waveform reaches its peak value one-quarter of a cycle *before* the voltage waveform. **Thus, you could say that the current in a capacitor *leads* the voltage across the capacitor by 90° (assuming that the capacitor itself**

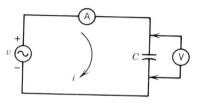

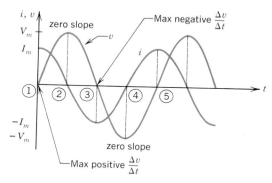

(a) Pure capacitance in an ac circuit

(b) Waveforms of current and voltage

FIGURE 23-1

Effect of a capacitor in an ac circuit.

has negligible series resistance, which is usually the case). This leading angle is referred to as the *phase angle*, because it shows by what angle the current and voltage waveforms are *out of phase*.

It is easy to remember that the current leads the voltage if you recall what happens in a capacitive dc circuit. When the voltage is first applied, the maximum charging current flows *immediately* (and gradually decreases to zero), while the capacitor voltage builds up *gradually* to maximum from a zero beginning.

23-2 CAPACITIVE REACTANCE

Since both the current and voltage waveforms in Fig. 23-1b are sinusoidal, you can obtain their rms values:

$$V = \frac{V_m}{\sqrt{2}} \qquad (15\text{-}17)$$

and

$$I = \frac{I_m}{\sqrt{2}} \qquad (15\text{-}16)$$

These are the values of the voltmeter and ammeter readings in Fig. 23-1a. The ratio of these two $\left(\frac{V}{I} \right)$ is called the opposition to current, or *capacitive reactance* (X_C). That is:

$$X_C = \frac{V_C}{I_C} \quad \text{ohms} \quad (23\text{-}1)$$

where: X_C is the capacitive reactance, in ohms (Ω)

V_C is the voltage across the pure capacitance, in volts (V)

I_C is the capacitor current, in amperes (A)

EXAMPLE 23-1

A sinusoidal voltage of peak-to-peak value 100 V is connected across a capacitor, as shown in Fig. 23-2. A series-connected ac milliammeter indicates 50 mA. What is the opposition to the current caused by the capacitor?

Solution

$$V = \frac{V_{p\text{-}p}}{2\sqrt{2}} \quad (15\text{-}19)$$

$$= \frac{100 \text{ V}}{2\sqrt{2}} \approx 35 \text{ V}$$

$$X_C = \frac{V_C}{I_C} \quad (23\text{-}1)$$

$$\approx \frac{35 \text{ V}}{50 \times 10^{-3} \text{ A}} \approx 700 \ \Omega$$

EXAMPLE 23-2

What is the capacitor current when a 120-V, 60-Hz source is applied to a capacitor whose reactance is 200 Ω?

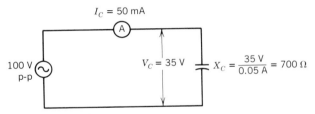

FIGURE 23-2

Circuit for Example 23-1, showing how $X_C = \dfrac{V_C}{I_C}$.

Solution

$$I_C = \frac{V_C}{X_C} \quad (23\text{-}1)$$

$$= \frac{120 \text{ V}}{200 \ \Omega} = \textbf{0.6 A}$$

EXAMPLE 23-3

What peak voltage occurs across a capacitor of reactance 2 kΩ when the capacitor current is 25 mA?

Solution

$$V_C = I_C X_C \quad (23\text{-}1)$$

$$= 25 \times 10^{-3} \text{ A} \times 2 \times 10^3 \ \Omega$$

$$= 50 \text{ V}$$

Peak capacitor voltage, $V_m = \sqrt{2} \times V_C \quad (15\text{-}18)$

$$= \sqrt{2} \times 50 \text{ V}$$

$$= \textbf{70.7 V}$$

23-3 FACTORS AFFECTING CAPACITIVE REACTANCE

The capacitor current depends upon the capacitance and the rate of change of capacitor voltage, as given by

$$i_C = C\left(\frac{\Delta v_C}{\Delta t}\right) \quad (22\text{-}1)$$

For a given supply voltage and frequency, if the capacitance (C) is *increased,* a larger current must flow. This means a *decrease* in capacitive reactance, so X_C is *inversely proportional* to C.

Now, consider an increase in the frequency (f) of the capacitor supply voltage. This means that the voltage across the capacitor must reach its *maximum* value in a *shorter* time (reduced period T). This leads to an *increase* in the maximum $\dfrac{\Delta v_C}{\Delta t}$ and an *increase* in the peak current (I_m). Once again, a reduction in X_C results, so X_C is *inversely proportional* to the frequency (f) as well as to C.

Combining these two effects and using the angular frequency ($\omega = 2\pi f$), you obtain:

$$X_C = \frac{1}{2\pi f C} = \frac{1}{\omega C} \quad \text{ohms} \quad (23\text{-}2)$$

where: X_C is the capacitive reactance, in ohms (Ω)
f is the frequency, in hertz (Hz)
C is the capacitance, in farads (F)
ω is the angular frequency, in radians per second (rad/s)

NOTE This equation is valid only for sinusoidal ac circuits. A full derivation of X_C is given in Appendix L.

EXAMPLE 23-4

What capacitive reactance will a 0.5-μF capacitor have at the following frequencies?
a. 60 Hz
b. 1 kHz

Solution

a. $X_C = \dfrac{1}{2\pi f C}$ (23-2)

$= \dfrac{1}{2\pi \times 60 \text{ Hz} \times 0.5 \times 10^{-6} \text{ F}}$

$= \mathbf{5305\ \Omega}$

b. $X_C = \dfrac{1}{2\pi f C}$ (23-2)

$= \dfrac{1}{2\pi \times 1 \times 10^3 \text{ Hz} \times 0.5 \times 10^{-6} \text{ F}}$

$= \mathbf{318\ \Omega}$

EXAMPLE 23-5

A capacitor draws 1.5 A from a 120-V, 60-Hz supply. Calculate the capacitance.

Solution

$X_C = \dfrac{V}{I}$ (23-1)

$= \dfrac{120 \text{ V}}{1.5 \text{ A}} = 80\ \Omega$

$X_C = \dfrac{1}{2\pi f C}$ (23-2)

$C = \dfrac{1}{2\pi f X_C}$

$= \dfrac{1}{2\pi \times 60 \text{ Hz} \times 80\ \Omega}$

$= 33 \times 10^{-6} \text{ F} = \mathbf{33\ \mu F}$

23-3.1   Comparison of Capacitance in DC and AC Circuits

Equation 23-2 and Example 23-4 should make clear that the capacitive reactance *decreases* as the frequency *increases*.

The graph in Fig. 23-3 shows that a capacitor approaches a *short circuit* (very low reactance) at *high* frequencies; at *low* frequencies, it approaches an *open circuit*. To put it another way, when $f = 0$, the capacitive reactance is *infinite*. This means that in a dc circuit ($f = 0$), no current flows, except for a momentary transient current when the voltage is first applied. Thus, a capacitor is said to *block* direct current but *allow* the flow of alternating current. The circuits in Fig. 23-4 show this.

In Fig. 23-4a, the reactance of the 50-μF capacitor at 60 Hz is only 53 Ω. This is negligible compared with the 576-Ω resistance of the lamp; therefore, the rated lamp current of 0.21 A flows, and the lamp burns brightly. But if a capacitor smaller than 50-μF is used, the reactance will be higher and the current in the circuit will be smaller; the lamp will burn less brightly. (The combined opposition to alternating current due to capacitance and resistance is called *impedance*. This is covered in Section 24-6.)

When a dc voltage is applied, as in Fig. 23-4b, no steady current flows, regardless of the size of the capacitor. The lamp does not light at all.

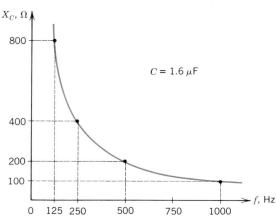

FIGURE 23-3
Decrease of X_C with increase in frequency f for a constant capacitance of 1.6 μF.

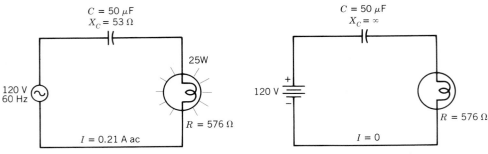

(a) With alternating current the low reactance of the capacitor allows a current of 0.21 A. The lamp lights.

(b) With direct current, there is an effective open circuit and there is no current. The lamp does not light.

FIGURE 23-4

Comparison of capacitance in ac and dc circuits.

23-3.2 Application as a Coupling Capacitor

An ac amplifier makes use of the capacitor's ability to block direct current and "pass" alternating current. When an amplified voltage is coupled from one stage of amplification to the next, a coupling capacitor may be used to block the direct current (from the biasing of the previous stage) and allow the ac component of signal voltage to be applied to the next stage. (See Fig. 23-5.)

If the coupling capacitor's reactance is small (compared with R), practically all of the ac input voltage appears across R for further amplification.

23-3.3 Application of Capacitors in Tone Controls*

An audio amplifier is usually designed to produce *equal* amplification of all frequencies over the nominal hearing range of 20 Hz to 20 kHz. This is called a *flat response*. However, the human ear is not equally responsive to all frequencies, especially when the volume of the sound is quite low. As a result, the low (bass) and high (treble) frequencies of music do not *seem* as loud as the mid-range

---
*This section may be omitted with no loss of continuity.

frequencies. Since these low and high frequencies (especially the *bass* frequencies) add depth and fullness to the music, it would be convenient if they could be accentuated (further amplified) without increasing the *overall* volume.

Tone controls provide selective amplification of various *bands* of frequencies. Figure 23-6 shows a three-position tone-control switch. In this circuit, the capacitor and resistor shown in Fig. 23-5 are *interchanged*. The circuit thus tends to allow the passage of low frequencies from V_i to V_o, but (depending on the size of the capacitor selected) shorts to ground, or *attenuates*, higher frequencies.

Consider the operation of the circuit when the switch is placed in the "flat" position, connected to the 0.01-μF capacitor. (See Fig. 23-6.) A voltage divider action takes place between the 5-kΩ resistor and the 0.01-μF capacitor. When the capacitive reactance is larger in comparison to the resistance, the output voltage (V_o) is practically equal to the input voltage (V_i). And, as shown in Table 23-1, for all frequencies from 100 Hz to almost 1000 Hz, $X_C \gg R$. For higher frequencies, where $X_C < R$, then $V_o < V_i$. (The way in which this voltage division takes place is covered, in equation form, in a later chapter.) Basically, at low frequencies, the capacitor is an *open circuit*, so no voltage drop takes place across the resistor ($V_o \approx V_i$). At higher frequencies, the low reactance of the capacitor causes a high-voltage drop across R (more current flows through the R and C combination), so V_o is small. The

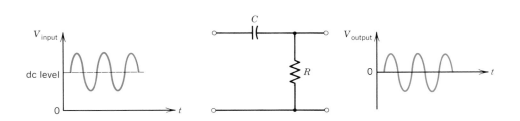

FIGURE 23-5

Blocking action of the capacitor couples the ac component of the input voltage to R but not the dc component.

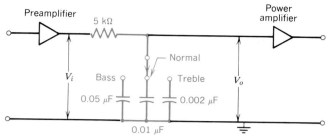

FIGURE 23-6

Three-position tone-control circuit.

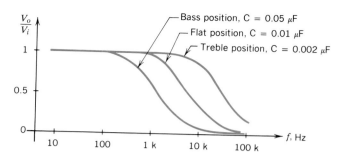

FIGURE 23-7

Frequency response curve for tone-control circuit of Fig. 23-6.

way in which the ratio of output voltage to input voltage (V_o/V_i) varies with frequency is shown in Fig. 23-7.

When the switch is placed in the "bass" position, the larger capacitor causes a shorting effect to occur at lower frequencies. As shown in Table 23-1, the capacitive reactance is now 3.2 kΩ at only 1000 Hz, rather than 5000 Hz. Thus, as Fig. 23-7 shows, only the relatively low frequencies are "passed on" to the power amplifier; the higher frequencies are attenuated. Since the higher frequencies are de-emphasized, this has the effect of *emphasizing* the low (bass) frequencies. Actually, as can be seen in the graphs of Fig. 23-7, frequencies of 100 Hz and lower are amplified approximately equally by the next stage, no matter what the position of the tone control. But when the control is in the bass position, the higher frequencies are attenuated, so the lower frequencies dominate.

In the "treble" position, the small (0.002-μF) capacitor requires the frequency to be much higher before a

shorting effect occurs. The high, or treble, frequencies will now be heard along with the low bass frequencies. (See Fig. 23-7.) Actually, with the switch set in the treble position, no capacitor is needed—just an open circuit. However, in some radios and other playback devices with small speakers that do not handle high frequencies very well, a small capacitance is usually introduced to reduce the "tinniness" of the sound from the small speaker.

An inexpensive tone control that provides continuous bass-to-treble control is shown in Fig. 23-8. When this control is moved upward to the bass position, the capacitor provides a shorting effect for the high frequencies (and thus, the *effect* of emphasizing the bass frequencies). In the treble position, the bypassing of high frequencies to ground is less effective than in the bass position. The high frequencies are thus amplified along with the bass frequencies.

A tone-control circuit with separate controls for bass

TABLE 23-1

Capacitive reactance and V_o/V_i ratio at various frequencies for each of the tone-control circuit capacitors in the circuit of Fig. 23-6

f, Hz	Bass 0.05 μF		Flat 0.01 μF		Treble 0.002 μF	
	X_C, kΩ	V_o/V_i	X_C, kΩ	V_o/V_i	X_C, kΩ	V_o/V_i
100	32	0.99	159	1.0	790	1.0
500	6.4	0.79	31.8	0.99	159	1.0
1,000	3.2	0.54	15.9	0.95	79	0.99
5,000	0.64	0.13	3.2	0.54	15.9	0.95
10,000	0.32	0.06	1.59	0.30	7.9	0.84
20,000	0.16	0.03	0.79	0.16	3.9	0.62

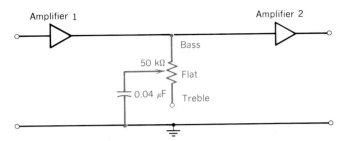

FIGURE 23-8
Continuous tone control.

"cut," the 50-kΩ resistor drops the high-frequency voltage fed to *M*, causing more attenuation.

Low- and mid-frequencies, which cannot pass through the 0.002-μF capacitor in the treble control, pass through the 12-kΩ resistor in the bass control. Mid-range frequencies pass through the 0.015-μF capacitor relatively unaffected. The low frequencies, however, can be boosted by the bass control being moved upward (+) or cut by the same control being moved downward (−). Again, a voltage dividing action is used.

and treble is shown in Fig. 23-9. Emphasis is provided for low or high frequencies, independently of each other. The "boosting" or "cutting" of these frequencies is accomplished with reference to the *flat response* (0) position on the two controls. Figure 23-10 shows the frequency response curve for this circuit, in decibels, with 0 dB as the reference level. A change of 3 dB corresponds to a doubling of sound intensity, so a 9-dB change means that intensity is eight times the initial level.

Note that, in Fig. 23-9, the output of amplifier 1 is fed through a 1-μF coupling capacitor to *both* the treble and bass controls. The outputs from these controls are "mixed" at point *M*, fed to the volume control, and, ultimately, to the power amplifier (amplifier 2).

If the treble control is moved toward "boost," the high frequencies can more easily pass directly through to point *M*, with less attenuation. If the control is moved toward

23-4 SERIES CAPACITIVE REACTANCES

When capacitors are *series-connected*, the total reactance is

$$X_{C_T} = X_{C_1} + X_{C_2} + X_{C_3} + \cdots + X_{C_n} \quad (23\text{-}3)$$

The current for a given applied voltage can be obtained from

$$I = \frac{V}{X_{C_T}} \quad (23\text{-}4)$$

and the voltage across each capacitor from $V_C = I X_C$.

NOTE When two or more capacitors are connected in series, the overall capacitance drops. However, this causes an overall increase in reactance (compared to one capacitor). This explains why the total reactance is *additive*, as in Eq. 23-3.

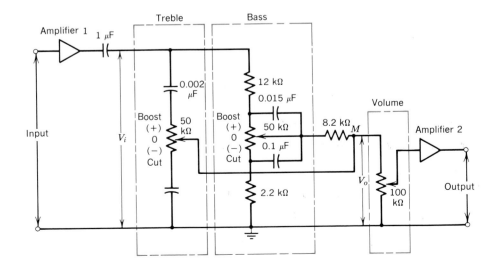

FIGURE 23-9
Tone-control circuit with separate bass and treble controls.

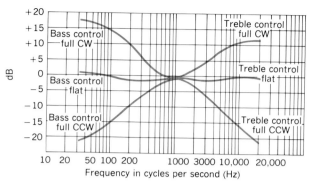

FIGURE 23-10

Frequency response curve for tone-control circuit in Fig. 23-9.

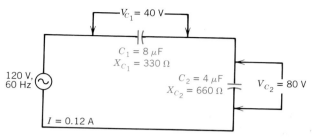

FIGURE 23-11

Circuit for Example 23-6.

Note that the voltage divides in such a way that the highest voltage (80 V) appears across the smallest capacitor (4 μF), as is the case with dc voltages.

EXAMPLE 23-6

Two capacitors of 8 and 4 μF are series-connected across a 120-V, 60-Hz supply as in Fig. 23-11. Calculate:

a. The total capacitive reactance.
b. The current drawn by the capacitors.
c. The voltage across each capacitor.

Solution

a. $X_{C_1} = \dfrac{1}{2\pi f C_1}$ (23-2)

$\quad = \dfrac{1}{2\pi \times 60\ \text{Hz} \times 8 \times 10^{-6}\ \text{F}}$

$\quad \approx 330\ \Omega$

$\quad X_{C_2} = \dfrac{1}{2\pi f C_2}$ (23-2)

$\quad = \dfrac{1}{2\pi \times 60\ \text{Hz} \times 4 \times 10^{-6}\ \text{F}}$

$\quad \approx 660\ \Omega$

$\quad X_{C_T} = X_{C_1} + X_{C_2}$ (23-3)

$\quad \approx 330\ \Omega + 660\ \Omega \approx \mathbf{990\ \Omega}$

b. $I = \dfrac{V}{X_{C_T}}$ (23-4)

$\quad = \dfrac{120\ \text{V}}{990\ \Omega} \approx \mathbf{0.12\ A}$

c. $V_{C_1} = I X_{C_1}$ (23-1)

$\quad = 0.12\ \text{A} \times 330\ \Omega \approx \mathbf{40\ V}$

$\quad V_{C_2} = I X_{C_2}$ (23-1)

$\quad = 0.12\ \text{A} \times 660\ \Omega \approx \mathbf{80\ V}$

23-5 PARALLEL CAPACITIVE REACTANCES

When capacitors are connected in parallel, the overall increase in capacitance causes a decrease in reactance.

Thus, similar to resistances in parallel, the total capacitive reactance is given by

$$\frac{1}{X_{C_T}} = \frac{1}{X_{C_1}} + \frac{1}{X_{C_2}} + \frac{1}{X_{C_3}} + \cdots + \frac{1}{X_{C_n}} \quad (23\text{-}5)$$

EXAMPLE 23-7

Two capacitors of 8 and 4 μF are connected in parallel across a 120-V, 60-Hz supply as in Fig. 23-12. Calculate:

a. The total capacitive reactance.
b. The total current drawn from the supply.

Solution

a. From Example 23-6, $X_{C_1} = 330\ \Omega$

$\quad\quad\quad\quad\quad\quad X_{C_2} = 660\ \Omega$

$\quad \dfrac{1}{X_{C_T}} = \dfrac{1}{X_{C_1}} + \dfrac{1}{X_{C_2}}$ (23-5)

Therefore, $X_{C_T} = \dfrac{X_{C_1} \times X_{C_2}}{X_{C_1} + X_{C_2}}$

$\quad = \dfrac{330\ \Omega \times 660\ \Omega}{330\ \Omega + 660\ \Omega} \approx \mathbf{220\ \Omega}$

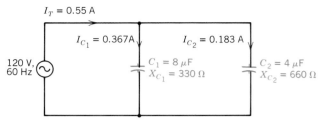

FIGURE 23-12
Circuit for Example 23-7.

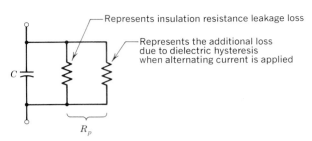

FIGURE 23-13
Ac equivalent circuit for a capacitor taking into account insulation resistance and dielectric hysteresis losses.

b. $I_T = \dfrac{V}{X_{C_T}}$ (23-4)

$= \dfrac{120 \text{ V}}{220 \text{ }\Omega} = \textbf{0.55 A}$

23-6 DISSIPATION FACTOR OF A CAPACITOR

You have learned that whenever a capacitor is charged by a dc source, the atoms of the dielectric become polarized in one direction. If alternating current is applied, the electrons of the dielectric are stressed—first in one direction, then in the other—due to the alternating electric field. This causes an energy loss in the dielectric in the form of heat. Since the effect is similar to the hysteresis loss in a magnetic core, it is called *dielectric hysteresis.*

Dielectric hysteresis can be represented in the equivalent circuit of a capacitor by connecting a resistor (an energy-dissipating element) in parallel with the capacitance. (See Fig. 23-13.)

Recall, however, that in Section 21-5.1, you considered a parallel resistance to represent the *insulation* resistance of a capacitor. The two resistance effects can be combined into a single resistance (R_p), as shown in Fig. 23-13. The dielectric hysteresis effect can be thought of as an *added ac* phenomenon, similar to ac resistance in a coil arising from magnetic hysteresis, skin effects, and so on.

Similar to the way in which you considered the *quality* (*Q*) or *storage* factor of a coil, you can obtain the *dissipation factor* (*D*) of a capacitor:

$$D = \frac{X_C}{R_p} = \frac{1}{2\pi f C R_p}$$ (23-6)

where: D is the dissipation factor of the capacitor (dimensionless)

X_C is the capacitive reactance of the capacitor, in ohms (Ω)

R_p is the total parallel ac resistance of the capacitor, in ohms (Ω)

As the name suggests, D is an indication of how much power is dissipated in a capacitor at a given frequency. A perfect, lossless capacitor would thus have a D equal to zero. If the capacitor is used only with direct current applied, of course, power dissipation would be due only to the relatively high insulation resistance (leakage current). Also, the additional power dissipated with alternating current applied depends upon the actual ac voltage across the capacitor.

The D factors for nonelectrolytic capacitors are in the range of 0.0005 for polystyrene and 0.02 for ceramics at 1 kHz. Electrolytic capacitors, however, may have D factors ranging from 0.15 at 475 WVDC up to 0.75 at 6 WVDC (specified for an ac component having a frequency of 60 Hz). It is especially important, when checking electrolytic capacitors, that the D factor be within allowable limits, since overheating can occur if there is any appreciable ac component of voltage across the capacitor. Electrolytic capacitors used in ac motor starting applications, for example, should be replaced if the dissipation factor exceeds 0.15 or 15%.

Generally, losses in capacitors (unlike those in inductors) are so small that they can be ignored, except in special cases or if problems develop.

Some capacitor-checking instruments show the *power factor* of the capacitor, rather than the dissipation factor. (For low-loss capacitors, these factors are approximately equal.) In this case, corresponding acceptable values of the power factor range from 0.15 to 0.6 for the same working voltages mentioned above. (See Section 25-6 for a discussion of the power factor.)

Finally, it should be mentioned that the *reciprocal* of D is the quality (Q) of the capacitor. Although this term is used primarily with coils, a high Q means a low-loss capacitor.

SUMMARY

1. When a sinusoidal voltage is applied across a capacitor, it causes a sinusoidal current that leads the voltage by 90°.
2. The opposition to alternating current offered by a capacitor is called the capacitive reactance; for a sinusoidal waveform, it is given by:

$$X_C = \frac{1}{2\pi f C}$$

3. Voltage across a capacitor in an ac circuit is given by $V_C = I_C X_C$.
4. A capacitor blocks direct current but passes alternating current, permitting it to be used for coupling purposes in an ac amplifier.
5. Tone-control circuits make use of the variable reactance of a capacitor at different frequencies to emphasize high- or low-frequency response.
6. A series connection of capacitors gives an increase in overall capacitive reactance, similar to series-connected resistors, with the largest voltage occurring across the smallest capacitor.
7. The equations for parallel-connected capacitive reactances are similar to those for parallel-connected resistors, with the largest current flowing in the branch with the highest capacitance.
8. When a capacitor is used in an ac circuit, there is an additional loss called dielectric hysteresis. This loss is caused by the repeated reversal of the alternating electric field.

SELF-EXAMINATION

Answer T or F or a, b, c, d, or e for multiple choice
(Answers at back of book)

23-1. In a sinusoidal ac circuit, the capacitor current is a maximum when the capacitor voltage passes through zero. _____

23-2. The phase angle between the capacitor voltage and the current is always 90°, no matter what else is connected in series with the capacitor, when the applied voltage is sinusoidal. _____

23-3. The only time the capacitor current is zero in a sinusoidal ac circuit is when the capacitor voltage is at its maximum values. _____

23-4. Capacitive reactance is the ratio of the rms capacitor current to the rms capacitor voltage. _____

23-5. A 1-μF capacitor, at a frequency of 500 Hz, has a capacitive reactance of:
a.   1000 Ω
b.   1000π Ω
c.   1000/π Ω

d. 300 Ω
e. any of the above
23-6. If the capacitor in Question 23-5 is connected to a 10-V, 500-Hz source, the current is approximately:
a. 10 mA
b. 31.4 mA
c. 3.14 mA
d. 0.314 A
e. 50 mA
23-7. If the frequency of the voltage applied to a capacitor is doubled, the current is reduced 50%. _____
23-8. If a very large capacitor is used in series with a lamp, the lamp will light no matter whether the applied voltage is alternating current or direct current. _____
23-9. The use of a capacitor to couple alternating current from one stage to another in an amplifier depends upon the capacitor's ability to block alternating current and pass direct current. _____
23-10. When two equal capacitors are series-connected across a 120-V ac source, the voltage across each capacitor is 60 V. _____
23-11. When two unequal capacitors are parallel-connected across a 120-V, 60-Hz source, the smaller capacitor will have the smaller current. _____
23-12. The same type of equation is used to combine series and parallel capacitive reactances as used for combining series and parallel capacitances. _____
23-13. Dielectric hysteresis is an effect that is considered only in ac applications. _____
23-14. The dissipation factor of a capacitor is a constant—independent of the frequency of operation. _____
23-15. If the dissipation factor of a capacitor is excessive, overheating of the capacitor can occur if used with a significant alternating voltage. _____

REVIEW QUESTIONS

1. If a capacitor consists of an insulator between two metal plates, how is it possible for a series-connected ac ammeter to show a steady ac current?
2. How does the phase angle between the capacitor current and voltage in an ac circuit differ from those for a purely inductive or purely resistive circuit? Why the difference?
3. a. Why is it not correct to characterize capacitive reactance as the effective ac resistance of a capacitor?
   b. Why should an *increase* in capacitance or frequency *increase* current in a capacitive circuit if the voltage is unchanged?
4. a. In what way can direct current be thought of as zero frequency alternating current as applied to a capacitor?
   b. How much opposition does a capacitor offer to direct current?
   c. What are at least two practical applications of this result?
5. Describe how you could use a voltmeter, ammeter, and ac voltage source of known frequency to measure the capacitance of a capacitor.

6. a. What do the tone controls described in this chapter rely on to provide variable frequency response?

b. What simple reason can you give for the circuit in Fig. 23-9 providing an increase in *both* bass and treble, compared with one *or* the other in the circuit of Fig. 23-6?

7. Why is the total capacitive reactance of capacitors in series given by the *sum* of the individual reactances?

8. Why is there no mutual reactance between capacitors as there may be with inductors?

9. Justify why series-connected capacitors cause the largest voltage to occur across the smallest capacitor.

10. a. What do you understand by the term *dielectric hysteresis?*

b. When is it a factor to take into account?

c. What are typical dissipation factors for electrolytic capacitors?

d. Why do you think electrolytics have higher D values than a paper capacitor?

PROBLEMS

(Answers to odd-numbered problems at back of book)

23-1. What capacitive reactance does a capacitor have if an applied emf of 230 V causes a current of 0.4 A?

23-2. A capacitor draws 20 mA when connected to a 10-V sinusoidal supply. What is its reactance?

23-3. What is the voltage across a 0.1-μF capacitor when the current is 5 mA at 400 Hz?

23-4. If a capacitor's current is 15 mA p–p at a frequency of 1.5 kHz, what does a voltmeter indicate across the 0.015-μF capacitor?

23-5. A 20-V p–p sinusoidal voltage at 50 Hz is connected across a capacitor. If a series-connected ammeter indicates 4 mA, calculate the capacitance.

23-6. A capacitor and a 2.2-kΩ resistor are series-connected across a 60-Hz supply. A voltmeter across the capacitor indicates 3.5 V when an oscilloscope across the resistor shows a peak value of 8 V. How much capacitance does the circuit have?

23-7. A 6.3-V sinusoidal voltage is applied to a 0.5-μF capacitor. How much current flows at the following frequencies?

a. 60 Hz

b. 6 kHz

c. 600 kHz

23-8. The following voltage is applied to a 1500-pF capacitor:

$$v = 50 \sin 10{,}000\pi t \text{ volts}$$

Determine the equation of the current waveform.

23-9. What is the maximum frequency that can be applied to a 4-μF capacitor if the capacitor current must not exceed 50 mA with an applied voltage of 24 V?

23-10. What is the minimum frequency that can be applied with a 10-μA constant-current, sine wave signal generator to a 0.002-μF capacitor to limit the peak capacitor voltage to 5 V?

23-11. Three capacitors—0.1 μF, 0.15 μF, and 0.2 μF—are series-connected across a 24-V, 400-Hz supply. Calculate:

    a.   The total capacitive reactance.

    b.   The current drawn by the capacitors.

    c.   The voltage across each capacitor.

    d.   The total capacitance of the circuit.

23-12. Three capacitors are connected in series across a 120-V, 60-Hz supply. The voltage across the first capacitor is 35 V; the capacitance of the second is 1.5 μF; and the current through the third is 15 mA. Calculate the capacitance of the first and third capacitors.

23-13. Three capacitors—0.1 μF, 0.15 μF, and 0.2 μF—are parallel-connected across a 48-V, 60-Hz supply. Calculate:

    a.   The total capacitive reactance.

    b.   The current drawn by each capacitor.

    c.   The total current drawn by the capacitors.

    d.   The total capacitance of the circuit.

23-14. Three capacitors are connected in parallel across a 240-V, 50-Hz supply, and draw a total current of 4 A. If the first capacitor is 20 μF, and the second capacitor has a current of 1.5 A, determine the capacitance of the second and third capacitors.

23-15. A 40-μF electrolytic capacitor dissipates a power of 1 W when used in a filter circuit with an ac voltage of 10 V at a frequency of 120 Hz. Calculate:

    a.   The effective parallel resistance R_p that determines the capacitor's losses.

    b.   The dissipation factor of this capacitor at this frequency.

CHAPTER 24

PHASORS IN ALTERNATING CURRENT CIRCUITS

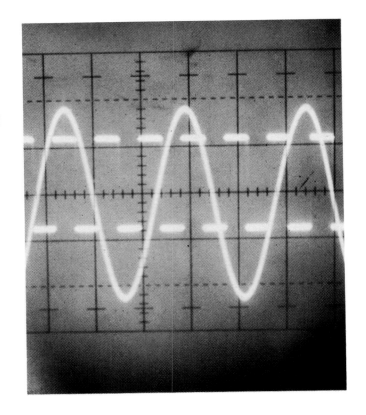

In this chapter, you will learn about the method of representing voltage and current waveforms using *phasor notation*. In this method, two phasors drawn at right angles to one another represent the phase angle relationships between current and voltage in pure *L* or *C* circuits.

When resistance and inductance are connected in series, their individual voltages must be added taking into account their phase angles. This leads to the expression for the total opposition to current, called *impedance* (*Z*), where R and X_L must be added using the square root of the sum of the squares of the individual oppositions. A similar expression may be used for a series *RC* circuit, except that the current *leads* the applied voltage in a capacitive circuit. In an inductive circuit, the current *lags* the applied voltage.

In a *series RLC* circuit, a *canceling* effect occurs between the inductive and capacitive reactances, causing an overall reduction of the total impedance. If $X_L = X_C$, the circuit displays a purely resistive property (a condition called *resonance*).

The *parallel* connection of R, L, and C is analyzed by obtaining the total current and obtaining the impedance by an application of "Ohm's law for an ac circuit" ($Z = V/I$). Finally, you will examine methods of measuring a circuit's phase angle between current and applied voltage. One method involves a *Lissajous* pattern on a single-beam oscilloscope; the other requires a dual-trace (dual-beam) oscilloscope.

24-1 REPRESENTATION OF A SINE WAVE BY A PHASOR

A convenient way to represent each waveform, when showing the phase angle relationship between the current and voltage in an ac circuit, is to use a line with an arrow, called a *vector* or *phasor*.* The length of the line represents the magnitude, while the direction of the line represents the phase angle with respect to some reference.

Consider the radius of a circle, initially in a horizontal position, then rotated counterclockwise through the full 360°. If you consider the radius at various angles (for example, every 30°), and project lines horizontally for each angle, you can plot a graph of the vertical distance from the end of this line to the horizontal axis against the angle (using some suitable scale for the x-axis in degrees). The result is shown in Fig. 24-1.

---

*A *vector* has both magnitude and direction, and is used to represent physical quantities, such as a force or velocity. A *phasor* also has magnitude and direction, but it represents a quantity that also *varies* in magnitude with *time* (such as an alternating voltage or current).

It should be clear, from right-angle trigonometry, that the vertical distances plotted are proportional to the *sine* of each angle involved. (Recall that $\sin\theta$ is the ratio of the opposite side to the hypotenuse in a right triangle. Since the hypotenuse in this case is always the same length—the radius of the circle—the opposite side is directly proportional to the sine of the angle θ.) Thus, the waveform produced by connecting each of the plotted points is a *sine wave*, with an amplitude or peak value equal to the length of the phasor (the radius of the circle). If very small angles, such as every 5°, are used for the projection, a very accurate sine wave can be drawn.

Therefore, you can use a horizontal line (phasor) to represent a sinusoidal waveform. The phasor is always assumed to rotate counterclockwise, at a speed equal to the frequency of the waveform. Strictly speaking, the length of the phasor should be the *peak* value of the waveform, but you can choose the length to represent the *rms* value of any given quantity, since this is generally what is used and measured in ac circuits.

Next, consider the waveform represented by a phasor drawn vertically upward and also assumed to rotate coun-

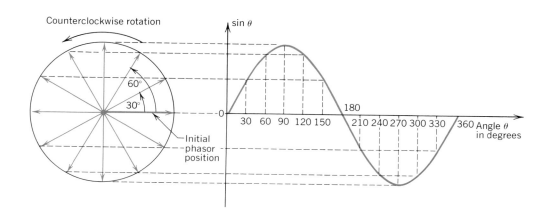

FIGURE 24-1

Showing how a sine wave can be constructed by using horizontal projections from a rotating radius vector.

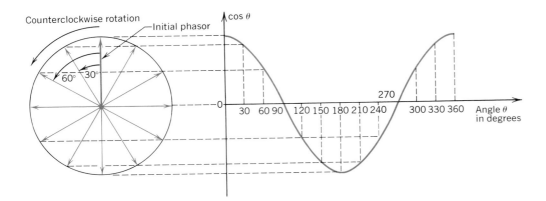

FIGURE 24-2
Showing how a vertical phasor represents a cosine curve.

terclockwise through a complete circle. Since the original (reference) phasor is assumed to start at 0°, this second waveform starts at its maximum vertical height. As the phasor rotates through 90°, the horizontal projections show a *decreasing* vertical height, reaching zero in one-quarter cycle. As shown in Fig. 24-2, the waveform now goes *negative* for 180°, then *positive* for the final quarter-cycle.

The resulting waveform is a *cosine* curve—a sine curve moved *ahead* by 90°, or *leading* a sine curve by one-quarter cycle. If two phasors (*A* and *B*) are drawn as a single diagram, with *A* ahead of *B* by 90°, the diagram can represent two waveforms that are 90° out of phase. (See Fig. 24-3a.)

Since the phasors are *always* assumed to rotate counterclockwise, it is obvious that *A* leads *B* by 90° and that *B* lags *A* by 90°. This *phase angle* relationship is shown in Fig. 24-3b, along with the waveforms represented by the phasors. Note the difference in magnitude and phase relations between the two waves. Both phasor and waveform representations are used: the phasor diagram for sinusoidal waves at the *same* frequency, and the waveforms for both

sinusoidal and nonsinusoidal waves at *any* frequency. Phasor diagrams are easier to draw and interpret, however, than the waveforms they represent.

24-2 VOLTAGE AND CURRENT PHASORS IN PURE *RLC* CIRCUITS

Figure 24-4 shows how phasors can be used to represent the voltage and current waveforms in sinusoidal ac circuits with pure resistance, inductance, or capacitance at a given frequency.

Since the same voltage is applied to each circuit, the *voltage* phasor has been drawn in the *horizontal* reference position for each. For pure resistance, the current and voltage are in phase, so the phasors are drawn ''in line'' with each other, indicating a phase angle of 0°. (See Fig. 24-4a.)

For a purely inductive circuit, current *lags* voltage by 90°, as the phasor diagram in Fig. 24-4b clearly shows.

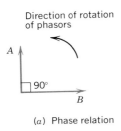

(a) Phase relation

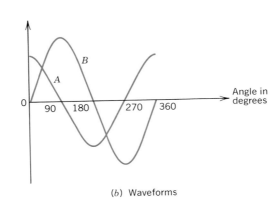

(b) Waveforms

FIGURE 24-3
Showing how two phasors *A* and *B*, with *A* leading *B* by 90°, represent two waveforms out of phase by 90°.

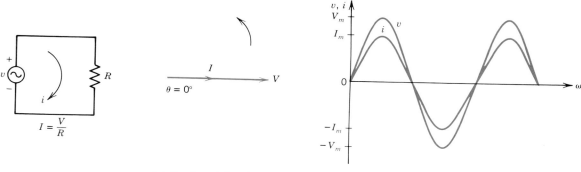

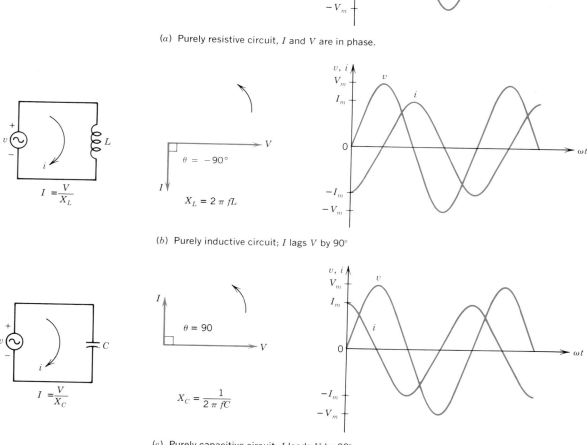

(a) Purely resistive circuit, *I* and *V* are in phase.

(b) Purely inductive circuit; *I* lags *V* by 90°

(c) Purely capacitive circuit, *I* leads *V* by 90°

FIGURE 24-4

Phasor representations and waveforms of *V* and *I* in pure resistance, inductance, and capacitance circuits.

That is, since the phasors are assumed to rotate counter-clockwise, *I* is *behind V* (or *lagging V*) by 90°. Alternatively, it could be said that *V* leads *I* by one-quarter cycle.

Similarly, for a purely capacitive circuit, current *leads* voltage by one-quarter cycle. This is shown by the phasor diagram in Fig. 24-4c, where the phasor *I* has been drawn

90° *ahead* of *V*. Alternatively, *V lags I* by 90°—phasor *V* is *behind* phasor *I*.

Note how much easier it is to describe the phase relationship between *V* and *I* by looking at the phasor diagram than the waveforms of *v* and *i*.

Also note that the current phasor has been drawn arbitrarily shorter than the voltage phasor. Except in cases where the phasors are drawn to scale in some graphical solutions, the lengths of the phasor lines are not important. They are usually indicated with rms values and identified with capital letters, as shown.

24-3 SERIES *RL* CIRCUIT

Now the principle of phasor representation can be applied to a series circuit containing pure resistance and pure inductance. (See Fig. 24-5.)

Since a series circuit is being considered, it is convenient to draw the current phasor (which is "common" to both resistor and inductor) in the horizontal reference po-

sition. Superimposed upon the current phasor is the voltage phasor across the resistor (V_R). This is done because, in a pure resistor, the current and voltage are always in phase with each other.

In the same way, the voltage phasor across the inductor (V_L) is drawn 90° ahead of, or *leading*, the current phasor. This is done because the current always lags the inductor voltage by 90° in a pure inductance.

Assume, for ease of discussion, that the resistance (R) is equal to the inductive reactance (X_L) of the coil at the frequency of the applied voltage. In Fig. 24-5a, $R = X_L = 50\ \Omega$. Then, since both components have the same current through them ($I = 2$ A in this example), they must have the same magnitude (size) of voltage across them. That is,

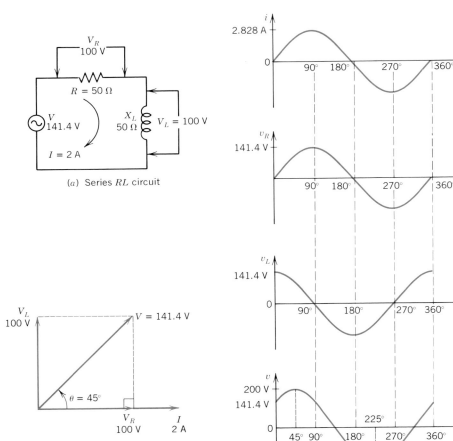

(a) Series *RL* circuit

(b) Phasor diagram

(c) Current and voltage waveforms

FIGURE 24-5

Series *RL* circuit with phasor diagram and waveforms.

$$V_R = I_R R \qquad \text{(15-15b)}$$
$$= 2 \text{ A} \times 50 \text{ }\Omega = 100 \text{ V}$$
$$V_L = I_L X_L \qquad \text{(20-1)}$$
$$= 2 \text{ A} \times 50 \text{ }\Omega = 100 \text{ V}$$

These two voltages, however, are 90° out of phase with each other. This means that the total voltage across the series combination cannot be obtained by simply adding V_R to V_L algebraically.* The *angle* between them must be taken into account. Kirchhoff's voltage law applies to this circuit as it does to any circuit. The applied voltage (V) is the (phasor) sum of V_R and V_L, with the phase angle also involved in the addition. This is known as *vector* or *phasor addition*.

This phasor addition can be carried out simply by constructing a parallelogram (a square, in this case) and drawing the diagonal. This is shown in Fig. 24-5b. Clearly, the *phasor* sum V is less than the *algebraic* sum of V_L and V_R. Also, because V is the hypotenuse of a right-angled triangle, it is given by

$$V = \sqrt{V_R^2 + V_L^2} \qquad \text{volts} \qquad \textbf{(24-1)}$$

where symbols are as above.
For the numerical values,

$$V = \sqrt{(100 \text{ V})^2 + (100 \text{ V})^2}$$
$$= \sqrt{10,000 + 10,000} \text{ V}$$
$$= \sqrt{20,000} \text{ V} = \textbf{141.4 V}$$

Thus the rms value of the applied voltage is 141.4 V, with a peak value of 200 V. (The waveforms in Fig. 24-5c show the peak values and how they would be displayed on an oscilloscope.)

The phase angle θ by which the current lags the applied voltage can now be obtained from the right triangle in Fig. 24-5b:

$$\tan \theta = \frac{V_L}{V_R} \qquad \text{(24-2)}$$

or

$$\theta = \tan^{-1} \frac{V_L}{V_R} \qquad \text{or arc tan } \frac{V_L}{V_R} \qquad \text{(24-2a)}$$

where symbols are defined as above.
The symbols $\tan^{-1}$ and arc tanV_L/V_R mean that θ is the

angle whose tangent is V_L/V_R.* It may be obtained with a scientific calculator using the INV and TAN buttons as follows:

For the numerical values,

$$\theta = \tan^{-1} \frac{100 \text{ V}}{100 \text{ V}}$$
$$= \tan^{-1} 1$$

Enter 1 into the calculator, press INV, then press TAN. The result is

$$\theta = \textbf{45°}$$

Note that in Fig. 24-5c, the applied voltage (V) reaches its peak value 45° before the current (I) does. This means that V leads I by 45°, or I lags V by 45°.

This result should be no surprise. If a circuit is purely resistive, the phase angle is 0°; if a circuit is purely inductive, the phase angle is 90°. Thus, if the circuit is made up of equal values of resistance and inductance, you should expect the angle to be midway between 0° and 90°, or **45°**.

EXAMPLE 24-1

A coil of negligible resistance and a 100-Ω resistor are series-connected across a 120-V, 60-Hz supply. A voltmeter connected across the resistor indicates 60 V. Calculate:

a. The reading of a voltmeter connected across the coil.
b. The phase angle between the applied voltage and the current.
c. The current in the circuit.
d. The inductance of the coil.

Solution

See Fig. 24-6.
a. $\quad V = \sqrt{V_R^2 + V_L^2} \qquad \text{(24-1)}$
$\quad V^2 = V_R^2 + V_L^2$
Therefore,
$$V_L = \sqrt{V^2 - V_R^2}$$
$$= \sqrt{(120 \text{ V})^2 - (60 \text{ V})^2}$$
$$= \sqrt{14,400 - 3600} \text{ V}$$
$$= \sqrt{10,800} \text{ V} = \textbf{104 V}$$

---

*You can, however, add the instantaneous values of v_R and v_L algebraically at *each instant of time* to obtain v in Fig. 24-5c.

---

*See Appendix F, inverse trigonometric functions.

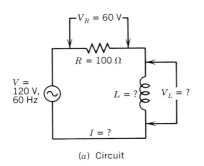

(a) Circuit

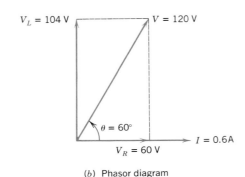

(b) Phasor diagram

FIGURE 24-6

Circuit and phasor diagram for Example 24-1.

b. $\theta = \tan^{-1} \dfrac{V_L}{V_R}$ (24-2a)

$= \tan^{-1} \dfrac{104\ \text{V}}{60\ \text{V}}$

$= \tan^{-1} 1.73 = \mathbf{60°}$

c. $I = \dfrac{V_R}{R}$ (15-15a)

$= \dfrac{60\ \text{V}}{100\ \Omega} = \mathbf{0.6\ A}$

d. $X_L = \dfrac{V_L}{I}$ (20-1)

$= \dfrac{104\ \text{V}}{0.6\ \text{A}} = 173\ \Omega$

$= 2\pi f L$ (20-2)

Therefore,

$$L = \dfrac{X_L}{2\pi f}$$

$$= \dfrac{173\ \Omega}{2\ \pi \times 60\ \text{Hz}} = \mathbf{0.46\ H}$$

24-4 IMPEDANCE OF A SERIES *RL* CIRCUIT

The total opposition to current in a series *RL* circuit is called the impedance (*Z*), and is the ratio of the total applied voltage (*V*) to the current (*I*). Impedance is measured in ohms (as are resistance and inductive reactance). But, as shown by the following, impedance is the *vector* sum of resistance and reactance. Consider the "voltage triangle" for a series *RL* circuit like the one shown in Fig. 24-7a. This is similar to the phasor diagram in Fig. 24-5b with V_L transferred to make a closed triangle.

Given $V = \sqrt{V_R^2 + V_L^2}$ (24-1)

and $V_R = IR, \qquad V_L = IX_L$

then $V = \sqrt{(IR)^2 + (IX_L)^2}$

$= \sqrt{I^2 R^2 + I^2 X_L^2}$

$= \sqrt{I^2 (R^2 + X_L^2)}$

$= I\sqrt{R^2 + X_L^2}$

and $\dfrac{V}{I} = \sqrt{R^2 + X_L^2}$

But $\dfrac{V}{I}$ is the impedance (*Z*).

Therefore, $Z = \sqrt{R^2 + X_L^2}$ ohms   (24-3)

where: *Z* is the impedance, in ohms (Ω)
        R is the resistance, in ohms (Ω)
        X_L is the inductive reactance, in ohms (Ω)

and $I = \dfrac{V}{Z}$ amperes   (24-4)

where: *I* is the circuit current, in amperes (A)
        V is the applied voltage, in volts (V)
        Z is the circuit impedance, in ohms (Ω)

Equation 24-4 is often referred to as "Ohm's law for an ac circuit," since it is the Ohm's law equivalent of $I = V/R$ for a dc circuit. As you will soon see, *Z* represents the total opposition to current for any combination of *RLC*, whether in series or parallel. The expression used to obtain *Z* will vary, however, according to the circuit.

Note that the impedance diagram (Fig. 24-7c) is drawn *without* arrows and is not referred to as a phasor diagram. This is because X_L, *R*, and *Z* do not vary with *time*, as do V_L, V_R, and *V*. Note also that the phase angle θ between the current (V_R) and voltage in Fig. 24-7a can now be obtained from the impedance triangle in Fig. 24-7c. This can be done because the two triangles are *similar*, and θ is the *same*.

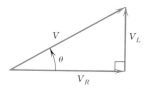

(*a*) Voltage triangle

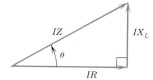

(*b*) Equivalent voltage triangle

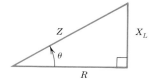

(*c*) Impedance triangle

FIGURE 24-7

Voltage and impedance triangles for a series *RL* circuit.

$$\tan\theta = \frac{X_L}{R} \qquad (24\text{-}5)$$

and
$$\theta = \tan^{-1}\frac{X_L}{R} \qquad (24\text{-}5a)$$

where: θ is the phase angle between the current and the applied voltage in a series *RL* circuit, in degrees or radians

X_L is the inductive reactance, in ohms (Ω)

R is the total circuit resistance, in ohms (Ω)

NOTE The phase angle is determined by inductive *reactance* compared with resistance, not by inductance alone. Even if the inductance and resistance remain constant, the phase angle can change if the frequency of the applied voltage changes.

As you can see from Fig. 24-7*c*, you can obtain the resistance and inductive reactance if the impedance and phase angle are known:

$$R = Z\cos\theta \qquad (24\text{-}6)$$
$$X_L = Z\sin\theta \qquad (24\text{-}7)$$

These equations will be useful when you consider practical coils, in which the coil resistance is not negligible.

EXAMPLE 24-2

A 30-mH coil of negligible resistance and a 200-Ω resistor are connected in series across a 10-V, 1-kHz sinusoidal supply. Determine:
a. The impedance of the circuit.

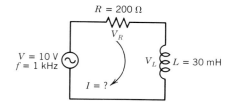

b. The current in the circuit.
c. The phase angle between the applied voltage and current.

Solution

See Fig. 24-8.
a. $X_L = 2\pi fL \qquad (20\text{-}2)$
$\quad = 2\pi \times 1 \times 10^3 \text{ Hz} \times 30 \times 10^{-3} \text{ H}$
$\quad = \textbf{188.5 } \Omega$

$Z = \sqrt{R^2 + X_L^2} \qquad (24\text{-}3)$
$\quad = \sqrt{200^2 + 188.5^2}\ \Omega$
$\quad = \textbf{274.8 } \Omega$

b. $I = \dfrac{V}{Z} \qquad (24\text{-}4)$

$\quad = \dfrac{10 \text{ V}}{274.8 \ \Omega} = \textbf{36.4 mA}$

c. $\theta = \tan^{-1}\dfrac{X_L}{R} \qquad (24\text{-}5a)$

$\quad = \text{arc tan } \dfrac{188.5 \ \Omega}{200 \ \Omega} = \textbf{43.3}°$

When a circuit contains both resistance and reactance, the relative value of the two determines the nature of the impedance. A series circuit is said to be predominantly *inductive* if the inductive reactance is in the order of 10 times as large as the resistance. For example, if $R = 1\ \Omega$ and $X_L = 10\ \Omega$,

$$Z = \sqrt{R^2 + X_L^2}$$
$$= \sqrt{1^2 + 10^2}\ \Omega$$
$$= \textbf{10.05 } \Omega$$

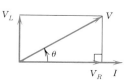

FIGURE 24-8

Circuit diagram, phasor diagram, and impedance triangle for Example 24-2.

and

$$\theta = \tan^{-1}\frac{X_L}{R}$$

$$= \arctan\frac{10}{1} = \mathbf{84°}$$

Similarly, if R is 10 times X_L, the circuit is considered to be predominantly *resistive*.

24-4.1 Measuring Inductance of a Practical Coil

So far, the resistance of a coil has been assumed to be negligible compared to X_L. Although this may be true at higher frequencies (since X_L increases with frequency), the coil's resistance may *not* be negligible at power line (60 Hz) frequencies, since X_L is small at such frequencies. In cases where the resistance is not negligible, the voltage across the coil is *not* 90° out of phase with the current through the coil. Further, it is impossible to make measurements across either the resistive portion or the inductive portion of the coil *alone*. To develop a voltage wave-

form proportional to the current, it is necessary to insert a small resistance in series with the coil. The resistor also allows a determination of the phase angle using a dual-beam (or "dual-trace") oscilloscope, as shown in Fig. 24-9.

If the dual-beam oscilloscope is connected as shown in Fig. 24-9a, channel 1 will display the applied voltage waveform, with a peak value, V_m. This is approximately equal to the coil voltage, because $v_{R'}$ is negligible (since R' is only a few ohms). Channel 2 displays a voltage waveform across R' that is proportional to the current in the circuit. That is,

$$I_m = \frac{V_{R'\max}}{R'}$$

The impedance of the whole circuit, which is approximately equal to the impedance of the coil, can now be obtained:

$$Z = \frac{V_m}{I_m}$$

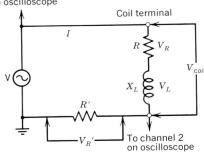

(a) Circuit of coil with small series resistor R'

(b) Phasor diagram

(c) Impedance diagram

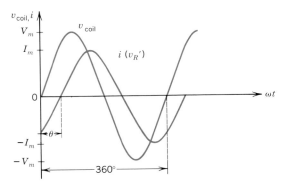

(d) Oscilloscope display of voltage waveforms across coil and R'

FIGURE 24-9

Voltage and current relations in a practical coil.

The phase angle (θ) can be determined from the oscilloscope by the direct ratio of the two horizontal distances representing θ and 360° in Fig. 24-9d. (The oscilloscope's horizontal sweep must be triggered by the same signal for both traces.) The resistive and inductive portions of the coil can now be obtained.

$$R = Z\cos\theta \qquad (24\text{-}6)$$
$$X_L = Z\sin\theta \qquad (24\text{-}7)$$

R is actually the ac resistance of the coil (R_{ac}), so the coil's Q may now be determined.

$$Q = \frac{X_L}{R_{ac}} \qquad (20\text{-}9)$$

and

$$L = \frac{X_L}{2\pi f} \qquad (20\text{-}2)$$

A method to find the phase angle using a *single-beam* oscilloscope is given in Section 24-8.

EXAMPLE 24-3

A coil and a 10-Ω resistor are series-connected across a 60-Hz supply. A dual-beam oscilloscope, connected as in Fig. 24-9a, displays a peak coil voltage of 18 V and a peak resistor voltage of 120 mV. A full cycle of the coil voltage occupies 5 cm and the beginning of the resistor voltage waveform is delayed 1 cm. Calculate:
a. The impedance of the circuit.
b. The phase angle between the applied voltage and current.
c. The ac resistance of the coil.
d. The inductive reactance of the coil.
e. The Q of the coil.
f. The inductance of the coil.

Solution

a. $I_m = \dfrac{V_{R'max}}{R'}$

$$= \frac{120 \times 10^{-3} \text{ V}}{10 \text{ Ω}} = 12 \text{ mA}$$

$$Z = \frac{V_m}{I_m}$$

$$= \frac{18 \text{ V}}{12 \times 10^{-3} \text{ A}} = \mathbf{1500 \ \Omega}$$

b. By direct ratio:

$$\frac{\theta}{360°} = \frac{1 \text{ cm}}{5 \text{ cm}}$$

therefore, $\theta = 360° \times \dfrac{1}{5} = \mathbf{72°}$

c. $R_{ac} = Z\cos\theta$ (24-6)

     $= 1500 \text{ Ω} \times \cos 72° = \mathbf{464 \ \Omega}$

d. $X_L = Z\sin\theta$ (24-7)

     $= 1500 \text{ Ω} \times \sin 72° = \mathbf{1427 \ \Omega}$

e. $Q = \dfrac{X_L}{R_{ac}}$ (20-9)

     $= \dfrac{1427 \text{ Ω}}{464 \text{ Ω}} = \mathbf{3.1}$

f. $L = \dfrac{X_L}{2\pi f}$ (20-2)

     $= \dfrac{1427 \text{ Ω}}{2\pi \times 60 \text{ Hz}} = \mathbf{3.8 \ H}$

NOTE $Z = 1500 \ \Omega$ is the impedance of the whole circuit so that $R_{ac} = 464 \ \Omega$ actually includes the 10-Ω resistance of R'. (A more accurate value for the coil's ac resistance is therefore 454 Ω.)

24-5 SERIES *RC* CIRCUIT

The phasor diagram for the voltage and current waveforms of a series *RC* circuit, similar to that used for the series *RL* circuit, is shown in Fig. 24-10.

Once again, the current phasor (I) has been drawn in the horizontal reference position, with the resistor voltage

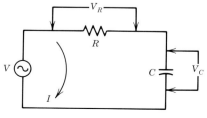

(*a*) Series *RC* circuit

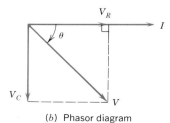

(*b*) Phasor diagram

FIGURE 24-10
Series *RC* circuit (*a*) with phasor diagram (*b*).

phasor (V_R) "in-line" with it. Lagging I by 90° is the capacitor voltage phasor (V_C), and V is the phasor sum:

$$V = \sqrt{V_R^2 + V_C^2} \qquad \text{volts} \qquad (24\text{-}8)$$

and

$$\theta = -\arctan\frac{V_C}{V_R} \qquad (24\text{-}9)$$

where: V is the total applied voltage, in volts (V)
V_R is the voltage across the resistor, in volts (V)
V_C is the voltage across the pure capacitance, in volts (V)
θ is the phase angle between the applied voltage and the circuit current

Note that the phase angle (θ) is negative in comparison with an inductive circuit. That is, the current *leads* the applied voltage in a capacitive circuit, but *lags* in an inductive circuit.

EXAMPLE 24-4

A capacitor and a 2.2-kΩ resistor are connected in series across a 60-Hz, sinusoidally alternating voltage. A series-connected ac ammeter indicates 15 mA, and an oscilloscope across the capacitor indicates 60 V peak-to-peak. Calculate:
a. The reading of a voltmeter connected across the resistor.
b. The rms voltage across the capacitor.
c. The applied voltage.
d. The phase angle between the current and applied voltage.
e. The amount of capacitance.

Solution

See Fig. 24-11.
a. $V_R = IR$ (15-15b)
    $= 15 \times 10^{-3}$ A $\times$ 2.2 $\times 10^3$ $\Omega =$ **33 V**

b. $V_C = \dfrac{V_{\text{p-p}}}{2\sqrt{2}}$ (15-17)

    $= \dfrac{60\text{ V}}{2\sqrt{2}} =$ **21.2 V**

c. $V = \sqrt{V_R^2 + V_C^2}$ (24-8)
    $= \sqrt{33^2 + 21.2^2}$ V $=$ **39.2 V**

d. $\theta = -\tan^{-1}\dfrac{V_C}{V_R}$

    $= -\arctan\dfrac{21.2\text{ V}}{33\text{ V}} =$ **−32.7°**

e. $X_C = \dfrac{V_C}{I}$ (23-1)

    $= \dfrac{21.2\text{ V}}{15 \times 10^{-3}\text{ A}} =$ 1.4 kΩ

    $C = \dfrac{1}{2\pi f X_C}$ (23-2)

    $= \dfrac{1}{2\pi \times 60\text{ Hz} \times 1.4 \times 10^3\ \Omega}$

    $=$ **1.9 μF**

The phasor diagram is shown in Fig. 24-11.

24-6 IMPEDANCE OF A SERIES *RC* CIRCUIT

It can be shown that, similar to an inductor, the total opposition to current in a series *RC* circuit is given by the impedance (*Z*). (See Fig. 24-12.)

$$Z = \sqrt{R^2 + X_C^2} \qquad \text{ohms} \qquad (24\text{-}10)$$

$$I = \frac{V}{Z} \qquad \text{amperes} \qquad (24\text{-}4)$$

and

$$\theta = -\tan^{-1}\frac{X_C}{R} \qquad (24\text{-}11)$$

where: Z is the impedance, in ohms (Ω)
R is the resistance, in ohms (Ω)
X_C is the capacitive reactance, in ohms (Ω)
θ is the phase angle between the current (I) and the applied voltage (V)

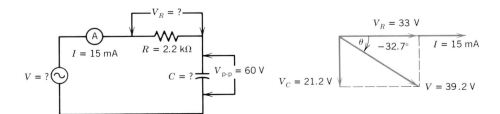

FIGURE 24-11
Circuit and phasor diagram for Example 24-4.

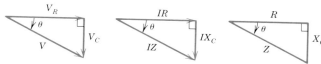

FIGURE 24-12

Voltage and impedance triangles for a series *RC* circuit.

EXAMPLE 24-5

a. How much resistance must be added in series with a 0.1-μF capacitor to limit the current to 5 mA when the series combination is connected to a 10-V, 1-kHz sine wave?

b. What is the phase angle between the applied voltage and current?

Solution

See Fig. 24-13.

a. $X_C = \dfrac{1}{2\pi fC}$ (23-2)

$\quad = \dfrac{1}{2\pi \times 1 \times 10^3 \text{ Hz} \times 0.1 \times 10^{-6} \text{ F}}$

$\quad = 1.6 \text{ k}\Omega$

$Z = \dfrac{V}{I}$ (24-4)

$\quad = \dfrac{10 \text{ V}}{5 \times 10^{-3} \text{ A}} = 2 \text{ k}\Omega$

$Z = \sqrt{R^2 + X_C^2}$ (24-10)

Therefore, $R = \sqrt{Z^2 - X_C^2}$

$\quad = \sqrt{(2 \times 10^3)^2 - (1.6 \times 10^3)^2}\ \Omega$

$\quad = \mathbf{1.2 \text{ k}\Omega}$

b. $\theta = -\tan^{-1}\dfrac{X_C}{R}$

$\quad = -\text{arc tan}\dfrac{1.6 \text{ k}\Omega}{1.2 \text{ k}\Omega} = \mathbf{-53.1°}$

The phasor diagram is shown in Fig. 24-13.

24-6.1 Voltage Division in a Series *RC* Circuit*

The concept of impedance is necessary when you consider the voltage developed across a capacitor that is in series with a resistor. (See Fig. 24-14*a*.)

Assume that an alternating voltage (V_i) is applied to the series circuit of *R* and *C*. What is an expression for V_o in terms of V_i, *R*, *C*, and *f*?

The current that flows due to V_i is

$$I = \frac{V_i}{Z}$$

where

$$Z = \sqrt{R^2 + X_C^2}$$

but

$$V_o = IX_C.$$

Therefore,

$$V_o = \frac{V_i}{Z} \times X_C$$

or,

$$V_o = V_i \times \frac{X_C}{\sqrt{R^2 + X_C^2}} \qquad (24\text{-}12)$$

Note the similarity of this voltage divider equation to that used for two resistors in series:

$$V_o = V_i \times \frac{R_o}{R + R_o}$$

where R_o takes the place of X_C.

Note also that the ratio V_o/V_i varies with frequency, since

$$X_C = \frac{1}{2\pi fC}$$

$$\frac{V_o}{V_i} = \frac{X_C}{\sqrt{R^2 + X_C^2}}$$

At low frequencies, X_C is much larger than *R*, and the impedance (denominator) is approximately equal to X_C, so V_o/V_i is approximately equal to 1.

As the frequency increases and X_C decreases, the ratio

---

*This section may be omitted with no loss of continuity.

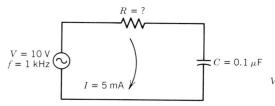

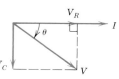

FIGURE 24-13

Circuit, phasor diagram, and impedance triangle for Example 24-5.

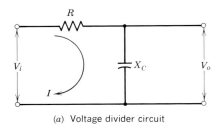

(a) Voltage divider circuit

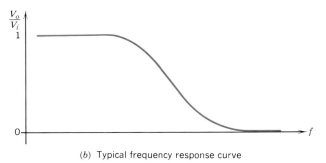

(b) Typical frequency response curve

FIGURE 24-14

Series *RC* circuit voltage division with low-frequency emphasis.

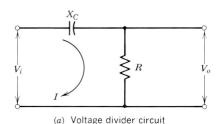

(a) Voltage divider circuit

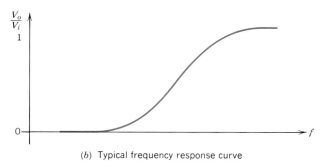

(b) Typical frequency response curve

FIGURE 24-15

Series *RC* circuit voltage division with high-frequency emphasis.

V_o/V_i approaches zero. A typical frequency response curve for the circuit is shown in Fig. 24-14b. This is the type of curve used to provide bass adjustment in the simple tone-control circuits discussed in Section 23-3.3.

If the locations of *R* and *C* are interchanged, as in Fig. 24-15a, the output voltage is given by

$$V_o = IR = \frac{V_i}{Z} \times R$$

or
$$V_o = V_i \times \frac{R}{\sqrt{R^2 + X_C^2}} \qquad (24\text{-}13)$$

The frequency response curve of Fig. 24-15b shows that this circuit attenuates the low frequencies but "passes" the highs. It has the effect of providing a treble control (high-frequency emphasis).

EXAMPLE 24-6

Given the circuit in Fig. 24-16a with $R = 5$ kΩ and $C = 0.05$ µF.

a. Determine the ratio of V_o to V_i at $f = 1$ kHz and several other frequencies to sketch a curve of V_o/V_i versus *f* from 10 Hz to 20 kHz.

b. How much larger is V_o at 1 kHz than at 10 kHz?

c. Repeat part (a) with the same size of components, but using the circuit in Fig. 24-17a.

d. How much larger is V_o at 1 kHz than at 100 Hz?

Solution

a. From Eq. 24-12.

$$V_o/V_i = \frac{X_C}{\sqrt{R^2 + X_C^2}}$$

At $f = 1$ kHz, $X_C = \dfrac{1}{2\pi f C}$ \qquad (23-12)

$$= \frac{1}{2\pi \times 1 \times 10^3 \text{ Hz} \times 0.05 \times 10^{-6} \text{ F}}$$

$$= 3.2 \text{ k}\Omega$$

$$Z = \sqrt{R^2 + X_C^2} \qquad (24\text{-}10)$$

$$= \sqrt{5^2 + 3.2^2} \text{ k}\Omega$$

$$= 5.9 \text{ k}\Omega$$

$$V_o/V_i = \frac{X_C}{\sqrt{R^2 + X_C^2}} \qquad (24\text{-}12)$$

$$= \frac{3.2 \text{ k}\Omega}{5.9 \text{ k}\Omega}$$

$$= \mathbf{0.54}$$

TABLE 24-1

Reactances and impedances for Example 24-6.

f, Hz	X_C, kΩ	$Z = \sqrt{R^2 + X_C^2}$, kΩ	X_C/Z	R/Z
10	320	320	1.00	0.02
100	32	32.4	0.99	0.15
500	6.4	8.1	0.79	0.62
1000	3.2	5.9	0.54	0.85
5000	0.64	5.0	0.13	1.0
10,000	0.32	5.0	0.06	1.0
20,000	0.16	5.0	0.03	1.0

Repeating the calculations for other frequencies, the results are shown in Table 24-1. The frequency response curve of V_o/V_i versus f is shown in Fig. 24-16b.

b. At $f = 10$ kHz, $V_o = 0.06V_i$.
At $f = 1$ kHz, $V_o = 0.54V_i$
Ratio of V_o at 1 kHz to V_o at 10 kHz is 0.54/0.06 = **9.**

c. For the circuit in Fig. 24-17a, with $R = 5$ kΩ and $C = 0.05$ μF,

$$V_o/V_i = \frac{R}{\sqrt{R^2 + X_C^2}} \qquad (24\text{-}13)$$

At $f = 1$ kHz, $X_C = 3.2$ kΩ

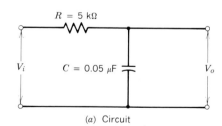

R = 5 kΩ

V_i $C = 0.05$ μF V_o

(a) Circuit

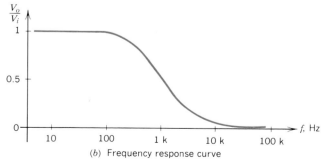

(b) Frequency response curve

FIGURE 24-16

Circuit and graph for Example 24-6A.

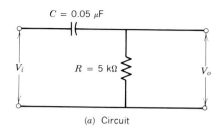

$C = 0.05$ μF

V_i $R = 5$ kΩ V_o

(a) Circuit

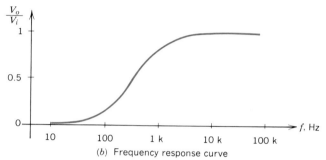

(b) Frequency response curve

FIGURE 24-17

Circuit and graph for Example 24-6C.

$$V_o/V_i = \frac{5 \text{ kΩ}}{\sqrt{5^2 + 3.2^2} \text{ kΩ}}$$

$$= \frac{5 \text{ kΩ}}{5.9 \text{ kΩ}}$$

$$= \mathbf{0.85}$$

Values for other frequencies are shown in Table 24-1, and the frequency response curve in Fig. 24-17b.

d. At $f = 1$ kHz, $V_o = 0.85 \, V_i$.
At $f = 100$ Hz, $V_o = 0.15 \, V_i$.

Ratio of V_o at 1 kHz to V_o at 100 Hz is $\dfrac{0.85}{0.15} = \mathbf{5.7.}$

24-7 SERIES *RLC* CIRCUIT

You have seen how the capacitance makes the current *lead* the applied voltage in a series *RC* circuit, and how the inductance makes the current *lag* in a series *RL* circuit. When *both* capacitance and inductance are in series, you should expect a canceling effect. This is evident when you consider the phasor and impedance diagrams in Fig. 24-18.

The inductive voltage (V_L) is 180° out of phase with the capacitive voltage (V_C) in Fig. 24-18b. If V_L is larger than V_C, the difference ($V_L - V_C$) would be indicated on a volt-

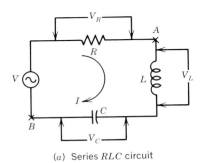

(a) Series *RLC* circuit

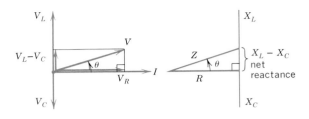

(b) Phasor diagram   (c) Impedance diagram

FIGURE 24-18
(*a*) Series *RLC* circuit;
(*b*) phasor diagram;
and (*c*) impedance
diagram.

meter connected between *A* and *B* in Fig. 24-18*a* (assuming the resistance of the coil is negligible). When $V_L - V_C$ is added to V_R vectorially, the result is the applied voltage (*V*). Thus,

$$V = \sqrt{V_R^2 + (V_L - V_C)^2} \quad \text{volts} \quad (24\text{-}14)$$

and

$$\theta = \tan^{-1}\frac{V_L - V_C}{V_R} \quad (24\text{-}15)$$

where: *V* is the applied voltage, in volts (V)
 V_R is the voltage across the total circuit resistance, in volts (V) (includes the resistance of the coil, if not negligible)
 V_L is the voltage across the inductance, in volts (V)
 V_C is the voltage across the capacitance, in volts (V)
 $V_L - V_C$ is the net reactive voltage, in volts (V)
 θ is the phase angle between the current and the applied voltage in degrees or radians

The impedance of a series *RLC* circuit can be obtained from the impedance diagram in Fig. 24-18*c*. The net reactance of the circuit is $X_L - X_C$, so that *Z* is given by

$$Z = \sqrt{R^2 + (X_L - X_C)^2} \quad \text{ohms} \quad (24\text{-}16)$$

and

$$\theta = \tan^{-1}\frac{X_L - X_C}{R} \quad (24\text{-}17)$$

and

$$I = \frac{V}{Z} \quad \text{amperes} \quad (24\text{-}4)$$

where: *Z* is the circuit impedance, in ohms (Ω)
 R is the circuit resistance, in ohms (Ω)
 X_L is the circuit inductive reactance, in ohms (Ω)
 $X_L - X_C$ is the net circuit reactance, in ohms (Ω)
 θ is the phase angle between the current and applied voltage, in degrees or radians

For the phasor diagram shown in Fig. 24-18*b*, the current *lags* the applied voltage (*V*), because the overall circuit is inductive. That is, X_L is greater than X_C, making V_L

larger than V_C. Clearly, if V_C is larger than V_L (X_C greater than X_L), then the current *leads* the applied voltage, and the overall circuit is said to be capacitive. Further, if $X_L = X_C$, then $V_L = V_C$, the phase angle is zero, and the circuit is purely resistive. This condition is called series *resonance;* it occurs when the inductive and capacitive reactances *cancel* each other. Resonance is a very important effect, and is considered in detail in Chapter 26.

EXAMPLE 24-7

A 30-mH coil (of negligible resistance), a 0.005-μF capacitor, and a 1-kΩ resistor are series-connected across a 2-V, 10-kHz sinusoidal supply. Calculate:

a. The circuit impedance.
b. The circuit current.
c. The voltage across each component.
d. The phase angle between the current and applied voltage.

Draw the circuit, the phasor and the impedance diagrams indicating all values.

Solution

a. $X_L = 2\pi f L \qquad\qquad (20\text{-}2)$
$\qquad = 2\pi \times 10 \times 10^3 \text{ Hz} \times 30 \times 10^{-3} \text{ H}$
$\qquad = 1885\ \Omega$

$\qquad X_C = \dfrac{1}{2\pi f C} \qquad\qquad (23\text{-}2)$

$\qquad\quad = \dfrac{1}{2\pi \times 10 \times 10^3 \text{ Hz} \times 0.005 \times 10^{-6} \text{ F}}$

$\qquad\quad = 3183\ \Omega$

$\qquad Z = \sqrt{R^2 + (X_L - X_C)^2}\ \Omega \qquad (24\text{-}16)$

$\qquad\quad = \sqrt{1000^2 + (1885 - 3183)^2}\ \Omega$

$\qquad\quad = \mathbf{1639\ \Omega}$

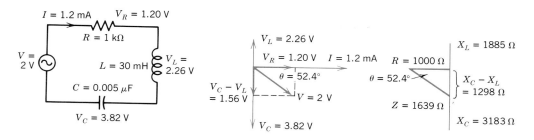

FIGURE 24-19

Series *RLC* circuit, phasor diagram, and impedance diagram for Example 24-7.

b. $I = \dfrac{V}{Z}$ (24-4)

$= \dfrac{2\ V}{1639\ \Omega} = \textbf{1.2 mA}$

c. $V_R = IR$ (15-15b)

$= 1.2 \times 10^{-3}\ A \times 1000\ \Omega = \textbf{1.20 V}$

$V_L = IX_L$ (20-1)

$= 1.2 \times 10^{-3}\ A \times 1885\ \Omega = \textbf{2.26 V}$

$V_C = IX_C$ (23-1)

$= 1.2 \times 10^{-3}\ A \times 3183\ \Omega = \textbf{3.82 V}$

d. $\theta = \tan^{-1} \dfrac{X_L - X_C}{R}$ (24-17)

$= \tan^{-1} \dfrac{1885 - 3183\ \Omega}{1000\ \Omega}$

$= \text{arc tan} - 1.3 = \textbf{-52.4°}$

The diagrams are shown in Fig. 24-19.

NOTE The voltages across the capacitor and inductor in Example 24-7 are each larger than the input voltage. This is a common phenomenon in series *RLC* circuits when the inductive and capacitive reactances are of the same order of magnitude. This is because the overall impedance is *lower* than if the capacitance or the inductance alone is in the circuit with the resistance. This means that a relatively high current flows, producing a relatively high voltage across the reactances. Note too that because X_C is greater than X_L, the overall circuit is capacitive. Also, V_C is greater than V_L and the current *leads* the applied voltage by 52.4°.

24-7.1 Application of a Series *RLC* Circuit

As shown in Example 24-7, the total impedance of a series *RLC* circuit may be less than if the circuit contained only resistance and inductive reactance. This is due to the cancelling effect the capacitor has on the overall reactance. The result is that less voltage is needed to cause the same current to flow through the original inductive circuit if a capacitor is connected in series. There is a very useful practical application of this fact. Assume that you have an inductive load (such as a relay, solenoid, or small motor) that requires 120 V for its proper operation. But, due to an excessive voltage drop in the connecting lines, only 110 or 105 V is available. By connecting a capacitor of the proper size in series with the load, the impedance can be reduced so that the proper current flows, even at the lower voltage. (As illustrated in Example 24-8, 120 V *does* appear across the load itself.)

NOTE The capacitor used must be suitable for ac operation.

EXAMPLE 24-8

A solenoid that requires a current of 200 mA to operate normally at 120 V, 60 Hz, has an ac resistance of 200 Ω. The solenoid must be operated at 95 V by inserting a capacitor in series with it. Calculate:

a. The size of the capacitor needed.
b. The voltage across the capacitor.
c. The voltage across the solenoid.
Draw the phasor diagram.

Solution

a. When operating normally, the solenoid's impedance is:

$$Z = \frac{V}{I} = \frac{120\ V}{0.2\ A} = 600\ \Omega$$

The solenoid's reactance may now be found:

$$Z = \sqrt{R^2 + X_L^2} \qquad (24\text{-}3)$$
$$X_L = \sqrt{Z^2 - R^2}$$
$$= \sqrt{600^2 - 200^2}\ \Omega$$
$$= 565.7\ \Omega$$

When the capacitor is connected in series with the solenoid, the total impedance must be:

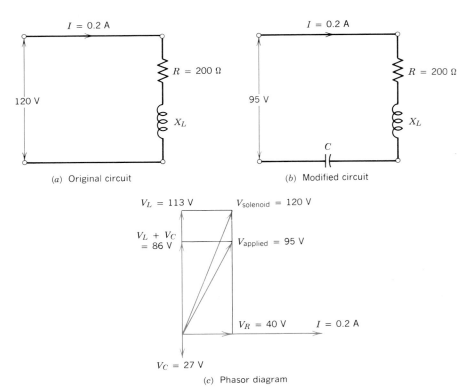

(a) Original circuit   *(b)* Modified circuit

(c) Phasor diagram

FIGURE 24-20
Circuits and phasor diagram for Example 24-8.

$$Z = \frac{V}{I} = \frac{95 \text{ V}}{0.2 \text{ A}} = 475 \text{ } \Omega$$

$$Z = \sqrt{R^2 + (X_L - X_C)^2} \qquad (24\text{-}16)$$

$$X_L - X_C = \sqrt{Z^2 - R^2}$$

$$= \sqrt{475^2 - 200^2} \text{ } \Omega$$

$$= 430.8 \text{ } \Omega$$

Therefore, $X_C = X_L - 430.8$

$$= 565.7 - 430.8 \text{ } \Omega$$

$$= 134.9 \text{ } \Omega$$

But $X_C = \dfrac{1}{2\pi fC}$ (23-2)

Therefore, $C = \dfrac{1}{2\pi fX_C}$

$$= \frac{1}{2\pi \times 60 \text{ Hz} \times 134.9 \text{ } \Omega}$$

$$= 19.7 \times 10^{-6} \text{ F} = \mathbf{19.7 \text{ } \mu F}$$

b. $V_C = IX_C$ (23-1)

$$= 0.2 \text{ A} \times 134.9 \text{ } \Omega$$

$$= \mathbf{27 \text{ V}}$$

c. $V_{\text{solenoid}} = IZ_{\text{solenoid}}$

$$= 0.2 \text{ A} \times 600 \text{ } \Omega$$

$$= \mathbf{120 \text{ V}}$$

(See Fig. 24-20.)

24-8   OSCILLOSCOPE MEASUREMENT OF PHASE ANGLES

Although the phase angle between two voltages can be observed most easily on a dual-trace oscilloscope, there is a useful method of displaying this information on a single-beam oscilloscope. The method involves applying one of the voltages to the vertical (*Y*) input and the other to the horizontal (*X*) input. Since both waveforms are at the same frequency, the phase relationship between the two is constant. The smaller of the two voltages is usually connected to the vertical input, since that input is more sensitive. To apply the other voltage to the horizontal deflection plates,

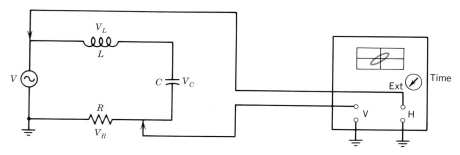

(a) Circuit connections to measure the phase angle

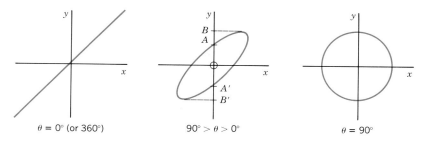

(b) Lissajous figures for various phase angles

FIGURE 24-21
Showing how to make phase angle measurements using Lissajous figures. (*Note*: The oscilloscope must be adjusted for equal vertical and horizontal deflections to obtain above figures, and in some oscilloscopes, the figures may slope the opposite way.)

it is usually necessary to move the time-base selector switch to "external" (EXT).

An example of the way in which the phase angle is measured in a series *RLC* circuit is shown in Fig. 24-21, along with typical *Lissajous figures.*

Since the frequency of the two voltages is the same, a stationary pattern results, as in Fig. 24-21b. In general, the figure produced is an ellipse with an overall height BB' and intercepts on the *Y*-axis at AA'. It can be shown that the sine of the phase angle θ between the two voltages applied to the oscilloscope is given by

$$\sin\theta = \frac{OA}{OB} = \frac{AA'}{BB'} \qquad (24\text{-}18)$$

$$\theta = arc\ sin\ \frac{OA}{OB} = arc\ sin\ \frac{AA'}{BB'} \qquad (24\text{-}18a)$$

Note that AA' and BB' are *distances* taken from the oscilloscope screen. It is *not* necessary that the oscilloscope be calibrated vertically or horizontally, but the Lissajous pattern *must* be centered horizontally.

In the case of a *circle*, AA' and BB' are equal, so $\theta =$

arc sin 1 = 90°. For a *straight line*, $AA' = 0$, so $\theta =$ arc sin 0 = 0°. If $\theta = 45°$, $\dfrac{OA}{OB} = \sin 45° = 0.7$.

Whether this is a leading or lagging phase angle cannot be determined from the Lissajous display alone.*

In the circuit of Fig. 24-21a, the phase angle determined is between V_R and V. But since V_R is proportional to (and in phase with) the current (I), the measured phase angle is between the current and the applied voltage. However, not *all* phase angle measurements necessarily give the phase angle between the current and the applied voltage. For example, what meaning would the phase angle determined from the oscilloscope have if R and C were interchanged in Fig. 24-21a? Recall that, in many cases, one side of the supply is grounded, and so is the low side on the oscilloscope. Care must be taken to determine what voltages are being compared. (See Problem 24-30.)

---

*But a dual-trace presentation clearly shows which waveform is leading or lagging. The advantage of the Lissajous method is that the phase angle is determined quite accurately.

EXAMPLE 24-9

An oscilloscope is used to make a phase angle measurement between the current and applied voltage in a series RLC circuit, as in Fig. 24-21a. The resulting ellipse is as shown in Fig. 24-21b, with $OA = 2$ cm and $OB = 3$ cm. Voltmeters across L and C indicate 8 and 5 V, respectively. Calculate:

a. The phase angle between the current and applied voltage.
b. The resistor voltage.
c. The applied voltage.
d. The reading of a voltmeter connected across the series LC combination.

Draw the phasor diagram.

Solution

a.　$\theta = \sin^{-1}\dfrac{OA}{OB}$　　　　　　　(24-18a)

　　$= \text{arc sin }\dfrac{2 \text{ cm}}{3 \text{ cm}} = \textbf{41.8°}$

b.　$\tan \theta = \dfrac{V_L - V_C}{V_R}$　　　　　(24-15)

　　Therefore, $V_R = \dfrac{V_L - V_C}{\tan \theta}$

　　　　　$= \dfrac{8 \text{ V} - 5 \text{ V}}{\tan 41.8°} = \textbf{3.36 V}$

c.　$V = \sqrt{V_R^2 + (V_L - V_C)^2}$　　　(24-14)

　　$= \sqrt{3.36^2 + (8 - 5)^2}$ V $= \textbf{4.5 V}$

d.　Voltage across LC combination $= V_L - V_C$

　　　　　　　　　　$= 8 - 5$ V $= \textbf{3 V}$

The phasor diagram is shown in Fig. 24-22.

24-9 PARALLEL RL CIRCUIT

When a resistor and an inductor are connected in parallel across a sinusoidal supply, the total current drawn (by Kirchhoff's current law) is the *phasor sum* of each branch current. Since voltage is common to both branch elements, V is used as the horizontal reference phasor, as in Fig. 24-23b.

The current drawn by each branch is given by

$$I_R = \frac{V}{R} \qquad (15\text{-}15a)$$

and

$$I_L = \frac{V}{X_L} \qquad (20\text{-}1)$$

Therefore, the total current I_T is

$$I_T = \sqrt{I_R^2 + I_L^2} \qquad \text{amperes} \qquad (24\text{-}19)$$

$$Z = \frac{V}{I_T} \qquad (24\text{-}4)$$

$$\theta = -\tan^{-1}\frac{I_L}{I_R} \qquad (24\text{-}20)$$

In this case, the negative phase angle shows that the total circuit current *lags* the applied voltage, as it does in a series RL circuit.

An expression may be obtained for the total impedance of the parallel-connected components using Eq. 24-19.

$$I_T = \sqrt{I_R^2 + I_L^2}$$

But　$I_R = \dfrac{V}{R}$ and $I_L = \dfrac{V}{X_L}$

Therefore,　$I_T = \sqrt{(V/R)^2 + (V/X_L)^2}$

$$= V\sqrt{\frac{1}{R^2} + \frac{1}{X_L^2}}$$

$$= V\sqrt{\frac{X_L^2 + R^2}{R^2 X_L^2}}$$

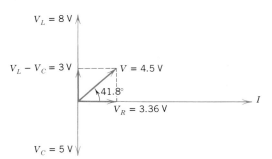

FIGURE 24-22
Phasor diagram for Example 24-9.

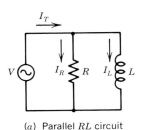

(a) Parallel RL circuit

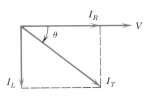

(b) Phasor diagram

FIGURE 24-23
(a) Parallel RL circuit and (b) phasor diagram.

But
$$Z = \frac{V}{I_T} = \sqrt{\frac{R^2 X_L^2}{R^2 + X_L^2}}$$

or
$$Z = \frac{R X_L}{\sqrt{R^2 + X_L^2}} \qquad (24\text{-}21)$$

Note the similarity to the equation for two resistors in parallel:

$$R_T = \frac{R_1 R_2}{R_1 + R_2} = \frac{\text{Product}}{\text{Sum}}$$

The numerator of Eq. 24-21 consists of the product of the two components, as for resistors. But in the denominator, you cannot merely add resistance directly to reactance; you must use the combination of resistance and reactance according to the series impedance equation:

$$Z = \sqrt{R^2 + X_L^2}$$

EXAMPLE 24-10

A 200-Ω resistor, and a coil of negligible resistance and 400-Ω reactance, are connected in parallel across a 40-V sinusoidal source. Calculate:
a. The total circuit current.
b. The impedance of the circuit.
c. The phase angle between the circuit current and applied voltage.

Solution

a. $I_R = \dfrac{V}{R}$ (15-15a)

$= \dfrac{40\text{ V}}{200\ \Omega} = 0.2\text{ A}$

$I_L = \dfrac{V}{X_L}$ (20-1)

$= \dfrac{40\text{ V}}{400\ \Omega} = 0.1\text{ A}$

$I_T = \sqrt{I_R^2 + I_L^2}$ (24-19)

$= \sqrt{0.2^2 + 0.1^2}\text{ A} = \textbf{0.224 A}$

b. $Z = \dfrac{V}{I_T}$ (24-4)

$= \dfrac{40\text{ V}}{0.224\text{ A}} = \textbf{179 }\boldsymbol{\Omega}$

Alternatively, $Z = \dfrac{R X_L}{\sqrt{R^2 + X_L^2}}$ (24-21)

$= \dfrac{200\ \Omega \times 400\ \Omega}{\sqrt{200^2 + 400^2}\ \Omega}$

$= \dfrac{80{,}000\ \Omega}{447.2} = \textbf{179 }\boldsymbol{\Omega}$

c. $\theta = -\tan^{-1}\dfrac{I_L}{I_R}$ (24-20)

$= -\text{arc tan}\ \dfrac{0.1\text{ A}}{0.2\text{ A}} = \textbf{-26.6°}$

Note that the overall circuit is considered to be more resistive than inductive because the resistive current is larger than the inductive current.

24-10 PARALLEL *RC* CIRCUIT

The phasor diagram for the current and voltage in a parallel *RC* circuit is shown in Fig. 24-24. It should be obvious that the total circuit current is given by

$$I_T = \sqrt{I_R^2 + I_C^2} \qquad \text{amperes (A)} \qquad (24\text{-}22)$$

$$Z = \frac{V}{I_T} \qquad (24\text{-}4)$$

$$\theta = \tan^{-1}\frac{I_C}{I_R} \qquad (24\text{-}23)$$

and
$$Z = \frac{R X_C}{\sqrt{R^2 + X_C^2}} \qquad (24\text{-}24)$$

EXAMPLE 24-11

A 200-Ω resistor and a 100-Ω capacitive reactance are connected in parallel with a 40-V sinusoidal supply. Calculate:
a. The total circuit current.
b. The impedance of the circuit.

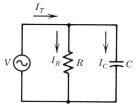

(*a*) Parallel *RC* circuit

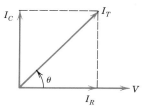

(*b*) Phasor diagram

FIGURE 24-24
(*a*) Parallel *RC* circuit and (*b*) phasor diagram.

c. The phase angle between the circuit current and applied voltage.

Solution

a. $I_R = \dfrac{V}{R}$ (15-15a)

     $= \dfrac{40 \text{ V}}{200 \text{ }\Omega} = 0.2 \text{ A}$

    $I_C = \dfrac{V}{X_C}$ (23-1)

     $= \dfrac{40 \text{ V}}{100 \text{ }\Omega} = 0.4 \text{ A}$

    $I_T = \sqrt{I_R^2 + I_C^2}$ (24-22)

     $= \sqrt{0.2^2 + 0.4^2} \text{ A} = \mathbf{0.447 \text{ A}}$

b. $Z = \dfrac{V}{I_T}$ (24-2)

     $= \dfrac{40 \text{ V}}{0.447 \text{ A}} = \mathbf{89.5 \text{ }\Omega}$

Alternatively, $Z = \dfrac{R X_C}{\sqrt{R^2 + X_C^2}}$ (24-24)

     $= \dfrac{200 \text{ }\Omega \times 100 \text{ }\Omega}{\sqrt{200^2 + 100^2} \text{ }\Omega}$

     $= \dfrac{20,000}{223.6} \text{ }\Omega = \mathbf{89.4 \text{ }\Omega}$

c. $\theta = \tan^{-1} \dfrac{I_C}{I_R}$ (24-23)

     $= \text{arc tan} \dfrac{0.4 \text{ A}}{0.2 \text{ A}} = \mathbf{63.4°}$

In this case, the circuit is more capacitive than resistive (I_T leads V), because I_C is larger than I_R. This is a result of X_C being *smaller* than R. (Note how this is opposite to a series *RC* circuit.)

24-11 PARALLEL *RLC* CIRCUIT

You should expect some canceling effect in a way similar to a series *RLC* circuit, when inductance and capacitance are both connected in parallel across an ac voltage source. In the circuit shown in Fig. 24-25, it is assumed that X_C is *less* than X_L so that I_C is *greater* than I_L. The resulting total current *leads* the applied voltage, and the circuit is said to be *capacitive* overall.

Conversely, if X_L is less than X_C, then I_L is greater than I_C and the circuit is considered to be *inductive*. But if $X_L = X_C$, the total current is in phase with the applied voltage (θ is zero), and the circuit is purely *resistive*. This condition is called *parallel resonance* and is discussed further in Section 26-6.

The total current, impedance, and phase angle of the parallel *RLC* circuit are given by:

$$I_T = \sqrt{I_R^2 + (I_C - I_L)^2} \quad \text{amperes} \quad (24\text{-}25)$$

$$Z = \dfrac{V}{I_T} \quad\quad\quad\quad (24\text{-}4)$$

and $\theta = \tan^{-1} \dfrac{I_C - I_L}{I_R}$ (24-26)

EXAMPLE 24-12

The resistor, inductor, and capacitor from the previous two examples ($R = 200 \text{ }\Omega$, $X_L = 400 \text{ }\Omega$, and $X_C = 100 \text{ }\Omega$) are connected in parallel with the same voltage source of $V = 40 \text{ V}$. Calculate:
a. The total circuit current.
b. The circuit impedance.

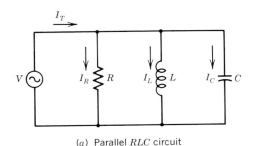

(a) Parallel *RLC* circuit

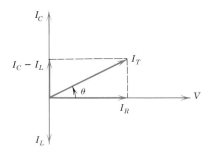

(b) Phasor diagram

FIGURE 24-25
(a) Parallel *RLC* circuit and (b) phasor diagram.

c. The phase angle between the circuit current and applied voltage.

Solution

a. As before,

$I_R = 0.2$ A

$I_L = 0.1$ A

$I_C = 0.4$ A

$I_T = \sqrt{I_R^2 + (I_C - I_L)^2}$ (24-25)

   $= \sqrt{0.2^2 + (0.4 - 0.1)^2}$ A $= \mathbf{0.36\ A}$

b. $Z = \dfrac{V}{I_T}$ (24-4)

   $= \dfrac{40\ V}{0.36\ A} = \mathbf{111\ \Omega}$

c. $\theta = \tan^{-1} \dfrac{I_C - I_L}{I_R}$ (24-26)

   $= \arctan \dfrac{0.4 - 0.1\ A}{0.2\ A} = \mathbf{56.3°}$

There are some important points to note in this example. First, observe that the *total* circuit current (0.36 A) is *less* than the capacitor branch current (0.4 A). It is not unusual, in some parallel *RLC* circuits where X_L and X_C are equal or close to equal in value, for a very large current (greater than the total current) to circulate in the *tank* circuit consisting of *L* and *C*.

Second, note that the total impedance of 111 Ω is *not* less than the smallest reactance of $X_C = 100$ Ω. That is, in a parallel *RL* or *RC* circuit, the total impedance (*Z*) *is* less than the smallest branch resistance or reactance. But the combined effect of $X_L = 400$ Ω and $X_C = 100$ Ω actually produces a *combined reactance* of 133 Ω (40 V/0.3 A). Obviously, parallel reactances and resistances cannot be combined in the same way that a purely resistive parallel circuit can be treated.

You will consider other combinations, such as series-parallel reactances, in Chapter 27, when you are better prepared to use complex numbers.

SUMMARY

1. A sine wave may be represented by a horizontal line called a *phasor*. The phasor is assumed to rotate counterclockwise at the frequency of the voltage supply.
2. A cosine curve is represented by a vertical phasor, and leads a sine curve by 90°.
3. Phasors at right angles can be used to represent the rms values of the current and voltage in pure *L* and *C* circuits.
4. A series *RL* circuit has a total applied voltage given by

$$V = \sqrt{V_R^2 + V_L^2}$$

and a phase angle between current and applied voltage given by

$$\theta = \tan^{-1} \frac{V_L}{V_R} = \tan^{-1} \frac{X_L}{R}$$

5. The total opposition to the current in a series *RL* circuit is called the impedance and is given by

$$Z = \sqrt{R^2 + X_L^2}$$

6. The current in an ac circuit is given by the equation:

$$I = \frac{V}{Z}$$

7. Practical coils (those in which resistance is *not* negligible) have a phase angle between the current and coil voltage of less than 90°.
8. A series *RC* circuit has a total applied voltage given by

$$V = \sqrt{V_R^2 + V_C^2}$$

and a phase angle between the current and applied voltage given by

$$\theta = -\tan^{-1}\frac{V_C}{V_R} = -\tan^{-1}\frac{X_C}{R}$$

9. The impedance of a series RC circuit is given by

$$Z = \sqrt{R^2 + X_C^2}$$

10. The impedance of a series RLC circuit is given by

$$Z = \sqrt{R^2 + (X_L - X_C)^2}$$

and

$$\theta = \tan^{-1}\frac{X_L - X_C}{R} = \tan^{-1}\frac{V_L - V_C}{V_R}$$

11. A single-beam oscilloscope display of two voltages in a stationary Lissajous figure can be used to obtain the phase angle between those two voltages using:

$$\theta = \sin^{-1}\frac{OA}{OB}$$

12. A parallel RLC circuit has a total line current:

$$I_T = \sqrt{I_R^2 + (I_C - I_L)^2}$$

impedance:

$$Z = \frac{V}{I_T}$$

and phase angle:

$$\theta = \tan^{-1}\frac{I_C - I_L}{I_R}$$

SELF-EXAMINATION

Answer T or F or a, b, c, d, or e for multiple choice
(Answers at back of book)

24-1. A phasor is a line that represents a quantity that varies with time. _____

24-2. Phasors are assumed to rotate clockwise at the same frequency as the voltage or current they represent. _____

24-3. Two phasors drawn at right angles can represent the voltage and current wave-forms in a purely capacitive or purely inductive circuit. _____

24-4. The phasor sum of the resistive and inductive voltages in a series RL circuit is *always* less than the algebraic sum. _____

24-5. If the resistive and inductive voltages in a series RL circuit are each 5 V, the total applied voltage and phase angle are
   a.   7.07 V, 90°
   b.   10 V, 45°
   c.   7.07 V, $\pi/4$ rad
   d.   10 V, 90°
   e.   none of the above

24-6. If the inductive and resistive voltages in a series RL circuit are equal, then the inductive reactance is equal to the resistance. _____

24-7. A series RL circuit has a resistance of 30 Ω and an inductive reactance of 40 Ω. The impedance is

     a. 50 Ω
     b. 70 Ω
     c. 10 Ω
     d. 35 Ω
     e. none of the above

24-8. A series RC circuit has a resistance of 50 Ω and a capacitive reactance of 120 Ω and is connected to a 130-V supply. The current drawn is
     a. 2.2 A
     b. 1.1 A
     c. 0.1 A
     d. 1 A
     e. none of the above

24-9. The voltage across a coil is always 90° out of phase with the current through the coil. _____

24-10. If a series RC circuit has a leading phase angle of 30°, the circuit is more capacitive than resistive. _____

24-11. If a series RC circuit has $X_C = 2R$, the phase angle between the current and applied voltage is
     a. 30°
     b. 63.4°
     c. 26.6°
     d. 60°
     e. 45°

24-12. The phase angle of a series RLC circuit is determined only by the values of R, L, and C. _____

24-13. If a capacitor is added in series with a series RL circuit, the result is always a decrease in the impedance. _____

24-14. If a capacitor of the correct reactance is connected in series with a series RL circuit, the result may be a purely resistive circuit. _____

24-15. The determination of a circuit's phase angle requires an oscilloscope with accurately calibrated vertical and horizontal inputs. _____

24-16. If a parallel RL circuit contains a resistance of 100 Ω and an inductive reactance of 1000 Ω, the circuit is considered predominantly inductive. _____

24-17. The total current leads the applied voltage in a parallel RL circuit whereas it lags in a series RL circuit. _____

24-18. The total line current in a parallel RLC circuit must, because it obeys Kirchhoff's current law, always be larger than any one of the branch currents.
_____

24-19. The total impedance of a parallel RLC circuit can be *larger* than the smallest branch resistance or reactance.

REVIEW QUESTIONS

1. Distinguish between a vector and a phasor, giving an example of each.
2. In what way would the waveform represented by a horizontal phasor be different if the phasor were assumed to rotate clockwise instead of counterclockwise?
3. Why, strictly speaking, should phasors be drawn or labeled with *peak* values and not *rms* values?

4. How would the waveforms of current and voltage for pure R, L, or C circuits be different (in Fig. 24-4) if the *current* phasor is drawn in the horizontal reference position?

5. What simple proof can you give, using voltage and impedance triangles, to show that the phase angle between the current and voltage in a series RL or RC circuit is determined by the ratio of the reactance to the resistance?

6. What is the difference between ac resistance, reactance, and impedance?

7. How would Eq. 24-3 be modified to obtain the total impedance of a series RL circuit in which the inductor's resistance R_L is not negligible?

8. Why is it not proper to refer to an impedance diagram as a phasor diagram?

9. Describe a method by which you could use a dual-beam oscilloscope to determine the resistance, inductance, and Q of a coil.

10. How would the impedance of a series RC circuit change, compared with an RL circuit, if the frequency of the applied voltage were increased?

11. How would you expect that the impedance and phase angle would change if you varied the frequency from direct current to some high value in a series RLC circuit? Sketch graphs of each.

12. How is it possible to develop larger voltages across the capacitor or inductor than the input voltage in a series RLC circuit? When is this effect most noticeable?

13. Sketch the circuit connections to a single-trace oscilloscope if you wanted to measure the phase angle between the current and applied voltage in a series RC circuit if one side of the supply and the low side of the oscilloscope are both grounded.

14. Give two ways in which a parallel RLC circuit differs from a parallel resistive circuit.

15. Show why $\tan \theta = -R/X_L$ for a *parallel* RL circuit, but $\tan \theta = X_L/R$ for a *series* RL circuit.

PROBLEMS

(Answers to odd-numbered problems at back of book)

24-1. Construct accurately a graph of a sine wave using the method of a rotating line as in Fig. 24-1. Use a 1-in. radius circle and angles every 15°. Make the horizontal axis of the graph approximately 2 in. long.

24-2. Repeat Problem 24-1 to construct a cosine curve as in Fig. 24-2.

24-3. Draw two vectors, A and B, at right angles to each other as in Fig. 24-3. Make them 1 in. long and construct the two waveforms they represent (as in Problems 24-1 and 24-2).

   Find the *phasor* sum, $A + B$, by constructing the diagonal as in Fig. 24-6b. Now use this line to construct the *waveform* represented by this phasor. Does it coincide with the *instantaneous* sums of the A and B *waveforms* taken every 30°?

24-4. Repeat Problem 24-3 but make A 1 in. long, and B 2 in. long.

24-5. A series RL circuit is connected to an ac supply. Voltmeters across the resistor and inductor (of negligible resistance) indicate 40 and 80 V, respectively. Calculate:
   a. The applied voltage.
   b. The phase angle between the current and applied voltage.
   c. The relative value of X_L compared with R.

24-6. A 24-V sinusoidal emf is applied to a series *RL* circuit. If a voltmeter across the resistor indicates 12 V, calculate:
   a.   The voltage across the inductance.
   b.   The phase angle between the applied voltage and current.
   c.   The relative value of X_L compared with R.

24-7. A coil of negligible resistance and a 2.2-kΩ resistor are series-connected across a 24-V, 400-Hz supply. If a voltmeter connected across the coil indicates 10 V, calculate:
   a.   The reading of a voltmeter connected across the resistor.
   b.   The phase angle between the applied voltage and current.
   c.   The current in the circuit.
   d.   The inductance of the coil.

24-8. The phase angle between current and applied voltage in a series *RL* circuit is known to be 30°. If the voltage across the 4.7-kΩ resistor is 15 V, calculate:
   a.   The inductor voltage.
   b.   The applied voltage.
   c.   The current in the circuit.
   d.   The inductive reactance of the coil.

24-9. A series *RL* circuit draws a current of 20 mA from a 24-V, 60-Hz source. If the inductive reactance is known to be twice as large as the resistance, calculate:
   a.   The circuit impedance.
   b.   The circuit resistance.
   c.   The circuit's inductive reactance.
   d.   The resistor voltage.
   e.   The inductor voltage.
   f.   The inductance of the coil.

24-10. Repeat Problem 24-9 with $R = 2X_L$.

24-11. A 200-μH coil of negligible resistance and a 47-Ω resistor are series-connected across a 20-V peak-to-peak, 10-kHz, sinusoidal signal generator. Calculate:
   a.   The circuit impedance.
   b.   The current in the circuit.
   c.   The phase angle between the current and applied voltage.

24-12. A 100-mH coil of unknown resistance is connected in series with a 220-Ω resistor and an 18-V peak-to-peak sine wave at 500 Hz. If a series-connected milliammeter indicates 12 mA, calculate:
   a.   The circuit impedance.
   b.   The resistance of the coil.
   c.   The voltage across the resistor.
   d.   The impedance of the coil.
   e.   The voltage across the coil.

24-13. A coil of unknown inductance and resistance is connected in series with a 4.7-Ω resistor and a 6.3-V, 60-Hz sinusoidal supply. A dual-beam oscilloscope, connected as in Fig. 24-9*a*, indicates a peak-to-peak resistor voltage of 80 mV. A full cycle of the resistor voltage occupies 10 cm and is delayed 2.25 cm compared with the coil voltage. Calculate:
   a.   The circuit impedance.
   b.   The phase angle between the current and applied voltage.
   c.   The ac resistance of the coil.
   d.   The inductive reactance of the coil.

    e.   The Q of the coil.

    f.   The inductance of the coil.

24-14. A coil with a Q of 2 is connected in series with a resistor and a 60-Hz supply. If the coil voltage is 100 V and the resistor voltage is 60 V, calculate the supply voltage. Draw the phasor diagram.

24-15. A capacitor and a 33-Ω resistor are connected in series across a 60-Hz sinusoidal supply. A series-connected ac ammeter indicates 1.65 A, and an oscilloscope across the capacitor indicates 150 V peak. Calculate:

    a.   The applied voltage.

    b.   The phase angle between the current and applied voltage.

    c.   The amount of capacitance.

24-16. The current in a series RC circuit is 40 mA. If the phase angle between the current and applied voltage is $-50°$ and the resistor is 330 Ω, calculate:

    a.   The applied voltage.

    b.   The impedance of the circuit.

    c.   The size of the capacitor if the frequency is 60 Hz.

24-17. a.   How much resistance must be added in series with a 200-pF capacitor to limit the current to 80 μA when the series combination is connected to a 150-mV, 1-MHz sine wave?

    b.   What is the circuit's phase angle between the applied voltage and current?

24-18. a.   How much capacitance must be added in series with a 50-Ω resistor to limit the current to 1.5 A when the series combination is connected to a 340-V peak-to-peak, 60-Hz sine wave?

    b.   What is the circuit's phase angle between the applied voltage and current?

24-19. For the circuit of Fig. 24-26, calculate V_o/V_i at 1 kHz and 10 kHz. Sketch a curve of V_o/V_i versus f.

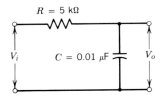

FIGURE 24-26

Circuit for Problems 24-19 and 24-20.

24-20. Repeat Problem 24-19 using $C = 0.002$ μF.

24-21. Given the circuit in Fig. 24-27, derive an equation for V_o/V_i in terms of R_1, R_2, and X_C.

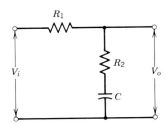

FIGURE 24-27

Circuit for Problem 24-21.

24-22. Given the circuit in Fig. 24-28:

a. Derive an equation for V_o/V_i in terms of R_1, R_2, and X_C.

b. Show that at very high frequencies, $\dfrac{V_o}{V_i} \approx \dfrac{R_2}{R_1 + R_2}$ for both circuits (Figs. 24-27 and 24-28).

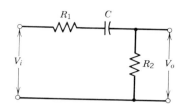

FIGURE 24-28
Circuit for Problem 24-22.

24-23. An 8-H coil (of negligible resistance), a 0.0035-μF capacitor, and a 2.2-kΩ resistor are series-connected across a 3.5-V, variable-frequency sinusoidal signal generator. For a frequency setting of 600 Hz, calculate:

a. The circuit impedance.

b. The circuit current.

c. The voltage across each component.

d. The phase angle between the current and applied voltage.

Draw the circuit, phasor, and impedance diagrams indicating all values.

24-24. Repeat Problem 24-23 using a frequency of 951 Hz. Compare the results with Problem 24-23.

24-25. Repeat Problem 24-23 using a frequency of 1200 Hz. Compare the results with Problems 24-23 and 24-24.

24-26. A relay coil has a resistance of 300 Ω and requires a current of 150 mA for its operation. When operated from a 60-Hz source, the required voltage is 180 V. Calculate the value of the capacitance in series with the relay coil that will allow its operation from a 120-V, 60-Hz source.

24-27. A relay coil is designed to operate at 24 V, 400 Hz, drawing a current of 100 mA. If the coil has an ac resistance of 100 Ω, calculate:

a. The series capacitance required to allow the relay to operate with 18 V applied.

b. The voltage across the capacitor.

c. The minimum voltage it is possible to use to operate the relay if the proper size capacitor is selected.

d. The size of capacitor for the operation stated in c.

24-28. A solenoid that requires a current of 250 mA to operate at 120 V, 60 Hz has an ac resistance of 300 Ω. Calculate the size of series capacitance needed to allow the combination to operate at 240 V, 60 Hz.

24-29. An oscilloscope is used to make a phase angle measurement between the current and applied voltage in a series RLC circuit. The resulting ellipse has an overall height of 6 cm and intercepts on the y-axis 5 cm apart. If a voltmeter connected across the series combination of L and C indicates 30 V, calculate:

a. The phase angle between the current and applied voltage.

b. The resistor voltage.

c. The applied voltage.

Draw a phasor diagram given that the circuit is capacitive.

24-30. Determine the oscilloscope display when it is connected as in the circuit of Fig. 24-29 to show a Lissajous figure. Justify your answer.

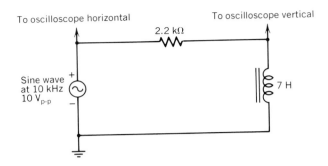

FIGURE 24-29
Circuit for Problem 24-30.

24-31. A 0.5-H inductor of negligible resistance and a 1.2-kΩ resistor are connected in parallel across a 48-V, 400-Hz source. Calculate:
    a.   The total current drawn from the source.
    b.   The impedance of the circuit.
    c.   The phase angle between the total current and applied voltage.
24-32. Repeat Problem 24-31 with the inductor replaced by a 0.2-μF capacitor.
24-33. A parallel *RLC* circuit consists of a 0.5-H coil (of negligible resistance), a 2.2-kΩ resistor, and a 0.03-μF capacitor connected across a 5-V, 1-kHz source. Calculate:
    a.   The total circuit current.
    b.   The circuit impedance.
    c.   The phase angle between the circuit current and applied voltage.
    Draw a phasor diagram showing all values.
24-34. a.   Calculate the size of the capacitor to be added in parallel with the circuit of Problem 24-33 to make the phase angle between the current and applied voltage equal zero.
    b.   Determine the new line current and impedance after adding the capacitor. Compare the results with Problem 24-33.

CHAPTER 25

POWER IN AC CIRCUITS

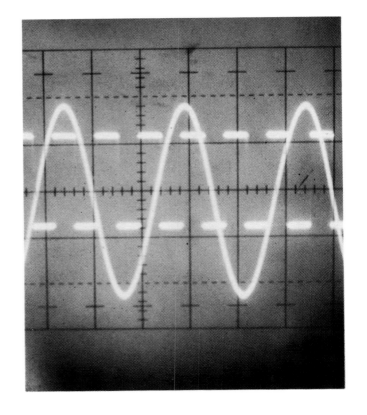

There are three types of power in an ac circuit: *true power, reactive power,* and *apparent power.* The first of these, true power, is the power dissipated in a resistance when electrical energy is converted to a different form. True power is measured in watts. Reactive power is power that is stored in capacitors and inductors and then returned to the system. It is measured in volt-amperes reactive (vars). The third type, apparent power, is simply the product of the total applied voltage and the total circuit current. It is measured in volt-amperes (VA).

The three types of power are related, and can be represented in a right-angled power triangle, from which the *power factor* can be obtained as the ratio of true power to apparent power. For inductive loads, the power factor is called *lagging,* to distinguish it from the *leading* power factor in a capacitive load.

The power factor of a circuit determines how much current is needed from the source to deliver a given true power. Since a *low* power factor requires a higher current than a unity power factor circuit, methods of correcting the power factor are examined in this chapter. This correction is usually accomplished by connecting a large capacitance across the line to offset the naturally inductive loads of motors and fluorescent lamps.

Finally, you will learn how to measure and calculate power in a three-phase, three-wire system, and how to achieve maximum power transfer to a load from an ac source with a given internal impedance.

25-1 POWER IN PURE RESISTANCE

As you learned in Chapter 15, current and voltage are in phase with each other in a purely resistive circuit, or in any *portion* of an ac circuit where only pure resistance is considered. Such an in-phase relationship can be represented by the phasor diagram in Fig. 25-1*b* or the waveforms in Fig. 25-1*c*.

Note that the power varies at twice the source frequency (as was shown in Section 15-3.1), and that it has an average value given by

$$P = I_R^2 R = V_R I_R = V_R^2/R \text{ watts} \quad (15\text{-}12 \text{ to } 15\text{-}14)$$

To distinguish the power *dissipated in a resistance* (in the form of heat) from other types of power, the term *real power* or *true power* is used. The symbol P (with no subscript) and the unit of measure watt (W) are used exclusively for real or true power. *True* power describes the conversion of electrical energy into another form, such as heat, light, rotating mechanical power, and so forth. The

true power in watts used by a dc or ac circuit is indicated by a properly connected wattmeter. The wattmeter will indicate power only if the circuit contains some power-dissipating element.

EXAMPLE 25-1

A wattmeter, connected in an ac circuit, indicates 50 W. An ammeter, connected in series with the only resistance in the whole circuit, reads 1.5 A. Determine the resistance of the ac circuit.

Solution

$$P = I_R^2 R \quad (15\text{-}12)$$
$$R = \frac{P}{I_R^2}$$
$$= \frac{50 \text{ W}}{(1.5 \text{ A})^2} = \textbf{22.2 } \boldsymbol{\Omega}$$

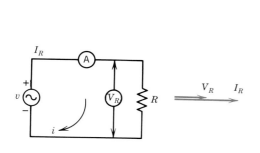

(a) Purely resistive circuit   (b) Phasor diagram

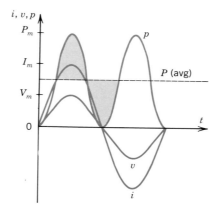

(c) Current, voltage, and power waveforms

FIGURE 25-1

Power in a pure resistance.

25-2 POWER IN PURE INDUCTANCE

If an ac circuit contains only inductance, the voltage and current will be 90° out of phase, as shown by the phasor diagram in Fig. 25-2*b*. Multiplying the *v* and *i* waveforms results in a power curve that has a frequency twice that of the source (Fig. 25-2*c*). However, over a complete cycle of input voltage, the power curve has an average value of *zero*. That is, the curve shows an equal alternation of positive and negative power above and below the zero time axis.

What do the terms *positive* and *negative* power mean? In Fig. 25-2*c*, the shaded portion of the power curve *above* the zero axis represents energy being delivered *to* the inductor (or load) *from* the source. This *positive* power actually represents a storage of energy in the magnetic field of the inductance. The shaded portion of the power curve *below* the zero axis represents energy returned *to* the source *from* the inductor. This *negative* power indicates that a flow of energy is taking place in the *opposite* direction (from load to source) when the coil's magnetic field collapses.

In a pure inductor with zero resistance, the power *returned* during a given quarter-cycle is the same as the power *delivered* during the preceding quarter-cycle. It is clear then that the *average true power (P) is zero* in a *pure inductance*. This simply means that if a wattmeter is connected to the circuit in Fig. 25-2*a*, it will read zero. Further, the load coil in Fig. 25-2*a* would never become warm, because no power is dissipated (converted to another form or "used up") in that coil.

The source, however, must be capable of delivering power for a quarter of a cycle, even though this power will be returned during the next quarter-cycle. This stored or transferred power is called *reactive* power (P_q).

In the case of a purely inductive circuit, the reactive power is given by

$$P_q = V_L I_L \qquad \text{volt-amperes reactive} \qquad (25\text{-}1)$$

where: P_q is the reactive (or quadrature) power,* in volt-amperes reactive (vars)

V_L is the voltage across the inductance, in volts (V)

I_L is the current through the inductance, in amperes (A)

Since $V_L = I_L X_L$ (20-1)

then, $P_q = I_L^2 X_L$ vars   (25-2)

and, $P_q = \dfrac{V_L^2}{X_L}$ vars   (25-3)

where X_L is the inductive reactance, in ohms.

Note that the equations for reactive power are similar to

---

*The subscript *q* in P_q represents quadrature power. It is the product of current and voltage that are one *quarter* of a cycle or 90° out of phase with each other. Some texts use the letter *Q* instead of P_q to represent reactive power. However, *Q* is also used for the quality of a coil. To avoid confusion, this text will use P_q for reactive power.

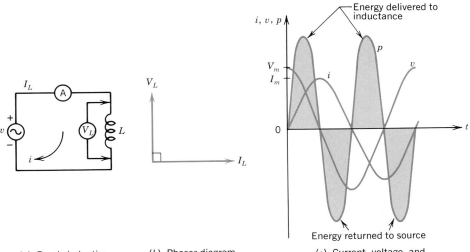

(*a*) Purely inductive circuit

(*b*) Phasor diagram

(*c*) Current, voltage, and power waveforms

FIGURE 25-2

Power in a pure inductance.

those for true power, with X_L used in place of R. You must remember to use vars for the unit of reactive power, however, rather than watts.

EXAMPLE 25-2

Calculate the reactive power of a circuit that has an inductance of 4 H when it draws 1.4 A from a 60-Hz supply.

Solution

$$X_L = 2\pi fL \tag{20-2}$$
$$= 2\pi \times 60 \text{ Hz} \times 4 \text{ H}$$
$$= 1508 \ \Omega$$
$$P_q = I_L^2 X_L \tag{25-2}$$
$$= (1.4 \text{ A})^2 \times 1508 \ \Omega$$
$$= 2956 \text{ vars} = \textbf{2.96 kvar}$$

Note that 1 kvar = 1 kilovar = 1000 vars.

The importance of reactive power will be seen in a later section. Note the basic difference between *true* and *reactive* power:

True power, in watts, is the product of the voltage and current that are *in phase* with each other.

Reactive power, in vars, is the product of the voltage and current that are 90° *out of phase* with each other.

Also, because the current *lags* the voltage in an inductive circuit, inductive reactive power is often called *lagging* reactive power.

25-3 POWER IN PURE CAPACITANCE

For pure capacitance, the voltage and current are once again 90° out of phase with each other, with the current *leading*, as shown by the phasor diagram in Fig. 25-3b.

The product of v and i gives a power curve as shown in Fig. 25-3c. Again, the energy that is *delivered* to the capacitor and stored in the electric field is represented as a *positive* quantity. One-quarter cycle later, *all* of this energy is *returned* to the source as the capacitor discharges. Thus the *average true power* (P) is *zero* in a *true capacitance*. This means that no power is dissipated in a pure capacitor and no heat develops.

However, as was the case with the coil, reactive power (P_q) is drawn by the capacitor, and the source must be able to supply this power.

For a purely capacitive circuit, the reactive power is given by:

$$P_q = V_C I_C \qquad \text{volt-amperes reactive} \tag{25-4}$$

where: P_q is the reactive power, in volt-amperes reactive (vars)

V_C is the voltage across the capacitance, in volts (V)

I_C is the current through the capacitance, in amperes (A)

Since $$V_C = I_C X_C \tag{23-1}$$
then, $$P_q = I_C^2 X_C \text{ vars} \tag{25-5}$$
and $$P_q = \frac{V_C^2}{X_C} \text{ vars} \tag{25-6}$$

where X_C is the capacitive reactance, in ohms.

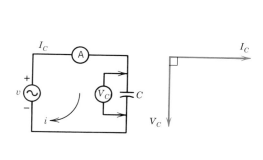

(a) Purely capacitive circuit

(b) Phasor diagram

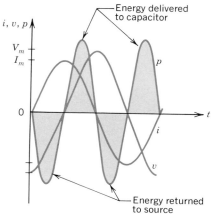

(c) Current, voltage and power waveforms

FIGURE 25-3

Power in a pure capacitance.

Again, the equations for reactive and true power are similar, with X_C used in place of R. But you must use vars, not watts, for reactive power.

EXAMPLE 25-3

A reactive power of 100 vars is drawn by a 10-μF capacitor due to a current of 0.87 A. Calculate the frequency.

Solution

$$P_q = I_C^2 X_C \qquad (25\text{-}5)$$

$$X_C = \frac{P_q}{I_C^2}$$

$$= \frac{100 \text{ vars}}{(0.87 \text{ A})^2} = 132 \ \Omega$$

$$X_C = \frac{1}{2\pi f C} \qquad (23\text{-}2)$$

$$f = \frac{1}{2\pi X_C C}$$

$$= \frac{1}{2\pi \times 132 \ \Omega \times 10 \times 10^{-6} \text{ F}} = \textbf{120 Hz}$$

Earlier, it was noted that inductive reactive power is referred to as lagging reactive power. In the same way, capacitive reactive power is known as *leading* reactive power because the current *leads* the voltage in a capacitive circuit. Reactive power can be measured by using a *varmeter* (a wattmeter used in conjunction with a phase-shifting circuit to provide a voltage 90° out of phase from the true load voltage). The product of the meter's current and voltage now gives an indication of the volt-amperes reac-

tive. The varmeter cannot, however, give a direct indication of whether the current is leading or lagging. A method of doing this is considered in Section 25-6.

25-4 POWER IN A SERIES *RL* CIRCUIT

As you have seen, inductance is always accompanied by resistance. Thus, coils in motors, generators, or similar devices always contain both resistance and inductance. When an ac voltage is applied, the current (I) is neither *in* phase nor 90° *out* of phase with the applied voltage (V), as shown in Fig. 25-4b.

This means that (unlike pure resistance and pure reactance) the product of the voltmeter and ammeter readings in Fig. 25-4a is a combination of true and reactive (quadrature) power. **The product of total V and total I is called *apparent power* (P_s). Since it is neither true power in watts nor reactive power in vars, a new unit called the volt-ampere (VA) is used to measure apparent power.**

$$P_s = V_T I_T \qquad \text{volt-amperes} \qquad (25\text{-}4)$$

where: P_s is the apparent power,* in volt-amperes (VA)
V_T is the total applied voltage, in volts (V)
I_T is the total circuit current, in amperes (A)

Note the use of the subscript T to emphasize that apparent power is the product of the *total* applied voltage and current. That is, multiplying the voltmeter reading in volts

---

*Some texts use the letter S to describe apparent power. We shall use the older but more descriptive notation P_s.

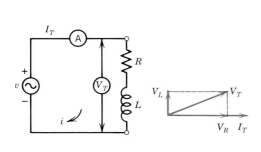

(a) Series *RL* circuit   (b) Phasor diagram

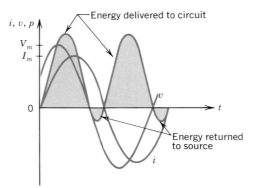

(c) Current, voltage, and power waveforms

FIGURE 25-4
Power in a series RL circuit.

by the ammeter reading in amperes gives the total apparent circuit power in volt-amperes. It will later be shown that P_s is the vector sum of P_q and P. (See Section 25-5.)

But
$$I_T = \frac{V_T}{Z} \qquad (24\text{-}4)$$

Thus
$$P_s = \frac{V_T^2}{Z} \qquad (25\text{-}8)$$

and
$$P_s = I_T^2 Z \qquad (25\text{-}9)$$

where: Z is the circuit impedance, in ohms.

Once again, the equations for true power and apparent power are similar, with Z used in place of R and volt-amperes substituted for watts.

EXAMPLE 25-4

A current of 2.6 A flows through a series circuit of 300-Ω resistance and 400-Ω inductive reactance. Calculate the apparent power drawn by the circuit.

Solution

$$Z = \sqrt{R^2 + X_L^2} \qquad (24\text{-}3)$$
$$= \sqrt{(300\ \Omega)^2 + (400\ \Omega)^2} = 500\ \Omega$$
$$P_s = I_T^2 Z$$
$$= (2.6\ \text{A})^2 \times 500\ \Omega$$
$$= 3380\ \text{VA} = \mathbf{3.38\ kVA}$$

Note that 1000 VA = 1 kilovolt-ampere = 1 kVA.

25-4.1 Measuring the Inductance of a Coil

Figure 25-4c shows a power curve that is more positive than negative. This means that there is some net true power delivered to the circuit and dissipated in the resistance. This true power must be indicated on a wattmeter connected as shown in Fig. 25-5.

If a voltmeter and ammeter are also connected as shown, it is possible (from the three meter readings) to determine the coil's resistance and inductance, if the frequency is known. (See Example 25-5.)

EXAMPLE 25-5

A coil is connected to the 120-V, 60-Hz line as in Fig. 25-5. The wattmeter indicates 525 W and the ammeter indicates 5 A. Calculate the inductance of the coil.

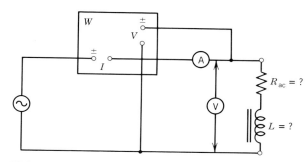

FIGURE 25-5

Measuring the inductance of a coil.

Solution

$$Z = \frac{V}{I} \qquad (24\text{-}4)$$
$$= \frac{120\ \text{V}}{5\ \text{A}} = 24\ \Omega$$

$$R = \frac{P}{I_R^2} \qquad (15\text{-}12)$$
$$= \frac{525\ \text{W}}{(5\ \text{A})^2} = 21\ \Omega$$

$$Z = \sqrt{R^2 + X_L^2} \qquad (24\text{-}3)$$
$$X_L = \sqrt{Z^2 - R^2}$$
$$= \sqrt{(24\ \Omega)^2 - (21\ \Omega)^2} = 11.6\ \Omega$$

$$X_L = 2\pi f L \qquad (20\text{-}2)$$

$$L = \frac{X_L}{2\pi f}$$
$$= \frac{11.6\ \Omega}{2\pi \times 60\ \text{Hz}} = \mathbf{30.8\ mH}$$

NOTE The calculated resistance (21 Ω) is the total ac resistance of the coil, including hysteresis and eddy current losses if a magnetic core is present.

Note also that this method of measuring the coil's inductance (and resistance) is limited to power line frequencies because of the frequency limitation of most wattmeters.

25-5 THE POWER TRIANGLE

The three types of power in an ac circuit have thus far been identified: true power in watts, reactive (quadrature) power in vars, and apparent power in volt-amperes. Is there a simple relationship among the three? Refer to Fig. 25-6.

You know that (see Fig. 25-6b)

$$V_T = \sqrt{V_R^2 + V_L^2} \qquad (24\text{-}1)$$

Therefore, $V_T I_T = \sqrt{(V_R I_T)^2 + (V_L I_T)^2}$

But

$$V_T I_T = P_s \quad \text{VA} \quad (25\text{-}7)$$

$$V_R I_T = P \quad \text{watts} \quad (15\text{-}14)$$

$$V_L I_T = P_q \quad \text{vars} \quad (25\text{-}1)$$

Therefore,

$$P_s = \sqrt{P^2 + P_q^2} \quad \textbf{volt-amperes} \quad (25\text{-}10)$$

where: P_s is the apparent power, in volt-amperes (VA)
   P is the true power, in watts (W)
   P_q is the reactive power, in volt-amperes reactive (vars)

This relation can be represented in a *power triangle,* as shown in Fig. 25-6c. The apparent power is represented by the hypotenuse of the right triangle. The true power is the product of the current and voltage in phase with each other and is drawn horizontally. The out-of-phase product of V_L and I_T gives the reactive power and is drawn vertically *downward.* This is a convention used to show a *lagging* inductive reactive power, corresponding to a lagging current. (A capacitive reactive power is drawn vertically *upward,* corresponding to a *leading* current.)

Note that no arrows are shown on the power triangle because the power varies at *twice* the frequency of the voltage and current in the circuit. The triangle is used to show the Pythagorean relationship of Eq. 25-10.

EXAMPLE 25-6

A series circuit of 300-Ω resistance and 400-Ω inductive reactance draws a current of 2.6 A. Calculate:
a. The true power.
b. The inductive reactive power.
c. The apparent power.

Solution

a. $P = I_R^2 R$ (15-12)
   $= (2.6 \text{ A})^2 \times 300 \ \Omega = \textbf{2028 W}$

b. $P_q = I_L^2 X_L$ (25-2)
   $= (2.6 \text{ A})^2 \times 400 \ \Omega = \textbf{2704 vars}$

c. $P_s = \sqrt{P^2 + P_q^2}$ (25-10)
   $= \sqrt{(2028 \text{ W})^2 + (2704 \text{ vars})^2}$
   $= 3380 \text{ VA} = \textbf{3.38 kVA}$

EXAMPLE 25-7

A coil connected across the 120-V, 60-Hz line draws a current of 5 A. A series-connected wattmeter indicates 525 W. Calculate:
a. The apparent power.
b. The reactive power.

Solution

a. $P_s = V_T I_T$ (25-7)
   $= 120 \text{ V} \times 5 \text{ A} = \textbf{600 VA}$

b. $P_s = \sqrt{P^2 + P_q^2}$ (25-10)
   $P_q = \sqrt{P_s^2 - P^2}$
   $= \sqrt{(600 \text{ VA})^2 - (525 \text{ W})^2} = \textbf{290 vars}$

25-5.1 Power Triangle for a Parallel *RL* Circuit

When a pure resistance and a pure inductance are parallel-connected across an ac supply, as in Fig. 25-7a, the total current is (see Fig. 25-7b):

$$I_T = \sqrt{I_R^2 + I_L^2} \quad (24\text{-}7)$$

Therefore, $V_T I_T = \sqrt{(V_T I_R)^2 + (V_T I_L)^2}$

and $P_s = \sqrt{P^2 + P_q^2}$ (25-10)

where all terms are as already defined.

Again, the power triangle can be drawn as in Fig. 25-7c. Evidently, Eq. 25-10 can be used for either a series or parallel *RL* combination.

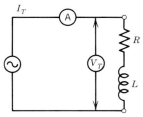

(a) Series *RL* circuit

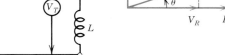

(b) Phasor diagram

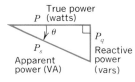

(c) Power triangle

FIGURE 25-6
Power triangle relations for a series *RL* circuit.

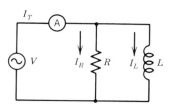

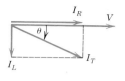

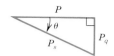

(a) Parallel *RL* circuit (b) Phasor diagram (c) Power triangle

FIGURE 25-7

FIGURE 25-7

Power triangle relations for a parallel *RL* circuit

EXAMPLE 25-8

A 300-Ω resistor and an inductive reactance of 400 Ω are parallel-connected across a 120-V, 60-Hz supply. Calculate:

a. The true power.
b. The reactive power.
c. The apparent power.
d. The current drawn from the supply.

Solution

a. $P = \dfrac{V_R^2}{R}$ (15-13)

   $= \dfrac{(120\ \text{V})^2}{300\ \Omega} = $ **48 W**

b. $P_q = \dfrac{V_L^2}{X_L}$ (25-3)

   $= \dfrac{(120\ \text{V})^2}{400\ \Omega} = $ **36 vars**

c. $P_s = \sqrt{P^2 + P_q^2}$

   $= \sqrt{(48\ \text{W})^2 + (36\ \text{vars})^2} = $ **60 VA**

d. Since $P_s = VI_T$

   $I_T = \dfrac{P_s}{V} = \dfrac{60\ \text{VA}}{120\ \text{V}} = $ **0.5 A**

25-5.2 Power Triangle for a Series or Parallel *RC* Circuit

Using the relationship

$$V_T = \sqrt{V_R^2 + V_C^2} \qquad (24\text{-}8)$$

for a series *RC* circuit (Fig. 25-8a) and

$$I_T = \sqrt{I_R^2 + I_C^2} \qquad (24\text{-}19)$$

for a parallel circuit (Fig. 25-8b), it can be shown that

$$P_s = \sqrt{P^2 + P_q^2} \qquad (25\text{-}10)$$

where all terms are as previously defined.

Note that the power triangle in Fig. 25-8c has the leading capacitive reactive power drawn vertically *upward*. This corresponds to a current that *leads* the applied voltage, for both series and parallel *RC* circuits.

In summary, it can be said that the Pythagorean power equation (Eq. 25-10) can be used for all circuit combinations, series or parallel, *RL* or *RC*.

25-5.3 Power Triangle for a Series or Parallel *RLC* Circuit

When a circuit contains both inductance and capacitance, there is a transfer of reactive power back and forth be-

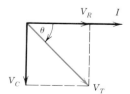

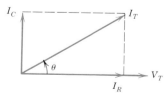

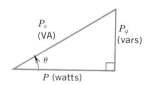

(a) Phasor diagram for a series *RC* circuit

(b) Phasor diagram for a parallel *RC* circuit

(c) Power triangle for either a series or parallel *RC* circuit

FIGURE 25-8

Phasor diagrams and power triangle for series and parallel *RC* circuits.

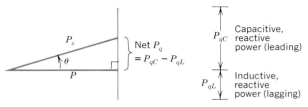

FIGURE 25-9
Power triangle for a series or parallel *RLC* circuit.

tween the inductance and the capacitance. This can be seen by examining Fig. 25-2c and Fig. 25-3c. During the first quarter-cycle, energy is being *delivered* to the inductance (positive power) at the same time that the capacitor is *returning* energy to the source (negative power). If both the capacitor and the inductor are in the same circuit (in series *or* parallel), this exchange of reactive power results in a *reduction* in the reactive power that must be supplied by the source. (See the power triangle in Fig. 25-9.)

The power triangle is now represented by the following equation:

$$P_s = \sqrt{P^2 + (P_{qC} - P_{qL})^2} \qquad \text{volt-amperes} \quad \text{(25-11)}$$

where: P_s is the apparent power, in volt-amperes (VA)
   P is the true power, in watts (W)
   P_{qC} is the capacitive reactive power, in volt-amperes reactive (vars)
   P_{qL} is the inductive reactive power, in volt-amperes reactive (vars)

Figure 25-9 has been drawn arbitrarily with $P_{qC} > P_{qL}$. This results in a net reactive power that is *capacitive*. If $P_{qC} < P_{qL}$, the circuit draws a net *inductive* reactive power. Also, if $P_{qC} = P_{qL}$, *no* reactive power is supplied by the source. This is an important condition, and is used in power factor correction. (See Section 25-8.)

EXAMPLE 25-9

A coil and a 12-μF capacitor are connected in parallel across a 240-V, 60-Hz line. The coil has an inductance of 0.25 H and a resistance of 75 Ω. Calculate:

a.   The true power.
b.   The net reactive power.
c.   The apparent power.
d.   The total line current.
Draw the circuit diagram and the power triangle.

Solution

a.   The coil's inductive reactance

$$X_L = 2\pi fL \qquad \text{(20-2)}$$
$$= 2\pi \times 60 \text{ Hz} \times 0.25 \text{ H}$$
$$= 94 \ \Omega$$

The coil's impedance

$$Z = \sqrt{R^2 + X_L^2} \qquad \text{(24-3)}$$
$$= \sqrt{(75 \ \Omega)^2 + (94 \ \Omega)^2} = 120 \ \Omega$$

The current through the coil

$$I = \frac{V}{Z} \qquad \text{(24-4)}$$
$$= \frac{240 \text{ V}}{120 \ \Omega} = 2 \text{ A}$$

The true power

$$P = I_R^2 R \qquad \text{(15-12)}$$
$$= (2 \text{ A})^2 \times 75 \ \Omega$$
$$= \textbf{300 W}$$

b.   The capacitor's reactance

$$X_C = \frac{1}{2\pi fC} \qquad \text{(23-2)}$$
$$= \frac{1}{2\pi \times 60 \text{ Hz} \times 12 \times 10^{-6} \text{ F}}$$
$$= 221 \ \Omega$$

Capacitive reactive power

$$P_{qC} = \frac{V_C^2}{X_C} \qquad \text{(25-6)}$$
$$= \frac{(240 \text{ V})^2}{221 \ \Omega} = 261 \text{ vars}$$

Inductive reactive power

$$P_{qL} = I_L^2 X_L \qquad \text{(25-2)}$$
$$= (2 \text{ A})^2 \times 94 \ \Omega$$
$$= 376 \text{ vars}$$

Net reactive power $= P_{qL} - P_{qC}$
$$= 376 - 261 \text{ vars}$$
$$= \textbf{115 vars} \text{ (inductive)}$$

c.   Apparent power

$$P_s = \sqrt{P^2 + (P_{qC} - P_{qL})^2} \qquad \text{(25-11)}$$
$$= \sqrt{(300 \text{ W})^2 + (261 \text{ vars} - 376 \text{ vars})^2}$$
$$= \textbf{321 VA}$$

d. $P_s = V_T I_T \qquad \text{(25-7)}$

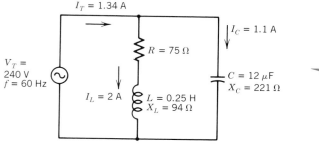
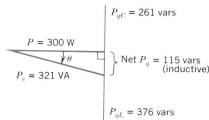

FIGURE 25-10
Circuit and power triangle for Example 25-9.

Therefore, $I_T = \dfrac{P_s}{V_T}$

$$= \dfrac{321 \text{ VA}}{240 \text{ V}} = \mathbf{1.34\ A}$$

The circuit and power triangle are shown in Fig. 25-10.

NOTE Example 25-9 provides a method of finding the total line current *without* a phasor addition of the coil and capacitor currents. This phasor addition is not simple because the coil current is neither 90° nor 180° out of phase with the capacitor current.

25-5.4 Application of a Capacitor as a Voltage Dropping Device*

A leading manufacturer of elapsed-time indicators (devices that record the length of time a piece of electrical equipment has been energized and operating) makes a basic unit that operates on 12-V ac. For use with higher voltages, the manufacturer uses a *series resistor* for 24 V, but a *capacitor* for 120-V operation. The reactance of the capacitor limits the current to the proper value without consuming any power, thus maintaining the low power consumption of the basic timer. This concept is illustrated in Example 25-10.

EXAMPLE 25-10

An elapsed-time indicator that requires 12 V and 60 Hz for its operation at 50 mA has a power consumption of 0.1 W. Calculate:

---

*This section may be omitted with no loss of continuity.

a. The series resistor and its power dissipation so the indicator can run on 24-V ac.
b. The size of series capacitor that would be required so the indicator could run on 24-V ac.
c. The size of series capacitor required and its voltage so the indicator can run on 120-V ac.
d. The size of a series resistor and its power dissipation if the indicator is to run on 120-V ac.

Solution

a. Impedance of indicator

$$Z = \dfrac{V}{I} \qquad (24\text{-}4)$$

$$= \dfrac{12 \text{ V}}{0.05 \text{ A}} = 240 \ \Omega$$

The resistance of the indicator may be found:

$$P = I^2 R_L \qquad (15\text{-}12)$$

$$R_L = \dfrac{P}{I^2} = \dfrac{0.1 \text{ W}}{(0.05 \text{ A})^2} = 40 \ \Omega$$

The reactance of the indicator may now be found:

$$Z = \sqrt{R_L^2 + X_L^2} \qquad (24\text{-}3)$$

$$X_L = \sqrt{Z^2 - R_L^2}$$

$$= \sqrt{(240 \ \Omega)^2 - (40 \ \Omega)^2} = 237 \ \Omega$$

Impedance of a series circuit operating at 24 V (see Fig. 25-11a) is

$$Z = \dfrac{V}{I} = \dfrac{24 \text{ V}}{0.05 \text{ A}} = 480 \ \Omega$$

$$Z = \sqrt{(R + R_L)^2 + X_L^2} \qquad (24\text{-}3)$$

$$R + R_L = \sqrt{Z^2 - X_L^2}$$

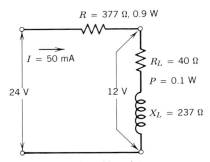

(a) 24-V operation with resistor

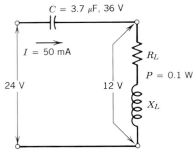

(b) 24-V operation with capacitor

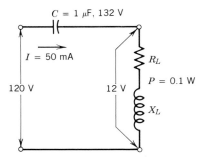

(c) 120-V operation with capacitor

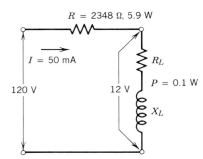

(d) 120-V operation with resistor

FIGURE 25-11

Circuits for Example 25-10.

$$= \sqrt{(480 \ \Omega)^2 - (237 \ \Omega)^2}$$
$$= 417 \ \Omega$$
$$R = 417 \ \Omega - R_L$$
$$= 417 \ \Omega - 40 \ \Omega$$
$$= \mathbf{377 \ \Omega}$$
$$P_R = I^2R \qquad (15\text{-}12)$$
$$= (0.05 \ \text{A})^2 \times 377 \ \Omega$$
$$= \mathbf{0.9 \ W}$$

b.   When a capacitor is used, the total impedance must still be (see Fig. 25-11b)

$$Z = \frac{V}{I} = \frac{24 \ \text{V}}{0.05 \ \text{A}} = 480 \ \Omega$$

where $\qquad Z = \sqrt{R_L^2 + (X_C - X_L)^2} \qquad (24\text{-}16)$

$$X_C - X_L = \sqrt{Z^2 - R_L^2}$$
$$= \sqrt{(480 \ \Omega)^2 - (40 \ \Omega)^2}$$
$$= 478 \ \Omega$$
$$X_C = 478 \ \Omega + X_L$$
$$= 478 \ \Omega + 237 \ \Omega$$
$$= 715 \ \Omega$$
$$X_C = \frac{1}{2\pi fC} \qquad (23\text{-}2)$$

Therefore, $\quad C = \dfrac{1}{2\pi fX_C}$

$$= \frac{1}{2\pi \times 60 \ \text{Hz} \times 715 \ \Omega}$$
$$= 3.7 \times 10^{-6} \ \text{F} = \mathbf{3.7 \ \mu F}$$

Although this capacitor would work, it is a relatively large value and more expensive than the 377-Ω, 0.9-W resistor (nearest standard value 383-Ω, 2-W). For this reason, the manufacturer chose to use a resistor, rather than a capacitor, even though the total power dissipation of the indicator is now 0.1 + 0.9, or 1 W.

c.   At 120 V, the total impedance must be (see Fig. 25-11c)

$$Z = \frac{V}{I} = \frac{120 \ \text{V}}{0.05 \ \text{A}} = 2400 \ \Omega$$

where $\qquad Z = \sqrt{R_L^2 + (X_C - X_L)^2} \qquad (24\text{-}16)$

$$X_C - X_L = \sqrt{Z^2 - R_L^2}$$
$$= \sqrt{2400^2 - 40^2} \ \Omega$$
$$\approx 2400 \ \Omega$$
$$X_C = 2400 \ \Omega + X_L$$
$$= 2400 \ \Omega + 237 \ \Omega$$
$$= 2637 \ \Omega$$
$$X_C = \frac{1}{2\pi fC} \qquad (23\text{-}2)$$

$$C = \frac{1}{2\pi f X_C}$$

$$= \frac{1}{2\pi \times 60 \text{ Hz} \times 2637 \text{ }\Omega}$$

$$= 1 \times 10^{-6} \text{ F} = \mathbf{1 \text{ }\mu F}$$

Voltage across this capacitor is

$$V_C = I X_C$$

$$= 0.05 \text{ A} \times 2637 \text{ }\Omega$$

$$= \mathbf{132 \text{ V}}$$

d. With a resistor at 120 V, the total impedance is still 2400 Ω (see Fig. 25-11d).

$$Z = \sqrt{(R + R_L)^2 + X_L^2} \qquad (24\text{-}3)$$

$$R + R_L = \sqrt{Z^2 - X_L^2}$$

$$= \sqrt{2400^2 - 237^2} \text{ }\Omega$$

$$= 2388 \text{ }\Omega$$

$$R = 2388 \text{ }\Omega - R_L$$

$$= 2388 \text{ }\Omega - 40 \text{ }\Omega$$

$$= \mathbf{2348 \text{ }\Omega}$$

$$P_R = I^2 R$$

$$= (0.05 \text{ A})^2 \times 2348 \text{ }\Omega$$

$$= \mathbf{5.9 \text{ W}}$$

In this case, the manufacturer chooses to use a 1-μF capacitor operating at 132 V for an indicator power consumption of only 0.1 W, rather than a 2350-Ω resistor, for a total power consumption of 6 W.

Note, in Fig. 25-11c, that the apparent power required by the circuit is 120 V × 0.05 A or 6 VA, but the true power drawn is only 0.1 W.

25-6 POWER FACTOR

The ratio of the true power delivered to an ac circuit, compared to the apparent power that the source must supply, is called the *power factor* of the load.

If you examine *any* power triangle, as in Fig. 25-12, you will see that the ratio of the true power to the apparent power is the *cosine* of the angle θ.

$$\text{power factor} = \frac{P}{P_s} = \cos\theta \qquad (25\text{-}12)$$

where: cos θ is the symbol for power factor (dimensionless)

  P is the true power dissipated in the load, in watts (W)

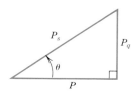 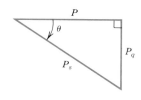

(a) Power triangle for a series or parallel *RC* circuit

(b) Power triangle for a series or parallel *RL* circuit

FIGURE 25-12

Power triangles showing that $P/P_s = \cos\theta$.

P_s is the apparent power drawn by the load, in volt-amperes (VA)

θ is the phase angle between current and voltage

EXAMPLE 25-11

A wattmeter connected to an ac circuit indicates 1 kW. If a voltmeter across the supply indicates 250 V when a series-connected ammeter reads 5 A, determine the power factor of the circuit.

Solution

$$\text{Apparent power } P_s = V_T I_T \qquad (25\text{-}7)$$

$$= 250 \text{ V} \times 5 \text{ A} = 1250 \text{ VA}$$

$$\text{Power factor} = \frac{P}{P_s} \qquad (25\text{-}12)$$

$$= \frac{1000 \text{ W}}{1250 \text{ VA}} = \mathbf{0.8}$$

If Eq. 25-12 is rearranged, you obtain

$$P = P_s \times \cos\theta$$

or $$P = V_T I_T \cos\theta \qquad \text{watts} \qquad (25\text{-}12a)$$

Equation 25-12a is very important. It shows that the power factor (cos θ) is simply the factor by which the apparent power ($V_T I_T$) must be multiplied to obtain the true power in watts.

This statement can be verified by considering the phasor diagrams in Fig. 25-13.

In Fig. 25-13a, the phasor diagram for a series *RL* circuit shows that

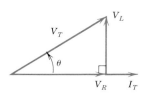

(a) Phasor diagram for a
series *RL* circuit

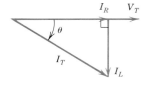

(b) Phasor diagram for a
parallel *RL* circuit

FIGURE 25-13
Phasor diagrams for (a) series and (b) parallel *RL* circuits.

$$\frac{V_R}{V_T} = \cos \theta$$

Therefore, $\qquad V_R = V_T \cos \theta$

But $\qquad P = V_T I_T \cos \theta \qquad$ (25-12a)

Rearranging, we obtain $\quad P = I_T(V_T \cos \theta)$

and therefore, $\qquad P = I_T V_R \qquad$ (15-14)

This equation gives the true power, because it is the product of the current and the *component* of the applied voltage that is *in phase* with the current.

Similarly, for a parallel circuit, as in Fig. 25-13*b*,

$$\frac{I_R}{I_T} = \cos \theta$$

Therefore, $\qquad I_R = I_T \cos \theta$

But $\qquad P = V_T (I_T \cos \theta) \qquad$ (25-12a)

and therefore, $\qquad P = V_T I_R$

This equation gives the true power, because it is the product of the applied voltage and the *component* of the total current that is *in phase* with the voltage.

To summarize, the equation $P = V_T I_T \cos \theta$ gives the true power in *any* single-phase ac circuit. It gives, in watts, the product of the *voltage and current components that are in phase*. This is what a wattmeter is designed to measure: the component of voltage that is in phase with current, or vice-versa.

In a similar manner, it can be shown that the equation

$$P_q = V_T I_T \sin \theta \qquad \text{vars} \qquad (25\text{-}13)$$

gives the product of the 90° *out-of-phase* components of voltage and current, resulting in a circuit's *reactive* power. $\sin \theta = P_q/P_s$ is called the *reactive factor*.

It should be clear from the preceding discussion that the

power factor angle (θ) between P and P_s is the *same* as the phase angle (θ) between the current and voltage. Since θ is determined by the circuit elements, the power factor is determined by the nature of the load:

1. If the circuit is purely resistive ($\theta = 0°$), the power factor, $\cos \theta$, is 1 and Eq. 25-12*a* reduces to $P = V_T I_T$. The apparent power is equal to the true power. Also, $\sin \theta = 0$ and Eq. 25-13 gives $P_q = 0$; there is no reactive power.

2. If the circuit is purely reactive ($\theta = 90°$), $\cos \theta$ is 0, and Eq. 25-12a gives $P = 0$. Now, $\sin \theta = 1$ and Eq. 25-13 gives $P_q = V_T I_T$; all the apparent power is reactive power.

EXAMPLE 25-12

A 240-V, 1-hp single-phase ac motor is running at full load and drawing a current of 6 A. An oscilloscope measurement determines a 41° phase angle between current and applied voltage. Calculate:
a. The power factor of the motor.
b. The apparent power drawn by the motor.
c. The true power delivered to the motor.
d. The reactive power drawn by the motor.
e. The efficiency of the motor.

Solution

a. Power factor $= \cos \theta \qquad$ (25-12)
$\qquad = \cos 41° = \textbf{0.75}$

b. Apparent power
$$P_s = V_T I_T \qquad (25\text{-}7)$$
$\qquad = 240 \text{ V} \times 6 \text{ A} = \textbf{1440 VA}$

c. True power $P = V_T I_T \cos \theta \qquad$ (25-12a)
$\qquad = 1440 \text{ VA} \times 0.75 = \textbf{1080 W}$

d. Reactive power
$$P_q = V_T I_T \sin \theta \qquad (25\text{-}13)$$
$\qquad = 1440 \text{ VA} \times \sin 41° = \textbf{945 vars}$

e. Efficiency $= \dfrac{P_{\text{out}}}{P_{\text{in}}}$

$\qquad = \dfrac{1 \text{ hp} \times 746 \text{ W/hp}}{1080 \text{ W}} \times 100\%$

$\qquad = \textbf{69\%}$

Note that the determination of efficiency compares useful output power to *true* input power in watts.

FIGURE 25-14
One type of power factor meter. This meter incorporates a solid-state power factor transducer and a direct-current measuring mechanism. The instrument is suitable for three-phase, three-wire operation using a potential and a current transformer. (Courtesy of Westinghouse Electric Corp.)

Even though the power factor is a *positive* dimensionless quantity, it is useful to distinguish between a capacitive load and an inductive load. Since inductive loads have current *lagging* applied voltage, an inductive circuit can be said to have a *lagging power factor.* Similarly, a capacitive load has a *leading power factor.* In each case, the power factor can be anywhere from 0 to 1, depending on the phase angle between the current and voltage.

Power factor *meters,* like the one shown in Fig. 25-14, indicate what the power factor *is,* and whether it is *leading* or *lagging.* Most industrial installations have a lagging power factor because they use a large number of ac induction motors. Such motors are inherently inductive.

EXAMPLE 25-13

Using the results of Example 25-9, find:
a. The power factor of the parallel combination of the coil and capacitor.
b. The phase angle between the applied voltage and current.

Solution

a. True power $P = 300$ W
   Net reactive power $P_q = 115$ vars (inductive)
   Apparent power $P_s = 321$ VA

   $$\text{Power factor} = \frac{P}{P_s} \qquad (25\text{-}12)$$

   $$= \frac{300 \text{ W}}{321 \text{ VA}}$$

   $$= \textbf{0.93 lagging}$$

b. $\cos\theta = 0.93$
   $\theta = \text{arc cos } 0.93$
   $= \textbf{21.6°}$

Note how an almost-unity power factor implies a voltage and current that are almost in phase with each other.

25-6.1 Detrimental Effect of a Low Power Factor

To show the important effect of the power factor, consider a 120-V, 60-Hz, 1-hp motor. Assume that it is 100% efficient, so that it draws a true power of 746 W. Such a motor has a typical power factor of 0.75, lagging.

To deliver 746 W from 120 V at a power factor of 0.75 requires a current of

$$I = \frac{P}{V \times \cos\theta} \qquad (25\text{-}12a)$$

$$= \frac{746 \text{ W}}{120 \text{ V} \times 0.75} = \textbf{8.3 A}$$

Now, assume that the motor can be modified in some way to make the power factor unity (1). The current now required is

$$I = \frac{P}{V \times \cos\theta} \qquad (25\text{-}12a)$$

$$= \frac{746 \text{ W}}{120 \text{ V} \times 1} = \textbf{6.2 A}$$

Evidently, it requires a higher current to deliver a given

quantity of true power if the power factor of the load is less than unity. This higher current means that more energy is wasted in the feeder wires serving the motor. In fact, if an industrial installation has a power factor less than 85% (0.85) overall, a ''power factor penalty'' is assessed by the utility company. For this reason, *power factor correction* is needed in large installations.

25-7 POWER FACTOR CORRECTION**

Load current is higher when the power factor is less than unity because some of the current is needed to supply the reactive power. Figure 25-15 shows the power triangle and phasor diagrams for the 1-hp motor described in Section 25-6.1 *before* power factor correction.

In Fig. 25-15b, which has been drawn with the voltage phasor as reference, the 8.3-A motor current has been resolved into two components. Component $I_x = 6.2$ A is in phase with the applied voltage and is responsible for delivering the *useful* true power (P), equal to 746 W. Component $I_y = 5.4$ A supplies the inductive reactive power (P_q), equal to 653 vars, which serves *no* useful purpose.

---

*Although it is possible to achieve unity power factor by connecting a capacitor in *series* with the load, power factor correction is *not* carried out this way because:
1. A reduction of the overall impedance (due to the canceling of the reactances) leads to an *increase* in line current, not a decrease.
2. The increased line current increases the voltage across the inductive load to a level *higher* than the normal operating value.

Now, consider adding a capacitor in parallel with the motor, as in Fig. 25-16a.* Further, the capacitor will be one whose reactance will draw a leading current of 5.4 A, as in Fig. 25-16b.

The total line current (supplied by the source) is now given by $I_T = I_x = 6.2$ A, since $I_C + I_y = 0$ (equal and opposite currents).

$$X_C = \frac{V_C}{I_C} \tag{23-1}$$

$$= \frac{120 \text{ V}}{5.4 \text{ A}} = 22 \text{ } \Omega$$

But $$X_C = \frac{1}{2\pi f C} \tag{23-2}$$

Therefore $$C = \frac{1}{2\pi f X_C}$$

$$= \frac{1}{2\pi \times 60 \text{ Hz} \times 22 \text{ } \Omega} = \textbf{120 } \boldsymbol{\mu}\textbf{F}$$

The function of the capacitor is to provide the inductive portion of the motor with the reactive power it needs instead of drawing this power from the source. In other words, the reactive power is transferred back and forth between the capacitor and the inductance. In fact, for unity power factor (cos θ = 1), the capacitor can be chosen to draw the *same* reactive power as the inductance:

Inductive reactive power,

$$P_{qL} = V_T I_T \sin \theta \tag{25-13}$$
$$= 120 \text{ V} \times 8.3 \text{ A} \times \sin 41° = 653 \text{ vars}$$

**This section may be omitted with no loss of continuity.

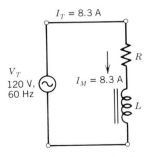
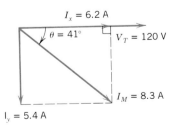
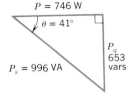

(a) Equivalent circuit of motor   (b) Phasor diagram with voltage used as reference   (c) Power triangle

FIGURE 25-15

Phasor diagram and power triangle for a 1-hp motor before power factor correction.

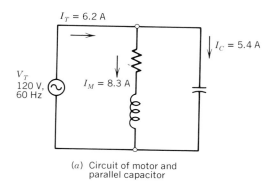

(a) Circuit of motor and
parallel capacitor

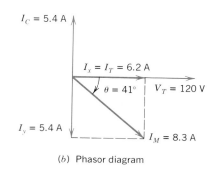

(b) Phasor diagram

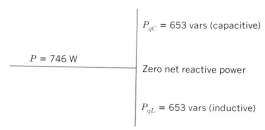

(c) Power diagram

FIGURE 25-16

Phasor and power diagrams for 1-hp motor after power factor correction.

Capacitive reactive power $P_{qC} = \dfrac{V_C^2}{X_C}$ (25-6)

Therefore, $X_C = \dfrac{V_C^2}{P_{qC}}$

For unity power factor correction,

$$P_{qC} = P_{qL} = 653 \text{ vars}$$

Therefore, $X_C = \dfrac{(120 \text{ V})^2}{653 \text{ vars}} = 22 \text{ }\Omega$

and $C = \textbf{120 } \boldsymbol{\mu}\textbf{F}$

It should be noted that if a wattmeter is connected to measure the power delivered to the motor (before power factor correction), it would indicate 746 W. This occurs because the wattmeter's operation takes into account the phase angle of 41° between the voltage and current. After power factor correction, the wattmeter still reads 746 W.*

---

*It should be noted that the power factor of a motor varies considerably with the variation in mechanical load connected to it. When the motor is idling or supplying a very light load, the power factor is very low; when the motor is running at the rated load, the power factor rises to approximately 0.7 or 0.8. Capacitors for power factor correction are usually chosen on the basis of full load. There are also *power factor controllers* that monitor the load current and automatically reduce the applied voltage when the current is light, improving the power factor and reducing losses.

A series-connected ammeter, however, will decrease from 8.3 to 6.2 A following power factor correction.

This illustrates an important precaution to observe when using a wattmeter in highly reactive circuits: if the power factor is very close to zero, the meter deflection will be very small (even though a very *large* current may be passing through the meter). Standard practice calls for using a series-connected ammeter to ensure that the wattmeter's current rating is not knowingly exceeded.

Similarly, a transformer's rating is given in kilovolt-amperes (kVA) because, if an inductive or capacitive load is connected, an appreciable current can flow, even though very little *true* power is delivered. Thus, a 1-kVA transformer rated at 250 V can deliver a maximum of 4 A. If the load power factor is 0.7, the true power is only 700 W, but the transformer would be working at its limit of 1000 VA. The transformer must not be asked to deliver 1000 W unless the load is purely resistive (has a power factor of unity).

Although it is possible to correct the power factor of each inductive load (see Fig. 25-17), large industrial installations often involve use of banks of parallel capacitors connected across the line where the service enters the building. Also, it is not usually economical to correct to unity power factor. Therefore, just sufficient capacitance is added to improve the power factor so that it is above the penalty level. It is also possible to improve the power fac-

FIGURE 25-17

The rectangular enclosure in the foreground contains power factor correction capacitors for the three-phase 480-V motor in the background. The unit contains six capacitor cells to provide a terminal-to-terminal capacitance of 144 μF rated at 25 kVAR. (Courtesy of Aerovox® Inc.)

tor by using large *synchronous* motors that can be adjusted to run at a leading current to offset the inductive load. These motors can then be used to provide some useful work, as well as to help correct the power factor.

EXAMPLE 25-14

A 15-W desk-type fluorescent lamp has an effective resistance of 200 Ω when operating. It is in series with a ballast that has a resistance of 80 Ω and an inductance of 0.9 H. The lamp and ballast are operated at 120 V, 60 Hz. Calculate:

a. The current drawn by the lamp.
b. The apparent power.
c. The power in the fluorescent lamp.
d. The power dissipated in the ballast.
e. The overall power factor.

f.   The reactive power.
g.   The size of the capacitor to provide unity power factor.
h.   The current drawn after power factor correction.

Solution

a. $Z = \sqrt{R_T^2 + X_L^2}$ (24-3)
$R_T = R_{lamp} + R_{ballast} = 200\ \Omega + 80\ \Omega = 280\ \Omega$
$X_L = 2\pi f L$ (20-2)
$= 2\pi \times 60\ \text{Hz} \times 0.9\ \text{H} = 339\ \Omega$
Therefore, $Z = \sqrt{(280\ \Omega)^2 + (339\ \Omega)^2} = 440\ \Omega$

$$I = \frac{V}{Z} \qquad (24\text{-}4)$$

$$= \frac{120\ \text{V}}{440\ \Omega} = \textbf{0.27 A}$$

See Fig. 25-18.

b. $P_s = V_T I_T$ (25-7)
$= 120\ \text{V} \times 0.27\ \text{A} = \textbf{32.4 VA}$

c. $P_{lamp} = I_R^2 R_{lamp}$ (15-12)
$= (0.27\ \text{A})^2 \times 200\ \Omega = \textbf{14.6 W}$

d. $P_{ballast} = I_R^2 R_{ballast}$ (15-12)
$= (0.27\ \text{A})^2 \times 80\ \Omega = \textbf{5.8 W}$

e.   Power factor $= \dfrac{P}{P_s}$ (25-12)

$$= \frac{20.4\ \text{W}}{32.4\ \text{VA}} = \textbf{0.63}$$

f. $P_q = V_T I_T \sin\theta$ (25-13)
$\cos\theta = 0.63$
$\theta = \cos^{-1} 0.63 = 51°$

Therefore, $P_{qL} = 32.4\ \text{VA} \times \sin 51° = \textbf{25.2 vars}$

g. $P_{qC} = P_{qL} = 25.2\ \text{vars}$

$$= \frac{V_C^2}{X_C} \qquad (25\text{-}6)$$

so that $X_C = \dfrac{V_C^2}{P_{qC}}$

$$= \frac{(120\ \text{V})^2}{25.2\ \text{vars}} = 571\ \Omega$$

$$X_C = \frac{1}{2\pi f C} \qquad (23\text{-}2)$$

Therefore, $C = \dfrac{1}{2\pi f X_C}$

$$= \frac{1}{2\pi \times 60\ \text{Hz} \times 571} = \textbf{4.6 μF}$$

h. $P = V_T I_T \cos\theta$ (25-12a)
The total power has not changed,

$$P = 14.6\ \text{W} + 5.8\ \text{W}$$

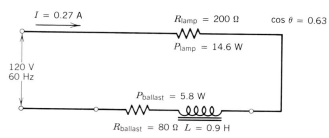

(a) Fluorescent lamp circuit before power factor correction

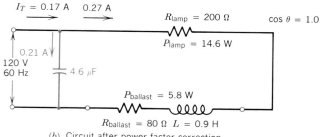

(b) Circuit after power factor correction

FIGURE 25-18

Circuits for Example 25-14.

$$= 20.4 \text{ W}$$
$$\cos \theta = 1$$

Therefore, $I_T = \dfrac{P}{V_T \cos \theta}$

$$= \frac{20.4 \text{ W}}{120 \text{ V} \times 1} = \mathbf{0.17 \ A}$$

25-8 POWER IN THREE-PHASE CIRCUITS

The power delivered to a load (balanced or unbalanced, that is, having equal or unequal line currents) in a three-phase, three-wire system can be measured using two wattmeters connected as in Fig. 25-19.

The total power is given by the sum of the two wattmeter readings

$$W_T = W_1 + W_2 \qquad (25\text{-}14)$$

If the wattmeters are connected symmetrically with respect to the marked terminals, but one of the meters reads *backwards,* reverse the current coil connections on that meter and treat the resulting reading as a negative quantity in Eq. 25-14.

If the three-phase load is balanced—if there are equal line currents and voltages—the total true power can be calculated using

$$P = \sqrt{3} \ VI \cos \theta \qquad \text{watts} \qquad (25\text{-}15)$$

where: P is the total true power in the load, in watts (W)
V is the line-to-line voltage, in volts (V)
I is the line current, in amperes (A)
θ is the phase angle between V and I

EXAMPLE 25-15

Two wattmeters are connected as in Fig. 25-19 to measure the power delivered to a three-phase, delta-con-

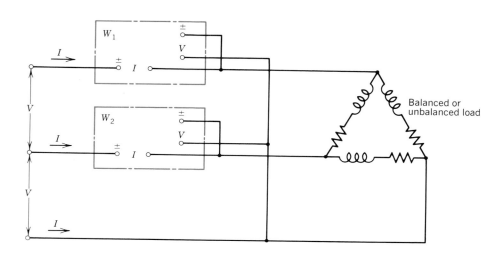

FIGURE 25-19

Measuring total power in a three-phase three-wire system using two wattmeters.

nected motor. One wattmeter reads 432 W and the other reads 1092 W. If the motor draws 2.5 A on each line at 440 V, calculate the power factor of the motor.

Solution

$$P = W_T = W_1 + W_2 \qquad (25\text{-}14)$$
$$= 432 \text{ W} + 1092 \text{ W} = 1524 \text{ W}$$

$$P = \sqrt{3} \, VI \cos \theta \qquad (25\text{-}15)$$

Therefore, $\cos \theta = \dfrac{P}{\sqrt{3} \, VI}$

$$= \frac{1524 \text{ W}}{\sqrt{3} \times 440 \text{ V} \times 2.5 \text{ A}} \approx 0.8$$

25-9 MAXIMUM POWER TRANSFER

In Chapter 9, you learned that maximum power is transferred to a load from a dc source when the load resistance (R) equals the internal resistance of the source (r).

For an ac source, you must consider the internal *impedance* of the source to determine what load receives maximum power. (Refer to Fig. 25-20.)

The total impedance of the circuit is at a minimum if the *capacitive* reactance of the load equals the *inductive* reactance of the source (or vice-versa, if the internal impedance of the source happens to be capacitive). Maximum current flows, and if $R_{\text{load}} = r$, maximum power transfer occurs (as in a dc circuit).

A load that has the same resistance as the source and the opposite (but equal) reactance is said to be the *conjugate* **of the source impedance. Thus, for maximum power transfer in an ac circuit, the load impedance must be the conjugate of the source impedance.**

If a transformer is used to match a load to a source (as in Section 18-1.4), Eq. 18-10 must be modified to the more general form:

$$Z_p = a^2 Z_L \qquad \text{ohms} \qquad (25\text{-}16)$$

where: Z_p is the impedance reflected into the primary, in ohms (Ω)

Z_L is the impedance of the load on the secondary, in ohms (Ω)

a is the turns ratio of the transformer $\left(\dfrac{N_p}{N_s}\right)$, dimensionless

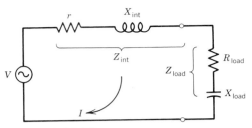

FIGURE 25-20
Load impedance must be the conjugate of the internal impedance for maximum power transfer.

Thus, if a load has a resistance of 5 Ω in series with an inductive reactance of 2 Ω, and if the transformer turns ratio is 4:1, the impedance reflected into the primary consists of a resistance of $4^2 \times 5 \ \Omega = 80 \ \Omega$ in series with an inductive reactance of $4^2 \times 2 \ \Omega = 32 \ \Omega$. The necessary source impedance to provide maximum power transfer would consist of 80 Ω of resistance in series with 32 Ω of capacitive reactance. (This neglects the resistance and reactance of the transformer itself.) See Problem 25-28.

EXAMPLE 25-16

A 10-V, 60-Hz supply has an internal resistance of 10 Ω and an inductance of 10 mH. A relay coil is to be connected whose inductance is known to be 50 mH. Determine:

a. The resistance of the coil for maximum power transfer.
b. The size of capacitor to be connected in series with the coil to provide maximum power transfer.
c. The power delivered to the coil with the capacitor.
d. The power delivered to the coil without the capacitor.

Solution

a. Coil resistance $R_{\text{load}} = r = $ **10 Ω**
b. The capacitive reactance of the capacitor must equal the *total* inductive reactance of the source and load to minimize the impedance.

$$X_L = 2\pi f L_T \qquad (20\text{-}2)$$
$$= 2\pi \times 60 \text{ Hz} \times (10 + 50) \times 10^{-3} \text{ H}$$
$$= 22.6 \ \Omega$$

$$X_C = \frac{1}{2\pi f C} \qquad (23\text{-}2)$$

Therefore, $C = \dfrac{1}{2\pi f X_C}$

$= \dfrac{1}{2\pi \times 60 \text{ Hz} \times 22.6 \text{ }\Omega}$

$= \mathbf{117 \text{ }\mu F}$

c. Since the reactances of the circuit have canceled each other, the impedance of the circuit is now purely resistive and equal to 20 Ω.

$I = \dfrac{V}{Z} \qquad\qquad (24\text{-}4)$

$= \dfrac{10 \text{ V}}{20 \text{ }\Omega} = 0.5 \text{ A}$

$P_{\text{coil}} = I_R^2 R \qquad\qquad (15\text{-}12)$

$= (0.5 \text{ A})^2 \times 10 \text{ }\Omega = \mathbf{2.5 \text{ W}}$

d. $Z = \sqrt{R_T^2 + X_L^2} \qquad\qquad (24\text{-}3)$

$= \sqrt{(20 \text{ }\Omega)^2 + (22.6 \text{ }\Omega)^2} = 30.2 \text{ }\Omega$

$I = \dfrac{V}{Z} \qquad\qquad (24\text{-}4)$

$= \dfrac{10 \text{ V}}{30.2 \text{ }\Omega} = 0.33 \text{ A}$

$P_R = I_R^2 R \qquad\qquad (15\text{-}12)$

$= (0.33 \text{ A})^2 \times 10 \text{ }\Omega = \mathbf{1.1 \text{ W}}$

Example 25-16 illustrates an important effect called *series resonance*. When the capacitive and inductive reactance cancel, the circuit becomes purely resistive and maximum current flows. For any given capacitor and inductor, this condition exists at only one frequency, so the circuit is said to be *frequency-selective*. Such a circuit can be used to "tune in" a desired frequency, as explained in the next chapter.

SUMMARY

1. Power dissipated in resistance is called true power, and is measured in watts. It can be calculated using the equations:

$$P = I_R^2 R = \dfrac{V_R^2}{R} = V_R I_R$$

2. Power delivered to a pure inductance is called reactive power, and is measured in vars. Reactive power is returned to the source, and is not dissipated in the inductance. Inductive reactive power can be calculated using the equations:

$$P_q = I_L^2 X_L = \dfrac{V_L^2}{X_L} = V_L I_L$$

3. Reactive power delivered to a capacitor during one quarter-cycle is returned to the source during the next quarter-cycle. Capacitive reactive power, in vars, can be calculated using:

$$P_q = I_C^2 X_C = \dfrac{V_C^2}{X_C} = V_C I_C$$

4. The apparent power delivered to a series or parallel RL or RC circuit is the product of the applied voltage and total circuit current, and is measured in volt-amperes (VA). Apparent power can be calculated using

$$P_s = I_T^2 Z = \dfrac{V_T^2}{Z} = V_T I_T$$

5. True power, reactive power, and apparent power can be represented in a power triangle with the inductive reactive power drawn vertically downward and the capacitive reactive power drawn vertically upward. The Pythagorean power equation, $P_s^2 = P^2 + P_q^2$, applies to *any* ac circuit combination.

6. If an ac circuit contains both inductance and capacitance, there is an exchange of reactive power between the two so that the source must supply only the net reactive power

$$P_{qC} - P_{qL}$$

7. The power factor of a circuit is defined as the ratio of the true power in the circuit to the apparent power, and is given by

$$\cos \theta = P/P_s$$

8. The true power in a single-phase ac circuit can always be obtained by using

$$P = V_T I_T \cos \theta \qquad \text{watts}$$

and the reactive power can be obtained by using

$$P_q = V_T I_T \sin \theta \qquad \text{vars}$$

9. Inductive loads have a lagging power factor; capacitive loads have a leading power factor.
10. A low power factor means that a higher current is required to deliver a given amount of true power than if the power factor is near unity.
11. Power factor correction requires that a capacitor be connected in parallel to draw the same amount of leading capacitive reactive power as the initial lagging inductive reactive power of the load.
12. The total power delivered to a three-phase, three-wire system can be measured using two wattmeters and finding the algebraic sum of their readings.
13. In a balanced three-phase load, the total true power can be calculated using

$$P = \sqrt{3} \, VI \cos \theta$$

14. Maximum power transfer occurs in an ac circuit when the load impedance is the conjugate of the internal source impedance.

SELF-EXAMINATION

Answer true or false or a, b, c, or d for multiple choice
(Answers at back of book)

25-1. Only resistance causes true power to be dissipated. _____
25-2. A wattmeter always shows the true power dissipated in an ac circuit, even if the circuit contains inductance and capacitance. _____
25-3. A purely resistive ac circuit only involves positive power whereas a purely inductive or capacitive circuit shows equal positive and negative power alternations. _____
25-4. The apparent power delivered to a purely inductive circuit is zero. _____
25-5. If a purely inductive circuit has a reactance of 120 Ω, the reactive power drawn by the circuit from a 120-V, 60-Hz supply is
    a. 14,400 vars
    b. 14.4 kvars
    c. 120 vars
    d. either part (a) or (b)
25-6. If a capacitor of 240-Ω reactance draws a current of 2 A, the reactive power supplied to the capacitor is

    a. 960 vars

    b. 480 vars

    c. 60 vars

    d. 240 vars

25-7. If a circuit has an impedance of 800 Ω and draws a current of 10 A, the apparent power is

    a. 8000 VA

    b. 80 kVA

    c. 8 kVA

    d. either part (a) or (c)

25-8. If a wattmeter indicates 300 W and a varmeter indicates 400 vars, the apparent circuit power is

    a. 2500 VA

    b. 25,000 VA

    c. 500 VA

    d. 700 VA

25-9. The Pythagorean power equation only applies to series and parallel *RL* circuits. _____

25-10. When an ac circuit contains both inductance and capacitance, in series or parallel, the net reactive power is always given by $P_q = P_{qC} - P_{qL}$. _____

25-11. If a circuit has more inductive reactive power than capacitive reactive power, the circuit has a lagging power factor. _____

25-12. The power factor is the factor by which the reactive power must be multiplied to obtain the true power. _____

25-13. True power is the product of the in-phase components of voltage and current while reactive power is the product of the 90° out-of-phase components of voltage and current. _____

25-14. If a wattmeter indicates 400 W, a series-connected ammeter indicates 2 A, and a parallel-connected voltmeter indicates 250 V, the power factor is

    a. 0.2

    b. 0.25

    c. 0.4

    d. 0.8

25-15. The major problem with low power factor circuits is the relatively high current necessary to deliver a given true power. _____

25-16. Power factor correction (to unity power factor) requires the addition of a capacitor in parallel with the circuit to make the applied voltage and circuit current in phase with each other. _____

25-17. When an ac motor has had its power factor corrected, the motor is more efficient. _____

25-18. If, in trying to correct a load's power factor, too large a capacitor is connected, there may be no reduction in the line current and in fact the current may increase. _____

25-19. Two wattmeters can be used to measure the total power in a three-phase, three-wire system only if the three line currents are equal. _____

25-20. If an ac source has an impedance consisting of 100 Ω resistance and a capacitive reactance of 50 Ω, the required load impedance for maximum power transfer is

    a. $R = 100\ \Omega$, $X_L = 100\ \Omega$

    b. $R = 100\ \Omega$, $X_C = 50\ \Omega$

c. $R = 50 \ \Omega, X_C = 50 \ \Omega$
d. $R = 100 \ \Omega, X_L = 50 \ \Omega$

REVIEW QUESTIONS

1. Distinguish, in your own words, between true power and reactive power.
2. Explain why a wattmeter has zero deflection when connected in a purely inductive or capacitive circuit.
3. What do you understand by "positive" power and "negative" power?
4. Why can't we use watts to describe reactive and apparent power?
5. Under what conditions is the apparent power equal to
   a. True power?
   b. Reactive power?
6. Describe a method to measure the inductance of a coil that requires a wattmeter.
   a. Why is this method limited to power line frequencies?
   b. If you measured the resistance of the coil with an ohmmeter, would you expect the reading to be higher or lower than the resistance obtained from the wattmeter and ammeter readings? Why?
7. a. How does a power triangle drawn for an inductive load differ from one drawn for a capacitive load?
   b. Why are no arrows shown on power triangles?
   c. Does a power triangle drawn for a series *RL* circuit differ from one drawn for a parallel *RL* circuit? Why or why not?
   d. Under what conditions does a power diagram for a series or parallel *RLC* circuit *not* result in a triangle?
8. a. Describe in your own words what is meant by "power factor."
   b. Describe two methods of measuring a circuit's power factor.
   c. What is the range of possible values for the power factor of a circuit?
9. How must the apparent power quantity be modified to obtain:
   a. True power?
   b. Reactive power?
10. Explain how the equation $P = V_T I_T \cos \theta$ gives the product of the *in-phase* components of current and voltage for true power in any single-phase ac circuit (series or parallel).
11. What is meant by leading and lagging power factors?
12. a. Why is a low power factor considered detrimental when supplying ac power from a source?
    b. Give two reasons why a low power factor can be costly both to the consumer and the supplier of ac power.
13. a. What is the general principle used to select a capacitor to provide unity power factor correction?
    b. What is the primary effect of correcting the power factor?
    c. How can a motor *and* its supply system, including feeder wires, be considered more *efficient* following power factor correction if the true power delivered to the motor is unchanged?
14. a. Draw a circuit diagram showing how to connect two wattmeters to measure the total power delivered to a load in a three-phase, three-wire system.
    b. What would you do to obtain the total power if one of the wattmeters read backward?

15. a. What is the maximum power transfer theorem as applied to an ac circuit?

b. What is meant by the term *conjugate?*

PROBLEMS

(Answers to odd-numbered problems at back of book)

25-1. A wattmeter, connected in an ac circuit, indicates 240 W. An ammeter, connected in series with the wattmeter, reads 4.5 A. Determine the resistance of the ac circuit.

25-2. An oscilloscope indicates 340 V p–p across the only resistor in a series ac circuit that draws a current of 600 mA. Calculate:

a. The resistance of the circuit.

b. The reading of a wattmeter connected to the circuit.

25-3. A 2-H coil of negligible resistance is connected across the 120-V, 60-Hz line. Calculate:

a. The coil's reactive power.

b. The coil's apparent power.

25-4. A coil of 100-Ω resistance and 0.8-H inductance draws a current of 400 mA from a 60-Hz supply. Determine the reading of a varmeter connected to the circuit.

25-5. A bank of capacitors used for power factor correction draws a reactive power of 20 kvars from a 550-V, 60-Hz, single-phase supply. Determine the amount of capacitance.

25-6. How much reactive power is supplied to a 50-μF motor-starting capacitor operating at 220 V, 60 Hz?

25-7. What is the apparent power supplied to a circuit that draws a current of 300 mA at 230 V?

25-8. What maximum current can be supplied by a 7.5-kVA transformer with a 240-V secondary?

25-9. A series circuit of 250-Ω resistance and 150-Ω inductive reactance draws a current of 1.8 A. What is the apparent power drawn by the circuit?

25-10. A coil is connected to the 220-V, 60-Hz line in series with an ammeter and wattmeter. If the ammeter indicates 1.25 A and the wattmeter indicates 80 W, calculate:

a. The inductance of the coil.

b. The quality of the coil.

25-11. A series circuit of 220-Ω resistance and 0.4-H inductance draws a current of 700 mA from a 60-Hz supply. Calculate:

a. The true power.

b. The reactive power.

c. The apparent power.

25-12. A relay coil connected across a 24-V, 50-Hz supply draws a current of 200 mA. If a series-connected wattmeter indicates 3 W, calculate:

a. The apparent power.

b. The reactive power.

25-13. A 470-Ω resistor and a 150-mH coil of negligible resistance are connected in parallel across a 100-V, 400-Hz supply. Calculate:

a. The true power.

b. The reactive power.

c. The apparent power.
d. The circuit's power factor.

25-14. Repeat Problem 25-13 assuming that the coil has a resistance of 500 Ω.

25-15. A 1-μF capacitor and a 100-Ω resistor are series-connected across a 20-V, 1-kHz source. Calculate:
a. The true power.
b. The reactive power.
c. The apparent power.
d. The circuit's power factor.

25-16. Repeat Problem 25-15 with the two components connected in *parallel* across the source.

25-17. A 50-μF capacitor, a 47-Ω resistor and a 100-mH coil of negligible resistance are connected in parallel across the 120-V, 60-Hz line. Calculate:
a. The true power.
b. The net reactive power.
c. The apparent power.
d. The total line current.
e. The power factor of the circuit.

25-18. Repeat Problem 25-17 assuming that the coil has a resistance of 40 Ω. Draw a power triangle.

25-19. A circuit drawing a current of 8.5 A from a 220-V, 60-Hz supply is known to have a lagging power factor of 0.6. Calculate:
a. The phase angle between the applied voltage and current.
b. The apparent power.
c. The true power.
d. The reactive power.

25-20. A 220-V, 1½-hp, single-phase ac motor is running at full load and drawing a current of 9 A. The phase angle between the applied voltage and the current is measured to be 35°. Calculate:
a. The power factor of the motor.
b. The apparent power.
c. The true power.
d. The reactive power.
e. The efficiency of the motor.

25-21. How much current is drawn by a 120-V, 60-Hz, 75% efficient, 1-hp, single-phase motor when operating at the following power factors:
a. 0.7 lagging.
b. 0.9 lagging.
c. 1.0.
d. 0.9 leading?

25-22. Repeat Problem 25-21 assuming the use of a three-phase motor.

25-23. For the circuit of Problem 25-19 calculate:
a. The capacitance required to provide unity power factor.
b. The line current following power factor correction.

25-24. For the circuit of Problem 25-20 calculate:
a. The capacitance required to provide unity power factor at 60 Hz.
b. The line current following power factor correction.

25-25. Repeat Example 25-14 assuming that the ballast is replaced by one that has a resistance of 50 Ω and an inductance of 1.1 H. What visual effect would be noticed with the new ballast?

25-26. What capacitance must be connected across a 120-V, 60-Hz, 75% efficient, 1-hp, single-phase motor operating at a 0.7 lagging power factor to produce an overall 0.9 leading power factor?

25-27. Two wattmeters, connected to measure the total power delivered to a 5-hp, three-phase, delta-connected motor, indicate 2860 and 1440 W, respectively. If the motor draws 5.2 A on each line at 550 V, calculate:

    a.   The motor's power factor.

    b.   The motor's efficiency.

25-28. A 40-V p–p sinusoidal source at 1 kHz has an internal impedance consisting of 100-Ω resistance in series with a 4-μF capacitor. A transformer is to be selected to maximize power transfer to a 4-Ω resistive load. Calculate:

    a.   The necessary turns ratio of the transformer.

    b.   The amount of inductance that must be in series with the 4-Ω resistance load to provide maximum power transfer.

    c.   The power delivered to the load with the transformer and inductance, assuming that the transformer is 100% efficient.

    d.   The power delivered to the load without the inductance.

    e.   The power delivered to the load if connected directly across the source.

25-29. A transmitter operating at 2 MHz delivers maximum power to an antenna when the antenna is "tuned" to represent a load of 150-pF capacitance in series with a resistance of 150 Ω. Determine the internal impedance of the transmitter.

CHAPTER 26

RESONANCE

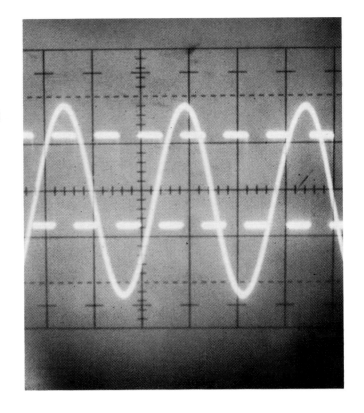

The antenna of a radio or TV receiver has voltages induced in it by the radio frequency signals from many different stations and TV channels. This chapter discusses how it is possible to select a desired *band* of frequencies and reject other signals.

Both series and parallel *RLC* circuits can be tuned to *resonate* at a frequency equal to that assigned to the desired broadcasting station. The resonant circuit, in addition to being able to receive a single *center* or *carrier* frequency, is *sensitive* to some *band* of frequencies emitted by the transmitter. By proper selection of components, the circuit's *bandwidth* can be made to receive all the transmitted information while rejecting the frequencies of adjacent stations. Stations must operate at specific frequencies assigned by the Federal Communications Commission and adjust their bandwidths so that there is no overlap.

In a *series RLC* circuit, there is a *resonant rise* of voltage across the reactive components that is (if the Q of the circuit is large) many times *larger* than the *input* voltage. Similarly, a resonant rise of the *tank* current occurs in a parallel *RLC* circuit, which can be much larger than the *line* current. Since both these effects are frequency *selective,* it is possible to *discriminate* against all other undesired frequencies by *tuning in* a desired station.

26-1 SERIES RESONANCE

In Section 24-7, you examined the basic relationship among the current, voltage, and impedance in a series *RLC* circuit. Now, the way in which such a circuit reacts to a *variable frequency source* will be examined. (See Fig. 26-1.)

You know that

$$Z = \sqrt{R^2 + (X_L - X_C)^2} \qquad (24\text{-}16)$$

$$I = \frac{V}{Z} \qquad (24\text{-}4)$$

and $$\theta = \arctan \frac{X_L - X_C}{R} \qquad (24\text{-}17)$$

The way in which X_L and X_C vary as the frequency increases is shown in Fig. 26-1b. The inductive reactance (X_L) is a positive *increasing* quantity; the capacitive reactance (X_C) is a positive *decreasing* quantity. Clearly, there is some frequency (f_r) where the difference of X_L and X_C, $(X_L - X_C)$ is zero. That is, the net reactance of the circuit is zero.

Thus, if the frequency of the source is adjusted to f_r, the reactances cancel and the impedance of the circuit is a *minimum* and equal to R (see Eq. 24-16). Also, from Eq. 24-4, when the impedance Z is a *minimum*, the current I is a *maximum*, and equal to V/R. From Eq. 24-17, the phase angle (θ) is zero. That is, the cur-

rent and applied voltage are in phase. All of these conditions occur at what is called the *resonant frequency* (f_r), and the series circuit is said to be in a condition of *series resonance.*

The conditions in the circuit at frequencies below, at, and above the resonant frequency are illustrated in Example 26-1.

EXAMPLE 26-1

A series *RLC* circuit consists of $R = 10\ \Omega$, $L = 200\ \mu H$, and $C = 50$ pF. A 100-mV variable frequency supply is connected to the circuit. Determine the following for frequencies of (1) 1 MHz, (2) 1.59 MHz, and (3) 2.2 MHz and present in tabular form. Also draw a phasor diagram for each frequency.
a. The impedance.
b. The phase angle.
c. The current.
d. V_R, V_L, and V_C.

Solution

1. a. At $f = 1 \times 10^6$ Hz,
$$X_L = 2\pi fL \qquad (18\text{-}2)$$
$$= 2\pi \times 1 \times 10^6\ Hz \times 200 \times 10^{-6}\ H$$
$$= 1257\ \Omega$$

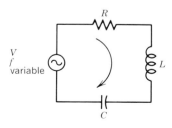

(a) Series *RLC* circuit

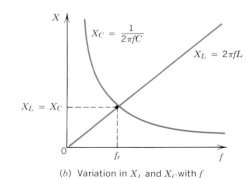

(b) Variation in X_L and X_C with f

FIGURE 26-1
Effect of a variable frequency on a series *RLC* circuit.

$$X_C = \frac{1}{2\pi fC} \tag{23-2}$$

$$= \frac{1}{2\pi \times 1 \times 10^6 \text{ Hz} \times 50 \times 10^{-12} \text{ F}}$$
$$= 3183 \ \Omega$$

$$Z = \sqrt{R^2 + (X_L - X_C)^2} \tag{24-16}$$
$$= \sqrt{(10 \ \Omega)^2 + (1257 \ \Omega - 3183 \ \Omega)^2}$$
$$= \mathbf{1926 \ \Omega} \quad \text{(capacitive)}$$

b. $$\theta = \arctan \frac{X_L - X_C}{R} \tag{24-17}$$
$$= \arctan \frac{1257 \ \Omega - 3183 \ \Omega}{10 \ \Omega} = \mathbf{-89.7°}$$

c. $$I = \frac{V}{Z} \tag{24-4}$$
$$= \frac{100 \text{ mV}}{1.926 \text{ k}\Omega} = \mathbf{51.9 \ \mu A}$$

d. $$V_R = IR \tag{13-15b}$$
$$= 51.9 \times 10^{-6} \text{ A} \times 10 \ \Omega$$
$$= \mathbf{0.5 \text{ mV}}$$
$$V_L = IX_L \tag{20-1}$$
$$= 51.9 \times 10^{-6} \text{ A} \times 1257 \ \Omega$$
$$= \mathbf{65.2 \text{ mV}}$$
$$V_C = IX_C \tag{23-1}$$
$$= 51.9 \times 10^{-6} \text{ A} \times 3183 \ \Omega$$
$$= \mathbf{165.2 \text{ mV}}$$

2. a. At $f = 1.59 \times 10^6$ Hz,
$$X_L = 2\pi fL \tag{18-2}$$
$$= 2\pi \times 1.59 \times 10^6 \text{ Hz} \times 200 \times 10^{-6} \text{ H}$$
$$\approx 2 \text{ k}\Omega$$
$$X_C = \frac{1}{2\pi fC} \tag{23-2}$$
$$= \frac{1}{2\pi \times 1.59 \times 10^6 \text{ Hz} \times 50 \times 10^{-12} \text{ F}}$$
$$\approx 2 \text{ k}\Omega$$

$$Z = \sqrt{R^2 + (X_L - X_C)^2} \tag{24-16}$$
$$= \sqrt{(10 \ \Omega)^2 + (2 \text{ k}\Omega - 2 \text{ k}\Omega)^2}$$
$$= \mathbf{10 \ \Omega} \quad \text{(resistive)}$$

b. $$\theta = \arctan \frac{X_L - X_C}{R} \tag{24-17}$$
$$= \arctan \frac{2 \text{ k}\Omega - 2 \text{ k}\Omega}{10 \ \Omega} = \mathbf{0°}$$

c. $$I = \frac{V}{Z} \tag{24-4}$$
$$= \frac{100 \text{ mV}}{10 \ \Omega} = \mathbf{10 \text{ mA}}$$

d. $$V_R = IR \tag{13-15b}$$
$$= 10 \times 10^{-3} \text{ A} \times 10 \ \Omega = \mathbf{100 \text{ mV}}$$
$$V_L = IX_L \tag{20-1}$$
$$= 10 \times 10^{-3} \text{ A} \times 2 \times 10^3 \ \Omega = \mathbf{20 \text{ V}}$$
$$V_C = IX_C \tag{23-1}$$
$$= 10 \times 10^{-3} \text{ A} \times 2 \times 10^3 \ \Omega = \mathbf{20 \text{ V}}$$

3. Similarly, at $f = 2.2$ MHz,
a. $X_L = 2764 \ \Omega$, $X_C = 1447 \ \Omega$,
    $Z = \mathbf{1317 \ \Omega}$ (inductive)
b. $\theta = \mathbf{89.6°}$
c. $I = \mathbf{75.9 \ \mu A}$
d. $V_R = \mathbf{0.76 \text{ mV}}$, $V_L = \mathbf{209.8 \text{ mV}}$,
    $V_C = \mathbf{109.8 \text{ mV}}$

See Table 26-1.
The phasor diagrams for the three frequencies are shown in Fig. 26-2.

In Example 26-1, the inductive and capacitive reactances are equal when the frequency is 1.59 MHz, the circuit is purely resistive, and the maximum current flows. Thus the resonant frequency of the circuit must be 1.59 MHz.

TABLE 26-1
Circuit Conditions in a Variable Frequency Circuit

Frequency	X_L, Ω	X_C, Ω	$X_L - X_C$, Ω	Z, Ω	θ	I, mA	V_R	V_L	V_C
1 MHz	1257	3183	-1926	1926	-89.7° (Capacitive)	0.05	0.5 mV	65 mV	165 mV
1.59 MHz	2000	2000	0	10	0° (Resistive)	10.0°	100 mV	20 V	20 V
2.2 MHz	2764	1447	1317	1317	89.6° (Inductive)	0.08	0.8 mV	210 mV	110 mV

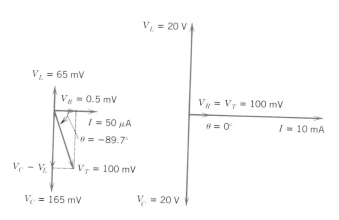

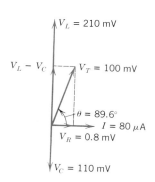

(a) $f < f_r$,
circuit is capacitive.

(b) $f = f_r$,
circuit is resistive.

(c) $f > f_r$,
circuit is inductive.

FIGURE 26-2

Phasor diagrams for series *RLC* circuit of Example 26-1 (not drawn to scale.)

26-2 RESONANT FREQUENCY

You can find the resonant frequency of any series circuit by noting that resonance occurs when $X_L = X_C$:

$$X_L = X_C$$

$$2\pi f_r L = \frac{1}{2\pi f_r C}$$

$$f_r^2 = \frac{1}{4\pi^2 LC}$$

Therefore, $\qquad f_r = \dfrac{1}{2\pi\sqrt{LC}} \qquad$ hertz $\qquad$ (26-1)

where: f_r is the resonant frequency of a series circuit, in hertz (Hz)

L is the inductance of the circuit, in henrys (H)

C is the capacitance of the circuit, in farads (F)

Note that the resonant frequency of a series circuit is independent of the circuit's resistance.

EXAMPLE 26-2

Calculate the resonant frequency of the circuit in Example 26-1, where $R = 10\ \Omega$, $L = 200\ \mu H$, and $C = 50$ pF.

Solution

$$f_r = \frac{1}{2\pi\sqrt{LC}} \qquad (26-1)$$

$$= \frac{1}{2\pi\sqrt{200 \times 10^{-6}\ H \times 50 \times 10^{-12}\ F}}$$

$$= 1.59 \times 10^6\ Hz$$

$$= \mathbf{1.59\ MHz}$$

EXAMPLE 26-3

What value of capacitance is necessary to provide series resonance with a 30-mH coil at 1 kHz?

Solution

$$f_r = \frac{1}{2\pi\sqrt{LC}} \qquad (26-1)$$

Therefore, $\qquad 4\pi^2 f_r^2 LC = 1$

and $\quad C = \dfrac{1}{4\pi^2 f_r^2 L}$

$$= \frac{1}{4\pi^2 \times (1 \times 10^3\ Hz)^2 \times 30 \times 10^{-3}\ H}$$

$$= \mathbf{0.84\ \mu F}$$

Example 26-1 showed the characteristics of a series circuit at three specific frequencies. If calculations for Z, I, and θ are made at other frequencies and plotted on a graph, the effect above and below the resonant frequency is as shown in Fig. 26-3.

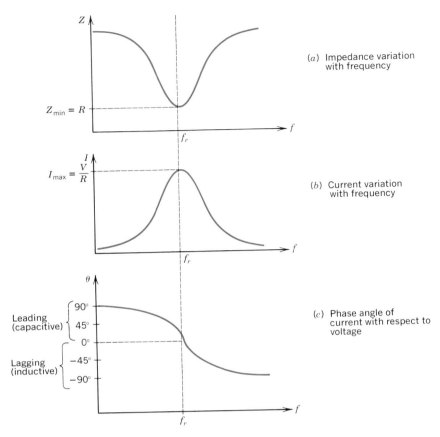

(a) Impedance variation with frequency

(b) Current variation with frequency

(c) Phase angle of current with respect to voltage

FIGURE 26-3

Effect of varying frequency on (a) impedance, (b) current, and (c) phase angle in a series *RLC* circuit.

Note that at the resonant frequency (f_r):

1. The circuit *impedance* is a *minimum,* equal to the resistance (R).
2. The circuit *current* is a *maximum,* equal to V/R.
3. The circuit reactance is zero, since $X_C = X_L$.
4. The circuit current is in phase with the applied voltage; θ is zero.

26-3 SELECTIVITY OF A SERIES CIRCUIT

The graph of *I* versus *f* in Fig. 26-3*b* shows that the circuit is *selective* in its *response* to a certain *band* of frequencies around the resonant frequency. That is, at frequencies that are much higher or lower than the resonant frequency (f_r),

the current in the circuit is very low. The circuit is said to have a low *response,* or is not very *sensitive* to these frequencies. The frequency *selectivity* of a resonant circuit is its most important characteristic. In order to compare the selectivity of one circuit with another, we refer to the *bandwidth* of the resonant circuit. (See Fig. 26-4.)

Consider the two frequencies f_1 and f_2 in Fig. 26-4. These have been selected where the circuit current is 0.707 times the maximum current at the resonant frequency (f_r). The bandwidth (Δf) is defined as the band of frequencies between f_1 and f_2, or:

$$\Delta f = f_2 - f_1 \qquad \text{hertz} \qquad (26\text{-}2)$$

where: Δf is the bandwidth of a resonant circuit, in hertz (Hz)

f_1 is the upper half-power frequency, in hertz (Hz)

f_2 is the lower half-power frequency, in hertz (Hz)

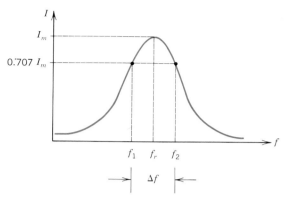

FIGURE 26-4
Bandwidth of a resonant circuit.

Note that f_1 and f_2 have been referred to as half-power frequencies.* At these frequencies, the power (true power) delivered to the circuit is one-half the power delivered at the resonant frequency (f_r). That is,

At $\quad f = f_r, \quad P = I_m^2 R$

At $\quad f = f_1, \quad P = (0.707 I_m)^2 R$
$$= 0.5 I_m^2 R$$
$$= \tfrac{1}{2} \times \text{power delivered at } f_r$$

Since a 50% change in the power level is not considered large (a 3-dB change in the power output from an audio amplifier is barely noticeable), the response of the circuit is considered relatively *flat* between f_1 and f_2. Beyond f_1 and f_2, the power drops off rapidly, and the circuit is considered as *rejecting* these frequencies. Thus, by the proper selection of the L and C components, a series resonant circuit can be made to "tune in" a given band of frequencies. This band will be centered at the circuit's resonant frequency, and adjacent frequencies will be rejected (but not completely, however). In most high-frequency communication circuits, f_1 and f_2 are symmetrically placed above and below f_r. Thus the "edge frequencies" (f_1 and f_2) can be obtained as follows:

$$f_1 = f_r - \frac{\Delta f}{2} \qquad (26\text{-}2a)$$

---

*f_1 and f_2 are also referred to as the lower and upper 3-dB frequencies. The notation dB stands for decibels, which are units used for comparison between two power levels (P_1 and P_2), according to the equation dB $= 10 \log P_2/P_1$. If $P_2 = 2P_1$, dB $= 10 \log 2 = 10 \times 0.3 = 3$ dB. Thus a 3-dB ratio refers to a 2:1 power level, such as occurs between f_1 and f_r, or f_2 and f_r.

$$f_2 = f_r + \frac{\Delta f}{2} \qquad (26\text{-}2b)$$

where terms are as above.

EXAMPLE 26-4

A series circuit has a resonant frequency of 400 kHz and a bandwidth of 12 kHz. Determine the "edge" or upper and lower half-power frequencies.

Solution

$$f_1 = f_r - \frac{\Delta f}{2} \qquad (26\text{-}2a)$$
$$= 400 \text{ kHz} - \frac{12 \text{ kHz}}{2} = \textbf{394 kHz}$$
$$f_2 = f_r + \frac{\Delta f}{2} \qquad (26\text{-}2b)$$
$$= 400 \text{ kHz} + \frac{12 \text{ kHz}}{2} = \textbf{406 kHz}$$

It can be shown that the bandwidth and the resonant frequency are related by the equation:

$$\Delta f = \frac{f_r}{Q} \qquad \text{hertz} \qquad (26\text{-}3)$$

where: Δf is the bandwidth of the circuit, in hertz (Hz)
  f_r is the resonant frequency of the circuit, in hertz (Hz)
  Q is the quality of the circuit, dimensionless

Recall, from Section 20-8, that the quality of a coil was defined as

$$Q = \frac{X_L}{R_{ac}} \qquad (20\text{-}9)$$

where R_{ac} is the total circuit resistance for a series RLC circuit. Thus, although the resistance does not determine a series circuit's resonant frequency, it does affect its bandwidth.

If Q is very large, the bandwidth Δf is very narrow, and we have a highly selective circuit. If Δf is too narrow, the circuit will not respond to all the incoming frequencies (as from a radio transmitter) so that some information is lost. Conversely, if Q is very small, the bandwidth is very broad, and interference from adjacent frequencies (or "channels") can occur because the circuit is not sufficiently selective. Each communication system has its own resonant frequency and bandwidth requirements.

For example, the intermediate frequency amplifiers in an AM (amplitude modulation) broadcast band receiver operate at an intermediate frequency of 455 kHz and require a bandwidth of approximately 10 kHz. An FM (frequency modulation) receiver operates at an intermediate frequency of 10.7 MHz and requires a 200-kHz bandwidth.

EXAMPLE 26-5

Determine the necessary Q values of circuits used in AM and FM intermediate amplifiers, using the data above.

Solution

$$\Delta f = \frac{f_r}{Q} \tag{26-3}$$

Therefore, $\quad Q = \dfrac{f_r}{\Delta f}$

For AM: $\quad Q = \dfrac{455 \text{ kHz}}{10 \text{ kHz}} = \mathbf{45.5}$

For FM: $\quad Q = \dfrac{10.7 \text{ MHz}}{0.2 \text{ MHz}} = \mathbf{53.5}$

The Q values in Example 26-5 are quite easy to attain commercially. In fact, it is because these Q values *are* practical that the commercially used intermediate frequencies of 455 kHz and 10.7 MHz were chosen.

EXAMPLE 26-6

Find the bandwidth of the circuit in Example 26-1 with L = 200 μH, R = 10 Ω, and C = 50 pF.

Solution

$$f_r = \frac{1}{2\pi \sqrt{LC}} \tag{26-1}$$

$$= \frac{1}{2\pi \sqrt{200 \times 10^{-6} \text{ H} \times 50 \times 10^{-12} \text{ F}}}$$

$$= 1.59 \text{ MHz}$$

$$Q = \frac{X_L}{R_{ac}} \tag{20-9}$$

$$= \frac{2\pi \times 1.59 \times 10^6 \text{ Hz} \times 200 \times 10^{-6} \text{ H}}{10 \text{ Ω}}$$

$$= 200$$

$$\Delta f = \frac{f_r}{Q} \tag{26-3}$$

$$= \frac{1.59 \times 10^6 \text{ Hz}}{200} \approx \mathbf{8 \text{ kHz}}$$

26-4 THE Q OF A SERIES-RESONANT CIRCUIT

We can determine another expression for the Q of a circuit to show how it determines the bandwidth.

For *any* resonant circuit, the Q can be defined as the ratio of the coil or the capacitor's reactive power to the true power.

$$Q = \frac{P_q}{P} \tag{26-4}$$

$$= \frac{I^2 X_L}{I^2 R_{ac}}$$

For a series circuit, the same current flows through the coil and resistor.

Therefore, $\quad Q = \dfrac{X_L}{R_{ac}} = \dfrac{2\pi f_r L}{R_{ac}} = 2\pi f_r \dfrac{L}{R_{ac}} \tag{20-9}$

But at resonance, $\quad f_r = \dfrac{1}{2\pi \sqrt{LC}} \tag{26-1}$

Therefore, $\quad 2\pi f_r = \dfrac{1}{\sqrt{LC}}$

Thus $\quad Q = \dfrac{1}{\sqrt{LC}} \dfrac{L}{R_{ac}}$

and $\quad Q = \dfrac{1}{R_{ac}} \sqrt{\dfrac{L}{C}} \tag{26-5}$

where: Q is the quality of a series-resonant circuit, dimensionless

R_{ac} is the total ac resistance of the circuit, in ohms (Ω)

L is the inductance, in henrys (H)

C is the capacitance, in farads (F)

EXAMPLE 26-7

Determine the Q of a series-resonant circuit in which R = 10 Ω, L = 200 μH, and C = 50 pF.

Solution

$$Q = \frac{1}{R_{ac}} \sqrt{\frac{L}{C}} \qquad (26\text{-}5)$$

$$= \frac{1}{10 \ \Omega} \sqrt{\frac{200 \times 10^{-6} \text{ H}}{50 \times 10^{-12} \text{ F}}}$$

$$= \frac{1}{10 \ \Omega} \sqrt{4 \times 10^6 \ \Omega^2} = \mathbf{200}$$

You can examine Eq. 26-5 to see the reason for its effect on the bandwidth. Note that if R_{ac} alone is reduced, Q is increased and the bandwidth is decreased without affecting the resonant frequency. Thus, if in Example 26-7, R is reduced from 10 to 5 Ω, Q is doubled to 400. The reason for the reduction in bandwidth can be seen in Fig. 26-5a. The current at resonance must now be double its initial value, causing a *steeper* rise in current around the resonant frequency (f_r).

Similarly, we see that an increase in the L/C ratio in Eq. 26-5 increases Q. Thus, when $L = 200$ μH, and $C = 50$ pF, $L/C = 4 \times 10^6$. If L is increased to 400 μH and C is reduced to 25 pF, the L/C ratio equals 16×10^6, so that Q is now double its initial value. (Note that the resonant frequency for this example is unchanged, since the product LC is still the same.)

The steeper rise in the I versus f curve in Fig. 26-5b shows that a narrower bandwidth results with a higher L/C ratio. The reason for this is found in the graphs of X_L and

X_C versus f in Fig. 26-1. A larger L/C ratio implies a larger value of L and a smaller value of C. This leads to *steeper* slopes of the X_L and X_C graphs. This means that as the frequency increases, the point of resonance (where X_L equals X_C) is reached more *rapidly*. If L were small and C were large, the two graphs would be relatively "flat," and a large change in f would be necessary to make X_L equal X_C for resonance.

Thus Q is the one term that combines the effects of all three components in determining the shape of the resonance curve of a series RLC circuit.

26-5 RESONANT RISE OF VOLTAGE IN A SERIES *RLC* CIRCUIT

One of the most useful features of a series RLC circuit is the *resonant rise of voltage* across the capacitor and inductor as resonance is approached. This is seen in Table 26-1 where, at the resonant frequency of 1.59 MHz, both the capacitor and inductor voltages are 20 V, with a supply voltage of only 100 mV.

Although the net reactance in the circuit is zero, allowing a large current to circulate, the individual reactances are quite large. The product of a large current (10 mA) and a large reactance (2 kΩ) produces a high voltage (20 V). Actually, because the capacitive reactance decreases

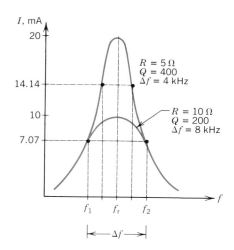

(a) A reduction in R increases Q and decreases the bandwidth.

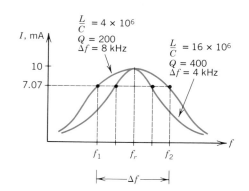

(b) An increase in the L/C ratio increases Q and decreases the bandwidth.

FIGURE 26-5

Effect of changes in R and L/C on the Q and bandwidth of a series resonant circuit.

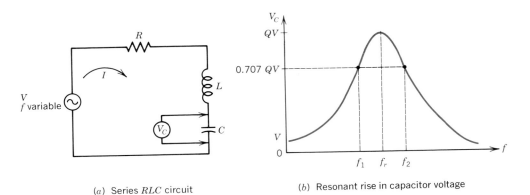

(a) Series *RLC* circuit   (b) Resonant rise in capacitor voltage

FIGURE 26-6

Effect on the capacitor voltage of varying the frequency of a series resonant *RLC* circuit.

as frequency increases, the product IX_C reaches a maximum at a frequency slightly *lower* than the resonant frequency (f_r). For high Q circuits, however, this difference is negligible and you can think of the capacitor (and inductor) as having a frequency response curve that peaks at f_r and has a bandwidth Δf given by Eq. 26-3. (See Fig. 26-6b.)

Note that the curve in Fig. 26-6b is not symmetrical. At very low frequencies, the capacitor voltage approaches the supply voltage (V), or an *open* circuit. At very high frequencies, V_C approaches zero, because X_C and I also approach zero. However, for high Q, $(Q > 10)$ high-frequency circuits, f_1 and f_2 can be considered to be symmetrically located above and below f_r.

An expression for the capacitor (and inductor) voltage at resonance can be obtained as follows:

$$V_C = IX_C \qquad (23\text{-}1)$$

But, at resonance:

$$I = \frac{V}{R} \qquad (15\text{-}15a)$$

and

$$X_C = X_L$$

Therefore,

$$V_C = \frac{V}{R}X_L$$

$$= \frac{X_L}{R}V$$

and

$$V_C = QV \quad \text{volts} \qquad (26\text{-}6)$$

where: V_C is the capacitor voltage at resonance, in volts (V)

Q is the total circuit quality, dimensionless

V is the voltage applied to the circuit at resonance, in volts (V)

Because Q is usually much larger than unity, we see a significant resonant rise in the capacitor voltage at resonance. A similar (but 180° out-of-phase) voltage rise occurs across the inductor. However, because of its resistance, the coil's voltage is not given simply by Eq. 26-6, although the two are close in value for high Q circuits.

EXAMPLE 26-8

A coil of 8-H inductance and 400-Ω resistance is series-connected with a variable capacitor and an ammeter across the 120-V, 60-Hz line. Calculate:

a. The capacitor voltage when the capacitor is adjusted for maximum circuit current.

b. The necessary capacitor for the condition in part (a).

Solution

a.   At resonance, the circuit current is a maximum, so f_r = 60 Hz.

$$Q = \frac{X_L}{R} = \frac{2\pi f L}{R} \qquad (20\text{-}9)$$

$$= \frac{2\pi \times 60\ \text{Hz} \times 8\ \text{H}}{400\ \Omega} = 7.5$$

$$V_C = QV \qquad (26\text{-}6)$$

$$= 7.5 \times 120\ \text{V} = \textbf{900 V}$$

b. $f_r = \dfrac{1}{2\pi \sqrt{LC}} \qquad (26\text{-}1)$

$$C = \frac{1}{4\pi^2 f^2 L}$$

$$= \frac{1}{4\pi^2 \times (60\ \text{Hz})^2 \times 8\ \text{H}} = \textbf{0.88 μF}$$

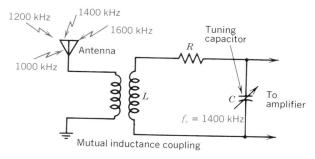

(a) Series *RLC* circuit tuned to 1400 kHz by *C* receives only one band of frequencies centered at 1400 kHz.

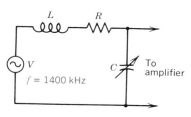

(b) Equivalent circuit.

FIGURE 26-7

Resonant rise of capacitor voltage in a series *RLC* circuit is used to select a desired band of frequencies.

Note, in Example 26-8, how it is possible for dangerously high voltages to be produced at resonance in a series *RLC* circuit.

Figure 26-7 shows how a capacitor's resonant rise of voltage is used to advantage in a frequency selective circuit, such as in a radio receiver or television set.

The antenna, as shown in Fig. 26-7a, "picks up" many different frequencies from the electromagnetic waves radiated by neighboring radio transmitters. These signals, in the order of several millivolts for an AM receiver, but only several microvolts for an FM receiver, must be "tuned in" by the resonant circuit. Assume that the circuit is tuned to a resonant frequency of 1400 kHz (the carrier frequency of a local AM station, for example) by adjusting *C*, as in Fig. 26-7b. Only the information contained in the band of frequencies associated with *that* resonant frequency will appear across the capacitor for amplification and further processing. If the circuit's bandwidth is correct, all other adjacent signals will be rejected, or "tuned out." The adjustment of *C* to a different value makes the circuit *sensitive* to a different band of frequencies, and a different station may be tuned in.

EXAMPLE 26-9

What is the necessary range of a variable capacitor to use with a 200-μH coil in a series circuit to tune in the standard radio broadcast band from 535 to 1605 kHz.

Solution

$$f_r = \frac{1}{2\pi \sqrt{LC}} \qquad (26\text{-}1)$$

$$C = \frac{1}{4\pi^2 f_r^2 L}$$

For $f_r = 535$ kHz,

$$C = \frac{1}{4\pi^2 \times (535 \times 10^3 \text{ Hz})^2 \times 200 \times 10^{-6} \text{ H}}$$
$$= \textbf{442 pF}$$

For $f_r = 1605$ kHz,

$$C = \frac{1}{4\pi^2 \times (1605 \times 10^3 \text{ Hz})^2 \times 200 \times 10^{-6} \text{ H}}$$
$$= \textbf{49 pF}$$

26-6 RESONANCE IN A THEORETICAL, PARALLEL *RLC* CIRCUIT

Now, consider the effect of varying frequency in a *parallel RLC* circuit, as in Fig. 26-8a.

The circuit shown is theoretical because we have assumed an ideal inductor having zero resistance. (Later, in Section 26-8, the effect of the coil's resistance will be taken into account.)

You know that at any frequency,

$$I_T = \sqrt{I_R^2 + (I_C - I_L)^2} \qquad (24\text{-}25)$$

$$I_T = \frac{V}{Z} \qquad (24\text{-}4)$$

and

$$\theta = \arctan \frac{I_C - I_L}{I_R} \qquad (24\text{-}26)$$

Also,

$$I_C = \frac{V}{X_C} \qquad (23\text{-}1)$$

and

$$I_L = \frac{V}{X_L} \qquad (20\text{-}1)$$

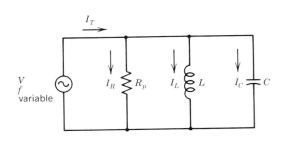

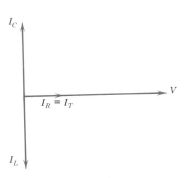

(a) Theoretical parallel resonant circuit.
(Inductor has no resistance.)

(b) Phasor diagram.

FIGURE 26-8

Resonance in a theoretical parallel resonant circuit.

If the source frequency is adjusted until $X_L = X_C$, then $I_L = I_C$. This means that $(I_L - I_C) = 0$; also, the total circuit current I_T is a *minimum* and equal to I_R. But if I_T is a *minimum*, the impedance Z must be a *maximum*. (See Eq. 24-4.)

$$Z = \frac{V}{I_T} = \frac{V}{I_R} = R_P$$

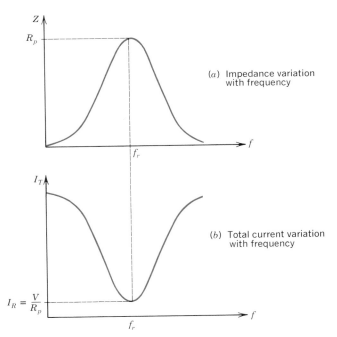

(a) Impedance variation with frequency

(b) Total current variation with frequency

FIGURE 26-9

Variation of (a) impedance and (b) total current in a parallel RLC circuit with a variable frequency source.

Thus the circuit is purely resistive ($\theta = 0°$) and the circuit current and applied voltage are in phase with each other. (The L and C components apparently act as an open circuit, with infinite impedance, as far as the source is concerned.)

This condition is called *parallel resonance*. It is also sometimes referred to as "antiresonance" because the effect on circuit current and impedance is opposite to the series resonant effect, as shown in Fig. 26-9.

Since parallel resonance occurs in *this* circuit when $X_L = X_C$, as it does in a series *RLC* circuit, the resonant frequency is given by the same equation:

$$f_r = \frac{1}{2\pi\sqrt{LC}} \quad (26\text{-}1)$$

EXAMPLE 26-10

A 100-μH coil of negligible resistance, a 50-pF capacitor, and a 100-kΩ resistor are connected in parallel across a variable frequency 50-mV signal generator. Calculate:
a. The resonant frequency of the circuit.
b. Each branch current when the signal generator is adjusted to the resonant frequency.
Show the currents in a circuit diagram.

Solution

a. $f_r = \dfrac{1}{2\pi \sqrt{LC}}$ (26-1)

$$= \frac{1}{2\pi\sqrt{100 \times 10^{-6}\ \text{H} \times 50 \times 10^{-12}\ \text{F}}}$$

$$= 2.25 \times 10^6\ \text{Hz} = \textbf{2.25 MHz}$$

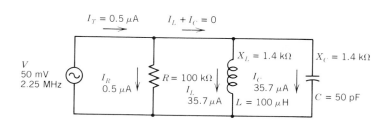

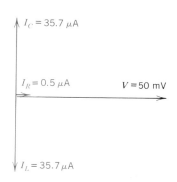

FIGURE 26-10

Circuit currents for Example 26-10.

b. $I_R = \dfrac{V}{R}$ (15-15a)

$= \dfrac{50\ mV}{100\ k\Omega} = \mathbf{0.5\ \mu A}$

$X_L = 2\pi fL$ (20-2)

$= 2\pi \times 2.25 \times 10^6\ Hz \times 100 \times 10^{-6}\ H$

$= 1414\ \Omega$

$I_L = \dfrac{V}{X_L}$ (20-1)

$= \dfrac{50\ mV}{1.4\ k\Omega} = \mathbf{35.7\ \mu A}$

$X_C = \dfrac{1}{2\pi fC}$ (23-2)

$= \dfrac{1}{2\pi \times 2.25 \times 10^6\ Hz \times 50 \times 10^{-12}\ F}$

$= 1414\ \Omega$

$I_C = \dfrac{V}{X_C}$ (23-1)

$= \dfrac{50\ mV}{1.4\ k\Omega} = \mathbf{35.7\ \mu A}$

See Fig. 26-10.

Note that the inductor and capacitor currents are equal in magnitude and 180° out of phase with each other. Thus, $I_L + I_C$ in Fig. 26-10 is zero, even though the individual currents are much larger than the total line current. This shows how the theoretical parallel *LC* circuit is an *open* circuit, since it draws no current from the source.

It is interesting to note what would happen if an ideal parallel *LC* circuit could be constructed. Consider the capacitor charged to some initial voltage and allowed to dis-charge through the inductor. The initial energy, $\frac{1}{2}CV^2$, stored in the electric field of the capacitor, would be transferred to the inductor in the form of a magnetic field having energy $\frac{1}{2}LI^2$. But when the capacitor is fully discharged, the magnetic field around the inductor would begin to collapse and, by Lenz's law, induce the opposite polarity of voltage across its terminals. This would cause the capacitor to become charged to the opposite polarity from its initial state. The cycle then would repeat itself, producing a sinusoidal waveform of voltage across the *LC* combination. The circuit is said to *oscillate* (or "ring"), producing a continuous sinusoidal voltage, at a frequency given by $f = \dfrac{1}{2\pi\sqrt{LC}}$.

In a practical circuit, where the inductor has resistance, the oscillations are "damped" and decay due to energy being dissipated in the resistance. Some *oscillator circuits*, as in some signal generators, rely on this tank circuit "flywheel" effect to produce a sinusoidal output voltage from a dc supply. An active device, such as a transistor, is used to supply energy at the right instants to make up for the energy dissipated in the coil's resistance.

The type of signal generator referred to in Example 26-10 is called a radio frequency (RF) generator. An RF generator produces signals from approximately 100 kHz to several hundred megahertz. The output voltage is much smaller than from an audio generator, sometimes as small as 0.1 V.

A typical RF generator is shown in Fig. 26-11. In addition to producing a sine wave output, some produce an *amplitude-modulated* output, in which the amplitude of the radio frequency signal is varied (modulated) at an *audio* rate, such as 400 Hz. This type of signal is useful for checking the operation of communications receivers.

FIGURE 26-11
A radio-frequency signal generator. (Courtesy of Dynascan Corporation, B & K-Precision Product Group.)

26-6.1 The Q of a Theoretical, Parallel Resonant Circuit

You have seen (in Section 26-4) that the Q of a resonant circuit is given by the following equation:

$$Q = \frac{P_q}{P} \tag{26-4}$$

For the theoretical parallel RLC resonant circuit, with no resistance in the inductor,

$$P_q = \frac{V^2}{X_L} \tag{25-3}$$

and

$$P = \frac{V^2}{R_p} \tag{15-13}$$

Thus

$$Q = \frac{V^2/X_L}{V^2/R_p}$$

and

$$Q = \frac{R_p}{X_L} \tag{26-7}$$

where: Q is the quality of a parallel RLC circuit, dimensionless

R_p is the resistance of the circuit, in ohms (Ω)

X_L is the inductive reactance of the *zero resistance* inductor, in ohms (Ω)

You have seen (Example 26-10) that R_p is usually much larger than X_L, so $Q > 1$ as in a series RLC circuit.

Note that the impedance of the parallel circuit at resonance can now be expressed in terms of Q:

$$Z = R_p = QX_L \tag{26-7a}$$

where terms are as previously given. Thus a high Q circuit has a high impedance at resonance.

The currents also can be expressed in terms of Q:

$$V = I_L X_L = I_R R_p$$

Therefore,

$$I_L = \frac{R_p}{X_L} \times I_R$$

But

$$\frac{R_p}{X_L} = Q \tag{26-7}$$

and

$$I_R = \text{line current} = I_T$$

Thus, considering only the *magnitudes* of the currents:

$$I_L = I_C = Q \times I_T \tag{26-8}$$

where symbols are as already stated.

Equation 26-8 states that there is a *resonant rise of current* in the *tank circuit* consisting of the parallel LC combination. This corresponds to the resonant rise of *voltage* in a *series* circuit. Very large currents can circulate in the tank circuit as energy is transferred back and forth between the coil and the capacitor, even if the line current is very low, because the Q of the circuit is high. (See Eq. 26-8.)

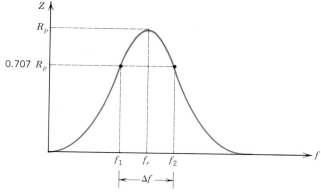

FIGURE 26-12

Bandwidth of a parallel resonant circuit.

EXAMPLE 26-11

For the circuit in Example 26-10, determine:
a. Q.
b. The inductor and capacitor currents at resonance.
c. The impedance at resonance.

Solution

a. $Q = \dfrac{R_p}{X_L}$ (26-7)

   $= \dfrac{100 \text{ k}\Omega}{1.4 \text{ k}\Omega} = \mathbf{71.4}$

b. $I_L = I_C = Q \times$ line current   (26-8)

      $= 71.4 \times 0.5 \text{ }\mu\text{A} = \mathbf{35.7 \text{ }\mu\text{A}}$

c. $Z = QX_L$ (26-7a)

    $= 71.4 \times 1.4 \text{ k}\Omega = \mathbf{100 \text{ k}\Omega}$

26-7 SELECTIVITY OF A PARALLEL RESONANT CIRCUIT

The bandwidth of a parallel resonant circuit (theoretical or practical) is the band of frequencies between f_1 and f_2 on the resonance curve, where the total impedance drops to 0.707 of the parallel resistance (R_p). See Fig. 26-12.

Again, it can be shown that

$$\Delta f = \frac{f_r}{Q} \tag{26-3}$$

so that a high Q results in a very selective circuit, as in the case of a series resonant circuit.

EXAMPLE 26-12

Determine the bandwidth of the circuit in Example 26-10.

Solution

$$\Delta f = \frac{f_r}{Q} \tag{26-3}$$

$$= \frac{2.25 \times 10^6 \text{ Hz}}{71.4} = \mathbf{31.5 \text{ kHz}}$$

To see the usefulness of a parallel circuit to select certain frequencies, you must now include the internal resistance of the source. This internal resistance is purposely made high for the reasons described below. (See Fig. 26-13.)

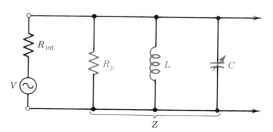

a. Parallel resonant circuit with high internal resistance source

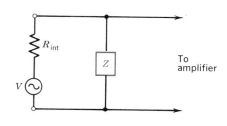

b. Equivalent circuit. At resonance, $Z = R_p$

FIGURE 26-13

Selectivity of a parallel resonant circuit is provided by a high internal resistance in the source.

If you represent the impedance of the parallel *RLC* circuit at *any* frequency by *Z*, as in Fig. 26-13*b*, you obtain a series voltage-divider circuit, where

$$V_{\text{out}} = V \times \frac{Z}{Z + R_{\text{int}}} \qquad (5\text{-}6)$$

At resonance:

$$V_{\text{out}} = V \times \frac{R_p}{R_p + R_{\text{int}}} \qquad (5\text{-}6a)$$

That is, at resonance, when *Z* is a maximum and equal to R_p, the maximum voltage is developed across the tank circuit. At frequencies above or below resonance, the impedance *Z* decreases, as in Fig. 26-12, so the output voltage also decreases. If R_{int} were zero or small, the voltage across the tank circuit would be the same as the source voltage at all frequencies, and no selectivity would result. For this reason, R_{int} is deliberately increased so that high selectivity is produced in the range of the resonant frequency.

EXAMPLE 26-13

A parallel *RLC* circuit has $R_p = 100$ kΩ. The circuit is connected to a 10-μV source with an internal resistance of 50 kΩ. Calculate:
a. The output voltage across the tank circuit when the capacitor is adjusted to provide resonance.
b. The output voltage for a frequency far from resonance where the tank circuit's impedance is 10 kΩ.

Solution

a. $V_{\text{out}} = V \times \dfrac{R_p}{R_p + R_{\text{int}}} \qquad (5\text{-}6a)$

$= 10 \text{ μV} \times \dfrac{100 \text{ kΩ}}{100 \text{ kΩ} + 50 \text{ kΩ}}$

$= \textbf{6.7 μV}$

b. $V_{\text{out}} = V \times \dfrac{Z}{Z + R_{\text{int}}} \qquad (5\text{-}6)$

$= 10 \text{ μV} \times \dfrac{10 \text{ kΩ}}{10 \text{ kΩ} + 50 \text{ kΩ}} = \textbf{1.7 μV}$

NOTE The addition of *Z* and R_{int} in part (b) above should actually be done *vectorially*, since, off resonance, the impedance is no longer pure resistance. But the error is small, since $R_{\text{int}} > Z$ and the total circuit is primarily resistive.

The selectivity of a parallel resonant circuit is used to produce sine wave voltages from *nonsinusoidal* pulses. These repetitive pulses have some *fundamental* sinusoidal frequency and many *harmonics*. (A harmonic is some multiple of the fundamental frequency.) Thus, if a circuit is resonant to only one frequency, it will reject the others and produce a pure sine wave. This property is used in Class C radio frequency amplifiers that conduct for less than half a cycle of a sine wave and produce pulses at some basic frequency. (This mode of operation produces a highly efficient amplifier for large power outputs.)

A parallel resonant circuit can also be used as a frequency doubler or *multiplier* if the circuit is tuned to the second harmonic rather than the fundamental. Finally, the output of an RF signal generator may not be a good sine wave until it is used with some resonant circuit (series or parallel) to reject the harmonics.

26-8 RESONANCE IN A PRACTICAL, TWO-BRANCH PARALLEL *RLC* CIRCUIT

In this section, you will consider the effect of the coil's resistance on the resonant frequency of a two-branch parallel *RLC* circuit. (See Fig. 26-14.)

Assume that the frequency of the source has been adjusted until $X_L = X_C$. Because of the (series) resistance of the coil (R_s), there is *less* current through the coil than through the capacitor. This means that the reactive power in *L* and *C* are *not* equal, so we do *not* have resonance. That is, for resonance (in series or parallel) the following essentially identical conditions are required:

1. $P_{qL} = P_{qC}$
2. $\cos \theta = 1$
3. $\theta = 0°$
4. $Z = R$

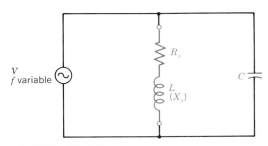

FIGURE 26-14
Practical parallel *RLC* circuit.

To make $P_{qL} = P_{qC}$, the frequency must be *lowered* slightly, reducing X_L so that I_L *increases* until $I_L^2 X_L = I_C^2 X_C$. When this occurs, it can be shown that the resonant frequency is given by

$$f_r = \frac{1}{2\pi\sqrt{LC}} \times \sqrt{\frac{Q^2}{1 + Q^2}} \quad \text{hertz} \quad (26\text{-}9)$$

where: f_r is the resonant frequency of a practical parallel circuit, in hertz (Hz)
   L is the inductance of the coil, in henrys (H)
   C is the capacitance of the circuit, in farads (F)
   Q is the quality of the coil, dimensionless

Note that

$$f_r < \frac{1}{2\pi\sqrt{LC}}$$

as expected, so the coil's resistance *does* determine the circuit's resonant frequency. However, if $Q \geq 10$, the resonant frequency (f_r) is given by $\frac{1}{2\pi\sqrt{LC}}$ with a maximum error of only 0.5%. Thus, Eq. 26-1 gives the resonant frequency of series *RLC* circuits *and* high Q, parallel *RLC* circuits.

Note also that the Q in Eq. 26-9 is specified as the Q of the coil itself. The reason for this can be seen in the equivalent circuit of Fig. 26-15b.

By the use of network analysis techniques, the practical two-branch parallel circuit in Fig. 26-15a can be converted to the equivalent three-branch theoretical circuit in Fig. 26-15b, where

$$R_p = \frac{R_s^2 + X_s^2}{R_s} = \frac{Z_s^2}{R_s} \quad (26\text{-}10)$$

$$X_p = \frac{R_s^2 + X_s^2}{X_s} = \frac{Z_s^2}{X_s} \quad (26\text{-}11)$$

where the symbols are as previously stated.

But you know that, for the theoretical circuit:

$$Q = \frac{R_p}{X_p} \quad (26\text{-}7)$$
$$= \frac{R_s^2 + X_s^2}{R_s} \div \frac{R_s^2 + X_s^2}{X_s}$$

Therefore,

$$Q = \frac{X_s}{R_s} = \frac{2\pi fL}{R_s} \quad (26\text{-}12)$$

where: Q is the quality of the coil (and the whole parallel circuit, if there is no additional parallel resistance), dimensionless
   L is the inductance of the coil, in henrys (H)
   R_s is the series resistance of the coil itself, in ohms (Ω)

Equation 26-9 can now be expressed in the following form:

$$f_r = \frac{1}{2\pi\sqrt{LC}} \times \sqrt{1 - \frac{CR_s^2}{L}} \quad \text{hertz} \quad (26\text{-}13)$$

where symbols are as already stated. Note that if $R_s = 0$, Eq. 26-13 reduces to Eq. 26-1,

$$f_r = \frac{1}{2\pi\sqrt{LC}}$$

EXAMPLE 26-14

A 10-μH coil having an ac resistance of 5 Ω is connected in parallel with a 0.01-μF capacitor across a radio frequency signal generator. Calculate:
a. The resonant frequency of the two-branch parallel circuit.
b. The Q of the circuit.

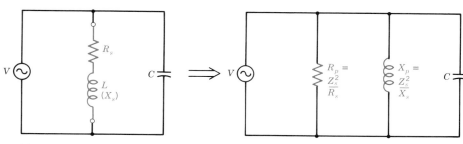

(a) Practical parallel *RLC* circuit
(b) Equivalent theoretical circuit

FIGURE 26-15
(a) A practical parallel resonant circuit may be converted to (b) an equivalent theoretical parallel circuit.

c. The bandwidth of the circuit.

d. The impedance of the circuit when the generator is tuned to the resonant frequency.

Solution

a. $f_r = \dfrac{1}{2\pi\sqrt{LC}} \times \sqrt{1 - \dfrac{CR_s^2}{L}}$ (26-13)

$= \dfrac{1}{2\pi\sqrt{10 \times 10^{-6}\,\text{H} \times 0.01 \times 10^{-6}\,\text{F}}}$

$\times \sqrt{1 - \dfrac{0.01 \times 10^{-6} \times 5^2}{10 \times 10^{-6}}}$

$= 503\,\text{kHz} \times 0.987 = \textbf{496 kHz}$

b. $Q = \dfrac{X_s}{R_s} = \dfrac{2\pi f L}{R_s}$ (26-12)

$= \dfrac{2\pi \times 496 \times 10^3\,\text{Hz} \times 10 \times 10^{-6}\,\text{H}}{5\,\Omega}$

$= \dfrac{31\,\Omega}{5\,\Omega} = \textbf{6.2}$

c. $\Delta f = \dfrac{f_r}{Q}$ (26-3)

$= \dfrac{496\,\text{kHz}}{6.2} = \textbf{80 kHz}$

d. $Z = R_p = \dfrac{R_s^2 + X_s^2}{R_s}$ (26-10)

$= \dfrac{(5\,\Omega)^2 + (31\,\Omega)^2}{5\,\Omega}$

$= \textbf{197}\,\boldsymbol{\Omega}$ (pure resistance)

Note that, since $Q = 6.2$ is less than 10, the resonant frequency is a little lower than that given by Eq. 26-1.

Also note the low impedance (resistance) of the circuit at resonance. This is due to the circuit's very low Q:

$Z = R_p = \dfrac{R_s^2 + X_s^2}{R_s}$

$= R_s + \dfrac{X_s^2}{R_s}$

$= R_s\left(1 + \dfrac{X_s^2}{R_s^2}\right)$

$= R_s(1 + Q^2)$ ohms (26-14)

$= 5\,\Omega(1 + 6.2^2) = \textbf{197}\,\boldsymbol{\Omega}$

26-8.1 Damping a Parallel Resonant Circuit

Introducing resistance to purposely reduce the Q of a tuned parallel circuit is called *damping*. This can be achieved for the practical two-branch circuit in two ways.

1. The series resistance of the coil (R_s) in Fig. 26-14 can be *increased*, as seen in Eq. 26-12, reducing the Q. This also has an interesting effect on the resonant frequency as shown by Eq. 26-13. If R_s is increased until $\dfrac{CR_s^2}{L} \geq 1$, there is *no* resonant frequency. This is called *critical* damping and is used in high-frequency circuits (such as TV receivers) to keep the coils and their distributed capacitances from *ringing* (oscillating at some undesired frequency).

2. Resistance can be added in *parallel* with the circuit. This *reduces* the effective value of R_p in Eq. 26-7, so the Q is also reduced. This is called *shunt* damping and is the preferred method to maintain a symmetrical resonant response, although both series and parallel damping combined is also used.

Damping is sometimes used in communications receivers (see Fig. 26-16) to broaden the bandwidth when looking for a weak signal. That is, a weak signal can easily be "tuned through" without detection by a circuit that has a very narrow bandwidth. However, once the weak signal is found when "tuned in," a narrow bandwidth would be an asset to block out competing, stronger adjacent signals.

EXAMPLE 26-15

In the parallel two-branch circuit of Example 26-14, with the 10-μH coil having a resistance of 5 Ω, and the 0.01-μF capacitor, determine:

a. The minimum resistance to be *added* in series with the coil to make sure the circuit has no resonant frequency.

b. The resistance to be added in *series* with the coil to broaden the bandwidth to 100 kHz and the new resonant frequency.

c. The shunting resistance to be connected in *parallel* with the coil to broaden the bandwidth to 100 kHz.

Solution

a. To avoid resonance: $\dfrac{CR_s^2}{L} \geq 1$

Therefore, minimum resistance

$R_s = \sqrt{\dfrac{L}{C}}$

$= \sqrt{\dfrac{10 \times 10^{-6}\,\text{H}}{0.01 \times 10^{-6}\,\text{F}}} = 31.6\,\Omega$

FIGURE 26-16

A communications receiver that uses parallel resonant circuits. (Courtesy of Trio-Kenwood Communications, Inc.)

Since the coil's resistance is 5 Ω, the *minimum* resistance to be added = 31.6 − 5 Ω = **26.6 Ω.**

b.   For a bandwidth of 100 kHz:

$$Q = \frac{f_r}{\Delta f} \qquad (26\text{-}3)$$

$$= \frac{496 \text{ kHz}}{100 \text{ kHz}} = 4.96$$

$$f_r = \frac{1}{2\pi\sqrt{LC}} \times \sqrt{\frac{Q^2}{1 + Q^2}} \qquad (26\text{-}9)$$

$$= 503 \text{ kHz} \times \sqrt{\frac{4.96^2}{1 + 4.96^2}}$$

$$= 503 \text{ kHz} \times 0.98 = \textbf{493 kHz}$$

Thus the new Q must be 4.93 to produce a bandwidth of 100 kHz at f_r = 493 kHz.

$$Q = \frac{X_s}{R_s} = \frac{2\pi f L}{R_s} \qquad (26\text{-}12)$$

$$R_s = \frac{2\pi f_r L}{Q}$$

$$= \frac{2\pi \times 493 \times 10^3 \text{ Hz} \times 10 \times 10^{-6} \text{ H}}{4.93}$$

$$= 6.3 \ \Omega$$

The resistance to be added in series with the coil = 6.3 − 5 Ω = **1.3 Ω.**

c.   For the equivalent parallel circuit with R_p:

$$Q = \frac{R_p}{X_p} \qquad (26\text{-}7)$$

Initially, with Δf = 80 kHz, the *equivalent* R_p = 197 Ω. For a Δf of 100 kHz, the *effective* R_s = 6.3 Ω:

$$R_p = \frac{R_s^2 + X_s^2}{R_s} \qquad (26\text{-}10)$$

$$= \frac{(6.3 \ \Omega)^2 + (31 \ \Omega)^2}{(6.3 \ \Omega)} = 159 \ \Omega$$

We need a resistance in parallel with 197 Ω to produce 159 Ω:

$$R_T = \frac{R_1 \times R_2}{R_1 + R_2} \qquad (6\text{-}10)$$

$$159 = \frac{197 \times R_2}{197 + R_2}$$

$$R_2 = \textbf{824 } \Omega$$

Example 26-15 shows how the resonant frequency of a parallel *RLC* circuit can be changed (3 kHz, in this case) by adding resistance in parallel with the overall circuit, or in series with the coil itself.

26-9 OPERATING PRINCIPLE OF A SUPERHETERODYNE AM RECEIVER

Series and parallel resonant circuits are widely used in AM, FM, and TV receivers. Figure 26-17 shows a block diagram of an AM radio receiver. (The principle of operation applies to FM receivers, as well, but with different frequencies.)

You have already seen (in Fig. 26-7) how a series resonant circuit can be tuned to select just one of the many bands of frequencies arriving at the antenna. This antenna signal is fed to a radio frequency amplifier, which resonantly tunes to the incoming signal whenever the tuning dial is turned. The incoming signal is very weak (a few microvolts) and must be amplified thousands of times before it is audible. This amplification is carried out in stages.

In the early days of radio, each stage had to be tuned to the signal being received, so that each stage had to be tunable over the full band, from 535 to 1605 kHz. This type of radio receiver was known as a "tuned radio frequency" (TRF) receiver; it had many disadvantages.

Modern receivers use *frequency conversion* or *heterodyne* circuits to produce a signal whose *radio* frequency is always the same (455 kHz), but which contains the *audio* frequencies of whichever station is tuned in. This allows all further stages of radio frequency amplification to take place at the *one fixed* frequency called the *intermediate* frequency (IF).

Assume that you want to listen to an AM radio station broadcasting at 1400 kHz. When you turn the dial to this frequency, the radio frequency amplifier is tuned to 1400 kHz. Also, "ganged" together, a *local oscillator* produces an RF sine wave equal to 1400 kHz + 455 kHz, or 1855 kHz. The 1400-kHz signal (containing the "intelligence" or desired audio) and the unmodulated 1855-kHz signal are applied to the *mixer*. As the name suggests, the mixer or *converter* produces an output signal that (among other things) contains frequencies of the *sum* and *difference* of the input signals. That is, frequencies of 1855 kHz + 1400 kHz = 3255 kHz and 1855 kHz − 1400 kHz = 455 kHz are available at the output. Each of these frequencies (3255 kHz and 455 kHz) contains the same audio as the initial signal. For practical considerations, however, only the *difference* frequency (455 kHz) is used.

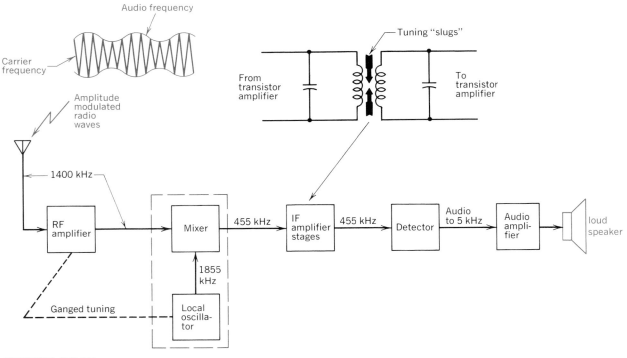

FIGURE 26-17

Block diagram of a superheterodyne AM radio receiver.

Thus, all further intermediate frequency amplifier stages are tuned to this frequency, since it is the *only* frequency produced, regardless of which station is tuned in. These stages use parallel *LC* circuits with inductive coupling from one stage to the next, as shown in Fig. 26-17. The coupling can be adjusted by using ferrite "slugs" to *align* the receiver at the proper frequency and provide the correct bandwidth of approximately 10 kHz. This bandwidth is necessary to prevent overlap between stations. As a result,

the highest audio frequency permitted is 5 kHz. When the carrier (1400 kHz in the example) is amplitude-modulated by a frequency of 5 kHz, *sideband* frequencies are produced that are 5 kHz *above* and *below* the carrier. All of these frequencies must be retrieved, except in a single sideband receiver, to reproduce the original sound. This process of *demodulation* is carried out in the detector stage and is followed by audio amplification to drive a loudspeaker.

SUMMARY

1. When a series circuit is in resonance, the circuit's impedance is a minimum (equal to R), the current is a maximum (equal to V/R), and the applied voltage and current are in phase with each other.

2. The resonant frequency of a series *RLC* circuit is given by

$$f_r = \frac{1}{2\pi\sqrt{LC}}$$

3. The bandwidth of a resonant circuit is the band of frequencies (Δf) between the two frequencies (f_1 and f_2) at which the power delivered to the circuit is one-half the power delivered at the resonant frequency.

$$\Delta f = \frac{f_r}{Q}$$

4. The Q of a series resonant circuit can be obtained using:

$$Q = \frac{P_q}{P} = \frac{X_L}{R_{ac}} = \frac{1}{R_{ac}}\sqrt{\frac{L}{C}}$$

5. At resonance in a series *RLC* circuit, there is a resonant rise of voltage across the capacitor and inductor given by

$$V_L = V_C = QV$$

6. At resonance in a theoretical *parallel RLC* circuit, the circuit *impedance* is a *maximum* (equal to R_p), the line *current* is a *minimum* (equal to V/R_p), and the circuit is purely resistive with the applied voltage and current in phase with each other.

7. The resonant frequency of a theoretical parallel *RLC* circuit is given by

$$f_r = \frac{1}{2\pi\sqrt{LC}}$$

8. The Q of a theoretical parallel *RLC* circuit is given by

$$Q = \frac{R_p}{X_L}$$

9. There is a resonant rise of current in the tank circuit of a parallel *RLC* circuit given by

$$I_L = I_C = Q \times I_R$$

10. The bandwidth of a parallel resonant circuit is the band of frequencies between f_1 and f_2, where the circuit impedance drops to 0.707 of the parallel resistance (R_p) and is given by

$$\Delta f = \frac{f_r}{Q}$$

11. The selectivity of a parallel resonant circuit depends upon the source having a high internal resistance to cause a voltage-divider effect that maximizes the output voltage at resonance.
12. Because of a coil's resistance, the resonant frequency of a practical *RLC* circuit is slightly less than the theoretical, unless $Q \geq 10$.
13. A parallel circuit can be damped to reduce Q and increase the bandwidth by adding series resistance to the coil or parallel resistance to the whole circuit.
14. A superheterodyne receiver uses a frequency mixer to produce a difference or intermediate frequency so that amplification can take place at a fixed frequency.

SELF-EXAMINATION

Answer true or false or a, b, c, or d for multiple choice
(Answers at back of book)

26-1. Series resonance always occurs where $X_L = X_C$. _____
26-2. At a series circuit's resonant frequency, the impedance is a maximum and the current is a minimum. _____
26-3. A series circuit behaves, as far as the source is concerned, as if it is a pure resistance when tuned to resonance. _____
26-4. The capacitor voltage can never be larger than the applied voltage in a series *RLC* circuit or Kirchhoff's voltage law would be violated. _____
26-5. The resonant frequency of a series *RLC* circuit is not determined by the resistance at all. _____
26-6. If a series circuit has $R = 2\ \Omega$, $L = 20$ mH, and $C = 2\ \mu$F, the approximate value of the circuit's resonant frequency is
    a. 8 kHz   c. 8 MHz
    b. 800 Hz   d. 80 kHz
26-7. The approximate value of Q in Question 26-6 is
    a. 8   c. 32
    b. 16   d. 2
26-8. The bandwidth of the circuit in Question 26-6 is approximately
    a. 50 Hz   c. 250 Hz
    b. 4 kHz   d. 100 Hz
26-9. The Q of a series circuit can be increased only by reducing the circuit's resistance. _____
26-10. If a 10-V_{p-p}, sine wave signal generator is connected to the series circuit in Question 26-6 and adjusted for resonance, the peak-to-peak voltage across the capacitor is
    a. 16 V   c. 160 V
    b. 80 V   d. 20 V
26-11. At resonance, the same voltage occurs across the inductor as across the capacitor for all conditions. _____

26-12. The resonant frequency of a theoretical parallel circuit is calculated using the same equation as for a series *RLC* circuit. _____

26-13. At parallel resonance the circuit has its maximum impedance and is purely reactive. _____

26-14. A 10-μH coil of negligible resistance, a 10-pF capacitor, and a 100-kΩ resistor are connected in parallel. The circuit's approximate resonant frequency is
   a. 1.6 MHz   c. 16 MHz
   b. 160 kHz   d. 160 MHz

26-15. The *Q* and bandwidth of the circuit in Question 26-14 are
   a. 50, 800 kHz   c. 10, 16 kHz
   b. 100, 160 kHz   d. 20, 80 kHz

26-16. The principal difference between the practical and theoretical parallel circuits is the resistance of the inductor. _____

26-17. A practical parallel circuit has a resonant frequency slightly higher than that given by the equation for the theoretical circuit. _____

26-18. A parallel resonant circuit can be damped either by adding resistance in series with the coil or in parallel with the coil. _____

26-19. A coil's series resistance can be increased to a point where the parallel *RLC* circuit will not resonate. _____

26-20. The advantage of a superheterodyne receiver is that most of the radio frequency amplification takes place at the fixed intermediate frequency. _____

REVIEW QUESTIONS

1. What are the conditions that characterize a series resonant circuit?

2. Why is the resonant frequency of a practical series circuit not affected by resistance whereas a practical parallel circuit's resistance does affect the resonant frequency?

3. a. What do you understand by the term *selectivity of a circuit?*
    b. How is selectivity measured?
    c. What is the origin of the term *half-power frequencies?*
    d. Why are the half-power frequencies sometimes referred to as 3-dB frequencies?
    e. What is meant by ''edge frequencies''?

4. a. What is the *general* definition of the *Q* of a circuit?
    b. What equation shows that *Q* is not determined only by *L* and *R*?
    c. Justify why a reduction in *R* or an increase in *L/C* increases *Q*.

5. a. How is it possible for the voltage across the capacitor in a series resonant circuit to be larger than the input voltage?
    b. How is this fact used to select only certain frequencies from an antenna?

6. a. Why is parallel resonance sometimes called antiresonance?
    b. What is theoretical about a pure *RLC* circuit?
    c. If it were possible to make a coil having zero resistance, what would be the impedance of a parallel coil and capacitor at resonance?
    d. What would be the line current and tank current for this condition?

7. Explain how it is necessary for the source, connected to a parallel resonant circuit, to have a high resistance to produce a frequency selective function.

8. Justify why the resonant frequency of a practical parallel circuit is *less* than that for a theoretical circuit having the same L and C values.

9. Why do you think the addition of sufficient resistance in series with a coil used in a parallel RLC circuit could prevent the circuit from having a resonant frequency?

10. a. What do you understand by the term *damping?*
  b. How can damping be achieved?

11. a. What range of frequencies must the local oscillator in Fig. 26-17 be able to produce to tune in the whole AM broadcast band?
  b. If an FM tuner uses an intermediate frequency of 10.7 MHz, what range of frequencies must be generated in the local oscillator to receive incoming frequencies from 88–108 MHz?
  c. What is the major difference between a superheterodyne receiver and a tuned radio frequency receiver?
  d. What does it mean to *align* a receiver?
  e. Why does an AM receiver not require a wider bandwidth than 10 kHz?

PROBLEMS

(Answers to odd-numbered problems at back of book)

26-1. A series RLC circuit consists of a 150-μH coil having a resistance of 12 Ω, and a 75-pF capacitor. A 20-mV radio frequency signal generator is connected to the circuit. For frequencies of (1) 1.485 MHz, (2) 1.500 MHz, and (3) 1.515 MHz, determine:
  a. The impedance. c. The current.
  b. The phase angle. d. V_R, V_L, and V_C.

26-2. Repeat Problem 26-1 using a 300-μH coil (12-Ω resistance) and a 37.5-pF capacitor.

26-3. Calculate the resonant frequency of the circuit in Problem 26-1.

26-4. Calculate the resonant frequency of the circuit in Problem 26-2.

26-5. What size of capacitance is required to provide series resonance with a 1-mH coil at 10-kHz?

26-6. What value of inductance will cause series resonance with a 0.0015-μF capacitor at 455 kHz?

26-7. A series circuit has a resonant frequency of 250 kHz and a bandwidth of 20 kHz. If 6 mW of power is delivered to the circuit at 250 kHz, calculate:
  a. The edge frequencies.
  b. The power delivered to the circuit when operating at the edge frequencies.
  c. The Q of the circuit.

26-8. The power delivered to a series RLC circuit is monitored at different frequencies. The power is noted to be a maximum, and then two frequencies are found, at 800 and 900 Hz, where the power drops to one-half of maximum. Determine:
  a. The bandwidth of the circuit.
  b. The resonant frequency.
  c. The Q of the circuit.

26-9. For the circuit in Problem 26-1, calculate:
  a. Q.
  b. The circuit's bandwidth.

26-10. For the circuit in Problem 26-2, calculate:
  a. Q.
  b. The circuit's bandwidth.

26-11. Determine the Q of a series resonant circuit in which $R = 15\ \Omega$, $L = 50\ \mu$H, and $C = 75$ pF.

26-12. A circuit must have a resonant frequency of 10.7 MHz and a bandwidth of 150 kHz. If an 8-Ω resistance coil is to be used, what is the required L/C ratio?

26-13. A series RLC circuit is resonant at 5 kHz when a 50-mH coil having a resistance of 150 Ω is used. If a 10-V p–p signal is applied to the circuit calculate:
  a. The capacitor voltage at resonance.
  b. The circuit capacitance.
  c. The circuit's bandwidth.
  d. The power delivered to the circuit at resonance.

26-14. An oscilloscope is connected across a capacitor in a series RLC circuit. The oscilloscope shows a maximum peak-to-peak voltage of 150 V when the input voltage of 24 V is adjusted in frequency to 400 Hz. If a series-connected wattmeter indicates 20 W, calculate:
  a. The Q of the circuit.
  b. The circuit resistance.
  c. The inductance.
  d. The capacitance.

26-15. The local oscillator frequency in an AM receiver must always be 455 kHz above the incoming frequency that ranges from 535 to 1605 kHz. If a 100-μH coil is used in the oscillator, calculate the range of a variable capacitor to tune-in the incoming frequencies.

26-16. If the coil in Problem 26-15 has a resistance of 15 Ω, calculate:
  a. The Q of the oscillator circuit at each end of the tuning range.
  b. The bandwidth of the oscillator circuit at each end of the tuning range.

26-17. A 0.5-mH coil of negligible resistance, a 0.0025-μF capacitor, and a 50-kΩ resistor are connected in parallel across a 10-mV, radio frequency signal generator. Calculate:
  a. The circuit's resonant frequency.
  b. Each branch current at resonance.
  c. The total line current at resonance.

26-18. Repeat Problem 26-17 using a 50-μH coil.

26-19. For the circuit of Problem 26-17, calculate:
  a. The Q of the circuit at resonance.
  b. The impedance of the circuit at resonance.
  c. The tank current at resonance.
  d. The circuit's bandwidth.

26-20. For the circuit of Problem 26-18, calculate:
  a. The Q of the circuit at resonance.
  b. The impedance of the circuit at resonance.
  c. The tank current at resonance.
  d. The circuit's bandwidth.

26-21. Assume now that the signal generator of Problem 26-17 has an internal resistance of 40 kΩ. Calculate the output voltage across the tank circuit at
  a. Resonance.
  b. A frequency where the impedance of the circuit has dropped to 5 kΩ (assume resistive).

26-22. A 10-mV signal generator of unknown internal resistance is connected in parallel with a 20-kΩ resistor, a capacitor, and a coil of negligible resistance. The frequency is adjusted until, at 100 kHz, maximum power transfer occurs *and* maximum circulating tank current occurs with a value of 20 μA. Calculate:
   a.   The internal resistance of the source.
   b.   The Q of the parallel *RLC* circuit.
   c.   The inductance of the coil.
   d.   The amount of capacitance.

26-23. A practical parallel resonant circuit is known to have a Q of 4 and a theoretical resonant frequency given by

$$f_r = \frac{1}{2\pi\sqrt{LC}}$$

of 150 kHz. What is the circuit's true resonant frequency?

26-24. A practical parallel *RLC* circuit is *measured* to have a resonant frequency of 200 kHz but has a *calculated* resonant frequency of 210 kHz when the equation

$$f_r = \frac{1}{2\pi\sqrt{LC}}$$

is used. Determine the Q of the circuit.

26-25. A 25-μH coil having an ac resistance of 4 Ω is connected in parallel with a 0.05-μF capacitor across a radio frequency signal generator. Calculate:
   a.   The resonant frequency of the circuit.
   b.   The Q of the circuit.
   c.   The circuit's bandwidth.

26-26. Repeat Problem 26-25 using 1 Ω for the coil's ac resistance.

26-27. For the parallel circuit of Problem 26-25, calculate:
   a.   The minimum resistance to be added in series with the coil to ensure that the circuit has no resonant frequency.
   b.   The resistance to be added in *series* with the coil to increase the bandwidth to 40 kHz, and the new resonant frequency.

26-28. For Problem 26-27, find the resistance to be connected in *parallel* with the coil to increase the bandwidth to 50 kHz.

26-29. The local oscillator frequency in an FM receiver must always be 10.7 MHz above the incoming frequency that ranges from 88–108 MHz. If a 1-μH coil is used in a parallel *RLC* circuit in the oscillator, calculate the range of a variable capacitor to tune-in the incoming frequencies.

CHAPTER 27

COMPLEX NUMBERS

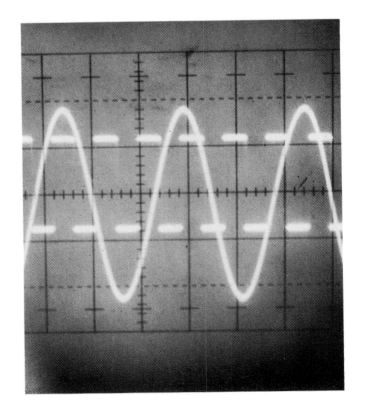

You have seen how ac circuits, because of their reactive components, have phase angles associated with current, voltage, and impedance. In this chapter, you will examine two *coordinate systems* that include the angular information as well as the magnitude.

The *polar system* directly indicates both the *magnitude* and the *phase angle*. The magnitude is always a positive quantity that corresponds to the reading of a meter (such as a voltmeter or ammeter). The phase angle, which may be either positive or negative, is measured from a (positive) horizontal zero-degree reference position. This system is particularly useful for multiplying and dividing phasor quantities, since it yields answers that include both magnitude and phase angle information.

The *rectangular coordinate system* consists of a *real* axis and an *imaginary j* axis. A *j* "operator" is used to distinguish between one phasor and another phasor (90° out of phase with the first), producing a *complex number* in *rectangular form*. Complex numbers in rectangular form are especially suitable for addition and subtraction, although multiplication and division are also possible (if the laws of algebra are observed).

A complex number in rectangular form can be used to represent a series circuit containing resistance and reactance. Then, after a rectangular-to-polar *conversion,* the equation $I = V/Z$ can be used to obtain current in a circuit. Similarly, parallel circuits may be solved using the concept of *admittance* (Y) and *susceptance* (B), which are the reciprocals (respectively) of impedance and reactance. Finally, complex number theory will be applied to a series-parallel circuit, and the use of Thévenin's theorem to solve ac circuits will be shown.

27-1 POLAR COORDINATES

You saw, in Section 24-3, that the total voltage applied to a series *RL* circuit is the *phasor* sum of V_R and V_L, as shown in Fig. 27-1. That is, the addition of two or more voltages that are not in phase with each other must take into account the phase angle between them. For two voltages that are 90° out of phase (as in Fig. 27-1*b*), the *magnitude* of the total voltage is

$$|V_T| = \sqrt{V_R^2 + V_L^2} \qquad (24\text{-}1)$$

and the phase angle for V_T with respect to I is

$$\theta = \text{arc tan} \frac{V_L}{V_R} \qquad (24\text{-}2a)$$

These two pieces of information for V_T, the *magnitude* and *phase angle,* may be combined into a *single notation,* called the *polar form:*

$$V_T = \sqrt{V_R^2 + V_L^2} \bigg/ \text{arc tan} \frac{V_L}{V_R} \qquad \text{volts} \quad (27\text{-}1)$$

where the symbols are as just shown.

EXAMPLE 27-1

Given $V_R = 30$ V, $V_L = 40$ V, find the polar form of the total voltage applied to the series *RL* circuit in Fig. 27-1.

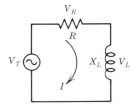

(a) Circuit diagram

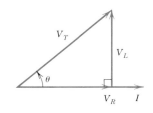

(b) Voltage triangle

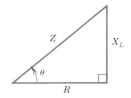

(c) Impedance triangle

FIGURE 27-1
Voltage and impedance triangles for a series *RL* circuit to show the use of polar coordinates.

Solution

The magnitude of V_T is given by

$$V_T = \sqrt{V_R^2 + V_L^2} \qquad (24\text{-}1)$$
$$= \sqrt{(30\ V)^2 + (40\ V)^2} = \textbf{50 V}$$

The phase angle is

$$\theta = \text{arc tan } \frac{V_L}{V_R} \qquad (24\text{-}2a)$$

$$= \text{arc tan } \frac{40\ V}{30\ V} = \textbf{53.1}°$$

The polar form of V_T is

$$V_T = \textbf{50 V} \underline{/+53.1°}$$

The answer to Example 27-1 is read as: "V_T equals 50 V *at an angle of* plus 53.1°."

The positive sign on the angle is sometimes used to distinguish the phasor from circuits in which the phase angle is negative. A polar coordinate specifies a quantity (such as voltage, impedance, or current) in terms of a *positive* magnitude, and either a positive or negative angle.

A *positive* angle is measured *counterclockwise* from the 0° horizontal reference line; a *negative* angle is measured *clockwise* from the same reference line.

As a further example, the impedance of the series *RL* circuit, in polar form (see Fig. 27-1c) is given by

$$Z = \sqrt{R^2 + X_L^2} \underline{/ \text{arc tan } \frac{X_L}{R}} \quad \text{ohms} \quad (27\text{-}2)$$

This is because the magnitude of the impedance is

$$|Z| = \sqrt{R^2 + X_L^2} \qquad (24\text{-}3)$$

and the phase angle is

$$\theta = \text{arc tan } \frac{X_L}{R} \qquad (24\text{-}5a)$$

EXAMPLE 27-2

A series *RL* circuit has a resistance of 4 Ω and an inductive reactance of 3 Ω. Find the circuit's impedance in polar form.

Solution

$$|Z| = \sqrt{R^2 + X_L^2} \qquad (24\text{-}3)$$

$$= \sqrt{(4\ \Omega)^2 + (3\ \Omega)^2} = \textbf{5 }\Omega$$

$$\theta = \text{arc tan } \frac{X_L}{R} \qquad (24\text{-}5a)$$

$$= \text{arc tan } \frac{3\ \Omega}{4\ \Omega} = \textbf{36.87}°$$

The polar form of Z is

$$Z = \textbf{5 }\Omega \underline{/36.9°}$$

Therefore, Z equals 5 Ω at an angle of plus 36.9°.

Similarly, for a series *RC* circuit, the total voltage and impedance in polar form are

$$V_T = \sqrt{V_R^2 + V_C^2} \underline{/ - \text{arc tan } \frac{V_C}{V_R}} \quad \text{volts (V)} \quad (27\text{-}3)$$

$$Z = \sqrt{R^2 + X_C^2} \underline{/ - \text{arc tan } \frac{X_C}{R}} \quad \text{ohms } (\Omega) \quad (27\text{-}4)$$

Thus, as an example, $V_T = 47\ V \underline{/-41°}$, $Z = 15.6\ \Omega \underline{/-64°}$.

In a similar way, the *current* in a circuit can be expressed in polar form, such as $I_1 = 3.5\ A \underline{/-72°}$ or $I_T = 56\ mA \underline{/+12°}$.

At this point, you may wonder about the reason for expressing an electrical quantity in polar coordinates. Is it merely a convenient notation?

Polar coordinates provide mathematical methods of *multiplying* and *dividing* electrical quantities that are very easy to perform. The resulting answers give magnitudes *and* phase angles. Also, the polar form is the reading you obtain from a meter. An ammeter or voltmeter reading *is* the magnitude of the polar form of the quantity.

Some instruments provide *both* magnitude *and* phase angle. Two such instruments, the vector impedance meter and the vector voltmeter, are shown in Fig. 27-2.

27-2 MULTIPLICATION AND DIVISION USING POLAR COORDINATES

In general, the polar form of any quantity can be expressed as

$$M\underline{/\theta}$$

where: M is the magnitude

θ is the phase angle

You know that M may have units of volts, ohms, and

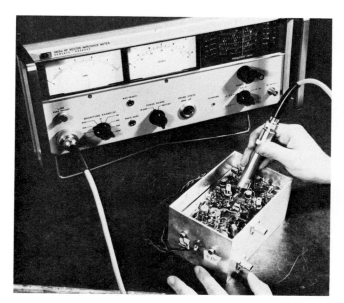

(a) Vector impedance meter measures the magnitude and phase angle of circuit impedance using a single probe

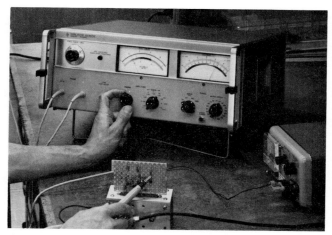

(b) Vector voltmeter measures the magnitude of, and the phase difference between, two voltage vectors from 1-1000 MHz

FIGURE 27-2

Meters to measure polar impedance and voltage. (Courtesy Hewlett-Packard Company.)

so on, and that θ may be positive or negative. Assume that you have two phasors given in polar form:

$$M_1\underline{/\theta_1}, \qquad M_2\underline{/\theta_2}$$

See Fig. 27-3.

It can be shown that the *product* of two phasors given in polar form is found as follows:

$$M_1\underline{/\theta_1} \times M_2\underline{/\theta_2} = M_1M_2\underline{/\theta_1 + \theta_2} \qquad (27\text{-}5)$$

That is, to multiply two phasors given in polar form, you should *multiply* their magnitudes and algebraically *add* their phase angles.

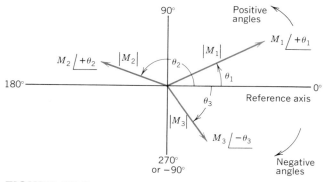

FIGURE 27-3

Three phasors given in polar form.

Similarly, for division,

$$\frac{M_1\underline{/\theta_1}}{M_2\underline{/\theta_2}} = \frac{M_1}{M_2}\underline{/\theta_1 - \theta_2} \qquad (27\text{-}6)$$

That is, to divide two phasors given in polar form, you should *divide* their magnitudes and algebraically *subtract* their phase angles.

EXAMPLE 27-3

Given three phasors: $A = 7.5\underline{/53°}$, $B = 3\underline{/12°}$, and $C = 5\underline{/-60°}$, find:
a. $A \times B$
b. $A \div B$
c. $A \times C$
d. $A \div C$
e. $C \div B$

Solution

a. $A \times B = 7.5\underline{/53°} \times 3\underline{/12°}$
$\qquad\qquad = 7.5 \times 3\underline{/53° + 12°} \qquad (27\text{-}5)$
$\qquad\qquad = \mathbf{22.5\underline{/65°}}$

b. $A \div B = \dfrac{7.5\underline{/53°}}{3\underline{/12°}}$

$$= \frac{7.5}{3} \underline{/53° - 12°} \qquad (27\text{-}6)$$

$$= 2.5\underline{/41°}$$

c. $A \times C = 7.5\underline{/53°} \times 5\underline{/-60°}$

$$= 7.5 \times 5\underline{/53° + (-60°)} \qquad (27\text{-}5)$$

$$= 37.5\underline{/-7°}$$

d. $A \div C = \dfrac{7.5\underline{/53°}}{5\underline{/-60°}}$

$$= \frac{7.5}{5}\underline{/53° - (-60°)} \qquad (27\text{-}6)$$

$$= 1.5\underline{/113°}$$

e. $C \div B = \dfrac{5\underline{/-60°}}{3\underline{/12°}}$

$$= \frac{5}{3}\underline{/-60° - 12°} \qquad (27\text{-}6)$$

$$= 1.67\underline{/-72°}$$

NOTE The result of multiplying (or dividing) two phasors has no simple *geometric* meaning with respect to the original phasors. In other words, the result cannot be found *graphically*. However, the result has a simple interpretation when applied to a circuit, as shown by Example 27-4.

27-2.1 Application of Polar Coordinates to Series Circuits

You can apply the polar coordinate system to the alternating current theory of Chapter 24 as shown in the following examples.

EXAMPLE 27-4.

A circuit consists of a 50-Ω resistor in series with a capacitive reactance of 120 Ω. A 120-V, 60-Hz supply is connected across the circuit. Determine the current and the phase angle it makes with the applied voltage.

Solution

$$Z = \sqrt{R^2 + X_C^2}\ \underline{/-\arctan \frac{X_C}{R}} \qquad (27\text{-}4)$$

$$= \sqrt{50^2 + 120^2}\ \Omega\ \underline{/-\arctan \frac{120}{50}}$$

$$= 130\ \Omega\underline{/-67.4°}$$

$$I = \frac{V}{Z} \qquad (24\text{-}4)$$

$$= \frac{120\ V\underline{/0°}}{130\ \Omega\underline{/-67.4°}} = \mathbf{0.92\ A\underline{/+67.4°}}$$

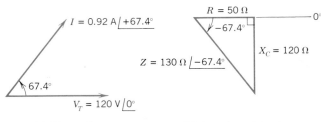

(a) Phasor diagram   (b) Impedance triangle

FIGURE 27-4
Diagrams for Example 27-4.

This states that the circuit current is 0.92 A and it *leads* the applied voltage by 67.4°. See Fig. 27-4.

Note that the polar form of 120 V is 120 V$\underline{/0°}$, since the applied voltage is made the reference phasor in this problem for convenience.

EXAMPLE 27-5.

Use polar coordinates to determine the magnitude and phase angle of the voltage in Example 27-4 across:
a. The resistor.
b. The capacitor.

Solution

a. $V_R = IR \qquad (15\text{-}15b)$

$$= 0.92\ A\underline{/+67.4°} \times 50\ \Omega\underline{/0°}$$

$$= \mathbf{46\ V\underline{/67.4°}}$$

b. $V_C = IX_C \qquad (23\text{-}1)$

$$= 0.92\ A\underline{/+67.4°} \times 120\ \Omega\underline{/-90°}$$

$$= \mathbf{110\ V\underline{/-22.6°}}$$

Note that the resistance has a polar form of 50 $\Omega\underline{/0°}$ (along the reference axis) and X_C has a polar form of 120 $\Omega\underline{/-90°}$. The phasor diagram in Fig. 27-5b shows how the V_R and V_C phasors are 90° out of phase, as required.

Note, however, that the phasor diagram has been rotated counterclockwise (67.4°) compared to the diagram in Fig. 24-11b. This has been done because the diagram in Fig. 27-5b uses the applied *voltage* in the reference position of 0°, whereas Fig. 24-11b uses *current* as the reference. Current is generally used as a reference for *series* circuits, because it is the same in *all* components.

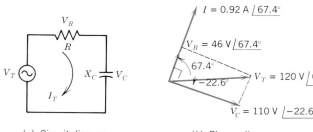

(a) Circuit diagram

(b) Phasor diagram

FIGURE 27-5

Diagrams for Example 27-5.

EXAMPLE 27-6

A series circuit with an applied emf of 48 V has a current of 6.4 A that lags the applied voltage by 40°. Determine:
a. The type of series circuit.
b. The amount of impedance in the circuit, using polar coordinate notation.
c. The resistance and reactance of the circuit.
d. The voltages throughout the circuit.

Solution

a. Since the current *lags* the voltage, the circuit must be **inductive.**

b. $Z = \dfrac{V}{I}$ (24-4)

$\quad = \dfrac{48 \text{ V}\underline{/0°}}{6.4 \text{ A}\underline{/-40°}} = \textbf{7.5 } \boldsymbol{\Omega\underline{/40°}}$

c. $R = Z \cos \theta$ (24-6)
$\quad = 7.5 \ \Omega \times \cos 40° = \textbf{5.75 } \boldsymbol{\Omega}$
$\quad X_L = Z \sin \theta$ (24-7)
$\quad = 7.5 \ \Omega \times \sin 40° = \textbf{4.8 } \boldsymbol{\Omega}$

d. $V_R = IR$ (15-15b)
$\quad = 6.4 \text{ A}\underline{/-40°} \times 5.75 \ \Omega\underline{/0°}$
$\quad = \textbf{36.8 V}\underline{/-40°}$
$\quad V_L = IX_L$ (20-1)
$\quad = 6.4 \text{ A}\underline{/-40°} \times 4.8 \ \Omega\underline{/90°}$
$\quad = \textbf{30.7 V}\underline{/50°}$

See the phasor diagram and impedance triangle in Fig. 27-6.

Note how the inductive reactance has a polar form of 4.8 $\Omega\underline{/+90°}$, and how V_R, in phase with I, is 90° out of phase with V_L.

27-3 THE *j* OPERATOR

For performing multiplication and division, the polar coordinate system is ideal. However, addition and subtraction are not directly possible in the polar form. Conversion to another coordinate system is necessary before these operations can be readily accomplished.

Consider a polar coordinate given by $1\underline{/90°}$. From Eq. 27-5, you know that

$$1\underline{/90°} \times 1\underline{/90°} = 1\underline{/180°}$$

But $1\underline{/180°} = -1$

and $1\underline{/90°} \times 1\underline{/90°} = (1\underline{/90°})^2$

Therefore, $(1\underline{/90°})^2 = -1$

Taking the square root of both sides:

$$1\underline{/90°} = \sqrt{-1}$$

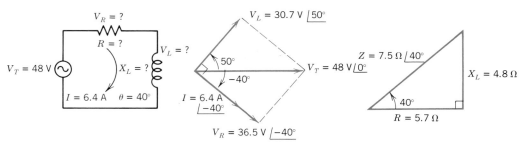

(a) Circuit diagram   (b) Phasor diagram   (c) Impedance triangle

FIGURE 27-6

Diagrams for Example 27-6.

You know from mathematics that $\sqrt{-1}$ is an *imaginary* number (there is no *real* number that, when squared, is equal to -1). Mathematicians use the letter i to represent this imaginary number ($\sqrt{-1}$). However, to avoid confusion with the symbol for current, the letter symbol j is used in this book to represent $\sqrt{-1}$. But more important, you can now see that

$$j = \sqrt{-1} = 1\underline{/90°} \quad \text{or} \quad j1 = j \times 1 = 1\underline{/90°}$$

This means that if you write $j2$ (or $j \times 2$), you are representing $2\underline{/90°}$. The effect of placing a j in front of a real number is to *cause a vector representing that real number to be rotated counterclockwise through an angle of +90°*.

For this reason, j is referred to as *an operator*. It *operates on* or *modifies* a number by rotating it through a positive angle of 90°.

Thus you can think of an *imaginary axis* with divisions of $j1$, $j2$, $j3$ (and so on), drawn vertically *upward* from the *real axis*, as in Fig. 27-7. This is the *positive* imaginary axis.

Now consider the effect of multiplying j by j.

$$j^2 = j \times j = \sqrt{-1} \times \sqrt{-1} = -1$$

Similarly, $j^2 \times 2 = -2$, and so on.

Thus, multiplying a positive imaginary number by j operates on the number to rotate it a further 90° to the negative real axis.

Now, if you multiply -1 by j, you obtain

$$-1 \times j = 1\underline{/180°} \times 1\underline{/90°} = 1\underline{/270°} = -j$$

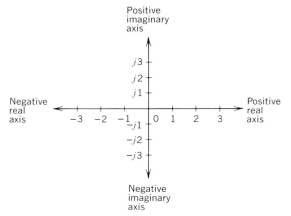

FIGURE 27-7
Real and imaginary axes make up the complex number plane or the rectangular coordinate system.

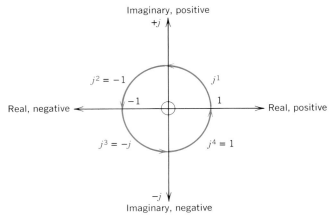

FIGURE 27-8
Showing how i operates on *any* number, real or imaginary, to rotate that number through an angle of +90°.

This establishes a *negative* imaginary axis due to the additional 90° rotation. Note that

$$-1 \times j = j^2 \times j = j^3 = -j$$

Finally, $$-j \times j = -j^2 = -(-1) = 1$$

or $$j^3 \times j = j^4 = 1$$

See Fig. 27-8.

27-4   RECTANGULAR COORDINATE SYSTEM

You now have a method of distinguishing between a phasor that is horizontal and another phasor that is 90° out of phase. Simply placing a j in front of the quantity indicates that a phase angle of +90° is associated with that number. Similarly, a $-j$ in front of a quantity indicates that a phase angle of −90° is associated with that number. For example, if a voltage is the *sum* of two voltages, one 90° out of phase with the other, you can show this sum with an addition sign if you use the j term.

Thus, for a series *RL* circuit, where V_L leads the resistor voltage V_R by 90°, the total applied voltage is given, in *rectangular* form, by

$$V_T = V_R + jV_L \tag{27-7}$$

Similarly, $$Z = R + jX_L \tag{27-8}$$

See Fig. 27-9.

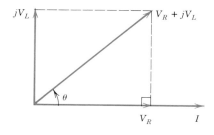

(a) Phasor diagram

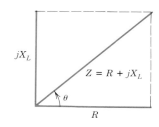

(b) Impedance diagram

FIGURE 27-9
Rectangular coordinate representations for a series *RL* circuit.

It is conventional practice to write the *real* number *first*, followed by the imaginary number. Such a number, because it is the combination of a real and an imaginary number, is called a *complex number*.

For a series *RC* circuit, it follows that the *rectangular coordinate* system gives

$$V_T = V_R - jV_C \qquad (27\text{-}9)$$

and

$$Z = R - jX_C \qquad (27\text{-}10)$$

Thus, the impedance of a series *RC* circuit may appear as $Z = 20 - j50\ \Omega$. This simply means that the circuit consists of a 20-Ω resistance in series with a capacitor whose reactance is 50 Ω.

Also, an impedance of $Z = 4.7 + j2\ \text{k}\Omega$ represents a series circuit with a 4.7-kΩ resistance and an inductive reactance of 2 kΩ.

For a series *RLC* circuit,

$$Z = R + j(X_L - X_C) \qquad (27\text{-}11)$$
$$V_T = V_R + j(V_L - V_C) \qquad (27\text{-}12)$$

See Fig. 27-10.

Again, it may seem all that has been gained with rectangular coordinates is a compact system of combining components that are 90° or 180° out of phase. The real value, however, is the mathematical manipulation that this system permits for solving complex circuit problems.

27-5 ADDITION AND SUBTRACTION USING RECTANGULAR COORDINATES

Addition and subtraction of phasors is especially simple when the phasors are given in rectangular form. (Recall that polar quantities do *not* lend themselves to addition and subtraction.)

Assume two phasors given in rectangular coordinates:

$$a + jb, \qquad c + jd$$

The sum of the two phasors is found by algebraically adding together first the real parts, then the imaginary parts (respectively).

$$(a + jb) + (c + jd) = (a + c) + j(b + d) \quad (27\text{-}13)$$

The geometric result of the addition of two phasors is shown in Fig. 27-11a. This applies whether the phasors are given in polar or rectangular coordinates. Similarly, the subtraction of two phasors requires the algebraic subtraction of their real and imaginary parts:

$$(a + jb) - (c + jd) = (a - c) + j(b - d) \quad (27\text{-}14)$$

Figure 27-11b shows that subtraction can be thought of as adding the negative of the phasor to be subtracted. That is, if $V_1 = a + jb$, then $-V_1 = -a - jb$. Adding $-V_1$ to V_2 results in $V_2 - V_1$.

(a) Series *RC* circuit $R = 20\ \Omega$ $X_C = 50\ \Omega$ $Z = 20 - j50\ \Omega$

(b) Series *RL* circuit $R = 4.7\ \text{k}\Omega$ $X_L = 2\ \text{k}\Omega$ $Z = 4.7 + j2\ \text{k}\Omega$

(c) Series *RLC* circuit $R = 40\ \Omega$ $X_L = 80\ \Omega$ $X_C = 30\ \Omega$ $Z = 40 + j(80 - 30)\ \Omega = 40 + j50\ \Omega$

FIGURE 27-10
Impedance of series circuits in rectangular coordinates.

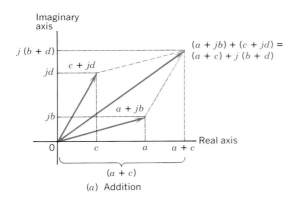

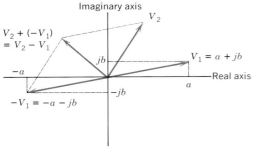

(a) Addition

(b) Subtraction

FIGURE 27-11

(a) Addition and (b) subtraction of complex numbers in rectangular coordinates.

EXAMPLE 27-7

Given three complex phasors: $A = 3 + j7$, $B = 4 - j5$, and $C = -6 + j8$. Find:

a. $A + B$.
b. $A - B$.
c. $B + C$.
d. $B - C$.

Solution

a. $A + B = 3 + j7 + 4 - j5$
$= (3 + 4) + j(7 - 5)$　　　　(27-13)
$= 7 + j2$

b. $A - B = (3 + j7) - (4 - j5)$
$= (3 - 4) + j(7 - (-5))$　　　(27-14)
$= -1 + j12$

　or $A - B = A + (-B)$
$= 3 + j7 + (-4 + j5)$
$= (3 - 4) + j(7 + 5)$
$= -1 + j12$

c. $B + C = (4 - j5) + (-6 + j8)$
$= (4 - 6) + j(-5 + 8)$　　　(27-13)
$= -2 + j3$

d. $B - C = (4 - j5) - (-6 + j8)$
$= (4 + 6) + j(-5 - 8)$　　　(27-14)
$= 10 - j13$

27-5.1　Application of Rectangular Coordinates to Series Circuits

The total impedance of a series circuit is the sum of the impedances throughout the circuit.

$$Z_T = Z_1 + Z_2 + Z_3 + \cdots + Z_n \quad (27\text{-}15)$$

Although Eq. 27-15 applies to polar *and* rectangular coordinates, it is the rectangular form that is most useful, as shown by Example 27-8.

EXAMPLE 27-8

Two coils and a capacitor are connected in series. The first coil has a resistance of 20 Ω and a reactance of 50 Ω; the second coil has a resistance of 15 Ω and a reactance of 35 Ω; the capacitor has a reactance of 10 Ω. Find the total series impedance.

Solution

$$Z_T = Z_1 + Z_2 + Z_3 \quad (27\text{-}15)$$
$$= (20 + j50) + (15 + j35) - j10$$
$$= (20 + 15) + j(50 + 35 - 10)$$
$$= 35 + j75$$

See Fig. 27-12.
Note that the original circuit is *equivalent* to a series circuit of 35-Ω *resistance* and 75-Ω *inductive* reactance.

You know that Kirchhoff's voltage law must be obeyed in *any* series circuit. Thus

$$V_T = V_1 + V_2 + V_3 + \cdots + V_n \quad (27\text{-}16)$$

as shown by Example 27-9.

EXAMPLE 27-9

Three loads are series-connected across the 120-V line. The first load has a voltage given by $40 + j30$ V; the

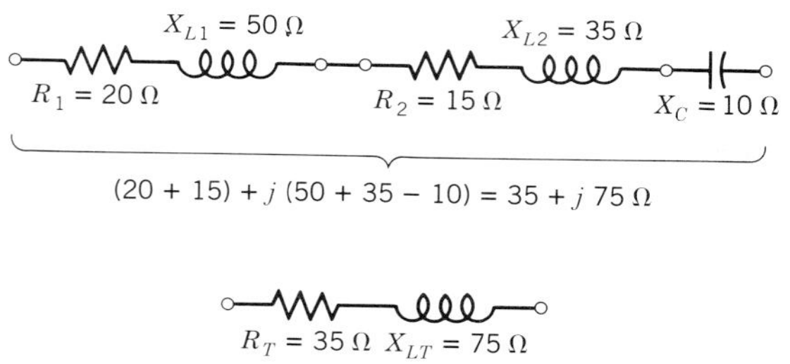

FIGURE 27-12
Addition of impedances in rectangular coordinates for Example 27-8.

second load has a voltage given by $25 - j90$ V. Find the rectangular form of the voltage across the third load.

Solution

$$V_T = V_1 + V_2 + V_3 \qquad (27\text{-}16)$$
$$120 + j0 = (40 + j30) + (25 - j90) + V_3$$

Therefore,

$$V_3 = (120 + j0) - (40 + j30) - (25 - j90) \text{ V}$$
$$= (120 - 40 - 25) + j(0 - 30 - (-90)) \text{ V}$$
$$= \mathbf{55 + j60 \text{ V}}$$

Note that the answer given for V_3 in Example 27-9 is *not* what a voltmeter would read. A voltmeter indicates the *polar* form, not the rectangular form. You will learn how to *convert* from rectangular to polar in Section 27-7.

27-6 MULTIPLICATION AND DIVISION USING RECTANGULAR COORDINATES

Although the processes of multiplication and division are *most easily* carried out using polar coordinates, these processes are not difficult using the rectangular form.
Given two complex numbers

$$a + jb, \qquad c + jd$$

the product is given by

$$(a + jb)(c + jd) = ac + jad + jbc + j^2bd$$
$$= ac + j^2bd + j(ad + bc)$$

But $$j^2 = -1$$

Therefore,

$$(a + jb)(c + jd) = (ac - bd) + j(ad + bc)$$

This is not to be used as a formula. It shows that multiplication in rectangular form follows the *distributive law* of algebra. The result is a complex number with a real and an imaginary part.

Similarly, the division of two complex numbers given in rectangular form also follows the laws of algebra, requiring a *rationalization* of the denominator.
Given two complex numbers

$$a + jb, \qquad c + jd$$

their quotient is given by

$$\frac{a + jb}{c + jd} = \frac{a + jb}{c + jd} \times \frac{c - jd}{c - jd}$$
$$= \frac{ac - jad + jbc - j^2bd}{c^2 - jcd + jcd - j^2d^2}$$
$$= \frac{(ac + bd) + j(bc - ad)}{c^2 + d^2}$$
$$= \frac{ac + bd}{c^2 + d^2} + j\frac{bc - ad}{c^2 + d^2}$$

Note that $c - jd$ is the *conjugate* of $c + jd$. Multiplying the numerator and denominator by $c - jd$ does not change the original fraction. It does, however, provide a real number in the final denominator because the imaginary components cancel. (See Example 27-10.)

EXAMPLE 27-10

Given two complex numbers, $A = 5 + j8$, $B = 7 - j4$, find:

a. $A \times B$.
b. A/B.

Solution

a. $A \times B = (5 + j8)(7 - j4)$
$$= 35 - j20 + j56 - j^2 32$$
$$= (35 + 32) + j(56 - 20)$$
$$= \mathbf{67 + j36}$$

b. $A/B = \dfrac{5 + j8}{7 - j4}$
$$= \frac{5 + j8}{7 - j4} \times \frac{7 + j4}{7 + j4}$$
$$= \frac{35 + j20 + j56 + j^2 32}{49 + j28 - j28 - j^2 16}$$
$$= \frac{(35 - 32) + j(20 + 56)}{(49 + 16) + j0}$$
$$= \frac{3 + j76}{65}$$

$$= \frac{3}{65} + j\frac{76}{65}$$
$$\approx 0.05 + j1.17$$

As you shall see in Section 27-8, an alternative to the procedures shown in Example 27-10 requires the conversion of the rectangular form *into* polar coordinates. However, if the result is required in rectangular form, another conversion must be made *from* the polar form after multiplying or dividing. In this case, the procedures shown in Example 27-10 may be easier.

27-6.1   Application of Rectangular Coordinates to Parallel Circuits

If two impedances (Z_1 and Z_2) are connected in parallel, their total impedance is

$$Z_T = \frac{Z_1 \times Z_2}{Z_1 + Z_2} \qquad (27\text{-}17)$$

where all impedances are given in complex numbers.

EXAMPLE 27-11

A coil of 5-Ω resistance and 8-Ω reactance is connected in parallel with a second coil having a resistance of 4 Ω and a reactance of 6 Ω. Find the total impedance.

Solution

$$Z_T = \frac{Z_1 \times Z_2}{Z_1 + Z_2} \qquad (27\text{-}17)$$
$$= \frac{(5 + j8)(4 + j6)}{(5 + j8) + (4 + j6)}\ \Omega$$
$$= \frac{20 + j30 + j32 + j^2 48}{(5 + 4) + j(8 + 6)}\ \Omega$$
$$= \frac{(20 - 48) + j(30 + 32)}{9 + j14}\ \Omega$$
$$= \frac{-28 + j62}{9 + j14}\ \Omega$$
$$= \frac{-28 + j62}{9 + j14} \times \frac{9 - j14}{9 - j14}\ \Omega$$
$$= \frac{-252 + j392 + j558 - j^2 868}{81 + 196}\ \Omega$$
$$= \frac{(868 - 252) + j(392 + 558)}{81 + 196}\ \Omega$$
$$= \frac{616 + j950}{277}\ \Omega$$

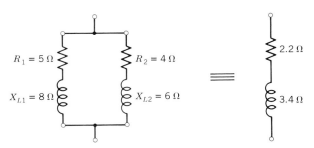

FIGURE 27-13
Circuits for Example 27-11.

$$= \frac{616}{277} + j\frac{950}{277}\ \Omega$$
$$\approx 2.2 + j3.4\ \Omega$$

See Fig. 27-13.

NOTE The total parallel impedance is equivalent to a *series* combination of a 2.2-Ω resistance and a 3.4-Ω inductive reactance. **A complex number in rectangular coordinates is *always* interpreted as a *series* impedance.**

EXAMPLE 27-12

A coil having a resistance of 4 Ω and an inductive reactance of 2 Ω is connected in parallel with a capacitor whose reactance is 3 Ω. Find the total impedance.

Solution

$$Z_T = \frac{Z_1 \times Z_2}{Z_1 + Z_2} \qquad (27\text{-}17)$$
$$= \frac{(4 + j2)(-j3)}{(4 + j2) + (0 - j3)}\ \Omega$$
$$= \frac{-j12 - j^2 6}{4 - j1}\ \Omega$$
$$= \frac{6 - j12}{4 - j1}\ \Omega$$
$$= \frac{6 - j12}{4 - j1} \times \frac{4 + j1}{4 + j1}\ \Omega$$
$$= \frac{24 + j6 - j48 - j^2 12}{16 - j^2 1}\ \Omega$$
$$= \frac{36 - j42}{17}\ \Omega$$
$$= \frac{36}{17} - j\frac{42}{17}\ \Omega$$
$$\approx 2.1 - j2.5\ \Omega$$

See Fig. 27-14.

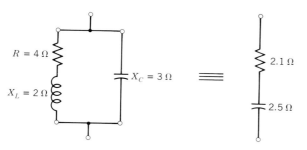

(a) Original parallel circuit   (b) Equivalent series circuit.

FIGURE 27-14
Circuits for Example 27-12.

Note that in this case, the equivalent circuit is capacitive. If more than two impedances are parallel-connected, they can be combined two at a time, as was the situation with parallel resistances. (See Section 6-4.)

Finally, note that you now have a method of obtaining the impedance of a parallel circuit *directly*. You do not have to assume a voltage, find the individual currents and their sum, and finally obtain Z_T from V/I_T, as you did in Chapter 24.

27-7 CONVERSION FROM RECTANGULAR TO POLAR FORM

It may be more convenient to work with a complex number in *polar* form, especially for multiplication and division. A phasor can be converted from rectangular to polar coordinates using the Pythagorean relationship, as shown in Fig. 27-15.

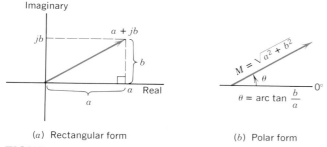

(a) Rectangular form   (b) Polar form

FIGURE 27-15
Conversion from rectangular to polar form.

Given a complex number $a + jb$, it can be converted to the form $M\underline{/\theta}$ as follows:

$$a + jb = \sqrt{a^2 + b^2}\underline{\bigg/ \arctan \frac{b}{a}} \qquad (27\text{-}18)$$

Care must be taken if a real number is negative, as shown by Example 27-13.

EXAMPLE 27-13.

Convert the following to polar form:
a. $3 + j4$.
b. $5 - j12$.
c. $-4 + j6$.
d. $-4 - j6$.

Solution

a. $3 + j4 = \sqrt{3^2 + 4^2}\underline{/\arctan \frac{4}{3}}$ (27-18)
   $= 5\underline{/53.1°}$

b. $5 - j12 = \sqrt{5^2 + 12^2}\underline{/\arctan -\frac{12}{5}}$ (27-18)
   $= 13\underline{/-67.4°}$

c. Make a sketch of the complex number as in Fig. 27-16a.
   $$M = \sqrt{4^2 + 6^2} = 7.2$$
   $$\phi = \arctan \frac{6}{4} = 56.3°$$
   $$\theta = 180° - 56.3° = 123.7°$$
   Therefore: $-4 + j6 = \mathbf{7.2\underline{/123.7°}}$

d. Make a sketch as in Fig. 27-16b.
   $M = 7.2$
   $\theta = -(180° - 56.3°) = -123.7°$
   Therefore, $-4 - j6 = \mathbf{7.2\underline{/-123.7°}}$
   or $-4 - j6 = 7.2\underline{/180° + \phi}$
   $= 7.2\underline{/180° + 56.3°}$
   $= \mathbf{7.2\underline{/+236.3°}}$

Note that the polar coordinate *angle* can be given as a positive or negative number, but the *magnitude* is always positive. Generally, if the positive angle is more than 180°, it is conventional to express the coordinates with a negative angle.

Most scientific calculators have a coded program for rectangular/polar conversions that avoid the difficulties arising from numbers being in the second and third quadrants. These programs vary, but the one that follows ap-

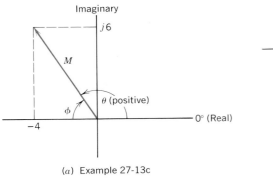

(a) Example 27-13c

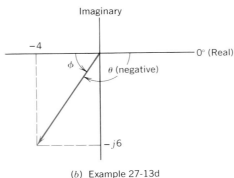

(b) Example 27-13d

FIGURE 27-16
Sketches for Example 27-13.

plies for one of the Texas Instruments type of calculators.*
Convert $-4 - j6$ to polar coordinates.

Enter	Press	Display	Comments
-4	x:y	0	Enters real part
-6	INV 2nd 18	$-123.7°$	Displays θ
	x:y	7.2	Displays M

Therefore, $-4 - j6 = 7.2\underline{/-123.7°}$

EXAMPLE 27-14

A coil having an ac resistance of 20 Ω and an inductive reactance of 50 Ω is connected in parallel with a capacitor that has an ac resistance of 10 Ω and a capacitive reactance of 20 Ω. Calculate the total impedance.

Solution

$$Z_T = \frac{Z_1 \times Z_2}{Z_1 + Z_2} \qquad (27\text{-}17)$$

$$= \frac{(20 + j50)(10 - j20)}{20 + j50 + 10 - j20} \ \Omega$$

$$= \frac{53.9\underline{/68.2°} \times 22.4\underline{/-63.4°}}{30 + j30} \ \Omega$$

$$= \frac{1207.4\underline{/4.8°}}{42.4\underline{/45°}}$$

$$= \mathbf{28.5} \ \Omega\underline{/-40.2°}$$

---

*In some calculators (such as the TI-55 shown in Fig. 1-1), press INV 2nd P → R or the button marked → P, instead of pressing INV 2nd "18" (the code for rectangular-to-polar conversion).

27-8 CONVERSION FROM POLAR TO RECTANGULAR FORM

The answer to Example 27-14 suggests a total impedance that is capacitive. To determine how much is capacitive and how much is resistive, however, requires a *polar-to-rectangular* conversion. That is, you can only identify an impedance in terms of series resistance and reactance when the impedance is given in rectangular coordinates.

The conversion from polar to rectangular can be done using right-angled trigonometry, as shown in Fig. 27-17.

Given a complex number in polar form $M\underline{/\theta},$ it can be converted to the form $a + jb$ as follows:

$$M\underline{/\theta} = M\cos\theta + jM\sin\theta \qquad (27\text{-}19)$$

This equation applies to all angles, positive or negative, with no restrictions.

EXAMPLE 27-15

Determine the components of the equivalent series impedance in Example 27-14, where $Z_T = 28.5$ $\Omega\underline{/-40.2°}$.

Solution

$$Z_T = 28.5 \ \Omega\underline{/-40.2°}$$

$$= 28.5\cos(-40.2°) + j28.5\sin(-40.2°)$$

$$= 28.5 \times 0.76 + j28.5 \times (-0.65)$$

$$= \mathbf{21.7 - j18.5} \ \Omega$$

Thus the total impedance is equivalent to a resistance

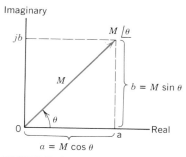

FIGURE 27-17
Conversion from polar to rectangular form.

of 21.7 Ω in series with a capacitor having a reactance of 18.5 Ω.

It is wise, when making conversions, to make a sketch of the initial coordinates (whether polar or rectangular) to check that the converted number falls into the correct quadrant.

EXAMPLE 27-16

Two coils and a capacitor are connected in series across a sinusoidal supply. The voltages across the coils are 50 V$\underline{/20°}$ and 30 V$\underline{/50°}$. The capacitor's voltage is 75 V$\underline{/-80°}$. Calculate the supply voltage.

Solution

By Kirchhoff's voltage law,

$$V_T = V_1 + V_2 + V_3 \tag{27-16}$$
$$V_1 = 50 \text{ V}\underline{/20°} = 50 \cos 20°$$
$$+j50 \sin 20° = 47 + j17.1 \text{ V}$$
$$V_2 = 30 \text{ V}\underline{/50°} = 30 \cos 50°$$
$$+j30 \sin 50° = 19.3 + j23 \text{ V}$$
$$V_3 = 75 \text{ V}\underline{/-80°} = 75 \cos (-80°)$$
$$+j75 \sin (-80°) = 13 - j73.9 \text{ V}$$
$$V_T = (47 + 19.3 + 13) + j(17.1 + 23 - 73.9) \text{ V}$$

Therefore, $V_T = 79.3 - j33.8$ V = **86.2 V$\underline{/-23.1°}$**

Note that it is necessary to convert to the polar form to obtain the total voltage that would be indicated by a voltmeter.

A polar-to-rectangular conversion (in a way similar to the rectangular/polar conversion in Section 27-7) can be made on some scientific calculators as follows*:

Convert 5$\underline{/30°}$ to rectangular coordinates.

Enter	Press	Display	Comments
5	x:y	0	Enters magnitude
30°	2nd 18	2.5	Value of imaginary
	x:y	4.3	Value of real

Therefore, 5$\underline{/30°}$ = 4.3 + j2.5

Note that for the TI calculator, the imaginary part of the complex number is displayed first. A rough sketch of 5$\underline{/30°}$ would make it clear that the smaller number must be the imaginary component.

EXAMPLE 27-17

A parallel circuit has two branches and draws a total current of 58 mA$\underline{/35°}$ from a 120-V, 60-Hz source. If the current in one branch is 35 mA$\underline{/-20°}$, calculate:

a. The current in the other branch, as indicated by an ammeter.
b. The total impedance of the parallel circuit.
c. The total true power dissipated in the whole circuit.
d. The total reactive power in the whole circuit.
e. The power factor of the whole circuit.

Solution

a. By Kirchhoff's current law,

$$I_T = I_1 + I_2 \tag{6-1}$$

Therefore,

$$I_2 = I_T - I_1$$
$$= 58\underline{/35°} - 35\underline{/-20°} \text{ mA}$$
$$= (47.5 + j33.3) - (32.9 - j12) \text{ mA}$$
$$= 14.6 + j45.3 \text{ mA}$$
$$= \textbf{47.6 mA}\underline{/72.1°}$$

b. $Z_T = \dfrac{V}{I_T}$ $\tag{24-4}$

---

*In some calculators (such as the TI-55 in Fig. 1-1), press the button marked $P \to R$ or $\to R$, instead of pressing "18."

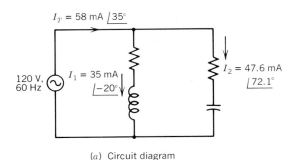

(a) Circuit diagram

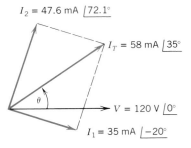

(b) Phasor diagram

FIGURE 27-18

(a) Circuit and (b) phasor diagrams for Example 27-17.

$$= \frac{120\ V\underline{/0°}}{58\ mA\underline{/35°}}$$

$$= 2.1\ k\Omega\underline{/-35°}$$

c. $Z_T = 2.1\ k\Omega\underline{/-35°}$

$$= 1.7 - j1.2\ k\Omega$$

This means that the circuit is equivalent to a current of 58 mA flowing through a resistance of 1.7 kΩ as far as total true power is concerned.

$P_T = I_R^2 R$ (15-12)

$$= (58 \times 10^{-3}\ A)^2 \times 1.7 \times 10^3\ \Omega$$

$$= \textbf{5.7 W}$$

d. $P_q = I_C^2 X_C$ (25-5)

$$= (58 \times 10^{-3}\ A)^2 \times 1.2 \times 10^3\ \Omega$$

$$= \textbf{4 vars}$$

e. $P_s = I^2 Z$ (25-9)

$$= (58 \times 10^{-3}\ A)^2 \times 2.1 \times 10^3\ \Omega$$

$$= \textbf{7.1 VA}$$

or, $P_s = VI_T$ (25-7)

$$= 120\ V \times 0.058\ A$$

$$= \textbf{7.1 VA}$$

f. Power factor $= \cos\theta = \dfrac{P_T}{P_s}$ (25-12)

$$= \frac{5.7\ W}{7.1\ VA} = \textbf{0.8}$$

or,   Power factor $= \cos\theta$ (25-12)

$$= \cos 35°$$

$$= \textbf{0.8 (leading)}$$

See Fig. 27-18.

27-9 PHASE-SHIFTING CIRCUITS*

In Chapter 24, you saw how the circuit of Fig. 27-19 provided an output that decreased in magnitude as the fre-

*This section may be omitted with no loss of continuity.

quency increased. (Refer to Section 24-6.1, Fig. 24-16.) In this section, you will learn about a method, using complex numbers, to determine how the *phase angle* between V_o and V_i varies with frequency. This has an application in controlling the instant at which a semiconductor switch will begin conducting (thus varying the average current flowing through a lamp or motor). The action of these phase-shifting circuits is illustrated in Example 27-18.

EXAMPLE 27-18

Given the *RC* circuit in Fig. 27-19.

a. Determine an expression for V_o/V_i in terms of R and X_C that gives both the magnitude and the phase angle.

b. Find the numerical values of the magnitude and phase angle for V_o/V_i at various frequencies, from 10 Hz to 20 kHz.

c. Calculate the frequency where the phase angle between V_o and V_i is $-45°$.

d. Determine the magnitude of V_o/V_i, where the phase angle is $-45°$.

e. Draw two graphs of V_o/V_i versus *f*, showing how magnitude and phase angle vary with frequency.

Solution

a. $I = \dfrac{V_i}{Z}$ (24-4)

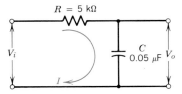

FIGURE 27-19

Circuit for Example 27-18.

$$= \frac{V_i}{R - jX_C}$$

$$V_o = I \times (-jX_C)$$

$$= \frac{V_i}{R - jX_C} \times -jX_C$$

$$\frac{V_o}{V_i} = \frac{-jX_C}{R - jX_C}$$

Multiply numerator and denominator by j.

$$\frac{V_o}{V_i} = \frac{X_C}{X_C + jR}$$

$$\frac{V_o}{V_i} = \frac{X_C}{\sqrt{X_C^2 + R^2}\underline{/+\text{arc tan } R/X_C}}$$

$$\frac{V_o}{V_i} = \frac{X_C\underline{/-\text{arc tan } R/X_C}}{\sqrt{X_C^2 + R^2}}$$

Note that the magnitude $\left|\dfrac{V_o}{V_i}\right| = \dfrac{X_C}{\sqrt{R^2 + X_C^2}}$ is the

same as obtained in Eq. 24-12, but the use of complex numbers has also provided a way of obtaining the phase angle, $\theta = -\text{arc tan } R/X_C$.

b. At $f = 1$ kHz,   (23-2)

$$X_C = \frac{1}{2\pi fC}$$

$$= \frac{1}{2\pi \times 1 \times 10^3 \text{ Hz} \times 0.05 \times 10^{-6} \text{ F}}$$

$$= 3.2 \text{ k}\Omega$$

$$\frac{V_o}{V_i} = \frac{X_C}{\sqrt{R^2 + X_C^2}}\underline{/-\text{arc tan } R/X_C}$$

$$= \frac{3.2 \text{ k}\Omega}{\sqrt{5^2 + 3.2^2} \text{ k}\Omega}\underline{/-\text{arc tan }} \frac{5 \text{ k}\Omega}{3.2 \text{ k}\Omega}$$

$$= \frac{3.2 \text{ k}\Omega}{5.9 \text{ k}\Omega}\underline{/-\text{arc tan } 1.563}$$

$$= 0.54\underline{/-57.4°}$$

Values of V_o/V_i at other frequencies are shown in Table 27-1.

c. For the phase angle between V_o and V_i to be $-45°$ requires

$$\text{arc tan } R/X_C = 45°$$

or $R/X_C = \tan 45° = 1$

Thus X_C must equal R:

$$\frac{1}{2\pi fC} = R$$

$$f = \frac{1}{2\pi RC}$$

TABLE 27-1

Magnitudes and Phase Angles of V_o/V_i for Example 27-18.

f, Hz	X_C, kΩ	$\dfrac{X_C}{\sqrt{R^2 + X_C^2}}$	R/X_C	$-\text{arc tan } R/X_C$
10	320	1.00	0.016	$-0.9°$
100	32	0.99	0.16	$-8.9°$
500	6.4	0.79	0.78	$-38.0°$
637	5.0	0.71	1.00	$-45.0°$
1000	3.2	0.54	1.56	$-57.4°$
5000	0.64	0.13	7.81	$-82.7°$
10,000	0.32	0.06	15.63	$-86.3°$
20,000	0.16	0.03	31.25	$-88.2°$

$$= \frac{1}{2\pi \times 5 \times 10^3 \text{ }\Omega \times 0.05 \times 10^{-6} \text{ F}}$$

$$= \textbf{636.6 Hz}$$

d. The magnitude of V_o/V_i is

$$\left|\frac{V_o}{V_i}\right| = \frac{X_C}{\sqrt{R^2 + X_C^2}}$$

When the phase angle is $-45°$, $X_C = R$; therefore,

$$\left|\frac{V_o}{V_i}\right| = \frac{R}{\sqrt{R^2 + R^2}}$$

$$= \frac{R}{\sqrt{2R^2}} = \frac{R}{\sqrt{2}R} = \frac{1}{\sqrt{2}} = \textbf{0.707}$$

e. Graphs showing how $\left|\dfrac{V_o}{V_i}\right|$ and the phase angle vary with frequency are shown in Fig. 27-20.

Note that the expression obtained for V_o/V_i in Example 27-18 may be written in a different form:

$$\frac{V_o}{V_i} = \frac{X_C}{X_C + jR}$$

$$= \frac{1}{1 + jR/X_C}$$

But $X_C = \dfrac{1}{2\pi fC}$ (23-2)

Therefore, $\dfrac{V_o}{V_i} = \dfrac{1}{1 + j2\pi fRC}$

When f is small, $2\pi fRC << 1$, and $V_o/V_i \approx 1$. That is, V_o

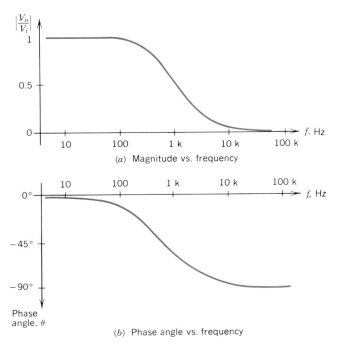

(a) Magnitude vs. frequency

(b) Phase angle vs. frequency

FIGURE 27-20
Graphs of magnitude and phase angle versus frequency for Example 27-18.

is approximately equal to V_i in magnitude and phase. When f is large, $2\pi fRC \gg 1$, and

$$\frac{V_o}{V_i} \approx \frac{1}{j2\pi fRC}$$

$$\approx \frac{1}{2\pi fRC\underline{/90°}}$$

$$\approx \frac{1}{2\pi fRC}\underline{/-90°}$$

That is, V_o is small compared to V_i and lags behind V_i by

almost 90°. For the special case of $2\pi fRC = 1$ or $f = \frac{1}{2\pi RC}$:

$$\frac{V_o}{V_i} = \frac{1}{1 + j1}$$

$$= \frac{1}{\sqrt{2}\underline{/45°}} = 0.707\underline{/-45°}$$

The phasor representations for these three cases are shown in Fig. 27-21.

The circuit of Fig. 27-19 is used at a fixed frequency of 60 Hz with the R variable (typically 1–100 kΩ, $C = 0.5$ μF) to provide a *phase-shifting* function. That is, V_o can be made to lag behind V_i by 0° to 90° by varying R. This "delayed" voltage is applied to the gate of a silicon-controlled rectifier (SCR) or a TRIAC to vary the point at which these solid-state switches will conduct and allow current to flow through a load. This, in turn, may vary the brightness of a lamp in a dimming circuit or the speed of a motor. (See Problem 27-35.)

It is also possible to provide a phase shift where the output voltage leads the input voltage. This is accomplished by interchanging the locations of R and C in Fig. 27-19. In this case, V_o (although small) leads V_i by almost 90° at low frequencies, dropping to 0° at high frequencies where V_o and V_i are approximately equal in magnitude. (See Problem 27-36.)

27-10 APPLICATION OF COMPLEX NUMBERS TO SERIES-PARALLEL CIRCUITS

In Section 6-7, you solved for current and voltage drops in a series-parallel *resistance* circuit. You can now apply complex number theory to any ac circuit using an approach

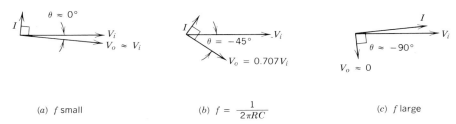

(a) f small

(b) $f = \dfrac{1}{2\pi RC}$

(c) f large

FIGURE 27-21
Phasor representations of V_o and V_i for Example 27-18.

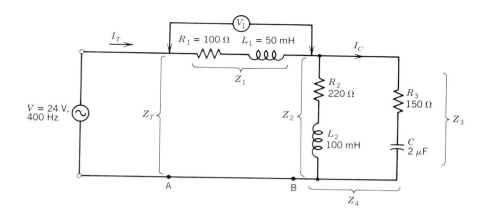

FIGURE 27-22

Circuit for Example 27-19.

similar to that used in Section 6-7, except that you must include phase angles. Voltage-divider and current-divider rules apply equally well to dc resistance and ac circuits when complex numbers are used, as shown in Example 27-19.

EXAMPLE 27-19

Given the circuit in Fig. 27-22, calculate:
a. The total impedance of the whole circuit.
b. The total supply current.
c. The reading of a voltmeter connected across Z_1.
d. The current through the capacitor.

Solution

a. $X_{L1} = 2\pi f L_1$ (20-2)

$\quad\quad = 2\pi \times 400\ \text{Hz} \times 50 \times 10^{-3}\ \text{H}$

$\quad\quad = 125.7\ \Omega$

$\quad Z_1 = R_1 + jX_{L1} = 100 + j125.7\ \Omega$

$\quad X_{L2} = 2\pi f L_2 = 251.4\ \Omega$

$\quad Z_2 = R_2 + jX_{L2} = 220 + j251.4\ \Omega$

$\quad X_C = \dfrac{1}{2\pi f C}$ (23-2)

$\quad\quad = \dfrac{1}{2\pi \times 400\ \text{Hz} \times 2 \times 10^{-6}\ \text{F}}$

$\quad\quad = 198.9\ \Omega$

$\quad Z_3 = R_3 - jX_C = 150 - j198.9\ \Omega$

$\quad Z_T = Z_1 + Z_4$

$\quad\quad = Z_1 + Z_2 \parallel Z_3$

$\quad\quad = Z_1 + \dfrac{Z_2 \times Z_3}{Z_2 + Z_3}$

$\quad\quad = 100 + j125.7$

$\quad\quad\quad + \dfrac{(220 + j251.4) \times (150 - j198.9)}{(220 + j251.4) + (150 - j198.9)}\ \Omega$

$\quad\quad = 100 + j125.7$

$\quad\quad\quad + \dfrac{334.1\underline{/48.8^\circ} \times 249.1\underline{/-53^\circ}}{370 + j52.5}\ \Omega$

$\quad\quad = 100 + j125.7 + \dfrac{83{,}224\underline{/-4.2^\circ}}{373.7\underline{/8.1^\circ}}\ \Omega$

$\quad\quad = 100 + j125.7 + 222.7\underline{/-12.3^\circ}\ \Omega$

$\quad\quad = 100 + j125.7 + 217.6 - j47.4\ \Omega$

$\quad\quad = 317.6 + j78.3\ \Omega$

$\quad\quad = \mathbf{327.1\ \Omega\underline{/13.8^\circ}}$

b. $I_T = \dfrac{V}{Z}$ (24-4)

$\quad\quad = \dfrac{24\ \text{V}\underline{/0^\circ}}{327.1\ \Omega\underline{/13.8^\circ}} = 0.0734\ \text{A}\underline{/-13.8^\circ}$

$\quad\quad = \mathbf{73.4\ mA\underline{/-13.8^\circ}}$

c. $V_1 = I_T Z_1$ (24-4)

$\quad\quad = 0.0734\ \text{A}\underline{/-13.8^\circ} \times (100 + j125.7\ \Omega)$

$\quad\quad = 0.0734\ \text{A}\underline{/-13.8^\circ} \times 160.6\ \Omega\underline{/51.5^\circ}$

$\quad\quad = \mathbf{11.8\ V\underline{/37.7^\circ}}$

d. $I_C = I_{Z3} = I_T \times \dfrac{Z_2}{Z_2 + Z_3}$ (6-12)

$\quad\quad = 73.4\ \text{mA}\underline{/-13.8^\circ}$

$\quad\quad\quad\quad \times \dfrac{220 + j251.4}{(220 + j251.4) + (150 - j198.9)}$

$\quad\quad = 73.4\ \text{mA}\underline{/-13.8^\circ} \times \dfrac{334.1\ \Omega\underline{/48.8^\circ}}{373.7\ \Omega\underline{/8.1^\circ}}$

$\quad\quad = 73.4\ \text{mA}\underline{/-13.8^\circ} \times 0.894\underline{/40.7^\circ}$

$\quad\quad = \mathbf{65.6\ mA\underline{/26.9^\circ}}$

566

27-11 CONDUCTANCE, SUSCEPTANCE, AND ADMITTANCE

There is an alternate method for solving parallel ac circuits. It involves the reciprocals of resistance, reactance, and impedance; these reciprocals are called conductance, susceptance, and admittance, respectively. Refer to Fig. 27-23.

You can express the total current in a pure parallel circuit in rectangular form:

$$I_T = I_R - jI_L + jI_C$$

or

$$\frac{V}{Z_T} = \frac{V}{R} - j\frac{V}{X_L} + j\frac{V}{X_C}$$

Therefore,

$$\frac{1}{Z_T} = \frac{1}{R} - j\frac{1}{X_L} + j\frac{1}{X_C}$$

In Eq. 2-5, $1/R$ was defined as *conductance* (G), measured in siemens (S). Conductance is the ability of a pure resistance to pass electric current.

Now, $1/X_L$ can be defined as the inductive *susceptance* (B_L) and $1/X_C$ as the capacitive *susceptance* (B_C), measured in siemens (S). Susceptance is the ability of a pure inductance or pure capacitance to pass alternating current.

Similarly, $1/Z$ can be defined as *admittance* (Y), measured in siemens (S). Admittance is the *overall* ability of a circuit to pass alternating current. Note that G, B, and Y are all measured in siemens:

$$B_L = \frac{1}{X_L} \quad \text{siemens} \quad (27\text{-}20)$$

$$B_C = \frac{1}{X_C} \quad \text{siemens} \quad (27\text{-}21)$$

$$Y = \frac{1}{Z} \quad \text{siemens} \quad (27\text{-}22)$$

where all terms are as previously defined.

You now can write, for a parallel *RLC* circuit,

$$Y = G - jB_L + jB_C \quad \text{siemens} \quad (27\text{-}23)$$

where: Y is the total admittance of a pure parallel circuit, in siemens (S)
G is the conductance, in siemens (S)
B_L is the inductive susceptance, in siemens (S)
B_C is the capacitive susceptance, in siemens (S)

Note that, in contrast to a series impedance expression ($Z = R + jX_L - jX_C$), inductive *susceptance* is preceded by a *negative j* and capacitive susceptance is preceded by a *positive j*.

Once the total admittance of a parallel circuit has been obtained, the following information is available:

1. The impedance can be found from

$$Z_T = \frac{1}{Y_T} \quad \text{ohms} \quad (27\text{-}22)$$

2. The components of an *equivalent parallel* circuit can be identified if the admittance is given in rectangular form. (This corresponds to the equivalent *series* components for *impedance* given in rectangular form.)

For example, if the admittance of a complex ac circuit, series or parallel, is reduced to

$$Y_T = 0.01 + j0.02 \text{ S}$$

you can interpret this as an equivalent parallel circuit with $G = 0.01$ S and $B_C = 0.02$ S. That is, the equivalent circuit would consist of a resistance

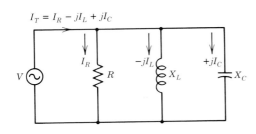

(a) Circuit diagram

(b) Phasor diagram

FIGURE 27-23
Complex numbers applied to a parallel circuit.

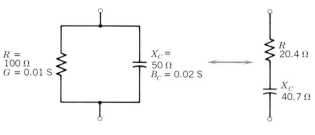

(a) $Y = 0.01 + j\,0.02$ S

(b) $Z = 20.4 - j\,40.7\ \Omega$

FIGURE 27-24

Equivalent parallel and series circuits.

$$R = \frac{1}{G} = \frac{1}{0.01\ \text{S}} = 100\ \Omega$$

in parallel with a capacitor having a capacitive reactance of

$$X_C = \frac{1}{B_C} = \frac{1}{0.02\ \text{S}} = 50\ \Omega$$

See Fig. 27-24a.

You can also find the polar form of an admittance and then find the equivalent impedance. For example,

$$Y_T = 0.01 + j0.02\ \text{S}$$
$$= 0.022\ \text{S}\underline{/63.4°}$$

$$Z_T = \frac{1}{Y_T} \qquad (27\text{-}22)$$

$$= \frac{1}{0.022\ \text{S}\underline{/63.4°}} = 45.5\ \Omega\underline{/-63.4°}$$

$$= 20.4 - j40.7\ \Omega$$

This represents a *series* circuit of 20.4-Ω resistance and 40.7-Ω capacitive reactance. (See Fig. 27-24b.) The two

circuits in Fig. 27-24 are equivalent to each other in all *ac* applications. This is the method by which a practical resonant circuit was converted to a theoretical parallel resonant circuit in Section 26-8. (See Fig. 26-15 and Eqs. 26-10 and 26-11.)

It should be noted that in any parallel circuit,

$$Y_T = Y_1 + Y_2 + Y_3 + \cdots + Y_n \qquad \text{siemens} \qquad (27\text{-}24)$$

as shown in Example 27-20.

EXAMPLE 27-20

For the parallel circuit given in Fig. 27-25a, calculate the following component values using the method of admittances:

a. An equivalent parallel circuit.
b. An equivalent series circuit.

Solution

a. $Z_1 = 20 + j50\ \Omega = 53.9\ \Omega\underline{/68.2°}$

$$Y_1 = \frac{1}{Z_1} \qquad (27\text{-}22)$$

$$= \frac{1}{53.9\ \Omega\underline{/68.2°}} = 0.0185\ \text{S}\underline{/-68.2°}$$

$$= 0.0069 - j0.0172\ \text{S}$$

$Z_2 = 25 + j40\ \Omega = 47.2\ \Omega\underline{/58°}$

$$Y_2 = \frac{1}{Z_2} \qquad (27\text{-}22)$$

$$= \frac{1}{47.2\ \Omega\underline{/58°}} = 0.0212\ \text{S}\underline{/-58°}$$

$$= 0.0112 - j0.0180\ \text{S}$$

$Z_3 = 40 - j40\ \Omega = 56.6\ \Omega\underline{/-45°}$

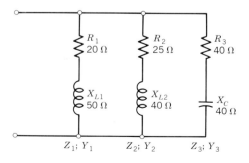

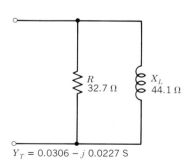

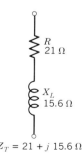

$Z_1; Y_1 \qquad Z_2; Y_2 \qquad Z_3; Y_3$

(a) Original circuit

$Y_T = 0.0306 - j\,0.0227$ S

(b) Equivalent parallel circuit

$Z_T = 21 + j\,15.6\ \Omega$

(c) Equivalent series circuit

FIGURE 27-25

Circuits for Example 27-20.

$$Y_3 = \frac{1}{Z_3} \qquad (27\text{-}22)$$

$$= \frac{1}{56.6\ \Omega \underline{/-45^\circ}} = 0.0177\ \text{S}\underline{/45^\circ}$$

$$= 0.0125 + j0.0125\ \text{S}$$

$$Y_T = Y_1 + Y_2 + Y_3 \qquad (27\text{-}24)$$

$$= (0.0069 - j0.0172) + (0.0112 - j0.0180)$$

$$+ (0.0125 + j0.0125)\ \text{S}$$

$$= 0.0306 - j0.0227\ \text{S} = 0.0381\ \text{S}\underline{/-36.6^\circ}$$

For an equivalent parallel circuit:

$$Y_T = G - jB_L$$

$$R = \frac{1}{G} \qquad (2\text{-}5)$$

$$= \frac{1}{0.0306\ \text{S}} = \mathbf{32.7\ \Omega}$$

$$X_L = \frac{1}{B_L} \qquad (27\text{-}20)$$

$$= \frac{1}{0.0227\ \text{S}} = \mathbf{44.1\ \Omega}$$

See Fig. 27-25*b*.

b. For an equivalent series circuit:

$$Z_T = \frac{1}{Y_T} \qquad (27\text{-}22)$$

$$= \frac{1}{0.0381\ \text{S}\underline{/-36.6^\circ}} = 26.2\ \Omega\underline{/36.6^\circ}$$

$$= 21 + j15.6\ \Omega$$

$$R = \mathbf{21\ \Omega}, \quad X_L = \mathbf{15.6\ \Omega}$$

See Fig. 27-25*c*.

27-12 APPLICATION OF COMPLEX NUMBERS TO THÉVENIN'S THEOREM

In Section 10-4, you saw how Thévenin's theorem could be used to convert a complicated circuit containing electromotive forces and resistances into a simple equivalent series circuit of one electromotive force and one resistance.

Thévenin's theorem can now be restated as it applies to an ac circuit containing emfs and impedances:

Any linear two-terminal network consisting of fixed impedances and sources of emf can be replaced by a single source of emf (V_{Th}) in series with a single

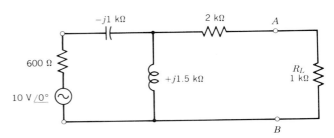

FIGURE 27-26

Circuit for Example 27-21.

impedance (Z_{Th}) whose values are given by the following:

1. V_{Th} is the open-circuit voltage at the terminals of the original circuit.
2. Z_{Th} is the impedance looking back into the original network from the two terminals with all sources of emf shorted out and replaced by their internal impedances, and all current sources opened.

EXAMPLE 27-21

Given the circuit in Fig. 27-26, use Thévenin's theorem to find the current through the 1-kΩ load resistor.

Solution

First remove the load (R_L) to find the open-circuit voltage between *A* and *B*, as in Fig. 27-27.

Since no current can now flow through the 2-kΩ resistor:

$$V_{\text{Th}} = V_{AB} = V_{CB}$$

$$V_{\text{Th}} = 10\ \text{V}\underline{/0^\circ} \times \frac{j1.5\ \text{k}\Omega}{0.6 - j1 + j1.5\ \text{k}\Omega}$$

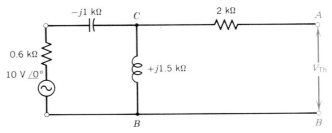

FIGURE 27-27

Obtaining the Thévenin equivalent voltage for Example 27-21.

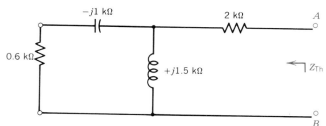

FIGURE 27-28

Obtaining the Thévenin equivalent impedance for Example 27-21.

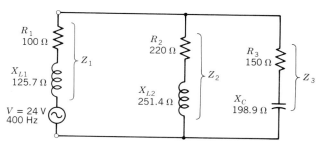

FIGURE 27-30

Circuit for Example 27-22.

$$= 10 \text{ V}\underline{/0°} \times \frac{j1.5}{0.6 + j0.5}$$

$$= 10 \text{ V}\underline{/0°} \times \frac{1.5\underline{/90°}}{0.78\underline{/39.8°}}$$

$$= 10 \text{ V}\underline{/0°} \times 1.92\underline{/50.2°}$$

$$= \textbf{19.2 V}\underline{\textbf{/50.2°}}$$

To find the Thévenin impedance, short the voltage source but leave its 600-Ω internal resistance in place. Find the impedance (Z_{Th}) between A and B looking back into the circuit, as in Fig. 27-28:

$$Z_{Th} = 2 \text{ k}\Omega + (j1.5 \text{ k}\Omega) \parallel (0.6 - j1 \text{ k}\Omega)$$

$$= 2 + \frac{j1.5 \times (0.6 - j1) \text{ k}\Omega}{0.6 + j1.5 - j1}$$

$$= 2 + \frac{1.5 + j0.9}{0.6 + j0.5} \text{ k}\Omega$$

$$= 2 + \frac{1.75\underline{/30.96°}}{0.78\underline{/39.81°}} \text{ k}\Omega$$

$$= 2 + 2.24\underline{/-8.85°} \text{ k}\Omega$$

$$= 2 + 2.21 - j0.34 \text{ k}\Omega$$

$$= \textbf{4.21} - \textbf{j0.34 k}\Omega$$

The Thévenin equivalent circuit is shown in Fig. 27-29.

The load current can now be found:

$$I = \frac{V_{Th}}{Z_{Th} + R_L}$$

$$= \frac{19.2 \text{ V}\underline{/50.2°}}{4.21 - j0.34 + 1 \text{ k}\Omega}$$

$$= \frac{19.2 \text{ V}\underline{/50.2°}}{5.21 - j0.34 \text{ k}\Omega}$$

$$= \frac{19.2 \text{ V}\underline{/50.2°}}{5.22 \text{ k}\Omega\underline{/-3.7°}} = \textbf{3.68 mA}\underline{\textbf{/53.9°}}$$

EXAMPLE 27-22

Given the circuit in Fig. 27-30, use Thévenin's theorem to find the current through the capacitor. (Note that this circuit is the same as that in Fig. 27-22, with Z_1 the internal impedance of the source.)

Solution

First remove the load, Z_3, to find the *open-circuit* voltage between A and B, as in Fig. 27-31.

$$V_{Th} = V \times \frac{Z_2}{Z_1 + Z_2} \qquad (5\text{-}5)$$

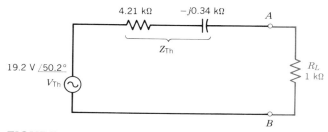

FIGURE 27-29

Thévenin equivalent circuit with load reconnected for Example 27-21.

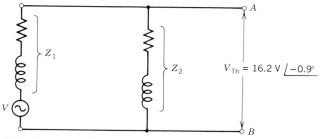

FIGURE 27-31

Obtaining the equivalent Thévenin voltage for Example 27-22.

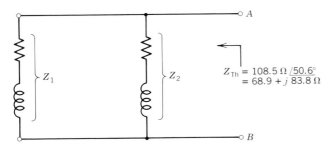

FIGURE 27-32
Obtaining the equivalent Thévenin impedance for
Example 27-22.

$= 24\ V\underline{/0°}$

$\times \dfrac{220 + j251.4\ \Omega}{(100 + j125.7) + (220 + j251.4)\ \Omega}$

$= 24\ V\underline{/0°} \times \dfrac{334.1\underline{/48.8°}\ \Omega}{320 + j377.1\ \Omega}$

$= 24\ V\underline{/0°} \times \dfrac{334.1\ \Omega\underline{/48.8°}}{494.6\ \Omega\underline{/49.7°}}$

$= 24\ V\underline{/0°} \times 0.675\underline{/-0.9°}$

$= \mathbf{16.2\ V\underline{/-0.9°}}$

To find the Thévenin impedance, short the voltage
source, V, and replace it with its internal impedance, Z_1,
as in Fig. 27-32.
This places Z_1 and Z_2 in parallel with each other.

$Z_{Th} = Z_1 \parallel Z_2$

$= \dfrac{Z_1 \times Z_2}{Z_1 + Z_2}$ (27-17)

$= \dfrac{(100 + j125.7) \times (220 + j251.4)}{(100 + j125.7) + (220 + j251.4)}\ \Omega$

$= \dfrac{160.6\underline{/51.5°} \times 334.1\underline{/48.8°}}{320 + j377.1}\ \Omega$

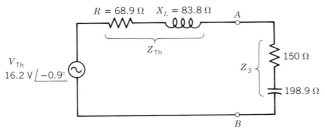

FIGURE 27-33
Thévenin equivalent circuit for Example 27-22.

$= \dfrac{53,656\underline{/100.3°}}{494.6\underline{/49.7°}}\ \Omega$

$= \mathbf{108.5\ \Omega\underline{/50.6°}} = \mathbf{68.9 + j83.8\ \Omega}$

The Thévenin equivalent circuit is shown in Fig. 27-33.

$I = \dfrac{V_{Th}}{Z_{Th} + Z_3}$ (24-4)

$= \dfrac{16.2\ V\underline{/-0.9°}}{68.9 + j83.8 + 150 - j198.9}$

$= \dfrac{16.2\ V\underline{/-0.9°}}{218.9 - j115.1\ \Omega}$

$= \dfrac{16.2\ V\underline{/-0.9°}}{247.3\ \Omega\underline{/-27.7°}}$

$= 0.0655\ A\underline{/26.8°} = \mathbf{65.5\ mA\underline{/26.8°}}$

The solution for this problem should be compared with
that for Example 27-19. Although the amount of work is
similar, the Thévenin approach is preferred if the current
is to be calculated for a number of different loads. Then,
only the last calculation for I must be repeated.

EXAMPLE 27-23

Given the circuit in Fig. 27-34, determine, using a Thévenin equivalent circuit, what the impedance of Z_L must
be to have maximum power delivered to the load, and
what this power is.

Solution

Remove the load impedance to obtain the Thévenin
open circuit voltage, as in Fig. 27-35.
The current I that would circulate can be found:

$I = \dfrac{V_2 - V_1}{Z_T}$

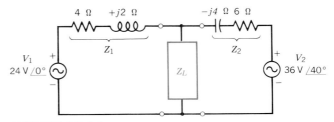

FIGURE 27-34
Circuit for Example 27-23.

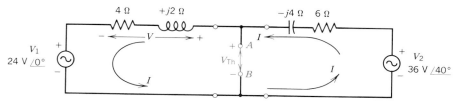

FIGURE 27-35
Obtaining the Thévenin open-circuit voltage for Example 27-23.

$$= \frac{36 \text{ V}\underline{/40°} - 24 \text{ V}\underline{/0°}}{4 + j2 + 6 - j4 \text{ } \Omega}$$

$$= \frac{27.6 + j23.1 - 24 \text{ V}}{10 - j2 \text{ } \Omega}$$

$$= \frac{3.6 + j23.1 \text{ A}}{10 - j2}$$

$$= \frac{23.4\underline{/81.1°}}{10.2\underline{/-11.3°}} \text{ A} = 2.3 \text{ A}\underline{/92.4°}$$

The voltage drop that occurs across Z_1 is:

$$V = IZ_1 \qquad (24\text{-}4)$$
$$= 2.3 \text{ A}\underline{/92.4°} \times (4 + j2 \text{ } \Omega)$$
$$= 2.3 \text{ A}\underline{/92.4°} \times 4.5 \text{ } \Omega\underline{/26.6°}$$
$$= 10.4 \text{ V}\underline{/119°}$$

Therefore, the Thévenin open-circuit voltage between A and B is given by

$$V_{Th} = V + V_1$$
$$= 10.4 \text{ V}\underline{/119°} + 24 \text{ V}\underline{/0°}$$
$$= -5.0 + j9.1 + 24 \text{ V}$$
$$= 19 + j9.1 \text{ V}$$
$$= \mathbf{21.1 \text{ V}\underline{/25.6°}}$$

To find the Thévenin impedance, short the voltage sources V_1 and V_2 as in Fig. 27-36:

$$Z_{Th} = Z_1 \| Z_2$$
$$= \frac{Z_1 \times Z_2}{Z_1 + Z_2} \qquad (27\text{-}17)$$

$$= \frac{(4 + j2)(6 - j4)}{(4 + j2) + (6 - j4)}$$

$$= \frac{24 - j16 + j12 + 8}{10 - j2}$$

$$= \frac{32 - j4}{10 - j2}$$

$$= \frac{(32 - j4) \times (10 + j2)}{(10 - j2) \times (10 + j2)}$$

$$= \frac{320 + j64 - j40 + 8}{100 + j20 - j20 + 4}$$

$$= \frac{328 + j24}{104}$$

$$= \frac{328}{104} + j\frac{24}{104}$$

$$= \mathbf{3.2 + j0.2 \text{ } \Omega}$$

The Thévenin equivalent circuit is shown in Fig. 27-37. For maximum power transfer, the load impedance must be the conjugate of the source impedance. Thus, load impedance required is a 3.2-Ω resistance in series with a capacitor that has a reactance of 0.2 Ω. That is $\mathbf{Z_L}$ $\mathbf{= 3.2 - j0.2 \text{ } \Omega.}$

With the load impedance in place, the load current can be obtained:

$$I = \frac{V}{Z_T} \qquad (24\text{-}4)$$

$$= \frac{21.1 \text{ V}\underline{/25.6°}}{3.2 + j0.2 + 3.2 - j0.2 \text{ } \Omega}$$

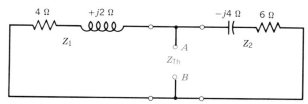

FIGURE 27-36
Obtaining the Thévenin impedance for Example 27-23.

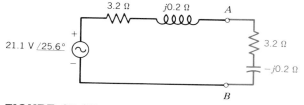

FIGURE 27-37
Thévenin equivalent circuit with load to produce maximum power transfer for Example 27-23.

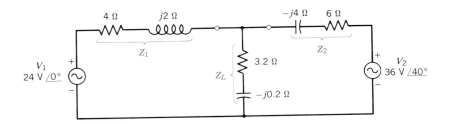

FIGURE 27-38
Circuit for Example 27-24.

$$= \frac{21.1 \text{ V}\underline{/25.6°}}{6.4 \text{ }\Omega} = 3.3 \text{ A}\underline{/25.6°}$$

Load power $= I^2R_L = (3.3 \text{ A})^2 \times 3.2 \text{ }\Omega = \mathbf{34.8 \text{ W}}$

27-13 APPLICATION OF COMPLEX NUMBERS TO NORTON'S THEOREM

As was the case with Thévenin's theorem, Norton's theorem can be restated to include impedances.

Any linear two-terminal network consisting of fixed impedances and sources of emf can be replaced with a single current source (I_N) in parallel with a single impedance (Z_N), whose values are given by the following:

1. I_N is the short-circuit current between the terminals of the original circuit.
2. Z_N is the impedance looking back into the original network from the two terminals with all sources of emf shorted out and replaced by their internal impedances, and any current sources opened.

EXAMPLE 27-24

Given the circuit in Fig. 27-38 (the same as Fig. 27-34 in Example 27-23, but with $Z_L = 3.2 - j0.2 \text{ }\Omega$), calculate the load current and the power delivered to the load Z_L.

Solution

The short circuit current is found by removing the load and shorting terminals A and B, as in Fig. 27-39:

$$I_N = I_1 + I_2$$
$$= \frac{V_1}{Z_1} + \frac{V_2}{Z_2}$$
$$= \frac{24 \text{ V}\underline{/0°}}{4 + j2 \text{ }\Omega} + \frac{36 \text{ V}\underline{/40°}}{6 - j4 \text{ }\Omega}$$
$$= \frac{24 \text{ V}\underline{/0°}}{4.5 \text{ }\Omega\underline{/26.6°}} + \frac{36 \text{ V}\underline{/40°}}{7.2 \text{ }\Omega\underline{/-33.7°}}$$
$$= 5.3 \text{ A}\underline{/-26.6°} + 5 \text{ A}\underline{/73.7°}$$
$$= 4.7 - j2.4 + 1.4 + j4.8 \text{ A}$$
$$= 6.1 + j2.4 \text{ A}$$
$$= \mathbf{6.6 \text{ A}\underline{/21.5°}}$$

The Norton impedance Z_N is found by shorting the voltage sources as in Fig. 27-40, placing Z_1 and Z_2 in parallel:

$$Z_N = Z_1 \parallel Z_2$$
$$= \frac{Z_1 \times Z_2}{Z_1 + Z_2} \qquad (27\text{-}17)$$
$$= \mathbf{3.2 + j0.2 \text{ }\Omega}$$

as calculated in Example 27-23.

The Norton equivalent circuit is shown in Fig. 27-41 with the load reconnected.

The load current can now be obtained by the current divider rule:

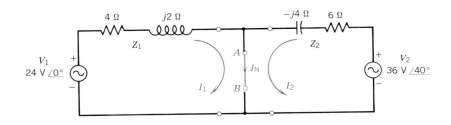

FIGURE 27-39
Obtaining the Norton short-circuit current for Example 27-24.

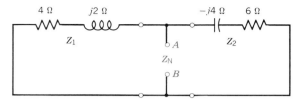

FIGURE 27-40
Obtaining the Norton equivalent impedance for
Example 27-24.

$$I_L = I_N \times \frac{Z_N}{Z_N + Z_L} \quad (6\text{-}12)$$

$$= 6.6 \underline{/21.5°} \times \frac{3.2 + j0.2 \ \Omega}{3.2 + j0.2 + 3.2 - j0.2 \ \Omega}$$

$$= 6.6 \text{ A} \underline{/21.5°} \times \frac{3.2 + j0.2}{6.4}$$

$$= 6.6 \text{ A} \underline{/21.5°} \times \frac{3.2 \underline{/3.6°}}{6.4}$$

$$= 6.6 \text{ A} \underline{/21.5°} \times 0.5 \underline{/3.6°}$$

$$= \mathbf{3.3 \text{ A} \underline{/25.1°}}$$

$$\text{Load power} = I_L^2 R_L$$

$$= (3.3 \text{ A})^2 \times 3.2 \ \Omega$$

$$= \mathbf{34.8 \text{ W}}$$

In this case, obtaining the Norton short-circuit current was easier than finding the Thévenin open-circuit voltage in Example 27-23. This means that the Thévenin equivalent voltage could have been found more easily from the Norton equivalent by multiplying I_N by Z_{Th} (or Z_N):

$$V_{Th} = I_N \times Z_{Th}$$

$$= 6.6 \text{ A} \underline{/21.5°} \times (3.2 + j0.2 \ \Omega)$$

$$= 6.6 \text{ A} \underline{/21.5°} \times 3.2 \ \Omega \underline{/3.6°}$$

$$= \mathbf{21.1 \text{ V} \underline{/25.1°}}$$

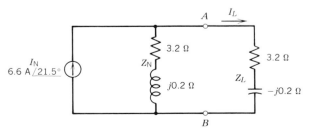

FIGURE 27-41
Norton equivalent circuit with load reconnected for
Example 27-24.

which is essentially the same as V_{Th} in Example 27-23. Thus, it is possible to convert from a Norton equivalent to a Thévenin equivalent, and vice-versa, as was done in Chapter 10 with dc circuits.

Similarly, it should be noted that *all* of the network analysis methods covered in Chapter 10 (branch currents, loop or mesh currents, superposition, and nodal analysis) can be applied to ac circuits—as you have done with Norton's and Thévenin's equivalents—provided complex numbers are used to represent the voltage, current and impedance.

For example, the circuit in Fig. 27-38 of Example 27-24 could be solved using the superposition theorem. If V_2 is shorted, a total impedance is presented to V_1 given by

$$Z_{T_1} = Z_1 + Z_L \| Z_2$$

This will allow the determination of current I_1 supplied by V_1 through Z_1:

$$I_1 = \frac{V_1}{Z_1}$$

The load current due to V_1 can be obtained by the current divider equation:

$$I_{L_1} = I_1 \times \frac{Z_2}{Z_2 + Z_L}$$

Now, if V_1 is shorted, the total impedance seen by V_2 is given by

$$Z_{T_2} = Z_2 + Z_L \| Z_1$$

The current supplied by V_2 alone is given by

$$I_2 = \frac{V_2}{Z_{T_2}}$$

and the load current due to V_2 is:

$$I_{L_2} = I_2 \times \frac{Z_1}{Z_1 + Z_L}$$

Superimposing the two load currents gives the total load current with V_1 and V_2 in place:

$$I_L = I_{L_1} + I_{L_2}$$

However, as shown above, the calculations are long and involved. And, in the case of branch currents, loop currents, and nodal analysis, equations with two or more unknowns must be solved simultaneously using complex numbers. The Thévenin (and/or Norton) approach, on the other hand, is relatively simple and will allow most ac networks to be solved more easily.

SUMMARY

1. The polar form of a complex number consists of a positive magnitude and a phase angle that may be positive or negative, such as $5 \text{ V}\underline{/-15°}$.

2. To multiply two complex numbers given in polar form, multiply their magnitudes and algebraically *add* their phase angles.

3. To divide two complex numbers given in polar form, divide their magnitudes and algebraically *subtract* their phase angles.

4. When the voltage and impedance are given in polar form, the equation $I = V/Z$ gives the magnitude and phase angle of the current with respect to the voltage.

5. The polar form of the current or voltage gives the reading of an ammeter or voltmeter, respectively.

6. The j operator, equal to $\sqrt{-1}$ or $1\underline{/90°}$, rotates a number through an angle of $+90°$.

7. The rectangular coordinate system consists of a *real* axis and an *imaginary* axis.

8. A complex number in rectangular form consists of a real part and an imaginary part, such as $Z = R + jX$.

9. When an impedance is given in rectangular form, the real part is the resistance portion and the imaginary part is the series reactive portion (inductive if positive, capacitive if negative).

10. The sum or difference of two complex numbers given in rectangular form is found by algebraically adding or subtracting the real parts, then the imaginary parts (respectively).

11. The multiplication of complex numbers in rectangular form follows the distributive law of algebra, whereas division requires a rationalization of the denominator. Both involve the fact that $j^2 = -1$.

12. The total impedance of a series circuit is given by $Z_T = Z_1 + Z_2 + Z_3$, and so on. The total impedance of a parallel circuit is given by

$$Z_T = \frac{Z_1 \times Z_2}{Z_1 + Z_2}$$

for two branches, with all impedances given in complex number form.

13. A complex number in rectangular form can be converted to polar form using:

$$a + jb = \sqrt{a^2 + b^2}\underline{\bigg/ \text{arc tan } \frac{b}{a}}$$

14. A complex number in polar form can be converted to rectangular form using:

$$M\underline{/\theta} = M\cos\theta + jM\sin\theta$$

15. Admittance (Y) is the reciprocal of impedance; susceptance (B), the reciprocal of reactance, and conductance (G), the reciprocal of resistance. All are measured in siemens (S).

16. If the admittance of a circuit is given in rectangular form, $Y = G + j(B_C - B_L)$, the resistance and reactive components of an equivalent parallel circuit can be identified, as well as the circuit's impedance, $Z = \dfrac{1}{Y}$.

17. The total admittance of a parallel circuit is given by $Y_T = Y_1 + Y_2 + Y_3$, and so on.

18. Thévenin's and Norton's theorems can be applied to an ac circuit if complex numbers are used to represent the impedances, voltage, and current.

SELF-EXAMINATION

Match the expressions in the right-hand column with those in the left-hand column. (Answers at back of book)

27-1. $30 - j40$ V =   a.   0.25 S
27-2. $R = 10\ \Omega,\ X_L = 20\ \Omega;\ Z =$ b. $30\ \text{V}\underline{/5°}$
27-3. $V_R = 60$ V, $V_L = 80$ V; $V_T =$ c. $10 + j20\ \Omega$
27-4. $I_R = 5$ A, $I_L = 6$ A, $I_C = 7$ A; $I_T =$ d. $2\ \text{A}\underline{/60°}$
27-5. $I = 3\ \text{A}\underline{/-40°},\ Z = 10\ \Omega\underline{/45°};\ V =$ e. $72\underline{/-45°}$
27-6. $V = 120\ \text{V}\underline{/10°},\ Z = 60\ \Omega\underline{/-50°};\ I =$ f. $50\ \text{V}\underline{/-53.1°}$
27-7. $24\underline{/-20°} \times 3\underline{/-25°}$ g. $7 - j8$
27-8. $3\underline{/-25°} \div 24\underline{/-40°}$ h. $5 + j1$ A
27.9 $5\underline{/43°} \times 3.4\underline{/47°}$ i. -4
27-10. $6\underline{/95°} \div 1.5\underline{/-85°}$ j. $-7 - j1$
27-11. $(3 + j7) + (4 - j15)$ k. $0.125\underline{/15°}$
27-12. $(-14 - j6) - (-7 - j5)$ l. $8 + j9$
27-13. $(6 - j4) \times (6 + j4)$ m. $j17$
27-14. $(3 - j5)^2$ n.   52
27-15. $(43 - j6) \div (2 - j3)$ o. $50 + j50$
27-16. $Z = 50\ \Omega\underline{/-53.1°};\ Y =$ p. $-16 - j30$
27-17. $70.7\underline{/45°} =$ q. $0.02\ \text{S}\underline{/53.1°}$
27-18. $8 + j9$: conjugate =   r.   4 S
27-19. $X_C = 0.25\ \Omega;\ B_C =$ s. $100\ \text{V}\underline{/53.1°}$
27-20. $X_L = 4\ \Omega;\ B_L =$ t. $8 - j9$

REVIEW QUESTIONS

1. a. When an impedance is given in polar form, how can you tell whether it is inductive or capacitive?
    b. How can you determine if there is more resistance than reactance?
2. If an impedance is given in rectangular form, how do you determine the resistive and reactive components?
3. What are two advantages of the polar notation?
4. For what mathematical operations is the rectangular notation especially suited?
5. What are the polar forms for j^2 and j^3?
6. The voltage across a pure capacitor in rectangular form is given by $-j150$ V. Is there anything *imaginary* about this voltage? Explain.
7. Although it is possible to multiply and divide phasors given in rectangular form (without conversion), how can addition and subtraction be performed directly (without conversion) in polar form?
8. a. What do you understand by the term *conjugate?*
    b. Give two applications in which the conjugate is used.
9. A circuit contains both resistance and inductance. How could you express the apparent power in terms of the true and reactive power using the rectangular form?
10. a. Given the admittance of a circuit in rectangular form, how do you interpret the real and imaginary parts?
    b. How would you determine the equivalent *series* circuit components?

11. How is Thévenin's theorem different when applied to an ac circuit compared with a dc circuit?

12. An ac bridge consists of a Wheatstone bridge circuit with an ac voltage source across one set of diagonally opposite corners and an ac null indicator across the other set of corners. If the resistance symbols are replaced by impedances, with Z_1 and Z_4 opposite each other, and Z_2 and Z_3 opposite each other, show that balance of the ac bridge occurs when the general bridge equation is satisfied:

$$Z_1 \times Z_4 = Z_2 \times Z_3$$

PROBLEMS

27-1. A series RL circuit has 80 V across the resistance and 40 V across the inductor. Determine the total applied voltage in rectangular and polar form.

27-2. A series RC circuit has 30 V across the resistance and 90 V across the capacitor. Determine the total applied voltage in rectangular and polar form.

27-3. A series RC circuit has 25 Ω of resistance and 40 Ω of reactance. Determine the impedance in rectangular and polar form.

27-4. A series RL circuit has 470 Ω of resistance and 300 Ω of reactance. Determine the impedance in rectangular and polar form.

27-5. A 68-Ω resistor and a 30-mH inductor of negligible resistance are series-connected across a 400-Hz sinusoidal supply. Determine the circuit impedance in rectangular and polar form.

27-6. A 2.2-kΩ resistor and a 0.1-μF capacitor are series-connected across a 1-kHz sinusoidal supply. Determine the impedance in rectangular and polar form.

27-7. Given $A = 50\underline{/60°}$ and $B = 4\underline{/-25°}$, find:
  a. $A \times B$
  b. $A \div B$
  c. $B \div A$

27-8. Given $C = 2 \times 10^{-3}\underline{/-51.2°}$ and $D = 4 \times 10^{-6}\underline{/-30.8°}$, find:
  a. $C \times D$
  b. $C \div D$
  c. $D \div C$

27-9. For the circuit of Problem 27-3, assume an applied voltage of 120 V and find, in magnitude and phase:
  a. The current.
  b. The resistor voltage.
  c. The capacitor voltage.
  Draw a phasor diagram.

27-10. For the circuit of Problem 27-4, assume an applied emf of 230 V and find, in magnitude and phase:
  a. The current.
  b. The resistor voltage.
  c. The inductor voltage.
  Draw a phasor diagram.

27-11. For the circuit of Problem 27-5, assume an applied emf of 48 V and find, in magnitude and phase:
  a. The current.

    b.   The resistor voltage.

    c.   The inductor voltage.

Draw a phasor diagram.

27-12. For the circuit of Problem 27-6, assume an applied emf of 20 V peak-to-peak and find, in magnitude and phase, rms values of:

    a.   The current.

    b.   The resistor voltage.

    c.   The capacitor voltage.

Draw a phasor diagram.

27-13. A series circuit with an applied emf of 60 V has a current of 250 mA that leads the applied voltage by 30°. Determine:

    a.   The type of series circuit.

    b.   The circuit impedance in polar form.

    c.   The resistance and reactance of the circuit.

27-14. A current of 40 mA flows when an 80-V dc source is applied to a series circuit. When a 120-V, 60-Hz source is applied, a current of 30 mA flows. Determine:

    a.   The circuit impedance in rectangular form.

    b.   The circuit impedance in polar form.

    c.   The inductance of the circuit.

27-15. Given $A = 3 - j7$ and $B = 8 + j11$, find:

    a. $A + B$

    b. $A - B$

    c. $B - A$

27-16. Given $C = -4 - j6$, $D = 4 - j9$, and $E = 11 + j10$, find:

    a. $C + D + E$

    b. $D - E - C$

    c. $2C + 3(D + E)$

27-17. A coil having a resistance of 50 Ω and a reactance of 150 Ω is connected in series with a capacitor whose reactance is 50 Ω. Find the total impedance in rectangular and polar form.

27-18. Three loads are series-connected across the 230-V line. The first load has a voltage given by $50 + j70$ V; the second load has a voltage given by $100 - j150$ V. Find the voltage across the third load in rectangular and polar form.

27-19. Given $A = 2 - j6$ and $B = 7 + j9$, find:

    a. $A \times B$

    b. $A \div B$

    c. $B \div A$, in rectangular form

27-20. Given $C = -4 - j5$, $D = 3 - j10$ and $E = 12 + j2$, find:

    a. $C \times D \times E$

    b. $C \div D$

    c. C^2, in rectangular form.

27-21. Two coils, one of 10-Ω resistance and 25-Ω reactance, the other of 15-Ω resistance and 30-Ω reactance, are connected in parallel across a 24-V, 60-Hz source. Calculate:

    a.   The total circuit impedance in rectangular and polar form.

    b.   The total current.

    c.   Each branch current.

27-22. Repeat Problem 27-21 with a 100-μF capacitor connected in parallel with the coils.

27-23. A coil having a resistance of 400 Ω and a reactance of 1 kΩ is connected in parallel with a capacitor whose reactance is 1.2 kΩ. When connected to a 100-mV source, calculate:

a.   The total circuit impedance.

b.   The total current.

c.   Each branch current.

27-24. Repeat Problem 27-23 assuming that the capacitor has a resistance of 1 kΩ connected in series with it.

27-25. Convert the following to polar form:

a. $6 + j10$

b. $3 - j7$

c. $-8 - j6$.

27-26. Convert the following to polar form:

a. $10 j$

b. -5

c. $3j^3$

27-27. Find the total impedance of the two parallel-connected coils in Problem 27-21 using rectangular to polar conversion.

27-28. What rectangular impedance must be connected in parallel with $5 + j4$ Ω to produce a total impedance of $2 - j1$ Ω? (*Hint:*   Let the impedance be $a + jb$ and equate $1/(2 - j1)$ to $1/(a + jb) + 1/(5 + j4)$.)

27-29. Convert the following to rectangular form:

a. $65\underline{/125°}$.

b. $12\underline{/-72°}$.

c. $20\underline{/200°}$.

27-30. A coil having an ac resistance of 8 Ω and an inductance of 200 μH is connected in parallel with a 1-μF capacitor that has an ac resistance of 4 Ω. Determine the components of the equivalent *series* impedance at a frequency of 10 kHz.

27-31. Repeat Problem 27-30 for a frequency of 5 kHz.

27-32. Two capacitors, a coil, and a resistor are connected in series. The voltages across the capacitors are $12 \text{ V}\underline{/-85°}$ and $15 \text{ V}\underline{/-90°}$; the coil voltage is $20 \text{ V}\underline{/45°}$; and the resistor voltage is $10 \text{ V}\underline{/0°}$. Calculate the supply voltage.

27-33. A parallel circuit has two branches and draws a total current of $120 \text{ mA}\underline{/-50°}$ from a 48-V, 400-Hz source. If the current in one branch is $30 \text{ mA}\underline{/30°}$, calculate:

a.   The reading of an ammeter connected in the other branch.

b.   The total impedance of the circuit.

c.   The total power dissipated in the whole circuit.

27-34. A coil and a pure capacitance, connected in parallel across a $120\text{-V}\underline{/0°}$, 60-Hz supply, draw a total current of $4 \text{ A}\underline{/0°}$. If the capacitor draws a current of $2 \text{ A}\underline{/90°}$, calculate:

a.   The reading of a wattmeter connected to measure the total power in the circuit.

b.   The current through the coil.

c.   The coil's resistance.

d.   The coil's inductance.

e.   The capacitance of the capacitor.

27-35. Given the phase-shifting circuit in Fig. 27-42, calculate:
  a. V_o/V_i in magnitude and phase with R minimum.
  b. V_o/V_i in magnitude and phase with R maximum.
  c. The value of R to provide a phase shift of 45°, and the magnitude of V_o at this value of R.

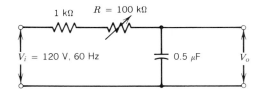

FIGURE 27-42

Circuit for Problem 27-35.

27-36. Given the circuit in Fig. 27-43:
  a. Determine an expression for V_o/V_i in terms of R and X_C that gives both the magnitude and the phase angle.

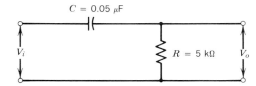

FIGURE 27-43

Circuit for Problem 27-36.

  b. Find the numerical values of the magnitude and phase angle for V_o/V_i at various frequencies from 10 Hz to 20 kHz.
  c. Calculate the frequency at which the phase angle between V_o and V_i is 45°.
  d. Determine the magnitude of V_o/V_i where the phase angle is 45°.
  e. Sketch two graphs of V_o/V_i versus f showing magnitude and phase angle.
27-37. Repeat Example 27-19, using a frequency of 1 kHz.
27-38. Repeat Example 27-19 with an impedance equal to Z_1 connected between A and B.
27-39. For the parallel circuit given in Fig. 27-44, calculate the values of:
  a. An equivalent parallel circuit.
  b. An equivalent series circuit, using the method of admittances, for a frequency of 1 kHz.

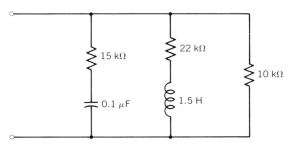

FIGURE 27-44

Circuits for Problems 27-39 through 27-42.

27-40. Repeat Problem 27-39, using a frequency of 100 Hz.
27-41. A 50-mV, 1-kHz sinusoidal source with an internal impedance of 12-kΩ resistance and 1-H inductance is connected to the parallel circuit of Fig. 27-44. Use Thévenin's theorem to calculate the current through the 1.5-H coil.
27-42. Repeat Problem 27-41, but find the current through the 10-kΩ resistor.

27-43. Given the circuit in Fig. 27-45 with $\omega = 1 \times 10^3$ rad/s, solve for the current through the 4-mH coil, using a Thévenin equivalent circuit as seen by the coil.

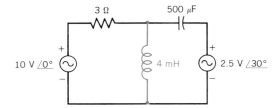

FIGURE 27-45
Circuit for Problems 27-43 through 27-45.

27-44. Repeat Problem 27-43, using a Norton equivalent circuit.
27-45. Given the circuit in Fig. 27-45, solve for the current through the 4-mH coil, using the principle of superposition. (First short one source and determine the load current, then short the other source and again determine the load current. Finally, combine the load currents.)
27-46. For the bridged-T network in Fig. 27-46, calculate the current through the 4 + $j4$-kΩ load, using a Thévenin or Norton equivalent for the rest of the circuit. (Note that this problem cannot be done by using series-parallel circuit theory.)

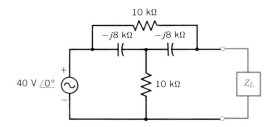

FIGURE 27-46
Circuit for Problem 27-46.

27-47. Given the circuit in Fig. 27-47, use a Norton equivalent circuit to find the current through the $-j10$ Ω load.
27-48. Repeat Problem 27-47, using Thévenin's theorem.

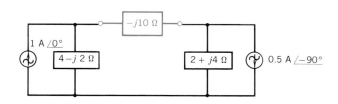

FIGURE 27-47
Circuit for Problems 27-47 and 27-48.

27-49. The ac bridge circuit of Fig. 27-48 balances when Z_1 is a 2.2-kΩ resistor, Z_2 is a 0.1-μF capacitor in series with a 100-kΩ resistor, Z_3 is a 5.1-kΩ resistor, and $f = 1$ kHz. Given that balance occurs when $Z_1 Z_4 = Z_2 Z_3$, determine the series components that make up Z_4.
27-50. The ac bridge circuit of Fig. 27-48 balances when Z_2 is a 15-kΩ resistor, Z_3 is a 6.8-kΩ resistor, Z_4 consists of a 0.015-μF capacitor in parallel with a 2.2-MΩ resistor, and $f = 1$ kHz. Determine:
   a. The series components that make up Z_1.
   b. The parallel components that make up Z_1.

27-51. An ac bridge used to determine accurately the inductance of a coil is called a Maxwell bridge. It has the form shown in Fig. 27-48, where Z_1 consists of a variable resistor (R_1) in parallel with a capacitor (C_S), Z_2 is a variable resistor (R_2), Z_3 is a fixed resistor (R_3), and Z_4 is the unknown inductance (L_x) in series with the unknown resistance (R_x). Given that at balance $Z_1Z_4 = Z_2Z_3$, derive the equations for R_x and L_x in terms of the known components. (*Hint:* Use $Z_4 = Y_1Z_2Z_3$, then equate real components to each other and imaginary components to each other.)

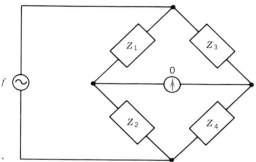

FIGURE 27-48
Circuit for Problems 27-49 through 27-51.

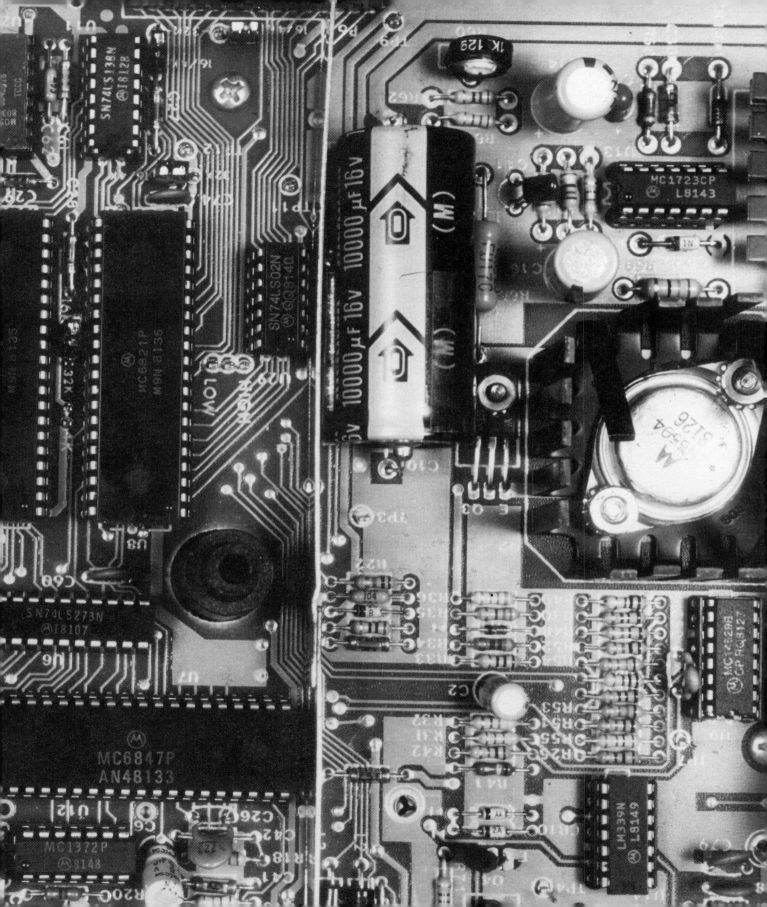

PART III

BASIC ELECTRONICS

CHAPTER 28

DIODES, RECTIFICATION, AND FILTERING

A diode is a device that permits current to flow through it in only one direction. The direction of flow depends on the way the diode is connected with respect to the polarity of the applied voltage. Diodes are extensively used in power supplies to convert alternating current to direct current. In this application, they are known as *rectifiers*. There are a number of nonrectifying uses for diodes, as well, in radio and television sets, computers, and industrial electronics. One example is the *light-emitting diode* (LED) that emits light when current passes through it.

Most modern solid-state diodes are made from the semiconductor material *silicon*. When an extremely pure wafer of silicon is modified by the addition of selected impurities, a *PN junction* is formed. When the PN junction is *forward-biased* by an external voltage (positive to *P,* negative to *N*), current will flow easily through the diode. The diode is then said to be *forward-biased* and *conducting.* Under *reverse-bias* conditions (positive to *N,* negative to *P*), almost no current will flow through the diode.

The conversion of alternating current to direct current can be accomplished in either *half-wave* or *full-wave* diode rectifying circuits. But since this leaves an appreciable *ripple* in the output voltage, *filtering* is usually added. This may be accomplished by connecting a large capacitor in parallel with the load to smooth the output voltage, or by using some combination of inductance and capacitance to produce an even-smoother dc output voltage.

28-1 PURE SILICON

As shown in Fig. 28-1, an atom of pure silicon has 14 electrons in orbit around its positive nucleus. The outer (valence) shell has four electrons. If this valence shell could obtain four more electrons, it would be complete, and strong *valence electron bonding* would result. This is how a crystal of pure silicon is structured, by the process called *covalent bonding*, as shown in Fig. 28-1b.

The central silicon atom in Fig. 28-1b is shown as *sharing* electrons from each of four neighboring silicon atoms. Each of the four neighboring atoms, in turn, is sharing one of the electrons of the central atom's valence shell for bonding purposes. Thus, none of the atoms *exclusively* has eight electrons in its valence shell, but shares electrons with neighboring atoms to fill its valence shell and form covalent bonds. The result is a crystal of pure silicon.

28-1.1 Pure Silicon at Room Temperature

If a silicon crystal were at absolute zero ($-273°C$), all the valence electrons would remain in the outer (valence) shell. Since no electron would have enough thermal energy to escape and become a *free electron*, the material would be a perfect insulator.

At room temperature, however, some of the valence electrons gain enough thermal energy to escape and break the covalent (shared) bond. They become *free* electrons, and are thus available to provide conduction (much like the free electrons in a copper wire). The number of free electrons in silicon is very small, however (only about 1 for every 10^{12} atoms), so pure silicon has a very high resistivity of 2300 Ω-m at room temperature. (For comparison, the ρ for copper, a good conductor, is 1.72×10^{-8} Ω-m at the same temperature.)

When an electron breaks a bond, it leaves a *vacancy* behind. Such an incomplete (broken and unshared) covalent bond is called a *hole*. This is shown in Fig. 28-2, where the hole is given a positive sign ($+$) to show the *absence* of an electron.

The important point to remember about holes is that they can contribute to current much as the free electrons can. For every electron that is liberated by thermal energy (becomes "free"), a hole is produced. You could refer to this combination as a *thermally generated electron-hole*

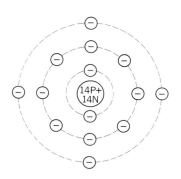

(*a*) Bohr concept of an atom of silicon.

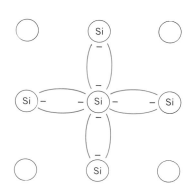

(*b*) Covalent bonding of silicon atoms
  form part of a crystal of pure silicon.

FIGURE 28-1
Diagrams representing atomic and crystal structure of silicon.

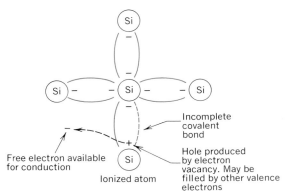

(a) Production of electron-hole pairs in pure silicon at room temperature

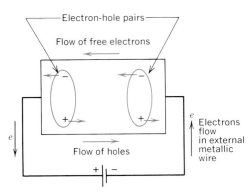

(b) Total current in pure silicon is due to the sum of electron and hole flow arising from thermally generated electron-hole pairs and applied voltage.

FIGURE 28-2
Pure silicon at room temperature.

pair. As the temperature increases, so does the number of electron-hole pairs.

28-1.2 Current in Pure Silicon

When a dc voltage is connected across a piece of pure silicon, a force acts on *both* the free electrons and the holes, as shown in Fig. 28-2*b*.

It is fairly easy to visualize the flow of free electrons in the direction shown, but how can a hole flow in the opposite direction?

A valence electron in a neighboring atom can leave its covalent bond quite easily to fill a hole. When the free electron leaves its bond, a hole is left behind—the hole has effectively moved in the *opposite* direction. This process continues, with the holes moving to the right (in the illustration) as electrons continue to move to the left. This mechanism does *not* involve the movement of *free* electrons; it is represented as the movement of positive charge or holes *only*. (This is supported by the Hall effect described in Chapter 11.) This motion of charge *(hole flow)* is in *addition* to that due to free electrons. As a result, the *total* current flowing in pure silicon is the *sum* of the free electron flow *and* the hole flow.

External to the silicon itself (when the silicon is placed in a circuit), only electrons can flow through the copper conducting wires. The electrons alone must then provide as much charge flow as the sum of the free electrons and hole flow in the silicon. Also, the current that flows through pure silicon is very temperature-de-

pendent, increasing with a rise in temperature, and vice-versa.

28-2 *N*-TYPE SILICON

Pure (intrinsic) silicon has very low conductivity, and thus is practically an insulator. The conductivity can be increased significantly, making the silicon a semiconductor, by the addition of certain materials known as *impurities*. The silicon is then called *extrinsic,* or an impure *(doped)* semiconductor. The added impurities are either *pentavalent* or *trivalent* atoms.

Phosphorus is a *pentavalent* material—it has *five* electrons in its outer valence shell. When a small percentage of phosphorus is added to pure silicon, the phosphorus atoms displace some of the silicon atoms, forming the arrangement shown in Fig. 28-3*a*. Since only four of the five electrons are needed for covalent bonding, one electron is *donated* to the crystalline structure. At room temperature, this fifth (unbonded) electron is easily detached from its parent atom to become a free electron. The phosphorus impurity is called a *donor N*-type *impurity* material. The atom itself is ionized, acquiring a bound positive charge that is not free to move.

The doped silicon is called *N* type *(negative* type), since it now has many more free electrons than were available in the pure silicon. Also, *N*-type silicon has many more free electrons than holes. Its resistivity is much lower, and it is capable of generating many more free electrons at room temperature than pure silicon.

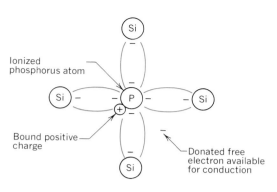

(a) Portion of a silicon crystal with a phosphorus atom substituted for a silicon atom to donate free electrons.

(b) Current in N-type silicon is due to electrons (majority carriers) and holes (minority carriers).

FIGURE 28-3

N-type silicon resulting from doping with pentavalent phosphorus impurity atoms.

28-2.1 Current in *N*-Type Silicon

Figure 28-3b represents the large increase in the number of free electrons available for conduction as a result of doping. Thermal agitation still causes electron-hole pairs, just as it does in pure silicon. But because of the abundance of donor electrons, the recombination of holes and electrons occurs rapidly (holes are filled by free electrons). Thus, there are even fewer holes in N-type silicon than in pure silicon.

The silicon is called N type (for negative) because the free electron density is predominant (over holes), even though the crystal as a whole is still electrically neutral. (Recall that *every* atom is electrically neutral, with balanced positive and negative charges. The pure silicon and pure phosphorus atoms are *each* electrically neutral.)

When an electromotive force is connected across the N-type silicon, both electrons and holes flow as before. But this time, the current is due primarily to the free donor electrons, so they are called the *majority current carriers*. The holes are termed the *minority current carriers*, since hole flow (in the opposite direction to electron flow) due to valence electron motion is very small by comparison.

28-3 *P*-TYPE SILICON

Now, consider the addition of aluminum, a *trivalent* impurity, to silicon. Since there are only three electrons in the aluminum atom's valence shell, the substitution of an aluminum atom for a silicon atom leaves an *incomplete* bond. This bond is usually completed by a neighboring

silicon atom giving up an electron. The aluminum atom thus acquires a bound negative charge, and a hole is left where the electron came from. (See Fig. 28-4a.)

Since the aluminum atom has created holes that can accept electrons, aluminum is known as an *acceptor* impurity. **The aluminum impurity has caused the *P*-type silicon to have many more holes than free electrons.**

28-3.1 Current in *P*-Type Silicon

Since holes predominate in P-type silicon, they are now the *majority* current carriers. The relatively small number of free valence electrons generated by thermal agitation are now the *minority* current carriers. The holes, being positive, account for the name "P-type" silicon. The motion of these current carriers when an emf is applied is shown in Fig. 28-4b. The total current is the sum of the majority hole flow and the minority electron flow. Like N-type silicon, P-type silicon is electrically neutral.

28-4 *PN* JUNCTIONS

Most modern PN junction diodes are manufactured by a diffusion process. They are made by placing a wafer of N-type silicon and an oxide of the impurity material (such as aluminum) in a furnace and raising the temperature to several hundred degrees Celsius. At high temperature, the impurity material becomes a gas and diffuses into the N-type silicon. This forms a thin P-type layer, resulting in a PN junction. Metal ohmic contacts are applied to the P-type

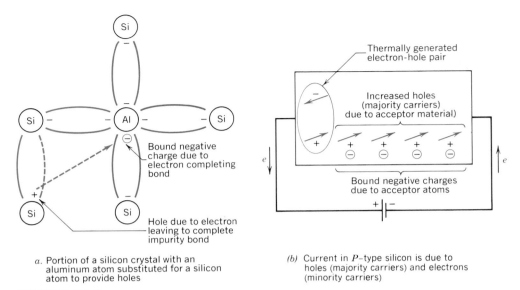

a. Portion of a silicon crystal with an aluminum atom substituted for a silicon atom to provide holes

(b) Current in *P*-type silicon is due to holes (majority carriers) and electrons (minority carriers)

FIGURE 28-4

P-type silicon resulting from doping with trivalent aluminum impurity atoms.

and *N*-type areas of the diode, and some form of encapsulation is provided to protect the relatively brittle silicon wafer.

The *PN* junction is *not* formed by joining together pieces of *P*- and *N*-type materials. However, to better understand what takes place at the *PN* junction, imagine that you have placed an *N*-type semiconductor and a *P*-type semiconductor side-by-side to "make" a *PN* junction. Since both the *P*- and *N*-type materials are electrically neutral, the *PN* junction is neutral, as well. A *PN* junction is shown in Fig. 28-5.

Note that each free electron in the *N*-type region neutralizes the bound positive charge of the impurity donor ions; in the *P*-type region, holes neutralize the bound negative charges.

At the moment of "joining together," some of the free electrons in the *N*-type region diffuse across the junction to the *P*-type region and combine with holes. Similarly, holes diffuse from the *P*-type to the *N*-type region and combine with electrons. This process establishes *unneutralized* bound negative and positive ions on opposite sides of the *junction*. These charges, referred to as *uncovered* charges, mean that the *P*-type region has acquired a net negative charge; the *N*-type region an equal net positive charge. This is shown in Fig. 28-5c.

Because of the diffusion of carriers across the junction, a potential barrier of about 0.7 V is established. Equilibrium develops because the *N*-type region, having become

positive, will prevent any further diffusion of holes from the *P*-type region. Similarly, the *P*-type region, with a negative charge, will repel any further diffusion of free electrons. A very narrow region (about 0.5 μm thick) that is depleted of mobile charges now exists around the *PN* junction. This is referred to as the *depletion* or *space charge* region. The *PN* junction, as you shall see, is actually a *diode* that permits easy conduction when forward-biased and practically no conduction when reverse-biased.

28-5 FORWARD-BIASED *PN* JUNCTION

Now, consider the application of an emf across the *PN* junction or *diode* with the positive side to the *P* type and the negative side to the *N* type. This is called *forward-bias*. (Refer to Fig. 28-6.) Forward bias tends to lower the potential on the *N* side and raise it on the *P* side, thus *lowering* the potential barrier. This allows the holes to *drift* across the junction once more from *P* to *N*, and the free electrons to drift in the reverse direction, from *N* to *P*.

The current across the junction is due primarily to these majority carriers, the holes in the *P* type and the electrons in the *N* type. The total current is due to the sum of the two flows. There is also a very small additional current due to the minority carriers (electrons in the *P* type and

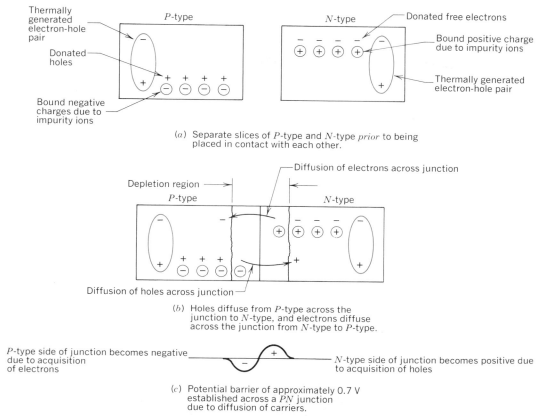

FIGURE 28-5

Development of a potential barrier when a *PN* junction is formed.

holes in the *N* type). This minority current is small because of the tiny number of minority carriers. Under forward bias, the depletion region also narrows somewhat—an important effect if the capacitance of the junction is considered.

The diode is considered to have a *low* **resistance under** *forward-bias* **conditions.** If the forward voltage is increased, the potential barrier is further lowered, and the depletion region narrows still further, allowing even more current to flow. The *conventional* direction of forward current (+ to −) is shown in Fig. 28-6*b* to be the *same* as the arrow in the symbol for the diode.

28-6 REVERSE-BIASED *PN* JUNCTION

When the external voltage is applied with opposite polarity to the diode (positive to *N* type, negative to *P* type), a condition of *reverse bias* is said to exist. Figure 28-7 clearly shows how the majority carriers in each side are pulled away from the junction. Since this uncovers more bound positive and negative ions, there is an *increase* in the potential barrier and the depletion region *widens*.*

At first, it would seem that *no* current would flow at all under a reverse-bias condition. Certainly, no *majority* carriers can cross the junction. The *minority* carriers, however, are not affected by the potential barrier. In fact, so far as the minority carriers are concerned, the *PN* junction is *forward* biased! A small reverse current (described as a *leakage* current) flows across the junction.

The reverse-biased minority current is strongly temper-

---

*The charges stored on opposite sides of the depletion zone cause a *capacitive* effect. This *junction capacitance* varies with the reverse voltage connected across the *PN* junction and is used to advantage in *varicap* or *varactor* diodes in communications tuning circuits.

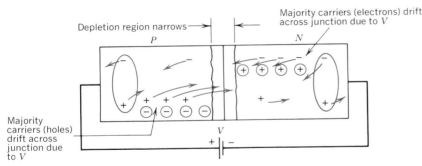

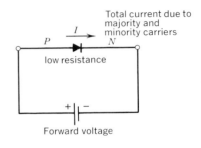

(a) Due to V lowering the potential barrier, the majority carriers can drift across the junction.

(b) Direction of conventional current for a forward-biased junction showing the symbol used for a diode

FIGURE 28-6
Forward-biased *PN* junction diode.

ature-dependent, because the minority carriers are due to thermally generated electron-hole pairs. This leakage current is on the order of only a few microamperes for low-current diodes. **Thus the diode is said to have a *high* resistance when *reverse* biased. The diode effectively *blocks* current compared with forward bias.**

28-7   DIODE *VI* CHARACTERISTICS

The forward- and reverse-bias properties of a diode are best shown in a plot of their current-voltage characteristics. A typical curve for a silicon diode is shown in Fig. 28-8.

As shown in Fig. 28-8, a very low, forward-bias voltage of approximately 0.6 V is necessary in the forward direction before the diode begins to conduct. This is often

referred to as the *cut-in* (or *knee* or *threshold*) voltage. It is the voltage needed to reduce the potential barrier sufficiently to allow majority carriers to begin migrating across the junction.

Beyond the cut-in voltage, the current increases sharply or exponentially, even with very small voltage increases. At high-current levels with relatively low voltages, the curve becomes a straight line, approximating the voltage drop across a resistor. In this region (above the cut-in voltage), the forward-biased diode can be thought of as a closed switch, having a low resistance, with approximately a 0.7-V drop across it.

When reverse biased, even with high reverse voltages, the diode is effectively an open switch, with a high resistance, and a leakage current of only a few microamperes. In most rectifier diodes, this current tends to increase *slightly* with a large increase in the reverse voltage, due to surface currents flowing across the *PN* junction.

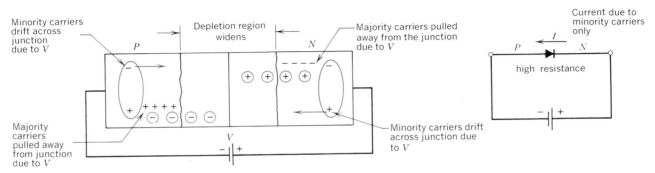

(a) Minority carriers drift across reverse-biased junction constituting a small leakage current.

(b) Direction of leakage current in reverse-biased junction.

FIGURE 28-7
Reverse-biased *PN* junction diode.

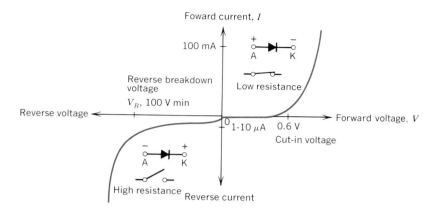

FIGURE 28-8

Forward and reverse *VI* characteristics of a silicon diode.

However, there is a value of reverse voltage at which there will be a significant increase in reverse current. This is called the *reverse breakdown* voltage (V_B). If the diode is operated beyond this point, it is said to have broken down, since it is no longer effectively blocking the current.

One explanation of breakdown is found in the *avalanche* effect. This takes place when the minority carriers that constitute the reverse current gain enough kinetic energy to knock bound electrons out of the covalent bonds. These electrons, in turn, collide with other atoms and dislodge other electrons, increasing the number of carriers available for reverse conduction.

This reverse voltage capability of a diode is often referred to as the peak inverse voltage (PIV) of the diode. Depending upon the type of diode, the peak inverse voltage may range from a minimum of about 100 V to several thousand volts.

In rectifier applications, the PIV must not be exceeded or permanent damage to the diode can result due to overheating of the *PN* junction. If no damage has occurred, the diode can be brought back to normal blocking operation by reducing the reverse bias below the breakdown level. If the reverse current is limited to a nondestructive value, the breakdown voltage can be used to advantage in *Zener* diodes.

28-7.1 Zener Diodes

In some diodes, breakdown can be made to occur very abruptly at accurately known values ranging from 2 to 200 V. (See Fig. 28-9a.) These are known as *Zener* diodes. Since their breakdown (or Zener) voltage is maintained over a wide range of current, they can be used in voltage regulator circuits, as shown in Fig. 28-9b.

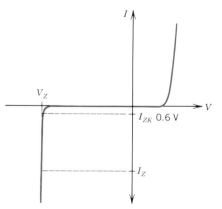

(a) Zener diode *VI* characteristic

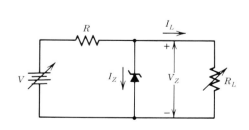

(b) Use of a Zener diode in a voltage-regulator circuit

FIGURE 28-9

Zener diode characteristics and application as a voltage regulator.

In this example, the reverse Zener diode voltage (V_Z) is also the voltage across the load (R_L). If either the input voltage or the load current changes, a different amount of current (I_Z) will flow through the Zener diode. But since V_Z changes very little when I_Z changes, a nearly constant load voltage will result. The product of I_Z and V_Z must be limited to the power dissipation capability of the particular Zener diode. Also, I_Z must not be allowed to drop below some minimum value (I_{ZK}), in order to maintain voltage regulation. Note that the Zener diode is used only in its reverse breakdown region and is not used for rectification.

28-7.2 Checking a Diode with an Ohmmeter

The terminals of a diode are often referred to as the *anode* and the *cathode*. These terms are carryovers from the electron tube diode, and refer to the P side and the N side, respectively. Figure 28-10a shows *four* types of diode encapsulation or marking used to determine diode polarity. In all cases, the P-type side is the anode and the N-type side is the cathode.

An ohmmeter is used to determine a diode's polarity,

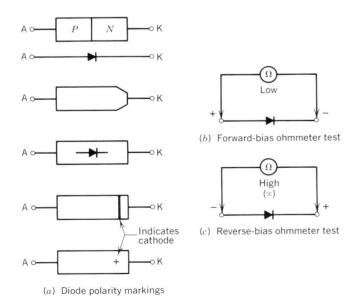

(a) Diode polarity markings

(b) Forward-bias ohmmeter test

(c) Reverse-bias ohmmeter test

FIGURE 28-10

Showing how to determine the anode and cathode from the physical appearance of four types of diodes, and how an ohmmeter can be used to check a good diode.

as well as checking the device to determine whether it is good, open, or shorted. The ohmmeter is essentially a voltage source in series with a resistor and an ammeter. A low resistance is indicated on the meter when forward-biasing the diode (Fig. 28-10b). A reading of close to infinity should result when the ohmmeter leads produce reverse bias (Fig. 28-10c). If *both* readings are low, the diode is shorted; if both are high, the diode is open.

Two precautions must be observed when using an ohmmeter with diodes. First, if you are using the meter to check the polarity of the diode, make sure that you know the polarity of the ohmmeter leads. Many VOMs, in the ohmmeter mode, use the COMMON as the negative and the OHMS as a positive terminal. The way that the batteries are inserted, however, may reverse this polarity (making COMMON positive and OHMS negative). You can check the ohmmeter lead polarity with a voltmeter, if in doubt.

The second precaution is related to the ohmmeter function of some electronic multimeters. Quite often, when the meter is set on the X1 range, a "good" diode will produce a reading indicating that it is open. This occurs because the cut-in voltage has not been exceeded. For a proper check, the meter should be switched to a higher (middle) ohms range.

Diodes are available in a wide range of voltage and current ratings. The silicon diodes shown in Fig. 28-11 range from low-current (1 A) types to the stud-mounted and hockey puck types capable of handling several hundred amperes.

28-8 THE GERMANIUM DIODE

Although most diodes used for rectification are made from silicon, some applications still require the use of germanium diodes. Germanium is a semiconductor with a resistivity lower than that of silicon. It has four electrons in its valence shell, so doping is achieved in ways similar to those used for silicon. Pentavalent materials such as arsenic and antimony are used for N-type germanium; trivalent materials such as gallium and indium are used for P-type germanium.

The *VI* characteristics for a germanium diode have the same general shape as those for silicon, but with two important differences. Germanium, with a potential barrier of 0.2–0.4 V, has a cut-in voltage of 0.2 V (compared to 0.6 V for silicon). Germanium also has a much higher (a minimum of 10 times greater) reverse current than a silicon diode. Both are affected by temperature in the same man-

FIGURE 28-11

Selection of silicon diodes showing low current, stud, and hockey puck types. (Courtesy of International Rectifier.)

ner, but silicon diodes can be operated with their junctions as high as 175°C, compared to a maximum of 85 to 100°C for germanium.

28-9 HALF-WAVE RECTIFICATION

Most electronic equipment must operate from a dc power source. Although batteries can be used in portable equipment, an electronic *power supply* is often included to allow operation from the 120-V, 60-Hz supply. The power supply converts the 120-V ac into direct current at a voltage suitable for use by the equipment. Since a diode is sensitive to polarity, it can be used to rectify the alternating voltage waveform into a unidirectional waveform. The average value of this waveform is the dc voltage value.

The simplest form of rectifier involves use of a single diode, as shown in Fig. 28-12a. Since many semiconductor (transistor) circuits operate at a relatively low voltage (6–20 V), a *transformer* is used to step down the ac voltage. As an example, assume that the secondary voltage is 6.3 V rms. Here, R_L represents the resistance of the load to which dc power is to be delivered, such as a radio, tape recorder, or amplifier. The input voltage to the diode is represented by a sine wave of peak value given by $V_m = 6.3\sqrt{2}$ V ≈ 9 V, shown in Fig. 28-12b.

On the positive half-cycle of V_s, diode D_1 is forward-biased and conducts, allowing current to flow through the load (R_L). As a simplification for this example, ignore the cut-in voltage of the diode and its resistance. In this case,

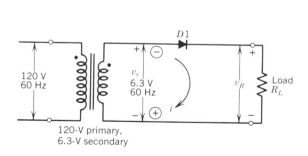

(a) Half-wave rectifier circuit. PIV = V_m.

FIGURE 28-12

Half-wave rectification.

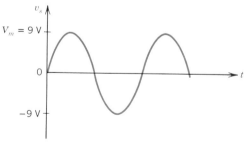

(b) Ac input to rectifying circuit.

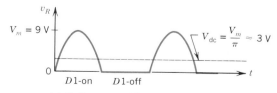

(c) Half-wave rectified output across load.

the voltage across the load will be a half sine wave with a peak value of 9 V, as shown in Fig. 28-12c.

When the input voltage *reverses* in polarity (shown by the circled polarity signs), the diode is *reverse-biased*. This blocking action means that no current flows and the voltage across the load is zero for one-half cycle. (Again, neglect the small reverse current that flows.) The load voltage will not increase toward the 9-V peak again until the input voltage reverses again.

This half-wave rectified output across the load is sometimes called *unidirectional pulsating* voltage. (See Fig. 28-12c.) Although the waveform is not smooth, it contains direct current and is satisfactory for some applications (such as battery chargers and some dc motors).

The average value of the half-wave rectified waveform can be shown to be

$$V_{dc} = \frac{V_m}{\pi} \qquad \text{volts} \qquad (28\text{-}1)$$

where: V_{dc} is the average or dc voltage, in volts (V)

V_m is the peak of the input voltage, in volts (V)

Note that V_{dc} is the reading of a dc voltmeter connected across the load (R_L). If a dc ammeter is connected in series with the load, it will indicate the average value of the current waveform that has the same wave shape as the voltage. This current is given by

$$I_{dc} = \frac{I_m}{\pi} \qquad \text{amperes} \qquad (28\text{-}2)$$

or by

$$I_{dc} = \frac{V_{dc}}{R_L} \qquad \text{amperes} \qquad (28\text{-}3)$$

where: I_{dc} is the average or dc current, in amperes (A)

I_m is the peak of the load current, in amperes (A)

R_L is the load resistance, in ohms (Ω)

EXAMPLE 28-1

A half-wave rectifier has an input voltage of 6.3 V ac, and a load of 220 Ω. Assuming an ideal diode (neglecting diode losses), calculate:

a. The dc voltage across the load.
b. The peak current through the load.
c. The reading of a dc ammeter in series with the load.
d. The dc power delivered to the load.

Solution

a. $V_m = 6.3 \text{ V} \times \sqrt{2} = 8.9 \text{ V}$ (15-18)

$$V_{dc} = \frac{V_m}{\pi} \qquad (28\text{-}1)$$

$$= \frac{8.9 \text{ V}}{\pi} = \textbf{2.8 V}$$

b. $I_m = \dfrac{V_m}{R_L}$ (15-17)

$$= \frac{8.9 \text{ V}}{220 \ \Omega} = \textbf{40.5 mA}$$

c. $I_{dc} = \dfrac{I_m}{\pi}$ (28-2)

$$= \frac{40.5 \text{ mA}}{\pi} = \textbf{12.9 mA}$$

d. $P_{dc} = V_{dc}I_{dc}$ (3-5)

$$= 2.8 \text{ V} \times 0.0129 \text{ A} = \textbf{0.036 W}$$
$$= \textbf{36 mW}$$

Note that when the diode is reverse-biased, it is subjected to a peak inverse voltage of V_m or 8.9 V in Example 28-1A. In high-voltage power supplies, this is an important item to check when selecting the diode.

28-9.1 Application of a Half-Wave Rectifier*

An interesting application of a half-wave rectifier is marketed as a life extender for incandescent bulbs. The device, in the form of a 1-in.-diameter disc, uses a microchip diode to convert ac to dc when placed in the lamp socket. (Also available are incandescent lamps with the diode built-in.) It is claimed that the life of the lamp is extended 10 to 100 times normal by reducing the two primary causes of failure: filament shock and bulb heat.

This is achieved by the half-wave rectified output delivering up to 35% less power (rms) to the lamp. As a result, the average temperature of the filament is reduced from about 2500°C to about 1700 to 1800°C. In addition, the filament is subjected to peak power dissipation only 60 times per second with the dc, compared to 120 times per second with the ac.

Of course, the reduced power delivered to the lamp results in less light output, so a larger lamp may be necessary to maintain the same light level. At first, it might seem that the rms power delivered to the lamp with half-wave rectification should be one-half the power delivered with a full sine wave. But the lower operating temperature of the filament means that the filament's resistance is also

---

*This section can be omitted with no loss of continuity.

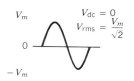

(a) Sine wave

(b) Half-wave rectified sine wave

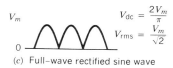

(c) Full-wave rectified sine wave

FIGURE 28-13
Rms and dc values for various waveforms.

lower. This tends to increase the lamp's current, so power dissipation is reduced less than 50%. Note that operation of the lamp at lower temperature, while extending lamp life, does not make the lamp more efficient. Less light (as measured in lumens) is produced per watt at a lower power level.

It should be noted that although the current flowing through the lamp is dc (there is no reversal of polarity), the total power delivered to the lamp (thus producing heat and light) is *rms* power and not just dc power. That is, 41% of the ac power delivered to a half-wave rectifier is converted to dc power; the rest exists in the load as ac power. To obtain the total rms power, it is necessary to know that the rms value of a *half* sine wave is given by $V_{rms} = V_m/2$. This is the reading that a true rms voltmeter will indicate when connected across a half-wave rectified sine wave. Recall that the rms voltage is an indication of the total heating effect of a waveform; the dc voltage is simply an average of the waveform. (See Fig. 28-13.)

EXAMPLE 28-2

A diode is connected in series with a 60-W incandescent bulb and a 120-V (rms), 60-Hz supply.

a. Calculate the normal hot resistance of the 60-W lamp when operating at 120 V.

b. Determine the rms voltage of the half-wave rectified output across the lamp.

c. Assuming the hot resistance of the filament is directly proportional to the rms voltage across the lamp, determine the resistance of the lamp when operating with the diode in place.

d. Calculate the rms power delivered to the lamp when operating with the diode and find the percentage reduction in power.

e. Determine the peak inverse voltage to which the diode is subjected.

Solution

a. $P = \dfrac{V^2}{R}$ (15-13)

$R = \dfrac{V^2}{P}$

$= \dfrac{(120\ V)^2}{60\ W} = \mathbf{240\ \Omega}$

b. $V_{rms} = \dfrac{V_m}{2}$

$= \dfrac{120\ V \times \sqrt{2}}{2} = \mathbf{85\ V}$

c. At 120 V, $R = 240\ \Omega$

At 85 V, $R = 240\ \Omega \times \dfrac{85\ V}{120\ V}$

$= \mathbf{170\ \Omega}$

d. $P_{rms} = \dfrac{V_{rms}{}^2}{R}$

$= \dfrac{(85\ V)^2}{170\ \Omega} = \mathbf{42.5\ W}$

See Fig. 28-14.

Percentage reduction in power $= \dfrac{60 - 42.5\ W}{60\ W} \times 100\%$

$= \dfrac{17.5}{60} \times 100\% = \mathbf{29\%}$

e. $P.I.V. = V_m$

$= 120\ V \times \sqrt{2}$

$= \mathbf{170\ V}$

Note that the light output in Example 28-2 would be reduced significantly. If, however, a 75-W lamp were used with the diode, the power delivered to the lamp would be 53 W, providing only 10% less light (approximately) than the 60-W lamp, but extending the life considerably, as well as reducing power consumption by 7 W.

It is interesting to observe, in Fig. 28-14, that both circuits draw a current of 0.5 A rms. However, the peak current is only 0.707 A with the ac supply, but 1 A with the half-wave rectifier. Similarly, the peak power supplied to the lamp's filament is 120 W with a pure sine wave, but

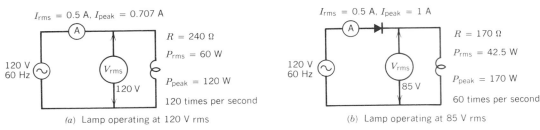

$I_{rms} = 0.5$ A, $I_{peak} = 0.707$ A

A

120 V
60 Hz

V_{rms}

120 V

$R = 240 \ \Omega$

$P_{rms} = 60$ W

$P_{peak} = 120$ W

120 times per second

(a) Lamp operating at 120 V rms

$I_{rms} = 0.5$ A, $I_{peak} = 1$ A

A

120 V
60 Hz

V_{rms}

85 V

$R = 170 \ \Omega$

$P_{rms} = 42.5$ W

$P_{peak} = 170$ W

60 times per second

(b) Lamp operating at 85 V rms

FIGURE 28-14
Circuit diagrams for Example 28-2.

170 W with the half-wave rectified sine wave. This is because somewhat more than half of the 60 W (42.5 W) must be supplied in only half the time with the diode in use. In this sense, filament ''shock'' is not reduced; however, the peaks occur only 60 times a second with the diode, compared to 120 times a second without.

Finally, note that the rms power (60 W) delivered to the circuit in Fig. 28-14a *is* the product of the applied rms voltage (120 V) and rms current (0.5 A), but this is *not* the case in Fig. 28-14b. Power is the product of voltage and current only if the two are in phase with each other. But you know that the effect of the diode is to prevent any current flowing for half a cycle of the applied sine wave of voltage, so the current and voltage are not in phase for a *complete cycle*. Across the lamp itself, however, the voltage waveform with an rms value of 85 V *is* in phase with the current waveform with an rms value of 0.5 A, so the product gives an rms power of 42.5 W.

28-10 FULL-WAVE RECTIFICATION

Figure 28-15a shows how a second diode can be combined with the first to provide full-wave rectification. This circuit also requires a transformer with a *center-tapped* secondary for a common return path. The line-to-line secondary voltage in Fig. 28-15a is 12.6 V, so that the line-to-center-tap voltage is 6.3 V, as shown. This circuit is a combination of two half-wave rectifier circuits, with one diode rectifying each half-cycle of the input voltage.

On the positive half-cycle of v_s, D_1 conducts and D_2 is reverse-biased and open. On the *negative* half-cycle (polarity signs shown circled), D_1 is open and D_2 conducts. The voltage waveform across the load is shown in Fig. 28-15c, where once again, ideal diodes are assumed.

It should be evident that the *average* of the full-wave rectified output voltage is now doubled, compared to half-wave rectification:

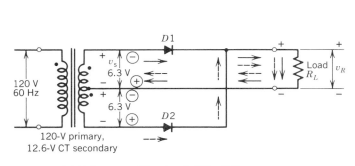

(a) Full-wave rectifier circuit. PIV = 2 V_m.

FIGURE 28-15
Full-wave rectification.

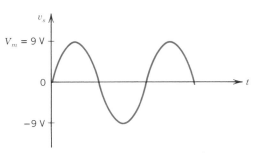

(b) Ac input to rectifying circuit.

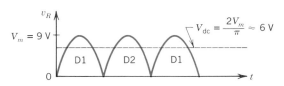

(c) Full-wave rectified output across load.

$$V_{dc} = \frac{2V_m}{\pi} \quad \text{volts} \qquad (28\text{-}4)$$

$$I_{dc} = \frac{2I_m}{\pi} \quad \text{amperes} \qquad (28\text{-}5)$$

where: V_{dc} is the average or dc voltage, in volts (V)
V_m is the peak of the input voltage, in volts (V)
I_{dc} is the dc current in the load, in amperes (A)
I_m is the peak of the load current, in amperes (A)

EXAMPLE 28-3

A 120-V primary, 12.6-V center-tapped secondary transformer is used in a two-diode, full-wave rectifier with a 220-Ω load. Neglecting diode losses (assume ideal diodes), calculate:

a. The dc (average) voltage across the load.
b. The peak current through the load.
c. The reading of a dc ammeter in series with the load.
d. The dc power delivered to the load.

Solution

a. $V_m = 6.3 \text{ V} \times \sqrt{2} = 8.9 \text{ V}$ (15-18)

$V_{dc} = \frac{2V_m}{\pi}$ (28-4)

$= \frac{2 \times 8.9 \text{ V}}{\pi} = \textbf{5.7 V}$

b. $I_m = \frac{V_m}{R_L}$ (15-7)

$= \frac{8.9 \text{ V}}{220 \text{ }\Omega} = \textbf{40.5 mA}$

c. $I_{dc} = \frac{2I_m}{\pi}$ (28-5)

$= \frac{2 \times 40.5 \text{ mA}}{\pi} = \textbf{25.8 mA}$

d. $P_{dc} = V_{dc}I_{dc}$ (3-5)
$= 5.7 \text{ V} \times 0.0258 \text{ A} = \textbf{0.147 W}$

These answers should be compared to those for Example 28-1, which uses the same value of R_L and ac input voltage.

The peak inverse voltage of each diode in the full-wave rectifier is $2V_m$ or 17.8 V in Example 28-3. This is because the peak voltage of V_m across the load occurs at the same time that the input voltage to the nonconducting diode is a peak. The sum of these two polarity-aiding volt-

ages is the diode's peak reverse voltage. (That is, the PIV is the peak voltage across the full secondary).

28-10.1 Bridge Rectifier

If *four* diodes are used in a *bridge* configuration, there is no need for a *center-tapped* (CT) transformer. This arrangement, shown in Fig. 28-16a, produces the same waveform as the full-wave two-diode CT transformer rectifier (Fig. 28-15c). Note that the line-to-line transformer secondary voltage is only 6.3 V. Conduction alternates between diodes D_1 and D_3 for one half-cycle of the input and diodes D_2 and D_4 for the other half-cycle. This causes a PIV for each diode of only V_m, again, the peak voltage across the secondary.

For convenience, commercial bridge rectifiers contain *four* diodes encapsulated in the same package, with two terminals provided for connecting the ac input and two terminals for the dc output. (See Fig. 28-17.)

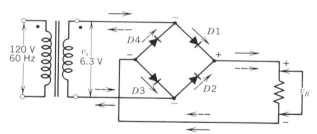

(a) Full-wave bridge rectifier circuit. PIV = V_m.

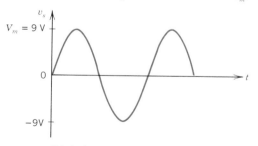

(b) Ac input voltage to rectifying circuit.

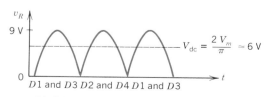

(c) Full-wave rectified output across load.

FIGURE 28-16
Full-wave rectification with a bridge rectifier.

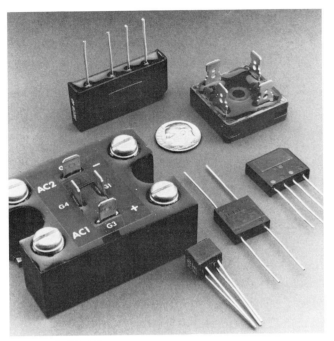

FIGURE 28-17
Encapsulated bridge rectifier packages. (Courtesy of International Rectifier.)

One advantage of the bridge rectifier is greater use of the transformer secondary winding in delivering the current to the load, as compared to either of the other two rectifiers. Therefore, for a given voltampere (apparent power) rating of the transformer, this circuit can deliver *more* dc power to the load than either of the two previous circuits.

That is, a 1-kVA transformer can deliver 287 W of dc power in a half-wave rectifying circuit with no filtering, 693 W in a two-diode *full-wave* rectifying circuit, and 812 W in a *bridge* rectifier circuit.

EXAMPLE 28-4

A 120-V primary, 6.3-V secondary transformer is used in a full-wave bridge rectifier with a 220-Ω load. Neglecting diode losses, calculate:
a. The dc voltage across the load.
b. The dc current through the load.
c. The dc power delivered to the load.
d. The PIV of each diode.

Solution

a. $V_m = 6.3 \times \sqrt{2} = 8.9 \text{ V}$ (15-18)

$$V_{dc} = \frac{2V_m}{\pi}$$ (28-4)

$$= \frac{2 \times 8.9 \text{ V}}{\pi} = \textbf{5.7 V}$$

b. $I_{dc} = \dfrac{V_{dc}}{R_L}$ (3-1a)

$$= \frac{5.7 \text{ V}}{220 \text{ }\Omega} = \textbf{25.9 mA}$$

c. $P_{dc} = V_{dc} I_{dc}$ (3-5)

$$= 5.7 \text{ V} \times 0.0259 \text{ A} = \textbf{0.147 W}$$

d. PIV = V_m = **8.9 V**

These answers should be compared to those from Example 28-3. (Numerical differences are due to rounding-off.)

28-10.2 Three-Phase, Full-Wave Rectifier

An important application of the rectifier is in the electrical system of modern automobiles. Since a three-phase alternator is used to supply the auto's electrical needs, an ac-to-dc conversion circuit is required. Also, because there are three phases, *six* diodes are required for full-wave rectification, as shown in Fig. 28-18*a*. The diodes are usually visible mounted on the alternator frame, which functions as a heat sink to cool the diodes.

Consider the interval from X to Y in Fig. 28-18*c*. In this case, v_{s_1} is positive, which means that current will flow from A, through D1 and the load, and back through D5 into B (v_{s_2} is negative). From Y to Z, v_{s_3} is near its maximum negative value, which means that A is positive with respect to C. Consequently, the current again flows out of A, through D1 and the load, and back through D6 into C. Due to these conduction periods of only 60°, the output voltage never drops below $\dfrac{\sqrt{3}\,V_m}{2}$.

It is apparent from Fig. 28-18*c* that the output voltage has a *higher* average value than any of the previous circuits, and can be shown to be

$$V_{dc} = \frac{3V_m}{\pi}$$ (28-6)

and $$I_{dc} = \frac{3I_m}{\pi}$$ (28-7)

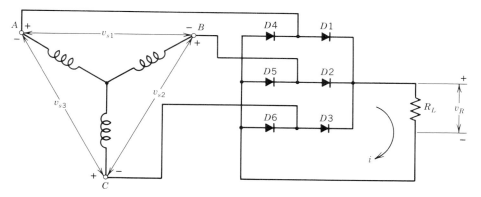

(a) Circuit of a three-phase, full-wave rectifier showing the six diodes.

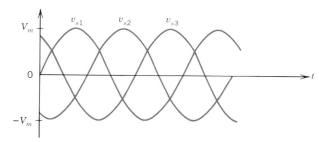

(b) Three-phase input voltage, each phase separated by 120°.

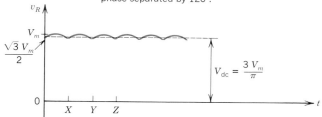

(c) Load voltage waveform showing dc output.

FIGURE 28-18
Circuit and waveforms for a three-phase full-wave rectifier.

Also, the PIV of each diode is only V_m, as in the bridge rectifier.

> **NOTE** The three-phase voltages could be obtained from the secondary windings of three single-phase transformers whose primary and secondary windings would be connected in a wye as shown, or in a delta.

EXAMPLE 28-5

A three-phase, full-wave rectifier must develop a dc output voltage of 14.5 V (typical in an automobile).

a. Determine the necessary rms output voltage from the alternator.

b. What is the peak current through the diodes when the alternator is supplying (after rectification) 30-A dc to the load?

c. Determine the PIV of the diodes.

Solution

a. $V_{dc} = \dfrac{3V_m}{\pi}$ (28-6)

Therefore, $V_m = \dfrac{\pi}{3}V_{dc}$

TABLE 28-1
Comparison of Rectifier Circuits. (No filter connected.)

	Half-Wave	Full-Wave	Bridge	Three-Phase
Secondary voltage line-to-line ($V_m = \sqrt{2}V$)	V	$2V$	V	V
Number of diodes	1	2	4	6
PIV	V_m	$2V_m$	V_m	V_m
V_{dc}	$\dfrac{V_m}{\pi}$	$\dfrac{2V_m}{\pi}$	$\dfrac{2V_m}{\pi}$	$\dfrac{3V_m}{\pi}$
Output waveform				
Ripple frequency	f	$2f$	$2f$	$6f$

$$= \frac{\pi \times 14.5}{3} \text{ V} = 15.2 \text{ V}$$

$$V_{rms} = V_m \times 0.707 \qquad\qquad (15\text{-}17)$$

$$= 15.2 \text{ V} \times 0.707 = \textbf{10.7 V}$$

b. $I_{dc} = \dfrac{3I_m}{\pi}$ (28-7)

Therefore, $I_m = \dfrac{\pi}{3} I_{dc}$

$$= \frac{\pi}{3} \times 30 \text{ A} = \textbf{31.4 A}$$

c. PIV $= V_m = \textbf{15.2 V}$

Characteristics of the four rectifier circuits are compared in Table 28-1.

28-11 CAPACITOR FILTERING

The dc output resulting from single-phase (and even from three-phase) rectification contains too much ac "ripple" for most electronics applications. In an audio circuit, this would show up as a 120-Hz "hum" from a full-wave rectifier circuit (120 Hz is the ripple frequency in the output for a 60-Hz input). A half-wave rectifier circuit has a 60-Hz ripple frequency, but this circuit is seldom used where a very "smooth" output voltage is required.

A simple way to filter out much of the unwanted alternating current is to use a large electrolytic capacitor in parallel with the load. This method, suitable for only light loads, is shown in Fig. 28-19.

The capacitor charges up to the *peak* value of the input voltage and tries to maintain this value (V_m). As the full-wave rectified input drops to zero, the capacitor discharges through R_L until the input voltage again increases to a value greater than the capacitor voltage. At this point, a diode will again be forward-biased and there will be a pulse of current through it, recharging the capacitor.

It is apparent from the output waveform in Fig. 28-19*b* that there is a *ripple voltage* that is said to be "riding on a dc level." For light loads (those with a large R_L), the dc voltage will remain near V_m. However, as the load increases (a decrease in R_L), the discharge of C will be greater, resulting in more ripple and a lower dc output voltage.

28-11.1 Ripple Factor

The success of a filter is measured by the ripple factor (r), which is defined as

$$r = \frac{V_{r,\text{rms}}}{V_{dc}} \qquad\qquad (28\text{-}8)$$

where: r is the ripple factor, dimensionless
$V_{r,\text{rms}}$ is the rms value of ripple voltage, in volts (V)
V_{dc} is the dc output voltage, in volts (V)

For a capacitor filter, the ripple is approximately triangular, with an rms value of

$$V_{r,\text{rms}} = \frac{V_{r,\text{p-p}}}{2\sqrt{3}} \qquad \text{volts} \qquad (28\text{-}9)$$

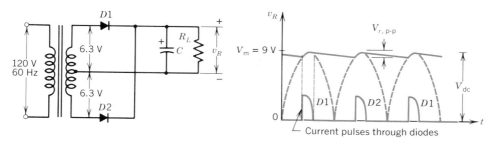

(a) Full-wave rectifier with capacitor filter

(b) Filtered output voltage showing peak-to-peak ripple voltage

FIGURE 28-19

Circuit and waveforms for simple capacitor filter.

where: $V_{r,\text{rms}}$ is the rms value of ripple voltage, in volts (V)

$V_{r,\text{p-p}}$ is the peak-to-peak value of the ripple voltage, in volts (V)

EXAMPLE 28-6

A full-wave rectifier supplies a load of 100 Ω with a filter capacitor of 1000 μF. The output is observed on an oscilloscope and found to have a peak of 9 V and a peak-to-peak ripple of 0.8 V. Calculate:

a. The rms value of the ripple voltage.
b. The dc output voltage.
c. The ripple factor.

Solution

a. $V_{r,\text{rms}} = \dfrac{V_{r,\text{p-p}}}{2\sqrt{3}}$ (28-9)

$= \dfrac{0.8\ \text{V}}{2\sqrt{3}} = \mathbf{0.23\ V}$

b. $V_{\text{dc}} = V_m - \left(\dfrac{V_{r,\text{p-p}}}{2}\right)$

$= 9\ \text{V} - \left(\dfrac{0.8\ \text{V}}{2}\right) = \mathbf{8.6\ V}$

c. $r = \dfrac{V_{r,\text{rms}}}{V_{\text{dc}}}$ (28-8)

$= \dfrac{0.23\ \text{V}}{8.6\ \text{V}} = \mathbf{0.027}$

or expressed as a percent,

$r = \mathbf{2.7\%}$

If the capacitor is recharged in a short time, compared with half of the period, the theoretical value of the ripple factor is given by

$$r = \frac{1}{4\sqrt{3}fR_LC} = \frac{2410}{CR_L} \qquad (28\text{-}10)$$

where: r is the theoretical ripple factor (full-wave), dimensionless

C is the capacitance, in microfarads (μF)

R_L is the load resistance, in ohms (Ω)

f is the supply frequency, 60 Hz

EXAMPLE 28-7

Calculate the theoretical ripple factor from the information given in Example 28-6 and compare it with the measured value.

Solution

$$r = \frac{2410}{CR_L} \qquad (28\text{-}10)$$

$= \dfrac{2410}{1000\ \mu\text{F} \times 100\ \Omega} = 0.024\ \text{or}\ \mathbf{2.4\%}$

This compares favorably with the measured value of 2.7%.

Note that Eq. 28-10 involves the *time constant* R_LC of the circuit in the denominator, so that an increase in C or R_L (or both) reduces the ripple.

Two precautions should be observed in a capacitor filter. First, the large amount of capacitance required means that a polarized capacitor (such as an electrolytic type) must be used. It must be connected with the proper polarity.

Second, any attempt to decrease the ripple simply by increasing the capacitance must be done carefully. A larger capacitor stores more energy, so it needs a larger pulse of current to recharge it. Also, this current must flow during an even shorter interval if there is less ripple. Taken together, these factors mean that the diodes must be able to handle the larger *peak* charging currents and must be selected accordingly.

28-12 INDUCTOR FILTERING

An alternative to using a *capacitor* in *parallel* with the load is to use an *inductor* in *series* with the load. Since the characteristic of an inductor is to oppose any change in current, the coil will have a *smoothing* effect, as shown in Fig. 28-20.

If you disregard the resistance of the inductor, the average or dc output voltage will be $\frac{2V_m}{\pi}$ as before. If the resistance is taken into account, the voltage will be reduced using the voltage divider equation.

The ripple factor can be shown to be

$$r = \frac{1}{6\pi\sqrt{2}f(L/R_L)} = \frac{R_L}{1600L} \qquad (28\text{-}11)$$

where: r is the theoretical ripple factor, (full-wave)
R_L is the load resistance, in ohms (Ω)
L is the inductance, in henrys (H)
f is the supply frequency, 60 Hz

EXAMPLE 28-8

A full-wave rectifier feeds a 100-Ω load. If a 6-H choke is used for filtering, calculate:
a. The theoretical ripple factor.
b. The dc output voltage if $V_m = 9$ V and choke resistance is neglected.
c. The dc output voltage if $V_m = 9$ V and the choke has a resistance of 25 Ω.

Solution

a. $r = \dfrac{R_L}{1600L}$ $\qquad\qquad (28\text{-}11)$

$= \dfrac{100}{1600 \times 6} = $ **0.01 or 1%**

b. $V_{dc} = \dfrac{2V_m}{\pi}$ $\qquad\qquad (28\text{-}4)$

$= \dfrac{2 \times 9 \text{ V}}{\pi} = $ **5.7 V**

c. $V_{dc} = 5.7 \text{ V} \times \dfrac{R_L}{R + R_L}$ $\qquad (5\text{-}5)$

$= 5.7 \text{ V} \times \dfrac{100 \ \Omega}{25 + 100 \ \Omega} = $ **4.6 V**

Note once again that the time constant (L/R_L) is in the denominator of Eq. 28-11, so a larger time constant is still desirable for effective filtering. However, for inductor filtering, this requires a large value of L or a *small* value of R_L (or both). Contrast this to capacitor filtering, where a *large* value of R_L is better. Inductor filtering is considered effective only where load current is large (R_L is small), whereas capacitor filtering is considered effective only where load current is small (R_L is large).

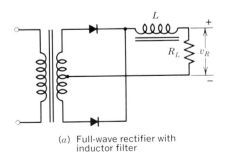

(*a*) Full-wave rectifier with inductor filter

(*b*) Filtered output voltage showing average dc voltage

FIGURE 28-20
Circuit and waveforms for simple inductor filter.

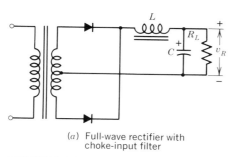 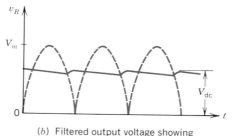

(a) Full-wave rectifier with choke-input filter

(b) Filtered output voltage showing average dc voltage

FIGURE 28-21
Circuit and waveforms for choke-input or *LC* filter.

28-13 CHOKE INPUT OR LC FILTER

The amount of ripple in the previous two filter circuits was determined in part by the load resistor; as the load changed, the amount of ripple also changed. If both a capacitor and an inductor are combined in one filter, the ripple is independent of the load. This is shown in Fig. 28-21. Neglecting the choke's resistance, the dc output voltage is $\frac{2V_m}{\pi}$. The ripple factor can be shown to be

$$r = \frac{0.83}{LC} \qquad (28\text{-}12)$$

where: r is the theoretical ripple factor (full-wave), dimensionless, for $f = 60$ Hz
L is the inductance, in henrys (H)
C is the capacitance, in microfarads (μF)

EXAMPLE 28-9

A full-wave rectifier uses an *LC* filter with $L = 6$ H and $C = 1000\ \mu$F to feed a 100-Ω load. Calculate, neglecting the choke's resistance:
a. The theoretical ripple factor.
b. The dc output voltage if $V_m = 9$ V.

Solution

a. $r = \dfrac{0.83}{LC}$ (28-12)

    $= \dfrac{0.83}{6\ \text{H} \times 1000\ \mu\text{F}} = \textbf{0.00014 or 0.014\%}$

b. $V_{dc} = \dfrac{2V_m}{\pi}$ (28-4)

    $= \dfrac{2 \times 9\ \text{V}}{\pi} = \textbf{5.7 V}$

TABLE 28-2
Summary of Filter Circuits

Type of Filter	Capacitor	Inductor	LC
Component values ($R_L = 100\ \Omega$)	$C = 1000\ \mu$F	$L = 6$ H	$L = 6$ H, $C = 1000\ \mu$F
Ripple factor, r (C in μF, L in H, R_L in Ω, $f = 60$ Hz)	$\dfrac{2410}{CR_L}$ (2.7%)	$\dfrac{R_L}{1600\ L}$ (1%)	$\dfrac{0.83}{LC}$ (0.014%)
DC output voltage (neglecting choke resistance)	$V_m - \dfrac{V_{r,p-p}}{2}$ (8.6 V)	$\dfrac{2V_m}{\pi}$ (5.7 V)	$\dfrac{2V_m}{\pi}$ (5.7 V)

The answers for this example should be compared with those from Examples 28-7 and 28-8. In all three cases, the load is the same, but the *LC* filter has produced the lowest ripple of all. The capacitor filter provided the highest dc output voltage. This is summarized in Table 28-2. See Fig. 28-22 for a 5-V dc power supply that can provide an output current of 6 A at 40°C. Note the large 5000-μF electrolytic capacitor used for filtering. This power supply also incorporates a voltage-regulating circuit.

It should be noted that in some very light load applications, the inductor in the *LC* filter can be replaced by a resistor. The value of the resistor should be equal to the inductive reactance of the coil calculated at the ripple frequency of 120 Hz.

Finally, it should be noted that the ripple factor in a three-phase, full-wave rectified output (as in Fig. 28-18*c*) is only 5.5%, *without* any filtering. This is comparable to the output of dc generators formerly used in automobiles. In some industrial applications, where the large amounts of dc power required would make filtering impractical, transformers can be interconnected to give 6-phase, 9-phase, or even 12-phase operation, resulting in less ripple when rectified.

FIGURE 28-22

A 5-V dc power supply with an output of 6 A at 40°C. Note the large electrolytic capacitor used for filtering. (Courtesy of Lambda Electronics, Division of Veeco Instruments, Inc.)

SUMMARY

1. Pure silicon is a semiconductor having four valence electrons. A semiconductor crystal is formed by the covalent bonding of atoms.
2. At room temperature, pure silicon has a small number of thermally generated electron-hole pairs and a relatively high resistivity.
3. A hole is an incomplete bond due to an electron vacancy. A hole can contribute to the current by being filled with a valence electron. Holes move in a direction opposite that of electrons.
4. The current in pure silicon is the sum of the free electron flow and hole flow, and it is very temperature-dependent.
5. An *N*-type semiconductor is formed by adding a pentavalent impurity such as phosphorus, causing an increase in free electrons and a reduction in resistivity.
6. A *P*-type semiconductor is formed by adding a trivalent impurity, such as aluminum, causing an increase in holes and a reduction in resistivity.
7. The majority current-carriers in an *N*-type semiconductor are electrons, and the minority carriers are holes. The reverse is true for *P*-type semiconductors.
8. When a *PN* junction is formed, carriers diffuse across the junction to establish a potential barrier of about 0.7 V (for silicon).
9. A forward-biased *PN* junction diode has a low resistance. This is obtained by connecting the positive terminal of a dc source to the *P*-type side and the negative to the *N*-type side. The reverse is true for a reverse-biased diode.
10. The current through a forward-biased *PN* junction consists primarily of majority carriers. The small reverse current (when reverse-biased) is due entirely to thermally generated minority carriers.

11. The diode *VI* characteristics show that a cut-in voltage of only 0.6 V is needed for forward current to flow through a silicon diode.

12. Every diode has a peak inverse voltage above which the diode breaks down, with an increase in reverse current.

13. An ohmmeter with known polarity can be used to determine the anode and cathode terminals of a diode.

14. A simple dc power supply can involve half-wave or full-wave rectification with single-phase or three-phase supplies, as summarized in Table 28-1.

15. A filter circuit is used to smooth the output from a rectifier circuit and produce a more "pure" direct current containing less ripple. This is done using a capacitor, inductor, or *LC* circuit.

16. The ripple factor is a measure of how well a filter circuit has removed the ac component from the output of a dc power supply. It is the ratio of the output ac ripple voltage to the dc output voltage.

17. The ripple in a capacitor filter increases as the load resistance decreases, while the opposite is true in an inductor filter. An *LC* filter has a ripple independent of load. These factors are summarized in Table 28-2.

SELF-EXAMINATION

Answer true or false
(Answers at back of book)

28-1. Both silicon and germanium are semiconductors with a valence of four. _____

28-2. In pure silicon there are just as many holes as there are free electrons. _____

28-3. If the temperature of pure silicon increases, the number of free electrons increase and the number of holes decrease. _____

28-4. A hole may be thought of as a positive charge carrier. _____

28-5. A hole is the result of a broken or incomplete bond. _____

28-6. When current flows in pure silicon, there are two components: one due to free electron flow, the other to hole flow. _____

28-7. Because free electrons and holes flow in the same direction, the total current is the sum of the two. _____

28-8. The current in pure silicon goes up as the temperature decreases due to a reduction in resistance. _____

28-9. A donor impurity must be pentavalent to form an *N*-type semiconductor. _____

28-10. An *N*-type semiconductor is negatively charged. _____

28-11. If the majority carriers in a semiconductor are electrons, the material must be *N*-type. _____

28-12. In a *P*-type material the minority carriers are holes. _____

28-13. An acceptor impurity is so called because, being trivalent, it has an incomplete bond that can accept a valence electron and create a hole. _____

28-14. A *P*-type material, although it has a large number of positive holes, is still neutral. _____

28-15. A potential barrier is established across a *PN* junction because of the diffusion of carriers across the junction uncovering bound ion charges. _____

28-16. The potential barrier makes the *P*-side positive and the *N*-side negative. _____

28-17. The application of a forward bias reduces the potential barrier while a reverse bias increases the barrier. _____

28-18. Forward bias involves connecting the positive to the *P*-type and the negative to the *N*-type. _____

28-19. The resistance of a forward-biased diode is low because majority carriers can easily cross the *PN* junction. _____

28-20. Absolutely no current flows through a reverse-biased diode because the barrier is made so large. _____

28-21. Because minority carriers are thermally generated, there is a high temperature dependence of the reverse current in diodes. _____

28-22. The cut-in voltage of 0.6 V for a silicon diode is a direct result of the potential barrier across a *PN* junction. _____

28-23. The *VI* characteristics can be approximated by an open switch when forward-biased and a closed switch when reverse-biased. _____

28-24. The PIV is the maximum reverse voltage a diode can withstand before an appreciable current starts to flow. _____

28-25. The reverse breakdown voltage of a Zener diode can be used to advantage in voltage regulation because of the small change in voltage across the diode when the current through it changes. _____

28-26. If an ohmmeter indicates infinity when the positive side is connected to a diode's anode and the negative side is connected to the cathode, the diode is good. _____

28-27. A half-wave rectifier produces a positive half-cycle of voltage in the output for each cycle of ac input. _____

28-28. A full-wave rectifier circuit uses a minimum of two diodes and can use as many as four or six. _____

28-29. If a two-diode, full-wave, rectifier circuit uses a transformer with the same line-to-line secondary voltage as a half-wave rectifier, the output voltage will be twice that for the half-wave.

28-30. In a bridge rectifier one side of the dc output voltage is common with one side of the transformer secondary. _____

28-31. The amount of dc power delivered to a load in a full-wave bridge rectifier is four times as large as the dc power in a half-wave rectifier utilizing the same ac input voltage on the transformer secondary. _____

28-32. The ripple frequency in a three-phase rectifier is six times the supply frequency and accounts for the very low ripple factor. _____

28-33. A capacitor filter relies upon the energy stored in a capacitor to be discharged at a slow rate for good filtering. _____

28-34. A larger capacitor will reduce ripple without having any effect on the diodes. _____

28-35. Capacitor filtering is best suited for light loads whereas an inductor filter is best with heavy loads. _____

28-36. For a given amount of ripple, a choke-input filter requires a smaller capacitor than a pure capacitor filter. _____

28-37. Since the property of an inductor is to oppose ac current due to inductive reactance, the dc resistance of the choke has no effect on the dc output voltage in an inductor or *LC* filter. _____

REVIEW QUESTIONS

1. What do you understand by the term *covalent bonding?*
2. What is an electron-hole pair?
3. Explain why an increase in temperature increases the conductivity in a semiconductor but decreases it in most metallic conductors.
4. Explain why the total current flowing in pure silicon is the sum of the free electron flow and the hole flow.
5. When a pentavalent phosphorus atom replaces a silicon atom at room temperature, what is the effect on the following:
   a. The number of free electrons in the overall silicon.
   b. The number of electron-hole pairs.
6. For the condition in Question 5, why doesn't the liberation of the fifth electron from the phosphorus atom create as many holes as it does free electrons?
7. a. What are the majority carriers in *N*-type silicon?
   b. What constitutes the minority carriers?
   c. Is the material still neutral from a charge standpoint?
   d. What is meant by ''bound positive charges?''
8. a. What acceptor impurity is used with silicon for *P*-type doping?
   b. How does this affect the electron-hole pairs?
   c. What are the majority and minority carriers?
9. What prevents the migration of all the electrons from the *N*-type material across the *PN* junction to the *P*-type side at the moment a *PN* junction is formed?
10. a. Which *PN* junction, silicon or germanium, has the higher potential barrier to the flow of majority carriers?
    b. How does this show up in the *VI* characteristics of the two diodes?
11. Explain how there can be *any* current flowing across a reverse-biased *PN* junction and why this current is very temperature-dependent.
12. A germanium atom has 32 electrons in orbit around its nucleus, silicon only 14. Can you suggest why the reverse current of a *PN* junction diode is larger when constructed from germanium instead of silicon?
13. Explain in your own words why a *PN* junction diode has a peak inverse voltage.
14. Explain in your own words how you would determine the anode and cathode of a diode, given an ohmmeter. What must you be sure of?
15. Describe the indications of an ohmmeter when checking diodes that are
    a. Shorted.
    b. Open.
    c. Good.
16. If a diode is checked with an ohmmeter in the forward-biased connection, explain why the readings are not constant as you change from a high range to lower ranges. Will the resistance readings get larger or smaller? Why?
17. If a diode is forward biased in a circuit, explain how, with a dc voltmeter, you could determine if it is germanium or silicon.
18. Consider the full-wave rectifier circuit of Fig. 28-15a. Describe the effects of:
    a. Reversing both diode directions.
    b. Reversing the direction of one diode.
19. Consider the bridge rectifier circuit of Fig. 28-16a. Describe the effects of:
    a. Reversing all the diodes.
    b. Reversing diode *D*1.

c.   Reversing diodes $D1$ and $D2$.
20. Explain why the peak inverse voltage of a diode in a half-wave rectifier is 2 V_m when a capacitor filter is used.
21. Explain why an increase in the time constant in a capacitor filter circuit decreases ripple.
22. Explain why repetitive peak diode current is increased as ripple is decreased in a capacitor filter circuit.
23. Describe how you would use an oscilloscope to experimentally determine the ripple factor of a dc power supply.
24. Suggest a reason why an inductor filter is more effective with a heavier load.
25. How does the ripple factor vary with load for the following filters:
     a.   Inductor.
     b.   Capacitor.
     c.   Choke input.
26. What advantage does an LC filter have over a capacitor filter? What disadvantage does it have?

PROBLEMS

(Answers to odd-numbered problems at back of book)
28-1. A 120-V primary, 24-V secondary transformer is used to supply a half-wave rectifier and a 470-Ω load. Neglecting diode losses, calculate:
       a.   The dc voltage across the load.
       b.   The peak diode current.
       c.   The dc load current.
       d.   The dc power delivered to the load.
       e.   The PIV across the diode.
       Sketch the waveforms of the load voltage and current.
28-2. Repeat Problem 28-1, using a 48-V secondary.
28-3. If a load requires a dc voltage of 9 V from a half-wave rectifier, determine:
       a.   The rms input voltage to the rectifier.
       b.   The load resistance if the peak load current through the diode must not exceed 500 mA.
28-4. Repeat Problem 28-3 using a dc voltage of 12 V.
28-5. A 120-V, 60-Hz voltage is applied to the primary of a 1:2 step-up transformer whose secondary is center-tapped, allowing a load of 1 kΩ to be connected to a full-wave rectifier utilizing two diodes. Neglecting diode losses, determine:
       a.   The dc voltage across the load.
       b.   The dc current through the load.
       c.   The dc power delivered to the load.
       d.   The PIV across each diode.
       e.   The ripple frequency of the output.
28-6. Repeat Problem 28-5 using a 1:4 step-up transformer.
28-7. a.   Specify what secondary transformer voltage is necessary to produce 18-V dc across a 50-Ω load connected to a two-diode, full-wave rectifier.
       b.   What will be the peak current through each diode?
28-8. Repeat Problem 28-7 for 30-V dc across a 100-Ω load.

28-9. Repeat Problem 28-5 using a bridge rectifier connected across the *full* secondary of the transformer.

28-10. a. What secondary voltage is necessary to be connected to a bridge rectifier to obtain a dc voltage of 100 V?

b. What will be each diode's PIV for this situation?

28-11. A 208-V, 60-Hz, three-phase system is applied to a full-wave rectifier circuit that supplies dc power to a 50-Ω load. Determine:

a. The dc output voltage across the load.

b. The dc power in the load.

c. Each diode's PIV.

d. The ripple frequency of the output.

28-12. Repeat Problem 28-11 using a 277-V, 400-Hz, three-phase system.

28-13. a. A filtered dc power supply is known to have a 10% ripple factor. If the dc output voltage is 12 V, what is the rms value of ripple voltage in the output?

b. Assuming that this ripple is sinusoidal in nature, what is its peak-to-peak voltage?

c. Sketch the waveform that a dc-coupled oscilloscope would indicate when connected across this power supply.

28-14. Repeat Problem 28-13 assuming that the ripple voltage is triangular in nature.

28-15. The output of a filtered dc power supply is observed on an oscilloscope and found to have an approximately sinusoidal ripple of 4 V_{p-p} "riding" on a dc level of 16 V.

a. Sketch this voltage waveform.

b. Determine the percent of ripple.

28-16. Repeat Problem 28-15 assuming a triangular ripple voltage.

28-17. A 12.6-V, center-tapped transformer operating at 60 Hz is used in a two-diode, full-wave rectifier with a 100-μF capacitor to provide filtering for a 1-kΩ resistive load. Calculate:

a. The theoretical percent ripple factor.

b. The dc output voltage.

28-18. Repeat Problem 28-17 using a 200-μF capacitor.

28-19. What minimum size of filter capacitor is necessary with a full-wave rectifier operating at 60 Hz to make sure that the ripple factor does not exceed 1% with a load of 200 Ω?

28-20. Repeat Problem 28-19 using a frequency of 400 Hz.

28-21. What value of inductance should be used in an inductor filter connected to a full-wave rectifier operating at 60 Hz if the ripple factor is not to exceed 3% for a 120-Ω load?

28-22. Calculate the dc output voltage for Problem 28-21 if a bridge circuit is used with a line-to-line voltage of 50 V, assuming that

a. The choke has negligible resistance.

b. The choke's resistance is 30 Ω.

28-23. An *LC* filter is to be used to provide a dc output with 0.5% ripple when operating from a full-wave rectifier operating at 60 Hz. To conserve the size of choke, a ratio for $L/C = 0.01$ is recommended (*L* in henrys, *C* in microfarads). Determine the required values of *L* and *C*.

28-24. The filter in Problem 28-23 is used with a 10-kΩ load and $V_m = 20$ V. Determine:

a. The dc output voltage neglecting the coil's resistance.
b. What value of resistance could be used to replace the inductor and still provide the same ripple.
c. The approximate dc output voltage with the resistor as in part (b).

28-25. Given the half-wave rectifier circuit in Fig. 28-23:

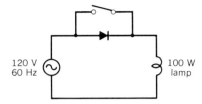

FIGURE 28-23
Circuit for Problem 28-25.

a. With the switch closed, determine the hot resistance of the 100-W lamp, and the rms value of the current.
b. With the switch open, determine the rms voltage of the half-wave rectified output across the lamp.
c. Calculate the dc voltage across the lamp.
d. Assuming the hot resistance of the filament is directly proportional to the rms voltage across the lamp, calculate the resistance of the lamp with the switch open.
e. Calculate the rms power delivered to the lamp when operating with the switch open and find the percentage reduction in power.
f. Calculate the dc power in the lamp with the switch open.

CHAPTER 29

INTRODUCTION TO TRANSISTORS, AMPLIFIERS, AND OSCILLATORS

The junction transistor, discovered in 1947 by a team of Bell Telephone Laboratory scientists, has revolutionized electronics. It has replaced the bulky, slower, heat-producing electron tube in almost every type of electronic equipment. Modern solid-state television sets consume a small amount of power (only a few tens of watts), compared to the hundreds of watts used by earlier tube-operated receivers.

Essentially, a transistor consists of two *PN* junction diodes back-to-back or front-to-front. Since both electrons *and* holes are involved in the operation of a junction transistor, these devices are often referred to as *bipolar junction transistors* (BJTs). This distinguishes them from the more recently developed *field effect transistors* (FETs), in which only electrons *or* holes flow (depending upon the transistor type). Both BJTs and FETs are transistors.

Transistors can be used to *amplify* signals. That is, they increase the magnitude of the input signal current or voltage (or both) to a higher-voltage, current, or power level by controlling a power-supply voltage. This can be done using one of three possible configurations: common emitter (CE), common base (CB), or common collector (CC). These names are derived from whichever of the three transistor terminals (emitter, base, collector) is common to both the input and output of the amplifier.

Virtually all commercial amplifiers use some form of *negative feedback.* This involves feeding some of the amplifier's output back to the input to reduce the input signal. Although this results in some loss of gain, many beneficial effects occur. These include increased stability, increased bandwidth, and a reduction in distortion of the amplifier's output. If *positive* feedback is supplied to the amplifier, the voltage gain increases and a condition of *oscillation* may occur. The amplifier now becomes an oscillator with a sinusoidal output, whose frequency is dependent upon the elements in the feedback circuit.

29-1 PHYSICAL CHARACTERISTICS OF TRANSISTORS

29-1.1 Types

There are two types of bipolar junction transistors (BJTs), referred to as *NPN* and *PNP* transistors. Figure 29-1 shows an *NPN* transistor, which consists of a layer of *P*-type material (the *base*), sandwiched between two *N*-type regions called the *emitter* and the *collector*.

The base is very thin, making up only about 1 to 2% of

the total length of the transistor. However, it is sufficient to produce two distinct *PN* junctions, represented in Fig. 29-1*b* as two diodes shown pointing "outward." The symbol in Fig. 29-1*c* uses an arrow, also pointing outward, to distinguish the emitter from the collector. The reasons for these names will become clear as you learn about the operation of the transistor.

If the transistor is constructed with an *N*-type base and *P*-type material for the emitter and collector, the result is a *PNP* transistor. Its symbol, along with a simple two-diode equivalent circuit, is shown in Fig. 29-2. Note that the arrow on the emitter points "inward."

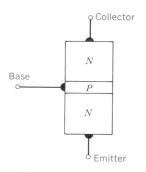

(*a*) Diagrammatic representation of an NPN sandwich

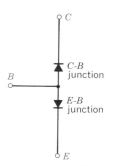

(*b*) Two-diode equivalent circuit of NPN transistor

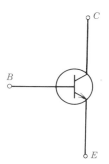

(*c*) Symbol used for NPN transistor

FIGURE 29-1
Different representations of an *NPN* transistor.

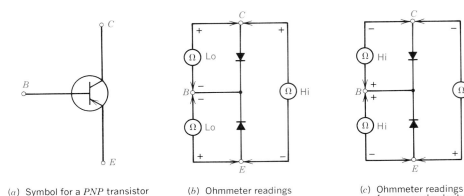

(a) Symbol for a *PNP* transistor

(b) Ohmmeter readings for given polarity

(c) Ohmmeter readings for reversed polarity

FIGURE 29-2

Checking a *PNP* transistor with an ohmmeter. (It is assumed that only one ohmmeter is connected to the transistor at any time.)

29-1.2 Ohmmeter Checking of a Transistor

Since a transistor is normally *soldered* into a circuit, it cannot be readily removed for testing. (Plug-in sockets, like those employed for electron tubes, are *not* generally used because of the large effect that oxidation of the contacts would have on the small voltages—only 0.6 V in some cases—found in transistor circuits.) A method of checking transistors without removing them from the circuit uses an ohmmeter, as shown in Figs. 29-2*b* and 29-2*c*.

Depending upon whether a junction is forward- or reverse-biased by the ohmmeter (no power is applied to the circuit otherwise, of course), a low or high reading will be obtained. A high reading is to be expected in *both* directions between *C* and *E,* except when punch-through has taken place. Care must be taken to include the effects of any other components connected across the transistor terminals before deciding that the transistor is bad. Also, when used on the *low-ohms* range, some digital multimeters may show a good junction to be open in both directions.

Care should be taken to use the middle ohmmeter ranges, since low ranges can produce excessive current and high ranges can produce excessive voltage that may destroy the transistor junctions.

29-1.3 Construction

Although the earliest transistors were made from germanium, practically all current transistors are manufactured

from silicon. Many different methods were used to construct transistors in the past, but the common method today is the *gaseous diffusion* technique, closely related to integrated circuit (IC) construction.

Figure 29-3 shows a simplified cross-sectional view of an *NPN* silicon *planar* transistor. Manufacturing begins with a block of *N*-type silicon covered with a layer of silicon dioxide. The base-collector junction is then etched by a photolithographic technique to expose an area of *N*-type material. Next, the silicon wafer is placed in a vessel (called a "boat") with an oxide of the appropriate doping impurity, such as boron, indium, or aluminum. The boat is slowly passed through a furnace where temperatures are accurately controlled to some point between 800°C and 1200°C (depending upon the type of junction desired). The

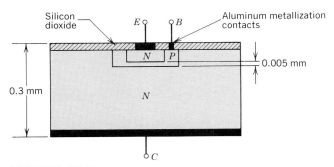

FIGURE 29-3

Typical construction of an *NPN* silicon planar transistor manufactured using the gaseous diffusion technique.

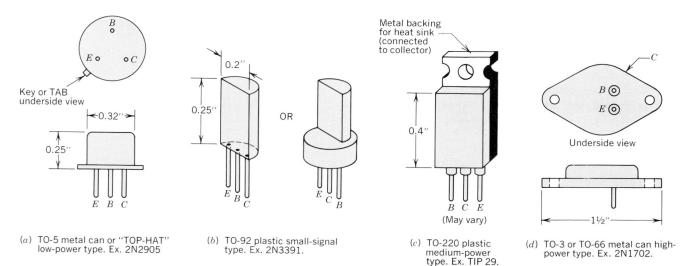

(a) TO-5 metal can or "TOP-HAT" low-power type. Ex. 2N2905

(b) TO-92 plastic small-signal type. Ex. 2N3391.

(c) TO-220 plastic medium-power type. Ex. TIP 29.

(d) TO-3 or TO-66 metal can high-power type. Ex. 2N1702.

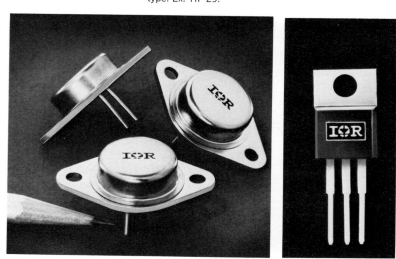

FIGURE 29-4

Typical transistor packages and lead identification. (Photographs in (e) courtesy of International Rectifier.)

(e) Medium and high-power types

impurity atoms, in gaseous form, diffuse into the area of N-type material that was exposed by etching, forming a thin layer of P-type material. This becomes the *base* region of the transistor.

The process is then repeated, using an N-type doping impurity to diffuse the *emitter* into the base. A final layer of silicon dioxide is applied, then aluminum metallization contacts (terminals) are added. They extend through the final layer of silicon dioxide to make contact with the emitter, base, and collector regions of the transistor.

It should be clear from Fig. 29-3 that the collector is a large region and that the base is a very thin layer between the collector and emitter. The transistor is *not* symmetrical, as may have been suggested by the representation in

Fig. 29-1. Also, the base is very lightly doped, in comparison to the emitter and collector. These are very important characteristics that will help to explain the operation of the transistor.

29-1.4 Packaging and Lead Identification

The silicon wafer from which the transistor is made is very brittle and must be protected by some form of encapsulation. Figure 29-4 shows four typical packages used for transistors.

In the medium-power and high-power types of transistors, the collector is often connected to the metal housing or backing. This helps to dissipate heat from the transistor to the metal chassis (heat sink) on which it is mounted. Where the chassis is used as a common electrical connection (such as ground), a mica washer may sometimes be used to insulate the collector from the chassis. Silicone grease is applied to improve heat transfer from transistor to chassis.

Power ratings for the small-signal type of transistors range from 75 to 300 mW, and from 5 to 150 W for power transistors at ambient temperatures of 25°C.

Designations for transistors registered with the Electronic Industries Association (EIA) have a 2N prefix, such as 2N3391 or a five-digit number such as 40340. Whether a transistor is *NPN* or *PNP* can only be determined (without testing) from a transistor catalog. The catalog also provides information on lead identification.

Many foreign transistor types use a prefix of 2SA or 2SB for *PNP* (such as 2SA 497) and 2SC or 2SD for *NPN* (such as 2SC 497). Often, because of lack of space, the 2S is omitted from the number.

29-2 BASIC COMMON-EMITTER AMPLIFYING ACTION

As already mentioned, a transistor amplifies an alternating input voltage signal by controlling the voltage of the power supply to which it is connected. To better understand how the transistor performs this function, refer to Fig. 29-5.

If the emitter-base (EB) junction is forward-biased by V_{BB}, a current will flow as in any *PN* junction diode. In a silicon diode, the voltage from base to emitter (V_{BE}) is between 0.5 and 0.6 V for a base current (I_B) of 0.2 mA. (See Fig. 29-5a.)

Now consider that a dc voltage supply, $V_{CC} = 12$ V, is connected from collector to emitter with the base open. Clearly, this *reverse* biases the collector-base junction, so that only a minority (leakage) current flows. This is referred to as I_{CEO} (the current from collector to emitter with the base open). Because this minority current is due to the thermally generated *minority* carriers (holes in the *N*-type material), the current is very temperature dependent. (See Fig. 29-5b.) Although I_{CEO} is very small in a silicon transistor (about 1 µA at 25°C), it causes instability at higher temperatures. (One method to reduce this effect will be examined in Section 29-6.)

Now consider the effect of simultaneously applying forward bias to the emitter-base (EB) junction and reverse bias to the collector-base (CB) junction, as shown in Fig. 29-5c. The result is a *large* current flowing in the collector. This current is typically 50 to 200 times as large as the base current in small-signal transistors. (Collector resistor R_C is used to limit the collector current to a safe value.) How is this possible if the collector-base junction is reverse biased?

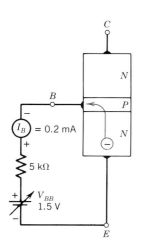

(a) Forward-biased emitter-base junction causes base current.

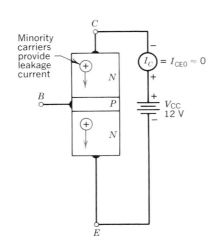

(b) Reverse-biased collector-base junction causes only small leakage current to flow in collector.

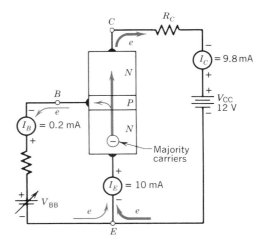

(c) Small base current causes large collector current to flow.

FIGURE 29-5

Current paths in an *NPN* transistor under different conditions of forward and reverse bias.

The answer lies in these three facts:

1. The base is very thin.
2. The base is very lightly doped, compared to the emitter and collector.
3. The emitter-base (EB) junction is forward biased.

The forward biasing of the emitter-base junction allows electrons (the *majority* carriers) in the *N*-type emitter to cross over into the base. Once these electrons are in the base, they come under the influence of the *N*-type collector region, which is connected to the positive side of V_{CC}. Almost all (98%) of the electrons emitted by the emitter are swept across the base and captured by the positive collector. Only a small base current flows because not many electrons (about 2% or less) recombine with holes in the base. This is due to the thinness of the base. The electrons do not spend much time in this region on their way to the collector. Also, the low availability of holes in the lightly doped base do not encourage very many recombinations.

Thus, a small base current (used to forward bias the emitter-base junction) can *control* a much larger collector current drawn from the supply (V_{CC}). And if the forward bias is increased, the collector current also increases. That is, the base current acts as a control valve or gate.

How much collector current actually *will* flow depends on how well the emitter-base junction has been *turned on*. When suitable bias currents are used, the control of the base current over the collector current is quite linear until a saturation effect occurs. This is shown in Fig. 29-6, using conventional current directions.

The base current does *not* generate or create collector current. The collector current is furnished by the collector supply (V_{CC}). The base current simply controls the conductivity between the collector and emitter (CE) terminals. Also, if the base current is zero, the transistor conductivity from the collector to the emitter essentially drops to zero. From this standpoint, it is easy to see how a transistor can be thought of either as a *valve* or as a *switch*. The latter is a very important application in digital logic devices.

29-3 CE COLLECTOR CHARACTERISTICS

The way in which the base current controls the collector current can also be shown in a *graphical* form, called *collector characteristic curves*. Figure 29-7a shows the circuit that can be used to obtain these curves experimentally, point by point. Alternatively, a *curve-tracer* will display this whole set of curves on a CRT screen, which can then be photographed. (See Fig. 29-8.) The characteristic curves in Fig. 29-7b show the influence of the collector-to-emitter voltage (V_{CE}). But for a given collector voltage, the family of curves shows essentially equal separations for equal increments in base current. This effect shows the *linear* nature of the transistor over *most* of its characteristics when used as an ac amplifier.

The collector curves are very useful in transistor amplifier design and in comparing transistors of the same type if matched characteristics are required in certain amplifier

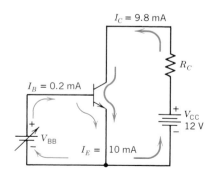

(a) A base current of 0.2 mA causes a collector current of 9.8 mA.

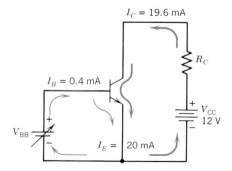

(b) A base current of 0.4 mA causes a collector current of 19.6 mA.

FIGURE 29-6

Conventional current directions in an *NPN* transistor showing how base current determines collector current.

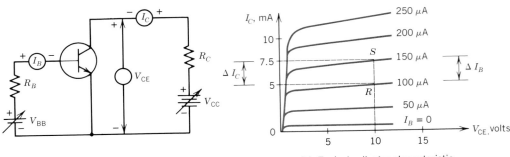

(a) Circuit to obtain collector characteristic curves for an *NPN* transistor in CE connection

(b) Typical collector characteristic curves for a small-signal type of transistor in the CE connection

FIGURE 29-7

Test circuit and CE collector characteristics for an *NPN* transistor.

applications. This is especially important in complementary symmetry amplifiers, where a matched pair of *NPN* and *PNP* transistors is used together.

29-3.1 Forward-Current Transfer Ratio (β)

You can also use the collector characteristic curves to calculate the "current gain" of the transistor (β), also known as the *forward-current transfer ratio*. Beta can be defined in terms of dc current or small-signal ac current. It is an indication of the control that the base current has over the collector current.

For dc:
$$\beta_{dc} = h_{FE} = \frac{I_C}{I_B} \qquad (29\text{-}1)$$

Evaluating β_{dc} (also known as h_{FE} in transistor manuals) at points R and S in Fig. 29-7b, you obtain

At R, $\beta_{dc} = \dfrac{I_C}{I_B} = \dfrac{5 \text{ mA}}{0.1 \text{ mA}} = \mathbf{50}$ \qquad ($V_{CE} = 10$ V)

At S, $\beta_{dc} = \dfrac{I_C}{I_B} = \dfrac{7.5 \text{ mA}}{0.15 \text{ mA}} = \mathbf{50}$ \qquad ($V_{CE} = 10$ V)

Instead of evaluating β at specific points on the collector characteristic curve, you can evaluate it as the ratio of a *small* change in the collector current for a given change in base current (ΔI_B), or

$$\beta_{ac} = h_{fe} = \frac{\Delta I_C}{\Delta I_B}, \qquad V_{CE} \text{ constant} \qquad (29\text{-}2)$$

Evaluating β_{ac} between points R and S in Fig. 29-7b, you find that

$$\beta_{ac} = \frac{\Delta I_C}{\Delta I_B} = \frac{7.5 \text{ mA} - 5 \text{ mA}}{0.15 \text{ mA} - 0.1 \text{ mA}} = \frac{2.5 \text{ mA}}{0.05 \text{ mA}} = \mathbf{50}$$

Over a large region of the characteristic curves, it is not unusual to find that $\beta_{dc} = \beta_{ac}$ and is fairly constant. As a result, the ac and dc subscripts are often dropped, and you can merely refer to the beta (β) of a transistor.

Values for β vary from 50–500 for *small-signal* transistors, but may be only 10–50 for high-power transistors. However, even among transistors of the *same type*, it is not unusual for a 3–1 spread to exist between the minimum and maximum β values. Furthermore, the β of a given transistor is extremely sensitive to, and varies with, the temperature.

Note that you now can obtain the collector current for a transistor if you know its β, from the following equation:

$$I_C = \beta I_B + I_{CEO} \qquad (29\text{-}3)$$

where: I_{CEO} is the leakage (minority carrier) current
I_B is the base current
I_C is the collector current
β is the transistor's forward-current transfer ratio

EXAMPLE 29-1

A silicon small-signal transistor has a leakage current of 1 μA and β = 200. Determine the transistor's collector current when the base current is

a. 0.

b. 50 μA.

c. 100 μA.

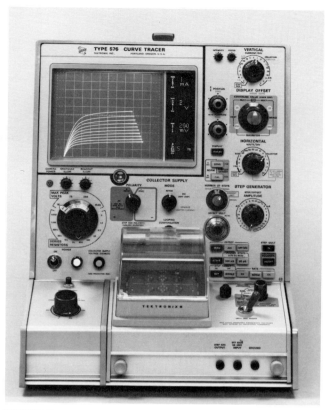

FIGURE 29-8
Type 576 curve tracer. (Courtesy of Tektronix, Inc.)

Solution

a. $I_C = \beta I_B + I_{CEO}$ (29-3)
    $= 200 \times 0 + 1\ \mu A$
    $= \mathbf{1\ \mu A}$

b. $I_C = B I_B + I_{CEO}$ (29-3)
    $= 200 \times 50\ \mu A + 1\ \mu A$
    $= 10\ mA + 1\ \mu A$
    $\approx \mathbf{10\ mA}$

c. $I_C = \beta I_B + I_{CEO}$ (29-3)
    $= 200 \times 100\ \mu A + 1\ \mu A$
    $= 20\ mA + 1\ \mu A$
    $\approx \mathbf{20\ mA}$

Example 29-1 shows that the reverse leakage current can be ignored, except at extremely small values of base current.

Note that β has no units, since it is the ratio of two currents having the same units.

29-4 CB COLLECTOR CHARACTERISTIC CURVES

If the transistor is connected with the base terminal common to both input and output, it is said to be operating in the *common base* mode. In this case, the *emitter* is the *input* terminal. Refer to Fig. 29-9a, where (for circuit variation), a *PNP* transistor is used in the common base connection.

The voltage (V_{EE}) forward-biases the emitter-base (EB) junction so that holes in the *P*-type emitter flow into the base region. Once again, approximately 98% of these charge carriers are swept through the narrow, lightly doped base into the *P*-type collector region. So, the base current is small as in the common emitter. Note that V_{EE} is necessary to forward-bias the emitter-base junction, but it is the "series" action of V_{EE} and V_{CC} that provides the emitter and collector current. If V_{CC} acts alone, with $I_E = 0$, only a leakage collector current flows, called I_{CBO}. This is the current from *collector* to *base* with the emitter *open*, and it is much smaller than I_{CEO} in a CE configuration.

Using the circuit in Fig. 29-9a, you can obtain the collector characteristic curves shown in Fig. 29-9b. The lines for each emitter current are practically horizontal, showing that the collector-base voltage (V_{CB}) has little effect on the collector current.

29-4.1 Forward-Current Transfer Ratio (α)

This is a measure of how much collector current flows, compared with the input emitter current. If you apply Kirchhoff's current law to the transistor, you obtain:

$$I_E = I_C + I_B \qquad (29-4)$$

It is clear then that the collector current is less than the emitter current by some small amount equal to the base current. In a way similar to that used for the CE connection, you can define the ratio of collector current to emitter current in terms of dc or ac quantities.

$$\alpha_{dc} = h_{FB} = \frac{I_C}{I_E} \qquad (29-5)$$

$$\alpha_{ac} = h_{fb} = \frac{\Delta I_C}{\Delta I_E}, \qquad V_{CB} \text{ constant} \qquad (29-6)$$

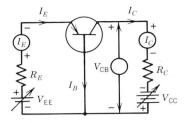

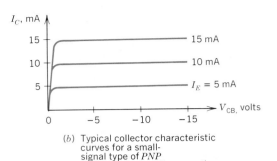

(a) Circuit to obtain collector characteristic curves for *PNP* transistor in CB connection

(b) Typical collector characteristic curves for a small-signal type of *PNP* transistor in the CB connection

FIGURE 29-9

Test circuit and CB collector characteristics for a *PNP* transistor.

These would be difficult to evaluate from the characteristic curves because I_C and I_E differ by very small amounts. However, the dc and ac values are so close to each other, you can simply refer to the transistor's alpha (α).

EXAMPLE 29-2

Dc current measurements for a transistor in the CB connection show an emitter current of 6 mA and a base current of 120 µA. Determine the transistor's alpha.

Solution

$$I_C = I_E - I_B \tag{29-4}$$
$$= 6 \text{ mA} - 0.12 \text{ mA}$$
$$= 5.88 \text{ mA}$$

$$\alpha = \frac{I_C}{I_E} \tag{29-5}$$
$$= \frac{5.88 \text{ mA}}{6 \text{ mA}} = \mathbf{0.98}$$

Note that the current gain in the CB connection is always slightly less than unity (one). That is, the output current is always slightly less than the input current. How, then, can a CB connection be useful in an amplifier if it does not amplify the current?

If the output current is made to flow through a collector resistor (R_C) that is significantly larger than the input resistance (R_{in}), it is possible to have a significant voltage gain. In other words, the output voltage is larger than the input voltage. The word *transistor,* in fact, is derived from the words *trans*fer-re*sistor* because of the ability of the device to transfer essentially the same current from a low-resistance input to a high-resistance output.

29-4.2 Relation Between α and β

It was shown above that the determination of α could not be done accurately because it involved only small differences in currents. However, β can be determined much more accurately, because of the large differences in collector and base currents. If a transistor's β is known, it can be shown that its α can be obtained quite accurately from

$$\alpha = \frac{\beta}{1 + \beta} \tag{29-7}$$

Solving Eq. 29-7 for β yields:

$$\beta = \frac{\alpha}{1 - \alpha} \tag{29-8}$$

Equation 29-8 shows that for a high β, a transistor must have α as close to unity as possible. This implies that high-gain transistors must have a very thin, lightly doped base to reduce the base current to a very low value. Under these conditions, the collector current is practically equal to the emitter current and α is practically unity.

EXAMPLE 29-3

Use the information in Example 29-2 to obtain:
a. The transistor's β.
b. The transistor's α.

(Use $I_E = 6$ mA, $I_B = 120$ µA.)

Solution

a. $I_C = I_E - I_B \tag{29-4}$
$= 6 \text{ mA} - 0.12 \text{ mA}$
$= 5.88 \text{ mA}$

$$\beta = \frac{I_C}{I_B} \tag{29-1}$$

$$= \frac{5.88 \text{ mA}}{0.12 \text{ mA}} = \mathbf{49}$$

b. $\alpha = \dfrac{\beta}{1 + \beta}$ (29-7)

$$= \frac{49}{1 + 49} = \mathbf{0.98}$$

Note that the collector current for a CB transistor connection can also be obtained from

$$I_C = \alpha I_E + I_{\text{CBO}} \tag{29-9}$$

But since I_{CBO}, which is a thermally dependent current, is typically only 0.025 μA at 25°C, the collector current in a CB configuration is much less affected by temperature than is a CE connection.

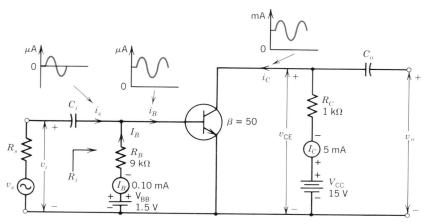

(a) CE ac amplifier circuit with two batteries

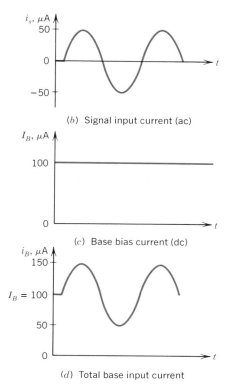

(b) Signal input current (ac)

(c) Base bias current (dc)

(d) Total base input current

FIGURE 29-10

Common emitter ac amplifier circuit and input current waveforms.

29-5 BASIC COMMON-EMITTER AMPLIFIER

You have seen how a small dc base current can cause a larger dc collector current to flow through the transistor. If the base current is varied, the collector current will *also* vary. And, if this collector current is made to flow through a load resistor, an output voltage will be developed.

Figure 29-10a shows V_{BB} (the base supply voltage) causing a dc-base bias current (I_B) of 100 μA. This, in turn, causes a dc collector current (I_C) of 5 mA. (Assuming that β = 50.) A sinusoidal signal voltage (v_s) with its internal resistance (R_s) is connected from base to emitter

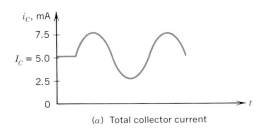

(a) Total collector current

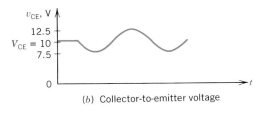

(b) Collector-to-emitter voltage

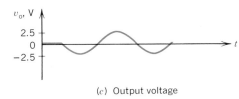

(c) Output voltage

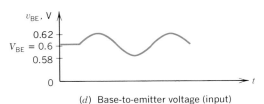

(d) Base-to-emitter voltage (input)

FIGURE 29-11
Current and voltage waveforms for a CE amplifier circuit.

through a coupling capacitor (C_i). This capacitor simply prevents V_{BB} from sending dc current through the applied signal. However, capacitor C_i allows the ac signal current (i_s) to be superimposed on the dc bias current (I_B) to produce a total base input current of i_B. These three current waveforms are shown in Figs. 29-10b, 29-10c, and 29-10d, where it is assumed that the ac signal current is 100 μA peak-to-peak.

This base current variation is amplified 50 times, causing a 5 mA peak-to-peak variation in the collector current superimposed on a dc level of 5 mA, referred to as i_C. (See Fig. 29-11a.)

When this varying collector current flows through the 1-kΩ collector resistor R_C, the collector-to-emitter voltage varies between 7.5 and 12.5 V. (See Fig. 29-11b.) That is, as the collector current reaches its peak of 7.5 mA, v_{CE} *drops* to 7.5 V; when i_C is at its *minimum* of 2.5 mA, v_{CE} *rises* to 12.5 V. This is because v_{CE} is equal to V_{CC} *minus* the voltage drop across R_C and is given by

$$v_{CE} = V_{CC} - i_C R_C \qquad (29\text{-}10)$$

Capacitor C_o blocks the 10-V dc level in v_{CE}, producing an output voltage (v_o) of 5 V peak-to-peak. (See Fig. 29-11c.) Since this voltage is an inverted sine wave compared with the input voltage (v_{BE}), the CE amplifier is said to cause *phase inversion*. (See Figs. 29-11c and 29-11d.)

29-5.1 Waveforms

From Fig. 29-10d, you can see that a minimum value of dc bias current (I_B) into the base is necessary to prevent the base current from dropping too low when an ac signal is applied. This prevents the transistor from being cut off, which would introduce distortion into the output current.

Similarly, the dc bias voltage of 0.6 V from base to emitter (Fig. 29-11d) is maintained by V_{BB}. In this example, a change of only 0.04 V peak-to-peak from the input signal is needed to vary the base and collector currents as shown.

If there is no distortion, the variations in current and voltage will be sinusoidal. This means that the dc readings of I_B, I_C, and V_{CE} will remain at 0.10 mA, 5 mA, and 10 V, respectively, whether a signal is applied or not.

29-5.2 Current, Voltage, and Power Gains

The ac current, voltage, and power amplifications can be defined as follows:

Current gain, $\quad A_i = \dfrac{i_c}{i_s}$ $\quad$ (29-11)

where: A_i is the current gain, dimensionless

$\quad\quad i_c$ is the ac component of the output collector current (i_C)

$\quad\quad i_s$ is the ac component of the input signal current

Voltage gain, $\quad A_v = \dfrac{v_o}{v_i}$ $\quad$ (29-12)

where: A_v is the voltage gain, dimensionless

$\quad\quad v_o$ is the ac component of the output collector voltage (v_{CE})

$\quad\quad v_i$ is the ac component of the input signal voltage from base to emitter (v_{BE})

(If v_s is used instead of v_i, the *open-circuit* voltage gain (A_{vs}) results.)

Alternatively, $\quad A_v = A_i \dfrac{R_o}{R_i} \approx -\beta \dfrac{R_o}{R_i}$ $\quad$ (29-13)

where: A_v is the voltage gain, dimensionless

$\quad\quad A_i$ is the current gain, dimensionless (approximately equal to $-\beta$ for a CE single-stage amplifier)

$\quad\quad R_o$ is the total ac load resistance on the amplifier (R_C in this case), in ohms (Ω)

$\quad\quad R_i$ is the ac input resistance to the amplifier, in ohms, given by $\dfrac{v_i}{i_s}$

Power gain, $\quad A_p = \dfrac{P_o}{P_i} = A_v \times A_i$ $\quad$ (29-14)

where: A_p is the amplifier power gain, dimensionless

$\quad\quad P_o$ is the ac output power

$\quad\quad P_i$ is the ac input power

Sometimes, power gain (A_p) is given in terms of *decibels* (dB), defined as

$$\text{power gain in dB} = 10\log\dfrac{P_o}{P_i} = 10\log A_p \quad (29\text{-}15)$$

EXAMPLE 29-4

Using the data in Figs. 29-10 and 29-11, calculate the following ac characteristics of the amplifier.

a. Current gain.

b. Voltage gain.

c. Input resistance.

d. Power gain.

Solution

a. $A_i = \dfrac{i_c}{i_s}$ $\quad$ (29-11)

$\quad = \dfrac{5\text{ mA p–p}}{0.1\text{ mA p–p}} = \mathbf{50}$

b. $A_v = \dfrac{v_o}{v_i}$ $\quad$ (29-12)

$\quad = \dfrac{5\text{ V p–p}}{0.04\text{ V p–p}} = \mathbf{125}$

c. $R_i = \dfrac{v_i}{i_s}$

$\quad = \dfrac{0.04\text{ V p–p}}{0.1 \times 10^{-3}\text{ A p–p}} = \mathbf{400\ \Omega}$

$\quad A_v = A_i \dfrac{R_o}{R_i}$ $\quad$ (29-13)

$\quad = 50 \times \dfrac{1000\ \Omega}{400\ \Omega} = \mathbf{125}$

d. $A_p = A_v \times A_i$ $\quad$ (29-14)

$\quad = 125 \times 50 = \mathbf{6250}$

$\quad \text{power gain in dB} = 10\log A_p$

$\quad\quad\quad = 10\log 6250$

$\quad\quad\quad = 10 \times 3.8 = \mathbf{38\ dB}$

Decibels are used to *linearize* power level *changes* to show, for example, what sound levels the human ear will notice. If an amplifier's output is doubled from 4 to 8 W, this corresponds to an increase of $10\log 2 = 3$ dB. If the output is doubled again from 8 W to 16 W, the increase is again:

$$10\log\dfrac{16\text{ W}}{8\text{ W}} = 3\text{ dB}$$

The doubling of power will sound the same in either case (even though in one case there is an increase of 8 W and in the other, there is an increase of only 4 W). Thus, the mathematical description of 3 dB in each case has indicated the same change in hearing response, (at a given frequency, for a given person) regardless of the sound levels. The decibel level change is now a linear response. Incidentally, a change of 3 dB can be noticed, but it is *not* an appreciable change in power to the human ear.

29-5.3 CE Amplifier with a Single Supply Voltage

The common emitter has two valuable features. One is the ability to produce a high *power* gain because it develops

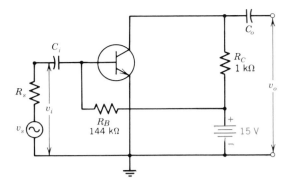

(a) *CE* ac amplifier operating from a single supply

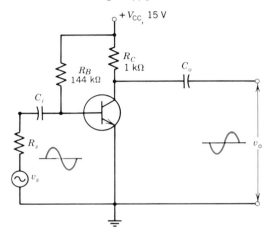

(b) *CE* ac amplifier circuit redrawn

FIGURE 29-12
CE amplifer operates with fixed bias from a single supply voltage.

both current and voltage gains. (The CB and CC amplifiers only amplify voltage or current, respectively, but not *both*.) The second valuable feature of the CE amplifier is the need for only *one* power supply. Figure 29-12*a* shows how the 15-V collector supply can be used to provide base bias if R_B is increased from 9 kΩ with $V_{BB} = 1.5$ V to 144 kΩ with $V_{CC} = 15$ V. $R_B = (15$ V $- 0.6$ V$)/0.1$ mA $= 144$ kΩ. This is possible because both supplies have their negative sides tied to the common emitter. (Note that this is not possible to do directly in a CB amplifier. See Fig. 29-9*a*.)

Figure 29-12*b* indicates the preferred way of drawing the CE amplifier circuit. This is especially convenient when a number of CE amplifiers, all operating from the same supply voltage, are connected in cascade. This

means that the output of one amplifier is connected, through a coupling capacitor, to the input of the next. In this case, the overall voltage gain is given by the *product* of the individual voltage gains of each amplifier. (If the individual stage gains are given in decibels, the total power gain is the *sum* of the individual gains.)

29-6 PRACTICAL CE AC AMPLIFIER WITH BIAS STABILIZATION

You have seen that the collector current of a transistor in the CE connection is given by

$$I_C = \beta I_B + I_{CEO} \qquad (29\text{-}3)$$

Both β and the minority leakage current (I_{CEO}) are temperature dependent. Also, as temperature *increases*, the input resistance *decreases*, causing an increase in I_B. Therefore, the collector current goes up with the temperature. This increase in the collector current causes more heat to be dissipated within the transistor, raising its temperature further and causing still another increase in collector current. Thus, any rise in the ambient temperature (that is, in the medium surrounding the transistor) can lead to *thermal runaway*. This means that the amplifier's operating point changes, causing distortion, and—in large power transistors—could even lead to transistor damage. The solution lies in arranging the circuit so that it provides temperature stability. This is called *bias stabilization*.

One method of providing bias stabilization is to make any *increase* in the average value of the collector current (I_C) cause a *decrease* in the base current (I_B). A simple way of doing this is to use a resistor in the emitter lead, as shown in Fig. 29-13.

The voltage now responsible for forward biasing the base-emitter junction is given by Kirchhoff's voltage law:

$$\begin{aligned} V_{BE} &= V_{R_2} - V_{R_e} \qquad (29\text{-}16) \\ &= V_{R_2} - I_E R_e \\ &\approx V_{R_2} - I_C R_e \end{aligned}$$

The voltage across R_2 (i.e., V_{R_2}), resulting from the voltage divider of R_1 and R_2, tends to *forward bias* the base-emitter junction. But the dc portion of the emitter current (which is practically equal to the collector current) develops a voltage across the emitter resistor (R_e) that tends to *reverse bias* the base-emitter junction. Furthermore, this reverse bias increases whenever the collector current increases. This produces a reduction in forward bias, causing I_B to drop and keeping I_C nearly constant

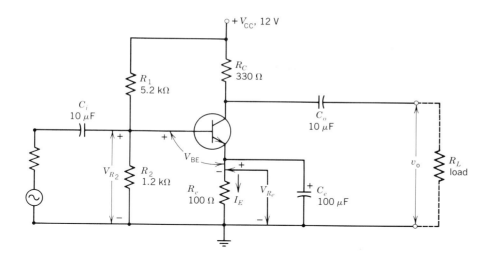

FIGURE 29-13

Typical CE ac amplifier circuit with emitter and voltage-divider bias to provide bias stabilization.

whenever I_C tries to increase due to a rise in temperature. (See Eq. 29-3.)

The resistor R_e is bypassed by the capacitor C_e so that the ac component of the emitter current will not pass through the emitter resistor. The ac component of current will take the easy path through the capacitor, bypassing the resistor. If this were not done, the ac voltage developed across R_e would reduce the ac input voltage between the base and the emitter of the transistor itself. This would decrease the ac voltage gain of the amplifier.

It should be noted that this bias stabilization is also useful in cases where a transistor is replaced by one of the same type, but with a very different β value. The circuit shown in Fig. 29-13 makes the gain virtually independent of β and more dependent on the values of bias resistors selected.

Although a CB amplifier is the most *stable* configuration and has no need of bias stabilization, the CE amplifier has the advantage of high-power gain, so the emitter feedback circuit of Fig. 29-13 is almost universally used.

29-7 FREQUENCY RESPONSE OF A CE AMPLIFIER

An important characteristic of an amplifier is how constant its voltage (and power) gain is over the band of frequencies it is designed to handle. To show this information, a frequency response curve can be plotted, either in terms of the voltage gain or the power gain variation in decibels. (See Fig. 29-14.)

The frequencies at which the voltage gain drops to

70.7% of the gain at some midfrequency (such as 400 Hz or 1 kHz) are called the upper and lower half-power frequencies (f_1 and f_2). Also, Δf, the band of frequencies between f_1 and f_2, is called the half-power or 3-dB bandwidth. Since a 3-dB change is not very noticeable to the human ear, the response is considered essentially "flat" over this bandwidth.

For acceptable audio (high-fidelity) amplifiers, f_1 is usually around 20 Hz and f_2 around 20 kHz to encompass the hearing range of most people. In some amplifiers that require a transformer in the output (to match a low-impedance speaker), the response may show a resonant rise due to distributed capacitance and inductance in the transformer.

The reason for the drop-off in response (Fig. 29-14) at *low* frequencies is the comparatively high capacitive reactance of C_i, C_o, and sometimes C_e in the circuit of Fig. 29-13. This causes a decrease in the base-emitter input voltage because of the voltage drop across C_i and a reduction in the output voltage across R_L due to the voltage drop across C_o.

The reduction in gain at *high* frequencies is due to the capacitive effects of the reverse-biased PN junctions in the transistor itself. These junction reactances (essentially X_C) become very low at high frequencies and tend to "short" the input signal to ground.

A transistor's high-frequency capability is indicated in the manufacturer's specifications by its small-signal, cut-off frequency (f_T). This is the frequency at which the short-circuit CE current gain drops to unity. Audio transistors have typical values of $f_T = 1$ MHz, while RF (radio-frequency) transistors have f_T values around 1 GHz.

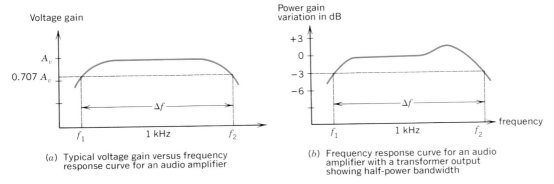

(a) Typical voltage gain versus frequency response curve for an audio amplifier

(b) Frequency response curve for an audio amplifier with a transformer output showing half-power bandwidth

FIGURE 29-14

Amplifier frequency response curves showing half-power bandwidths in terms of voltage gain and power gain in decibels.

The type of transistor configuration connection also determines the value of f_2. A CB amplifier is much better than a common emitter at high frequencies and often is found in the front end of an FM receiver operating at frequencies around 100 MHz.

29-8 NEGATIVE FEEDBACK

The circuit of Fig. 29-13 operated on the principle of emitter feedback to stabilize the transistor in the event of a temperature rise. This is a form of negative feedback.

The principle of negative feedback is a very important one and is widely used in amplifiers, communications circuits, and control systems. Basically, it involves feeding a signal (voltage or current) back from the output of an am-

plifier or system to the input of the system. The general concept is shown in Fig. 29-15.

If the switch in the feedback path is open, there is *no* feedback, and the voltage gain of the amplifier is

$$A_v = \frac{V_o}{V_i} \qquad (29\text{-}12)$$

With the switch in Fig. 29-15 closed, a portion of the *new* output, equal to kV'_o is added to the input. This changes the *new* input to the amplifier to $v_i + kV'_o$. Whether this increases or decreases the input depends upon the phase relationships or the signs (positive or negative) of k and A_v. In *negative feedback*, the input is deliberately decreased.

It can be shown that the overall gain of the amplifier with feedback is now given by

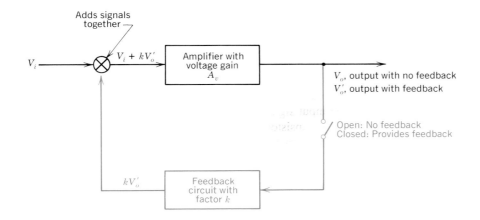

FIGURE 29-15

Block diagram showing how feedback can be introduced to an amplifier. Negative feedback results if kA_v is negative; positive feedback results if kA_v is positive.

$$A_{vf} = \frac{V'_o}{V_i} = \frac{A_v}{1 - kA_v} \qquad (29\text{-}17)$$

where: A_{vf} is the voltage gain with feedback
A_v is the voltage gain without feedback
k is the feedback circuit factor, usually between 0 and 1

Note that kA_v is the *loop gain*. For negative feedback, kA_v is negative, the denominator *increases* and A_{vf} decreases. But if kA_v is positive (*positive* feedback), the denominator of Eq. 19-17 decreases, and A_{vf} increases. (Such increases can convert the amplifier to an oscillator.)

This is clearly shown in Example 29-5.

EXAMPLE 29-5

An amplifier has a voltage gain of -40. Calculate the overall voltage gain when feedback is provided as follows:
a. 10% *negative* feedback.
b. 20% *negative* feedback.
c. 1% *positive* feedback.

Solution

a. When $k = 0.1$ (10%); $kA_v = 0.1 \times -40 = -4$

$$A_{vf} = \frac{A_v}{1 - kA_v} \qquad (29\text{-}17)$$

$$= \frac{-40}{1 - 0.1 \times (-40)} = \frac{-40}{1 - (-4)}$$

$$= \frac{-40}{1 + 4} = \mathbf{-8}$$

b. When $k = 0.2$ (20%); $kA_v = 0.2 \times -40 = -8$

$$A_{vf} = \frac{A_v}{1 - kA_v} \qquad (29\text{-}17)$$

$$= \frac{-40}{1 - 0.2 \times (-40)} = \frac{-40}{1 - (-8)}$$

$$= \frac{-40}{1 + 8} = \mathbf{-4.4}$$

Note that the gain has decreased because negative feedback has increased from 10 to 20%. (The denominator has increased.)

c. $k = -0.01$ (1%); $kA_v = -0.01 \times -40 = +0.4$

$$A_{vf} = \frac{A_v}{1 - kA_v} \qquad (29\text{-}17)$$

$$= \frac{-40}{1 - (-0.01)(-40)} = \frac{-40}{1 - 0.4} = \mathbf{-66.7}$$

Note that in the case of *positive* feedback, the denominator of Eq. 29-17 decreases, causing the overall gain (with feedback) to increase.

29-8.1 Reasons for Using Negative Feedback

It can be seen from the above example that *negative* feedback *reduces* the voltage gain, while *positive* feedback *increases* the voltage gain. Why then should anyone want to use negative feedback, if it is going to decrease voltage gain? The reduction in gain is the price that must be paid to obtain the following benefits with negative feedback:

1. A *reduction* in *distortion* of the amplifier's output.
2. A *widening* of the amplifier's *bandwidth*.
3. Often, an *increase* in *input resistance*.
4. Often, a *decrease* in *output resistance*.
5. Sometimes, a *decrease* in amplifier *noise*.
6. *Stabilized* voltage gain (effects of changes reduced in the active device itself).

Almost all commercial amplifiers, like the one shown in Fig. 29-16, use some form of negative feedback to achieve these benefits. These characteristics are *improved* by the same factor, $1 - kA_v$ (recall that the term kA_v is negative), as the voltage gain is *reduced*. That is,

$$R_{if} = R_i(1 - kA_v) \qquad (29\text{-}18)$$

where: R_{if} is the input resistance *with* feedback
R_i is the input resistance *without* feedback

$$BW_f = BW(1 - kA_v) \qquad (29\text{-}19)$$

where: BW_f is the bandwidth *with* feedback
BW is the bandwidth *without* feedback

$$D_f = \frac{D}{1 - kA_v} \qquad (29\text{-}20)$$

where: D_f is the distortion *with* feedback
D is the distortion *without* feedback

It can be shown that the *gain-bandwidth product* of an amplifier is a *constant* (for a given transistor) by multiplying Eqs. 29-17 and 29-19:

$$A_{vf} = \frac{V'_o}{V_i} = \frac{A_v}{1 - kA_v} \qquad (29\text{-}17)$$

$$BW_f = BW(1 - kA_v) \qquad (29\text{-}19)$$

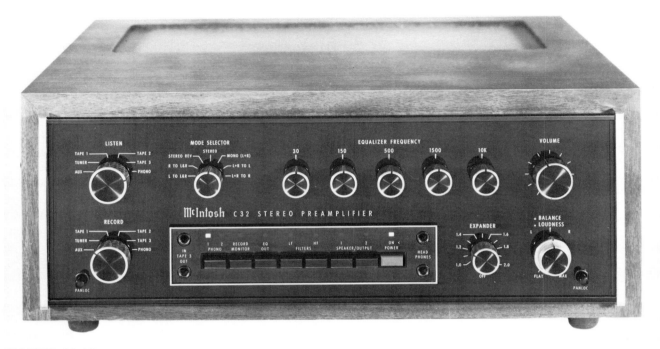

FIGURE 29-16

Commercial audio amplifier. (Courtesy of McIntosh Laboratory, Inc.)

$$A_{vf} \times BW_f = \frac{A_v}{(1 - kA_v)} \times BW(1 - kA_v)$$

or $\quad A_{vf} \times BW_f = A_v \times BW \qquad$ **(29-21)**

where: $A_{vf} \times BW_f$ is the gain-bandwidth product *with* feedback

$\quad A_v \times BW$ is the gain-bandwidth product *without* feedback

This important equation shows that the bandwidth can only be *increased* to a higher value (BW_f) at the expense of a corresponding *reduction* in gain to a lower value (A_{vf}).

EXAMPLE 29-6

A single-stage CE amplifier has an output of 6 V p–p when a 100-mV p–p input is applied. The amplifier's input resistance is 2 kΩ; its bandwidth is 10 kHz; and its distortion is 6%. After 15% negative feedback is introduced to the amplifier, calculate:

a.   The voltage gain.

b.   The input resistance.

c.   The bandwidth.

d.   The distortion.

e.   The gain-bandwidth product. Compare this with the gain-bandwidth product before feedback.

Solution

a. $A_v = \dfrac{v_o}{v_i}$ $\qquad\qquad$ **(29-12)**

$\quad = \dfrac{-6 \text{ V p–p}}{0.1 \text{ V p–p}} = -60$

(The negative sign indicates the inherent phase inversion of a single-stage CE amplifier.)

$\quad A_{vf} = \dfrac{A_v}{1 - kA_v}$ $\qquad\qquad$ **(29-17)**

$\qquad = \dfrac{-60}{1 - 0.15 \times (-60)} = \dfrac{-60}{1 + 9} = \mathbf{-6}$

b. $R_{if} = R_i(1 - kA_v)$ $\qquad\qquad$ **(29-18)**

$\qquad = 2 \text{ k}\Omega(1 - 0.15 \times (-60))$

$\qquad = 2 \text{ k}\Omega(1 + 9) = \mathbf{20 \text{ k}\Omega}$

c. $BW_f = BW(1 - kA_v)$ $\qquad\qquad$ **(29-19)**

$$= 10 \text{ kHz } (1 - 0.15 \times (-60))$$
$$= 10 \text{ kHz } (1 + 9) = \textbf{100 kHz}$$

d. $\quad D_f = \dfrac{D}{1 - kA_v}$ (29-20)

$$= \dfrac{6\%}{1 - 0.15(-60)} = \dfrac{6\%}{1 + 9} = \textbf{0.6\%}$$

e. Gain-bandwidth product $= A_{vf} \times BW_f$ (29-21)
$$= (-6) \times 100 \text{ kHz}$$
$$= \textbf{−600 kHz}$$

Before feedback, gain-bandwidth

$$\text{product} = A_v Bw$$
$$= (-60) \times 10 \text{ kHz}$$
$$= \textbf{−600 kHz}$$

29-8.2 Introduction of Negative Feedback in a CE Amplifier

Figure 29-17 shows how ac negative feedback can be introduced in a CE amplifier by not bypassing all of the emitter resistor. Note that this circuit is essentially the same as that of Fig. 29-12, with only a portion of the emitter resistor shunted by the capacitor. The unshunted portion of R_e is R_{e1} in Fig. 29-17.

Such shunting causes an ac voltage to be developed across R_{e1} instead of the entire emitter resistor R_e. (The circuit acts exactly as Fig. 29-13 using Eq. 29-16.) The voltage across R_{e1} directly subtracts from the input voltage (v_i), reducing the base-emitter input voltage and thus the output (v_o). That is,

$$v_{be} = v_i - v_{R_{e1}}$$ (29-22)

Note the similarity of Eqs. 29-22 and 29-16.

It can be shown that the amount of voltage fed back is given by

$$k = \dfrac{R_{e1}}{R_o}$$ (29-23)

where: k is the feedback factor, dimensionless
$\quad R_{e1}$ is the unbypassed emitter resistor, in ohms (Ω)
$\quad R_o$ is the total ac load resistance equal to $R_C \| R_L$, in ohms

Since the voltage gain (A_v) of a CE amplifier is negative (phase inversion), the product kA_v is negative, producing negative feedback in this case.

EXAMPLE 29-7

A CE amplifier with a 1.5-kΩ collector resistor has an 82-Ω resistor in the emitter circuit. When R_e is completely bypassed, the amplifier has an input resistance of 1.2 kΩ. Given that the transistor's β is 100, calculate:

a. The amplifier's voltage gain (without feedback).
b. The voltage gain when the emitter bypass capacitor is not connected (with feedback).
c. The new voltage gain, with feedback, when a 1-kΩ load, R_L, is added.

Solution

a. $\quad A_v = A_i \dfrac{R_o}{R_i} \approx -\beta \dfrac{R_o}{R_i}$ (29-13)

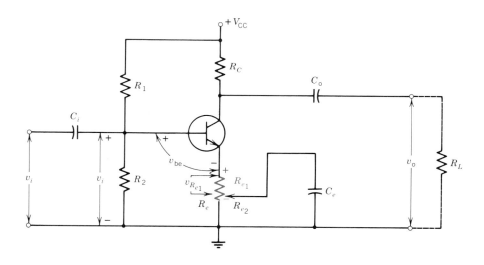

FIGURE 29-17
Showing how negative feedback can be introduced in a CE amplifier due to the unbypassed portion of the emitter resistor, R_{e1}.

$$R_o = R_C, \quad A_v \approx -100 \times \frac{1500 \ \Omega}{1200 \ \Omega} = -125$$

b.
$$k = \frac{R_e}{R_o} = \frac{R_e}{R_C} \qquad (29\text{-}23)$$

$$= \frac{82 \ \Omega}{1500 \ \Omega} = 0.055 \quad (5.5\%)$$

$$A_{vf} = \frac{A_v}{1 - kA_v} \qquad (29\text{-}17)$$

$$= \frac{-125}{1 - 0.055 \times (-125)} = \frac{-125}{1 + 6.88}$$

$$= -15.9$$

c. With $R_L = 1 \ \mathrm{k\Omega}$ added, the voltage gain without feedback is changed because R_o is now

$$R_o = R_C \parallel R_L = \frac{R_C \times R_L}{R_C + R_L} \qquad (6\text{-}10)$$

$$= \frac{1500 \ \Omega \times 1000 \ \Omega}{1500 \ \Omega + 1000 \ \Omega} = 600 \ \Omega$$

The new voltage gain *without* feedback is given by

$$A_v = A_i \frac{R_o}{R_i} \approx -\beta \frac{R_o}{R_i} \qquad (29\text{-}13)$$

$$\approx -100 \times \frac{600 \ \Omega}{1200 \ \Omega} = -50$$

The new feedback factor is

$$k = \frac{R_e}{R_o} \qquad (29\text{-}23)$$

$$= \frac{82 \ \Omega}{600 \ \Omega} = 0.137 \ (13.7\%)$$

The new voltage gain *with* feedback is

$$A_{vf} = \frac{A_v}{1 - kA_v} \qquad (29\text{-}17)$$

$$= \frac{-50}{1 - 0.137 \times (-50)} = \frac{-50}{1 + 6.85} = -6.4$$

29-9 POSITIVE FEEDBACK

Example 29-8 shows the effect of increasing the amount of positive feedback in an amplifier until the loop gain (kA_v) equals one.

EXAMPLE 29-8

An amplifier has a voltage gain of 40 without feedback. Determine the voltage gains when positive feedback of the following amounts is applied:

a. $k = 0.01$.

b. $k = 0.02$.

c. $k = 0.025$.

Solution

a. $A_{vf} = \dfrac{A_v}{1 - kA_v} \qquad (29\text{-}17)$

$$= \frac{40}{1 - 0.01 \times 40} = \frac{40}{0.6} = 66.7$$

b. $A_{vf} = \dfrac{A_v}{1 - kA_v} \qquad (29\text{-}17)$

$$= \frac{40}{1 - 0.02 \times 40} = \frac{40}{0.2} = 200$$

c. $A_{vf} = \dfrac{A_v}{1 - kA_v} \qquad (29\text{-}17)$

$$= \frac{40}{1 - 0.025 \times 40} = \frac{40}{0} = ?$$

Equation 29-17 suggests that the gain will be infinite when the loop gain reaches the critical value of $kA_v = +1$. Of course, the output voltage cannot be infinite—the amplifier will stop amplifying and start oscillating. That is, Eq. 29-17 no longer holds and an output voltage will develop without the need for any separate input. If the feedback path contains a frequency selective network, the requirement that $kA_v = 1$ is met at only one frequency. Sinusoidal oscillations result, and you have a sine wave oscillator.

29-9.1 Wien Bridge Oscillator Circuit

A circuit that can produce sine wave oscillations with a distortion of less than 1% is shown in Fig. 29-18.

The cascaded two-stage CE amplifier provides a 360° phase shift so that the output (V_o) is of the correct phase to feed back to the input. However, to ensure stable operation at a single frequency, the series-parallel RC network is used in the feedback path. It can be shown that V_2 is in phase with V_o only when $R = X_C$, at which point $V_2 = \frac{1}{3}V_o$. The amount of unbypassed emitter resistance in the first stage can be varied (through negative feedback) to provide the correct amount of overall gain so that kA_v is only slightly greater than one (for minimum distortion).

Variable frequency operation from 20 Hz to 200 kHz is possible through simultaneous adjustment of the two capacitors. The frequency of oscillation is given by

$$f = \frac{1}{2\pi RC} \qquad (29\text{-}24)$$

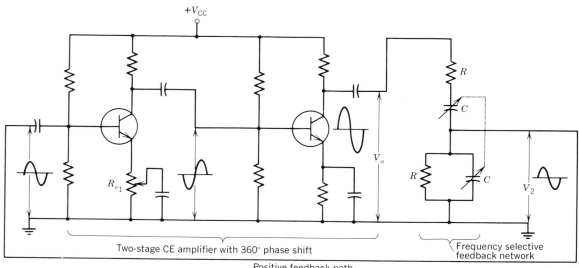

FIGURE 29-18

Variable frequency Wien bridge oscillator. An example of how positive feedback causes oscillations.

where: f is the frequency of oscillation, in hertz (Hz)
    R is the resistance, in ohms (Ω)
    C is the capacitance, in farads (F)

EXAMPLE 29-9

What is the frequency of oscillation in a Wien bridge oscillator in which $R = 10\ k\Omega$ and $C = 200\ pF$?

Solution

$$f = \frac{1}{2\pi RC} \qquad (29\text{-}24)$$

$$= \frac{1}{2\pi \times 10 \times 10^3\ \Omega \times 200 \times 10^{-12}\ F}$$

$$= \mathbf{79.6\ kHz}$$

There are many different forms of oscillators. Some, designed to produce radio frequencies, use resonant LC tank circuits. Examples of these are the Hartley and Colpitts oscillators found in communications transmitters and receivers.

Nonsinusoidal oscillators often produce a square wave output. They are referred to as multivibrators and usually involve two interconnected transistors that are switched alternately full-on, then full-off. Timing is accomplished by means of the discharging rate of a capacitive circuit.

It should be noted that many of the amplifying and oscillating functions mentioned in this chapter can be found in encapsulated integrated circuits. (Discrete components are used, however, where the power levels are very high.) Nevertheless, it is often necessary to connect external components to these ICs for proper operation. This makes a working knowledge of each part of an amplifier a good preparation for using ICs.

SUMMARY

1. Bipolar junction transistors may be of either the *PNP* or *NPN* type.
2. The emitter-base and collector-base junctions of transistors can be checked with an ohmmeter (the ohmmeter midranges should be used to avoid damaging the transistors).

3. In normal amplifying action, the emitter-base junction is always forward biased and the collector-base junction is always reverse biased.

4. In either a CB or CE connection, majority carriers are swept across the narrow, lightly doped base to the collector. The resulting collector current is controlled by a relatively small base current that is the result of the forward-biased, emitter-base junction.

5. The forward current transfer ratio in a CE configuration is called h_{fe} or β (beta) and is the ratio of the collector current to the base current; in a CB configuration, it is called the h_{fb} or α (alpha) and is the ratio of the collector current to the emitter current.

6. The ratio β ranges from 10 for power transistors up to 500 for small-signal transistors; α is usually only slightly less than one.

7. Alpha can be obtained from beta, using

$$\alpha = \frac{\beta}{1 + \beta}$$

8. A CE amplifier involves superimposing an ac signal current on a dc base bias current. The corresponding variations in collector current (magnified by an amount equal to β) cause an ac voltage to be developed across the collector resistor. This output voltage is 180° out of phase with the input signal.

9. A CE amplifier can be operated from a single power supply and provides both current and voltage gain to give the largest power gain.

10. The power gain can be calculated in decibels using

$$dB = 10 \log\frac{P_o}{P_i}$$

11. Most practical CE amplifiers use emitter self-bias and a voltage divider, forward-bias circuit to provide bias stabilization. This reduces the problems arising from temperature increases.

12. An amplifier's frequency response curve shows a drop in voltage gain at low frequencies because of circuit capacitances, and at high frequencies due to transistor junction capacitances.

13. The voltage gain of an amplifier that uses voltage feedback is given by

$$A_{vf} = \frac{A_v}{1 - kA_v}$$

In the above equation:
a. If kA_v is *negative*, the gain is *reduced* due to negative feedback.
b. If kA_v is *positive*, the gain is *increased* and may cause oscillations.

14. The beneficial effects of negative feedback include a reduction in distortion, greater stability, and an increase in bandwidth.

15. Negative feedback can be introduced in a CE amplifier by means of an unbypassed resistor in the emitter lead.

16. The frequency of oscillation of a Wien bridge oscillator can be varied by changing the two capacitors in the frequency selective network:

$$f = \frac{1}{2\pi RC}$$

At this point, the output is in phase with the input, thereby sustaining oscillations due to positive feedback.

SELF-EXAMINATION

Answer true or false.
(Answers at back of book)

29-1. The symbol for a *PNP* transistor has the arrow on the collector pointing inward. _____

29-2. The two-diode equivalent circuit for an *NPN* transistor has the base-emitter diode pointing out and the collector-base diode pointing in.

29-3. The base, emitter, and collector regions must be manufactured with exactly the same thicknesses. _____

29-4. For an *NPN* transistor an ohmmeter connected from collector to base, positive at the collector, will have a high reading. _____

29-5. A transistor manufactured using the gaseous diffusion technique has similar sizes of collector and emitter. _____

29-6. The reason for connecting the collector to the metal case in a power transistor is to help dissipate heat. _____

29-7. The location of the base terminal in a signal-type transistor is always in the middle between the collector and emitter. _____

29-8. In normal amplifying action, both the emitter-base junction and the collector-base junction must be forward biased. _____

29-9. If the base is open in a CE amplifier, the current that flows in the collector is the leakage current called I_{CEO}. _____

29-10. A larger emitter-base forward bias causes more minority carriers to flow from the emitter to the collector for either the *NPN* or *PNP*. _____

29-11. The characteristic curves for a common emitter consist of graphs of the collector current against the collector-to-emitter voltage with the base current as the control. _____

29-12. If, in a common emitter, the collector current is 15 mA when the base current is 50 μA, the transistor's β is 30. _____

29-13. If a transistor's α is 0.99, the base current will be 0.1 mA when the emitter current is 10 mA. _____

29-14. The collector characteristic curves for a common-base are essentially horizontal lines with the emitter and collector currents practically equal. _____

29-15. If a transistor's α = 0.99, its β = 101. _____

29-16. If a transistor's β = 99, its α = 0.99. _____

29-17. In a CE amplifier the base dc bias current must be at least equal to the amplitude of the input ac signal current. _____

29-18. In a CE amplifier, as the base input current increases, the collector current also increases. _____

29-19. A single supply can be used in a CE amplifier by connecting a resistor from the collector supply voltage to the base. _____

29-20. Input and output capacitors are used for blocking direct current. _____

29-21. Phase inversion occurs between the input and output voltages in a common emitter. _____

29-22. A CE amplifier provides current and power gain but no voltage gain. _____

29-23. A CB amplifier provides both current and voltage gain. _____

29-24. A bypassed emitter resistor provides bias stabilization due to dc negative feedback. _____

29-25. Removing the bypass capacitor in a CE amplifier introduces ac negative feedback but still provides dc bias stabilization. _____

29-26. A frequency response curve shows how constant an amplifier's output is over a band of frequencies. _____

29-27. A reduction in gain at low and high frequencies is due to capacitive effects either in the amplifier circuit or within the transistor itself. _____

29-28. The CE amplifier is noted for its high power gain and the CB amplifier for its high-frequency capability. _____

29-29. Negative feedback is used in an amplifier to increase its voltage gain. _____

29-30. The advantages of negative feedback include reduced distortion and reduced bandwidth. _____

29-31. An oscillator involves positive feedback in which a signal is fed back from the output, in proper phase, to the input. _____

29-32. Oscillators can be made to produce outputs at different frequencies by varying components in the feedback network. _____

REVIEW QUESTIONS

1. Draw the symbol for an *NPN* transistor and its two-diode equivalent circuit, indicating the readings of ohmmeters connected across the three pairs of terminals for each ohmmeter polarity.

2. What is the advantage of the collector region's being physically much larger in area than the emitter?

3. How is a *PNP* type distinguished from an *NPN* type in many foreign transistors?

4. a. In what manner are the emitter-base and collector-base junctions biased in a *PNP* transistor when used in an amplifier?
   b. Does it matter whether it is used in a common base or a common emitter?

5. a. Why are the leakage currents of a transistor very temperature-dependent?
   b. How are they designated?

6. Explain how it is possible for current to flow from the emitter to the collector if the collector-base junction is reverse biased.

7. Why does an increase in base current cause an increase in collector current in a common emitter?

8. Of what is β a measure? How is it related to α?

9. a. Sketch a CE amplifier circuit with a *PNP* transistor and indicate clearly the polarities of the base and collector supply voltages.
   b. Do the same with a CB amplifier, also using a *PNP* transistor.
   c. Do you see why a CE amplifier is preferred over a CB as far as supply voltages are concerned?

10. Explain in your own words why phase inversion occurs between the input and output voltages in a CE amplifier.

11. Compare the magnitudes of the input currents in a CB and CE (for the same emitter-base voltage) and suggest which has the higher input resistance.

12. Distinguish between the following notations:
   a. i_c, i_C, and I_C
   b. $(I_E = I_C + I_B)$ and $(i_E = i_C + i_B)$
   c. V_{BB} and V_{CC}

13. Describe in your own words the origin of the name "transistor."

14. Describe how you would experimentally obtain a transistor's

    a.   ac beta.

    b.   dc alpha.

15. If A_i is less than one in a CB amplifier, how is it possible to get any power gain?

16. a. Sketch the circuit for a *PNP* CE amplifier with emitter and voltage divider bias.

    b. Indicate the polarities of the dc voltages across R_2, R_e, and the base-emitter junction.

    c. Describe in your own words how this circuit is stabilized against any increases in ambient temperature.

17. In the circuit of Question 16, describe what happens to the collector current if the transistor is replaced with one having a much lower value of β.

18. What factors are responsible for the drop-off in voltage gain at low and high frequencies?

19. Why are f_1 and f_2 referred to as half-power frequencies in a frequency response curve?

20. a. Describe why negative feedback, which reduces the output voltage, *widens* the bandwidth of an amplifier.

    b. Sketch the typical frequency response curves of an amplifier before and after adding negative feedback.

21. Why do you think kA_v is referred to as *loop gain* in Eq. 29-17?

22. Refer to Fig. 29-17. What happens to the following characteristics as the variable contact on R_e is moved upward?

    a. Voltage gain.

    b. Output distortion.

    c. Frequency response.

    d. Input resistance.

23. a. If an amplifier has a positive feedback loop, what condition can cause oscillations?

    b. What are oscillations?

24. a. In Fig. 29-18, explain why V_o is in phase with the input to the first stage.

    b. If a second stage was not used, what would the feedback network have to do?

25. Assume that the circuit of Fig. 29-18 is providing oscillations.

    a. Describe what happens to V_o, in quantity and quality, as the moving contact on R_{e1} is moved downward.

    b. What extreme condition could result?

26. a. What effect will a reduction of capacitance have on the frequency of oscillation in the Wien bridge oscillator?

    b. Could this have been achieved by a similar reduction of resistance?

    c. Why is it preferable to reduce the capacitance and not the resistance?

PROBLEMS

(Answers to odd-numbered problems at back of book)

29-1. Determine the unknown current given the following:

    a. $I_B = 300$ μA, $I_C = 15$ mA, $I_E = $?

    b. $I_E = 10.1$ mA, $I_C = 9.9$ mA, $I_B = $?

    c. $I_B = 250$ μA, $I_E = 25$ mA, $I_C = $?

29-2. Determine the unknown current given the following:
  a. $I_E = 6.4$ A, $I_C = 5.9$ A, $I_B = ?$
  b. $I_B = 100$ mA, $I_C = 2.5$ A, $I_E = ?$
  c. $I_E = 250$ mA, $I_B = 2.5$ mA, $I_C = ?$

29-3. For the information given in the three parts of Problem 29-1, determine:
  a. The value of β for each part.
  b. The value of α for each part.

29-4. For the information given in the three parts of Problem 29-2, determine:
  a. The value of β for each part.
  b. The value of α for each part.

29-5. For the CE characteristic curves of a silicon NPN transistor given in Fig. 29-19, determine $\beta_{dc} = h_{FE}$ at the following points:
  a. $I_B = 20$ µA, $V_{CE} = 15$ V
  b. $I_B = 100$ µA, $V_{CE} = 15$ V
  c. $I_B = 180$ µA, $V_{CE} = 15$ V
  d. Sketch a graph of β_{dc} versus I_C, for the three calculated values of β_{dc}.

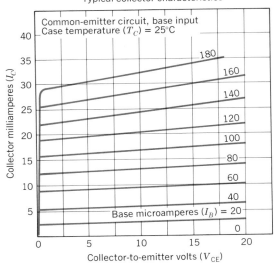

FIGURE 29-19
Collector characteristics for Problems 29-5 through 29-8.

29-6. Repeat Problem 5 using $V_{CE} = 2.5$ V instead of 15 V. Sketch the curve of β_{dc} versus I_C on the same axes as in Problem 29-5 and comment on the results.

29-7. Using the CE characteristic curves in Fig. 29-19 determine $\beta_{ac} = h_{fe}$ at $V_{CE} = 15$ V for various average collector currents from 2 to 35 mA. Sketch a curve of β_{ac} versus I_C.

29-8. Repeat Problem 29-7 using $V_{CE} = 2.5$ V instead of 15 V.

29-9. A transistor's $\beta = 150$. If its base current is 0.15 mA, calculate the collector and emitter currents.

29-10. A transistor's $\beta = 200$. If its collector current is 45 mA, calculate the base and emitter currents.

29-11. The emitter current in a transistor is 3 A. If the transistor's $\beta = 20$, calculate the base and collector currents.

29-12. Find the value of α for the transistors in:
  a. Problem 29-9.

b. Problem 29-10.

c. Problem 29-11.

29-13. A transistor in the CB connection has its emitter current increased from 15 to 20 mA. This caused an increase in its base current from 0.32 to 0.48 mA.

a. Calculate α_{ac}.

b. Determine β_{ac} in two ways.

29-14. A transistor in the CE connection has its base current increased from 15 to 20 μA, causing an increase in the collector current from 6 to 8 mA.

a. Calculate h_{fe}.

b. Determine h_{fb} in two ways.

29-15. a. A germanium transistor's collector current is 51 mA when the base current is 0.4 mA. If $\beta_{dc} = 125$, calculate the transistor's collector cutoff current I_{CEO}.

b. Why would this be an impractical method for determining a silicon transistor's I_{CEO}?

29-16. A silicon power transistor has $\beta_{dc} = 180$ and $I_{CEO} = 40$ mA. Calculate the collector current for the following conditions:

a. $I_B = 0$.

b. $I_B = 10$ mA.

c. $I_B = 20$ mA.

29-17. A suitably biased CE amplifier circuit, as in Fig. 29-12b, has a transistor with $\beta = 125$ and $R_C = 2.2$ kΩ. An input voltage of 50 mV peak to peak causes a 40-μA peak-to-peak variation in the base current. Its 3-dB bandwidth is 15 kHz with a distortion of 4%. Calculate:

a. The peak-to-peak variation in the collector current.

b. The current gain, A_i.

c. The peak-to-peak output voltage.

d. The voltage gain, A_v.

e. The input resistance, $R_i = \dfrac{\text{input voltage}}{\text{base current}}$.

f. The voltage gain, A_v, using Eq. 29-13. Compare with part (d).

g. The power gain, A_p, in dB.

29-18. Repeat Problem 17 using:

a. $R_C = 3.3$ kΩ.

b. $\beta = 100$ and $R_C = 1$ kΩ.

29-19. For the amplifier conditions in Problem 17, with $V_{BE} = 0.6$ V and $I_B = 40$ μA, determine:

a. V_{CE}.

b. V_{CB}.

c. The value of R_B if $V_{CC} = 30$ V.

29-20. Repeat Problem 29-19 using $R_C = 3.3$ kΩ.

29-21. To provide bias stabilization, a 100-Ω resistor (bypassed adequately with a capacitor) is placed in the emitter lead of the amplifier in Problem 29-17. Calculate the value of R_B to maintain $V_{BE} = 0.6$ V, $I_B = 40$ μA, with $V_{CC} = 30$ V. [Compare with Problem 19(c).]

29-22. If a voltage-divider, $R_1 - R_2$, is used instead of R_B to provide forward bias for the amplifier in Problem 29-21, what voltage must exist across R_2?

29-23. An amplifier has a power output of 20 W and a voltage gain of 200 at a frequency of 1 kHz. When operated at the extremes of its -3-dB bandwidth calculate:

    a.   The power output of the amplifier.

    b.   The voltage gain of the amplifier.

    c.   The rms output current if the input voltage is 50 mV.

29-24. Repeat Problem 29-23 using -6 dB instead of -3 dB.

29-25. An amplifier has a voltage gain of 150. Calculate the new voltage gain for the following conditions:

    a.   5% negative feedback introduced.

    b.   10% negative feedback introduced.

    c.   0.5% positive feedback introduced.

29-26. Repeat Problem 29-25 with voltage gain $= 100$.

29-27. For the amplifier in Problem 29-17 assume a 100-Ω unbypassed emitter resistor is added as in Fig. 29-17. Calculate:

    a.   The feedback factor, k.

    b.   The new voltage gain.

    c.   The new input resistance.

    d.   The new bandwidth.

    e.   The new distortion.

    f.   The gain-bandwidth product before and after adding the 100-Ω resistor.

    g.   The current gain, A_i.

    h.   The new power gain, in dB.

29-28. Repeat Problem 29-27 using $R_C = 3.3$ kΩ.

29-29. Repeat Problem 29-27 assuming a 1-kΩ load resistor, R_L, is added, as in Fig. 29-17.

29-30. If $C = 50$ pF is the lowest practical value to use in the Wien bridge oscillator circuit of Fig. 29-18, determine the value of R and the maximum value of C to provide a range of frequencies from 20 to 200 kHz.

29-31.   a.   Determine the frequency of oscillation of a Wien bridge oscillator with $R = 5.1$ kΩ and $C = 500$ pF.

    b.   If V_2 in Fig. 29-18 is 200 mV, what is the output of the oscillator, V_o?

    c.   What must be the minimum voltage gain of the two-stage CE amplifier in Fig. 29-18?

APPENDICES

APPENDIX
A

COMMON SYMBOLS USED IN SCHEMATIC DIAGRAMS

Conductor or wire

Crossing conductors not connected

Connected conductors

Shielded single conductor

Single cell

Multiple cell battery

Alternating current source

Momentary contact push button

Single-pole single-throw switch, SPST

Single-pole double-throw switch, SPDT

Double-pole double-throw switch, DPDT

Incandescent lamp

Neon lamp

Buzzer

Bell

Loudspeaker

Motor

Generator

Fuse

Circuit breaker

Ground (earth)

Chassis

Common return path at same potential

Antenna

Microphone

Nonpolarized connector (plug)

Resistor (fixed)

Adjustable resistor (rheostat)

Potentiometer

640

Capacitor (fixed)

Capacitor, variable

Polarized capacitor

Air-cored inductor or coil

Iron-cored inductor

Adjustable inductor

Variable powdered-iron core inductor

Air-cored transformer (high frequency)

Iron-cored transformer (power)

Meter

*Insert one of the following:
A   Ammeter
V   Voltmeter
OHM Ohmmeter
P   Wattmeter
O   Oscilloscope
G   Galvanometer

Amplifier

Headphones

Diode, rectifier

Light-emitting diode

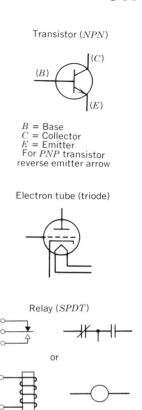

Transistor (*NPN*)

(B)   (C)

(E)

B = Base
C = Collector
E = Emitter
For *PNP* transistor
reverse emitter arrow

Electron tube (triode)

Relay (*SPDT*)

or

STANDARD RESISTANCE VALUES FOR RESISTORS WITH VARIOUS TOLERANCES

±1%	±2%	±5%	±10%	±20%	±1%	±2%	±5%	±10%	±20%
100	100	10	10	10	316	316			
102					324				
105	105				332	332	33	33	33
107					340				
110	110	11			348	348			
113					357				
115	115				365	365	36		
118					374				
121	121	12	12		383	383			
124					392		39	39	
127	127				407	407			
130		13			412				
133	133				422	422			
137					432		43		
140	140				442	442			
143					453				
147	147				464	464			
150		15	15	15	475		47	47	47
154	154				487	487			
158					499				
162	162	16			511	511	51		
165					523				
169	169				536	536			
174					549				
178	178				562	562	56	56	
182		18	18		576				
187	187				590	590			
191					604				
196	196				619	619	62		
200		20			634				

±1%	±2%	±5%	±10%	±20%	±1%	±2%	±5%	±10%	±20%
205	205				649	649			
210					665				
215	215				681	681	68	68	68
221		22	22	22	698				
226	226				715	715			
232					732				
237	237				750	750	75		
243		24			765				
249	249				787	787			
255					806				
261	261				825	825	82	82	
267					845				
274	274	27	27		866	866			
280					887				
287	287				909	909	91		
294					931				
301	301	30			953	953			
309					976				

NOTE The table shows the significant figures available for each tolerance. Resistors are available in multiples and sub-multiples, such as 2.2 Ω, 220 Ω, 2200 Ω, and so on.

APPENDIX C

AWG CONDUCTOR SIZES AND METRIC EQUIVALENTS

Gauge	Dia. (mil)	Resistance (Ω/1000 ft)	Dia. (mm)	Resistance (Ω/km)
0000	460	0.049	11.68	0.160
000	409.6	0.062	10.40	0.203
00	364.8	0.078	9.266	0.255
0	324.9	0.098	8.252	0.316
1	289.3	0.124	7.348	0.406
2	257.6	0.156	6.543	0.511
3	229.4	0.197	5.827	0.645
4	204.3	0.248	5.189	0.813
5	181.9	0.313	4.620	1.026
6	162	0.395	4.115	1.29
7	144.3	0.498	3.665	1.63
8	128.5	0.628	3.264	2.06
9	114.4	0.792	2.906	2.59
10	101.9	0.999	2.588	3.27
11	90.7	1.26	2.30	4.10
12	80.8	1.59	2.05	5.20
13	72	2	1.83	6.55
14	64.1	2.52	1.63	8.26
15	57.1	3.18	1.45	10.4
16	50.8	4.02	1.29	13.1
17	45.3	5.06	1.15	16.6
18	40.3	6.39	1.02	21.0
19	35.9	8.05	0.912	26.3
20	32	10.1	0.813	33.2
21	28.5	12.8	0.723	41.9
22	25.3	16.1	0.644	52.8
23	22.6	20.3	0.573	66.7
24	20.1	25.7	0.511	83.9
25	17.9	32.4	0.455	106
26	15.9	41	0.405	134
27	14.2	51.4	0.361	168
28	12.6	64.9	0.321	213

	Resistance		Resistance	
Gauge	Dia. (mil)	(Ω/1000 ft)	Dia. (mm)	(Ω/km)
29	11.3	81.4	0.286	267
30	10	103	0.255	337
31	8.9	130	0.227	425
32	8	164	0.202	537
33	7.1	206	0.180	676
34	6.3	261	0.160	855
35	5.6	329	0.143	1071
36	5	415	0.127	1360
37	4.5	523	0.113	1715
38	4	655	0.101	2147
39	3.5	832	0.090	2704
40	3.1	1044	0.080	3422

APPENDIX D

SOLVING LINEAR EQUATIONS USING DETERMINANTS

The process of solving a set of linear equations in two or more variables can be reduced to a purely mechanical process.

This process involves writing the coefficients of the unknowns in rows and columns to construct a *determinant*. The following is an example of a 2×2 determinant:

$$\begin{vmatrix} 4 & -5 \\ 3 & 7 \end{vmatrix}$$

A determinant can be "evaluated" as follows.

$$\begin{vmatrix} a_1 & b_1 \\ a_2 & b_2 \end{vmatrix} = a_1 \times b_2 - a_2 \times b_1$$

Thus $\begin{vmatrix} 4 & -5 \\ 3 & 7 \end{vmatrix} = 4 \times 7 - 3 \times (-5)$
$= 28 - (-15)$
$= 28 + 15 = \mathbf{43}$

Consider two equations in x and y with *known* coefficients, written in the following form:

$$a_1 x + b_1 y = c_1$$
$$a_2 x + b_2 y = c_2$$

Cramer's rule gives the solution for x and y in terms of determinants:

$$x = \frac{\begin{vmatrix} c_1 & b_1 \\ c_2 & b_2 \end{vmatrix}}{\begin{vmatrix} a_1 & b_1 \\ a_2 & b_2 \end{vmatrix}} \qquad y = \frac{\begin{vmatrix} a_1 & c_1 \\ a_2 & c_2 \end{vmatrix}}{\begin{vmatrix} a_1 & b_1 \\ a_2 & b_2 \end{vmatrix}}$$

Note that the *denominator* determinant is the same for both x and y, and is formed by writing down the coeffi-

cients of x and y in the same order that they appear in the original equations.

The *numerator* determinant is formed by repeating the denominator determinant, but with the coefficients for the variables being sought replaced by the constants. Thus, if you are solving for x, you replace a_1 and a_2 with c_1 and c_2, respectively. When solving for y, you replace b_1 and b_2 with c_1 and c_2, respectively.

EXAMPLE D-1

Solve the following equations from Example 10-1.

$$2.5I_B + 2.0I_G = 13.2$$
$$2.0I_B + 2.1I_G = 14.5$$

Solution

By Cramer's rule:

$$I_B = \frac{\begin{vmatrix} 13.2 & 2.0 \\ 14.5 & 2.1 \end{vmatrix}}{\begin{vmatrix} 2.5 & 2.0 \\ 2.0 & 2.1 \end{vmatrix}} = \frac{13.2 \times 2.1 - 2.0 \times 14.5}{2.5 \times 2.1 - 2.0 \times 2.0}$$

$$= \frac{27.72 - 29.0}{5.25 - 4.0}$$

$$= \frac{-1.28}{1.25} = \mathbf{-1.024}$$

$$I_G = \frac{\begin{vmatrix} 2.5 & 13.2 \\ 2.0 & 14.5 \end{vmatrix}}{1.25} = \frac{2.5 \times 14.5 - 2.0 \times 13.2}{1.25}$$

$$= \frac{36.25 - 26.4}{1.25}$$

$$= \frac{9.85}{1.25} = \textbf{7.88}$$

Cramer's rule applies to any number of unknowns. The following simple method of evaluating 3×3 determinants, however, does *not* apply to higher orders. Given a determinant:

$$\begin{vmatrix} a_1 & b_1 & c_1 \\ a_2 & b_2 & c_2 \\ a_3 & b_3 & c_3 \end{vmatrix}$$

its value can be found by repeating the first two columns and making the indicated products:

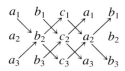

$$\text{Value} = a_1 \times b_2 \times c_3 + b_1 \times c_2 \times a_3 + c_1 \times a_2 \times b_3$$
$$- a_3 \times b_2 \times c_1 - b_3 \times c_2 \times a_1 - c_3 \times a_2 \times b_1$$

$$= 160 + 0 + 0 - 120 - 810 - 0$$
$$= 160 - 930 = \textbf{-770}$$

Denominator:
$$\begin{array}{ccccc} 3 & 0 & 2 & 3 & 0 \\ 2 & -4 & 9 & 2 & -4 \\ 0 & 9 & -4 & 0 & 9 \end{array}$$

$$\text{Value} = 3 \times (-4) \times (-4) + 0 + 2 \times 2 \times 9$$
$$-0 - 9 \times 9 \times 3 - 0$$
$$= 48 + 36 - 243 = \textbf{-159}$$

$$I_1 = \frac{-770}{-159} = \textbf{4.84}$$

For practice, solve for I_2 (-2.67) and I_3 (-2.26).

EXAMPLE D-2

Solve the following equations from Example 10-2 for I_1:

$$3I_1 + 0I_2 + 2I_3 = 10$$
$$2I_1 - 4I_2 + 9I_3 = 0$$
$$0I_1 + 9I_2 - 4I_3 = -15$$

Solution

$$I_1 = \frac{\begin{vmatrix} 10 & 0 & 2 \\ 0 & -4 & 9 \\ -15 & 9 & -4 \end{vmatrix}}{\begin{vmatrix} 3 & 0 & 2 \\ 2 & -4 & 9 \\ 0 & 9 & -4 \end{vmatrix}}$$

Numerator:
$$\begin{array}{ccccc} 10 & 0 & 2 & 10 & 0 \\ 0 & -4 & 9 & 0 & -4 \\ -15 & 9 & -4 & -15 & 9 \end{array}$$

$$\text{Value} = 10 \times (-4) \times (-4) + 0 \times 9$$
$$\times (-15) + 2 \times 0 \times 9$$
$$-(-15) \times (-4) \times 2 - 9 \times 9$$
$$\times 10 - (-4) \times 0 \times 0$$

APPENDIX E

EFFECT OF ANGLE ON THE VOLTAGE INDUCED IN A MOVING CONDUCTOR

Consider a conductor moving at velocity v through a magnetic field B at some angle θ with the field, as shown in Fig. E-1.

After a period of time Δt, the conductor has moved a distance Δs. The only part of this distance that contributes anything to the induced voltage, however, is the *horizontal* component ($\Delta s'$), which actually *cuts* the field. The relationship between $\Delta s'$ and Δs obviously has something to do with the angle θ. This relationship (See Appendix F) is given by

$$\frac{\Delta s'}{\Delta s} = \sin \theta$$

For example, if θ is a 30° angle, $\sin \theta = 0.5$. This simply means that $\Delta s'$ is one-half the distance Δs. Thus

$$\Delta s' = \Delta s \times \sin \theta$$

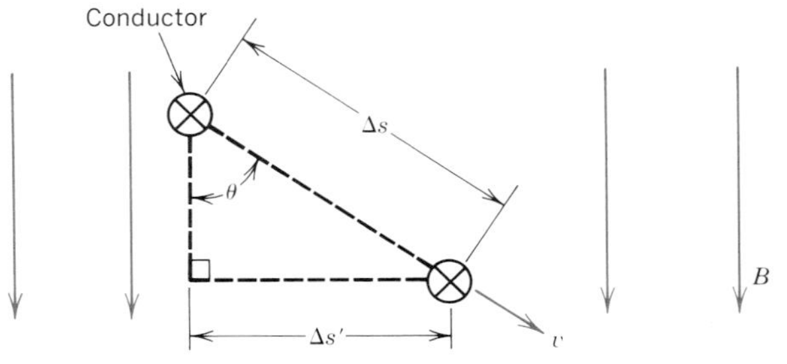

But
$$v_{\text{ind}} = B \times l \times \frac{\Delta s'}{\Delta t}$$
$$= B \times l \times \frac{\Delta s}{\Delta t} \times \sin \theta$$

And
$$\frac{\Delta s}{\Delta t} = v$$

Thus
$$V_{\text{ind}} = Blv \sin \theta = V_m \sin \theta \qquad (14\text{-}5)$$

FIGURE E-1
Conductor being moved at velocity *v* through a magnetic field *B* at an angle θ.

648

APPENDIX F

TRIGONOMETRIC RATIOS

For any right-angled triangle, no matter what its size, the ratio of one side to another has a definite value, depending only upon the angle θ.

Consider the triangle shown in Fig. F-1. The side opposite the right angle is called the *hypotenuse*. As far as the angle θ is concerned, side a is the *opposite* side and side b is the *adjacent* side.

The name *sine* θ is given to the ratio of the opposite side to the hypotenuse in terms of the angle θ.

$$\sin \theta = \frac{\text{opposite side}}{\text{hypotenuse}} = \frac{a}{c}$$

where sin θ is an abbreviation for sine θ.

Similarly, two other common ratios are named:

$$\cos \theta = \text{cosine } \theta = \frac{\text{adjacent side}}{\text{hypotenuse}} = \frac{b}{c}$$

$$\tan \theta = \text{tangent } \theta = \frac{\text{opposite side}}{\text{adjacent side}} = \frac{a}{b}$$

Each of these three ratios, called trigonometric ratios or functions, has its own numerical value for any given angle. Values of these ratios for angles between 0° and 90° can be found in trigonometric tables. Alternatively, they can be obtained on a scientific calculator similar to the one introduced in Chapter 1. Table F-1 shows values for some common angles.

This table can be interpreted as follows: If, in a right angle, θ is 30°, the side opposite θ is only half the length of the hypotenuse (since sin 30° = 0.5). Similarly, if θ is 45°, the adjacent and opposite sides are equal in length (since tan 45° = 1).

Angles between 0° and 90° are called *acute* angles. All their trigonometric ratios are positive. It is also possible to consider the trigonometric ratios of angles from 90° to 360°, called *obtuse* angles. Some of these ratios are negative and do not specifically refer to the side of a triangle. That is, although trigonometric functions have been defined in terms of the sides of a right triangle, alternative definitions include angles of any size. For angles greater than 360°, the trigonometric ratios are a repeat of those from 0° to 360°.

A scientific calculator gives the trigonometric ratios of

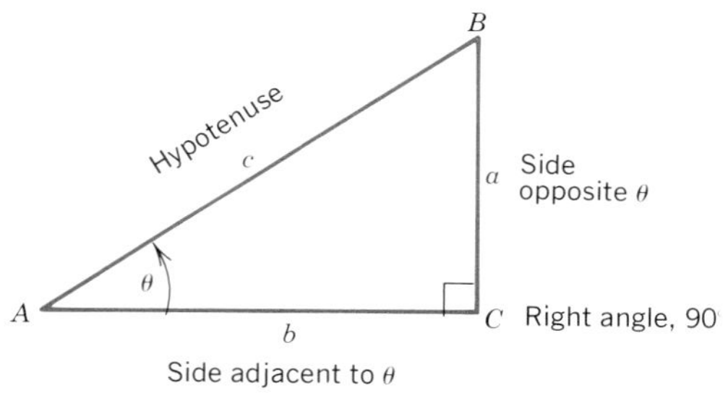

FIGURE F-1
Relation of sides to θ in a right triangle.

TABLE F-1
Trigonometric Values of Some Common Angles

Angle, θ		Sin θ	Cos θ	Tan θ
Degrees	Radians			
0	0	0	1.000	0
30	$\frac{\pi}{6}$	0.500	0.866	0.577
45	$\frac{\pi}{4}$	0.707	0.707	1.000
60	$\frac{\pi}{3}$	0.866	0.500	1.732
90	$\frac{\pi}{2}$	1.000	0	∞

any angle, given in radians or degrees, both in magnitude and sign (positive or negative). Thus

$$\sin 270° = -1$$
$$\cos 390° = 0.866$$
$$\tan 120° = -1.732$$
$$\sin 2 \text{ rad} = 0.909$$

INVERSE TRIGONOMETRIC FUNCTIONS

When the sine (or any trigonometric ratio) of an angle is known, it is possible to find the angle itself by using the *inverse* function.

For example, given $\sin \theta = 0.5$, you can say that θ is "an angle whose" sine is 0.5. Mathematically, you would express this statement for θ using the following notation:

$$\theta = \text{arc } \sin 0.5$$

where "arc" stands for "an angle whose."

Another notation used is

$$\theta = \sin^{-1} 0.5$$

where $\sin^{-1}$ also means "an angle whose."

Both of these notations refer to the *inverse sine function*. Note that $\sin^{-1}$ does *not* refer to any reciprocal.

Since, for this example, you know that the sine of 30° is 0.5, then

$$\theta = \text{arc } \sin 0.5 = \sin^{-1} 0.5 = 30°$$

In general, you can obtain the value of an inverse trigonometric function by using a scientific calculator.

For $\sin^{-1} 0.5$, enter 0.5 into the calculator, press INV, then press SIN. The result is 30° or 0.5236 rad. (Some calculators may have a SIN^{-1} button.) Note that the angle displayed is in the first quadrant. That is, $\sin 150°$ is also 0.5.

Other examples of inverse trigonometric functions are

$$\theta = \text{arc } \tan 1 = \tan^{-1} 1 = 45°$$
$$\theta = \text{arc } \cos 0.8 = \cos^{-1} 0.8 = 36.87°$$

APPENDIX G

FACTORS AFFECTING THE AMPLITUDE OF A GENERATOR'S SINE WAVE

Refer to Fig. 14-9.

Voltage induced in each conductor $= Blv \sin \theta$ (14-5)

For two conductors A and B in series, $v_{\text{ind}} = 2 Blv \sin \theta$.

We need an expression for the velocity, v:

$$v = \frac{\text{distance traveled}}{\text{time taken}}$$

$$= \frac{\text{distance traveled by one conductor in one revolution}}{\text{time for one revolution}}$$

Let $b =$ diameter of the coil.

distance traveled $=$ circumference of circle of diameter b

$$= \pi b$$

Let $n =$ the speed of rotation of the coil in rev/s.

$$\text{time for one revolution} = \frac{1}{n} \text{s}$$

Therefore,
$$v = \frac{\pi b}{1/n} = \pi bn$$

and
$$v_{\text{ind}} = 2 Blv \sin \theta$$

$$= 2 Bl\pi bn \sin \theta$$

$$= 2\pi nBA \sin \theta$$

since $bl =$ area of the coil, A.

If the coil has N turns, the total voltage induced by the coil is

$$v_{\text{ind}} = 2\pi nBAN \sin \theta = V_m \sin \theta \qquad (14\text{-}7)$$

APPENDIX H

INDUCED VOLTAGE IN TERMS OF FREQUENCY AND TIME

A sine wave of voltage can be expressed by the equation $v = V_m \sin \theta$. The angle θ can be expressed in terms of the frequency (f), which is a more readily known quantity, as follows.

Consider a generator with one pair of poles. You know that one cycle of voltage corresponds to one revolution or 360°. (This is also true for any number of pairs of poles if you use electrical degrees.)

The angle "swept out" or passed through by a coil generating a voltage of f cycles per second equals $360 f$ degrees every second.

If at $t = 0$, the angle (and voltage induced) is zero, then at some later time t (in seconds), the angle that the coil has passed through, $\theta = 360ft$ degrees. Thus, the induced voltage is now given by

$$v = V_m \sin (360ft)°$$

But $360° = 2\pi$ radians.

Thus,
$$v = V_m \sin 2\pi ft \qquad (15\text{-}3)$$

If you let $\omega = 2\pi f$ (the angular frequency in rad/s), then

$$v = V_m \sin \omega t \qquad (15\text{-}4)$$

APPENDIX I

DERIVATION OF THE INDUCTIVE REACTANCE FORMULA, X_L

Refer to Fig. I-1.

The voltage induced across the coil will be a maximum, V_m, when the current is changing at its maximum rate. Thus

since

$$v_L = L\left(\frac{\Delta i}{\Delta t}\right)$$

$$V_m = L\left(\frac{\Delta i}{\Delta t}\right)_{max} = L \times \text{initial slope of } i \text{ curve}$$

Let us find an expression for the initial slope of the current curve:

Given $i = I_m \sin 2\pi ft$, $i = 0$, when $t = 0$

Assume that a short time interval, Δt, elapses. Then

$$i = I_m \sin 2\pi f\Delta t$$

But if Δt is small (near zero), $2\pi f \Delta t$ is also small (near zero). For this condition, it can be shown that

$$\sin 2\pi f\Delta t \approx 2\pi f \Delta t \quad \text{in radians}$$

$$\left[\text{Try } \sin 1° = \sin\left(\frac{\pi}{180}\text{rad}\right) \approx \frac{\pi}{180}\right]$$

Thus, for a very small Δt, near zero,

$$i = I_m \sin 2\pi f\Delta t = I_m 2\pi f \Delta t$$

The initial slope of the i curve is given (see Fig. I-1) by

$$\text{slope} = \frac{\text{rise}}{\text{run}} = \frac{i}{\Delta t} = \frac{I_m 2\pi f \Delta t}{\Delta t} = 2\pi fI_m$$

$$V_m = L \times \text{initial slope of the } i \text{ curve}$$

Therefore, $V_m = L \times 2\pi fI_m = 2\pi fLI_m$

The effective voltage across the coil (equal to the supply voltage) is

$$V = \frac{V_m}{\sqrt{2}} = \frac{2\pi fLI_m}{\sqrt{2}}$$

also, $$I = \frac{I_m}{\sqrt{2}}$$

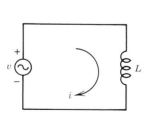

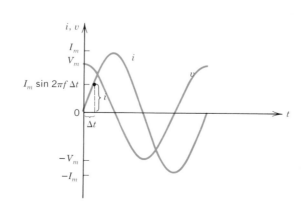

FIGURE I-1

Initial slope of current curve equals $i/\Delta t$.

Opposition to the ac current,

$$X_L = \frac{V}{I} = \frac{\dfrac{2\pi f L I_m}{\sqrt{2}}}{\dfrac{I_m}{\sqrt{2}}}$$

$$= \frac{2\pi f L I_m}{\sqrt{2}} \times \frac{\sqrt{2}}{I_m} = 2\pi f L$$

$$X_L = 2\pi f L = \omega L \qquad\qquad (20\text{-}2)$$

APPENDIX J

DERIVATION OF THE CAPACITANCE EQUATION WITH AIR DIELECTRIC

The electric field intensity between the two parallel plates of a charged capacitor is given by

$$E = \frac{V}{d} \quad \text{volts/meter (V/m)}$$

where: E is the electric field strength in volts per meter, V/m

V is the potential difference across the plates in volts, V

d is the separation of the plates in meters, m

The electric field intensity is also given by

$$E = \frac{\sigma}{\epsilon_0} \quad \text{newtons/coulomb, (N/C or V/m)}$$

where: E is the electric field strength in newtons per coulomb, N/C (or V/m)

σ is the charge per unit area on the plates in coulombs per square meter, C/m^2

ϵ_0 is the permittivity of free space $= 8.85 \times 10^{-12} \ C^2/Nm^2$

Since

$$\sigma = \frac{\text{charge on one plate}}{\text{area of one plate}} = \frac{Q}{A}$$

then

$$E = \frac{\sigma}{\epsilon_0} = \frac{Q}{\epsilon_0 A} \text{ and } Q = \epsilon_0 A E$$

Also, since

$$E = \frac{V}{d}, \quad V = dE$$

thus

$$C = \frac{Q}{V} = \frac{\epsilon_0 A E}{dE}$$

and

$$C = \epsilon_0 \frac{A}{d}$$

If the dielectric is a material other than air, having a dielectric constant K, the capacitance is

$$C = K\frac{\epsilon_0 A}{d} \qquad \text{(21-2)}$$

OBSERVATION OF A TRANSFORMER CORE'S *B-H* HYSTERESIS LOOP USING AN INTEGRATING CIRCUIT

A transformer core's *B-H* curve can be observed on an oscilloscope using the circuit shown in Fig. K-1.

The current in the transformer primary develops a voltage across R_2 that is proportional to the magnetizing intensity $\left(H = \dfrac{NI}{l}\right)$. This voltage, connected to the oscilloscope's horizontal input, deflects the beam proportionally to H, 60 times per second. Here, N is the number of turns in the primary and l is the length of the magnetic circuit of the transformer's core. If you know the value of R_2 and the peak voltage, the horizontal axis can be calibrated in ampere-turns/meter.

On the transformer secondary,

$$i = C\frac{\Delta v_C}{\Delta t} \tag{22-1}$$

$$= C\frac{dv_C}{dt} \quad \text{(using the differential form)}$$

Thus $\quad i\,dt = C\,dv_C \quad$ and $\quad dv_C = \dfrac{1}{C}i\,dt$

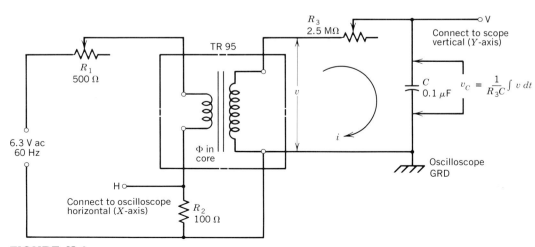

FIGURE K-1

Circuit to observe a Lissajous figure of a transformer's *B-H* hysteresis loop.

or
$$v_C = \frac{1}{C}\int i\,dt$$

The integral sign ($\int$) indicates the integrating process referred to in Section 22-9.1.

A minimum value of R_3 is approximately 1 MΩ. Since $X_C \approx 30$ kΩ, the secondary current is determined primarily by R_3:

$$i \approx \frac{v}{R_3} \quad \text{and} \quad v_C \approx \frac{1}{R_3 C}\int v\,dt$$

The capacitor voltage is the *integral* of the input voltage v.

But
$$v = N\frac{d\Phi}{dt} \qquad (14\text{-}2)$$

Therefore,
$$v_C \approx \frac{1}{R_3 C}\int N\frac{d\Phi}{dt}\times dt$$
$$= \frac{N}{R_3 C}\int d\Phi$$
$$= \frac{N\Phi}{R_3 C}$$

Since the secondary number of turns (N) is constant, as are R_3 and C, the capacitor voltage (v_C) is proportional to the core flux (Φ). Also, since A, the cross-sectional area of the core, is constant, the flux density $B = \frac{\Phi}{A}$ is proportional to v_C.

Thus the oscilloscope beam is deflected vertically proportional to the flux density (B), resulting in a Lissajous figure that is representative of the core's *B-H* characteristics.

If the horizontal and vertical axes are known for their volts/cm deflection, the two axes can be calibrated in At/m and in webers/m^2, respectively. Also, the core's permeability can be obtained quantitatively, using $\mu = \frac{B}{H}$.

APPENDIX L

DERIVATION OF THE CAPACITIVE REACTANCE FORMULA, X_C

The current will be a maximum, I_m, when the voltage is increasing at its maximum rate, at the origin (see Fig. L-1). Thus

since

$$i = C\left(\frac{\Delta v}{\Delta t}\right)$$

$$I_m = C\left(\frac{\Delta v}{\Delta t}\right)_{max} = C \times \text{initial slope of the } v \text{ curve}$$

Given $v = V_m \sin 2\pi ft$, $v = 0$ when $t = 0$

Assume a short time interval, Δt.

$$v = V_m \sin 2\pi f \Delta t$$

But if Δt is small (near zero), $2\pi f \Delta t$ is also small (near zero). For this condition, it can be shown that $\sin 2\pi f \Delta t \approx 2\pi f \Delta t$ in radians. Thus, for a very small Δt, near zero,

$$v = V_m \sin 2\pi f \Delta t = V_m 2\pi f \Delta t$$

The maximum initial slope of v curve is given by

$$\text{slope} = \frac{\text{rise}}{\text{run}} = \frac{v}{\Delta t} = \frac{V_m 2\pi f \Delta t}{\Delta t} = 2\pi f V_m$$

Therefore, $I_m = C \times$ initial slope of the v curve
$$= C \times 2\pi f V_m = 2\pi f C V_m$$

Effective capacitor current $I = \dfrac{I_m}{\sqrt{2}} = \dfrac{2\pi f C V_m}{\sqrt{2}}$

Also,
$$V = \frac{V_m}{\sqrt{2}}$$

Opposition to the ac current,

$$X_C = \frac{V}{I} = \frac{V_m/\sqrt{2}}{2\pi f C V_m/\sqrt{2}}$$

$$= \frac{V_m}{\sqrt{2}} \times \frac{\sqrt{2}}{2\pi f C V_m} = \frac{1}{2\pi f C}$$

Then
$$X_C = \frac{1}{2\pi f C} = \frac{1}{\omega C} \qquad (23\text{-}2)$$

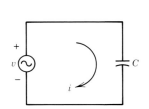

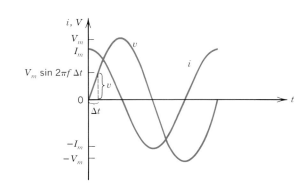

FIGURE L-1
Initial slope of voltage curve equals $v/\Delta t$.

GLOSSARY

TERM	DEFINITION	SYMBOL OR ABBREVIATION	BASIC UNIT
Admittance	A measure of the ability of a reactive circuit to permit the flow of current. The reciprocal of impedance.	Y	siemens
Alternating current	A current that continuously reverses direction because of the continuously reversing polarity of an alternating voltage.	ac	ampere
Alternator	An ac generator.		
Ammeter	An instrument used to measure electrical current.	—(A)—	
Ampere	The basic unit of electrical current (coulomb per second).	A	
Ampere-hour	The basic unit of charge storage capacity of a cell or battery. A measure of the capacity of a battery to supply current for a given period of time.	Ah	
Ampere-turn	The basic unit of magnetizing force (magnetomotive force) caused by current flowing through a coil.	At	
Ampere-turn per meter	The basic MKS unit of magnetizing intensity. Magnetomotive force per unit length of magnetic circuit.	At/m	

TERM	DEFINITION	SYMBOL OR ABBREVIATION	BASIC UNIT
Amplification	The process of increasing the voltage, current, or power of an electrical signal.		
Amplifier	A device used to increase the voltage, current, or power of an electrical signal by means of an active component, such as a transistor or electron tube.		
Amplitude	The maximum instantaneous value of an alternating voltage or current. Peak value.	V_m, I_m	volt, ampere
Analog	A measurable quantity that takes on a *continuous* set of values. Nondigital.		
Anode	The terminal of a device (such as a battery) that *loses* electrons into the external circuit during normal operation. The *negative* electrode of a battery is an anode.		
Apparent power	The power that *appears* to be delivered to a reactive circuit, found by multiplying the applied voltage and current.	P_s	VA Volt-ampere
Aquadag	A graphite coating on the inside of a cathode-ray tube for collecting secondary electrons emitted by the screen.		
Arc	A spark or flash caused by a high voltage ionizing a gas or vapor.		
Arc tangent	An inverse trigonometric function meaning "the angle whose tangent is." Also called *inverse* tangent or $\tan^{-1}$.	$\tan^{-1}$	

TERM	DEFINITION	SYMBOL OR ABBREVIATION	BASIC UNIT
Armature	In a relay, the moving part that is attracted to the electromagnet. In a generator, the windings in which a voltage is induced. In a motor, the rotating part or *rotor*.		
Atom	The smallest particle of an element that combines with similar particles of other elements to form molecules and compounds.		
Attenuation	The process of reducing the strength of a signal.		
Audio	The range of sound frequencies detectable by the human ear, nominally 20 Hz–20 kHz.		
Autotransformer	A transformer that uses one coil for both primary and secondary.		
Average power	Average rate of energy consumption.	P_{av}	watt
AWG	American Wire Gauge. A table of standard wire sizes based on their circular mil areas.		
Ayrton shunt	A series interconnection of resistors placed in parallel with a meter movement to provide a multiple-range ammeter.		
Bandwidth (3 dB)	The range of frequencies over which the output of a device does not drop by 70.7% (or 3 dB) below its maximum output.	$BW \; \Delta f$	Hz
Base	The central semiconductor region in a bipolar junction transistor.		
Battery	Two or more interconnected cells that use a chemical reaction to produce electrical energy.	$\dashv \! \vert \! \vert \! \vert \! \vdash$	

TERM	DEFINITION	SYMBOL OR ABBREVIATION	BASIC UNIT
Beta	The ratio of collector current to base current in a bipolar junction transistor.	β	dimensionless
B-H curve	A graph showing the relationship between flux density (*B*) and applied magnetizing intensity (*H*) for a magnetic material.		
Bias	The application of a voltage or current to an active device (such as a transistor) to produce a desired mode of operation.		
Bifilar winding	A double winding, consisting of two wires wound side-by-side to reduce inductive effects (in a resistor) or to produce close coupling between two coils.		
Bimetal strip	A strip made of two different metals welded together, each having a different temperature coefficient of expansion. The strip bends as the temperature changes.		
Bipolar junction transistor (BJT)	A three-terminal semiconductor device with a base sandwiched between a collector and an emitter. Used for amplification or switching. May be *PNP* or *NPN* construction.		
Bohr model	A planetary model of the atom in which the central nucleus, made up of neutrons and protons, is surrounded by orbiting electrons.		
Branch	One of the paths for current in a parallel circuit.		

TERM	DEFINITION	SYMBOL OR ABBREVIATION	BASIC UNIT
Bridge	Four components arranged in a diamond-shaped circuit, used for the precise measurement of resistance, etc. A voltage source is applied to one pair of opposite corners and a galvanometer connected to the other pair.		
Brush	A carbon or graphite structure used to conduct current between a stationary terminal and a rotating commutator or slip ring in a motor or generator.		
Capacitance	The property of a capacitor to store electrical charge and energy.	C	farad
Capacitive reactance	The opposition provided by capacitance to the flow of alternating current.	X_C	ohm
Capacitor	An electrical device consisting of two metal plates separated by an insulating material called a dielectric.		
Capacity	The ability of a battery to store electrical charge and energy.		Ah
Cell	A device that uses a chemical reaction to produce a dc voltage. The basic building block of a battery.		
Cemf	Counter (or back) electromotive force, produced in a motor or by an inductance, that opposes the applied voltage.		volt
Center tap	A connection at the midpoint of a transformer's primary or secondary winding.	CT	

TERM	DEFINITION	SYMBOL OR ABBREVIATION	BASIC UNIT
Charge	An electrical property of an electron (negative) or a proton (positive) that can cause a potential difference to exist between two points when excess charges accumulate.	Q	coulomb
Chassis	A metal case or frame upon which an electrical circuit or system is constructed.		
Choke	Another name for an inductor or coil, especially one that *chokes* or blocks high frequencies.	L	
Circuit	An interconnection of components to form a complete path for the flow of electrical current.		
Circuit breaker	A resettable thermal or electromagnetic switch that opens a circuit when the circuit current exceeds a selected value.		
Circular mil	The unit of the cross-sectional area of a wire, found by squaring the diameter given in mils (thousandths of an inch).	CM	
Clamp-on meter	An ammeter to measure current by clamping around a wire instead of physically breaking the circuit. Makes use of the magnetic field *around* the current-carrying wire for its operation.		
Closed circuit	A complete path that permits the flow of current.		
Coefficient of coupling	The portion of magnetic flux set up by one coil that links a second coil.	k	dimensionless
Coercivity	A measure of the demagnetizing intensity (coercive force) needed to reduce the residual magnetism of a material to zero.		At/m

TERM	DEFINITION	SYMBOL OR ABBREVIATION	BASIC UNIT
Coil	A common term for an inductor or the turns of wire making up the windings in a transformer, motor, or similar device.		
Common	A term sometimes used for a common reference point or ground (neutral) in an electrical circuit or system.		
Common-base (CB) amplifier	A junction transistor amplifier circuit in which the base is common to both the input and output.		
Common-emitter (CE) amplifier	A junction transistor amplifier circuit in which the emitter is common to both the input and output.		
Commutator	Copper segments on the rotor of a motor or generator that switch current at the correct instant to generate direct current or permit a motor to run from a source of direct current.		
Complex number	A combination of a real and an imaginary number that may be given in rectangular or polar form.	$a + jb$ $M\underline{/\theta}$	
Condenser	Older name used for a capacitor.		
Conductance	The ability of a circuit to allow the flow of current. Reciprocal of resistance.	G	siemens
Conductivity	A measure of the ease with which a material conducts electrical current. Reciprocal of resistivity.	σ	siemens/meter
Conductor	A material that allows electrical current to flow relatively easily, such as copper.		
Continuity	A complete path through which current can flow. A resistance or impedance less than infinity.		

TERM	DEFINITION	SYMBOL OR ABBREVIATION	BASIC UNIT
Core	A material (usually magnetic) inside a coil that concentrates the magnetic flux.		
Coulomb	The unit of electrical charge.	C	
Counter emf	See Cemf.		
CRT	Cathode Ray Tube.		
Curie temperature	The temperature at which a magnetic material loses all its magnetism.		
Current	The rate of flow of electrical charge in a circuit.	I	ampere
Current transformer	A transformer used to extend the range of an ac ammeter by stepping down the primary line current by a known ratio.		
Cycle	One complete positive and one complete negative alternation of a repetitive quantity, such as voltage or current.	$\sim$	
Decibel	A logarithmic unit used to show the power or voltage gain of an amplifier, or the attentuation of a filter.	dB	
Degree	A unit of angular measure equal to 1/360th of a complete circle or revolution.	°	
Delta	A method of connecting a three-phase system so that the line and phase voltages are equal. A configuration of three connected components, sometimes referred to as a π (pi) connection. Symbol meaning "a change in."	Δ	

TERM	DEFINITION	SYMBOL OR ABBREVIATION	BASIC UNIT
Diamagnetic materials	Materials that have a permeability slightly less than that of air and that slightly oppose the setting up of magnetic lines of force.		
Dielectric	An insulator; usually refers to the insulating material between the plates of a capacitor.		
Dielectric constant	The ratio of capacitance with a dielectric to the capacitance with air, all other variables being held constant. A measure of the ability of an insulator in a capacitor to induce an opposing electric field.	K	dimensionless
Dielectric strength	The electric field intensity required to cause an insulator to break down and conduct.		V/m
Differentiator	A circuit that produces an output proportional to the slope (or derivative) of the input.		
Digital	A nonlinear process in which a variable takes on discrete values.		
Diode	An electronic device that allows current to flow through it in only one direction.		
Direct current	An electric current that flows in only one direction.	dc	
Dissipation factor	An indication of how much power is dissipated (lost) in a capacitor or coil at a given frequency. The reciprocal of Q.	D	dimensionless
Distortion	An undesired change in output waveform compared with input, due to irregularities in amplitude, frequency, or phase.		
DMM	Digital multimeter.		

TERM	DEFINITION	SYMBOL OR ABBREVIATION	BASIC UNIT
Domains	Groups of adjacent atoms (10^{15}–10^{17}) that respond collectively to an external magnetic field.		
DVM	Digital voltmeter.		
Eddy current	An induced current that circulates in a conducting core that carries a varying magnetic field.		
Effective value	A measure of the heating effect of a varying voltage or current. Also known as the *rms* or root-mean-square value.	V_{eff} I_{eff}	volt ampere
Efficiency	The ratio of output power (or work) to input power (or work), generally expressed as a percentage.	η	%
Electric field (intensity)	A region near a charged body where a force is exerted on another electrically charged body.	E	V/m N/C
Electricity	A form of energy due to electric charges at rest (static electricity) or in motion (current) that can be transmitted through wires to produce effects of heat, light, motion, sound, etc.		
Electrode	A conductor or terminal, usually metallic, through which an electric current enters or leaves.		
Electrolyte	A solution or paste capable of conducting an electric current because of its dissociation into positive and negative ions, as in a voltaic cell (battery).		
Electromagnet	A magnet produced by passing current through a coil of wire wound on a soft iron core.		

TERM	DEFINITION	SYMBOL OR ABBREVIATION	BASIC UNIT
Electromotive force	The open circuit voltage produced by an electrical source (such as a battery or generator), that causes current to flow in a complete circuit.	emf	volt
Electron	Negatively charged particle of an atom.		
Electronic	Related to the control and movement of electrical charges through semiconductor and vacuum tube devices.		
Electrostatic force	The force of attraction or repulsion between stationary charged bodies.		
Element	A substance composed entirely of one type of atom.		
Emitter	One of the three terminals of a BJT that emits the majority carriers.	(E)	
Energy	The ability or capacity to do work.	W	joules ft lb
Equivalent circuit	A circuit that represents a more complicated circuit, but permits easier analysis.		
Exponent	The power of a number, indicating how many times that number should be multiplied by itself.		
Farad	The unit of capacitance.	F	
Feedback	The process of returning some of the output of a circuit (such as an amplifier) back to the input. May be positive or negative.		
Ferromagnetic materials	Materials that have a permeability hundreds of times greater than that of free space, resembling the magnetic properties of iron.		

TERM	DEFINITION	SYMBOL OR ABBREVIATION	BASIC UNIT
Field	A region in which a force is exerted on a static charge (electric field), or on a moving charge (magnetic field), or on a mass (gravitational field).		
Field winding	A coil in a motor or generator used to set up a magnetic field.		
Filter	A frequency selective circuit that passes certain frequencies and rejects all others.		
Fluorescent lamp	A device that produces a source of light that is the result of ultraviolet energy making a phosphor powder fluoresce (glow).		
Flux	The lines of force in a magnetic or electric field.		
Flux density	The amount of flux per unit cross-sectional area.		
Free electron	A valence electron that has acquired enough energy to break away from its parent atom, and thus is free to move at random through the atoms of the given material.		
Frequency	The number of times a varying quantity goes through a complete cycle of variation in one second.	f	hertz
Fuel cell	A chemical reaction, aided by a catalyst, in which electrical energy is produced directly, using a process that is the reverse of electrolysis.		
Full-wave rectifier circuit	A diode circuit that uses both positive and negative alternations of an alternating current to produce a direct current.		

TERM	DEFINITION	SYMBOL OR ABBREVIATION	BASIC UNIT
Function generator	A piece of electronic equipment that is capable of producing several waveforms, such as sine, square, pulse, and triangular.		
Fuse	A thermal protective device that opens, when current becomes excessive, to disconnect a circuit from the source.		
Gain	The ratio of the output of an amplifier to the input, in terms of voltage, current, or power.	A_v A_i A_p	dimensionless
Galvanometer	An electrical instrument (ammeter) used to detect (measure) very small amounts of direct current.		
Gauss	The cgs unit of magnetic flux density.	G	
Gaussmeter	An instrument (fluxmeter) used to measure magnetic flux density.		
Generator	A machine that converts mechanical energy into electrical energy (ac or dc).		
Germanium	A semiconductor material used to make diodes and transistors.	Ge	
Giga	A prefix used to designate 10^9.	G	
Ground	A common or reference point in an electrical circuit that can be a chassis or earth ground.		
Hall emf	The voltage produced across the edges of a semiconductor strip that is carrying a current and located in a magnetic field.		
Harmonic	An integral multiple of a fundamental frequency.		
Henry	The unit of inductance.	H	

TERM	DEFINITION	SYMBOL OR ABBREVIATION	BASIC UNIT
Hertz	The unit of frequency. One cycle per second.	Hz	
Heterodyne	The process of beating or mixing two signals of different frequencies.		
Hole	A positive charge carrier (majority) in a *P*-type semiconductor. The space left in an atom's valence shell by a departed electron.		
Horsepower	Unit of power in the English system. 1 hp = 550 ft lb/s = 746 W.	hp	
Hydrometer	A device used to measure specific gravity.		
Hypotenuse	The longest side of a right triangle that is opposite the 90° right angle.		
Hysteresis	A lagging of the magnetic flux in a magnetic material behind the magnetizing force that is producing it.		
Impedance	The total opposition to the flow of alternating current in a circuit containing resistance and reactance.	Z	ohm
Incandescent lamp	A device that produces light as a result of heating a filament to a white-hot incandescence.		
Induced voltage	Voltage induced in a coil as a result of a changing magnetic field.	V_{ind}	volt
Inductance	The property of an inductor to oppose any change of current through it.	L	henry
Induction motor	An ac motor in which a rotating magnetic field, produced by current in the stator windings, induces current in the rotor, resulting in a rotating motion.		
Inductive reactance	The opposition that inductance offers to an alternating current.	X_L	ohm

TERM	DEFINITION	SYMBOL OR ABBREVIATION	BASIC UNIT
Inductor	A coil or choke having the property of inductance.	⌇lℓℓℓℓ⌇	
Infinite	Having no bounds or limits.	∞	
Input	The voltage, current, or power applied to a circuit.		
Instantaneous value	The value of a variable quantity at a given instant of time.		
Insulator	A material that does not conduct electrical current.		
Integrated circuit	A circuit in which all components and their interconnections are fabricated on the same chip of semiconductor material.	IC	
Integrating circuit	A circuit that produces an output that is proportional to the mathematical integration of the input. A summation device.	⊳∫	
Internal resistance	The resistance contained within a voltage or power source, such as a power supply, battery, or amplifier.	r	ohm
Ion	An atom that has an excess or deficiency of electrons.		
Iron-vane meter	An instrument that uses a moving iron vane and a stationary coil to measure low-frequency ac voltage and current.		
Joule	The unit of work and energy in the MKS system.	J	
Kilo	A prefix used to designate 10^3.	k	
Kilowatt-hour	A common unit of energy used by power companies to compute electricity usage bills.	kWh	
Kirchhoff's laws	A set of laws that describes how voltage and current are distributed throughout an ac or dc circuit.		

TERM	DEFINITION	SYMBOL OR ABBREVIATION	BASIC UNIT
Knee	An abrupt change in slope between two relatively straight segments of a curve.		
Lag	A condition that describes the delay of one waveform behind another in time or phase.		
Laminated core	A core built up from thin sheets of metal (often insulated from each other) and used in transformers, relays, motors, and other devices operating from an ac supply.		
Lead	A condition that describes how one waveform is ahead of another in time or phase. Also, a wire or cable connection.		
Lenz's law	A law which states that the polarity of induced emf is always such as to oppose the cause that created it.		
Linear	A relationship between two variables in which one is directly proportional to the other. A straight-line relationship.		
Line of force	A line drawn to represent the direction of a force in an electric or magnetic field.		
Load	A device that converts electrical energy into another form, such as heat, light, or motion.		
Logarithm	The exponent to which a base number must be raised to equal a given number. ($\log_{10} 100 = \log_{10} 10^2 = 2$)	log	
Loop	A closed path in a circuit.		

TERM	DEFINITION	SYMBOL OR ABBREVIATION	BASIC UNIT
Magnet	A magnetized material that aligns itself with the earth's magnetic field and attracts magnetic materials.		
Magnetic	Related to magnetism; north and south poles and their associated lines of force.		
Magnetic circuit	A complete circuit for establishing magnetic lines of force.		
Magnetic core	A form or frame made from those materials (iron, nickel, cobalt, and certain alloys) that can be magnetized.		
Magnetic field	A region in which a force of attraction is exerted on a magnetic material or on a moving charge.		
Magnetic field strength	The magnetomotive force per unit length of magnetic circuit. Also known as magnetizing force, magnetic field intensity, or magnetizing intensity.	H	At/m
Magnetic flux	Magnetic lines of force.	Φ	webers
Magnetic flux density	Magnetic lines of force per unit area. Also known as magnetic induction.	B	Wb/m^2 or teslas
Magnetize	To make a magnetic material act as a magnet by aligning its magnetic domains.		
Magneto	A generator that produces alternating current using a permanent magnet.		
Magnetomotive force (mmf)	The magnetizing force produced by current flowing through a coil of wire wrapped around a magnetic core.	F_m	At
Magnitude	The size of some electrical quantity, such as voltage or current, without regard to its phase angle.		

TERM	DEFINITION	SYMBOL OR ABBREVIATION	BASIC UNIT
Maximum power transfer	A circuit condition in which the load resistance equals the source resistance (dc), or where the load impedance is the conjugate of the source impedance (ac), causing maximum power to be transferred from source to load.		
Maxwell	The cgs unit of magnetic flux.	Mx	
Mega	A prefix designating 10^6 (1 million).	M	
Megger	A portable, hand-operated high-voltage dc generator used as an ohmmeter to measure insulation and other high resistances.		
Megohm	One million ohms.	$M\Omega$	
Mesh	An arrangement of current loops in a circuit.		
Meter loading	A change in circuit conditions caused by the finite resistance of a meter when connected to that circuit to make measurements.		
Micro	A prefix designating 10^{-6} (1 millionth).	μ	
Milli	A prefix designating 10^{-3} (1 thousandth).	m	
Milliammeter	An ammeter that measures current in thousandths of an ampere.		
Modulation	The process of varying the amplitude (AM), frequency (FM), or phase (PM) of a carrier wave using a signal such as audio or video to convey information.		
Multimeter	An instrument designed to measure two or more electrical quantities. Often portable.		
Multiplier	A resistor used to extend the voltage range of a meter movement.	R_s	ohm

TERM	DEFINITION	SYMBOL OR ABBREVIATION	BASIC UNIT
Mutual inductance	The inductance between two separate coils that share a common magnetic field.	M	henry
Nano	A prefix designating 10^{-9}.	n	
Negative charge	A body or terminal that has an excess of electrons compared to a neutral state.		
Neutral	The common or grounded wire in a single or three-phase system.		
Neutron	The electrically neutral particle of an atom. It carries no electrical charge, but has a mass equal to that of a proton.		
Nodal analysis	A method of analyzing a network in which total current entering a node is equated to the total current leaving the node.		
Node	A junction in a circuit where current divides into different paths.		
Nonlinear	A scale that has unequal divisions for equal increments. A device that produces an output that is not directly proportional to the input.		
Norton equivalent circuit	A current source in parallel with a resistance (impedance) that is equivalent to a given linear network.		
Nucleus	The central part of an atom that consists mainly of protons and neutrons.		
Null	A zero or minimum.		
Oersted	The cgs unit of magnetizing intensity.	Oe	At/cm
Ohm	The unit of resistance.	Ω	
Ohmmeter	An instrument that measures resistance.		

TERM	DEFINITION	SYMBOL OR ABBREVIATION	BASIC UNIT
Open circuit	A circuit with an incomplete path for current. Infinite resistance.		
Oscillator	A circuit that uses positive feedback to produce an output without an external input signal.		
Oscilloscope	An instrument that displays on its screen the manner in which a waveform varies over time.	—⊙—	
Output	The voltage, current, or power developed by a circuit in response to an input.		
Overload	A load that draws more than the rated current or power.		
Parallel	A connection of two or more components that operate from the same voltage.		
Parallel resonance	A condition in a parallel *RLC* circuit in which the circuit impedance is resistive and a maximum, causing the applied voltage and current to be in phase.		
Paramagnetic materials	Materials that have a permeability slightly greater than that of free space (air) and slightly aid the setting up of magnetic lines of force.		
Peak value	The maximum value of a waveform, equal to the amplitude for symmetrical alternating waveforms.	V_p I_p	volt ampere
Period	The time required for a periodic waveform to complete a full cycle.	T	s
Periodic	A variation that repeats at definite time intervals.		
Permeability	A measure of the ease with which magnetic lines of force can be established in a material.	μ	Wb/At-m

TERM	DEFINITION	SYMBOL OR ABBREVIATION	BASIC UNIT
Permittivity	A measure of the ability of a material to permit an electric field to be established in that material.	ϵ	C^2/Nm^2
Phase angle	An angle that shows the displacement of a time-varying waveform from some reference.	θ, ϕ	degree radian
Phasor	A line drawn to represent a time-varying waveform in magnitude and phase angle.		
Photovoltaic cell	A silicon *PN* junction that produces voltage from light energy. A solar cell.		
Pico	A prefix designating 10^{-12}.	p	
Piezoelectric	The production of voltage by applying pressure to a crystal.		
Polarity	Refers to the negative and positive terminals in an electric circuit or the north and south poles of a magnet.		
Polarization	The collection of hydrogen gas around the electrode of a discharging cell, or the distortion of the path of orbiting electrons in the dielectric of a charged capacitor.		
Pole	The regions of a magnet where the flux lines are concentrated. An electrode of a battery.		
Polyphase	A circuit or supply that uses more than one phase of alternating current.		
Positive charge	A body or terminal that has a deficiency of electrons, compared to a neutral state.		
Potential	The ability to do work with respect to some reference.		volt
Potential difference	A measure of how much work may be done to move an electrical charge between two points.	*V*	volt

TERM	DEFINITION	SYMBOL OR ABBREVIATION	BASIC UNIT
Potentiometer	A three-terminal variable resistor device used to vary voltage. Also, an instrument used to make accurate voltage measurements.		
Power	The rate of doing work or converting energy from one form to another.	P	watt
Power factor	The ratio of a circuit's true power in watts to the apparent power in volt-amperes.	$\cos \theta$	dimensionless
Power factor correction	A process by which capacitors are connected in parallel with an inductive ac load to make total current drawn from the line more in phase with line voltage.		
Power supply	A piece of equipment that supplies voltage and current, often variable (ac or dc), from the 120-V, 60-Hz line or other available sources.		
Primary cell	A cell (battery) that cannot be recharged to its initial state.		
Primary winding	The winding of a transformer that is connected to the input.		
Proton	A positively charged particle in the nucleus of an atom.		
Pulsating direct current	A varying direct current that periodically returns to zero.		
Quality factor	The ratio of reactive power to true power in a coil or resonant circuit. Used to show the amount of inductive reactance in a coil, compared with its ac resistance.	Q	dimensionless
Radian	A unit of angular measurement. Approximately 57.3°.	rad	

TERM	DEFINITION	SYMBOL OR ABBREVIATION	BASIC UNIT
Range	The maximum value that a meter is capable of measuring.		
Reactance	The opposition to the flow of alternating current offered by capacitance or inductance (or both).	X	ohm
Reactive power	The power that is alternatively delivered to a reactive component or circuit and then returned to the source.	P_q	var
Reactor	Another name for an inductor.	⎯ⱸⱸⱸⱸ⎯	henry
Real power	The power that is dissipated as heat in a resistor or converted to another form. Also known as true power.	P	watt
Rectification	The process of converting alternating current to direct current.		
Rectifier	An electronic device or circuit that converts alternating current to direct current.	⎯▶⎯	
Regulation	The change in terminal voltage of a source from no load to full load, expressed as a percentage of full load voltage.	V_{reg}	%
Relative permeability	The ratio of the permeability of a material to the permeability of free space (air).	μ_r	dimensionless
Relay	A switching device that uses electromagnetism to move an armature that opens or closes one or more contacts; often controlled from a remote location.		
Reluctance	The opposition that a material offers to the establishing of magnetic flux.	R_m	At/Wb

TERM	DEFINITION	SYMBOL OR ABBREVIATION	BASIC UNIT
Residual magnetism	The magnetism that remains in a material after the magnetizing force has been removed.		Wb/m^2
Resistance	The opposition to the flow of current that converts electrical energy into heat.	R	ohm
Resistivity	The resistance of a specific quantity of a given material.	ρ	$\Omega \cdot m$
Resistor	An electrical component whose property of resistance opposes current.	—⋁⋁⋁—	
Resonance	A series or parallel RLC circuit that, at one frequency, has the applied voltage and current in phase with each other.		
Resonant frequency	The frequency that causes a series or parallel RLC circuit to be in a state of resonance.	f_r	hertz
Retentivity	The measure of the ability of a material to retain its magnetism after the magnetizing force is removed.		Wb/m^2 T
Rheostat	A two-terminal variable resistor used to vary current.		
Right angle	A 90° angle.		
Root-mean-square	The value of direct current that is as effective in producing heat as the given time-varying waveform of current or voltage. Also known as the effective value.	rms	ampere volt
Saturation	A condition in which an increase in the driving force produces little or no effect, as in the saturation of a magnetic material.		
Secondary cell	A cell (battery) that is capable of being repeatedly recharged to its initial state.		
Secondary winding	The output winding of a transformer that is connected to the load.		

TERM	DEFINITION	SYMBOL OR ABBREVIATION	BASIC UNIT
Selectivity	The ability of a tuned circuit to respond to a certain band of frequencies and reject others. Inversely related to bandwidth.		
Self-inductance	The property of a coil to induce a voltage within itself because of a changing current in the coil.	L	henry
Self-induction	The process of generating a voltage in a coil or circuit because of a changing magnetic field within the coil or circuit itself.		
Semiconductor	A material, such as silicon or germanium, that generally has four valence electrons and a resistivity between that of a conductor or an insulator (but closer to that of a conductor).		
Series	A connection of two or more components so that there is only one path for the flow of current.		
Series resonance	A condition in a series RLC circuit where, at that one frequency, the impedance is a minimum and pure resistance exists because of equal inductive and capacitive reactances.		
Short	A path of zero or very low resistance between two points.		
Shunt	A parallel-connected component used (as in an ammeter) to divert current around a meter movement.		
Siemen	The unit used for conductance, susceptance, and admittance. The reciprocal of an ohm.	S	
Signal	A low-level voltage or current that is usually time-varying.		

TERM	DEFINITION	SYMBOL OR ABBREVIATION	BASIC UNIT
Silicon	A semiconductor material used to make electronic devices, such as diodes, ICs, and transistors.	Si	
Sine	In a right triangle, the name given to the ratio of the side opposite a given acute angle to the hypotenuse.	sin	dimensionless
Sine wave	An alternating waveform whose instantaneous value is related to the trigonometric sine function of time or angle.		
Sinusoidal	A waveform having the shape of a sine wave.		
Skin effect	The effect that takes place at high frequencies because of induced voltage, causing an alternating current to flow near the surface of a conductor, which results in an effective increase in the resistance of the conductor.		
Slope	The steepness of a line or curve given by the ratio of rise to run of a tangent line drawn to the curve.	m $\dfrac{\Delta y}{\Delta x}$	
Solar cell	A semiconductor device that converts light energy directly into electrical energy.		
Solenoid	A coil with a movable plunger that operates because of electromagnetism.		
Source	A device that produces or converts energy.		
Specific gravity	The ratio of the weight of a given volume of a substance to the weight of the same volume of water.		
Standard cell	A cadmium sulfate cell that produces a precisely known voltage (1.01830 V) used for calibrating a potentiometer.		

TERM	DEFINITION	SYMBOL OR ABBREVIATION	BASIC UNIT
Static	A fixed, nonvarying condition; without motion.		
Steady state	A condition, following a transient period, during which sudden changes no longer take place.		
Superposition principle	For any given component in a linear multi-emf circuit, the total current (or voltage) is the algebraic sum of the currents (voltages) that result from each emf acting separately.		
Susceptance	The ability of a purely reactive component to permit the flow of alternating current. The reciprocal of reactance.	B	siemens
Switch	An electronic or electrical device used to control the flow of current in a circuit.		
Tangent	In a right triangle, the name given to the ratio of the side opposite a given acute angle to the adjacent side.	tan	dimensionless
Tank	A parallel resonant LC circuit.		
Temperature coefficient	A constant specifying the amount of change in the value of a temperature dependent variable per unit change of temperature.	α	per °C
Tesla	The MKS unit of magnetic flux density. $1\ T = 1\ Wb/m^2$	T	
Thermistor	A semiconductor device that provides a large change (usually a drop) in resistance when its temperature increases.		
Thermocouple	A pair of dissimilar metals joined together so that an emf is generated at the open ends when the junction is heated or cooled.		

TERM	DEFINITION	SYMBOL OR ABBREVIATION	BASIC UNIT
Theta	A Greek letter used to represent an angle.	θ	degree radian
Thévenin equivalent circuit	A voltage source in series with a resistance (impedance) that is equivalent to a given linear network.		
Three-phase	Three sinusoidal voltages or currents displaced from each other by 120 electrical degrees.	3ϕ	
Time constant	The time needed for the voltage or current in an RC or RL circuit to change 63.2% from the initial value to the final value.	τ	second
Tolerance	The upper and lower limits of variation from the component's nominal value.		
Tone control	A circuit, often using variable resistors with capacitors, that provides emphasis of high (treble) or low (bass) frequencies in an audio amplifier.		
Transducer	In electricity and electronics, a device that converts a variable (such as flow or pressure or temperature) to an electrical signal; or a device that converts an electrical signal to a different form of energy, such as sound.		
Transformer	An electrical device that makes use of electromagnetic induction to transfer electrical energy from a primary winding to a secondary winding, causing an increase or decrease in the ac voltage.		
Transient	A short-lived change in circuit conditions from one steady state to another.		

TERM	DEFINITION	SYMBOL OR ABBREVIATION	BASIC UNIT
Transistor	A three-terminal semiconductor device used for amplification or switching. May be of the bipolar or field effect type.		
Tuned circuit	A resonant circuit that responds to a narrow band of frequencies.		
Turns ratio	The ratio of the number of turns in a transformer's primary winding to the number of turns in the secondary winding.	a	dimensionless
Unity power factor	A condition in an ac circuit where the inductive and capacitive reactive powers are equal, causing the applied voltage and current to be in phase with each other.	$\cos \theta = 1$	dimensionless
Valence electron	An electron that is in the outermost shell (orbit) of an atom.		
Vector	A line drawn to represent a quantity in magnitude and direction.		
Volt	The unit of voltage, electromotive force, or electrical potential.	V	
Voltage	The potential difference between two points. A measure of the potential energy available to move charge or a measure of the amount of work that must be done to move charge between two points.	V	volt
Volt-ampere	The unit of apparent power in an ac circuit.	VA	
Volt-ampere reactive	The unit of reactive power in an ac circuit.	var	
Voltmeter	An electrical instrument used to measure the voltage between two points.	—(V)—	

TERM	DEFINITION	SYMBOL OR ABBREVIATION	BASIC UNIT
Voltmeter sensitivity	The ohms per volt rating of a voltmeter, equal to the ratio of the total resistance of a voltmeter on a given range to the FSD voltage of that range. Also equal to the reciprocal of the meter movement's full-scale current for analog instruments.	S	Ω/V
Watt	The MKS unit of power.	W	
Watt-hour	A unit of energy.	Wh	
Wattmeter	An electrical instrument used to measure power.	—(P)—	
Waveform	The shape of a wave obtained when instantaneous values of a varying quantity are plotted against time.		
Wavelength	The distance, usually expressed in meters, traveled by a wave during one complete cycle.	λ	meter
Weber	The MKS unit of magnetic flux.	Wb	
Wheatstone bridge	See Bridge.		
Winding	The coil or turns of wire on an inductor or transformer.	—ℓℓℓℓ—	
Wiper	The movable contact in a potentiometer or other variable device.		
Work	The product of an applied force and the distance moved by the force.		
Wye	A method of connecting a three-phase system, using a common point so that the line and phase currents are equal. A configuration of three connected components, sometimes referred to as a T (tee) connection.	Y	

TERM	DEFINITION	SYMBOL OR ABBREVIATION	BASIC UNIT
Zener diode	A diode used with reverse bias in the breakdown mode where voltage is relatively independent of current. Used for voltage regulation.		

ANSWERS TO SELF-EXAMINATIONS

CHAPTER 1

1. T   2. F   3. T   4. T   5. T   6. F   7. T   8. T
9. T   10. F   11. T   12. F   13. F   14. T   15. T
16. T   17. F   18. T   19. T   20. T   21. F   22. T
23. T   24. F   25. F   26. F   27. T   28. T

CHAPTER 2

1. T   2. F   3. T   4. T   5. F   6. T   7. T   8. T
9. F   10. T   11. T   12. T   13. T   14. T   15. F
16. T   17. F   18. T   19. T   20. T   21. F   22. T
23. T   24. F   25. T   26. T   27. T   28. F   29. T
30. T   31. T   32. F   33. T   34. T

CHAPTER 3

1. T   2. T   3. T   4. F   5. T   6. F   7. T   8. F
9. T   10. F   11. T   12. T   13. T   14. T   15. F
16. T   17. F   18. T   19. F   20. T   21. 10 Ω
22. 20 A   23. 50 V   24. 3 A   25. 220 V   26. 40 W
27. 22 Ω   28. 1.2 kWh   29. $2.00   30. YVYG

CHAPTER 4

1. d   2. b   3. T   4. F   5. b   6. a   7. c   8. T
9. F   10. b   11. T   12. T

CHAPTER 5

1. T   2. T   3. F   4. T   5. F   6. T   7. c   8. c
9. b   10. a   11. d   12. a   13. b   14. c   15. d
16. d   17. T   18. T   19. T   20. F

CHAPTER 6

1. T   2. F   3. T   4. F   5. c   6. a   7. b   8. d
9. c   10. c   11. c   12. d   13. b   14. a   15. b

CHAPTER 7

1. T   2. F   3. F   4. F   5. d   6. c   7. b   8. T
9. F   10. T   11. T   12. F   13. T   14. T   15. T
16. F

CHAPTER 8

1. F   2. c   3. b   4. T   5. T   6. F   7. b   8. d
9. F   10. c   11. F   12. T   13. c   14. a   15. d
16. T   17. T   18. T   19. F

CHAPTER 9

1. T   2. c   3. F   4. b   5. c   6. F   7. d   8. T
9. T   10. F   11. d   12. T   13. F   14. b   15. F
16. F

CHAPTER 10

1. T   2. T   3. F   4. F   5. T   6. T   7. T   8. T
9. T   10. T   11. F   12. T   13. F   14. T   15. T
16. F   17. T   18. T   19. T   20. F   21. F   22. T

CHAPTER 11

1. T   2. F   3. T   4. F   5. T   6. T   7. T   8. T
9. F   10. F   11. T   12. F   13. T   14. T   15. F
16. T   17. F   18. F   19. T   20. T   21. F   22. T
23. T   24. F   25. F   26. F   27. T   28. T   29. F
30. T

CHAPTER 12

1. T   2. F   3. T   4. F   5. T   6. T   7. F   8. c
9. T   10. T   11. T   12. F   13. T   14. b   15. T
16. T   17. T   18. a   19. c   20. T   21. c   22. F
23. T   24. b   25. d

CHAPTER 13

1. T 2. T 3. F 4. T 5. F 6. F 7. T 8. F 9. c 10. F 11. T 12. T 13. F 14. T 15. T 16. T 17. F 18. T 19. T 20. T 21. F 22. T

CHAPTER 14

1. F 2. T 3. F 4. T 5. T 6. F 7. T 8. c 9. c 10. b 11. c 12. F 13. T 14. T 15. F 16. a 17. T 18. T 19. F 20. T 21. d 22. b 23. b 24. T 25. F 26. T 27. b 28. T

CHAPTER 15

1. c 2. d 3. F 4. T 5. T 6. a 7. d 8. a 9. b 10. c

CHAPTER 16

1. T 2. T 3. F 4. F 5. T 6. T 7. F 8. F 9. T 10. b 11. c 12. T 13. F 14. T 15. F 16. T 17. F 18. T

CHAPTER 17

1. T 2. F 3. b 4. d 5. F 6. T 7. F 8. T 9. F 10. c 11. d

CHAPTER 18

1. T 2. T 3. F 4. a 5. c 6. F 7. b 8. T 9. F 10. F 11. F 12. T 13. F

CHAPTER 19

1. T 2. T 3. F 4. F 5. T 6. F 7. c 8. b 9. F 10. F 11. T 12. T 13. F 14. T 15. T

CHAPTER 20

1. T 2. T 3. T 4. F 5. d 6. a 7. T 8. T 9. F 10. T 11. F 12. F 13. T 14. F 15. T 16. T 17. T

CHAPTER 21

1. T 2. F 3. F 4. c 5. d 6. b 7. c 8. T 9. b 10. T 11. T 12. F 13. T 14. T 15. F 16. F 17. T 18. F 19. T 20. F 21. c 22. a 23. F 24. F 25. F

CHAPTER 22

1. T 2. T 3. F 4. F 5. b 6. c 7. e 8. a 9. T 10. c 11. F 12. T 13. F 14. d 15. T 16. F 17. T

CHAPTER 23

1. T 2. T 3. T 4. F 5. c 6. b 7. F 8. F 9. F 10. T 11. T 12. F 13. T 14. F 15. T

CHAPTER 24

1. T 2. F 3. T 4. T 5. c 6. T 7. a 8. d 9. F 10. F 11. b 12. F 13. F 14. T 15. F 16. F 17. F 18. F 19. T

CHAPTER 25

1. T 2. T 3. T 4. F 5. c 6. a 7. b 8. c 9. F 10. T 11. T 12. F 13. T 14. d 15. T 16. T 17. F 18. T 19. F 20. d

CHAPTER 26

1. T 2. F 3. T 4. F 5. T 6. b 7. a 8. d 9. F 10. b 11. F 12. T 13. F 14. c 15. b 16. T 17. F 18. T 19. T 20. T

CHAPTER 27

1. f 2. c 3. s 4. h 5. b 6. d 7. e 8. k 9. m 10. i 11. g 12. j 13. n 14. p 15. l 16. q 17. o 18. t 19. r 20. a

CHAPTER 28

1. T 2. T 3. F 4. T 5. T 6. T 7. F 8. F 9. T 10. F 11. T 12. F 13. T 14. T 15. T 16. F 17. T 18. T 19. T 20. F 21. T 22. T 23. F 24. T 25. T 26. F 27. T 28. T 29. F 30. F 31. T 32. T 33. T 34. F 35. T 36. T 37. F

CHAPTER 29

1. F 2. F 3. F 4. T 5. F 6. T 7. F 8. F 9. T 10. F 11. T 12. F 13. T 14. T 15. F 16. T 17. T 18. T 19. T 20. T 21. T 22. F 23. F 24. T 25. T 26. T 27. T 28. T 29. F 30. F 31. T 32 T

ANSWERS TO ODD-NUMBERED PROBLEMS

c. 45 mΩ

2-25 a. 0.667 mS

b. 0.213 μS

c. 22.2 S

CHAPTER 3

3-1 146.7 Ω

3-3 6 kΩ

3-5 2.2 mA

3-7 306 V

3-9 0.5 mA

3-11 1.2 kΩ

3-13 a. 3.6×10^5 J

b. 0.1 kWh

3-15 83.3%

3-17 1¢

3-19 a. 3.6×10^6 J

b. 2.78×10^{-7} kWh

3-21 a. 1.44 kW

b. 1.93 hp

3-23 117 V

3-25 39.8 A

3-27 7.07 mA

3-29 26.1 V

3-31 2 W

3-33 a. 33 kΩ ± 10%

b. 510 Ω ± 5%

CHAPTER 4

4-1 a. 5.43×10^{-2} Ω

b. 0.136 V

c. No. 16

4-3 2.92×10^3 m

4-5 9.69×10^{-6} m^2

No. 7 min. gauge

4-7 9.89 cm

4-9 a. 0.625 mm

CHAPTER 5

5-1 500 Ω

5-3 a. 180 mA

b. 27 V, 54 V, 9 V

5-5 a. 4.86 W, 9.72 W, 1.62 W

b. 16.2 W

5-7 a. 11.4 V

b. −8.6 V

5-9 4 kΩ

5-11 $R_4 = 156$ Ω

2-27 a. 25 Ω

b. 2.2 MΩ

c. 1 Ω

c. 2.2 MΩ ± 20%

d. 6.8 Ω ± 5%

e. 10 Ω ± 10%

f. 124 Ω ± 1%

g. 3010 Ω ± 2%

h. 523 kΩ ± 1%

3-35 a. Brn Brn Or Gold

b. Gray Red Brn Silver

c. Brn Grn Blue

d. Or Blk Gold Gold

e. Red Viol Yell Gold

f. Brn Wh Blu Blk Brn

g. Grn Brn Brn Silver Blk

h. Gray Red Grn Yell Brn

3-37 a. 15 kΩ

b. 270 kΩ

c. 5.6 MΩ

3-39 a. 237 kΩ

b. 365 Ω

c. 6.81 Ω

b. 0.0417 mm

4-11 0.239 Ω

4-13 18.7 Ω

4-15 363.8°C

4-17 a. 923 Ω

b. No cold inrush current

4-19 1003.6 Ω

$R_3 = 338$ Ω

$R_1 = 483$ Ω

5-13 1.25 kΩ, 1.25 kΩ, 2.5 kΩ, 1.25 kΩ

5-15 36.1 V

5-17 a. 0.6 Ω

b. 43.3 A

5-19 100-W lamp: 90 V, 56.25 W

60-W lamp: 150 V, 93.75 W

5-21 a. 30.1 mV

b. 0 V
c. 15.1 mV
5-23 6.02 μA
5-25 181 mW
5-27 6 V, 6 W
5-29 a. 15 V
    b. 90 Ω, 60 Ω
    c. -15 V

d. -9 V
5-31 a. 0.8 mA
    b. $+42$ V
    c. -38 V
5-33 a. 0.7 mA
    b. 10.8 mW
    c. -14.4 V

CHAPTER 6

6-1 56.88 Ω
6-3 4.03 kΩ
6-5 19.70 Ω
6-7 253 W
6-9 a. 18 V, 3 A, 2.25 A, 0.75 A
    b. 108 W
6-11 a. 48 Ω
    b. 6.9 Ω, 6 Ω
6-13 4.7 kΩ
6-15 8 kΩ, 8 kΩ
6-17 a. $I_{47\ k} = 6.24$ mA

      $I_{51\ k} = 5.76$ mA
    b. $P_{47\ k} = 1.83$ W (2 W)
      $P_{51\ k} = 1.69$ W (2W)
6-19 500 Ω, 25 V
6-21 $I_1 = 13.3$ mA
    $I_2 = 10$ mA
6-23 14.1 V
6-25 $I_1 = 1.707$ mA
    $I_2 = 0.776$ mA
    $I_3 = 0.517$ mA

CHAPTER 7

7-1 a. 7.02 kΩ
    b. 5.7 mA
    c. 2.28 mA
    d. 26.8 V
7-3 a. 2.21 mA
    b. 0.98 mA
    c. 4.88 mW
7-5 a. No. 5
    b. 541 W
    c. 6762 W
7-7 a. 680 Ω
    b. 529.4 Ω
    c. 600 Ω
7-9 a. 4.16 mA; 10.4 mW
    b. The first 800 Ω
7-11 3⅓ Ω
7-13 a. 0.6 A
    b. 36 W
    c. 120 W
7-15 a. 20 kΩ
    b. 0.5 W
    c. 40 V
    d. 1 mA
7-17 a. 10 Ω
    b. 10 Ω
7-19 a. 31.2 μA

    b. 7.76 V
    c. 5.6 μA
7-21 a. 30 Ω, 1.2 W
      (30 Ω, 5%, 2 W)
    b. 6.6 V
    c. 13.2 V
7-23 $R_B = 1.5$ MΩ, 1.5 W
    (1.5 MΩ, 5%, 2 W)
    $R_S = 83.3$ kΩ, 3 W
    (82 kΩ, 5%, 5 W)
    1.895 kV
7-25 a. 8.87 kΩ
      (9.1 kΩ, 5%)
    b. $P_B = 7.1$ W (10 W)
      $P_S = 10.2$ W (15 W)
    c. 322 V
7-27 a. 36 Ω, 5%
    b. 4 Ω
      (3.9 Ω, 5%)
    c. 2.25 W each
    d. 17.9 V
    e. 13.43 V
7-29 $R_S = 2.5$ kΩ, 4W
    (2.4 kΩ, 5%, 5 W)
    $R_1 = 8$ kΩ, 3.2 W)
    (8.2 kΩ, 5%, 5 W)

$R_B = 18 \text{ k}\Omega, 0.45 \text{ W}$
(18 kΩ, 5%, 1 W)
7-31 $R_1 = R_4 = 100 \ \Omega, 0.64 \text{ W}$
(100 Ω, 5%, 1 W)
$R_2 = R_3 = 400 \ \Omega, 0.36 \text{ W}$
(390 Ω, 5%, 1W)
7-33 $V = 19$ V
$V_1 = 9$ V

$V_{AD} = -10$ V
$V_{BF} = 12$ V
$V_{CD} = 6$ V
$V_{FD} = -3$ V
7-35 $I_{R1} = 12$ A
$I_{R2} = 8$ A
$V_N = 10$ V

CHAPTER 8

8-1 a. 9.38 lb
b. 10.62 lb
8-3 a. 16.5 h
b. 53 h
c. 87 h
8-5 a. 0.49 Ω, 283 W
b. 31.2 Ω

c. 5 min
d. No. Insufficient current
8-7 a. 30 A
b. 0.05 Ω
8-9 a. 6.2 A
b. 0.24 Ω

CHAPTER 9

9-1 a. 10 mA
b. 25 V
c. 75 V
d. 0.44 W
e. 40 mA
9-3 a. 0.25 Ω
b. 1.25 Ω
c. 0.5 Ω
9-5 a. 2.256 V
b. 0.256 Ω
9-7 20%
9-9 11.5 V
9-11 12 A
9-13 a. 24 V, 10 A, 0.04 Ω
b. 6 V, 40 A, 0.0025 Ω
c. 12 V, 20 A, 0.01 Ω

9-15 a. 12.68 V
b. 0.024 Ω
c. 2.92 V
d. 252 A
e. 3.81 kW, 2.54 kW
9-17 a. 0.02 Ω
b. 300 A
c. 1800 W
c. 1.7 hp
e. 1800 W
f. 6 V
g. 50%
9-19 a. 1.375 Ω, 46.625 Ω
b. 14.7%, 85.4%
c. 46.6 Ω

CHAPTER 10

10-1 a. 1.55 A recharging
b. 14.23 A
c. 12.68 A
d. 12.68 V
10-3 a. 180.4 A discharging
b. 176.8 A
c. 3.6 A opposite direction
d. 3.6 V opposite polarity
10-5 $I_{R_1} = 3\frac{1}{3}$ mA
$I_{R_2} = 3\frac{1}{3}$ mA
$I_{R_3} = 0$

$I_{R4} = 1\frac{2}{3}$ mA
$I_{R5} = 1\frac{2}{3}$ mA
10-7 a. $I_{R_1} = 2.04$ mA
$I_{R_2} = 2.51$ mA
$I_{R_3} = 1.39$ mA
$I_{R_4} = 3.16$ mA
$I_{R_5} = 0.65$ mA
b. 1.43 V
c. 3.3 kΩ
10-9 0 A
10-11 a. 3.04 kΩ

b. 4.03 kΩ

10-13 I_{R_1} = 3.33 mA

I_{R_2} = 3.33 mA

I_{R_3} = 0.01 mA ≈ 0

I_{R_4} = 1.66 mA

I_{R_5} = 1.66 mA

10-15 0 A

10-17 3.71 A

10-19 a. 1.52 A recharging

b. 14.2 A

c. 12.68 A

d. 12.68 V

10-21 3.12 mA

10-23 0.65 mA

10-25 3.08 mA

10-27 a. 0.186 mA

b. 0.243 mA

10-29 5.7 mA

10-31 66⅔ Ω, 37.5 W

10-33 V_{Th} = −1 V

R_{Th} = 0 Ω

10-35 V_{Th} = 0.77 V

R_{Th} = 3.8 kΩ

10-37 2.76 mA

10-39 0.706 A upwards

10-41 0.5 mA

10-43 0.15 V

10-45 All 30 kΩ

10-47 2.6 kΩ

10-49 0.5 mA, 150 mV

CHAPTER 11

11-1 0.6 T

11-3 1.6×10^{-4} Wb

11-5 2.43 A

11-7 2125 At/m

11-9 a. 1.35 T

b. 750 At

c. 1705 At/m

d. 7.92×10^{-4} Wb/At-m

e. 630

11-11 9.66 A

11-13 a. 1.07×10^6 At/Wb

b. 1.4×10^{-4} Wb

11-15 a. 3.75 N

b. 0.563 N · m

11-17 a. 0.5 A

b. 3 A

c. 6 V

d. 114 V

e. 78 W

f. 420 W

g. 282 W, 0.38 hp

h. 67%

11-19 384 turns

11-21 6.52×10^{-4} Wb

11-23 5158 turns

11-25 11,140 turns

11-27 0.54 A

CHAPTER 12

12-1 a. 40.8 Ω

b. 0.2 V

c. 40 Ω

12-3 a. 222.2 Ω, 40.8 Ω,

8.03 Ω, 2.002 Ω, 0.400 Ω

b. 0.2 V

c. 200 Ω, 40 Ω,

8 Ω, 2 Ω, 0.4 Ω

12-5 a. 0.01 Ω

b. 4.98 Ω

c. 0.5 W

12-7 R_1 = 1600 Ω, R_2 = 320 Ω,

R_3 = 80 Ω

12-9 a. R_1 = 142.1 Ω, R_2 = 15.8 Ω

b. 15.7 Ω

c. 15.1 Ω

d. 15 Ω

12-11 a. ±1 mA

b. 4 to 6 mA

14 to 16 mA

24 to 26 mA

c. 20%

6.7%

4%

12-13 a. 0.682 mA, 1.02 mW

b. 0.469 mA, 0.48 mW

c. 0.439 to 0.499 mA

d. 0.652 mA

e. 0.352 to 0.952 mA

12-15 a. 10 kΩ, 100 kΩ, 500 kΩ

b. 8 kΩ, 98 kΩ, 498 kΩ

c. 10 kΩ/V

d. 500 kΩ

12-17 a. 10 kΩ, 100 kΩ, 500 kΩ
 b. 8 kΩ, 90 kΩ, 400 kΩ
 c. 10 kΩ/V
 d. 500 kΩ

12-19 a. 40 μA
 b. 6 V
 c. 8.33 V
 d. 9.91 V

12-21 a. 4 V
 b. 3.88 V
 c. 0.92 V

d. 0.92 V
 e. 2.93 V

12-23 a. 4.902 V
 b. 4.898 V
 c. 4.46 V
 d. 4.46 V
 e. 4.85 V

12-25 a. 9.32 kΩ/V
 b. 13.87 to 16.37 V
 c. 245 μA
 d. 9%

CHAPTER 13

13-1 a. 1.253 V
 b. −0.047 V
 c. 3.1%

13-3 a. 113.24 Ω
 b. 1.1%

13-5 a. 19.9 kΩ
 b. 12.5%

13-7 a. 50.1 Ω
 b. 10%

13-9 0, 7.5 kΩ, 20 kΩ,
 45 kΩ, 120 kΩ, ∞

13-11 a. 18.7 kΩ
 b. 1.3 kΩ too high
 c. 0.2 V

13-13 a. 0
 b. 57 kΩ
 c. 0.25 A
 d. 125 mA, 25 μA
 e. 12.5 mA, 25 μA
 f. 36 Ω, 4 Ω

13-15 × 3 kΩ

13-17 a. 45.2 Ω
 b. ±0.6%
 c. 0.181 A

13-19 a. 104.5 kΩ
 b. ±3.1%
 c. 55 μA
 d. 320 μW

13-21 a. 104.5 kΩ
 b. ±3.1%
 c. 18.5 μA
 d. 35.8 μW

13-23 a. 1 ± 0.0003 V,
 ± 0.03%
 b. 1 ± 0.0012 V,
 ±0.12%

13-25 a. 2 ± 0.003 mA, ±0.15%
 b. 9 ± 0.01 mA, ±0.11%

13-27 a. ±0.264 kΩ
 b. ±0.032%

CHAPTER 14

14-1 a. Negative
 b. Positive

14-3 a. Negative
 b. Positive

14-5 a. Negative
 b. Positive

14-7 a. 3.3×10^{-3} Wb/s
 b. 0.5 Wb/s
 c. 0.9 Wb/s
 d. 4×10^{3} Wb/s

14-9 5×10^{-3} Wb/s

14-11 1 T

14-13 b. +1.5 Wb/s, 0,
 +2 Wb/s, −5 Wb/s
 c. 75 V, 0,
 100 V, −250 V

14-15 15 V

14-17 a. 7.5 V
 b. 10.6 V
 c. 14.8 V
 d. 15 V

14-19 a. 12.5 V

b. 12.5 V, − 11.9 V,
10.8 V
14-21 471 V
14-23 a. 60 Hz

b. 16.7 ms
14-25 a. 5.77 V
b. 199 V

CHAPTER 15

15-1 a. $v = 5 \sin 40\pi \times 10^3 t$ volts
b. 2.94 V
15-3 a. 70 V
b. 140 V
c. 50 rad/s
d. 7.96 Hz
e. 0.126 s
f. −18.4 V
15-5 a. 1.82 A
b. 400 Hz
c. $v = 60 \sin 800\pi t$ volts
d. $i = 1.82 \sin 800\pi t$ amperes
15-7 a. 109 W
b. 54.5 W
c. 800 Hz
15-9 a. 5.3 V
b. 4.95 mA

c. 7.07 mV
d. 12 A
15-11 a. 339 V
b. 679 V
c. 42.4 A
d. 4.5 μV
15-13 a. 127.6 V
b. 74 W
15-15 0.25 W
15-17 a. 11 mW
b. 2.27 kΩ
15-19 a. 5 Ω, 3.08 Ω, 1.47 Ω
b. 28.8 W, 17.7 W, 8.5 W
c. 22.9 V
15-21 a. 18.5 V
b. 17 kΩ
c. 12.3 kΩ

CHAPTER 16

16-1 a. 26.5 mV
b. 0.53 V
c. 13.25 V
16-3 a. 1.25 Hz
b. 312.5 Hz
c. 6.25 kHz
d. 125 kHz
16-5 600 Hz
16-7 a. 0.98 mA
b. 2.5 kHz
c. $v = 4.6 \sin 5000\pi t$ volts
d. $i = 0.98 \sin 5000\pi t$ mA

16-9 2.65 kΩ
16-11 a. 42.4 V
b. 120 V_{p-p}
c. 1.93 mA
d. 0.2 ms
16-13 a. 28 V
b. 9.33 kΩ
c. 5 kΩ
d. 3 kΩ
16-15 a. 2 mA
b. 3.43 mA, 0 V
c. 0 mA, 0 V, 48.6 V_{p-p}

CHAPTER 17

17-1 a. 14 A/s
b. 98 V
17-3 600 A/s
17-5 a. 0.13 mH
b. 23.4 mH
17-7 0.35 V, 0 V, −3 V
17-9 a. 120 mH

b. 10.9 mH
c. 20 mH in series with a
parallel combination of
40 and 60 mH
17-11 a. 21 H, 9H
b. 0.4
17-13 419 μH

CHAPTER 18

18-1 16.3 V
18-3 a. 32.6 A
      b. 7.1 Ω
18-5. a. 9.58
       b. 1.6 A
       c. 0.167 A
       d. 38.4 VA
18-7 a. 4.8 A
      b. 5.3 A
18-9 a. 2.83
      b. 17.7 V, 0.277 A
      c. 6.3 V, 0.782 A
      d. 4.9 W
      e. 1.9 W
18-11 a. 755 A
       b. 2523 A
       c. 5000 kW
18-13 a. 1210 W

b. 11 A
c. 10.1 A
d. 0.9 A
e. 8.2%
18-15 a. 40 A
       b. 1725 V
       c. 52.5 kW
       d. 69 kVA
18-17 a. 480 V
       b. 4.45 V
       c. 35.6 V
       d. 18.9 V
18-19 a. D to E, use V_{CF}
       b. B to C, D to F,
           use V_{AE}
       c. B to E, use V_{AF}
       d. B to D, C to E,
           use V_{AF}

CHAPTER 19

19-1 a. 1500 A/s
      b. 2 mA
      c. 1.33 μs
      d. 6.7 μs
19-3 a. 48 Ω
      b. 30 ms
      c. 1.44 H
19-5 a. 0.33 A
      b. 2.4 V
      c. 41.3 ms
      d. 17.5 ms
19-7 a. 0.324 A
      b. 2.2 V
      c. 40.2 ms
      d. 17.3 ms
19-9 0.563 J
19-11 b. 46 mA
       c. − 12 V
       d. 2.8 μs

19-13 b. 44.6 mA
       c. − 12.14 V
       d. 2.77 μs
19-15 a. 64 μs
       b. 11 mH
       c. 1.94 V
19-17 a. 2.5 A/s
       b. 2.42 ms
       c. 6.1 mA
       d. 12.1 ms
       e. 20 V
       f. 122 V
       g. 15.3 A/s
       h. 0.34 ms
       i. 1.7 ms
19-19 a. 9.32 A
       b. 13.38 V
       c. − 93.2 V
       d. 5.66 A, − 56.6 V

CHAPTER 20

20-1 2.4 kΩ
20-3 1.19 H
20-5 3.37 MHz
20-7 a. 0.8 mA
      b. $i = 1.13 \sin 2\pi \times 10^4 t$ mA
20-9 a. 12 V dc

b. 44 mV$_{p-p}$
20-11 a. 1 mA
       b. 10 V, 20 V, 30 V
20-13 a. 892 Ω
       b. 11.2 mA
       c. 4.3 V, 5.7 V

20-15 0.36 H, 0.72 H
20-17 a. 0.128 H
      b. 0.191 H
      c. 0.67

20-19 a. 5.48 kΩ
      b. 10.9 mA
20-21 a. 1.78 kΩ
      b. 580 Ω

CHAPTER 21

21-1 5 μF
21-3 20 V
21-5 a. 0.02 μF
    b. 1600 V
21-7 a. 1.51×10^{-3} m^2
    b. 330 V
21-9 5.08×10^{-3} m^2
21-11 6.7×10^6 V/m
21-13 a. 122 μF
     b. 16 V
     c. 1.95 mC

21-15 a. 6.25 μF
     b. 312.5 μC
     c. 12.5 V
     d. 100 V
21-17 a. 0.00857 μF
     b. 114.3 V, 85.7 V
21-19 60 μF
21-21 0.0156 J
21-23 32 μF
21-25 a. 484 pF
     b. 6 μF

CHAPTER 22

22-1 a. 5000 V/s
    b. 1 mA
    c. 10 μC
22-3 45 mA, 0, -37.5 mA
22-5 a. 1818 V/s
    b. 0.27 mA
    c. 6V
22-7 a. 25 s
    b. 0.2 V/s
    c. 3.15 V
    d. 125 s
    e. 50 mA
22-9 a. 205 μA
    b. 75.7 μA
    c. 167 V
    d. 440 s
22-11 a. 6 s
     b. 100 mA
     c. 15 ms
     d. 40 A
     e. 24 J
     f. 1600 W
22-13 a. 34.4 V
     b. 3.1 s
22-15 a. 131 V
     b. 4 mA

     c. 2.3 ms
     d. -2.5 mA
22-17 a. 120 μs
     b. 1.2 μF
     c. 833 Hz
22-19 a. 13.3 kHz
     b. 0.33 V/μs
22-21 a. 8.8 V
     b. 0.2 V
     c. 8.9 V, 0.1 V
22-23 a. 35.3 V
     b. 3.36 s
22-25 a. 132 V
     b. 3.8 mA
     c. 2.26 ms
     d. -2.5 mA
22-27 a. 7.28 V
     b. 2.72 mA
22-29 a. 5 mA
     b. 77.7 V
     c. 1.12 mA
     d. 7.77 mC
     e. 5.18 mA
     f. 0.5 s
     g. 51.8 V

23-1 575 Ω

23-3 19.9 V

23-5 1.8 μF

23-7 a. 1.19 mA

     b. 119 mA

     c. 11.9 A

23-9 82.9 Hz

23-11 a. 8.62 kΩ

      b. 2.78 mA

24-5 a. 89.4 V

     b. 63.4°

     c. 2:1

24-7 a. 21.8 V

     b. 24.6°

     c. 9.9 mA

     d. 0.4 H

24-9 a. 1.2 kΩ

     b. 537 Ω

     c. 1074 Ω

     d. 10.74 V

     e. 21.48 V

     f. 2.85 H

24-11 a. 48.7 Ω

      b. 145 mA

      c. 15°

24-13 a. 1050 Ω

      b. 81°

      c. 159.6 Ω

      d. 1037 Ω

      e. 6.5

      f. 2.75 H

24-15 a. 119 V

      b. $-62.8°$

      c. 41.3 μF

24-17 a. 1.7 kΩ

      b. $-25.1°$

25-1 11.9 Ω

25-3 a. 19.1 vars

     b. 19.1 VA

25-5 175.7 μF

25-7 69 VA

25-9 946 VA

25-11 a. 107.8 W

c. 11.1 V, 7.4 V, 5.5 V

d. 0.046 μF

23-13 a. 5.89 kΩ

      b. 1.81 mA, 2.71 mA, 3.61 mA

      c. 8.15 mA

      d. 0.45 μF

23-15 a. 100 Ω

      b. 0.33

24-19 0.954, 0.303

24-21

$$\frac{V_o}{V_i} = \sqrt{\frac{R_2^2 + X_C^2}{(R_1 + R_2)^2 + X_C^2}}$$

24-23 a. 45.7 kΩ

      b. 76.6 μA

      c. $V_R = 0.17$ V, $V_L = 2.31$ V, $V_C = 5.81$V

      d. $-87.2°$

24-25 a. 22.6 kΩ

      b. 154.9 μA

      c. $V_R = 0.34$ V, $V_L = 9.36$ V, $V_C = 5.87$ V

      d. 84.4°

24-27 a. 5.8 μF

      b. 6.85 V

      c. 10 V

      d. 1.8 μF, 21.8 V

24-29 a. 56.4°

      b. 19.9 V

      c. 36 V

24-31 a. 55.3 mA

      b. 868 Ω

      c. $-43.7°$

24-33 a. 2.36 mA

      b. 2.12 kΩ

      c. $-16°$

     b. 73.9 vars

     c. 130.7 VA

25-13 a. 21.3 W

      b. 26.5 vars

      c. 34 VA

      d. 0.63

25-15 a. 1.1 W

b. 1.8 vars
c. 2.1 VA
d. 0.52
25-17 a. 306 W
b. 111 vars (inductive)
c. 326 VA
d. 2.7 A
e. 0.94 (lagging)
25-19 a. 53.1°
b. 1870 VA
c. 1122 W
d. 1496 vars
25-21 a. 11.8 A
b. 9.2 A
c. 8.3 A
d. 9.2 A
25-23 a. 82 μF

b. 5.1 A
25-25 a. 0.248 A
b. 29.8 VA
c. 12.3 W
d. 3.1 W
e. 0.52
f. 25.5 vars
g. 4.7 μF
h. 0.13 A,
    Dimmer lamp
25-27 a. 0.87
b. 86.7%
25-29 150 Ω resistance in
    series with an inductive
    reactance of 531 Ω;
    (L = 42 μH)

CHAPTER 26

26-1 a. 31.8 Ω capacitive
    12.04 Ω resistive
    29.7 Ω inductive
b. −67.8°, −4.8°, 66.2°
c. 0.629 mA, 1.66 mA, 0.673 mA
d. 7.55 mV, 880 mV, 899 mV
    19.9 mV, 2.35 V, 2.35 V
    8.1 mV, 961 mV, 943 mV
26-3 1.5005 MHz
26-5 0.25 μF
26-7 a. 240 kHz, 260 kHz
b. 3 mW
c. 12.5
26-9 a. 118
b. 12.7 kHz
26-11 54.4
26-13 a. 37.1 V (rms)
b. 0.02 μF
c. 476 Hz

d. 83.3 mW
26-15 60 to 258 pF
26-17 a. 142.4 kHz
b. I_R = 0.2 μA
    I_L = I_C = 22.4 μA
c. 0.2 μA
26-19 a. 112
b. 50 kΩ
c. 22.4 μA
d. 1.27 kHz
26-21 a. 5.6 mV
b. 1.1 mV
26-23 145.5 kHz
26-25 a. 140.1 kHz
b. 5.5
c. 25.5 kHz
26-27 a. 18.4 Ω
b. 2.3 Ω, 136.7 kHz
26-29 1.8 to 2.6 pF

CHAPTER 27

27-1 80 + j40 V
    89.4 V $\underline{/26.6°}$
27-3 25 − j40 Ω
    47.2 Ω $\underline{/-58°}$
27-5 68 + j75.4 Ω
    101.5 Ω $\underline{/48°}$
27-7 a. 200 $\underline{/35°}$
b. 12.5 $\underline{/85°}$

c. 0.08 $\underline{/-85°}$
27-9 a. 2.54 A $\underline{/58°}$
b. 63.5 V $\underline{/58°}$
c. 101.6 V $\underline{/-32°}$
27-11 a. 0.47 A $\underline{/-48°}$
b. 32.2 V $\underline{/-48°}$
c. 35.7 V $\underline{/42°}$

27-13 a. Capacitive
    b. 240 Ω $\underline{/-30°}$
    c. 208 Ω, 120 Ω
27-15 a. $11 + j4$
    b. $-5 - j18$
    c. $5 + j18$
27-17 $50 + j100$ Ω,
    112 Ω $\underline{/63.4°}$
27-19 a. $68 - j24$
    b. $-0.308 - j0.462$
    c. $-1 + j1.5$
27-21 a. $6.1 + j13.7$ Ω,
    14.95 Ω $\underline{/66°}$
    b. 1.6 A $\underline{/-66°}$
    c. 0.89 A $\underline{/-68.2°}$,
    0.72 A $\underline{/-63.4°}$
27-23 a. 2.89 kΩ $\underline{/4.8°}$
    b. 34.6 μA $\underline{/-4.8°}$
    c. 92.8 μA $\underline{/-68.2°}$,
    83.3 μA $\underline{/90°}$
27-25 a. 11.7 $\underline{/59°}$
    b. 7.6 $\underline{/-66.8°}$
    c. 10 $\underline{/-143°}$
27-27 14.93 Ω $\underline{/66°}$
27-29 a. $-37.3 + j53.2$

    b. $3.7 - j11.4$
    c. $-18.8 - j6.8$
27-31 10.9 Ω, 127 μH
27-33 a. 118.5 mA
    b. 400 Ω $\underline{/50°}$
    c. 3.7 W
27-35 a. 0.983 $\underline{/-10.7°}$
    b. 0.052 $\underline{/-87°}$
    c. $R = 4.3$ kΩ
    $V_o = 85$ V
27-37 a. 382.7 Ω $\underline{/45.9°}$
    b. 62.7 mA $\underline{/-45.9°}$
    c. 20.7 V $\underline{/26.4°}$
    d. 63.1 mA $\underline{/-31.2°}$
27-39 a. 4.86 kΩ, 15.9 H
    b. 4.85 kΩ, 38.2 mH
27-41 0.57 μA $\underline{/-41.6°}$
27-43 1.74 A $\underline{/-105.2°}$
27-45 1.75 A $\underline{/-105.1°}$
27-47 0.67 A $\underline{/26.5°}$
27-49 $R = 232$ kΩ, $C = 0.043$ μF
27-51 $R_x = \dfrac{R_2 R_3}{R_1}$, $L_x = C_s R_2 R_3$

CHAPTER 28

28-1 a. 10.8 V
    b. 72.2 mA
    c. 23 mA
    d. 0.25 W
    e. 34 V
28-3 a. 20 V
    b. 56.6 Ω
28-5 a. 108 V
    b. 108 mA
    c. 11.7 W
    d. 339 V
    e. 120 Hz
28-7 a. 40 V
    b. 0.57 A
28-9 a. 216 V
    b. 216 mA
    c. 46.7 W
    d. 339 V
    e. 120 Hz

28-11 a. 281 V
    b. 1.58 kW
    c. 294 V
    d. 360 Hz
28-13 a. 1.2 V
    b. 3.4 V
28-15 b. 8.8%
28-17 a. 2.41%
    b. 8.55 V
28-19 1205 μF
28-21 2.5 H
28-23 1.3 H, 130 μF
28-25 a. 144 Ω, 0.83 A
    b. 84.9 V
    c. 54 V
    d. 102 Ω
    e. 70.7 W, 29.3%
    f. 28.6 W

CHAPTER 29

29-1 a. 15.3 mA
      b. 0.2 mA
      c. 24.75 mA
29-3 a. 50, 49.5, 99
      b. 0.98, 0.98, 0.99
29-5 a. 125
      b. 172
      c. 190
29-7 At $I_C = 2.5$ mA, $\beta_{ac} = 146$
      At $I_C = 17$ mA, $\beta_{ac} = 181$
      At $I_C = 32$ mA, $\beta_{ac} = 223$
29-9 22.5 mA, 22.65 mA
29-11 0.143 A, 2.86A
29-13 a. 0.968
      b. 30.3
29-15 a. 1 mA
      b. I_{CEO} is too small
29-17 a. 5 mA
      b. -125
      c. 11 V
      d. -220
      e. 1.25 kΩ
      f. -220
      g. 44.4 dB
29-19 a. 19 V
      b. 18.4 V
      c. 740 kΩ

29-21 722.5 kΩ
29-23 a. 10 W
      b. 141
      c. 1.42 A
29-25 a. 17.6
      b. 9.4
      c. 600
29-27 a. 0.045
      b. -20
      c. 13.75 kΩ
      d. 165 kHz
      e. 0.36%
      f. -3300 kHz
      g. -125
      h. 34 dB
29-29 a. 0.145
      b. -6.3
      c. 13.8 kΩ
      d. 165 kHz
      e. 0.36%
      f. -1035 kHz
      g. -125
      h. 29 dB
29-31 a. 62.4 kHz
      b. 600 mV
      c. 3

INDEX